IMPORTANT:

HERE IS YOUR REGISTRATION CODE TO ACCESS
YOUR PREMIUM McGRAW-HILL ONLINE RESOURCES.

For key premium online resources you need THIS CODE to gain access. Once the code is entered, you will be able to use the Web resources for the length of your course.

If your course is using **WebCT** or **Blackboard**, you'll be able to use this code to access the McGraw-Hill content within your instructor's online course.

Access is provided if you have purchased a new book. If the registration code is missing from this book, the registration screen on our Website, and within your WebCT or Blackboard course, will tell you how to obtain your new code.

Registering for McGraw-Hill Online Resources

TO gain access to your McGraw-Hill web resources simply follow the steps below:

1. USE YOUR WEB BROWSER TO GO TO: **http://www.mhhe.com/sverdrup/**

2. CLICK ON **FIRST TIME USER**.

3. ENTER THE REGISTRATION CODE* PRINTED ON THE TEAR-OFF BOOKMARK ON THE RIGHT.

4. AFTER YOU HAVE ENTERED YOUR REGISTRATION CODE, CLICK **REGISTER**.

5. FOLLOW THE INSTRUCTIONS TO SET-UP YOUR PERSONAL UserID AND PASSWORD.

6. WRITE YOUR UserID AND PASSWORD DOWN FOR FUTURE REFERENCE.
   KEEP IT IN A SAFE PLACE.

TO GAIN ACCESS to the McGraw-Hill content in your instructor's **WebCT** or **Blackboard** course simply log in to the course with the UserID and Password provided by your instructor. Enter the registration code exactly as it appears in the box to the right when prompted by the system. You will only need to use the code the first time you click on McGraw-Hill content.

Thank you, and welcome to your McGraw-Hill online Resources!

* YOUR REGISTRATION CODE CAN BE USED ONLY ONCE TO ESTABLISH ACCESS. IT IS NOT TRANSFERABLE.

0-07-247280-4 SVERDRUP/DUXBURY/DUXBURY: AN INTRODUCTION TO THE WORLD'S OCEANS, 7E

MCGRAW-HILL
ONLINE RESOURCES

REGISTRATION CODE

disruptive-49212951

How's Your Math?

Do you have the math skills you need to succeed?

Why risk not succeeding because you struggle with your math skills?

Get access to a web-based, personal math tutor:

- Available 24/7, unlimited use
- Driven by artificial intelligence
- Self-paced
- An entire month's subscription **for much less** than the cost of one hour with a human tutor

ALEKS is an inexpensive, private, infinitely patient math tutor that's accessible any time, anywhere you log on.

Log On for a
FREE 48-hour Trial

www.highedstudent.aleks.com

ALEKS is a registered trademark of ALEKS Corporation.

an introduction to the

World's Oceans

an introduction to the

World's Oceans

Seventh Edition

Keith A. Sverdrup

University of Wisconsin–Milwaukee

Alyn C. Duxbury

University of Washington

Alison B. Duxbury

Seattle Community College

Boston   Burr Ridge, IL   Dubuque, IA   Madison, WI   New York   San Francisco   St. Louis
Bangkok   Bogotá   Caracas   Kuala Lumpur   Lisbon   London   Madrid   Mexico City
Milan   Montreal   New Delhi   Santiago   Seoul   Singapore   Sydney   Taipei   Toronto

McGraw-Hill Higher Education 🜨

*A Division of The **McGraw-Hill** Companies*

AN INTRODUCTION TO THE WORLD'S OCEANS, SEVENTH EDITION

Published by McGraw-Hill, a business unit of The McGraw-Hill Companies, Inc., 1221 Avenue of the Americas, New York, NY 10020. Copyright © 2003, 2000, 1997 by The McGraw-Hill Companies, Inc. All rights reserved. No part of this publication may be reproduced or distributed in any form or by any means, or stored in a database or retrieval system, without the prior written consent of The McGraw-Hill Companies, Inc., including, but not limited to, in any network or other electronic storage or transmission, or broadcast for distance learning.

Some ancillaries, including electronic and print components, may not be available to customers outside the United States.

♻ This book is printed on recycled, acid-free paper containing 10% postconsumer waste.

International   1 2 3 4 5 6 7 8 9 0 VNH/VNH 0 9 8 7 6 5 4 3 2
Domestic   1 2 3 4 5 6 7 8 9 0 VNH/VNH 0 9 8 7 6 5 4 3 2

ISBN 0-07-247280-4
ISBN 0-07-119489-4 (ISE)

Publisher: *Margaret J. Kemp*
Sponsoring editor: *Thomas C. Lyon*
Senior developmental editor: *Donna Nemmers*
Executive marketing manager: *Lisa Gottschalk*
Project manager: *Sheila M. Frank*
Senior production supervisor: *Sandy Ludovissy*
Senior media project manager: *Tammy Juran*
Media technology producer: *Judi David*
Coordinator of freelance design: *Michelle D. Whitaker*
Cover designer: *Maureen McCutcheon*
Cover image: *Yngve Rakke/Getty Images*
Lead photo research coordinator: *Carrie K. Burger*
Photo research: *Karen Pugliano*
Supplement producer: *Brenda A. Ernzen*
Compositor: *Shepherd, Inc.*
Typeface: *10.5/12 Goudy*
Printer: *Von Hoffmann Press, Inc.*

The credits section for this book begins on page 497 and is considered an extension of the copyright page.

Library of Congress Cataloging-in-Publication Data

Sverdrup, Keith A.
    An introduction to the world's oceans.—7th ed. / Keith A. Sverdrup, Alyn C. Duxbury, Alison B. Duxbury.
        p.   cm.
    Rev. ed. of: An introduction to the world's oceans. 6th ed. / Alyn C. Duxbury, Alison B. Duxbury, Keith A. Sverdrup. ©2000.
    Includes bibliographical references and index.
    ISBN 0–07–247280–4 (acid-free paper)
    1. Oceanography. I. Duxbury, Alyn C., 1932– . II. Duxbury, Alison. III. Duxbury, Alyn C., 1932– . Introduction to the world's oceans. IV. Title.

GC11.2 .D89   2003
551.46—dc21   2002067854
                                                             CIP

INTERNATIONAL EDITION ISBN 0–07–119489–4
Copyright © 2003. Exclusive rights by The McGraw-Hill Companies, Inc., for manufacture and export. This book cannot be re-exported from the country to which it is sold by McGraw-Hill. The International Edition is not available in North America.

www.mhhe.com

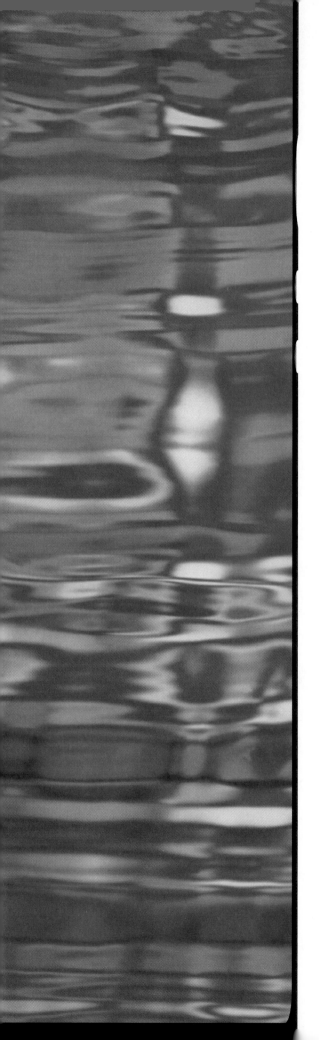

Dedicated to
George and Jean Sverdrup
and
Andrew, Jeannie,
and Alec Duxbury

brief contents

contents

preface

A Note to Students

Human beings have been curious about the oceans since they first walked along their shores. As people have learned more about the oceans, they have come to more fully understand and appreciate the tremendous influence these bodies of salt water have on our lives. The oceans cover over 70% of Earth's surface, creating a habitat for thousands of known species and countless others still to be discovered. The sea contains vast quantities of diverse natural resources in the water and on the sea floor; some are actively exploited today, and many more may be recovered in the future with improved technology and greater demand. Global climate and weather are strongly influenced by the oceans as they interact with the atmosphere through the transfer of moisture and heat energy. The ocean basins also serve as the location of great geological processes and features such as earthquakes, volcanoes, massive mountain ranges, and deep trenches, all of which are related to the creation and destruction of sea floor in the process of plate tectonics.

Much of what happens in the oceans and on the sea floor is hidden from direct observation. Although the *Hubble Space Telescope* can form images made from light that has traveled over 10 billion trillion kilometers, we can not see more than a few tens of meters below the ocean's surface even under the most favorable conditions because of the efficient scattering and absorption of light by seawater. Consequently, most of what we know about the oceans comes from indirect, or remote, methods of observation. With constantly improving technology and innovative applications of that technology, we continue to learn more about the geological, physical, chemical, and biological characteristics of the oceans.

Although careful scientific study of the oceans is often difficult and challenging, it is both necessary and rewarding. Our lives are so intimately tied to the oceans that we benefit from each new fact that we discover. Continued research and a better understanding of the oceans become increasingly important as the population of this planet grows ever larger. Early in the new millennium, there is both good news and bad news concerning global population growth. The rate of population increase has slowed with falling birth rates, and there is some indication that the human population will level off by the end of this century. But even if the human population does stabilize, it will not do so before there is an increase of several billion people over today's population. We clearly will continue to face difficult environmental decisions affecting the oceans as well as the land in the foreseeable future. Our best chance of dealing wisely and effectively with these challenges is to promote more widespread understanding of the oceans.

Although it is critical that we continue to train marine scientists to study the oceans, it is no less important for people in all walks of life to develop a basic understanding of how the oceans influence our lives and how our actions influence the oceans. In studying oceanography, you are preparing yourself to be an informed global citizen. It is likely that at some point in the future you will have the opportunity to voice your concern about the health of the oceans, either directly or through the governmental process. Your interest in and study of oceanography will help you participate in future discussions and decision-making processes in an informed manner.

The Online Learning Center at www.mhhe.com/sverdrup provides you with links to Internet addresses applicable to this text. To expand your knowledge of oceanography, there are Internet exercises for many of these sites. The exercises are found within the Online Learning Center.

A Note to Instructors

A major objective of this text is to stimulate student interest and curiosity by blending contemporary information and research with basic principles in order to present an integrated introduction to the many and varied sciences used in the study of the oceans. To do so, we have extensively reviewed and rewritten material from the sixth edition to produce this new seventh edition. In the face of constant and rapid change, we have added new material for both content and interest.

We realize that the students who will use this book come from diverse backgrounds and that for many of them this will be an elective course. The content continues to be reasonably rigorous, but we have chosen to use simple algebra rather than advanced mathematics. For instance, we use centrifugal force to explain tidal principles because most students do not have much background in vectors. More emphasis is placed on the physical and geological aspects of oceanography than on the chemical and geochemical because the latter two disciplines require more background knowledge.

An ecological approach and descriptive material are used to integrate the biological chapters with the other subject fields. We strive to emphasize oceanography as a cohesive and united whole rather than a collection of subjects gathered under a marine umbrella.

In order to understand the constant barrage of information concerning our planet and marine issues, students must have a basic command of the language of marine science in addition to mastering processes and principles. For this reason we maintain an emphasis on critical vocabulary. All terms are defined in the text; terms that are particularly important are printed in boldface. A list of important terms is at the end of each chapter, and a glossary is included at the end of the book. The Online Learning Center for this text also hosts flashcards of key terms for student study.

Summaries at the ends of chapters provide quick reviews of key concepts. Study Problems are included in many chapters, and Study Questions are at the end of each chapter. The Study Questions are not intended merely for review, but also to challenge students to think further about the lessons of the chapter.

This book may be used in a one-quarter or one-semester course. Because the experience and emphasis of faculty using this book will differ, it is expected that each instructor will emphasize and elaborate on some topic at the expense of other topics. We continue to make each chapter stand as independently as possible and encourage instructors to use the chapters in the order that best suits their purposes. Cross-references from one chapter to another indicate discussion of topics elsewhere in the text. Faculty wishing to expand quantitative material are encouraged to make use of Appendix C, Equations and Quantitative Relationships. The answers to the Study Questions and Study Problems from the text appear in the Instructor's Manual, in the password-protected instructor's area of the Online Learning Center.

Changes to the Seventh Edition

In addition to revisions and updates based on current research, this edition contains several new boxed readings.

The Prologue contains a significant amount of new material on the history of oceanography, as well as an updated section on oceanography of the recent past, present, and future. In Chapter 1 we have added material on the origin of the oceans. The discussions of the Earth's internal structure and the characteristics of plates and their boundaries have been extensively rewritten and many new figures have been added. Chapter 3's section describing the characteristics of marine sediments has been completely revised and a new section on gas hydrates has been added. Chapter 4 contains a new section on temperature and heat before the discussion of changes in the state of water. Chapter 5 now begins with a new section explaining the pH scale and the pH of seawater in order to form a foundation for the discussion of seawater chemistry. Both chapters 6 and 7 have been renamed and reorganized. Chapter 6 is now named "The Structure and Motion of the Atmosphere." It focuses exclusively on atmospheric processes critical to understanding the links between the atmosphere and the oceans. Chapter 7 is now entitled "Circulation and Ocean Structure." This chapter includes a revised discussion of ocean density structure and covers topics in vertical circulation, upwelling and downwelling, the layering of the oceans, and updated material on sampling methods and measurement techniques. Chapter 7 also contains two new special interest boxes covering topics including Arctic Ocean studies (with a discussion of the North Pole Environmental Observatory) and Ocean Gliders (independent, unmanned vehicles). The material in Chapter 8 has been reordered and sections on convergence and divergence have been rewritten to enhance clarity. Chapter 9's discussion of Energy from Waves and the boxed reading on Tsunami Warning Systems have both been updated. The tide and current tables have been updated in Chapter 10. The discussions of beach dynamics and beach types in Chapter 11 have been reordered for clarity and the material previously in chapter 12 on modifying beaches has been moved to chapter 11. In Chapter 12 the discussion of the Gulf of Mexico dead zone has been updated

and rewritten. Chapter 13 has a new name, "The Living Ocean"; the chapter has been completely revised and rewritten. More emphasis is given to ocean biology and there are new sections on ocean biology, groups of organisms and environmental zones. Symbiotic relationships are now included. Chapter 14 has new information on the roles of nitrogen and iron, and a new boxed reading on CalCOFI—Fifty Years of Coastal Ocean Data. Chapter 15 contains updated material on the krill harvest, more information on the microbial loop, and the section on harmful algal blooms has been revised and updated. There is revised and expanded information on skates and rays, marine mammal populations, and bathypelagic organisms in Chapter 16. In addition, all fisheries data has been updated. Chapter 17 has new and revised information on corals including bleaching, predation, disease, and the effect of human activities. New information on Lost City vents is included in Chapter 17 and all harvesting data has been updated. In addition, Chapter 17 hosts a new boxed reading on ice worms living on deep-sea ice hydrates.

Instructor Supplements

McGraw-Hill offers a variety of supplements to assist the instructor with both preparation and classroom presentation. Materials that are specific to this text and that are available to the instructor include a text-specific **Online Learning Center (OLC),** which contains many useful tools for the instructor. The OLC for this text can be found at www.mhhe.com/sverdrup. Within the password-protected Online Learning Center instructors will find the **Instructor's Manual,** which includes answers to the Study Questions and Study Problems from the text, as well as suggested test questions and answers for each chapter of the text. Also available through the Online Learning Center and on CD-ROM, instructors will have access to the **Digital Content Manager (DCM).** This multimedia collection of visual resources allows instructors to utilize artwork from the text in multiple

formats to create customized classroom presentations, visually-based tests and quizzes, dynamic course website content, or attractive printed support materials. The digital assets on this cross-platform CD-ROM are grouped by chapter within easy-to-use folders. On the CD-ROM version of the Digital Content Manager, instructors will have access to 34 **video segments from Scripps Institution of Oceanography,** grouped by chapter to go with the seventh edition of this text. The Scripps video segments are also available on videotape.

PowerWeb: Oceanography can be accessed through the Online Learning Center, and contains articles from current magazines, newspapers, and journals; weekly updates of current issues; web research tips; an online library of updated research links to help you find the right information; up-to-the-minute headlines from around the world including course-specific and general news; and online quizzing and assessments for your students.

A **Brownstone testing system** is available on a cross-platform CD-ROM to generate tests; and a set of **overhead transparencies** contains over 200 figures from the text.

For those schools offering a laboratory with their oceanography course, McGraw-Hill offers an exceptional lab manual entitled "Investigating the Ocean" by R. Leckie and R. Yuretich of University of Massachusetts-Amherst. Additional earth science supplements offered by McGraw-Hill which are appropriate for this course include the *Journey Through Geology* CD-ROM by the Smithsonian Institution; the JLM Visuals Physical Geology Photo CD; and a geoscience videotape library.

Contact your McGraw-Hill sales representative for details on these products.

Student Supplements

The Internet makes oceanographic information and data available to researchers and it also provides images and information in many forms to instructors and students. Public agencies and museums, universities and research laboratories, satellites and oceanographic projects, interest groups and individuals all over the planet provide information that can be publicly accessed. The text-specific **Online Learning Center (OLC)** website, which can be found at www.mhhe.com/sverdrup, provides chapter-sorted links to many websites that contain information pertinent to each chapter's content. In addition, web links are provided within the OLC for further information on many boxed readings and figures within each chapter. Wherever you see this icon web link in your textbook, you will find associated web links for the indicated figure or boxed reading on the OLC. The OLC also hosts a complete Student Study Guide, chapter quizzing, key term flashcards, and Internet exercises to help with chapter study. In addition, **PowerWeb** is a great way to get information you need quickly and easily! Through the OLC, students can access PowerWeb: Oceanography, which contains articles from current magazines, newspapers, and journals; weekly updates of current issues; web research tips; an online library of updated research links to help you find the right information; up-to-the-minute headlines from around the world including course-specific and general news; online quizzing and assessments to measure your understanding of course material, and more!

Acknowledgments

As a book is the product of many experiences, it is also the product of people other than the authors. We extend many thanks to our friends and colleagues who have graciously answered our questions and provided us with information and access to their photo files. We owe very special thanks to faculty and staff of the School of Oceanography, College of Ocean and Fishery Sciences, University of Washington, and to the scientists and staff of the National Oceanic and Atmospheric Administration's Northwest Regional Office, who have answered questions, supplied data and provided many of the illustrations in this edition. We are also grateful to Scripps Institution of Oceanography, who have allowed us the privilege of providing their videotape series as an instructor ancillary to this seventh edition of the text.

We would also like to acknowledge the valuable contributions of the following people:

Marcia McNutt
Monterey Bay Aquarium Research Institute

Charlie Eriksen
School of Oceanography, University of Washington

Suzan Huney
Polar Science Center, University of Washington

Charles Fisher and Erin McMullin
Mueller Laboratory, Pennsylvania State University

Steve Nathan and R. Mark Leckie
University of Massachusetts, Amherst

Reviewers for the Seventh Edition

Claude D. Baker
Indiana University Southeast

Douglas Biggs
Texas A&M University

Wayne V. Bloechl
Cabrillo College

Gregg Brooks
Eckerd College

Hans G. Dam
University of Connecticut

Cynthia Domack
Hamilton College

Nancy Glass
Baldwin Wallace College

Lawrence Krissek
Ohio State University

James McWhorter
Miami-Dade Community College

Keith H. Meldahl
MiraCosta College

A. Conrad Neumann
University of North Carolina at Chapel Hill

Dan E. Olson
Edgewood College

Mike Phillips
Illinois Valley Community College

Reed Scherer
Northern Illinois University

Guatam Sen
Florida International University

Kent M. Syverson
University of Wisconsin at Eau Claire

Margaret Vose
University of Southern Maine

We thank all members of the team at McGraw-Hill, without whose help, enthusiasm, and coordinated efforts this seventh edition could not have been completed.

A variety of tools within this textbook have been designed to assist with chapter review and critical analysis of chapter topics.

KEY TERMS

Key Terms are all boldfaced and defined within the text, and end-of-chapter key terms listings indicate the most important terms and their locations within each chapter.

farming, including methods for building the pond, selecting the stock, and harvesting it. In China, Southeast Asia, and Japan, fish farming in fresh and salt water has continued to the present as a practical and productive method of raising large quantities of fish such as carp, milkfish, *Tilapia*, and catfish. Most of these farms are family-run, labor-intensive operations. Fish farmers may practice **monoculture,** raising fish as a single species, or **polyculture,** raising more than one species as surface feeders and bottom feeders to make use of the total volume of the pond.

In the United States fish farming produces only 2% of fishery products, but that amount includes 50% of the catfish and nearly all the trout. To be successful and profitable in the United States, fish farming requires a proven market and large-scale operation combined with the development of technology to reduce costs. The species chosen for farming must reproduce easily in captivity, have juvenile forms that survive well under controlled conditions,

After the area has been restructured it must be monitored, and changes must be made as needed to maintain the desired environment. If these preproject and postproject studies are not made and followed by necessary corrections, the mitigation effort will not succeed. Mitigation may preserve some habitats and species, but it only approximates a lost environment; it cannot duplicate it.

Pressures to develop deep-sea areas for energy and mining projects are likely to increase in the future. Tropical OTEC plants (ocean thermal energy conversion; see chapter 7) will bring cold, nutrient-rich water to the sea surface, changing the surface productivity and potentially interfering with the survival of tropical species. Manganese nodule mining (see chapter 3) will increase deep-sea turbidity, changing the environment for bottom organisms in mined areas. How mitigation might be undertaken in the open ocean has not yet been considered, but the threat of deep-sea exploitation is already causing some scientists to call for the setting aside of marine reserves in the open ocean.

Summary

Marine ecology is the study of interactions between marine organisms and their environment. The organisms of the sea are classified for identification and relationship. Marine organisms are commonly divided into plankton, nekton, and benthos. All organisms are classified for identification and relationship into kingdoms and a new system that includes domains.

The marine environment is subdivided into zones. The major environments are the benthic and pelagic zones; there are numerous subdivisions of each of these zones.

Organisms living in the sea are buoyed and supported by the seawater. Adaptations for staying afloat include low-density body fluids, gas bubbles, gas-filled floats, swim bladders, oil and fat storage, and extended surface areas and appendages.

Most marine fish lose water by osmosis. They drink continually and excrete salt to prevent dehydration. Sharks have the same concentration of salt in their tissues as there is in seawater. They therefore do not have a water-loss problem. Salinity is a barrier to some organisms; others can adapt to large salinity changes.

Temperature affects density, viscosity, and the water's buoyancy, as well as the stability of the water column. The body temperature and metabolism of all marine organisms, except for birds and mammals, are controlled by the sea temperature. Some fish conserve heat in their body muscles and elevate the temperature in these muscles. Temperatures at depth are uniform; sea surface temperatures change with latitude and seasons.

Changes in pressure affect organisms that have gas-filled cavities. Marine mammals have a unique ability to undergo large pressure changes due to their physiology and body chemistry. The swim bladders of fish are affected by pressure changes, and the fish must change depth slowly.

The carbon dioxide—oxygen balance in the oceans influences the distribution of all organisms. The availability of

nutrients and light limits plant populations. The depth of light penetration in the oceans is controlled by the angle of the Sun's rays, the properties of the water, and the material in the water. Light limits plant life, but nutrients are also required. Some organisms produce chemical light, known as bioluminescence. Animals use color for concealment and camouflage and also to warn predators of poisonous flesh and bitter taste.

Winds, tides, and currents mix the water. Moving water carries food and oxygen, removes waste, and disperses organisms. Floating populations are not necessarily scattered but keep their place due to movements between surface currents and deeper currents. Upwellings supply nutrients and hold plants in the surface layers. Downwellings are regions of low plant growth.

Different substrates provide food, shelter, and attachment for different groups of organisms. Some live attached to the surface, some move across it, some burrow into the sea floor, and some modify the sea floor, providing habitats for others. Some animals are carnivores; others are herbivores. Large numbers of ocean organisms live in close associations such as commensalism, mutualism, and parasitism. Barriers for marine organisms include water properties, light intensity, zones of convergence and divergence, seafloor topography, and geography.

Development of marine areas, with the consequent loss of habitat and therefore populations, has led to the concept of mitigation. Mitigation requires developers to preserve or replace habitats in an effort to maintain and preserve species.

Key Terms

All key terms from this chapter can be viewed by term, or by definition, when studied as flashcards on this book's Online Learning Center at www.mhhe.com/sverdrup (click on this book's cover).

CHAPTER SUMMARY

The Summary in each chapter provides a quick review of key concepts.

Study Questions

1. The processes that form a stretch of flat, uniform coast depend on whether the land is rising or sinking. Discuss the processes that form a flat coastal area under these two conditions.
2. Are multiple berms more likely to be seen on a beach between March and August than between September and February?
3. Fjord coasts and drowned river valleys, or ria coasts, are primary coasts. Explain why their appearances are distinctly different. Give examples of each.
4. Describe the processes required to create a tombolo. Consider and discuss (a) the distribution of wave energy and (b) the longshore transport.
5. What conditions are required to maintain a beach with a constant profile and composition? Consider both a static and a dynamic environment.
6. Why does an eroding beach that is supplied with material from the cliffs of an old glacier deposit become an armored beach, while an eroding beach that is supplied by river sediments does not?
7. How does the profile of a sand beach change during alternating seasonal periods of storm waves and more gentle small waves?
8. Sketch a beach section that is stable with respect to the supply and removal of beach sediments. Indicate the source of the sediments and the final deposition of sediments in the system. If the longshore transport or wave energy distribution is altered by a barrier perpendicular to the beach, what changes will occur?
9. What is causing the apparent rise of sea level around the world? Is the rise similar in all places?
10. What processes move beach sediments in a coastal circulation cell?
11. Compare the circulation of a semienclosed basin at 30°N with that of an estuary located at 60°N. Which is less likely to accumulate waste products at depth? Why?
12. Estuaries are classified by their net circulation and salt distribution in this chapter. What other features could be used to describe and classify them?
13. Why are estuaries with short flushing times less apt to degrade when used as the receiving water for urban runoff than estuaries with long flushing times?
14. Compare the circulations and histories of San Francisco Bay and Chesapeake Bay. How do these estuaries differ; how are they similar? What do you see in the future for each?
15. In order to prevent erosion of his beach, your neighbor wants to build a groin. What effect might the groin have on your neighbor's beach? On your beach if you live on the upstream side of the longshore current? On your beach if you live on the downstream side of the longshore current?

Study Problems

1. Determine the flushing time of an estuary in which $T_o = 9 \times 10^7$ m³/day and the volume of water is 30×10^8 m³.
2. An estuary has a volume of 50×10^9 m³. This water is 5% fresh water, and the fresh water is added at the rate of 6×10^7 m³/day. Why does the rate of addition of fresh water from the estuary to the ocean equal the input of fresh water from the land to the estuary? Consider the average salinity of the estuary as constant. What is the residence time of the fresh water?
3. Water entering an estuary at depth has a salinity of 34.5‰. The water leaving the estuary has a salinity of 29‰. The river inflow is 20×10^5 m³/day. Calculate T_o, the seaward transport.
4. A bay with no freshwater input can flush only by tidal exchange. If on each tidal cycle (ebb and flood) 10% of the bay's water volume is exchanged with ocean water, how much of the original water will still be in the bay after four tidal cycles? Solve for (1) no mixing between bay water and ocean water; (2) complete mixing between bay water and ocean water.
5. What is the flushing time of the bay in problem 4 for each case? Which case best duplicates natural conditions?

Links to Related Websites

Visit the book's Online Learning Center at www.mhhe.com/sverdrup (click on the book's cover) to find live Internet links for additional topics related to this chapter's content.

• Coastal and marine geology
• Coastal management

Visit the book's Online Learning Center at www.mhhe.com/sverdrup (click on this book's cover) to find these additional chapter tools: Suggested Readings; links to further information on boxed readings, selected figures, and related chapter topics; and additional study aids.

STUDY QUESTIONS AND PROBLEMS

Study Problems and Study Questions serve not only as a concept review, but challenge students to think further about the lessons within each chapter.

LINKS TO RELATED WEBSITES

Find Internet links to each chapter's content, boxed readings, and figures within the Online Learning Center for this text at www.mhhe.com/sverdrup. Watch for the special web link icon to indicate Internet-linked material.

The Sea Bear

The sea and sea ice are essential components of the polar bear's environment. The polar bear is the top predator of the Arctic's marine food chains, and its scientific name is appropriate, *Ursus maritimus*, or "sea bear" (box fig. 1). Polar bears are long-lived (fifteen to twenty years or more), late-maturing (four to five years) carnivores with dense fur and a blubber layer for insulation. Adult males weigh 250–800 kg (550–1700 lb) and measure 2.5–3.0 m (8–10 ft) from tip of nose to tail; females weigh 100–300 kg (200–700 lb) and measure 1.8–2.5 m (6–8 ft) see (box fig. 2). In late October and November females make their dens in drifted snow. One or two cubs are born in December or January, and mother and cubs emerge in late March and early April. Most female bears keep their cubs with them until the cubs are about two and a half years old. The bears feed primarily on ringed seals and to a lesser extent on other seal species and whale and walrus carcasses. They use their keen sense of smell to locate seals and hunt by stalking the animals basking on the ice, lying in wait at breathing holes, and breaking into their birth-dens. Polar bears travel long distances over the ice, 30 km (19 mi) or more each day; they are also strong swimmers, swimming continuously for 100 km (62 mi).

web link **Box Figure 1** A polar bear on the ice in the central Arctic Ocean.

Polar bear distribution is circumpolar but not continuous. Populations that return to feeding and denning areas are kept apart by the sea ice, ice movements, and land barriers. The bears move extensively between seasons, depending on regional patterns of freezing and breakup of the sea ice. Polar bears depend on sea ice as a platform from which to hunt and feed and as a base on which to seek mates, breed, and travel long distances. About 68% of the polar ice region appears to be suitable habitat for polar bears.

The world's total population of polar bears is estimated at between 25,000 and 40,000. The Canadian population is about 15,000, or roughly half of the world's polar bears. Polar bears are also found in Denmark (Greenland), Norway (Svalbard Islands), and Russia. There are two Alaska polar bear populations, estimated at 3000 animals; one population is in the vicinity of the Chukchi and Bering Seas and the other is in the Beaufort Sea area.

In the 1950s and 1960s there was growing international concern for the polar bear populations as the number of bears being killed for their hides increased. Russia prohibited all hunting of polar bears in 1956, but poaching increased dramatically. The U.S. Marine Mammal Protection Act of 1972 prohibited the killing of all marine mammals including polar bears, except by native people for subsistence purposes. Norway prohibited all polar bear hunting in 1973, and Canada increased regulations on native peoples' hunting and allowed a limited sport hunt of 500 bears per year. Greenland harvests 100–150 bears annually in a tightly regulated hunt for residents only.

In October 2000 the United States and Russia signed a treaty covering the polar bears in northwest Alaska and the Chukota region of northeast Siberia; the treaty includes the islands of the Chukchi and Bering Seas. New U.S. and Russian commissions were formed in which native tribes participate in setting harvest quotas. For the first time quotas were placed on the number of bears hunted in both countries. Den areas were ruled off-limits; commercial hunting and killing females and cubs under one year are prohibited. No poison, traps, snares, aircraft, or large motorized vehicles and vessels are permitted in the hunt.

The polar bears still face an uncertain future. If climate change increases ice-free periods, the bears cannot return to the ice to hunt and mate. Other potential problems include toxic contaminants, poaching, disruption of the food chain by overfishing, human

activities related to oil and gas exploration and development, and increased tourism and human contacts.

To Learn More About the Sea Bear
Eliot, J. 1998. Polar Bears, Stalkers of the High Arctic. *National Geographic* 193(1): 52–71.

Internet References
Visit the book's Online Learning Center at www.mhhe.com/sverdrup (click on the book's cover) to explore links to further information on related topics.

Box Figure 2 This polar bear is being weighed, measured, and fitted with a tracking collar that will be followed by satellite. A measuring tape is lying on top of the bear. The scientists are members of the 1994 Arctic Section Expedition.

---

surface for edible refuse, including scraps from ships and fishing boats. Its primary foods are squid, crab, and surface fish. The smallest of the oceanic birds, Wilson's petrel, is a swallowlike bird that breeds in Antarctica and flies 16,000 km (10,000 mi) along the Gulf Stream to Labrador during the Southern Hemisphere winter, returning to the Southern Hemisphere for the southern summer, another 16,000 km.

Penguins (fig. 16.9a) do not fly but swim in flocks, using their wings for propulsion and steering with their feet. Their underwater swimming speed is almost 10 mph. They feed on fish, krill, squid, and shellfish. All but the Galápagos penguin are found in the Southern Hemisphere, and two species, the emperor and the Adelie, are found in Antarctica.

Pelicans and cormorants are large fishing birds with big beaks. They are strong fliers found mostly in coastal areas, but some venture far out to sea. A pelican (fig. 16.9b) has a particularly large beak from which hangs a pouch used in catching fish. White pelicans of North America fish in groups, herding small schools of fish into shallow water and then scooping them up in their large pouches. The Pacific's

brown pelican does a spectacular dive from up to 10 m (30 ft) above the water to capture its prey. Cormorants are black, long-bodied birds with snakelike necks and moderately long bills that are hooked at the tip. They settle on the water and dive from the surface, swimming primarily with their feet but also using their wings. Because they do not have the water-repellent feathers of other seabirds, cormorants must return to land periodically to dry out.

Terns (fig. 16.9c) and gulls are found all over the world except in the South Pacific between South America and Australia. Gulls are strong flyers and will eat anything; they forage over beach and open water. The terns are small, graceful birds with slender bills and forked tails. The Arctic tern breeds in the Arctic and each winter migrates south of the Antarctic Circle, a round-trip of 35,000 km (20,000 mi).

Puffins, murres, and auks are heavy-bodied, short-winged, short-legged diving birds. They feed on fish, crustaceans, squid, and some krill. All are limited to the North Atlantic, North Pacific, and Arctic areas, where most nest on isolated cliffs and islands. The great auk was a large, slow,

www.mhhe.com/sverdrup

This text-specific website hosts many useful tools for the instructor, as well as features to help students sharpen their study skills and make the grade in their oceanography course. A complete **Instructor's Manual,** as well as the **Digital Content Manager,** can be accessed within a special password-protected instructors' section of the Online Learning Center (OLC). The Digital Content Manager provides photos and artwork from the text in multiple formats to assist in creating customized classroom presentations, dynamic course website content, or attractive printed support materials. Students will find electronic flashcards of the key terms within each chapter, as well as chapter-grouped web links to provide additional information on chapter topics, individual figures, boxed readings, and interactive quizzes for each chapter.

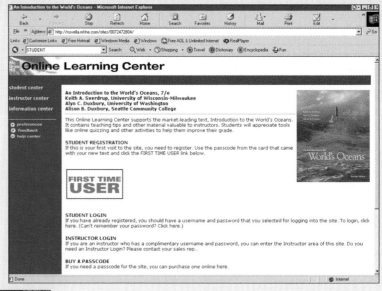

Students: Partner Up with Our Online Study Guide

The Online Learning Center hosts a complete **Study Guide** for each chapter, which includes Chapter Objectives, Key Concepts in lecture outline format, and a variety of self-test features to help you understand and retain the content within each chapter of your textbook.

Charge Up Your Oceanography Research with PowerWeb

Through our Oceanography Online Learning Center you may access **PowerWeb**—your personal research assistant. PowerWeb provides articles from current magazines, newspapers, and journals; weekly updates of current issues; web research tips; an online library of updated research links to help you locate the right information; up-to-the-minute headlines from around the world; plus online quizzing and assessments to measure your understanding of course material.

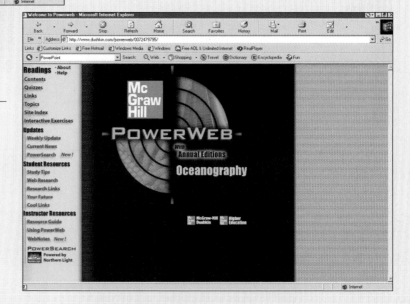

The History of Oceanography

Where are your monuments, your battles, martyrs?
Where is your tribal memory? Sirs, in that grey
vault. The sea. The sea has locked them up. The sea is History.

Derek Walcott
From *The Sea is History*

Navigational aides before electronics: sextant, pen, log book, and
map of Falmouth, England.

O ceanography is a broad field in which many sciences are focused on the common goal of understanding the oceans. Geology, geography, geophysics, physics, chemistry, geochemistry, mathematics, meteorology, botany, and zoology have all played roles in expanding our knowledge of the oceans. The field is so broad that oceanography today is usually broken down into a number of subdisciplines.

Geological oceanography includes the study of the Earth at the sea's edge and below its surface and the history of the processes that formed the ocean basins. Physical oceanography investigates how and why the oceans move; marine meteorology (the study of heat transfer, water cycles, and air-sea interactions) is often included in this discipline. Chemical oceanography studies the composition and history of the water, its processes, and its interactions. Biological oceanography concerns marine organisms and the relationship between these organisms and the environment of the oceans. Ocean engineering is the discipline that designs and plans equipment and installations for use at sea.

The study of the oceans was promoted by intellectual and social forces as well as by our needs for marine resources, trade and commerce, and national security. Oceanography started slowly and informally; it began to develop as a modern science in the mid-1800s and has grown dramatically, even explosively, in the last few decades. Our progress toward the goal of understanding the oceans has been uneven, and it has frequently changed direction. The interests and needs of nations as well as the scholarly curiosity of scientists have controlled the rate at which we study the oceans, the methods we use to study them, and the priority we give to certain areas of study. To gain perspective on the current state of knowledge about the oceans, we need to know something about the events and incentives that guided people's previous investigations of the oceans.

The Early Times

People have been gathering information about the oceans for millennia, accumulating bits and pieces of knowledge, passing it on by word of mouth. Curious individuals must have acquired their first ideas of the oceans from wandering the seashore, wading in the shallows, gathering food from the ocean's edges. During the Paleolithic period humans developed the barbed spear, or harpoon, and the gorge. The gorge was a double-pointed stick inserted into a bait and attached to a string. At the beginning of the Neolithic period the bone fishhook was developed and later the net. By 5000 B.C. copper fishhooks were in use (see traditional Native American implements in prologue fig. I).

As early humans moved slowly away from their inland centers of development, they were prepared to take advantage of the sea's food sources when they first explored and later settled along the ocean shore. The remains of shells and other refuse, in piles known as kitchen middens, have been found at the sites of ancient shore settlements. These remains show that our early ancestors gathered shellfish, and fish bones found in some middens suggest that they also began to use rafts or some type of boat for offshore fishing. Drawings on ancient temple walls show fishnets; on the tomb of the Egyptian Pharaoh Ti, Fifth Dynasty (5000 years ago), is a drawing of the poisonous pufferfish with a hieroglyphic description and warning. As long ago as 1200 B.C. or earlier, dried fish were traded in the Persian Gulf; in the Mediterranean the ancient Greeks caught, preserved, and traded fish, while the Phoenicians founded fishing settlements, such as "the fisher's town" Sidon, that grew into important trading ports.

Early information about the oceans was mainly collected by explorers and traders. These voyages left little in the way of recorded information. Using descriptions passed down from one voyager to another, early sailors piloted their way from one landmark to another, sailing close to shore and often bringing their boats up onto the beach each night.

Some historians believe that seagoing ships of all kinds are derived from early Egyptian vessels. The first recorded voyage by sea was led by Pharaoh Snefru about 3200 B.C. In 2750 B.C. Hannu led the earliest documented exploring expe-

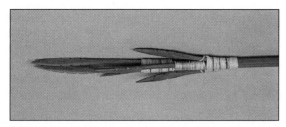

(a)

(b)

(c)

Prologue I Traditional fishing and hunting implements from coastal Native American cultures of the Pacific Northwest. (a) A duck spear made of cedar. (b) A bone harpoon point and lanyard made of sinew, hemp, and twine. (c) A fishhook made of bone and steam-bent cedar root. *Courtesy Thomas Burke Memorial Washington State Museum, catalog nos. BWSM 1-11433, 1-794, 1-158.*

web (link) **Prologue II** A traditional navigator from the Caroline Islands instructing others in the use of a "star structure" showing the relative points on the horizon where key stars rise and set.

dition from Egypt to the southern edge of the Arabian Peninsula and the Red Sea.

The Phoenicians, who lived in present-day Lebanon from about 1200 B.C. to 146 B.C., were well known as excellent sailors and navigators. While their land was fertile it was also densely populated, so they were compelled to engage in trade with others to acquire many of the goods they needed. They accomplished this by establishing land routes to the East and marine routes to the West. The Phoenicians were the only nation in the region at that time that had a navy. They traded throughout the Mediterranean Sea with the inhabitants of North Africa, Italy, Greece, France, and Spain. They also ventured out of the Mediterranean Sea to travel north along the coast of Europe to the British Isles and south to circumnavigate Africa in about 590 B.C. In 1999 the wreckage of two Phoenician cargo vessels circa 750 B.C. was explored using remotely operated vehicles (ROVs) that could dive to the

wreckage and send back live video images of the ships. The ships were discovered about 48 km (30 mi) off the coast of Israel at depths of 300–900 m (roughly 1000–3000 ft).

Extensive migration throughout the Southwestern Pacific may have begun by 2500 B.C. These early voyages were relatively easy because of the comparatively short distance between islands in the far Southwestern Pacific region. By 1500 B.C. the Polynesians had begun more extensive voyages to the east, where the distance between islands grew from tens of miles at the edge of the western Pacific to thousands of miles in the case of voyages to the Hawaiian Islands. They successfully reached and colonized the Hawaiian Islands sometime between A.D. 450 and 600. By the eighth century A.D., they had colonized every habitable island in a triangular region roughly twice the size of the United States bound by Hawaii on the north, New Zealand in the southwest, and Easter Island to the east.

A basic component of navigation throughout the Pacific was the careful observation and recording of where prominent stars rise and set along the horizon. Observed near the equator, the stars appear to rotate from east to west around the Earth on a north-south axis. Some rise and set farther to the north and some farther to the south, and they do so at different times. Navigators created a "star structure" by dividing the horizon into thirty-two points where the stars for which the points are named rise and set. These points form a compass and provide a reference for recording information about the direction of winds, currents, waves, and the relative positions of islands, shoals, and reefs (prologue fig. II). The Polynesians also navigated by making close observations of waves and cloud formations. Observations of birds and the distinctive smells of land such as flowers and wood smoke alerted them to possible landfalls. Once islands were discovered, their locations relative to one another and to the regular patterns of sea swell and waves bent around islands could be recorded with stick charts constructed of bamboo and shells (prologue fig. III).

Prologue III A navigational chart (*rebillib*) of the Marshall Islands. Sticks represent a series of regular wave patterns (swells). Curved sticks show waves bent by the shorelines of individual islands. Islands are represented by shells.

As early as 1500 B.C. Arabs of many different ethnic groups and regions were exploring the Indian Ocean. In the seventh century A.D. they were unified under Islam and began to control the trade routes to India and China and consequently the commerce in silk, spices, and other valuable goods (this monopoly wasn't broken until Vasco da Gama defeated the Arab fleet in 1502).

The Greeks called the Mediterranean "Thalassa" and believed that it was encompassed by land, which in turn was surrounded by the endlessly circling river Oceanus. In 325 B.C. Alexander the Great reached the deserts of the Mekran Coast, now a part of Pakistan. He sent his fleet down the coast in an apparent effort to probe the mystery of Oceanus. He and his troops had expected to find a dark, fearsome sea of whirlpools and water spouts inhabited by monsters and demons; they did find tides that were unknown to them in the Mediterranean Sea. His commander, Nearchus, took the first Greek ships into the ocean, explored the coast, and brought them safely to the port of Hormuz eighty days later. Pytheas (350–300 B.C.), a navigator, geographer, astronomer, and contemporary of Alexander, made one of the earliest recorded voyages from the Mediterranean to England. From there he sailed north to Scotland, Norway, and Germany. He navigated by the Sun, stars, and wind, although he may have had some form of sailing directions. He recognized a relationship between the tides and the Moon and made early attempts at determining latitude and longitude. These early sailors did not investigate the oceans; for them the oceans were only a dangerous road, a pathway from here to there, a situation that continued for hundreds of years. However, the information that they accumulated began to build into a body of lore to which sailors and voyagers added from year to year.

While the Greeks traded and warred throughout the Mediterranean, they observed and also asked themselves questions about the sea. Aristotle (384–322 B.C.) believed that the oceans occupied the deepest parts of the Earth's surface; he knew that the Sun evaporated water from the sea surface, which condensed and returned as rain. He also began to catalog marine organisms. The brilliant Eratosthenes (c. 264–194 B.C.) of Alexandria, Egypt, mapped his known world and calculated the circumference of the Earth to be 40,250 km, or 25,000 mi (today's measurement is 40,067 km, or 24,881 mi). Posidonius (c. 135–50 B.C.) reportedly measured an ocean depth to about 1800 m (6000 ft) near the island of Sardinia according to the Greek geographer Strabo (c. 63 B.C.–A.D. 21). Pliny the Elder (c. A.D. 23–79) related the phases of the Moon to the tides and reported on the currents moving through the Strait of Gibraltar. Ptolemy (c. A.D. 127–51) produced the first world atlas and established world boundaries: to the north the British Isles, Northern Europe, and the unknown lands of Asia; to the south an unknown land, "Terra Australis Incognita," including Ethiopia, Libya, and the Indian Sea; to the east China; and to the west the great Western Ocean reaching around the Earth to China on the other side (prologue fig. IV.) His atlas listed more than 8000 places by latitude and longitude, but his work contained a major flaw. He had accepted a value of 29,000 km (18,000 mi) for the Earth's circumference. This value shortened Earth distances and led Columbus, more than 1000 years later, to believe that he had reached the eastern shore of Asia when he landed in the Americas.

The Middle Ages

After Ptolemy, intellectual activity and scientific thought declined in northern Europe for about 1000 years. However, shipbuilding improved during this period; vessels became more seaworthy and easier to sail, so sailors could make longer voyages. The Vikings (Norse for *piracy*) were highly accomplished seamen who engaged in extensive exploration, trade, and colonization for nearly three centuries from about 793 to 1066 (prologue fig. V). They sailed to Iceland in 871, where it is believed that as many as 12,000 immigrants eventually settled. During this time they also journeyed inland on Russian rivers throughout central and eastern Europe and western Asia. Voyages into the Mediterranean Sea led them to Rome and Baghdad. Erik Thorvaldsson (known as Erik the Red) sailed west from Iceland in 982 and discovered Greenland. He lived there for three years before returning to Iceland to recruit more settlers. Icelander Bjarni Herjolfsson, on his way to Greenland to join the

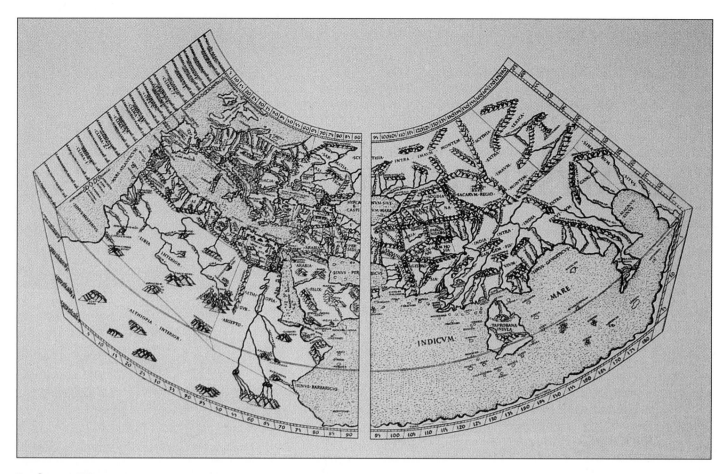

Prologue IV A chart from an Italian fifteenth-century edition of Ptolemy's *Geographia*.

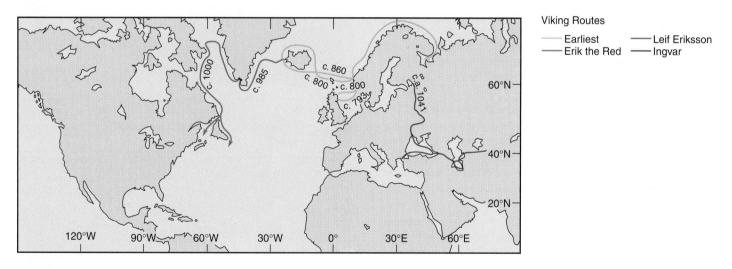

web link **Prologue V** Major routes of the Vikings to the British Isles, to Asia, and across the Atlantic to Iceland, Greenland, and North America.

colonists in 985–6, was blown off course, sailed south of Greenland, and is believed to have come within sight of Newfoundland before turning back and reaching Greenland. Leif Eriksson, son of Erik the Red, sailed west from Greenland in 1002 and reached North America roughly 500 years before Columbus.

To the south, in the region of the Mediterranean after the fall of the Roman Empire, the Arabs preserved Greek and Roman knowledge and they continued to build on it. The Arabic writer El-Mas'údé (d. 956) gives the first description of the reversal of the currents due to the seasonal monsoon winds. Using this knowledge of winds and currents, the

Marine Archaeology

More than 2000 years of exploration, war, and trading have left thousands of wrecks scattered across the oceans' floors. These wrecks are great storehouses of information for archaeologists and historians searching for information about ships, trade, warfare, and the details of human life in ancient times. Marine archaeologists use techniques developed for oceanographic research to find, explore, recover, and preserve wrecks and other artifacts lying under the sea. Sound beams that sweep the sea floor produce images that are viewed onboard ship, and an object of interest can be located very precisely with computer-controlled positioning systems. Magnetometers are also used in these searches to detect iron from sunken vessels. Divers can be sent down to verify a wreck in shallow water, and a research submersible (submarine) carrying observers can be used in deeper water. Unstaffed, towed camera and instrument sleds or remotely operated vehicles (ROVs) equipped with underwater video cameras explore in deep water or in areas that are difficult or unsafe for divers and submersibles. ROVs and submersibles also collect samples to help identify wrecks. (See the box titled "Undersea Robotic Technology" in chapter 2.)

The oldest known shipwreck, from the fourteenth century B.C., a Bronze Age merchant vessel, was discovered in 1983 more than 33 m (100 ft) down in the Mediterranean Sea off the Turkish coast. Divers have recovered thousands of artifacts from its cargo, including copper and tin ingots, pottery, ivory, and amber. Using these items, archaeologists have been able to learn about the life and culture of the period, trace the ship's trade route, and understand more about Bronze Age people's shipbuilding skills.

The waters around northern Europe have claimed thousands of wrecks since humans began to trade and voyage along these coasts. Two of these vessels, found in shallow water, have provided quantities of information for historians and archaeologists. In the summer of 1545 the English warship *Mary Rose* sank as it sailed out to engage the French fleet. The ship was studied in place and raised in 1982. More than 17,000 objects were salvaged from it, giving archaeologists and naval historians insights into the personal as well as the working lives of the officers and crew of a naval vessel at that time. The portion of the hull that was buried in the mud was preserved and is on display at Portsmouth, England (box fig. 1). The Swedish man-of-war *Vasa* sank in 1628 in Stockholm Harbor at the beginning of its maiden voyage. The ship was located in 1956 and raised in 1961. The *Vasa* was a great ship, over 200 ft long, and, like others of the period, it was fantastically decorated with carvings and statues. Divers searched the seabed every summer from 1963–67 and recovered sculptures and carved details that had adorned the ship. Because the water of the Baltic Sea is much less salty than the open ocean, there were no shipworms to destroy the wooden hull, and the *Vasa* is now on exhibit in Stockholm, Sweden (box fig. 2).

Another fighting ship of this period, the Spanish galleon *San Diego*, sank in 1600 as it engaged two Dutch vessels off the Philippine coast. An intensive two-year archaeological excavation began in 1992. Relics recovered include hundreds of pieces of intact Chinese porcelain, a bronze astrolabe for determining latitude, and a bronze and glass compass. Most of the hull had been destroyed by

web link **Box Figure 1** The hull of the *Mary Rose* in its display hall at Portsmouth Dockyard, England. To preserve and stabilize the remaining wood, it is constantly sprayed with a preservative solution.

web link **Box Figure 2** The *Vasa*, a seventeenth-century man-of-war, is on permanent display in Stockholm, Sweden. Most of the *Vasa* is original, including two of the three masts and parts of the rigging.

 web link **Box Figure 3** Surrounded by a cofferdam to keep out mud and water, archaeologists excavate amidships of the wreck of the *Belle*. Shop vacs and water hoses are used to expose the ship's structure and cargo, including casks and boxes of muskets.

Box Figure 4 The World War II German battleship *Bismarck* was sunk in 1941 and lies 4750 m (15,600 ft) below the surface in the North Atlantic. In 1989, the towed camera sled *Argo* photographed part of the ship's superstructure.

shipworms and currents; surviving pieces were measured and then covered with sand for protection.

Wrecks that lie in deep water are initially much better preserved than those in shallow water because these deep-water wrecks lie below the depths of the strong currents and waves that break up most shallow-water wrecks. Shallow-water wrecks are also often the prey of treasure hunters who destroy the history of the site while they search for adventure and items of market value. Over long periods, the cold temperatures and low oxygen content of deep water favor preservation of wooden vessels by slowing decomposition and excluding the marine organisms that bore into wood in shallow areas. Also, the muds and sands falling from above cover objects on the deep-sea floor much more slowly than they do in shallow water.

In 1685 the *Belle*, the ship of the explorer Robert Cavelier, Sieur de la Salle, sank in a storm in Matagordo Bay, along the Texas coast. In 1995 this wreck was located by scanning the bay for magnetic anomalies caused by iron in the wreckage. Reaching the vessel, under 4 m (12 ft) of water and tons of mud, required a cofferdam made of two concentric octagonal walls of interlocking steel plates. It was constructed 12 m (40 ft) below the bay floor and 18 m (6 ft) above sea level. The water and mud were then pumped out of the area within the cofferdam to allow access to the *Belle* (box fig. 3). The wreck yielded a rich harvest of artifacts including cannons, brass pots, candlesticks, coils of rope and brass wire, personal belongings, and trade goods such as rings, glass beads, and combs.

Robert Ballard of the Woods Hole Oceanographic Institution has led several expeditions using sophisticated electronic and robotic equipment to find both ancient and more modern vessels. He discovered the wreck of the *Titanic* in 1985, 4000 m (13,000 ft) down in the North Atlantic. In 1989 he searched 192 km^2 (120 mi^2) of ocean to find the German battleship *Bismarck*, sunk in 1941 after one of the most famous sea battles of World War II. The ship lies 4750 m (15,600 ft) below the surface (box fig. 4). Both ships were located using a towed underwater camera sled, controlled from the vessel at the surface. Once the wrecks were found, observers descended via submersible for direct observation and further inspection of the vessels. Photographic surveys

using ROVs that could be maneuvered to take pictures inside and outside the wrecks documented both vessels.

In 1997 a team of oceanographers, engineers, and archaeologists led by Robert Ballard discovered a cluster of five Roman ships, along with thousands of artifacts, on the bottom of the Mediterranean Sea. The ships lie in 762 m (2500 ft) of water beneath an ancient trade route, where they were lost sometime between about 100 B.C. and A.D. 400. Prior to this discovery, no major ancient shipwreck sites had been discovered and explored by archaeologists in water deeper than 61 m (200 ft). The ships were remarkably well preserved, having been protected from looting by divers and encrustation with coral by the deep water in which they sank. The vessels were spread over an area of about 32 km^2 (20 mi^2) and probably sank when they were caught in sudden, violent storms. The ships were initially located using the U.S. Navy's nuclear submarine *NR-1*. The *NR-1* is capable of searching over large areas for extended periods of time, and its long-range sonar can detect objects at a much greater distance than conventional sonar systems typically used by oceanographers. Once the site had been identified, the small ROV *Jason* was used for detailed mapping, observation, and the recovery of over 100 artifacts selected to help the archaeologists date the ancient wrecks.

These expeditions have moved marine archaeology from the shore into the deep sea. Today's oceanographic instrumentation makes all shipwrecks available to archaeologists.

To Learn More About Marine Archaeology

Ballard, R. D. 1985. How We Found *Titanic*. *National Geographic* 168 (6): 698–722.

Ballard, R. D. 1986. A Long Last Look at *Titanic*. *National Geographic* 170 (6): 698–727.

Ballard, R. D. 1989. Finding the *Bismarck*. *National Geographic* 176 (5): 622–37.

Ballard, R. D. 1998. High-Tech Search for Roman Shipwrecks. *National Geographic* 193 (4): 32–41.

Bass, G. F. 1987. Oldest Known Shipwreck Reveals Splendors of the Bronze Age. *National Geographic* 172 (6): 693–734.

Gerard, S. 1992. The Caribbean Treasure Hunt. *Sea Frontiers* 38 (3): 48–53.

Goddio, F. 1994. The Tale of the *San Diego*. *National Geographic* 186 (1): 35–56.

LaRoe, L. 1997. La Salle's Last Voyage. *National Geographic* 191 (5): 72–83.

Oceanus 28 (4) 1985–86. Winter issue devoted to the discovery of the *Titanic*.

Roberts, D. 1997. Sieur de la Salle's Fateful Landfall. *Smithsonian* 28 (1): 40–52.

Internet References

Visit the book's Online Learning Center at www.mhhe.com/sverdrup (click on the book's cover) to explore links to further information on related topics.

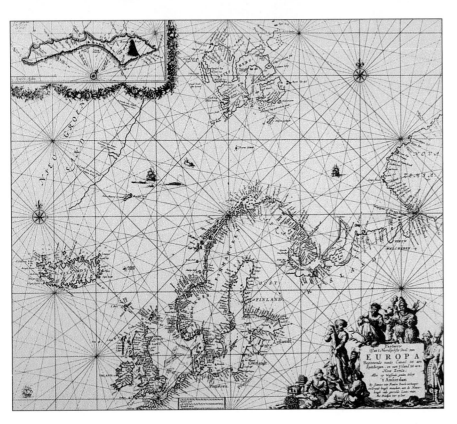

Prologue VI A navigational chart of northern Europe from Johannes van Keulen's *Sea-Atlas* of 1682–84.

Arabs established regular trade routes across the Indian Ocean. In the 1200s large Chinese junks with crews of 200 to 300 sailed the same routes (between China and the Persian Gulf) as the Arab dhows.

During the Middle Ages, while scholarship about the sea remained static, the knowledge of navigation increased. Harbor-finding charts, or *portolanos*, appeared. These charts carried a mileage scale and noted hazards to navigation, but they did not have latitude or longitude. With the introduction of the magnetic compass to Europe from Asia in the thirteenth century, compass directions were added. A Dutch navigational chart from Johannes van Keulen's *Great New and Improved Sea-Atlas or Water-World* of 1682–84 is shown in prologue figure VI. The compass course directions follow the pattern used in early fourteenth-century portolanos.

Although tides were not understood, the Venerable Bede (673–735) illustrated his account of the tides with data from the British coast. His calculations were followed in the tidal observations collected by the British Abbot Wallingford of Saint Alban's Monastery in about 1200. His tide table, titled "Flod at London Brigge," documented the times of high water. Sailors made use of Bede's calculations until the seventeenth century.

As scholarship was reestablished in Europe, Arabic translations of early Greek studies were translated into Latin and thus became available to European scholars. The study of tides continued to absorb the medieval scientists, and they were also interested in the saltiness of the sea. By the 1300s Europeans had established successful trade routes, including some partial ocean crossings. An appreciation of the importance of navigational techniques grew as trade routes were extended.

Voyages of Discovery

Early in the fifteenth century the Chinese organized seven voyages to explore the Pacific and Indian Oceans. More than 300 ships, one more than 122 m (400 ft) long, participated in these ventures to extend Chinese influence and demonstrate the power of the Ming dynasty. The voyages ended in 1433 when their explorations led the Chinese to believe that other societies had little to offer, and the government of China withdrew within its borders, beginning a 400-year period of isolation.

In Europe the desire for riches from new lands persuaded wealthy individuals, often representing their countries, to underwrite the costs of long voyages to all the oceans of the world. The individual most responsible for the great age of European discovery was Prince Henry the Navigator (1394–1460) of Portugal. In 1419, his father, King John, made him governor of Portugal's southernmost coasts. Prince Henry was keenly interested in and studied navigation and mapmaking. He established a naval observatory for the teaching of navigation, astronomy, and cartography about 1450. From 1419 until his death in 1460, Prince Henry sent expedition after expedition down the west coast of Africa to secure trade routes and to establish

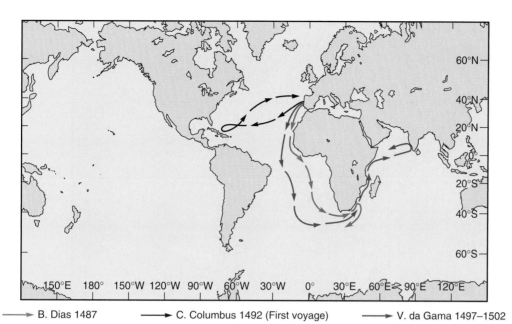

→ B. Dias 1487   → C. Columbus 1492 (First voyage)   → V. da Gama 1497–1502

colonies. These expeditions moved slowly due to the mariners' belief that waters at the equator were at the boiling point and that sea monsters would engulf ships. It wasn't until twenty-seven years after Prince Henry's death that Bartholomeu Dias (1450?–1500) braved these "dangers" and rounded the Cape of Good Hope in 1487 in the first of the great voyages of discovery (prologue fig. VII). Dias sailed in search of new and faster routes to the spices and silks of the East.

Portugal's slow progress down the west coast of Africa in search for a route to the east finally came to fruition with Vasco da Gama (1469–1524). In 1498 he followed Bartholomeu Dias's route to the Cape of Good Hope and then continued beyond along the eastern coast of the African continent. He successfully mapped a route to India but was challenged along the way by Arab ships. In 1502, da Gama returned with a flotilla of fourteen heavily armed ships and defeated the Arab fleet. By 1511, the Portuguese conquered the spice routes and had access to the Spice Islands. In 1513, Portuguese trade extended to China and Japan.

Christopher Columbus (1451–1506) made four voyages across the Atlantic Ocean in an effort to find a new route to the East Indies by traveling west rather than east. By relying on inaccurate estimates of the Earth's size he badly underestimated the distances involved and believed he had found islands off the coast of Asia when he reached the New World.

The Italian navigator Amerigo Vespucci (1454–1512) made several voyages to the New World (1499–1504) for Spain and Portugal, exploring nearly 10,000 km (6000 mi) of South American coastline. He accepted South America as a new continent not part of Asia, and in 1507 the German cartographer Martin Waldseemüller applied the name "America" to the continent in Vespucci's honor. Vasco Núñez de Balboa (1475–1519) crossed the Isthmus of Panama and found the Pacific Ocean in 1513, and in the same year Juan Ponce de León (1460?–1521) discovered Florida and the Florida Current. All claimed the new lands they found for their home countries. Although they had

sailed for fame and riches, not knowledge, they more accurately documented the extent and properties of the oceans, and the news of their travels stimulated others to follow.

Ferdinand Magellan (1480–1521) left Spain in September 1519 with 270 men and five vessels in search of a westward passage to the Spice Islands. The expedition lost two ships before finally discovering and passing through the Strait of Magellan and rounding the tip of South America in November 1520. Magellan crossed the Pacific Ocean and arrived in the Philippines in March 1521, where he was killed in a battle with the natives on April 27, 1521. Two of his ships sailed on and reached the Spice Islands in November 1521, where they loaded valuable spices for a return home. In an attempt to guarantee that at least one ship made it back to Spain the two ships parted ways. The *Victoria* continued sailing west and successfully crossed the Indian Ocean, rounded Africa's Cape of Good Hope, and arrived back in Spain on September 6, 1522, with eighteen of the original crew. This was the first circumnavigation of the Earth (prologue fig. VIII). Magellan's skill as a navigator makes his voyage probably the most outstanding single contribution to the early charting of the oceans. In addition, during the voyage he established the length of a degree of latitude and measured the circumference of the Earth. It is said that Magellan tried to test the mid-ocean depth of the Pacific with a hand line, but this idea seems to come from writings by a nineteenth-century German oceanographer; writings from Magellan's time do not support this story.

By the latter half of the sixteenth century, adventure, curiosity, and hopes of finding a trading shortcut to China spurred efforts to find a sea passage around North America. Sir Martin Frobisher (1535?–94) made three voyages in 1576, 1577, and 1578, and Henry Hudson (d. 1611) made four voyages (1607, 1608, 1609, and 1610), dying with his son when set adrift in Hudson Bay by his mutinous crew. The Northwest Passage continued to beckon, and in 1615 and 1616 William Baffin (1584–1622) made two unsuccessful attempts.

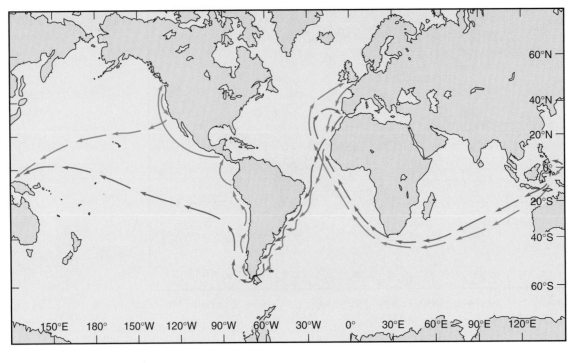

———————▶ F. Magellan 1519–22

———————▶ F. Drake 1577–80

 web link **Prologue VIII** The sixteenth-century circumnavigations by Ferdinand Magellan's ship *Victoria* and Sir Francis Drake and the *Golden Hind*.

While European countries were setting up colonies and claiming new lands, Francis Drake (1540–96) set out in 1577 with 165 crewmen and five ships to show the English flag around the world (prologue fig. VIII). He was forced to abandon two of his ships off the coast of South America. He was separated from the other two ships while passing through the Straits of Magellan. During the voyage Drake plundered Spanish shipping in the Caribbean and in Central America and loaded his ship with treasure. In June 1579, Drake landed off the coast of present-day California and sailed north along the coast to the present United States–Canadian border. He then turned southwest and crossed the Pacific Ocean in two months' time. In 1580 he completed his circumnavigation and returned home in the *Golden Hind* with a cargo of Spanish gold, to be knighted and treated as a national hero. Queen Elizabeth I encouraged her sea captains' exploits as explorers and raiders because, when needed, their ships and their knowledge of the sea brought military victories as well as economic gains.

The Beginnings of Earth Science

New ideas and new knowledge had stimulated the practical exploration of the oceans during the fifteenth and sixteenth centuries, but most of the thinking about the sea was still rooted in the ideas of Aristotle and Pliny. In the seventeenth century, although the practical needs of commerce, national security, and economic and political expansionism guided events at sea, scientists on land were beginning to show an interest in experimental science and the study of the specific properties of substances. Curiosity about the Earth flourished; scientists wrote pamphlets and formed societies in which to discuss their discoveries. The works of Johannes Kepler (1571–1630) on planetary motion and those of Galileo Galilei (1564–1642) on mass, weight, and acceleration would in time be used to understand the oceans. Although both Kepler and Galileo had theories of tidal motion, it was not until Sir Isaac Newton (1642–1727) wrote his *Principia* in 1687, which gave the world the unifying law of gravity, that the processes governing the tides were explained. Edmund Halley (1656–1742), the astronomer of comet fame and a contemporary and friend of Newton, also had an interest in the oceans; he made a voyage to measure longitude and study the variation of the compass in 1698 and suggested that the age of the oceans could be calculated by determining the rate at which rivers carry salt to the sea.

The Importance of Charts and Navigational Information

As colonies were established far away from their home countries, and as trade, travel, and exploration expanded, there was renewed interest in developing better charts and more accurate navigational techniques. The first hydrographic office dedicated to mapping the oceans was established in France in 1720 and was followed in 1795 by the British Admiralty's appointment of a hydrographer.

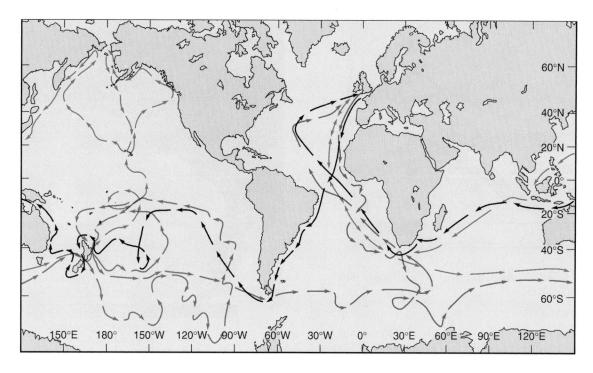

→ Cook's first voyage 1768–71
→ Cook's second voyage 1772–75
→ Cook's third voyage 1776–79

web link **Prologue X** The three voyages of Captain James Cook.

web link **Prologue IX** John Harrison's fourth chronometer. A copy of this chronometer was used by Captain James Cook on his 1772 voyage to the southern oceans.

As early as 1530 the relationship between time and longitude had been proposed by the Flemish astronomer Gemma Frisius, and in 1598 King Philip III of Spain had offered a reward of 100,000 crowns to any clockmaker building a clock that would keep accurate time onboard ship. In 1714 Queen Anne of England authorized a public reward for a practical method of keeping time at sea, and the British Parliament offered 20,000 pounds sterling for a seagoing clock that could keep time with an error not greater than two minutes on a voyage to the West Indies from England. A Yorkshire clockmaker, John Harrison, built his first chronometer (high accuracy clock) in 1735, but it was not until 1761 that his fourth model met the test, losing only fifty-one seconds on the eighty-one-day voyage. Harrison was awarded only a portion of the prize after his success in 1761, and it was not until 1775, at the age of eighty-three, that he received the remainder from the reluctant British government. In 1772 Captain James Cook took a copy of the fourth version of Harrison's chronometer (prologue fig. IX) to produce accurate charts of new areas and to correct previously charted positions.

Captain James Cook (1728–79) made his three great voyages to chart the Pacific Ocean between 1768 and 1779 (prologue fig. X). In 1768 he left England in command of the *Endeavour* on an expedition to chart the transit of Venus; he returned in 1771 having circumnavigated the globe and having explored and charted the coasts of New Zealand and eastern Australia. Between 1772 and 1775 he

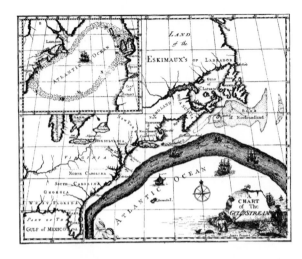

web link **Prologue XI** The Franklin-Folger map of the Gulf Stream, 1769.

commanded an expedition of two ships, the *Resolution* and the *Adventure,* to the South Pacific, charted many islands, explored the Antarctic Ocean, and by controlling his sailors' diet prevented vitamin C deficiency and scurvy, the disease that had decimated the crews of vessels that spent long periods of time at sea. Cook sailed on his third and last voyage in 1776 in the *Resolution* and *Discovery.* He spent a year in the South Pacific and then sailed north, discovering the Hawaiian Islands in 1778. He continued on to the northwest coast of North America and into the Bering Strait, searching for a passage to the Atlantic. He returned to Hawaii for the winter and was killed by the native islanders at Kealakekua Bay on the island of Hawaii in 1779. Cook takes his place not only as one of history's greatest navigators and seamen but also as a fine scientist. He made soundings to depths of 400 m (1200 ft) and accurate observations of winds, currents, and water temperatures. Cook's careful and accurate observations produced much valuable information and made him one of the founders of oceanography.

In the United States, Benjamin Franklin (1706–90) became concerned about the amount of time required for news and cargo to travel between England and America. With Captain Timothy Folger, his cousin and a whaling captain from Nantucket, he constructed the 1769 Franklin-Folger chart of the Gulf Stream current (prologue fig. XI), which, when published, encouraged captains to sail within the Gulf Stream en route to Europe and to avoid it on the return passage. Since the Gulf Stream carries warm water from low latitudes to high latitudes, it is possible to map its location with satellites that measure sea surface temperature. Compare the Franklin-Folger chart in prologue figure XI to a map of the Gulf Stream shown in prologue figure XII based on the average sea surface temperature during 1996. In 1802 Nathaniel Bowditch (1773–1838), another American, published the *New American Practical Navigator.* In this book Bowditch made the techniques of celestial navigation available for the first time to every competent sailor and set the stage for U.S. supremacy of the seas during the years of the Yankee Clip-

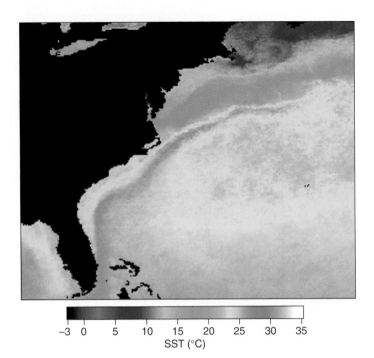

| -3 | 0 | 5 | 10 | 15 | 20 | 25 | 30 | 35 |

SST (°C)

web link **Prologue XII** Average sea surface temperature for the year 1996. The red-orange streak of 28° to 30°C water shows the Gulf Stream.

pers. When Bowditch died, his copyright was bought by the U.S. Navy, and his book continued in print. Its information was updated and expanded with each edition, serving generations of mariners and navigators.

In 1807 the U.S. Congress, at the direction of President Thomas Jefferson, formed the Survey of the Coast under the Treasury Department, later named the Coast and Geodetic Survey and now known as the National Ocean Survey. The U.S. Naval Hydrographic Office, now the U.S. Naval Oceanographic Office, was set up in 1830. Both were dedicated to exploring the oceans and producing better coast and ocean charts. In 1842 Lieutenant Matthew F. Maury (1806–73), seen in prologue figure XIIIa, who had worked with the Coast and Geodetic Survey, was assigned to the Hydrographic Office and founded the Naval Depot of Charts. He began a systematic collection of wind and current data from ships' logs. He produced his first wind and current charts of the North Atlantic in 1847. At the 1853 Brussels Maritime Conference, Maury issued a plea for international cooperation in data collection, and from the ships' logs he received he produced the first published atlases of sea conditions and sailing directions. His work was enormously useful, and as a result ships sailed more safely and were able to take days off their sailing times between major ports around the world. The British estimated that Maury's sailing directions took thirty days off the passage from the British Isles to California, twenty days off the voyage to Australia, and ten days off the sailing time to Rio de Janeiro. In 1855 he published *The Physical Geography of the Sea.* This work includes chapters on the Gulf Stream, the atmosphere, currents, depths, winds, climates, and storms, and the first bathymetric chart

(a)

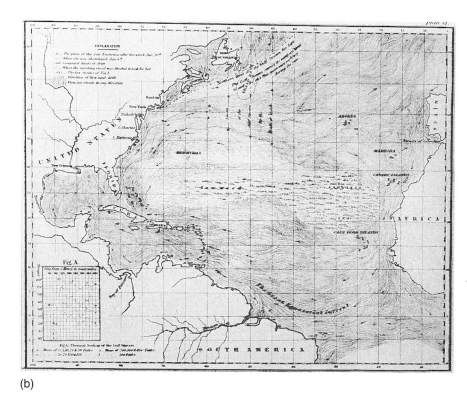

(b)

web link **Prologue XIII** (a) Lieutenant Matthew F. Maury of the U.S. Navy, 1806–73. (b) Maury's chart of the Gulf Stream and North Atlantic Ocean surface currents from *The Physical Geography of the Sea,* 1855.

of the North Atlantic with contours at 6000, 12,000, 18,000, and 24,000 ft. (See prologue fig. XIIIb for the Gulf Stream chart from this book and compare the change in detail and style to the Franklin-Folger chart in prologue fig. XI.) Many marine scientists consider Maury's book the first textbook of what we now call oceanography and consider Maury the first true oceanographer. Again, national and commercial interests were the driving forces behind the study of the oceans.

Ocean Science Begins

As charts became more accurate and as information about the oceans increased, the oceans captured the interest of naturalists and biologists. Baron Alexander von Humboldt (1769–1859) made observations on a five-year (1799–1804) cruise to South America; he was particularly fascinated with the vast numbers of animals inhabiting the current flowing northward along the western coast of South America, the current that now bears his name. Charles Darwin (1809–82) joined the survey ship *Beagle* and served as the ship's naturalist from 1831–36 (prologue fig. XIV). He described, collected, and classified organisms from the land and sea. His theory of atoll formation is still the accepted explanation.

At approximately the same time, another English naturalist, Edward Forbes (1815–54), began a systematic survey of marine life around the British Isles and the Mediterranean and Aegean Seas. He collected organisms in deep water and, on the basis of his observations, proposed a system of ocean depth

zones, each characterized by specific animal populations. However, he also mistakenly theorized that there was an azoic, or lifeless, environment below 550 m (1800 ft). His announcement is curious, because twenty years earlier the Arctic explorer Sir John Ross (1777–1856), looking again for the Northwest Passage, had taken bottom samples at over 1800 m (6000 ft) depth in Baffin Bay with a "deep-sea clamm," or bottom grab, and had found worms and other animals living in the mud. Ross's nephew, Sir James Clark Ross (1800–1862), took even deeper samples from Antarctic waters and noted their similarity to the Arctic species recovered by his uncle. Still, Forbes's systematic attempt to make orderly predictions about the oceans, his enthusiasm, and his influence make him another candidate as a founder of oceanography.

Christian Ehrenberg (1795–1876), the German naturalist, found the skeletons of minute organisms in seafloor sediments and also recognized that the same organisms were alive at the sea surface; he concluded that the sea was filled with microscopic life and that the skeletal remains of these tiny organisms were still being added to the sea floor. The investigation of the minute drifting plants and animals of the ocean was not seriously undertaken until the German scientist Johannes Müller (1801–58) began his work in 1846. He used an improved fine mesh tow net similar to that used by Charles Darwin to collect these organisms, which he examined microscopically. This work was continued by Victor Hensen (1835–1924), who improved the Müller net, introduced the quantitative study of these minute drifting sea organisms, and gave them the name *plankton* in 1887. A

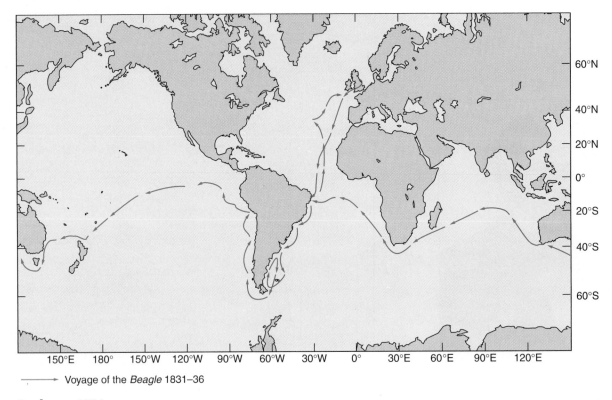

web ⚭ link **Prologue XIV**   The voyage of Charles Darwin and the survey ship HMS *Beagle*, 1831–36.

Prologue XV   This illustration of microscopic drifting animals is from volume 25 of *Fauna and Flora des Golfes von Neapel (Fauna and Flora of the Gulf of Naples)*, published in 1899.

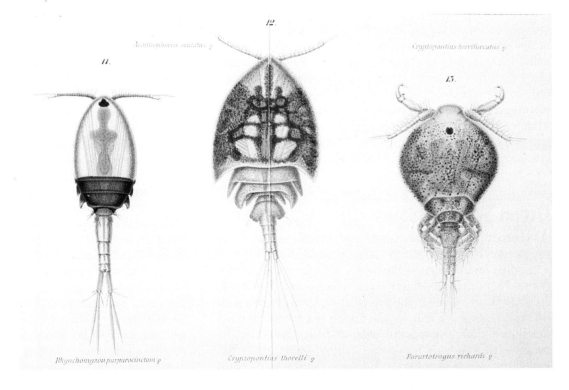

portion of a plate from an 1899 publication describing these organisms is seen in prologue figure XV.

Although science blossomed in the seventeenth and eighteenth centuries, there was little scientific interest in the sea except as we have seen for the practical reasons of navigation, tide prediction, and safety. In the early nineteenth century ocean scientists were still few and usually only temporarily attracted to the sea. Some historians believe that the subject and study of the oceans were so vast, requiring so many people and such large amounts of money, that it required the interest and support of government before the growth of oceanography as a science could occur. This did not happen until the nineteenth century in Great Britain.

In the last part of the nineteenth century the laying of transatlantic telegraph cables made a better knowledge of the deep sea a necessity. Engineers needed to know about seafloor conditions, including bottom topography, currents, and organisms that might dislodge or destroy the cables. The British began a series of deep-sea studies stimulated by the retrieval of a damaged cable from more than 1500 m (5000 ft) deep, well below Forbes's azoic zone. When the cable was brought to the surface it was found to be covered with organisms, many of which had never been seen before. In 1868 the *Lightning* dredged between Scotland and the Faroe Islands at depths of 915 m (3000 ft) and found many animal forms. The British Admiralty continued these studies with the *Porcupine* during the summers of 1869 and 1870, dredging up animals from depths of more than 4300 m (14,000 ft). Charles Wyville Thomson (1830–82), like Forbes, a professor of natural history at Edinburgh University, was one of the scientific leaders of these two expeditions. On the basis of their results he wrote *The Depths of the Sea*, published in 1873, which became very popular and is regarded by some as the first book on oceanography.

The *Challenger* Expedition

With public interest running high, the Circumnavigation Committee of the British Royal Society was able to persuade the British Admiralty to organize the most comprehensive single oceanographic expedition ever undertaken. The Society obtained the use of the naval corvette *Challenger*, a sailing vessel with auxiliary steam power. All but two of the corvette's guns were removed, and the ship was refitted with laboratories, winches, and equipment, including 232 km (144 mi) of sounding rope. The leadership was offered to Charles Wyville Thomson, and his assistant was a young geologist, John Murray (1841–1914). The *Challenger* sailed from Portsmouth, England, on December 21, 1872, for a voyage that was to last nearly three-and-a-half years, during which time the vessel logged 110,840 km (68,890 mi) (prologue fig. XVI). The first leg of the voyage took the vessel to Bermuda, then to the South Atlantic island of Tristan da Cunha, around the Cape of Good Hope, and east across the southernmost part of the Indian Ocean; this vessel was the first steamship to cross the Antarctic Circle. It continued on to Australia, New Zealand, the Philippines, Japan, and China. Turning south to the Marianas Islands, the vessel took its deepest sounding, 8180 m (26,850 ft). Sailing across the Pacific to Hawaii, Tahiti, and through the Strait of Magellan, it returned to England on May 24, 1876. Queen Victoria conferred a knighthood on Thomson, and the *Challenger* expedition was over.

The *Challenger* expedition's purpose was scientific research; during the voyage the crew took soundings at 361 ocean stations, collected deep-sea water samples, investigated deep-water motion, and made temperature measurements at all depths. Thousands of biological and sea-bottom samples were collected. The cruise brought back evidence of an ocean teeming with life at all depths and opened the way for the era of descriptive oceanography that followed.

Although the *Challenger* expedition ended in 1876, the work of organizing and compiling information continued for twenty years, until the last of the fifty-volume *Challenger Reports* was issued. John Murray (later Sir John Murray) edited the reports after Thomson's death and wrote many of them himself. He is considered the first geological oceanographer. William Dittmar (1833–92) prepared the information on seawater chemistry for the *Challenger Reports*. He identified the major elements present in the water and confirmed the findings of earlier chemists that in a seawater sample the proportion of the major dissolved elements to each other is constant. Oceanography as a modern science is usually dated from the *Challenger* expedition. The *Challenger Reports* laid the foundation for the science of oceanography.

The *Challenger* expedition stimulated other nations to mount ocean expeditions. Although their avowed purpose was the scientific exploration of the sea, in large measure prestige was at stake. Norway explored the North Atlantic with the *Voringen* in the summers of 1876–78; Germany studied the Baltic and North Seas in the SS *Pomerania* in 1871 and 1872 and in the *Crache* in 1881, 1882, and 1884. The French government financed cruises by the *Travailleur* and the *Talisman* in the 1880s. The Austrian ship *Pola* worked in the Mediterranean and Red Seas in the 1890s. The U.S. vessel *Enterprise* circumnavigated the Earth between 1883 and 1886, as did Italian and Russian ships between 1886 and 1889.

Oceanography as Science

During the late nineteenth century and the early twentieth century, intellectual interest in the oceans increased. Oceanography was changing from a descriptive science to a quantitative one. Oceanographic cruises now had the goal of testing hypotheses by gathering data. Theoretical models of ocean circulation and water movement were developed. The Scandinavian oceanographers were particularly active in the study of water movement. One of them, Fridtjof Nansen (1861–1930) (prologue fig. XVII), a well-known athlete, explorer, and zoologist, was interested in the current systems of the polar seas. This extraordinary man decided to test his ideas about the direction of ice drift in the Arctic by freezing a vessel into the polar ice pack and drifting with it to reach the North Pole. To do so he had to design a special vessel that would be able to survive the great pressure from the ice; the 39 m (128 ft) wooden *Fram* ("to push forward"), shown in prologue figure XVII, was built with a smoothly rounded hull and planking over 60 cm (2 ft) thick.

Nansen departed with thirteen men from Oslo in June 1893. The ship was frozen into the ice nearly 1100 km (700 mi) from the North Pole and remained in the ice for thirty-five months. During this period measurements were made through holes in the ice that showed that the Arctic Ocean was a deep-ocean basin and not the shallow sea that had been expected. Water and air temperatures were recorded, water chemistry was analyzed, and the great plankton blooms of the area were observed. Nansen became impatient with the slow rate of drift and, with F. H. Johansen, left the *Fram* locked in the ice some 500 km (300 mi) from the

continued

Prologue XVI The *Challenger* Expedition: December 21, 1872 to May 24, 1876. Engravings from the *Challenger Reports*, volume 1, 1885.
(a) "H.M.S. *Challenger*—Shortening Sail to Sound," decreasing speed to take a deep-sea depth measurement. (b) "H.M.S. *Challenger*." (c) "Dredging and Sounding Arrangement on board *Challenger*." Rigging is hung from the ship's yards to allow the use of the over-the-side sampling equipment. A biological dredge can be seen hanging outboard of the rail. The large cylinders in the rigging are shock absorbers. (d) Sieving bottom samples for organisms. (e) "Deep Sea Deposits." This plate shows the shells of microscopic organisms making up different kinds of muds and clays from the floor of the deep sea. (f) "H.M.S. *Challenger* at St. Paul's Rocks," in the equatorial mid-Atlantic. (g) "Zoological Laboratory on the Main Deck."
(h) "Chemical Laboratory." (i) A biological dredge used for sampling bottom organisms. Note the frame and skids that keep the mouth of the net open and allow it to slide over the sea floor.

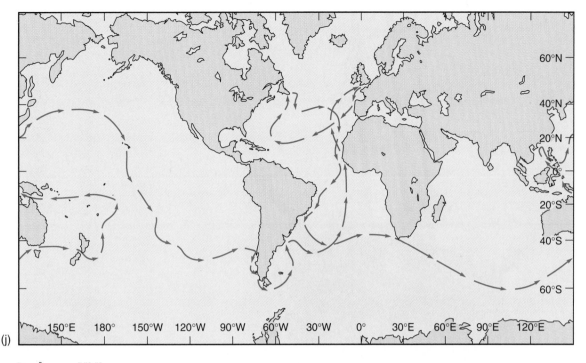

web link **Prologue XVI** *continued* (j) The cruise of the *Challenger*, 1872–76, the first major oceanographic research effort.

pole. They set off with dogsleds toward the pole, but after four-and-a-half weeks they were still more than 300 km (200 mi) from the pole, with provisions running low and the condition of their dogs deteriorating. The two men turned away from the pole and spent the winter of 1895–96 on the ice, living on seals and walrus. They were found by a British polar expedition in June 1896 and returned to Norway in August of that year. The crew of the *Fram* continued to drift with the ship until they freed the vessel from the ice, also in 1896, and returned home. Nansen's expedition had laid the basis for future Arctic work.

After the expedition's findings were published, Nansen continued to be active in oceanography, and his name is familiar today from the Nansen bottle, which he designed to isolate and collect water samples taken from deep water. In 1905 he turned to a career as a statesman, working for the peaceful separation of Norway from Sweden. During the years after World War I he worked with the League of Nations to resettle refugees, for which he received the 1922 Nobel Peace Prize. The well-designed *Fram* paid another visit to polar waters, carrying the Norwegian explorer Roald Amundsen (1872–1928) to the Antarctic continent on his successful 1911 expedition to the South Pole. It was also Amundsen who finally made a Northwest Passage entirely by water in the *Gjoa*, leaving Norway in 1903 and arriving in Nome, Alaska, three years later (prologue fig. XVIII).

Fluctuations in the abundance of commercial fish in the North Atlantic and adjacent seas, and the effect of these changes on national fishing programs, stimulated oceanographic research and international cooperation. As early as 1870, researchers began to realize their need for knowledge of ocean chemistry and physics in order to understand ocean biology. The study of the ocean and its fisheries required the crossing of national boundaries, and in 1902 Germany, Russia, Great Britain, Holland, and the Scandinavian countries formed the International Council for the Exploration of the Sea (ICES) to coordinate and sponsor research in the ocean and in fisheries.

Advances in theoretical oceanography sometimes could not be verified with practical knowledge until new instruments and equipment were developed. Lord Kelvin (1824–1907) invented a tide-predicting machine in 1872 that made it possible to combine tidal theory with astronomical predictions to produce predicted tide tables. Deep-sea circulation could not be systematically explored until approximately 1910, when Nansen's water-sampling bottles were combined with thermometers designed for deep-sea temperature measurements and an accurate method for salinity was devised by the chemist Martin Knudsen (1871–1949). The reliable and accurate measurement of ocean depths had to wait until the development of the echo sounder, which was given its first scientific use on the 1925–27 German cruise of the *Meteor*. Interestingly, although the *Meteor* expedition was supposedly sent out for purely oceanographic reasons, it was also an attempt by the German government to find an affordable way to separate dissolved gold from seawater. Although the expedition failed to find a cheap way to produce gold, it did accumulate a great deal of information about the South Atlantic.

Oceanography as Science   **17**

(b)

web ⦿ link **Prologue XVII** (a) Fridtjof Nansen, Norwegian scientist, explorer, and statesman (1861–1930), using a sextant to determine his ship's position. (b) The *Fram*, frozen in the ice. As the ice pressure increased, it lifted the specially designed and strengthened hull so that the ship would not be crushed.

(a)

U.S. Oceanography in the Twentieth Century

In the United States, government agencies related to the oceans proliferated during the nineteenth century. These agencies were concerned with gathering information to further commerce, fisheries, and the Navy. After the Civil War, the replacement of sail by steam lessened government interest in studying winds and currents and in surveying the ocean floor. Private institutions and wealthy individuals took over the support of oceanography in the United States.

Alexander Agassiz (1835–1910), mining engineer, marine scientist, and Harvard professor, financed a series of expeditions that greatly expanded knowledge of deep-sea biology. Agassiz served as the scientific director on the first ship built especially for scientific ocean exploration, the U.S. Fish Commission's *Albatross*, commissioned in 1882. He designed and financed much of the deep-sea sampling equipment that enabled the *Albatross* to recover more specimens of deep-sea fishes in one haul than the *Challenger* had collected during its entire three-and-a-half years at sea.

One of Agassiz's students, William E. Ritter, became a professor of zoology at the University of California–Berkeley.

(a)

(b)

web link **Prologue XVIII** (a) Roald Amundsen, Norwegian explorer of the Arctic and Antarctic (1872–1928). (b) The *Gjoa*, preparing for its journey through the Northwest Passage (1903).

From 1892–1903, Ritter conducted summer field studies with his students at various locations along the California coast. In 1903 a group of business and professional people in San Diego established the Marine Biological Association and invited Ritter to locate his field station in San Diego permanently. With financial support from members of the Scripps family, who had made a fortune in newspaper publishing, Ritter was able to do this. This was the beginning of the University of California's Scripps Institution of Oceanography (prologue fig. XIX*a*). The property and holdings of the Marine Biological Association were formally transferred to the University of California in 1912.

(a)

(b)

 web link **Prologue XIX** (a) The Scripps Institution of Oceanography in La Jolla, California. Established in 1903 by William Ritter, a zoologist at the University of California–Berkeley, with financial support from E.W. Scripps and his daughter Ellen Browning Scripps. The first permanent building was erected in 1910. (b) The Woods Hole Oceanographic Institution in Woods Hole, Massachusetts. In a rare moment all three WHOI research vessels are in port. *Knorr* is in the foreground with *Oceanus*, bow forward, and *Atlantis*, stern forward, on the opposite side of the pier.

Also, in the first twenty years of the century, the Carnegie Institute funded a series of exploratory cruises, including investigations of the Earth's magnetic field, and maintained a biological laboratory. In 1927, a National Academy of Sciences committee recommended that ocean science research be expanded by creating a permanent marine science laboratory on the East Coast. This led to the establishment of the Woods Hole Oceanographic Institution in 1930 (prologue fig. XIX*b*). It was funded largely by a grant from the Rockefeller Foundation. The Rockefeller Foundation allocated funds to stimulate other programs in marine research and construct additional laboratories at this same time, and oceanography began to move onto university campuses. Teaching oceanography required that the subject material be consolidated, and in 1942, *The Oceans*, by Harald U. Sverdrup, Martin W. Johnson, and Richard H. Fleming, was published. It captured between its covers nearly all the world's knowledge of oceanographic processes and was used to train a generation of ocean scientists.

Oceanography mushroomed during World War II, when practical problems of military significance had to be solved quickly. The United States and its allies needed to move personnel and materials by sea to remote locations, to predict ocean and shore conditions for amphibious landings, to know how explosives behaved in seawater, to chart beaches and harbors from aerial reconnaissance, and to use underwater sound to find submarines. Academic studies ceased as oceanographers pooled their knowledge in the national effort.

After the war, oceanographers returned to their classrooms and laboratories with an array of new, sophisticated instruments, including radar, improved sonar, automated wave detectors, and temperature-depth recorders. They also returned with large-scale government funding for research and education. The Earth sciences in general and oceanography in particular blossomed during the 1950s. The numbers of scientists, students, educational programs, research institutes, and professional journals all increased.

Major funding for applied and basic ocean research was supplied by both the Office of Naval Research (ONR) and the National Science Foundation (NSF). The Atomic Energy Commission (AEC) financed oceanographic work at the South Pacific atoll sites of atomic tests. During the 1950s the Coast and Geodetic Survey expanded its operations and began its seismic sea wave (tsunami) warning system. International cooperation brought about the 1957–58 International Geophysical Year (IGY) program, in which sixty-seven nations cooperated to explore the sea floor and made discoveries that completely revolutionized geology and geophysics. As a direct result of the IGY program, special research vessels and submersibles were built to be used by both federal agencies and university research programs.

The decade of the 1960s brought giant strides in programs and equipment. In 1963–64 another multinational endeavor, the Indian Ocean Expedition, took place. In 1965 a major reorganization of governmental agencies occurred. The Environmental Science Services Administration (ESSA) was formed by consolidating the Coast and Geodetic Survey and the Weather Bureau among others. Under ESSA, federal environmental research institutes and laboratories were established, and the use of satellites to obtain data became a major

(a)

(b)

web link **Prologue XX** (a)The *Glomar Challenger*, the Deep Sea Drilling Program drill ship used from 1968–1983. (b) The *JOIDES Resolution*, the Ocean Drilling Program drill ship in use since 1985.

focus of ocean research. In 1968 the Deep Sea Drilling Program (DSDP), a cooperative venture between research institutions and universities, began to sample the Earth's crust beneath the sea (prologue fig. XXa; discussed in more detail in chapter 2) using the specially built drill ship *Glomar Challenger*. It was finally retired in 1983 after fifteen years of extraordinary service. The *Glomar Challenger* was named after the ship used during the *Challenger* expedition of 1872–76. Electronics developed for the space program were applied to ocean research. Computers went aboard research vessels, and for the first time data could be sorted, analyzed, and interpreted at sea. This made it possible for scientists to adjust experiments while they were in progress. Government funding allowed large-scale ocean experiments. Fleets of oceanographic vessels from many institutions and nations carried scientists studying all aspects of the oceans.

In 1970 the U.S. government reorganized its Earth science agencies once more. The National Oceanic and Atmospheric Administration (NOAA) was formed under the Department of Commerce. NOAA combined several formerly independent agencies including the National Ocean Survey,

National Weather Service, National Marine Fisheries Service, Environmental Data Service, National Environmental Satellite Service, and Environmental Research Laboratories. NOAA also administers the National Sea Grant College Program. This program consists of a network of twenty-nine individual programs located in each of the coastal and Great Lakes states. Sea Grant encourages cooperation in marine science and education among government, academia, and industry.

The ten-year International Decade of Ocean Exploration (IDOE) occurred in the 1970s. The IDOE was a multinational effort to survey seabed mineral resources, improve environmental forecasting, investigate coastal ecosystems, and modernize and standardize the collection, analysis, and use of marine data.

At the end of the 1970s oceanography faced a reduction in funding for ships and basic research, but the discovery of deep-sea hot-water vents and their associated animal life and mineral deposits renewed the excitement over deep-sea biology, chemistry, geology, and ocean exploration in general. Instrumentation continued to become more sophisticated and expensive as deep-sea mooring, deep-diving submersibles, and the remote sensing of the ocean by satellite became possible. Increased cooperation among institutions led to the integration of research at sea between subdisciplines and resulted in large-scale, multifaceted research programs. Although collection of oceanographic data from vessels at sea expanded, data collected by satellites since the 1970s has increasingly presented researchers with the ability to observe sea surface changes on a global scale. Currents, eddies, plant production, sea-level changes, waves, thermal properties, and air-sea interactions are all monitored via satellite, allowing scientists to develop computerized prediction models and to test them against natural phenomena.

During these years of expanding programs, Earth scientists began to recognize the signs of global degradation and the need for policy and management of living and nonliving resources. Students were attracted to the programs, and ocean management courses were added to curricula. As more and more nations were turning to the sea for food and as technology was increasing our ability to harvest the sea, problems of resource ownership, dwindling fish stock, and the need for fishery management had to be faced.

In 1983 the Deep Sea Drilling Project became the Ocean Drilling Program (ODP). The objectives of the ODP included drilling into the thick sediments near continental margins. The ODP uses a larger drilling ship, the *JOIDES Resolution* (prologue fig. XXb), to replace the retired *Glomar Challenger*. The *JOIDES Resolution* is named after the HMS *Resolution*, used by Captain James Cook to explore the Pacific Ocean basin over 200 years ago.

NASA's *NIMBUS-7* satellite, launched in 1978, carried a sensor package called the Coastal Zone Color Scanner (CZCS) that detected multiband radiant energy from chlorophyll in sea and land plants. The sensor operated from 1978–86, when it finally failed. CZCS images can be used to determine levels of biological productivity in the oceans. They can also indicate how, and where, physical processes in the oceans influence the distribution and health of marine

Satellite Oceanography

With satellites, oceanographers can study the oceans as a global system. It is not possible to equip enough research vessels to study more than a small area of one ocean at one time, but each satellite makes millions of observations every day as it follows the changing conditions across the world's oceans. Oceanographic vessels linked to satellites via computer enable scientists to use immediate data to plan their sampling programs while at sea. The many satellites that now orbit our Earth provide huge amounts of digital information that is processed and manipulated by sophisticated computers to increase our knowledge of currents, waves, plant life, sea ice, storms, and even the sea floor. Throughout this book satellite images are presented to show you views of entire oceans and to help you understand the processes that act on them.

Satellites are equipped with specialized sensors that are either passive—recording information from the Earth, such as radiation energy—or active—sending out radar, microwave, or laser signals that echo back to the satellite (box fig. 1). Data are sent back to the Earth in the form of radio signals that are recorded, processed, and analyzed to create images from the satellite's observations. In the United States satellites are the responsibility of some private companies, the military services, the National Oceanographic and Atmospheric Administration (NOAA), and the

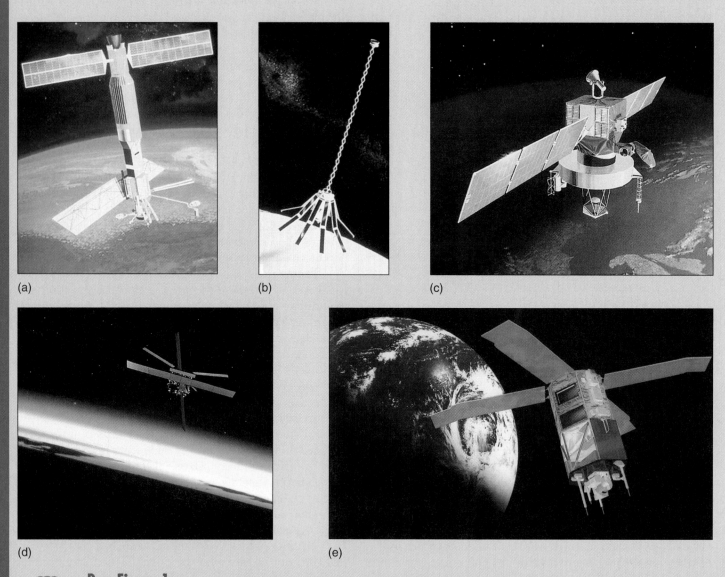

(a) (b) (c)

(d) (e)

web link **Box Figure 1** Satellites with Earth-sensing systems. (a) *SEASAT* operated between June and October 1978. (b) *GEOSAT* monitored the oceans between 1985 and 1990. (c) *TOPEX-Poseidon* is a joint program with NASA and the French space agency CNES. It began a three-year ocean observation program in 1993. (d) *ERS-2*, launched in 1995, is a European satellite; it carries radar and microwave systems to take measurements through clouds and darkness. (e) Since 1997 *SEASTAR* has carried a sensor to study the distribution of ocean plant life.

National Aeronautic and Space Administration (NASA). Some satellites are military intelligence gatherers; some are communication links for global telephone, radio, and television connections; some are navigation aids (see section 1.5 in chapter 1), and some gather data for Earth sciences research.

The satellite remains in its orbit because its orbit speed produces outward-directed forces that balance the Earth's gravity. Many combinations of orbit speed, height above the Earth, and the Earth's gravity allow this balance. When an eastward-traveling satellite is approximately 35,800 km (22,300 mi) above the Earth's surface and its orbit coincides with the Earth's equatorial plane, the satellite remains above one point on the Earth's surface. This satellite is called a geosynchronous satellite, and it circles the Earth's center every twenty-four hours. NOAA's series of weather satellites (Geostationary Operational Environmental Satellites, or GOES) are geosynchronous satellites in equatorial orbits; GOES-8 was launched in 1994 and GOES-J in 1995. Two additional satellites were launched in 1998, one in polar orbit called NOAA-K, and GOES-10 in geosynchronous orbit. GOES-10 was launched to replace the faltering four-year-old GOES-8. GOES-10 provides important information about severe weather such as tornadoes, flash floods, and hurricanes as well as sea and lake ice conditions for much of the United States, Latin America, and the Pacific Ocean. GOES-9, launched in 1995, continues to operate well, providing coverage of the eastern United States and parts of the Atlantic Ocean. The GOES data provide cloud images and movies used on television weather reports.

A satellite in a low-Earth orbit that is perpendicular to the equatorial plane circles the Earth approximately every ninety minutes. Such a satellite has an observational band, or swath, that covers a new band of Earth surface each time the satellite crosses the equator. Complete Earth surface coverage is accomplished in twelve hours with a swath width of 22 ½° longitude at the equator.

Polar orbits that are displaced westward with the Sun, as the Earth turns eastward, are Sun-synchronous. NOAA's Advanced Television Infrared Observation Satellite (TIROS) weather satellite, the first Earth-observing satellite, was a Sun-synchronous satellite. It was launched in 1960 and has been followed by a series of TIROS satellites in polar orbits at heights of 550–850 km (340–500 mi).

NASA's NIMBUS-7 satellite, launched in 1978, monitored back-radiation and reflection from the Earth's surface and carried a sensor package called the Coastal Zone Color Scanner that detected multiband radiant energy reflecting from chlorophyll in sea and land plants. Data from NIMBUS-7 were used to make the images of the world's production of marine plant life in prologue figure XXI; the data were collected between 1978 and 1986, when this scanner failed.

NASA's Sea Satellite (SEASAT) was a specialized oceanographic satellite that was launched in June 1978 and operated only until October of that year. This short-lived satellite carried a radar altimeter capable of measuring the distance between the satellite and the sea surface with an accuracy of about 5 cm (2 in) over a narrow observational width. Radar echoes returning to multiple receiving antennas on the satellite measured the scattering patterns caused by surface waves. These patterns gave information on wave height related to wind speed and scattering direction related to

wind direction. The altimeter detected elevation changes in the sea surface that were used to measure tides, currents, seafloor topography, and changes in sea level. See figure 3.3 for an ocean-floor image made in this way.

The U.S. Navy Geodynamic Experimental Ocean Satellite (GEOSAT), launched in 1985, was designed to collect high-resolution data for the military, but its orbit was changed to replace the failed SEASAT, and from 1986–90, it monitored sea-level topography, surface winds and waves, local gravity changes, and abrupt boundaries between water types. GEOSAT's measurements of month-to-month sea-level changes were within 4 cm (1.6 in) of tide gauge records. It was also able to detect major currents and water boundaries on the basis of water-level changes.

Many satellites carry an AVHRR, advanced very high resolution radiometer, which senses sea surface temperatures using infrared radiation. See the satellite images related to El Niño in figure 6.29.

In 1992 Japan's Japanese Earth Resource Satellite (JERS-1) began an environmental and resource observation program, and in 1993 the joint U.S.–French ocean TOPography EXperiment, TOPEX/Poseidon system began a program measuring global sea levels, tides, and surface currents; it has also monitored interaction between the atmosphere and ocean surface. Altimetry measurements that are repeated every ten days provide readings of sea surface height to within 1–2 cm (less than an inch).

The European Environmental Remote Sensing satellites, ERS-1 launched in 1991 and ERS-2 launched in 1995, carry all-weather radar and microwave systems, enabling data to be taken even when the satellite's view is obscured by clouds and darkness. ERS-2 also measures the levels of ozone and trace gases in the Earth's outer atmosphere. The ERS satellites and JERS-1 are able to track ocean vessels, monitor sea conditions, detect oil spills from changes in wave patterns, and provide improved weather forecasting to the Southern Hemisphere. ERS-1 and ERS-2 fly similar paths and can simultaneously observe the same Earth area, allowing stereo images to be produced. Another European satellite with similar capabilities, Environmental Satellite ENVISAT, is scheduled for a January 2002 launch.

The SEASTAR, launched in 1997, carries a color scanner known as SeaWiFS; or sea-viewing wide-field-of-view sensor. SeaWiFS monitors the distribution of chlorophyll and plant life at the ocean surface; this had not been possible since the 1986 loss of the coastal zone color scanner aboard the NIMBUS-7. NASA launched its first Earth Observing System (EOS) satellite, Terra, in December of 1999. The Total Ozone Mapping Spectrometer (TOMS) was also launched in 1999 and is scheduled to be replaced by an advanced ozone mapping satellite OMI in early 2004. These satellites will complement ongoing programs by monitoring the roles of clouds, radiation, water vapor, and precipitation in our weather systems; the gas exchange between the oceans and the atmosphere; and the role of polar ice.

Each new satellite is assigned new tasks but also continues the measurements made by its predecessors. Consequently, vast amounts of new data are being collected, and data storage facilities and management and distributions systems are needed to make the data available to researchers. A new data system is being developed with the EOS project to serve up to 10,000 scientists around the world. As scientists struggle to forecast global environmental cycles in their efforts to maintain a healthy global environment, satellite oceanography becomes more and more important. It is costly, but it is the only way to achieve a global view of the Earth, its oceans, and its atmosphere.

To Learn More About Satellite Oceanography

Clery, D. 1993. ERS-1 Gives Europeans New Views of the Oceans. *Science* 260 (5115): 1742–43.

Lyon, K. G. 1993. SeaStar: Ocean Color, Space Commercialization. *Sea Technology* 34 (10): 17–21.

Sea Technology 38 (8) 1997. Issue on environmental monitoring and remote sensing.

Internet References

Visit the book's Online Learning Center at www.mhhe.com/sverdrup (click on the book's cover) to explore links to further information on related topics.

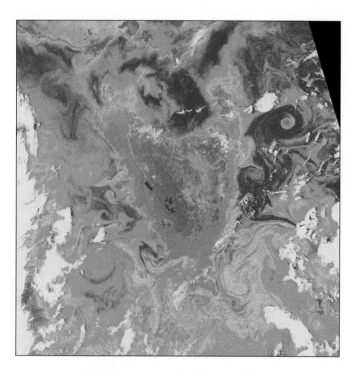

 web link **Prologue XXI** False color image of the oceans centered on the island of Tasmania. Tasmania is located south of the eastern coast of Australia. *Yellows* and *reds* indicate high concentrations of phytoplankton, *greens* and *blues* low concentrations, and *dark blue* and *purple* very low concentrations. The complex current interactions around the island have significant influence on the distribution of phytoplankton.

biological communities, particularly the small marine plants called phytoplankton, which are discussed in detail in chapter 15 (prologue fig. XXI).

In 1985 the U.S. Navy Geodynamic Experimental Ocean Satellite *(GEOSAT)* was launched. It was designed to collect data for military purposes but its orbit was changed to replace the failed *SEASAT*. From 1986–90 it monitored sea-level topography, surface winds and waves, local gravity changes, and abrupt "boundaries" in the ocean caused by changes in salinity and temperature.

Oceanography of the Recent Past, the Present, and the Future

Today's scientists see the Earth not as a single system but as a complex of systems and subsystems acting as a whole. Projects that emerged in the 1990s and continue in the 2000s require that scientists cross from one discipline to another and share information for common goals. Satellites are used for global observation. Earth and ocean scientists are able to manipulate online and archival data quickly by computer at sea or on land, and they share it rapidly over the Internet. In addition, successful integrated approaches to Earth studies require that governments, agencies, universities, and national and international programs agree to set common priorities and to share program results.

Several large-scale oceanographic programs have been developed to better understand the role of the oceans in processes of the atmosphere-ocean-land system. These programs provide data for models that scientists use to predict the evolution of the Earth's environment as well as the consequences of human-influenced changes. The World Ocean Circulation Experiment (WOCE) studies the world oceans using computer models and chemical tracers to model the present state of the oceans and then predict ocean evolution in relation to long-term changes in the atmosphere. This effort combines sampling by ship, satellites, and floating independent buoys with sensors. The U.S. Joint Global Ocean Flux Study (JGOFS) tracks the relationship between ocean plant production and solar radiation. Scientists are monitoring the worldwide abundance of plant life by ship and satellite to understand how carbon and other biologically active elements move between the ocean, atmosphere, and land. The Global Ocean Atmosphere-Land System (GOALS) studies the energy transfer between the atmosphere and the tropical oceans to better understand El Niño and its effects and to provide improved large-scale climate prediction.

Two large international programs, the Deep Ocean Drilling Program (DODP) and the Ridge Interdisciplinary Global Experiment (RIDGE) explore the Earth's ocean floors and the margins of the continents. They investigate the structure and history of the Earth and probe the ocean's great mountain range systems and their relationships to the chemistry of the oceans.

In August 1992, the satellite *TOPEX/Poseidon* was launched in a joint U.S.–French mission to explore ocean circulation and its interaction with the atmosphere. *TOPEX/Poseidon* measures sea level along the same path every ten days. This information is used to relate changes in ocean currents with atmospheric and climate patterns. The measurements obtained allow scientists to chart the height of the seas across ocean basins with an accuracy of less than 10 cm (4 in). *TOPEX/Poseidon*'s three-year prime mission ended in the fall of 1995 and is now in its extended observational phase. A major follow-on mission to continue these studies began in December, 2001 with the launch of *Jason-1*. *Jason-1*'s mission is the same as *TOPEX/Poseidon*'s but it is designed to acquire continuous data over longer periods of time in order to measure long-term circulation processes more accurately.

A meeting of representatives from twenty-four nations recommended the development of a Global Ocean Observing System (GOOS) to include satellites, buoy networks, and research vessels. The goal of this program is to enhance our understanding of ocean phenomena so that events such as El Niño and its impact on climate can be predicted more accurately and with greater lead time. The successful prediction of the 1997–98 El Niño six months in advance of its peak made it possible to plan for its arrival.

An integral part of the GOOS is a project called Argo, named after the mythical vessel used by the ancient Greek seagoing hero Jason. Argo is an international project that will deploy an array of 3000 independent instruments, or floats, throughout the oceans by the year 2005 (prologue fig. XXII*a*). The first of these floats are being deployed and tested now. Each float will be programmed to descend to a depth of 2000 m (6560 ft, or about 1.25 miles), where it will remain for about ten days (prologue fig. XXII*b*). It will then ascend to the surface, measuring temperature and salinity as it rises. When it reaches the surface it will relay the data to shore via satellite and then descend once again and wait for its next cycle. In this fashion the entire array will provide detailed temperature and salinity data of the upper 2000 m of the oceans every ten days.

The United Nations designated 1998 as the Year of the Ocean. Goals included (1) a comprehensive review of federal ocean policies and programs to ensure coordinated advancements leading to beneficial results and (2) raising public awareness of the significance of the oceans in human life and the impact that human life has on the oceans. Also in 1998, the National Research Council's Ocean Studies Board released a report highlighting three areas that are likely to be the focus of future research: (1) improving the health and productivity of the coastal oceans, (2) sustaining ocean ecosystems for the future, and (3) predicting ocean-related climate variations.

Up until the late nineteenth century the great oceanographic voyages were largely voyages of exploration. Explorers such as James Cook and scientists such as those

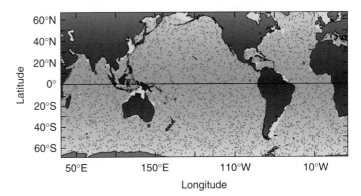

(a)

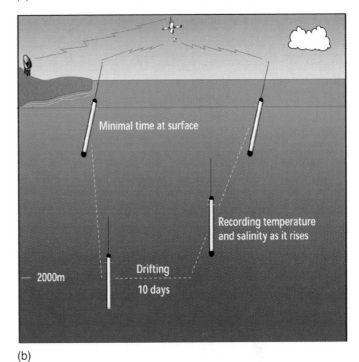

(b)

web link **Prologue XXII** (a) The Argo array will consist of 3000 autonomous floats spread throughout the oceans. (b) Every ten days the float will ascend to the surface, measuring temperature and salinity as it rises. When it reaches the surface it will transmit the data to shore via satellite and then descend to 2000 m (6560 ft) once again.

on the *Challenger* set sail into unknown waters to discover what they could find. With the beginning of the twentieth century modern oceanographic science matured to become a process of hypothesis testing. Modern-day oceanographers typically first form a new idea based on current knowledge and then carefully design an experiment to test it. The sense of exploration has largely been in the background in recent decades with a few clear exceptions, such as the discovery of hydrothermal vent systems on mid-ocean ridges. The Office of Ocean Exploration (OOE) at NOAA was created in 2001 to encourage and fund new exploratory missions in the oceans once again. The OOE will also fund the development of new technology to support underwater exploration, such as crewed or robotic submersibles and underwater-imaging systems.

Although large-scale, federally funded studies are presently in the forefront of ocean studies, it is important to remember that studies driven by the specific research interests of individual scientists are essential to point out new directions for oceanography and other Earth sciences. In the following chapters, you will follow the development of the ideas that have enabled us to build an understanding of the dynamic and complex systems that are the Earth's oceans.

Summary

Oceanography is a multidisciplinary field in which geology, geophysics, chemistry, physics, meteorology, and biology are all used to understand the oceans. Early information about the oceans was collected by explorers and traders such as the Phoenicians, the Polynesians, the Arabs, and the Greeks. Eratosthenes calculated the circumference of the Earth, and Ptolemy produced the first world atlas.

During the Middle Ages, the Vikings crossed the North Atlantic, while shipbuilding and chartmaking improved. In the fifteenth and sixteenth centuries the Chinese, Dias, Columbus, da Gama, Vespucci, and Balboa made voyages of discovery. Magellan's expedition became the first to circumnavigate the Earth. In the sixteenth and seventeenth centuries, some explorers searched for the Northwest Passage while others set up trading routes to serve developing colonies.

By the eighteenth century, national and commercial interests required better charts and more accurate navigation techniques. Cook's voyages of discovery to the Pacific produced much valuable information, and Franklin sponsored a chart of the Atlantic's Gulf Stream. A hundred years later, the U.S. Navy's Maury collected wind and current data to produce current charts and sailing directions and then wrote the first book on oceanography.

Ocean science began with the nineteenth-century expeditions and research of Darwin, Forbes, Müller, and others. The three-and-a-half-year *Challenger* expedition laid the foundation for modern oceanography with its voyage, which gathered large quantities of data on all aspects of oceanography. Exploration of the Arctic and Antarctic Oceans was pursued by Nansen and Amundsen into the beginning of the twentieth century.

In the twentieth century, private institutions played an important role in developing U.S. oceanographic research, but the largest single push came from the needs of the military during World War II. After the war, large-scale government funding and international cooperation allowed oceanographic projects that made revolutionary discoveries about the ocean basins. Development of electronic equipment, deep-sea drilling programs, research submersibles, and use of satellites continued to produce new and more detailed information of all kinds. At present, oceanographers are focusing their research on global studies and the management of resources as well as continuing to explore the interrelationships of the chemistry, physics, geology, and biology of the sea.

Study Questions

1. Eratosthenes estimated the circumference of the Earth at approximately 25,000 miles. Compare this estimate with the circumference used by Ptolemy. What difference would it have made to later voyages of discovery if Eratosthenes' measurement had been used rather than that of Ptolemy?
2. Who first assigned the name America to the "New World"? For whom was it named?
3. Why was there such great interest in finding and establishing a Northwest Passage?
4. Who first understood the tides and published an explanation of them?
5. What were Captain James Cook's contributions to our understanding of the oceans?
6. Why did Benjamin Franklin consider it so important to chart the Gulf Stream current?
7. Who was Matthew F. Maury, and why is he considered by many to be the "founder of oceanography"?
8. Why do you think Edward Forbes concluded that there was no life in the oceans below 1800 ft?
9. What did the engineers who laid the first transatlantic cable need to know about the oceans?
10. The *Challenger* and its expedition are often called unique. Why is this term used? What were the benefits of this expedition to the science of the oceans?
11. What was Fridtjof Nansen trying to prove by freezing the *Fram* into the polar ice?
12. The amount of ocean data has been expanding at an ever-increasing rate since the first years of ocean exploration. Why?
13. How has each of the following affected twentieth-century oceanography? (a) Economics. (b) Commerce and transportation. (c) Military needs.
14. In what ways have computers altered oceanography?
15. What are the reasons for the increased global interest in resources of the sea? What types of management do you think may be required for these resources in the future?

Links to Related Websites

Visit the book's Online Learning Center at www.mhhe.com/sverdrup (click on the book's cover) to find live Internet links for additional topics related to this chapter's content.

- Timeline of ocean exploration
- The early times: Polynesians

- The Middle Ages: The Vikings
- Voyages of discovery
- The importance of charts and navigational information
- The *Challenger* expedition
- Oceanography as science
- Oceanography in the twentieth century
- Oceanography of the recent past, the present, and the future
- Oceanographic institutions, museums, and aquaria

Visit the book's Online Learning Center at www.mhhe.com/sverdrup (click on this book's cover) to find these additional chapter tools: Suggested Readings; links to further information on boxed readings, selected figures, and related chapter topics; and additional study aids.

chapter 1

The Water Planet

Early morning cumulus cloud forming over the South China Sea.

One evening he asked the miller where the river went. "It goes down the valley," answered he, "and turns a power of mills—six score mills, they say, from here to Unterdeck—and it none the wearier, after all. And then it goes out into the lowlands, and waters the great corn country, and runs through a sight of the fine cities (so they say) where kings live all alone in great palaces, with a sentry walking up and down before the door. And it goes under bridges with stone men upon them, looking down and smiling so curious at the water, and living folks leaning their elbows on the wall and looking over too. And then it goes on and on, and down through marshes and sands, until at last it falls into the sea, where the ships are that bring parrots and tobacco from the Indies. Ay, it has a long trot before it as it goes over our weir, bless its heart!"

"And what is the sea?" asked Will.

"The sea!" cried the miller. "Lord help us all, it is the greatest thing God made! That is where all the water in the world runs down into a great salt lake. There it lies, as flat as my hand and as innocent-like as a child, but they do say when the wind blows it gets up into water-mountains bigger than any of ours, and swallows down great ships bigger than our mill, and makes such a roaring that you can hear it miles away upon the land. There are great fish in it five times bigger than a bull, and one old serpent as long as our river and as old as all the world, with whiskers like a man, and a crown of silver on her head."

Robert Louis Stevenson
From *The Merry Men*

Questions concerning the origin and the age of the planet we call the Earth are related to the origin of the universe and the beginnings of our solar system. Many theories have been suggested for these origins. In this chapter we consider the most widely accepted hypotheses for the birth of our universe and our planet as well as the methods now used to measure time and calculate the age of the Earth.

Water did not exist on the Earth in the beginning, but its formation on a planet that was neither too far from nor too close to the Sun, neither too cold nor too hot, changed the Earth and allowed the development of life. In this chapter we begin to investigate our water planet (fig. 1.1)—the cycling of water, its distribution, and its largest reservoirs, known as the oceans. We will begin to understand how scientists and mariners find their way about the oceans, and we will learn some of the basic principles required for the study of the oceans, which we call oceanography.

Beginnings 1.1

Origin of the Universe

For centuries, our concept of the nature of the universe was governed by visual observations from the Earth's surface. Our current understanding of its history and structure have been greatly enhanced by observations made with instruments that are sensitive to energy across the **electromagnetic spectrum,** from radio waves to gamma rays, as well as the operation of the optical *Hubble Space Telescope (HST)* (fig. 1.2).

The *HST* was deployed in April 1990, in low-Earth orbit at an altitude of 595 km (370 mi), where it circles the Earth every ninety-seven minutes. Because of its location above the Earth's atmosphere, the 2.4 m (94.5 in) reflecting telescope (the size of a reflecting telescope refers to the diameter of its mirror) has an optical resolution, or image clarity, that is about ten times better than the best ground-based telescopes can achieve. It can detect objects one-billionth as bright as the human eye is capable of detecting.

In recent years, observational data have provided increasing evidence that the universe originated in an event known as the **Big Bang.** The Big Bang model envisions all energy and matter in the universe as having initially been concentrated in an extremely hot, dense singularity much smaller than an atom. Roughly 13 billion years ago this singularity experienced a cataclysmic explosion that caused the universe to rapidly expand and cool as it grew larger. One second after the Big Bang, the temperature of the universe was about 10 billion K (roughly 1000 times the temperature of the Sun's interior). The Kelvin (K) temperature scale is an absolute temperature scale. On this scale, 0 K is absolute zero, the coldest possible temperature. At this temperature all atoms and molecules would stop moving. Room temperature is about 300 K (see appendix B for conversions from K to °C and °F). At this time the universe consisted mostly of elementary particles, light, and other forms of radiation. The elementary particles, such as protons and electrons, were too energetic to combine into atoms. One hundred seconds after the Big Bang the temperature had cooled to about 1 billion K (roughly the temperature in the centers of the hottest stars at the present time).

The universe was now cool enough for protons, the nuclei of hydrogen atoms, as well as nuclei of deuterium, helium, and lithium to begin to form. While the temperature was still very high, the universe was dominated by the radiation. Later, as the universe cooled, matter took over. Eventually, when the temperature had dropped to a few thousand degrees, electrons and nuclei would have started to combine to form atoms, and strong interactions between matter and radiation ceased. It was then possible for small perturbations in the distribution of matter to begin to grow gravitationally. Denser, cooler regions pulled in additional matter gravitationally, increasing their density even further. An extraordinary composite image was compiled from hundreds of millions of individual observations obtained by the *COBE (Cosmic Background Explorer)* satellite over a period of four years. The spatial variation in temperature, which reflects the clumping of mass in the universe, approximately 300,000 years after the Big Bang, can be seen clearly (fig. 1.3). The variation in mass density was very small, just 0.001% change from the highest density to the lowest. A billion years or so after the Big

web ⚭ link **Figure 1.1** The water planet. Earth as seen from space.

web ⚭ link **Figure 1.2** The *Hubble Space Telescope (HST)* is a joint venture between the European Space Agency and the National Aeronautics and Space Administration. It was first proposed in the 1940s, designed and constructed in the 1970s and 1980s, and began its operational life with its launch in 1990.

Bang, gravity began to pull matter into the structures we see in the universe today. The first stars probably formed many billions of years before the Sun.

The universe has a distinct structure. On a small scale, there are individual stars. Stars are responsible for the formation of elements heavier than lithium. Stars fuse hydrogen and helium in their interiors to form heavier elements such as carbon, nitrogen, and oxygen. The higher temperatures of more massive stars continue the nuclear fusion process to create elements as heavy as iron. These elements, so important in oceanographic processes, are created in stars and were not made by the Big Bang at the formation of the universe. Some of these stars are at the center of solar systems like our own, with planets that orbit them. Detecting the presence of planets orbiting other stars is extremely difficult because they are small and dark, and the brilliance of their parent stars commonly conceals their presence. Consequently, the first confirmation of extra-solar planets came from indirect observations. Both a large planet and the star it orbits will rotate around their common center of mass, causing a wobble in the star's motion. The existence of planets outside of our solar system was first demonstrated several years ago by the detection of just such a wobble in other stars. Astronomers have recently directly observed a number of faint points of light in the constellation Orion that they believe are planetlike objects a few times more massive than Jupiter in our solar system. These objects can only be seen because they are very young and still warm after the process of formation, and they do not orbit a nearby star that would mask the light they radiate.

Galaxies are composed of clumps of stars. Our galaxy, the Milky Way, is composed of about 200 billion stars. It is shaped like a flattened disk, with a thickness of about 1000 **light-years** and a diameter of about 100,000 light-years. A light-year is equal to the distance light travels in one year, which is 9.46×10^{12} km (5.87×10^{12} mi). The observable universe contains from 10 billion to 100 billion galaxies. Galaxies are preferentially found in groups called **clusters.** A single cluster may contain thousands of galaxies. Clusters typically have dimensions of 1 million to 30 million light-years. Individual clusters tend to

www.mhhe.com/sverdrup

group in long, stringlike or walllike structures called superclusters. Superclusters may contain tens of thousands of galaxies. The largest supercluster known is about 500 million light-years across. At very large scales, the universe looks something like a sponge, with galaxies arranged in interconnected lines and sheets interspersed with huge bubbles in which very few galaxies are seen.

Throughout the universe, some stars are burning out or exploding; others are still being formed, incorporating original matter from the Big Bang and recycling matter from older generations of stars. We do not yet have sufficient information to know the fate of our universe. Some theorists hold that the universe will expand indefinitely; others believe it will expand to a point and then collapse back to produce another Big Bang.

Origin of Our Solar System

Present theories attribute the beginning of our solar system to the collapse of a single, rotating interstellar cloud of gas and dust that included material that was produced within older stars and liberated into space when the old stars died. This rotating cloud, or **nebula,** appeared about 5 billion years ago. The shock wave from a nearby exploding star, or supernova, is thought to have imparted spin to the cloud, pushing it together and causing it to shrink from its own gravitational pull. As the nebula collapsed, its speed of rotation increased, and, heated by its own gravitational energy, its temperature rose. The gas and dust, spinning faster and faster, contracted parallel to the axis of spin, forming a disk. At the center of the disk a star, our Sun, was formed. Self-sustaining nuclear reactions kept the Sun hot, but the outer regions began to cool. In this cooler outer portion of the rotating disk, molecules of gas and dust began to collide, accrete (or stick together), and chemically interact. The collisions and interactions produced particles that grew from further accretion of other particles and became large enough

to have sufficient gravity to attract still other particles. The planets of our solar system had begun to form. After a few million years the Sun was orbited by nine planets (in order from the Sun): Mercury, Venus, Earth, Mars, Jupiter, Saturn, Uranus, Neptune, and Pluto.

If Mercury, Venus, Earth, and Mars are compared to Jupiter, Saturn, Uranus, and Neptune, the four planets closer to the Sun are seen to be much smaller in diameter and mass. (See table 1.1. Note use of metric units; see appendix B for further information.) These four inner planets are rich in metals and rocky materials. The four outer planets are cold giants, dominated by ices of water, ammonia, and methane. Their atmospheres are made up of helium and hydrogen; the planets

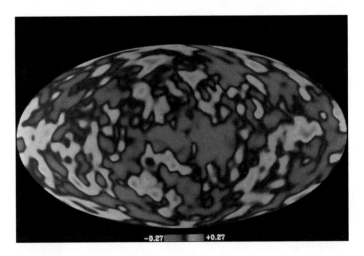

web link **Figure 1.3** Structure of the early universe about 300,000 years after the Big Bang, as seen by the *Cosmic Background Explorer (COBE)* satellite. The red spots are slightly warmer than the mean temperature, and the blue spots are slightly cooler. The difference in temperature between the two is about 0.001% of the mean temperature.

Table 1.1 Features of the Planets in the Solar System

Planet	Mean Distance from Sun (10^6 km)	Diameter (km)	Mass Relative to Earth Mass	Rotation Period[1] (hours, days)	Orbit Period (years)	Mean Temperature of Surface (°C)	Principal Atmospheric Gases[2]
Mercury	57.9	4,878	0.055	58.6 d	0.24	−170 night 430 day	Na
Venus	108.2	12,104	0.815	−243 d	0.62	−23 clouds 480 surface	CO_2, N_2
Earth	149.6	12,756	1.000	23.94 h	1.00	16	N_2, O_2
Mars	227.9	6,787	0.107	−24.62 h	1.88	−50 (average)	CO_2, N_2, Ar
Jupiter	778.3	142,800	317.8	9.93 h	11.86	−150	H_2, He, CH_4, NH_3
Saturn	1429	120,000	95.2	10.5 h	29.48	−180	H_2, He, CH_4, NH_3
Uranus	2875	50,800	14.5	−17.24 h	84.01	−210	H_2, He, CH_4
Neptune	4504	48,600	17.2	16 h	164.8	−220	H_2, CH_4, He?
Pluto	5900	2,245	0.002	6.4 d	247.71	−230	CH_4 (temporary)

1. Negative rotation period indicates rotation in a direction opposite that of the Earth's rotation.
2. Na = sodium; CO_2 = carbon dioxide; N_2 = nitrogen; O_2 = oxygen; Ar = argon; H_2 = hydrogen; He = helium; CH_4 = methane; NH_3 = ammonia.

Origin of the Oceans

The oldest sedimentary rocks found on Earth, rocks that formed by processes requiring liquid water at the surface, are about 3.9 billion years old. This indicates that there have been oceans on the Earth for approximately 4 billion years. Where did the water in the oceans come from? There are two possible sources for this water, the interior of the Earth and outer space.

Traditionally, scientists have suggested that the water in the oceans and atmosphere originated in the interior of the Earth in a region called the mantle (discussed in more detail in chapter 2, section 2.1) and was brought to the surface by volcanism, a process that continues to this day (box fig. 1). The rock that makes up the mantle is thought to be similar in composition to meteorites, which contain from 0.1%–0.5% water by weight. The total mass of rock in the mantle is roughly 4.5×10^{27} g, thus the original mass of water in the mantle would have been approximately 4.5×10^{24} to 2.25×10^{25} g. This is from three to sixteen times the amount of water currently in the oceans, so it is clear that the mantle is an adequate source for the water in the oceans; but is enough water brought to the surface through volcanism to actually fill the oceans? Magmas erupted by volcanoes contain dissolved gases that are held in the molten rock by pressure. Most magmas consists of 1%–5% dissolved gas by weight, most of which is water vapor. The gas that escapes from Hawaiian magmas is about 70% water vapor, 15% carbon dioxide, 5% nitrogen, and 5% sulfur, with the remainder consisting mostly of chlorine, hydrogen, and argon. It is estimated that thousands of tons of gas are ejected in volcanic eruptions each day. Undoubtedly the rate of volcanic eruptions on the Earth has varied with time, probably being much greater earlier in Earth's history when the planet was still very hot. However, if we conservatively assume that the present rate of ejection of water vapor by volcanism has been roughly constant over the last 4 billion years, then the volume of water expelled by volcanoes would have produced roughly 100 times the volume of water in the oceans.

The traditional view of the interior of the Earth serving as the source of ocean water has recently been challenged by a bold new suggestion that large volumes of water are continually being added

 Box Figure 1 Volcanic eruptions release gases including water vapor, carbon dioxide, and sulfur dioxide to the atmosphere and oceans. Even if it is not erupting, an active volcano can release thousands of tons of sulfur per day.

 Box Figure 2 A prelaunch photo of the *Dynamics Explorer 1 and 2 (DE-1/DE-2)* spacecraft stack before being covered by their fairings and mated with the Delta Launch vehicle. *DE-1* is on the bottom. *DE-1* was launched into a high-altitude elliptical orbit while *DE-2* was launched into a lower orbit.

to the Earth from outer space. Evidence for this idea comes from data collected by a polar-orbiting satellite called the *Dynamics Explorer 1 (DE-1)* (box fig. 2). The *DE-1* carried an ultraviolet photometer capable of taking pictures of Earth's **dayglow.** Dayglow is ultraviolet light, invisible to the naked eye, emitted by atomic oxygen in the upper atmosphere when it absorbs and reradiates electromagnetic energy from the Sun. In many of the dayglow images of Earth obtained by the satellite, there are distinct dark spots, roughly 48 km (30 mi) in diameter, that appear to move across the face of the Earth suggesting that they were caused by moving objects (box fig. 3). The direction of motion of the dark spots matches the direction of motion of meteoritic material as it approaches Earth. Astronomer Louis

Frank has suggested that these dark spots are created when small icy comets vaporize in the outer atmosphere, creating clouds of water vapor that absorb the ultraviolet radiation of Earth's dayglow over a small area, thus creating a dark spot in the bright ultraviolet background. The size of the spots implies that the average mass of the comets is about 10 kg (22 lb). He estimates that there are an average of twenty of these comets that enter the atmosphere each minute, or a staggering 10 million each year. If all of the water in these comets condensed to form a layer on the surface of the Earth it would be roughly 0.0025 mm (0.0001 in) deep. While this doesn't seem like a significant amount of water, over 4 billion years this rate of accumulation would fill the oceans two to three times.

There is still some debate about the role of comet impacts in the formation of the oceans. Additional study may give us further insight to the relative importance of volcanism and the impact of extraterrestrial objects in creating the oceans. It is very likely that both processes have contributed to their formation.

To Learn More About the Formation of the Oceans

Delsemme, A. H. 2001. An Argument for the Cometary Origin of the Biosphere. *American Scientist* 89 (5): 432–42.

Frank, L. A., J. B. Sigwarth, and J. D. Craven. 1986. On the influx of small comets into Earth's upper atmosphere. *Observations and Interpretations Geophysical Research Letters* 13 (4): 303–10.

Kasting, J. F. 1998. The Origins of Water on Earth. *Scientific American Presents: The Oceans* 9 (3): 16–21.

Vogel, S. 1996. Living Planet. *Earth* 5 (2): 26–35.

Weisburd, S. 1985. Atmospheric Footprints of Icy Meteors. *Science News* 128: 391.

Internet References

Visit the book's Online Learning Center at www.mhhe.com/sverdrup (click on the book's cover) to explore links to further information on related topics.

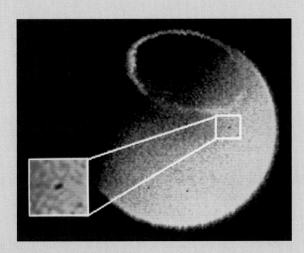

web link **Box Figure 3** Image of the Earth's dayglow at ultraviolet wavelengths taken from an altitude of 18,500 km (11,500 mi). The dayglow is due to the excitation of atomic oxygen by solar radiation. *Inset* shows a magnified view of a dark spot, or "hole," in the dayglow thought to be caused by the vaporization of small cometlike balls of ice.

located nearer the Sun lost these lighter gases because the higher temperature and intensity of solar radiation tend to push these gases out and away from the center of the solar system. If the mass of each planet in table 1.1 is divided by its volume, the results will show that the outer planets are composed of lighter, or less dense, materials than the inner planets.

Pluto, a little-known small planet at least 500 times less massive than the Earth, has an elliptic orbit that takes it inside the orbit of Neptune. It has a reflective surface that may be primarily composed of frozen methane. Because of its unusual orbit, it has been suggested that Pluto might at one time have been a satellite of Neptune.

Early Planet Earth

Before the beginning of the geologic record, during the first billion years of the Earth's existence, the Earth is thought to have been a mixture of silicon compounds, iron and magnesium oxides, and small amounts of other naturally occurring

elements. According to this model, the Earth formed originally from cold matter, but events occurred that raised the Earth's temperature and initiated processes that obliterated its earlier history and resulted in its present form. The early Earth was bombarded by particles of all sizes, and a portion of their energy was converted into heat on impact. Each new layer of accumulated material buried the material below it, trapping the heat and raising the temperature of the Earth's interior. At the same time, the growing weight of the accumulating layers compressed the interior, and the energy of compression was converted to heat, raising the Earth's internal temperature to approximately 1000°C. Atoms of radioactive elements, such as uranium and thorium, disintegrated by emitting subatomic particles that were absorbed by the surrounding matter, further raising the temperature.

Some time during the first few hundred million years after the Earth formed, its interior reached the melting point of iron and nickel. When the iron and nickel in the planet melted, they migrated toward the center. Frictional heat was generated, and lighter substances were displaced. In this way the temperature of the Earth was raised to an average of 2000°C. The less dense material from the partially molten interior moved upward and spread over the surface, cooling and solidifying. The melting and solidifying probably happened repeatedly, separating the lighter, less dense compounds from the heavier, denser substances in the interior of the planet. In this way the Earth became completely reorganized and differentiated into a layered system, which is explored in greater detail in chapter 2.

The Earth's oceans and atmosphere are probably both, at least in part, by-products of this heating and differentiation. As the Earth warmed and partially melted, water locked in the minerals as hydrogen and oxygen was released and brought to the surface along with other gases in volcanic eruptions. As the Earth's surface cooled, water vapor was able to condense to form the oceans. Another possible source of water for the oceans is from space objects, such as cometlike balls of ice or meteorites, that have collided with the Earth throughout its history. The origin of the oceans is discussed further in the box titled "Origin of the Oceans."

At first the Earth must have been too small and had too little gravity to have accumulated an atmosphere. It is generally believed that during the process of differentiation, gases released from the Earth's hot, chemically active interior formed the first atmosphere, which was primarily made up of water vapor, hydrogen gas, hydrogen chloride, carbon monoxide, carbon dioxide, and nitrogen. Any free oxygen present would have combined with the metals of the crust to form compounds such as iron oxide. Oxygen gas could not accumulate in the atmosphere until its production exceeded its loss by chemical reactions with the Earth's crust. Oxygen production did not exceed loss until life evolved to a level of complexity in which green plants could convert carbon dioxide and water with the energy of sunlight into organic matter and free oxygen. This process and its significance to life are discussed in chapters 5 and 14.

Age and Time   1.2

Age of the Earth
Over the centuries, people have asked the question, How old is the Earth? In the seventeenth century Archbishop Ussher of Ireland attempted to answer the question by counting the generations listed in the Bible; he determined that the first day of creation was Sunday, October 23, 4004 B.C. In the late 1800s the English physicist Lord Kelvin calculated the time necessary for molten rock to cool to present temperatures and dated the Earth as 20 million to 40 million years old. In 1899 another physicist, John Joly, used the amount of salt in the oceans to determine the Earth's age and calculated time for the rivers to wash the salt from the land to be 100 million years.

It was not until scientists understood radioactive decay, and a method was developed to apply it to the dating of rock samples, that reliable age data became available.

This method, known as **radiometric dating,** uses radioactive **isotopes** of certain elements.

An atom of radioactive isotope has an unstable nucleus. This unstable nucleus changes, or decays, and emits one or more particles plus energy. For example, the radioactive isotope carbon-14 decays or changes to nitrogen-14; uranium-235 decays to lead-207; and potassium-40 decays to argon-40. The time at which any single nucleus will decay is unpredictable, but if large numbers of atoms of the same radioactive isotope are present it is possible to predict that a certain fraction of the isotope will decay over a certain period of time. The time over which one-half of the atoms of a radioactive isotope decay, or over which the atom changes from one element (the parent element) to another element (the daughter product), is known as the isotope's **half-life** (fig. 1.4). The half-life of each radioactive isotope is characteristic and constant. For example, the half-life of carbon-14 is 5730 years; that of uranium-235 is 704 million years, and that of potassium-40 is 1.3 billion years. Therefore, if a substance is found that was originally made up only of atoms of uranium-235, in 704 million years, by a series of reactions, the substance is one-half uranium-235 and one-half lead-207. In another 704 million years three-quarters of the substance will be lead-207 and only one-quarter uranium-235. Because each radioactive isotopic system behaves uniquely in nature, data must be carefully tested, compared, and evaluated. The best data are those in which different radioactive isotopic systems give the same date.

The Earth is an active planet, and its original surface rocks no longer exist. The oldest rocks on the surface of the Earth have been dated at slightly more than 4.0 billion years old. The oldest Moon rocks are dated at a little over 4.4 billion years old. Meteorites that have survived the journey from space through the Earth's atmosphere have been dated between 4.5 billion and 4.6 billion years old. The substances found in many meteorites represent materials that condensed out of hot gases thought to be present at the beginning of the solar system. These ages agree with theoretical calculations made for the age of the Sun. This information sets the accepted age of the Earth at about 4.6 billion years.

Geologic Time
To refer to events in the history and formation of the Earth, scientists use geologic time (table 1.2). The principal divisions are the four eons: the Hadean (4.6 to 4.0 billion years ago), the Archean (4.0 to 2.5 billion years ago), the Proterozoic (2.5 billion to 570 million years ago), and the Phanerozoic (since 570 million years ago). The first three eons are collectively popularly known as the Precambrian. Fossils are known from other eons but are common only from the Phanerozoic. This eon is divided into three eras: the Paleozoic era of ancient life; the Mesozoic era of middle life (popularly called the Age of Reptiles); and the Cenozoic era of recent life (the Age of Mammals). Each of these eras is subdivided into periods and epochs; today, for instance, we live in the Holocene epoch of the Quaternary period of the Ceno-

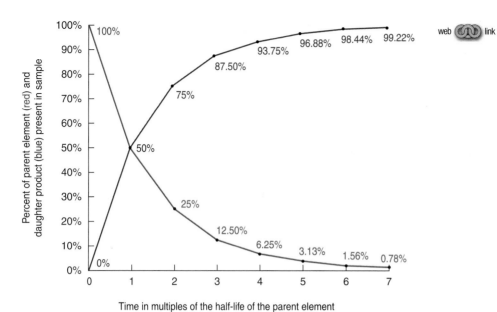

web ⊕ link **Figure 1.4** The half-life for a radioactive isotope is the time in which half of the parent element atoms decay to produce daughter product atoms. The radioactive decay curve is exponential. It displays the percentage of original parent element atoms and daughter product atoms as a function of time (measured in multiples of the half-life of the parent element).

zoic era. The appearance or disappearance of fossil types was used to set the boundaries of the time units before radiometric dating enabled scientists to set these timescale boundaries more accurately. The accurate calibration of radiometric dates and relative time determined from fossils is an ongoing process, and the exact dates defining the time units are constantly being adjusted with the acquisition of new data. Dating time units in the Paleozoic era and Precambrian eons is particularly difficult because of the lack of very old marine sediments and fossils in the modern ocean basins.

Very long periods of time are incomprehensible to most of us. We often have difficulty coping with time spans of more than ten years—What were you doing exactly ten years ago today? We have nothing with which to compare the 4.6-billion-year age of the Earth or the 500 million years since the first **vertebrates** (animals with a spinal column) appeared on this planet. In order to place geologic time in a framework we can understand, let us divide the Earth's age by 100 million. If we do so, then we can think of the Earth as being just forty-six years old. What has happened over that forty-six years?

There is no remaining record of events during the first three years. The earliest history preserved can be found in some rocks of Canada, Africa, and Greenland that formed forty-three years ago. Sometime between thirty-five and thirty-eight years ago the first primitive living cells of bacteria-type organisms appeared. Oxygen production by living cells began about twenty-three years ago, half the age of the planet. Most of this oxygen combined with iron in the early oceans and did not accumulate in the atmosphere. It took about eight years, or until roughly fifteen years ago, for enough oxygen to accumulate in the atmosphere to support significant numbers of complex oxygen-requiring cells. Oxygen reached its present concentration in the atmosphere approximately eleven years later. The first invertebrates (animals without backbones) developed seven years ago and two years later the first vertebrates (animals

with backbones) appeared. Primitive fish first swam in the oceans, and corals appeared just five years ago. Three years, eight and one-half months ago the first sharks could be found in the oceans and roughly five months later reptiles could be found on land. A massive extinction struck the planet just two and one-half years ago, killing 96% of all life. Following this catastrophic event the dinosaurs appeared just two years, three months, and ten days ago. Three and one-half weeks later the first mammals developed. A second major extinction occurred roughly two years ago. This event killed over half of the species on the Earth, leaving mostly dinosaurs on land. Just eighteen and one-half months ago the first birds flew in the air, and they would have seen the first flowering plants a little less than five months later. A third major extinction struck the Earth two hundred thirty-seven days ago, killing off the dinosaurs as well as many other species. Two hundred eleven days ago the mammals, birds, and insects became the dominant land animals. Our first human ancestors (the first identifiable member of the genus *Homo*) appeared just a little less than six days ago. About half an hour ago, modern humans began the long process we know as recorded civilization, and only one minute ago, the Industrial Revolution began changing the Earth and our relationship with it for all time.

Natural Time Periods

People first defined time by the natural motions of the Earth, Sun, and Moon. Later, people grouped natural time periods for their convenience, and still later, artificial time periods were created for people's special uses. Time is used to determine the starting point of an event, the event's duration, and the rate at which the event proceeds. An accurate measurement of time is required to determine location or position; this use of time is discussed in section 1.4.

The year is the time required by the Earth to complete one orbit about the Sun. The time required for this orbit is 365¼ days, adapted for convenience to 365 days with an

Table 1.2 The Geologic Timescale

Eon	Era	Period	Epoch	Began (millions of years ago)	Life Forms/Events
Phanerozoic	Cenozoic "Age of Mammals"	Quaternary	Holocene	0.01	Modern humans
			Pleistocene	1.6	Earliest humans
		Tertiary	Pliocene	5.3	
			Miocene	23.7	Earliest hominids
			Oligocene	36.6	Flowering plants
			Eocene	57.8	Earliest grasses Mammals, birds, and insects dominant
			Paleocene	65	

Cretacecus-Teritiary boundary: The extinction of dinosaurs and many other species at the end of the Mesozoic era (65 million years ago).

Eon	Era	Period	Epoch	Began (millions of years ago)	Life Forms/Events
	Mesozoic "Age of Reptiles"	Cretaceous		144	Earliest flowering plants (115) Dinosaurs in ascendence
		Jurassic		208	First birds (155) Dinosaurs abundant

Triassic - Jurassic boundary: The extinction of over 50% of all species on Earth, including the last of the mammallike reptiles, leaving mainly dinosaurs on land (208 million years ago).

Eon	Era	Period	Epoch	Began (millions of years ago)	Life Forms/Events
		Triassic		245	First turtles (210) First mammals (221) First dinosaurs (228) First crocodiles (240)

Permian-Triassic boundary: The greatest mass extinction of all time; 96% of all life on Earth perishes at the end of the Paleozoic era (245 million years ago).

Eon	Era	Period	Epoch	Began (millions of years ago)	Life Forms/Events
	Paleozoic	Permian	⎫ "Age of Amphibians"	286	Extinction of trilobites and many other marine animals
		Carboniferous	⎭	360	First reptiles (330) Large coal swamps Amphibians abundant
		Devonian	⎫ "Age of Fishes"	408	First seed plants (365) First sharks (370) First insect fossils (385) Fishes dominant
		Silurian	⎭	438	First vascular land plants (430)
		Ordovician	⎫ "Age of Invertebrates"	505	First land plants similar to lichen (470) First fishes (505) Earliest corals Marine algae
		Cambrian	⎭	570	Abundant shelled invertebrates Trilobites dominant
Proterozoic	⎫ Collectively these are popularly known as the Precambrian			2500	First invertebrates (700) Earliest shelled organisms (~ 750) Oxygen begins to accumulate in the atmosphere (1500) First fossil evidence of single-celled life with a cell nucleus: eukaryotes (1500)
Archean				4000	First evidence of by-products of eukaryotes (2700) Earliest primitive life, bacteria and algae: prokaryotes (3500–3800). These will dominate the world for the next 3 billion years. Oldest surface rocks (4030)
Hadean	⎭			4600	Oldest single mineral (4300) Oldest Moon rocks (4440) Oldest meteorites (4560)

From Fundamentals of Oceanography, 4th edition, Duxbury, Duxbury, and Sverdrup. Copyright 2000 The McGraw-Hill Companies. All rights reserved.

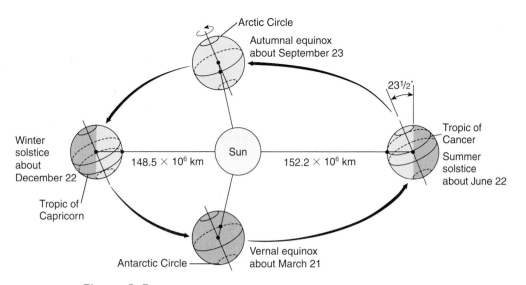

web link **Figure 1.5** The Earth's yearly seasons. The fixed orientation of the Earth's axis during its orbit of the Sun causes different portions of the Earth to remain in shadow at different seasons. The mean distance between the Earth and the Sun is 149×10^6 km, but note the change in this distance during the year.

extra day added every four years, except years ending in hundreds and not divisible by 400. As the Earth follows its orbit around the Sun, those who live in temperate zones and polar zones are very conscious of the seasons and of the differences in the lengths of the periods of daylight and darkness. The reason for these seasonal changes is seen in figure 1.5. The Earth moves along its orbit with its axis tilted 23½° from the vertical. As the Earth orbits the Sun during the year, the Earth's North Pole is sometimes tilting toward the Sun and sometimes tilting away from it. The Northern Hemisphere receives its maximum hours of sunlight when the North Pole is tilted toward the Sun; this is the Northern Hemisphere's summer. During the same period, the South Pole is tilted away from the Sun, so the Southern Hemisphere receives the least sunlight; this period is winter in the Southern Hemisphere (note in fig. 1.5 that summer in the Northern Hemisphere and winter in the Southern Hemisphere occur when the Earth is farthest from the Sun). When the Earth is closest to the Sun at the other side of its orbit, the North Pole is tilted away from the Sun, creating the Northern Hemisphere's winter, and the South Pole is inclined toward the Sun, creating the Southern Hemisphere's summer. Note also that during summer in the Northern Hemisphere it is light around the North Pole and dark around the South Pole; the opposite is true during the Northern Hemisphere's winter.

As we are carried along by the spinning Earth orbiting the Sun, we are not conscious of any movement. What we sense is that the Sun rises in the east and sets in the west daily and slowly moves up in the sky from south to north and back during one year. The periods of daylight in the Northern Hemisphere increase as the Sun moves north to stand above 23½°N, the **Tropic of Cancer.** It reaches this position at the **summer solstice,** on or about June 22, the day with the longest period of daylight and the beginning of summer in the Northern Hemisphere. On this day the Sun does not sink below the horizon above the **Arctic Circle,** 66½°N latitude, nor does it rise above the **Antarctic Circle,** 66½°S. Following the summer solstice the Sun appears to move southward until on or about September 23, the **autumnal equinox,** when it stands directly above the equator. On this day the periods of daylight and darkness are equal all over the world. The Sun continues its southward movement until about December 21, when it stands over 23½°S, the **Tropic of Capricorn;** this position marks the **winter solstice** and the beginning of winter in the Northern Hemisphere. On this day the daylight period is the shortest in the Northern Hemisphere; above the Arctic Circle the Sun does not rise, and above the Antarctic Circle the Sun does not set. The Sun then begins to move northward, and on about March 21, the **vernal equinox,** it stands again above the equator; spring begins in the Northern Hemisphere, and the periods of daylight and darkness are once more equal around the world. Follow figure 1.5 around again, checking the position of the South Pole, and note how the seasons of the Southern Hemisphere are reversed from those of the Northern Hemisphere.

The greatest annual variation in the intensity of direct solar illumination occurs in the temperate zones. In the polar regions the seasons are dominated by the long periods of light and dark, but the direct heating of the Earth's surface is small because the Sun is always low on the horizon. Between the Tropics of Cancer and Capricorn there is little seasonal change in solar radiation levels, because the Sun never moves beyond these boundaries.

Weeks and months, as they presently exist, modify natural time periods. It requires 27⅓ days for the Moon to orbit the Earth, but a period of 29½ days defines the **lunar month.** In the lunar month the Moon passes through four phases: new Moon, first quarter, full Moon, and last quarter. The four phases approximately match the four weeks of the month. Days are grouped into twelve months of unequal length in order to form one calendar year. The present arrangement is known as the Gregorian calendar after Pope Gregory XIII, who in the sixteenth century made the changes necessary to correct the old Julian calendar, adopted in 46 B.C. and named after Julius Caesar. The Gregorian calendar was adopted in the United States in 1752, by which time the Julian calendar was eleven days in error. In that year, by parliamentary decree, in both Great Britain and the United States, September 2 was followed by September 14. People rioted in protest, demanding their eleven days back. Also in 1752, the beginning of the calendar year was changed from the original date of the vernal equinox, March 21, to January 1; 1751 had no months of January and February.

The day is derived from the Earth's rotation on its axis as it orbits the Sun. The average time for the Earth to make one rotation relative to the Sun is twenty-four hours; this is the average, or mean, **solar day**—our clock day. Another measure of a day is the time required for the Earth to make a complete rotation with respect to a far-distant point in space. This is known as the **sidereal day** and is about four minutes shorter than the mean solar day; it gives the true rotational period of the Earth. The sidereal day is useful in astronomy and navigation.

Living organisms respond to these natural cycles. In temperate zones, flowers bloom and die back; forest trees lose their leaves, enter a period of dormancy, and then produce new leaves and buds as the length of the periods of daylight and darkness change and the temperatures increase or decrease; but tropical forests remain lush year-round. Some animals migrate and alternate periods of activity and hibernation or estivation with the seasons. Other animals set their internal clocks to the day-night pattern, hunting in the dark and sleeping in the light; still others do the reverse. Plants and animals of the sea also react to these rhythms, as do the physical processes that move the atmosphere and circulate the water in the oceans. An understanding of these cycles helps us understand processes that occur at the ocean surface and are discussed in later chapters: climate zones, winds, currents, vertical water motion, plant life, and animal migration.

Shape of the Earth   1.3

As the Earth cooled and turned in space, gravity and the forces of rotation produced its nearly spherical shape. The Earth sphere has a mean, or average, radius of 6371 km (3959 mi). It has a shorter polar radius (6356.9 km; 3950 mi) and a longer equatorial radius (6378.4 km; 3963 mi). This difference of 21.5 km, or about 13 mi, occurs because the Earth is not a rigid sphere. As the Earth spins it tends to bulge at the equator, much as a ball of potter's clay bulges when spun on a stick (fig. 1.6). Because the landmasses are presently concentrated in the middle region of the Northern Hemisphere and centered on the South Pole in the Southern Hemisphere, the Earth's surface is depressed slightly in these areas and is elevated at the North Pole and in the middle region of the Southern Hemisphere. This land distribution causes the Earth to have a very slight pear shape, about 15 m (50 ft) between depressions and elevations. The Earth is a nearly perfect sphere.

The Earth is also quite smooth. The top of the Earth's highest mountain, Mount Everest in the Himalayas, is about 8840 m (29,000 ft) above sea level; the deepest ocean depth, the bottom of the Challenger Deep in the Mariana Trench of the Pacific Ocean, is about 11,000 m (36,000 ft). If these measurements are divided by the mean radius of the Earth (6371 km, or 20,896,000 ft), the resulting elevation-to-radius ratios are 0.00139 for the mountain and 0.00173 for the

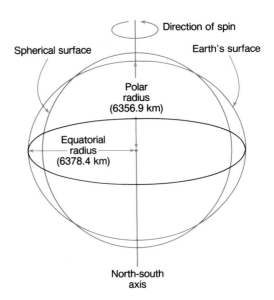

Figure 1.6 The rotation of the Earth on its axis makes it bulge outward at the equator. Notice that the equatorial radius is larger than the polar radius.

trench. On a scale model of the Earth with a radius of 50 cm (20 in), Mount Everest would be about 0.07 cm (0.027 in) high and the Challenger Deep would be about 0.086 cm (0.034 in) in depth. The Earth model's surface would feel rather like the skin of a grapefruit or the surface of a basketball. The topography of the Earth's surface, its high mountains and deep oceans, is minor compared to the size of the planet.

Location Systems   1.4

Latitude and Longitude

In order to find our way around on the surface of our planet we need a reference (or location) system. Most of us use such a system daily: city name, street name or number, and building number. Armed with a city map, we confidently navigate to areas never visited before. Most of the Earth's surface, however, is not provided with streets and building numbers, and we must use another system. To determine the location of a position on the Earth, we use a grid of reference lines that are superimposed on the Earth's surface and cross at right angles. These grid lines are called lines of **latitude** and **longitude.** Lines of latitude, also known as **parallels,** begin at the **equator.** The equator is created by passing a plane through the Earth halfway between the poles and at right angles to the Earth's axis. This process is much like cutting an orange in two pieces halfway between the depressions marking the stem and the navel. The equator is marked at 0° latitude, and other latitude lines are drawn around the Earth parallel to the equator, northward to 90°N, or the North Pole, and southward to 90°S, or the South Pole (fig. 1.7a). Notice that the parallels of latitude describe increasingly smaller circles as the poles are approached. Notice also that all parallels of latitude must be designated as an angle either north or south of the equator. The latitude value is determined by the internal angle (ϕ, or

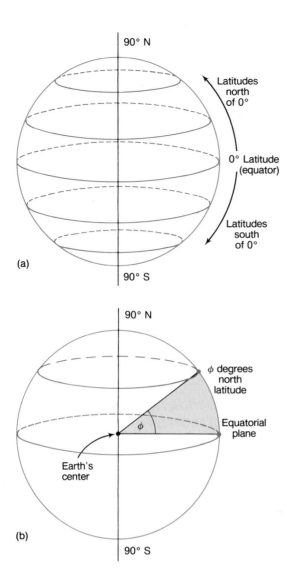

Figure 1.7 (a) Latitude lines are drawn parallel to the equatorial plane. (b) The value of a latitude line is expressed in angular degrees determined by the angle formed between the equatorial plane and the latitude line to the Earth's center. This is the angle ø (phi). The degree value of ø must be noted as north or south of the equator.

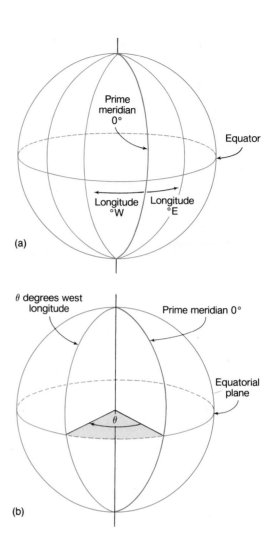

Figure 1.8 (a) Longitude lines are drawn with reference to the prime meridian. (b) The value of a longitude line is expressed in angular degrees determined by the angle formed between the prime meridian and the longitude line to the Earth's center. This is the angle θ (theta). The value of θ is given in degrees east or west of the prime meridian.

phi) between the latitude line and the equatorial plane at the Earth's center (fig. 1.7b). The previously mentioned Tropics of Cancer and Capricorn (see fig. 1.5) correspond to latitudes 23½°N and S, respectively. Latitudes 66½°N and S, respectively, correspond to the Arctic and Antarctic Circles.

Lines of longitude, or **meridians,** are formed at right angles to the latitude grid (fig. 1.8). Longitude begins at an arbitrarily chosen point: 0° longitude is a line on the Earth's surface extending from the North Pole to the South Pole and passing directly through the Royal Naval Observatory in Greenwich, England, just outside London. The 0° longitude line is shown outside the Greenwich Observatory in figure 1.9. On the other side of the Earth, 180° longitude is directly opposite 0°. The 0° longitude line is known as the **prime meridian.** The 180° longitude line approximates the **international date line.** Longitude lines are identified by their angular displace-

ment (*θ*, or theta) to the east and west of 0° longitude, as shown in figure 1.8b. Thus longitude may be reported as either 0°–180° east (also known as positive longitude) and 0°–180° west (also known as negative longitude) or 0°–360° east (always positive). In this manner –90°, 90°W, and 270° longitude all mark the same meridian. Note in figure 1.8a that meridians are the same size, much like the lines marking the segments of an orange. The meridians mark the intersection of the Earth's surface with a plane passing through the Earth's center at right angles to the parallels of latitude. Any circle at the Earth's surface with its center at the Earth's center is a **great circle.** All longitude lines form great circles; only the equator is a great circle of latitude. A great circle connecting any two points on the Earth's surface defines the shortest distance between them.

To identify any location on the Earth's surface, we use the crossing of the latitude and longitude lines; for example, 158°W, 21°N is the location of the Hawaiian Islands, and

Figure 1.9 The Royal Naval Observatory at Greenwich, England. The brass strip set into the courtyard marks the prime meridian, the division between east and west longitudes.

20°E, 33°S identifies the Cape of Good Hope at the southern tip of Africa. Because the distance expressed in whole degrees is large (1 degree of latitude equals 60 nautical miles), each degree of arc is divided into 60 minutes, each minute into 60 seconds, and each second into tenths of a second. One **nautical mile** is equal to one minute of arc length of latitude or longitude at the equator, or 1852 m (1.15 land mi; see appendix B). Positions on the Earth can be specified with great accuracy when this system is used.

Chart Projections

Charts and maps show the Earth's three-dimensional surface on a flat, or two-dimensional, surface. Maps usually show the Earth's land features or land and sea relationships, whereas charts depict the sea and sky. Any flat chart or map produces a distorted image of the curved surface of the Earth. The task of the mapmaker, or cartographer, is to produce the most accurate, most convenient, and least distorted picture for the task for which the map or chart is to be used.

Maps are made by projecting the Earth's features and its latitude-longitude system onto a surface; the resulting picture is a map or chart **projection.** Imagine a transparent globe with the continents and the latitude and longitude lines painted on its surface. Place a light in the center of the globe and let the light rays shine out through it. The light will project the shadows of the continents and the latitude and longitude lines onto a piece of paper held up to the outside of the globe. Different projections are obtained by varying the position of the light and the type of surface on which the projection is made. Many chart and map projections have been constructed, but most of them are modifications

of three basic types: cylindric, conic, and tangent plane, all shown in figure 1.10. In these projections, the surface that is to be the map is rolled around the globe as a cylinder (fig. 1.10*a*), made into a cone (fig. 1.10*b*), or laid flat (tangent) against the sphere (fig. 1.10*c*). Although the cylindric and conic surfaces may be placed around, over, or tangent to the Earth at any location, they are usually placed so that the cylinder touches the Earth at its equator and the cone is centered on the polar axis.

Compare the three parts of figure 1.10 and notice that distortion on a map or chart increases as the distance from the place of contact to the globe increases. Consider Greenland. In the tangent plane projection (fig. 1.10*c*) its size and shape are very close to its true form on the Earth. In the conic projection (fig. 1.10*b*) the island has grown larger, and in the cylindric projection (fig. 1.10*a*) both its size and its shape are greatly distorted. The traditional and familiar world map used in many books and school classrooms is the **Mercator projection,** an adjusted form of the cylindric type shown in figure 1.10*a*. Although distortion is great at high latitudes and poles cannot be shown, the Mercator projection, unlike other projections, has the advantage that a straight line as drawn on the map is a line of true direction or constant compass heading, and therefore the Mercator projection is useful in navigation. Each type of chart or map has its own characteristics. The user must select the projection with the least distortion and best properties for his or her purpose.

Maps that show lines connecting points of similar elevation on land (known as **contours** of elevation) show the Earth's **topography;** they are topographic maps. Charts of the ocean showing contour lines connecting points of the same depth below the sea surface depict the area's **bathymetry;** they are bathymetric charts (fig. 1.11). Color, shading, and perspective drawings may be added to indicate elevation changes and produce visual representations or bird's-eye views, called **physiographic maps.** See figure 1.12 for an example of a physiographic map and compare it to the bathymetric chart in figure 1.11. Today, computers use electronic ocean-depth measurements to form detailed bathymetric charts that are rapidly converted into three-dimensional, color images of the sea floor that can be viewed from any angle (fig. 1.13).

Measuring Latitude

Early maps show us that the first cartographers and navigators had considerable difficulty in precisely describing and locating the then known Earth features. When accurate measurement failed, artistic license appeared to fill the gaps (fig. 1.14). The major problem was that early navigators were not able to determine their position accurately. As navigational techniques improved, so did the maps. It was known by early navigators that the North Star, **Polaris,** appeared to hang in the sky above the North Pole and did not appreciably move from that spot. In the Northern Hemisphere, therefore, measuring the angle of elevation of Polaris above the horizon gave a good estimate of one's latitude.

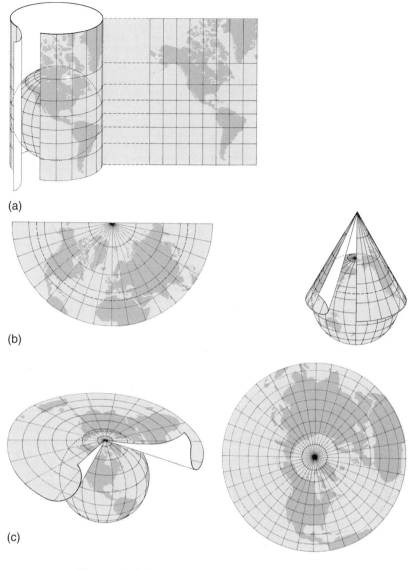

Once a ship was out of sight of land, it sailed north or south until reaching the desired latitude, then sailed east or west along that line of latitude until again reaching land. Adjustments north and south to reach the desired landfall were made close to shore and according to visible landmarks.

Longitude and Time

Determining longitude was a much more difficult task. Because the longitude lines rotate with the turning Earth, 360° in twenty-four hours, it becomes necessary to know the position of the Sun or the stars with time relative to one's longitude line. Although the theory for using time to determine longitude had been proposed by the Flemish astronomer Gemma Frisius in 1530, early clocks did not work satisfactorily on rolling ships, and precise longitude measurements were not possible until the construction of an accurate ship's chronometer in the eighteenth century. See the prologue for the history of this achievement.

If a clock is set to exactly noon when the Sun is at its **zenith,** or highest elevation above a reference longitude, and if that clock is then carried to a new location and the zenith time of the Sun is determined at this new location, the clock's time difference between the Sun's zenith at the reference longitude and at the new location is used to determine the longitude at the new location. When this technique is used, a position that is 15° of longitude west of the reference longitude is directly under the Sun one hour later, or at 1 P.M. by the clock, because the Earth has turned eastward 15° during that hour. A position 15° east of the reference longitude is directly under the Sun at 11 A.M., because it requires an hour to turn the 15° to bring the Sun to its zenith over the reference longitude (fig. 1.15).

web link **Figure 1.10** The three basic types of map projections. (a) An equatorial cylindric projection. (b) A simple polar conic projection. (c) A polar tangent plane projection.

Figure 1.11 A bathymetric, or contour, chart of the sea floor along a section of generalized coast. Changes in the pattern of and the spacing between contour lines indicate changes in depth.

Figure 1.12 A physiographic map of the area shown in figure 1.11. Shading and perspective have been added.

UCSB Geology 23-AUG-89   East Pacific Rise 13 N (looking SSW)

Figure 1.13 Three-dimensional computer-processed image of a section of the East Pacific Rise. Data for this image were obtained by using the Sea Beam echo sounder.

The reference longitude in use today is the prime meridian, or 0° longitude. The clock time is set to noon when the Sun is at its zenith above the prime meridian. This is **Greenwich Mean Time (GMT),** now **Universal Time** or **Zulu time** (zero meridian time). Because Sun time changes by one hour for each 15° of longitude, the Earth has been divided into time zones that are 15° of longitude wide. The time zones do not exactly follow lines of longitude; they follow political boundaries when necessary for the convenience of the people living in those zones (fig. 1.16).

Modern Navigational Techniques   1.5

Modern navigators still use chronometers and wait for clear skies to "shoot" the Sun or stars with a sextant to determine positions at sea, but such measurements are used primarily to check their modern electronic navigational equipment.

Chronometers are calibrated, or reset, by broadcast time signals. When vessels are near land, **radar** (radio detecting and ranging) bounces radio pulses off a target such as the shoreline or another vessel. The image on the radar screen is formed by radiation energy that is sent out by a transmitter, reflected from an object, returned to the antenna, and then displayed on a screen. **Loran** (long-range navigation) can be used farther out at sea. This electronic timing device measures the difference in arrival time of radio signals from pairs of land stations. The position of the receiving ship is plotted on a chart that shows the time delay lines for these stations. Recent improvements in loran include receivers with computers that can be programmed with the latitude and longitude of the desired destination. The loran receiver monitors the signals from the stations and directly reads out the course to be sailed and the distance to the destination. The computer can also monitor the signals and continuously calculate the latitude and longitude of the vessel, enabling the ship's personnel and ocean scientists to know their position at all times.

The **satellite navigation system** is an accurate and sophisticated navigational aid. Satellites orbiting the Earth emit signals of a precise frequency that are picked up by a receiver on the ship. The ship's receiver monitors the frequency shift of the signal as the satellite passes and determines the exact instant in time at which the frequency is correct. At this instant the ship's path and the satellite's orbit are at right angles. Given this information, a computer programmed with the satellite's orbital properties can determine the ship's position to within 30 m (100 ft) or less.

A more versatile and accurate method of finding one's position uses the U.S. Navstar **Global Positioning System (GPS).** GPS is a worldwide radio-navigation system consisting of twenty-four navigational satellites, twenty-one operational and three active spares, and five ground-based monitoring stations (fig. 1.17). The satellites orbit at an altitude of about 20,165 km (12,500 mi) and repeat the same track and relative configuration over any point about every twenty-four hours. At any given time, from five to eight satellites are visible from

Figure 1.14 A map of the Americas from Flemish geographer Abraham Ortelius's 1570 atlas, *Theatre of the World*.

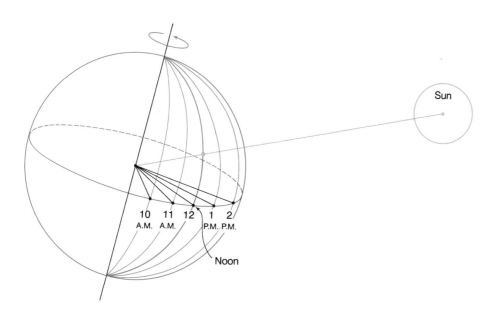

Figure 1.15 The time on a meridian relative to the Sun changes by one hour for each 15° change in longitude.

web link **Figure 1.16** Distribution of the world's time zones. Degrees of longitude are marked along the *bottom*. Time zones are positive (west zones) or negative (east zones). Time at Greenwich is determined by adding the zone number to the local time.

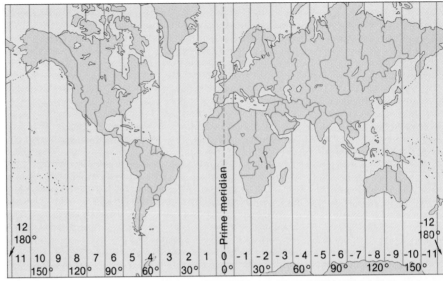

Time zones west — Time zones east

web link **Figure 1.17** The Global Positioning System (GPS) consists of (a) twenty-four satellites and (b) five monitoring stations. The main control station is located at Falcon Air Force Base in Colorado Springs.

(a)

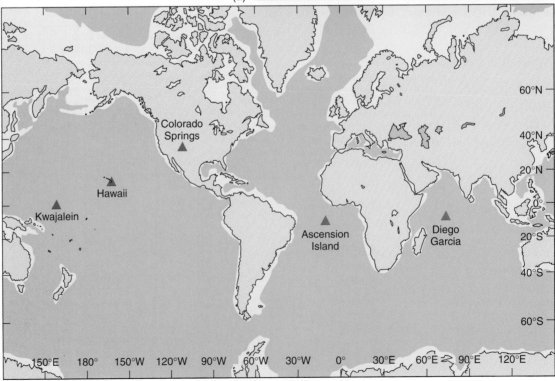

(b)

any point on the Earth. The system uses this constellation of satellites as reference points for calculating positions on the surface with an accuracy of a few meters for commercial and private users and to better than a centimeter with advanced forms of GPS that are currently classified for military purposes.

Each GPS satellite transmits a unique digital code that is sent as a radio signal to ground receivers. GPS receivers generate codes that are identical to those sent by the satellites at exactly the same time the satellites do. Because of the distance the satellite signals travel, there is a time lag between when the receiver generates a specific signal and when it receives the same signal generated at the same time by the satellite. This time delay is a function of the distance between the receiver and the satellite. The satellite signals travel at the speed of light, or approximately 300,000 km/s (186,000 mi/s), so measuring the arrival time of the signal accurately is critically important. The signal from a satellite directly overhead would reach a ground receiver in roughly 0.06 s. Precise measurements of distance between a receiver and four satellites will pinpoint the exact location of the receiver. In practice, there are a variety of sources of error in the measurements, so measurements are made between the receiver and as many satellites as it can detect to maximize location accuracy. GPS can be used to measure the velocity of a moving receiver as well as its location.

GPS is used in a wide variety of applications, including positioning and navigation, mapping land and sea surface features, monitoring tides and currents, measuring the motion of tectonic plates, and monitoring the state of stress along faults. It is even used to study the viscosity of the Earth's mantle (see chapter 2); it observes three-dimensional crustal velocities as regions previously depressed under heavy ice sheets during the last glacial period continue slowly to rise as the ice melts and releases its tremendous weight. Oceanographers are currently developing new techniques for using GPS in underwater applications. A major problem in the use of GPS underwater is that GPS signals are radio waves, which are quickly absorbed by water. Underwater GPS systems that are being tested involve surface buoys containing GPS receivers that communicate with underwater targets using acoustic (sound) energy. A further challenge is that it is difficult to rapidly transmit the large amounts of information that are contained in GPS radio signals in an acoustic signal. Solutions to these problems will likely be developed rapidly because of the enormous benefit of being able to use GPS in underwater applications.

The U.S. GPS system is similar to a global position system, GLONASS, employed by the former Soviet Union. The increasing cooperation between the United States and the Russia Federation is pointing to the development of receivers that will accept both systems, allowing increased positioning accuracy, better Earth coverage, and independent verifications of results.

Shipboard computers can now store an electronic atlas that includes surface charts and seafloor bathymetry. The ship's position is tracked and its position determined continuously. Oceanographic vessels use devices that draw charts showing the vessel's changing position and at the same time conduct a seafloor survey and keep track of water measurements (such as temperature and salt content) made automatically as the ship moves along. Today's oceanographers are able to return more and more accurately to the same place at sea to repeat measurements and follow changes in ocean processes. Scientists now have the ability to evaluate data as they are being taken, allowing changes to be made in a vessel's sampling pattern and thereby ensuring that the data satisfy a project's research requirements.

Earth: The Water Planet 1.6

Water on the Earth's Surface

As the Earth and the other planets of the solar system cooled, the Sun's energy gradually replaced the heat of planet formation in maintaining their surface temperatures. The Earth developed a nearly circular orbital path at a mean distance of 149 million km (149×10^6 km), or 93 million mi (93×10^6 mi) from the Sun. (If you are unfamiliar with scientific notation to express very large numbers, see appendix A.) Moving along this orbit the Earth is 152.2×10^6 km (94.5×10^6 mi) from the Sun in June and 148.5×10^6 km (92.2×10^6 mi) away in December, as shown in figure 1.5. At these distances from the Sun, the Earth's orbit keeps the annual heating and cooling cycle within moderate limits. The Earth's mean surface temperature is about 16°C, which allows water to exist as a gas, as a liquid, and as a solid.

The rotation of the Earth on its axis is also important in moderating temperature extremes. The Earth completes one rotation, turning from west to east, in twenty-four hours. If the Earth rotated more slowly, the side of the Earth toward the Sun would be exposed to the Sun's energy for a longer period than it is at present and would become very hot, while the side in darkness would lose heat and become very cold. Temperature changes from day to night would be large. By contrast, a shorter period of rotation would decrease the present variation from day to night.

The Earth's solar orbit, its rotation, and its blanket of atmospheric gases produce surface temperatures that allow the existence of liquid water. The atmosphere covering the Earth's surface acts as a protective shield between the Earth and the Sun. Without it, solar heating would evaporate water at a much higher rate. Compare the Earth's distance to the Sun, its period of rotation, and the time required for the Earth to complete one orbit of the Sun with those of other planets, as shown in table 1.1. Note the surface temperature of the Earth as compared to those of other planets.

The amount of water on the Earth's surface can be expressed in several ways. For example, the oceans cover 361 million km² (361×10^6 km²), or 139 million mi² (139×10^6 mi²). Because these numbers are so large, they do not convey a clear idea of size; therefore, an easier concept is to remember that 71% of the Earth's surface is covered by the oceans, and only 29% of the surface area is land above sea level.

The volume of water in the oceans is enormous: 1.33 billion km³ (1.33×10^9 km³, or 0.317×10^9 mi³). Another way to

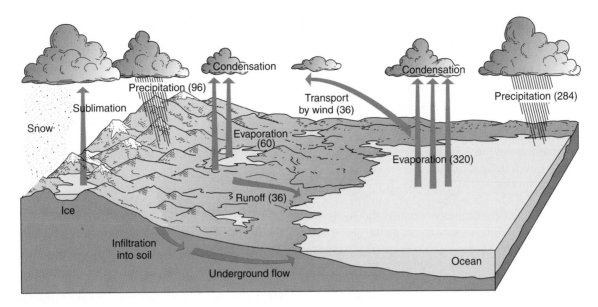

web ⟨∞⟩ link **Figure 1.18** The hydrologic cycle and annual transfer rates for the whole Earth. Precipitation transfer rate includes both snow and rain. Evaporation transfer rate from the continents includes evaporation of surface water, transpiration (the release of water to the atmosphere by plants), and sublimation (the direct change in state from ice to water vapor). Runoff from the continents includes both surface flow and underground flow. Annual transfer rates are in thousands of cubic kilometers ($10^3 km^3$).

Table 1.3 The Earth's Water Supply

Approximate Water Volume			
Reservoir	(km³)	(mi³)	Approximate Percent of Total Water
Oceans and sea ice	1,338,500,000	320,600,000	97.24
Ice caps and glaciers	29,289,000	7,000,000	2.14
Groundwater	8,368,000	2,000,000	0.61
Freshwater lakes	125,500	30,000	0.009
Saline lakes and inland seas	105,000	25,000	0.008
Soil moisture	67,000	16,000	0.005
Atmosphere	13,000	3,100	0.001
Rivers	1,250	300	0.0001
Total water volume	1,376,468,750	329,674,400	100

From Fundamentals of Oceanography, 4th edition, Duxbury, Duxbury, and Sverdrup. Copyright 2000 The McGraw-Hill Companies. All rights reserved.

express the oceanic volume is to think of a smooth sphere with exactly the same surface area as the Earth (510×10^6 km², or 197×10^6 mi²), uniformly covered with the water from the Earth's oceans. The ocean water would be 2686 m (8800 ft) deep, a depth of about 1.7 mi. If the water from all other sources in the world were added, the depth would rise 56 m to 2742 m (9000 ft). When water volumes are considered as depths over a smooth sphere, they are referred to as **sphere depths.** The ocean sphere depth is 2686 m, and the total water sphere depth of all the Earth's water is 2742 m.

Hydrologic Cycle

The Earth's water occurs as a liquid in the oceans, rivers, lakes, and below the ground surface; it occurs as a solid in glaciers, snow packs, and sea ice; it takes the form of water

droplets and gaseous water vapor in the atmosphere. The places in which water resides are called **reservoirs,** and each type of reservoir, when averaged over the entire Earth, contains a fixed amount of water at any one instant. But water is constantly moving into and out of reservoirs. This movement of water through the reservoirs, diagrammed in figure 1.18, is called the **hydrologic cycle.**

Water is taken out of the oceans and moved into the atmosphere by evaporation. Most of this water returns directly to the sea by precipitation, but air currents carry some water vapor over the continents. Precipitation in the form of rain and snow transfers this water from the atmosphere to the land surface, where it percolates into the soil; is taken up by plants; fills rivers, streams, and lakes; or remains for longer periods as snow and ice in some areas. Some of this

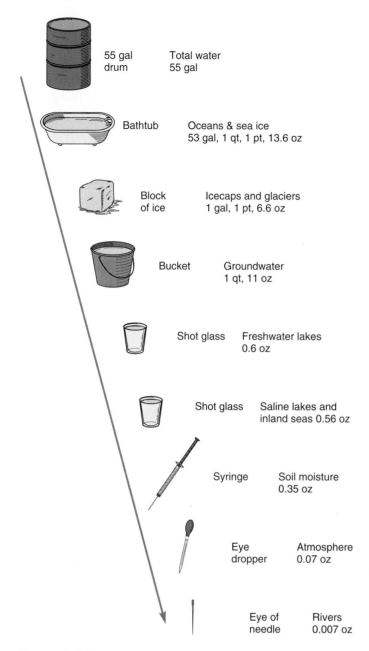

Figure 1.19 Comparison of the amount of the world's water supply held in each of the major water reservoirs. For purposes of illustration the Earth's total water supply has been scaled down to the volume of a 55 gal drum.

water returns to the atmosphere by evaporation of surface water, **transpiration,** the release of water by plants, and **sublimation,** the conversion of ice directly to water vapor. Melting snow and ice, rivers, groundwater, and land runoff move the water back to the oceans to complete the cycle and maintain the oceans' volume. For a comparison of the water stored in the Earth's reservoirs, see table 1.3 and figure 1.19.

In local areas, excess water may evaporate from the land and excess precipitation may occur over the sea, but on a worldwide average, the cycle operates with a net removal of ocean water by evaporation, a net gain of water on land as a result of precipitation, and a return of the excess water on

land to the sea by rivers and land drainage. The properties of climate zones are principally determined by their surface temperatures and their evaporation-precipitation patterns: the moist, hot equatorial regions; the dry, hot subtropical deserts; the cool, moist temperate areas; and the cold, dry polar zones. Differences in these properties, coupled with the movement of air between the climate zones, moves water through the hydrologic cycle from one reservoir to another at different rates. The transfer of water between the atmosphere and the oceans alters the salt content of the oceans' surface water, and, with the seasonal and latitudinal changes in surface temperature, determines many of the characteristics of the world's oceans. These characteristics are explored in chapters 7 to 9.

Reservoirs and Residence Time

Because the total amount of water on Earth is nearly constant, the hydrologic cycle must maintain a balance between the addition and removal of water from the Earth's water reservoirs. The rate of removal of water from a reservoir must equal the rate of addition to it, for if the balance is disturbed one reservoir will gain at the expense of another. The average length of time that a water molecule spends in any one reservoir is called the **residence time** for water in that reservoir. Water's residence time can be calculated by dividing the volume of water in the reservoir by the rate at which it is either added or removed. A large reservoir generally has a long residence time because of its large volume, while small reservoirs generally have a short residence time and the water in them can be replaced comparatively quickly. The size of the reservoir also determines how it reacts to changes in the rate at which the water is gained or lost. Large reservoirs show little effect from small rate changes, whereas small reservoirs may alter substantially when exposed to the same variation in rates of gain or loss. For example, if the ocean volume decreased by 6½% and that volume of water were added to the land ice, the result would be a 400% increase in the present volume of land ice but only a 250 m (820 ft) drop in sea level. This example reflects the changes that have occurred on Earth during the major ice ages.

About 380,000 km³ (90,820 mi³) of water move through the atmosphere each year. Because the atmosphere holds the equivalent of 13,000 km³ (3100 mi³) of liquid water at any one time, a little arithmetic shows that the water in the atmosphere can be replaced twenty-nine times each year. Atmospheric water has a very short residence time. The residence time for water in the other, larger reservoirs is much longer. For example, it would take 36,844 years to evaporate and pass the water from all of the oceans through the atmosphere, to the land as precipitation, and back to the oceans via rivers. Further study of water's movement shows us that annually 320,000 km³ (76,480 mi³) is evaporated from the oceans, and 60,000 km³ (14,340 mi³) is evaporated from land. When the water returns as precipitation, 284,000 km³ (67,876 mi³) is returned directly to the sea surface, and 96,000 km³ (22,944 mi³) returns to the land. However, the excess gained by the land (36,000 km³ or

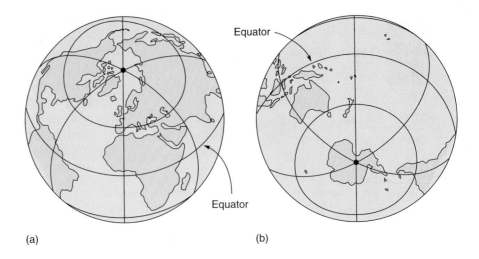

(a)   (b)

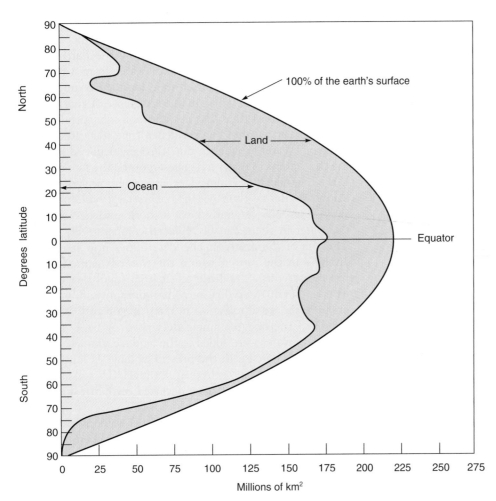

Figure 1.21 Distribution of land and ocean by latitude. In the Northern Hemisphere, middle-latitude areas of land and ocean are nearly equal. Land is almost absent at the same latitudes in the Southern Hemisphere. The areas are calculated on the basis of 5° latitude intervals.

8604 mi³) flows back to the oceans in rivers, streams, and groundwater (see fig. 1.18).

Distribution of Land and Water
To understand the present distribution of land and water on the Earth, consider the Earth when it is viewed from the north (fig. 1.20a) and from the south (fig. 1.20b). About 70% of the Earth's landmasses are in the Northern Hemisphere, and most of this land lies in the middle latitudes. The Southern Hemisphere is the water hemisphere, with its land located mostly in the tropical latitudes and in the polar region. The details of this land-water-latitude distribution are presented in figure 1.21.

Oceans
The distribution and shape of the continents divide the world ocean into at least three, and some would argue as many as five, individual oceans. Oceanographers all recognize the three major oceans—the Pacific, Atlantic, and Indian Oceans—that extend northward as three fingers from a common source of undivided ocean surrounding Antarctica. This circumpolar body of water can either be thought of as an additional ocean at latitudes above 50°S, commonly called the Southern Ocean or the Antarctic Ocean, or it can be divided along lines of longitude to become the southernmost extensions of the Pacific, Atlantic, and Indian Oceans. The Arctic Ocean is often considered an extension of the North Atlantic, but because of its size and relative isolation many people consider it to be an independent ocean basin. The four most commonly recognized oceans, the Pacific, Atlantic, Indian, and Arctic, are shown in figure 1.22. Each of these oceans has its own characteristic surface area, volume, and mean depth.

The Pacific Ocean has greater surface area, volume, and mean depth than any of the other oceans. The Pacific was

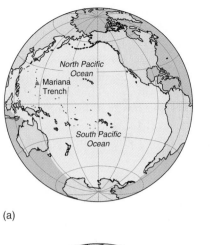

(a)

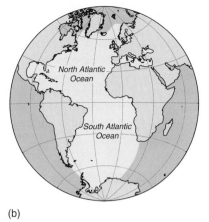

(b)

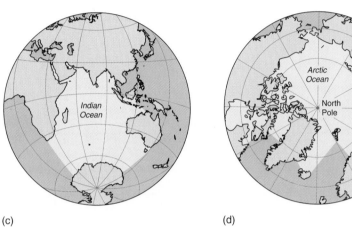

(c)

(d)

web link **Figure 1.22** The world's four major oceans in order of decreasing size (a) Pacific Ocean (the deepest spot in the oceans is located in the Mariana Trench); (b) Atlantic Ocean; (c) Indian Ocean; and (d) Arctic Ocean.

named by Ferdinand Magellan in 1520 for the calm weather he and his crew enjoyed while crossing it (*paci* = peace). It covers a little over a third of Earth's surface and just over half of the world ocean's surface. At its maximum width near 5°N, the Pacific stretches 19,800 km (12,300 mi) from Indonesia to Columbia. There are roughly 25,000 islands in the Pacific, the majority of which are south of the equator. This is more than the total number of islands in the rest of the oceans combined. There are a number of marginal seas along the edges of the Pacific, including the Celebes Sea, Coral Sea, East China Sea, Sea of Japan, Sulu Sea, and Yellow Sea.

The Atlantic Ocean is the second largest. Its name is derived from Greek mythology and means "Sea of Atlas" (Atlas was a Titan who supported the heavens by means of a pillar on his shoulders). The land area that drains into the Atlantic is four times larger than that of either the Pacific or Indian Oceans. There are relatively few islands given the size of the basin. The irregular coastline of the Atlantic includes a number of bays, gulfs, and seas. Some of the larger ones include the Caribbean Sea, Gulf of Mexico, Mediterranean Sea, North Sea, and Baltic Sea.

The Indian Ocean is primarily a Southern Hemisphere ocean; it is the third largest but is quite deep. The northernmost extent of the Indian Ocean is in the Persian Gulf at about 30°N. The Indian Ocean is separated from the Atlantic Ocean to the west by the 20°E meridian and from the Pacific Ocean to the east by the 147°E meridian. It is nearly 10,000

km (6200 mi) wide between the southern tip of Africa and Australia. For centuries it has had tremendous strategic importance as a trade route between Africa and Asia.

The Arctic Ocean is the smallest of the four, occupying a roughly circular basin over the North Pole region. It is connected to the Pacific Ocean through the Bering Strait and to the Atlantic Ocean through the Greenland Sea. Its floor is divided into two deep basins by an underwater mountain range. The major flow of water into and out of the Arctic Ocean is through the North Atlantic Ocean. Use table 1.4 to compare the four oceans.

Hypsographic Curve

Another method used by oceanographers to depict land-water relationships is shown in figure 1.23. This graph of depth or elevation versus the Earth's area is called a **hypsographic curve.** Find the line indicating sea level and note that the elevation of land above sea level is given in meters along the left margin; the depth below sea level is given in meters along the right margin. The scale across the top of the figure indicates total Earth area in 100 million km² (10^8 km²). The scales along the bottom of the figure indicate percentages of the Earth's surface area; note that the curve crosses sea level at the 29% mark, showing that 29% of the Earth's surface is above sea level and 71% is below sea level. The lower of the two scales gives land and ocean areas as separate percentages. Referring to the land area percentage scale, note that only 20% of all land areas are at elevations above 2 km. The ocean area scale shows that approximately 85% of the ocean area is below 2 km in depth. The hypsographic curve helps us to see not only that our Earth is 71% covered with water but also that the areas whose depths are well below the sea surface are much greater than the areas whose elevations are well above it; there are basins beneath the sea that are about four times greater in area than the area of land in mountains above sea level.

Mount Everest, the highest land peak, reaches 8.84 km (5.49 mi) above sea level, whereas the ocean's deepest trench descends 11.02 km (6.84 mi) below sea level.

Because the hypsographic curve is constructed as a plot of area versus height, an area of the diagram shows volume, because volume is the product of area times height. The mean elevation of the land is 840 m (2750 ft), and the entire land volume above sea level fits within a box 840 m high, covering 29% of the Earth's surface. The mean depth

Figure 1.23 The hypsographic curve displays the area of the Earth's surface at elevations above and below sea level.

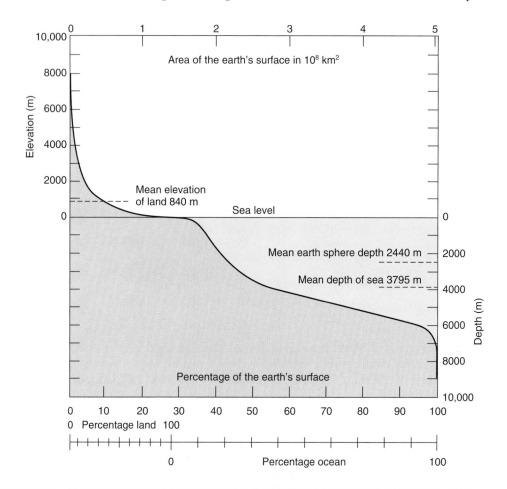

Table 1.4 Ocean Depths, Areas, and Volumes

Ocean	Average Depth	Area	Volume	Percent of Ocean Surface Area	Percent of Earth's Surface Area
Pacific	4028 m (13,216 ft)	179.7 × 10⁶ km² (69.3 × 10⁶ mi²)	723.8 × 10⁶ km³ (173.3 × 10⁶ mi³)	50.1	35.5
Atlantic	3332 m (10,932 ft)	93.7 × 10⁶ km² (36.1 × 10⁶ mi²)	312.2 × 10⁶ km³ (74.8 × 10⁶ mi³)	25.9	18.4
Indian	3897 m (12,786 ft)	73.6 × 10⁶ km² (28.4 × 10⁶ mi²)	286.8 × 10⁶ km³ (68.7 × 10⁶ mi³)	20.4	14.4
Arctic	1117 m (3665 ft)	14.1 × 10⁶ km² (5.4 × 10⁶ mi²)	15.7 × 10⁶ km³ (3.8 × 10⁶ mi³)	3.9	2.8
All Oceans	3795 m (12,451 ft)	361.1 × 10⁶ km² (139.3 × 10⁶ mi²)	1338.5 × 10⁶ km³ (320.6 × 10⁶ mi³)	100	70.8

From Fundamentals of Oceanography, 4th edition, Duxbury, Duxbury, and Sverdrup. Copyright 2000 The McGraw-Hill Companies. All rights reserved.

of the ocean is 3795 m (12,400 ft). In this case the ocean volume is the mean ocean depth times 71% of the Earth's surface area.

Refer to the discussion of sphere depths as a device for expressing water volumes and remember that, because humans are land-dwellers, their reference line is sea level. But if we are considering the hypsographic curve for the total Earth, a mean Earth elevation becomes more suitable for describing the location of the Earth's solid surface; this value is called the **mean Earth sphere depth.** This level is 2440 m (8003 ft) below the present sea level, as shown in figure 1.23. This value represents the level at which the volume of crust above and water volume below the Earth sphere depth are equal. To arrive at the mean Earth sphere depth, think of moving all the elevated land and some of the sea bottom down into the ocean depressions until the Earth is a perfectly smooth ball. When this is accomplished the mean Earth sphere depth reference level would stand at the 2440 m (8003 ft) depth. Such an operation would also result in the displacement of seawater upward, until a new sea level 246 m (807 ft) above the present sea level was reached. This depth of water is the **mean ocean sphere depth,** 2686 m (8810 ft). It is the depth that the oceans would have if the volume of ocean water were spread over an area equal to the total Earth area rather than over 71% of the Earth's area. The oceans swallow the land, with only a comparatively small rise in water level, emphasizing again that Earth is the water planet (or the ocean planet) and not the land planet.

Summary

The beginning of the expansion of the universe was followed by the first stars, the reactions that produced the atomic elements, and billions of galaxies. Our solar system is part of the Milky Way galaxy; it began as a rotating cloud of gas. A series of events produced nine planets orbiting the Sun, each planet having unique characteristics. Over approximately 1.5 billion years the Earth heated, cooled, changed, and collected a gaseous atmosphere and an accumulation of liquid water.

Reliable age dates for Earth rocks, meteorites, and Moon samples are obtained by radiometric dating. The accepted age of the Earth is 4.6 billion years. Geologic time is used to express the timescale of the Earth's history.

The distance between the Earth and the Sun, the Earth's orbit, its period of rotation, and its atmosphere protect the Earth from extreme temperature change and water loss. Because the Earth rotates, its shape is not perfectly symmetrical. Its exterior is relatively smooth. Natural time periods (the year, day, and month) are based on the motions of the Sun, Earth, and Moon. Because of the tilt of the Earth's axis as it orbits the Sun, the Sun moves annually between 23½°N and 23½°S, producing the seasons.

Latitude and longitude are used to form a grid system for the location of positions on the Earth's surface. Different types of map and chart projections have been developed to show the Earth's features on a flat surface. All of these projections distort the Earth's features to some extent. Bathymetric and physiographic charts and maps use elevation and depth contours to depict the Earth's topography.

In order to determine longitudinal position one must be able to measure time accurately. This need required the development of the seagoing clocks for celestial navigation.

Modern navigational techniques make use of radar, radio signals, computers, and satellites. A satellite network provides more accurate position readings and also maps storms, tides, sea level, and properties of surface waters.

The Earth is the water planet. Of the Earth's surface, 71% is covered by its oceans. There is a fixed amount of water on Earth. Evaporation and precipitation move the water through the reservoirs of the hydrologic cycle. Water's residence time varies in each reservoir and depends on the volume of the reservoir and the rate of addition and removal of the water.

The Earth's Northern Hemisphere is the land hemisphere; the Southern Hemisphere is the water hemisphere. The Earth has three large oceans extending north from the Southern Ocean. Each has a characteristic surface area, volume, and mean depth. The hypsographic curve is used to show land-water relationships of depth, elevation, area, and volume. It is also used to determine mean land elevation, mean ocean depth, Earth sphere depth, and ocean sphere depth.

Key Terms

All key terms from this chapter can be viewed by term, or by definition, when studied as flashcards on this book's Online Learning Center at www.mhhe.com/sverdrup (click on this book's cover).

electromagnetic spectrum, 29	longitude, 38
Big Bang, 29	parallel, 38
galaxy, 30	equator, 38
light-year, 30	meridian, 39
cluster, 30	prime meridian, 39
nebula, 31	international date line, 39
radiometric dating, 34	great circle, 39
isotope, 34	nautical mile, 40
half-life, 34	projection, 40
vertebrate, 35	Mercator projection, 40
Tropic of Cancer, 37	contour, 40
summer solstice, 37	topography, 40
Arctic Circle, 37	bathymetry, 40
Antarctic Circle, 37	physiographic map, 40
autumnal equinox, 37	Polaris, 40
Tropic of Capricorn, 37	zenith, 41
winter solstice, 37	Greenwich Mean Time, 42
vernal equinox, 37	Universal Time, 42
lunar month, 37	Zulu Time, 42
solar day, 38	radar, 42
sidereal day, 38	loran, 42
latitude, 38	satellite navigation system, 42

Study Questions

1. How and why have estimates of the age of the Earth changed over the past few hundred years? Do you think the present estimate of the Earth's age will change in the future?
2. Describe the distribution of water and land on the Earth.
3. Why does the Earth's average surface temperature differ from the surface temperature of other planets in the solar system?
4. Why is the twilight period at sunset shorter at low latitudes than it is at high latitudes?
5. The route of a ship sailing a constant compass course on a Mercator projection is indicated by a straight line that cuts all longitude lines at the same angle. This is a rhumb line. Discuss how this line appears (a) on a polar conic projection, (b) on a globe, and (c) on a tangent plane projection centered on the polar axis.
6. Discuss how the hypsographic curve is used to determine the mean depth and sphere depth of the oceans.
7. Why are the Arctic and Antarctic Circles displaced from the poles by 23½°, so that they are located at 66½°N and 66½°S?
8. What are some advantages of using satellites for oceanographic research? Are there any disadvantages?
9. How will the seasons change over a calendar year at each of these latitudes: (a) 10°N; (b) 70°N; (c) 30°S? Make a simple diagram for each latitude to show why the seasonal pattern occurs.
10. Explain why the Earth sustains a wide variety of life forms but the other planets of our solar system do not.
11. Trace several possible routes for a water molecule moving between a mountain lake and an ocean. In which reservoirs would the molecule spend the greatest amount of time and in which the least?
12. Use an atlas to find the appropriate latitudes and longitudes for each of the following:
    a. St. John, Newfoundland, and London, England
    b. Cape Town, South Africa, and Melbourne, Australia
    c. Anchorage, Alaska, and Moscow, Russia
    d. Strait of Gibraltar, Strait of Magellan, Straits of Florida
    e. Galápagos Islands, Tristan da Cunha, Reykjavík, Iceland
13. Although latitude and longitude were used on very early charts, navigators continued to use charts with many compass direction lines (portolano type) well into the seventeenth century. Why?
14. If the lunar month were used as the length of a month, what would happen to the calendar year relative to the Sun?
15. At what 1000 m depth interval in the world's oceans would the greatest change in ocean area occur? Use the hypsographic curve to determine your answer.

Study Problems

1. Determine the distance between two locations: 110°W, 38½°N and 110°W, 45°N. Express this distance in nautical miles and in kilometers.
2. The contour interval on a bathymetric chart is constant and equal to 100 m. Graph the slope of the sea floor across four evenly spaced contour lines if the distance between the first line and the fourth line is 2.5 km.
3. A plane leaves Tokyo, Japan, on June 6, at 0800 hours local Tokyo time and flies for nine hours, landing in San Francisco, California. Give the local time and date of arrival in San Francisco.
4. Show that the annual net evaporative loss of water from the world's oceans equals the annual net gain of water by precipitation on the land. Why does the ocean volume not decrease?
5. Use the volume of the oceans and the Earth's area to determine the sphere depth of the oceans.

Links to Related Websites

Visit the book's Online Learning Center at www.mhhe.com/sverdrup (click on the book's cover) to find live Internet links for additional topics related to this chapter's content.

- General astronomy
- Solar system
- Planet Earth
- Geologic time
- Natural cycles
- Latitude and longitude
- Time zones
- Map projections
- Nautical charts
- Navigation
- The hydrologic cycle
- The oceans

Visit the book's Online Learning Center at www.mhhe.com/sverdrup (click on this book's cover) to find these additional chapter tools: Suggested Readings; links to further information on boxed readings, selected figures, and related chapter topics; and additional study aids.

chapter 2

Plate Tectonics

I n its entire length, the basin of this sea (the Atlantic) is a long
trough, separating the Old World from the New, and extending
probably from pole to pole. This ocean-furrow was scored into the solid
crust of our planet by the Almighty hand, that there the waters which
"he called seas" might be gathered together, so as to "let the dry land
appear," and fit the Earth for the habitation of man. . . .

    Could the waters of the Atlantic be drawn off, so as to expose
to view this great sea-gash, which separates continents and extends
from the Arctic to the Antarctic, it would present a scene the most
rugged, grand, and imposing. The very ribs of the solid Earth, with the
foundations of the sea, would be brought to light, and we should have
presented to us at one view, in the empty cradle of the ocean, "a
thousand fearful wrecks," with that dreadful array of dead men's
skulls, great anchors, heaps of pearl and inestimable stones which, in
the poet's eye, lie scattered in the bottom of the sea, making it hideous
with sights of ugly death.

Matthew Fontaine Maury
From *The Physical Geography of the Sea*, 1855

The San Andreas Fault crossing the Carrizo Plain in California.

M any features of our planet have presented Earth scientists with contradictions and puzzles. For example, the remains of warm water coral reefs are found off the coast of the British Isles, marine fossils occur high in the Alps and the Himalayas, and coal deposits that were formed in warm, tropic climates are found in northern Europe, Siberia, and northeastern North America. In addition, similar patterns were observed in widely scattered places but for no known reason. Great mountain ranges divide the oceans, the volcanoes known as the ring of fire border both the east and west coasts of the Pacific Ocean, and deep-ocean trenches are found adjacent to long island arcs. No single coherent theory explained all these features, until the technology and scientific discoveries in the 1950s and early 1960s combined to trigger a complete reexamination of the Earth's history.

In this chapter we investigate the interior of the Earth, and we explore the history of the plate tectonic theory, as well as the evidence that allowed this theory to become accepted fact. We review the research that continues to provide us with new insights into the Earth's past, to comprehend and appreciate its present, and even to look forward into its future.

Interior of the Earth 2.1

Investigating Earth's Structure

Although we cannot directly observe the interior of the planet, scientists have been able to learn a great deal about the structure, composition, physical state, and behavior of the Earth's interior using indirect methods. The Earth's spherical shape, its mean (or average) radius, and its mass can be used to determine the average density of the Earth, 5.51 g/cm^3. **Density** is a measure of mass per unit volume and is usually given in grams per cubic centimeter, written g/cm^3. This calculated density is considerably greater than the average density of surface rocks, which is about 2.7 g/cm^3. The Earth's high average density requires that the material below the surface have a much greater density. Because the Earth wobbles only slightly as it rotates and the acceleration due to gravity over its surface is relatively uniform, the Earth's mass must be distributed fairly uniformly about its center as a series of concentric layers. Gravity, density, and the Earth's dimensions enable us to calculate the pressures within the Earth and also the temperatures that can be reached under these pressures. Because the Earth has a magnetic field, we conclude that its central part must include materials that produce magnetic fields.

Another clue to the Earth's structure is furnished by meteorites that occasionally hit the Earth and are considered to be the remains of planets. More than half of the meteorites that have been found are "stony" silicate or rocky lumps; another large group is made mainly of iron, nickel, and other metals; and a few are "stony-iron" with metal inclusions. Radiometric dating of meteorites gives a maximum age of 4.6 billion years, the same as the age of the solar system and Earth. These fragments allow us to directly analyze the density, chemistry, and mineralogy of the nickel-iron cores and stony shells of bodies that we believe to have a composition similar to that of Earth.

The most detailed information we have about the interior has come from roughly a century of recording and studying the passage of **seismic waves** through the body of the Earth. Geologists and geophysicists monitor recording stations all over the surface of the Earth that measure the type, strength, and arrival time of seismic waves generated by earthquakes, volcanic eruptions, and deliberately caused detonations.

Two basic kinds of seismic waves occur: surface waves travel relatively slowly along the surface of the Earth, and body waves travel at higher speeds through the Earth's interior (fig. 2.1). Most of the information we have about the Earth's interior comes from the study of body waves. There are two kinds of body waves: **P-waves,** or primary waves (so called because they travel faster than any other seismic waves and are the first to arrive at a recording station), and **S-waves,** or secondary waves (so called because they travel more slowly than P-waves and are the second waves to arrive at a station).

P-waves and S-waves produce different types of motion in the material they travel through. P-waves, also known as compressional waves, alternately compress and stretch the material they pass through, causing an oscillation in the same direction as they move. P-waves can travel through all three states of matter:

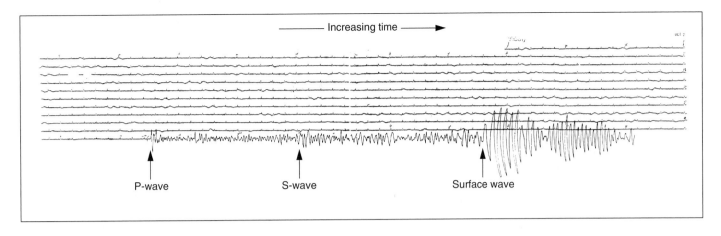

Increasing time

P-wave   S-wave   Surface wave

 web link **Figure 2.1** A seismogram of an earthquake that occurred in Taiwan recorded in Berkeley, California, 10,145 km (6300 mi) away. The faster body waves (P- and S-waves) arrive before the slower surface waves. Time increases from top to bottom and from left to right on the seismogram.

solid, liquid, and gas (sound propagates through the air and the oceans as a compressional wave). S-waves, also known as shear waves, oscillate at right angles to their direction of motion (similar to a plucked string). S-waves propagate only through solids. The motion generated in materials by these two types of waves is shown in figure 2.2.

The speed and direction of seismic waves depend on the characteristics of the material they travel through, including its chemistry, its density, and changes in the physical state of the material (solid, partially molten, or molten) caused by variations in pressure and temperature with depth. Detailed modeling of the paths taken by seismic waves, and their expected travel times through the Earth as a function of distance, has produced an Earth model consisting of four major layers: the inner core, the outer core, the mantle, and the crust (fig. 2.3). These are discussed in detail in the next section. As seismic body waves move through one layer and into another, their speeds change and the waves bend, or **refract,** as shown in figure 2.4*a, b.* The paths taken by P- and S-waves as they pass through the Earth provide information about the dimensions, structure, and physical properties of each of the internal layers. The outer core is a unique region where P-waves are strongly refracted (fig. 2.4*a*) and S-waves cannot propagate (fig. 2.4*b*), indicating that the core behaves like a liquid (this is discussed further on). The presence of the outer core creates both P- and S-wave shadow zones (distances along the surface of the Earth away from an earthquake where P- and S-waves generated by the event will not be recorded). The size of these shadow zones is determined by the depth to the top of the outer core. If the interior of the Earth had uniform properties, seismic body waves would follow straight lines, their speed would be constant, and there would be no shadow zones, as shown in figure 2.4*c.*

Internal Layers

At the planet's center is the **inner core;** its radius is 1222 km (759 mi). It is solid and nearly five times as dense as common surface rocks such as granite because of the tremendous pressure at that depth. The inner core is composed primarily of iron with lesser amounts of lighter elements that most likely include nickel, sulfur, and oxygen. The temperature of the inner core ranges from about 4000°–5500°C, far above the melting temperature of iron at the Earth's surface, but it remains solid because of the high pressure. The inner core is surrounded by a shell, 2258 km (1402 mi) thick, of similar composition and lower temperature (3200°C) and pressure. This shell is called the **outer core.**

As early as 1926 studies of the tidal deformation of the Earth made it clear that at least a portion of the core must behave like a fluid. In 1936, seismologist Inge Lehmann used earthquake data to establish the fact that there was a solid inner core surrounded by a "fluid" outer core that S-waves do not pass through. Although it behaves like a fluid, the outer core may not be completely molten. It would behave like a fluid even if as much as 30% of it were composed of suspended crystals, most probably of iron oxides and sulfides, that had formed from the surrounding liquid. Recent studies have determined that the inner core rotates about 1° per year faster than the mantle. This increase in eastward rotation is thought to be caused by motion in the fluid outer core. The fluid motion in the outer core moves at a speed that is probably on the order of kilograms per year, generating the Earth's magnetic field.

Detailed study of P-waves traveling in the vicinity of the core-mantle boundary has revealed that the upper surface of the core is not smooth but has peaks and valleys over its surface. These features extend as much as 11 km (7 mi) above and below the mean surface of the outer core. It is thought that the regions above a peak are areas where the mantle has excess heat and mantle material rises, drawing the core upward. Cooler, denser, and more viscous mantle material sinks to cause depressions in the core's surface. It is likely that these peaks and valleys last only as long as it takes a rising plume to lose its excess heat and sink back toward the core, perhaps a hundred million years.

Interior of the Earth **55**

The **mantle** makes up about 70% of the Earth's volume. The mantle is 2866 km (1780 mi) thick and is less dense and cooler (1100°–3200°C) than the core. It is composed of magnesium-iron silicates (rocky material rather than metallic like the core). Although the mantle is solid, some parts of it are weaker than others. Material in the mantle flows very slowly in response to variations in temperature, which create changes in density. Warmer, more buoyant material rises toward the surface while cooler, denser material sinks. The velocity of this motion is generally on the order of centimeters per year, which is much slower than the flow in the liquid outer core.

Data from an increasingly densely spaced and sophisticated array of seismic-recording stations and the computer capacity to analyze the travel times of the thousands of seismic waves generated by the world's earthquakes are used to produce three-dimensional maps of the interior of the Earth. This process, known as **seismic tomography,** is giving us a more detailed description of the Earth's interior layers and is demonstrating that these layers, especially the mantle, are less homogeneous than once thought. Three-dimensional tomographic images of the mantle show that there are large regions of the mantle with higher-than-average and lower-than-average seismic wave velocities, indicating regions of colder and warmer rock, respectively (fig. 2.5).

The Earth's outermost layer is the cold, rigid, thin surface layer called the **crust.** The boundary between the crust and the mantle is a chemical boundary; the rocks on either side of it have different chemical compositions. We know this by studying rocks that originated in the mantle and are now exposed at the surface. This boundary is called the **Mohorovicic discontinuity** in honor of its discoverer, Andrija Mohorovicic. It is more commonly known simply as the **Moho.** There are two kinds of crust, continental and oceanic. Continental crust is relatively light and averages about 40 km (25 mi) in thickness. Its composition and structure are highly variable, but it consists primarily of **granite**-type rock, which has a high content of sodium, potassium, aluminum, and silica. Oceanic crust is relatively dense, with an average thickness of about 7 km (4.3 mi). It is more homogeneous both chemically and structurally than continental crust and is composed primarily of **basalt**-type rock, which is low in silica and high in iron, magnesium, and calcium. See table 2.1 for a comparison of these layers and their properties.

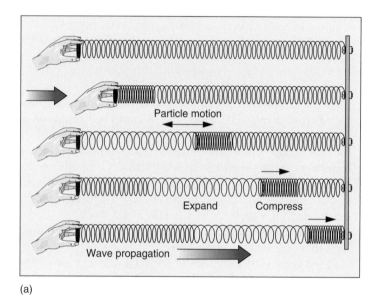

(a)

(b)

web link **Figure 2.2** Particle motion in seismic waves. (a) P-wave motion can be illustrated with a sudden push on the end of a stretched spring. Vibration is parallel to the direction of propagation. (b) S-wave motion can be illustrated by shaking a rope to transmit a deflection along its length. Vibration is perpendicular to the direction of propagation.

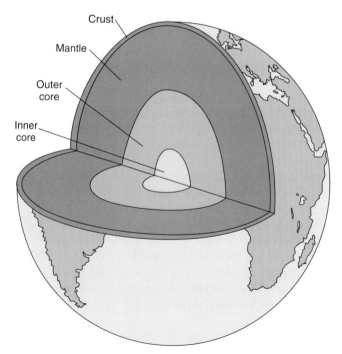

Figure 2.3 The layered structure of the Earth.

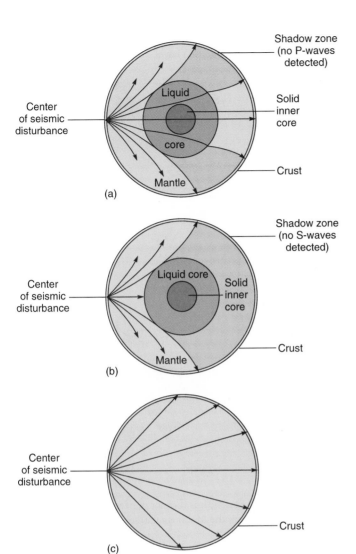

Figure 2.4 Movement of seismic waves through the Earth. (a) Refraction of P-waves and shadow zones produced by the Earth's interior structure. (b) Refraction of S-waves and shadow zones produced by the Earth's interior structure. (c) No refraction and no shadow zones occur in an Earth with a uniform structure.

Lithosphere and Asthenosphere 2.2

The Layers

More detailed study of the uppermost part of the Earth, the upper mantle and the crust, has shown that, independent of the sharp chemical boundary marked by the Moho, it is possible to identify a different layered structure characterized by changes in the mechanical properties of the rock from rigid to ductile behavior. Rocks that behave rigidly do not deform or change shape when a force is applied. Rocks that have ductile behavior will deform, or flow, in response to an applied force.

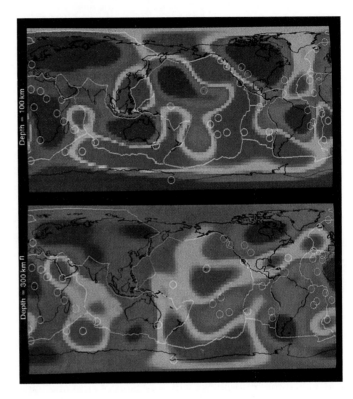

Figure 2.5 Color-coded map views of seismic velocities in the mantle at depths of 100 and 300 km (60 and 180 mi). *Blue* indicates high velocity due to lower temperatures. *Red* indicates low velocity due to higher temperatures. *White lines* outline tectonic plates and *white circles* are major hot spots.

Table 2.1 Layers of the Earth

Layer	Depth (km)	Thickness (km)	State	Composition	Density (g/cm³)	Temperature (°C)
Crust						
Continental	0–65	40 (average)	Solid	Silicates rich in sodium, potassium, and aluminum	2.67	−89–1000
Oceanic	0–10	7 (average)	Solid	Silicates rich in calcium, magnesium, and iron	3.0	0–1100
Mantle	Base of crust–2891	2866	Solid and mobile	Magnesium-iron silicates	3.4–5.6	1100–3200
Outer core	2891–5149	2258	Liquid	Iron, nickel	9.9–12.2	3200
Inner core	5149–6371	1222	Solid	Iron, nickel	12.8–13.1	4000–5500

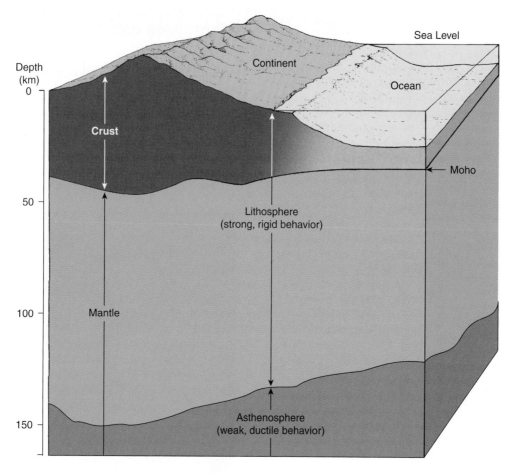

Figure 2.6 The lithosphere is formed from the fusion of crust and upper mantle. It varies in thickness; it is thinner in ocean basins and thicker in continental regions. The lithosphere rides on the weak, partially molten asthenosphere. Notice that the Moho is relatively close to the Earth's surface under the ocean's basaltic crust but is depressed under the granitic continents.

This has led to the identification of a strong, rigid surface shell called the **lithosphere,** which consists of crust and upper mantle material fused together. It is the lithosphere that comprises the plates of plate tectonics, discussed in detail in section 2.4. In oceanic regions the lithosphere thickens with increasing age of the sea floor. It reaches a maximum thickness of about 100 km (62 mi) at an age of 80 million years. In continental regions the thickness of the lithosphere is slightly greater than in the ocean basins, varying from about 100 km (62 mi) beneath the young marginal edges of continents to about 150 km (93 mi) beneath old continental crust. The base of the lithosphere corresponds roughly to the region in the mantle where temperatures reach 650°C ± 100°C. At the higher temperatures found at these depths, mantle rock begins to lose its strength.

The lithosphere is underlain by a weak, deformable region in the mantle called the **asthenosphere** where the temperature and pressure conditions lead to partial melting of the rock and loss of strength. The asthenosphere is often equated with a region of low seismic velocity called the low velocity zone (LVZ). Seismic waves travel more slowly through the asthenosphere, indicating that it may be as much as 1% melt. The asthenosphere behaves in a ductile

manner, deforming and flowing slowly when stressed. It behaves roughly the way hot asphalt does.

With increasing depth, the increase in pressure results in greater strength once more in the lower mantle, sometimes called the **mesosphere,** where the material is once again solid but will convect slowly, moving upward in some regions and downward in others, because of temperature gradients and density differences. The depth to the base of the asthenosphere remains a matter of scientific debate. It if does correspond roughly to the LVZ, it would extend from the base of the lithosphere to about 350 km (217 mi). Some scientists believe it may be as shallow as 200 km (122 mi), while others think it may extend to as much as 700 km (435 mi). The lithosphere and underlying asthenosphere are shown in figure 2.6 and are compared in table 2.2

Isostasy

The distribution of elevated continents and depressed ocean basins requires that a balance be kept between the internal pressures under the land blocks and those under the ocean basins. This is the principle of **isostasy.** The balance is possible because the greater thickness of low-density granitic crust in the continental regions is compensated for by the el-

Table 2.2 Upper Layers of the Earth, Based on Their Response to Applied Stress

Layer	Depth (km)	Thickness (km)	Characteristics
Lithosphere	0–100 (oceanic regions) 0 to 100–150 (continental regions)	0–150	Solid Rigid response
Asthenosphere	Base of lithosphere—350	200–350	Solid (~1% melt) Ductile response
Mesosphere	350—core-mantle boundary	4949	Solid Mobile

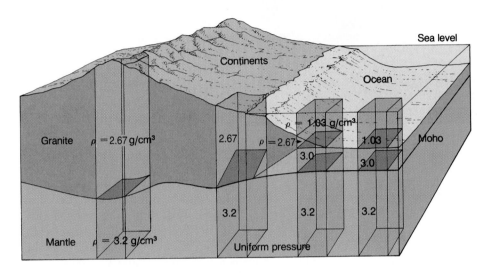

Figure 2.7 Isostasy. Columns of crustal material are unequal in height and density but generate the same pressure at the same depth within the mantle.

evated higher-density mantle material under the thinner crust of the oceans (fig. 2.7). The situation is often compared to the floating of an iceberg. The top of the iceberg is above the sea surface, supported by the buoyancy of the displaced water below the surface. The deeper the ice extends below the surface, the higher the iceberg reaches above the water. The less dense continental land blocks float on the denser mantle in the same way, with most of the continental volume below sea level.

In other words, the asthenosphere offers buoyant support to a section of lithosphere sagging under the weight of a mountain range. Because the lithosphere is cooler, it is more rigid and stronger than the asthenosphere. If a thick section of lithosphere has a large area and is mechanically strong, it depresses the asthenosphere slightly and is able to support a mountain range such as the Himalayas or the Alps. In other places where the crust is fractured and mechanically weak, a mountainous region must penetrate deeper into the mantle in order to provide buoyancy. The Andes are a mountain range with roots deep in the mantle.

If material is removed from or added to the continents, isostatic adjustment occurs. For example, parts of North America and Scandinavia continue to rise as the continents readjust to the lost weight of the ice sheets that receded at the close of the last ice age 10,000 years ago. Newly formed volcanoes protruding above the sea surface as islands often subside or sink back under the sea as their weight depresses the oceanic crust. The upper mantle gradually changes its

shape in response to weight changes in the overlying, more rigid crust. In the middle 1800s the concept of isostasy was firmly rooted in geological studies. It was thought that landmasses separated and oceans changed boundaries in response to changes in isostasy as sections of the Earth's crust moved up or down.

Movement of the Continents 2.3

History of a Theory: Continental Drift

As world maps became complete and more accurate, observant individuals were intrigued by the shapes of the continents on either side of the Atlantic Ocean. The possible "fit" of the bulge of South America into the bight of Africa was noted by the English scholar and philosopher Francis Bacon (1561–1626), the French naturalist George Buffon (1707–88), the German scientist and explorer Alexander von Humboldt (1769–1859), and others. In the 1850s the idea was expressed that the Atlantic Ocean had been created by a separation of two landmasses during some unexplained cataclysmic event early in the history of the Earth.

Thirty years later it was suggested that a portion of the Earth's continental crust had been torn away to form the Moon, creating the Pacific Ocean and triggering the opening of the Atlantic. As scientific studies of the Earth's crust continued, patterns of rock formation, fossil distribution, and mountain range placement began to show even greater

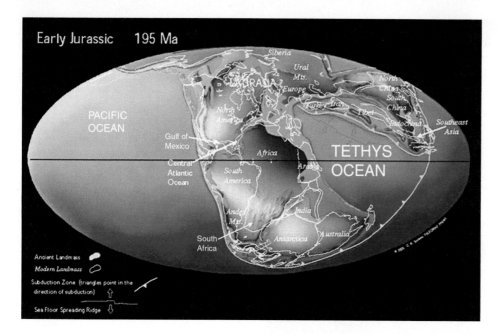

Early Jurassic   195 Ma

LAURASIA

PACIFIC
OCEAN

Siberia

Ural
Mts.

Europe

North
America

Iran

Tibet

North
China

South
China

Indochina

Southeast
Asia

Gulf of
Mexico

Central
Atlantic
Ocean

Africa

Arabia

TETHYS
OCEAN

South
America

Andes
Mts.

India

South
Africa

Antarctica

Australia

Ancient Landmass

Modern Landmass

Subduction Zone (triangles point in the
direction of subduction)

Sea Floor Spreading Ridge

web link **Figure 2.8** Pangaea in the early Jurassic, 195 Ma (million years before the present). Pangaea is composed of the two subcontinents Laurasia and Gondwanaland, which will split apart with the opening of the central Atlantic Ocean. There is one great ocean, Panthalassa, which was the ancestral Pacific Ocean, and a second small ocean, the Tethys, which would eventually close with the opening of the Indian Ocean.

similarities between the lands now separated by the Atlantic Ocean. In a series of volumes published between 1885 and 1909, an Austrian geologist, Edward Suess, proposed that the southern continents had been joined into a single continent he called **Gondwanaland.** He assumed that isostatic changes had allowed portions of the continents to sink and create the oceans between the continents. This idea was known as the subsidence theory of separation. At the beginning of this century, Alfred L. Wegener and Frank B. Taylor independently proposed that the continents were slowly drifting about the Earth's surface. Taylor soon lost interest, but Wegener, a German meteorologist, astronomer, and Arctic explorer, continued to pursue this concept until his death in 1930.

Wegener's theory, often called **continental drift,** proposed the existence of a single supercontinent he called **Pangaea** (fig. 2.8). He thought that forces arising from the rotation of the Earth combined with tidal forces began Pangaea's breakup. First, the northern portion composed of North America and Eurasia, which he called **Laurasia,** separated from the southern portion formed from Africa, South America, India, Australia, and Antarctica, for which he retained the earlier name Gondwanaland. The continents as we know them today then gradually separated and moved to their present positions. Wegener based his ideas on the geographic fit of the continents and the way in which some of their older mountain ranges and rock formations appeared to be related to each other when the landmasses were assembled to form Pangaea. He also noted that fossils more than 150 million years old collected on different continents were

remarkably similar, implying the ability of land organisms to move freely from one landmass to another. Fossils more recent than 150 million years old showed quite different forms in different places, suggesting that the continents and their evolving populations had separated from one another.

Wegener's theory provoked considerable debate in the 1920s. The most serious weakness in the theory was Wegener's inability to identify a mechanism that could cause the continents to break apart and drift through the ocean basins. When challenged on this point he suggested that there were two possible driving mechanisms, centrifugal force from the rotation of the Earth and the same tidal forces that raise and lower sea level. Both of these were quickly demonstrated to be too small to reasonably be capable of moving continental rock masses through the rigid basaltic crust of the ocean basins. With the lack of a viable driving mechanism, much of the scientific community remained skeptical of the continental drift theory and little attention was given to it after Wegener's death.

Evidence for a New Theory: Seafloor Spreading

Armed with new sophisticated instruments and technology developed during World War II, Earth scientists returned to their studies of the sea floor in the 1950s. Although the use of sound to examine the sea floor began in the 1920s, by the 1950s the equipment was greatly improved and much more readily available. (See chapter 3 for a discussion of echo sounders and depth recorders.) In the 1950s a worldwide effort was made to survey the sea floor, and for the first time, scientists were able to examine, in detail, the deep-ocean floor. These surveys discovered a series of mountain ranges through the ocean basins (fig. 2.9). These ranges are 65,000 km (40,000 mi) long; they rise 2–3 km (1.2–1.9 mi) above the adjacent sea floor and typically are between 1000 and 3000 km (600 and 1800 mi) wide. These are the deep seas' mid-ocean **ridge** and **rise** systems. If the slopes of these mountain ranges are steep and the width of the ranges is narrow, they are referred to as ridges (such as the Mid-Atlantic Ridge and the Mid-Indian Ridge); if the slopes are gentler and their width is broad, they are called rises (such as the East Pacific Rise). The ocean ridges typically have a central **rift valley,** a depression along the axis of the ridge, 50–3000 m (165–9850 ft) deep and 20–50 km (12.5–31 mi) wide. Ocean rises typically do not have a central rift valley; instead, they have an elevation (their shallowest depth) along the center. Along the axes of both ridges and rises is a narrow zone roughly 2 km (1.2 mi) wide that is volcanically active.

Another dominant feature of the ocean floor was found to be narrow, steep-sided ocean **trenches,**

6000–11,000 m (20,000–36,000 ft) deep, that are most characteristic of the Pacific Ocean. Some of these trenches are located seaward of chains of volcanic islands; Japan, Indonesia, the Philippines, and the Aleutian Islands are all associated with trenches. Other deep-sea trenches follow the edges of South and Central America (fig. 2.9). Ridges, rises, and trenches are discussed further in chapter 3. Theories current at the time of their discovery had not predicted the existence of such extensive mid-ocean features and could not explain them. However, a new theory was advanced that made the old idea of continental drift seem a possible and plausible explanation.

In the early 1960s, H. H. Hess of Princeton University promoted the concept that deep within the Earth's mantle there are currents of low-density molten material heated by the Earth's natural radioactivity. When these upward-moving mantle currents reach the lithosphere, they move along under it, cooling as they do so until they became cool enough and dense enough to sink down toward the core again. These patterns of moving mantle material are called **convection cells** (fig. 2.10). There are two proposed models of mantle convection. Some scientists believe that two sets of convection exist, cells one confined to the upper mantle above a depth of 700 km (435 mi) and the other in the lower mantle. Other scientists believe that convection occurs throughout the entire mantle to the core-mantle boundary; this model is called whole-mantle convection.

If the upward-moving mantle material, or magma, breaks through the lithosphere of the sea floor instead of continuing to flow underneath it, underwater volcanoes are produced and a ridge system forms along the crack in the crust. As the magma oozes out, it becomes lava, cools, hardens, and is added to the Earth's surface as new oceanic-type basaltic crust. If new crust is being produced in this manner, there must be some mechanism to remove old crust because there is no measurable change in the size of the Earth. The great, deep trenches of the Pacific were proposed as areas where the older, cooler, and denser oceanic lithosphere depresses and fractures the sea floor as it sinks back into the Earth's interior, eventually to be recycled into the mantle and the convection cell system. Figure 2.10 shows this process of producing new lithosphere at the ridges and losing old lithosphere at the trenches in a system driven by the motion of convection cells.

Although some of the ascending molten material breaks through the crust and solidifies, most of the rising material is turned aside under the rigid lithosphere and moves away toward the descending sides of the convection cells, dragging pieces of the lithosphere with it. This lateral movement of the oceanic lithosphere produces **seafloor spreading** (fig. 2.10). Areas in which new sea floor, and oceanic lithosphere, is formed above rising magma are **spreading centers;** areas of descending older oceanic lithosphere are **subduction zones.** The seafloor spreading mechanism provides the forces causing continental drift. The continents are not moving through the basalt of the sea floor; instead, they are being carried as passengers on the lithosphere, similar to boxes on a conveyor belt.

Hess's model of the sea floor riding atop mantle convection cells has been modified as additional forces acting on the lithosphere have been identified. These forces are discussed in more detail in section 2.5.

Evidence for Crustal Motion

Additional evidence was needed to support the idea of seafloor spreading. As oceanographers, geologists, and geophysicists explored the Earth's crust on land and under the oceans, evidence began to accumulate.

Epicenters of earthquakes were known to be distributed around the Earth in narrow and distinct zones. Epicenters are the points on the Earth's surface directly above the actual earthquake location, which is called the earthquake **focus** or **hypocenter.** These zones were found to correspond to the areas along the ridges, or spreading centers, and the trenches, or subduction zones. Earthquakes that are shallower than 100 km (60 mi) are prevalent in the ocean basins along ridges and rises and at trenches where the lithosphere bends and fractures as it is subducted into the mantle. These relatively shallow events are also found in continental regions that are undergoing significant deformation. Earthquakes deeper than 100 km are generally associated with the subduction of oceanic lithosphere. The deeper the earthquakes are the more they are displaced away from the trench beneath continents or island arcs (fig. 2.11).

Researchers sank probes into the sea floor to measure the heat from the interior of the Earth moving through the oceanic crust (fig. 2.12). The measured heat flow shows a pattern with a high degree of variability, even over closely spaced intervals. In part, this variation in the data is attributed to the seeping of seawater down through porous or fractured portions of the crust at one location along a ridge system and its rising as heated water at another. This circulation does not occur over regions of the sea floor that are sealed by a thick layer of loose particles, or **sediment.** In these areas, the measured heat flow shows a regular pattern. It is highest in the vicinity of the mid-ocean ridges over the ascending portion of the mantle convection cell, where the crust is more recent, and it decrease as the distance from the ridge center and the crustal age increases (fig. 2.13).

Radiometric dating of the age of rocks from the land and from the sea floor shows that the oldest rocks from the oceanic crust are only about 200 million years old, whereas the rocks from the land are much older. The sea floor formed by convection-cell processes at the spreading centers is young and short-lived, for it is lost at the subduction zones, where it plunges back down into the mantle.

Vertical, cylindric samples, or **cores,** were obtained by drilling through the sediments that cover the ocean bottom and into the rock of the ocean floor. (A detailed discussion of sediments is found in chapter 3.) Drilling cores through the average of 500–600 m (1600–2000 ft) of sediments that cover the deep-ocean floor and then drilling on into the ocean floor required the development of a new technology and a new kind of ship. In the late summer of 1968 the specially constructed drilling ship

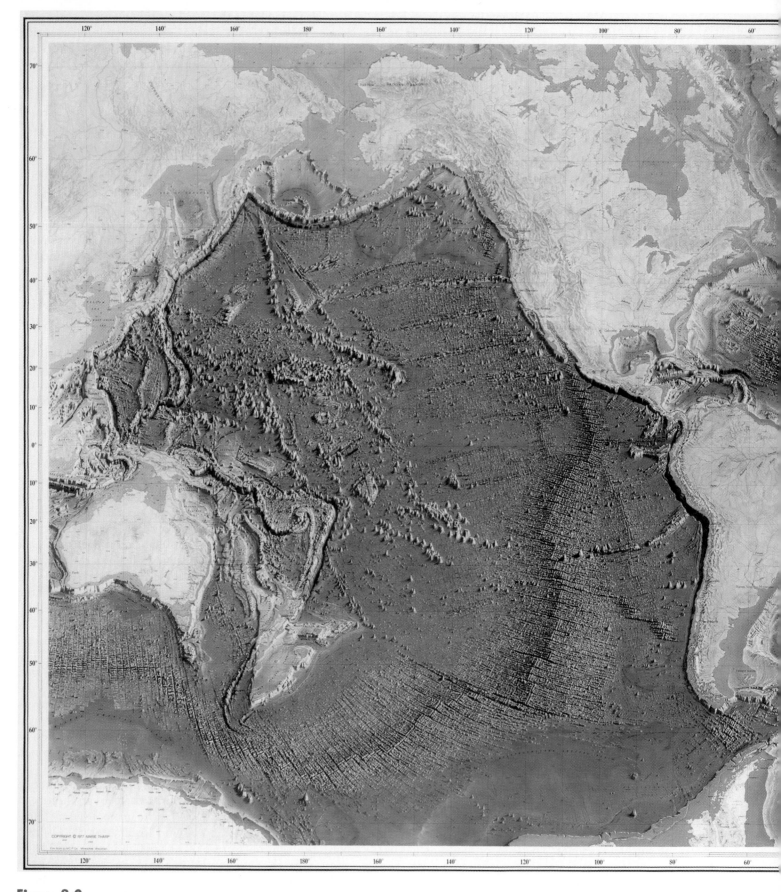

Figure 2.9 A physiographic chart of the world's oceans. *From World Ocean Floor, Bruce C. Heezen and Marie Tharp. © Marie Tharp 1977. Reproduced by permission of Marie Tharp, Washington Ave., South Nyack, NY 10960.*

WORLD OCEAN FLOOR

BY BRUCE C. HEEZEN AND MARIE THARP

Based on Research and Exploration Initiated and Supported by the

UNITED STATES NAVY · OFFICE OF NAVAL RESEARCH

1977

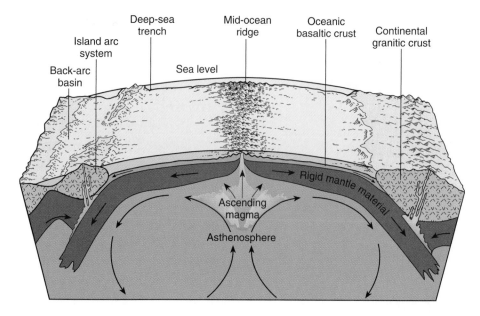

Figure 2.10 Seafloor spreading creates new crust at mid-ocean ridges and loses old crust in deep-sea trenches. This process is shown being driven by convection cells in the asthenosphere.

Glomar Challenger (see prologue fig. XX*a*) was used for a series of studies. This ship, 122 m (400 ft) long, with a beam of 20 m (65 ft) and a draft of 8 m (27 ft), displaces 10,500 tons when loaded. Its specialized bow and stern thrusters and its propulsion system respond automatically to computer-controlled navigation, using acoustic beacons on the sea floor. This setup enables the ship to remain for long periods of time in a nearly fixed position over a drill site in water too deep to anchor. An acoustic guidance system enables it to replace drill bits and reenter the same bore holes in water about 6000 m (20,000 ft) deep. This method is illustrated in figure 2.14. A more general discussion of cores and coring methods is found in chapter 3.

In 1983 the *Glomar Challenger* was retired, after logging 600,000 km (375,000 mi), drilling 1092 holes at 624 drill sites, and recovering a total of 96 km (60 mi) of deep-sea cores for study. A new deep-sea drilling program with a new drill vessel, the *JOIDES Resolution*, began in 1985. The *JOIDES Resolution* is about the same size as the *Glomar Challenger*, complete with the world's most sophisticated, state-of-the-art scientific and drilling equipment (fig. 2.15). The drill pipe comes in individual sections 9.5 m (31 ft) long that can be screwed together to make a single pipe up to 8200 m (27,000 ft) long. Information on current research being conducted by the *JOIDES Resolution* appears in section 2.7.

The cores taken by the *Glomar Challenger* provided much of the factual data needed to establish the existence of seafloor spreading. No ocean crust older than 180 million years was found, and sediment age and thickness were shown to increase with distance from the ocean ridge system (fig. 2.16). Note that the sediments closest to the ridge system are thin over the new crust, which has not had long

to accumulate its sediment load. The crust farther away from the ridge system is older and is more heavily loaded with sediments.

Although each of these pieces of evidence fits the theory that the Earth's crust produced at the ridge system is new and young, the most elegant proof for seafloor spreading came from a study of the magnetic evidence locked into the ocean's floors.

The Earth's familiar geographic North (90°N) and South (90°S) Poles mark the axis about which it rotates. The Earth also behaves as if it has a giant bar magnet embedded in its interior tilted roughly 11.5° away from its axis of rotation (fig. 2.17). The north magnetic pole is located near 79.3°N, 71.5°W in the Northwest Territories of Canada, and the south magnetic pole is almost directly opposite near 79.3°S, 108.5°E in the South Pacific Ocean. A magnetic field like the Earth's, with two opposite poles, is called a **dipole.** The Earth's magnetic field consists of invisible lines of magnetic force that are parallel to the Earth's surface at the magnetic equator and converge and dip toward the Earth's surface at the magnetic poles. A small, freely suspended magnet (such as a compass needle) will align itself with these lines of magnetic force, with the north-seeking end pointing to the north magnetic pole and the south-seeking end pointing to the south magnetic pole. In addition, it dips toward the Earth's surface by differing amounts depending on its distance from the magnetic equator. At the magnetic equator a small magnet would be horizontal, or parallel to the Earth's surface. In the north magnetic hemisphere the north-seeking end of the magnet would point downward at an increasing angle until it pointed vertically downward at the north magnetic pole. In the south magnetic hemisphere the north-seeking end of the

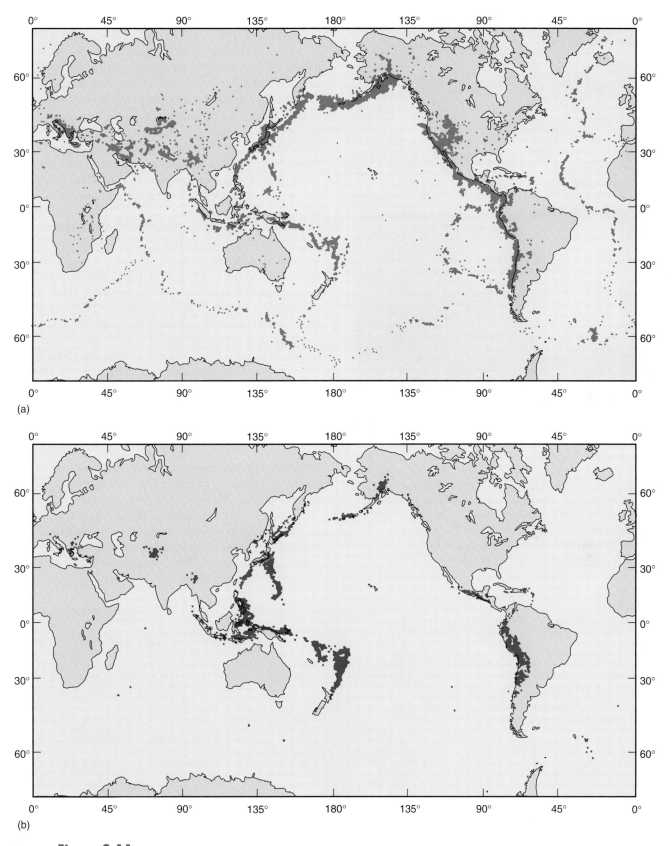

web link **Figure 2.11** Earthquake epicenters, 1961–67. (a) Epicenters of earthquakes with depths less than 100 km (60 mi) outline regions of crustal movement. (b) Epicenters of earthquakes with depths greater than 100 km are related to subduction.

Figure 2.12 Recovering a heat flow probe from the deep sea floor. The probe consists of a cylinder holding the measuring and recording instruments above a thin probe that penetrates the sediment. Thermisters (devices that measure temperature electronically) at the top and bottom of the probe measure the temperature difference over a fixed distance equal to the length of the probe. This temperature change with depth is used to compute the heat flow through the sediment. *Photo courtesy of Dr. Robert Kieckhefer.*

magnet would point upward at an increasing angle until it pointed vertically upward at the south magnetic pole.

Most igneous rocks contain particles of a naturally magnetic iron mineral called magnetite. Particles of magnetite act like small magnets. These particles are particularly abundant in basalt, the type of rock that makes up the oceanic crust. When basaltic magma erupts on the sea floor along ocean ridges it cools and solidifies to form basaltic rock. During this time the magnetite grains in the rock become magnetized in a direction parallel to the existing magnetic field at that time and place. When the temperature of the rock drops below a critical level called the **Curie temperature,** or Curie point (named in honor of the twice Nobel Prize–winning physicist Madame Marie Curie), roughly 580°C, the magnetic signature (magnetic strength and orientation) of these particles is "frozen" into the rock, creating a "fossil" magnetism that remains unchanged unless the rock is heated again to a temperature above the Curie temperature. In this manner, rocks can preserve a record of the strength and orientation of the Earth's magnetic field at the time of their formation. The investigation of fossil magnetism in rocks is the science of **paleomagnetism.**

Research on age-dated layers of volcanic rock found on land shows that the polarity, or north-south orientation, of the Earth's magnetic field reverses for varying periods of geologic time. Thus, at different times in the Earth's history the present north and south magnetic poles have changed places. Each time a layer of volcanic material cooled and solidified, it recorded the magnetic orientation and polarity of the time period in which it occurred. The dating and testing of samples taken through a series of volcanic layers have enabled scientists to build a calendar of these events (fig. 2.18). During these **polar reversals,** the Earth's magnetic field gradually decreases in strength by a factor of about ten over a period of a few thousand years. There is some evidence that during a

Figure 2.13 Heat flow through the Pacific Ocean floor. Values are shown against age of crust and distance from the ridge crest. *Data from J. G. Sclater and J. Crowe, "On the Variability of Oceanic Heat Flow Average" in* Journal of Geographical Research, *81:17 (June 1976), p. 3004. American Geophysical Union, Washington, D.C.*

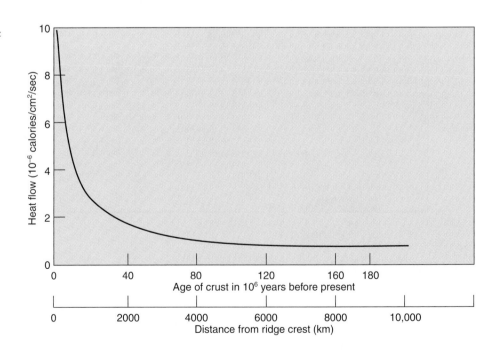

reversal the Earth's field may not remain a simple dipole. The collapse of the magnetic field during reversals enables more intense cosmic radiation to penetrate the planet's surface, and recent work has suggested a correlation between such periods and a decrease in populations of delicate, single-celled organisms that live in the surface layers of the ocean. Nearly 170 reversals have been identified during the last 76 million years, and the present magnetic orientation has existed for 780,000 years. The interval between reversals has varied, with some notable periods of constant orientation for long time intervals. During the Cretaceous period (from roughly 120 million to 80 million years before present) the field was stable and normally polarized. The field was stable and reversely polarized for about 50 million years in the Permian period, roughly 300 million years before present. The cause of reversals is unknown but is thought to be associated with changes in the motion of the magnetic material of the Earth's liquid outer core. How long our current polarity will last we do not know.

During the 1950s oceanographers began to frequently measure the strength of the magnetism of the oceanic crust during research cruises by towing marine magnetometers behind the ship, away from the magnetic noise of the vessel. From 1958–61 the results of several marine magnetic surveys were published as maps of high and low magnetic intensity that resembled the ridges and valleys of a topographic map. These alternating areas of high and low intensity came to be known as "magnetic stripes." The stripes varied in width from a few kilometers to many tens of kilometers and were as much as several thousand kilometers long. When the stripes were compared to marine bathymetry it was noted that the maps revealed a mirror image pattern of roughly parallel magnetic stripes on each side of the mid-ocean ridge (fig. 2.19). Their significance was not understood until 1963, when F. J. Vine and D. H. Matthews of Cambridge University proposed that these stripes represented a recording of the polar reversals of the vertical component of the Earth's magnetic field, frozen into the sea floor. As the molten basalt rose along the crack of the ridge system and solidified, it locked in the direction of the prevailing magnetic field. Seafloor spreading moved this material off on either side of the ridge, to be replaced by more molten materials. Each time the Earth's magnetic field reversed, the direction of the magnetic field was recorded in the new crust. Vine and Matthews proposed that if such were the case, there should be a symmetric pattern of magnetic stripes centered at the ridges and becoming older away from the ridges. Their ideas were confirmed; the polarity and age of these stripes corresponded to the same magnetic field changes found in the dated layers on land and dramatically demonstrated seafloor spreading.

The accumulation of magnetic stripe data across the world's oceans has been used to produce a map of the age of the sea floor (fig. 2.20). Seafloor age increases away from the oceanic spreading centers, providing another verification of seafloor spreading. The present ocean basins are not old but new, created during the past 200 million years, or the last 5% of the Earth's history.

Polar Wandering Curves

Paleomagnetic studies of fossil magnetism in continental rocks provide evidence of relative motion between tectonic plates and the Earth's magnetic poles through time. By carefully measuring the fossil magnetism in a rock, the location of the north magnetic pole at the time of the rock's formation can be calculated. This is done by determining the direction and distance to the pole's position. The horizontal component of the rock's magnetism points in the direction of the pole. The angle of dip of the magnetic signature reveals the magnetic latitude at which the rock formed, which

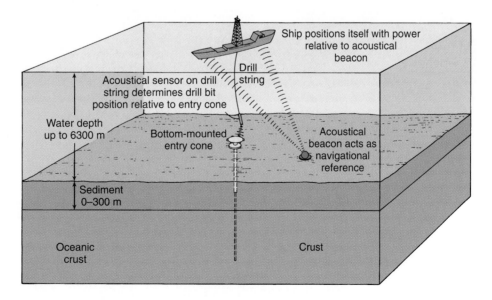

web link **Figure 2.14** Deep-ocean drilling technique. Acoustical guidance systems are used to maneuver the drilling ship over the bore hole and to guide the drill string back into the bore hole.

Figure 2.15 The deep-sea drilling vessel *JOIDES Resolution.*

Figure 2.16 Age and thickness of seafloor sediments.

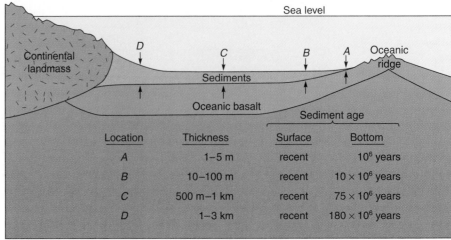

Location	Thickness	Surface	Bottom
A	1–5 m	recent	10^6 years
B	10–100 m	recent	10×10^6 years
C	500 m–1 km	recent	75×10^6 years
D	1–3 km	recent	180×10^6 years

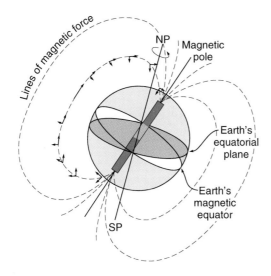

web 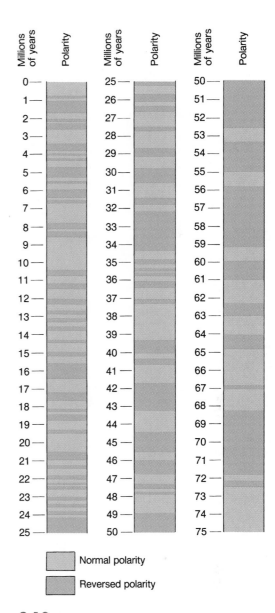 link

Figure 2.17
Lines of magnetic force surround the Earth and converge at the magnetic poles. A freely suspended magnet aligns itself with these lines of magnetic force, with the north-seeking end of the magnet pointing to the north magnetic pole and the south-seeking end pointing to the south magnetic pole. The magnet hangs parallel to the Earth's surface at the magnetic equator and dips at an increasing angle as it approaches the magnetic poles. *NP* and *SP* indicate the north and south geographic poles, respectively.

Figure 2.18 Polarity reversal timescale during the Cenozoic Era. Time is given in millions of years.

Figure 2.19 Reversals in the Earth's magnetic polarity cause the symmetrically striped pattern centered on the Mid-Atlantic Ridge. The age of the sea floor increases with the distance from the ridge. The spreading rate along the Mid-Atlantic Ridge is about 1 cm per year.

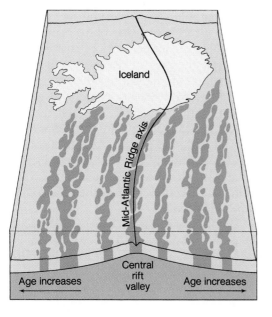

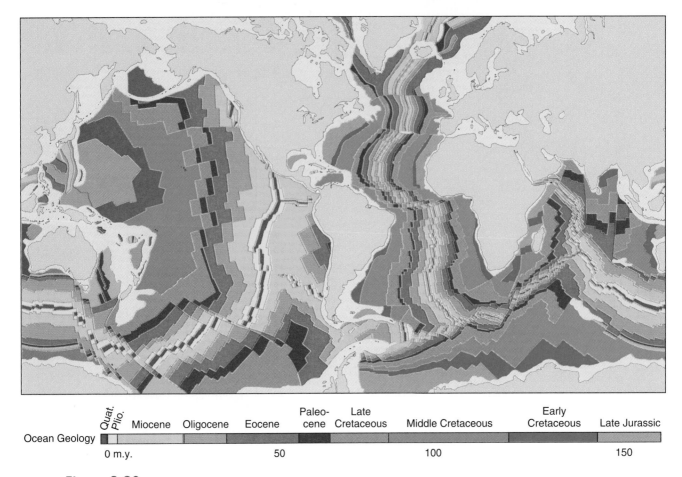

Ocean Geology	Quat. Plio.	Miocene	Oligocene	Eocene	Paleo-cene	Late Cretaceous	Middle Cretaceous	Early Cretaceous	Late Jurassic

0 m.y.   50   100   150

web link **Figure 2.20** The age of the ocean crust based on seafloor spreading magnetic patterns is shown in this color-shaded image. The age of the ocean floor increases with increasing distance from the mid-ocean ridge as a result of seafloor spreading. Oceanic fracture zones offset the ridge and the age patterns. *After* The Bedrock Geology of the World *by R.L. Larson, W.C. Pitman, III et al. W.H. Freeman.*

is directly related to the distance between the north magnetic pole and the rock's formation site.

By measuring the fossil magnetism in continental rocks of different ages that are all found on the same tectonic plate, it is possible to create a plot of the apparent location of the north magnetic pole through time. Such a plot is known as a **polar wandering curve** (fig. 2.21). A polar wandering curve is evidence that the relative positions of the continent, or the plate it is embedded in, and the north magnetic pole have changed with time. This may be either the result of the movement of the pole or the movement of the plate. While we know that the magnetic poles do move a little with time, their average location remains near the geographic poles. Consequently, a polar wandering curve records the motion of the plate with respect to a nearly stationary north magnetic pole.

Because all plates do not move along the same path, each plate produces its own polar wandering curve, which converges with all other polar wandering curves at the current location of the north magnetic pole. If we look at the polar wandering curves constructed for the North American and Eurasian Plates we see that they have the same general shape but diverge with increasing age of the rocks (fig. 2.21).

The divergence of these two polar wandering paths is a record of the opening of the North Atlantic Ocean and the drifting apart of the North American and European Plates. When these two polar wandering curves are superimposed, the movements of these plates reconstruct the formation of the two-lobed portion of Pangaea proposed by Wegener, with Laurasia to the north, Gondwanaland to the south, and the Sea of Tethys in between. On the basis of this method, the present landmasses join nicely at the edges of the continental blocks, at a boundary on the continental slope that is about 2000 m (6560 ft) below the present sea level. When the plates and their landmasses are moved together, major ancient mountain ranges and fault systems spanning several of our modern continents join; for example, the fault through the Caledonian Mountains of Scotland joins the Cabot Fault extending from Newfoundland to Boston. Also, countries of western Europe and Great Britain that are at present located in temperate and high latitudes were once found in the equatorial zone; their previous locations explain their fossil coral reefs and desert-type sand deposits.

Current research leads scientists to believe that the assumption that the rotational polar axis has not varied its location over time may not be correct. The rotating Earth

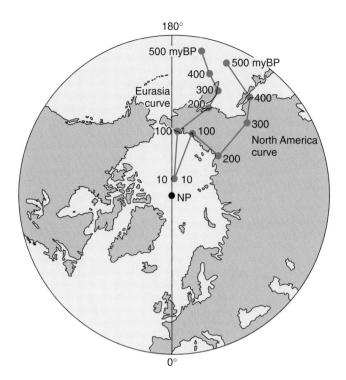

Figure 2.21 The positions of the north magnetic pole millions of years before the present (myBP), judged by the magnetic orientation and the age of the rocks of North America and Eurasia. The divergence of the two paths indicates that the landmasses of North America and Eurasia have been displaced from each other as the Atlantic Ocean opened.

From Fundamentals of Oceanography, *4th edition, Duxbury, Duxbury, and Sverdrup. Copyright 2000 The McGraw-Hill Companies. All rights reserved.*

Table 2.3 Approximate Plate Area (10^6 km²)	
Major Plates	
Pacific Plate	105
African Plate	80
Eurasian Plate	70
North American Plate	60
Antarctic Plate	60
South American Plate	45
Australian Plate	45
Smaller Plates	
Nazca Plate	15
Indian Plate	10
Arabian Plate	8
Philippine Plate	6
Caribbean Plate	5
Cocos Plate	5
Scotia Plate	5
Juan de Fuca Plate	2

possesses tremendous **inertia,** which keeps it spinning on its axis. However, as the lithospheric plates migrate about on the Earth's surface, the Earth may tend to roll about its center slightly, trying to adjust so that the majority of the Earth's crustal mass is centered about the equator. If the Earth does roll about its center, the location of the pole points marking the Earth's rotational axis relative to the crust may also shift.

Plate Tectonics 2.4

When the ideas of continental drift and seafloor spreading were joined, it led to the formation of a single unified concept of the fragmentation and movement of the outer rigid shell of the Earth; this concept is known as **plate tectonics.** This rigid shell is the layer we call the lithosphere (you may want to review section 2.2 and figure 2.6). The lithosphere is fragmented into seven major plates along with a number of smaller ones (fig. 2.22 and table 2.3). The plates are outlined by the major earthquake belts of the world (see fig. 2.11); this relation was first pointed out in 1965 by J. T. Wilson, a geophysicist at the University of Toronto.

Plates and Their Boundaries

Each lithospheric plate consists of the upper roughly 80–100 km (50–62 mi) of rigid mantle rock capped by either oceanic or continental crust. Lithosphere capped by oceanic

crust is often simply called oceanic lithosphere, and, in a similar fashion, lithosphere capped by continental crust is often referred to as continental lithosphere. Some plates, such as the Pacific Plate, consists entirely of oceanic lithosphere but most plates, like the South American Plate, consist of variable amounts of both oceanic and continental lithosphere with a gradual transition from one to the other along the margins of continents. The plates move with respect to one another, sliding on top of the ductile asthenosphere below. As the plates move, they interact with one another along their boundaries, producing the majority of the earthquake and volcanic activity that occurs on Earth.

There are three basic kinds of plate boundaries. These are defined by the type of relative motion between the plates they separate. Each of these boundaries is associated with specific kinds of geological features and processes (table 2.4 and fig. 2.23). Plates move away from each other along **divergent plate boundaries.** Ocean basins and the sea floor are created along marine divergent boundaries marked by the mid-ocean ridges and rises, while continental divergent boundaries are often marked by deep rift zones. Plates move toward one another and collide along **convergent plate boundaries.** Marine convergent boundaries are associated with deep-ocean trenches, the destruction of sea floor, and the closing of ocean basins, while continental convergent boundaries are associated with the creation of massive mountain ranges. The third type of plate boundary occurs where two plates are neither converging nor diverging but are simply sliding past one another. These are called **transform boundaries** and are marked by large faults called **transform faults.**

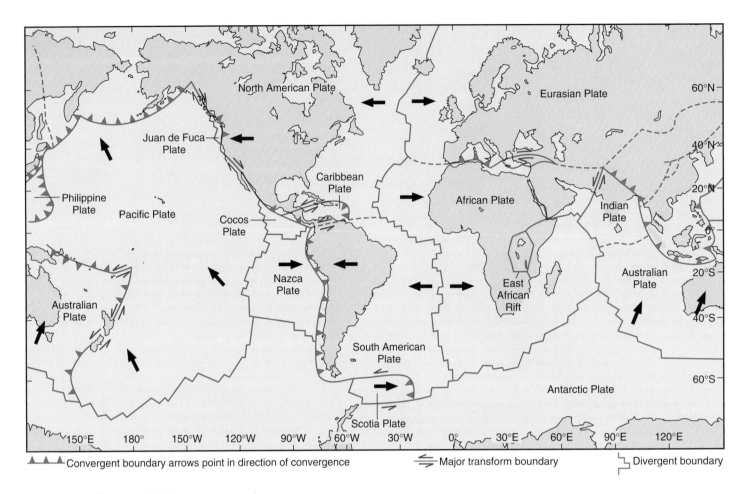

Convergent boundary arrows point in direction of convergence   Major transform boundary   Divergent boundary

web link **Figure 2.22** The Earth's lithosphere is broken into a number of individual plates. The edges of the plates are defined by three different types of plate boundaries. *Arrows* indicate direction of plate motion.

Table 2.4 Types and Characteristics of Plate Boundaries

Plate Boundary	Types of Lithosphere	Geologic Process	Geologic Feature	Earthquakes	Volcanism	Examples
Divergent (move apart)	Ocean–Ocean	New sea floor created, ocean basin opens	Mid-ocean ridge	Yes, shallow	Yes	Mid-Atlantic Ridge, East Pacific Rise
	Continent–Continent	Continent breaks apart, new ocean basin forms	Continental rift, shallow sea	Yes, shallow	Yes	East African Rift, Red Sea, Gulf of Aden, Gulf of California
Convergent (move together)	Ocean–Ocean	Old seafloor destroyed by subduction	Ocean trench	Yes, shallow to deep	Yes	Aleutian, Mariana, and Tonga Trenches (Pacific Ocean)
	Ocean–Continent	Old seafloor destroyed by subduction	Ocean trench	Yes, shallow to deep	Yes	Peru-Chile and Middle-America Trenches (Eastern Pacific Ocean)
	Continent–Continent	Mountain building	Mountain range	Yes, shallow to intermediate	No	Himalaya Mountains, Alps
Transform (slide past each other)	Ocean	Sea floor conserved (neither created nor destroyed)	Transform fault (offsets segments of ridge crest)	Yes, shallow	No	Mendocino and Clipperton (Eastern Pacific Ocean)
	Continent	Sea floor conserved (neither created nor destroyed)	Transform fault (offsets segments of ridge crest)	Yes, shallow	No	San Andreas Fault, Alpine fault (New Zealand), North and East Anatolian Fault (Turkey)

From Fundamentals of Oceanography, *4th edition, Duxbury, Duxbury, and Sverdrup. Copyright 2000 The McGraw-Hill Companies. All rights reserved.*

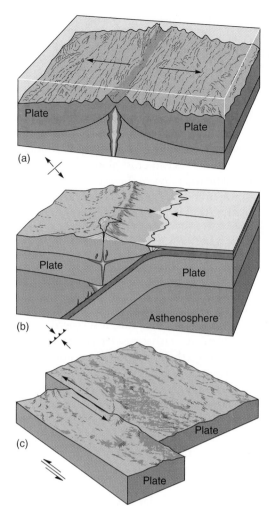

Divergent Boundaries

Ocean basins are created along divergent boundaries that break apart continental lithosphere (fig. 2.24). A series of successive divergent boundaries were responsible for the breakup of Pangaea, producing the individual continents and ocean basins we see today. Upwelling of hot mantle rock is thought to be responsible for the creation of divergent boundaries in continental lithosphere. Mantle upwelling heats the base of the lithosphere, thinning it and causing it to dome upward (fig. 2.24a). This weakens the lithosphere and produces extensional forces and stretching that lead to faulting and volcanic activity. Faulting along the boundary creates rifting and subsidence in the continental crust, form-ing a **rift zone** along the length of the boundary (fig. 2.24b). A present-day example of this early rifting is the continuing development of the East African Rift valley, stretching from Mozambique to Ethiopia (fig. 2.25). Volcanoes in the rift in-clude Kilimanjaro and Mount Kenya. Such a sunken rift zone is called a **graben.** The next "stage" in continental rifting can be seen further to the north, where two shallow seas, the Red Sea and the Gulf of Aden, have formed (see fig. 2.24c). If this rift system remains active in the future, East Africa will

web 🔗 link **Figure 2.23** The three basic types of plate boundaries include (a) divergent boundaries where plates move apart, (b) convergent boundaries where plates collide, and (c) transform boundaries where plates slide past one another.

Figure 2.24 (a) Continental rifting begins as rising magma heats the overlying continental crust, making it dome upward and thin as tensional forces cause it to stretch. (b) Stretching and pulling apart of the crust produce a rift valley with active volcanism. (c) Continued spreading results in the formation of new sea floor along the boundary and the creation of a young, shallow sea. (d) Eventually, a mature ocean basin and spreading ridge system are created.

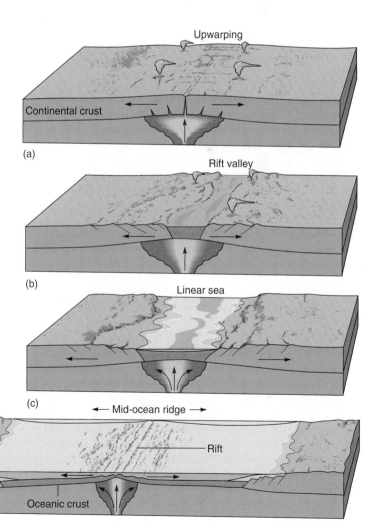

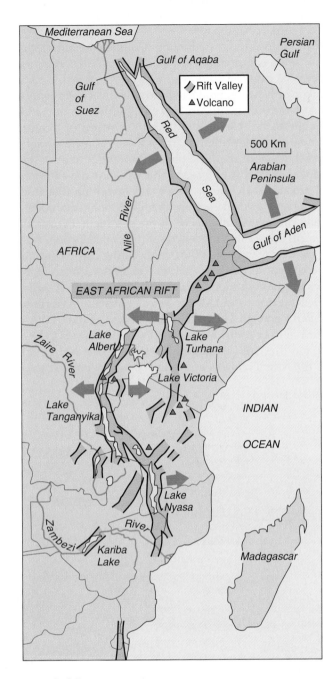

Figure 2.25 An active continent-continent divergent boundary, still in its early stage of development, has created the East African Rift system. This same boundary is more fully developed to the north, where it has created two shallow seas, the Red Sea and the Gulf of Aden.

up from the mantle and flow onto the ocean floor. In the early stages of a volcanic eruption, basaltic magma can flow rapidly onto the sea floor in relatively flat flows called sheet flows that may extend several kilometers away from their source. Basaltic magma has a low viscosity, allowing it to flow freely, because there is relatively little silicon dioxide in its chemistry. As the eruption rate decreases, the magma is often extruded more slowly onto the sea floor, creating rounded flows called pillow lavas, or **pillow basalts** (fig. 2.27). The magma solidifies to form new ocean crust along the edges of each diverging plate. The thickness of the oceanic crust remains relatively constant from the time of its formation at the ridge. However, as the two plates diverge, the mantle rock immediately beneath the crust cools, fuses to the base of the crust, and begins to behave rigidly, causing the thickness of the oceanic lithosphere to increase with age and distance from the ridge. Plates move apart at different rates along the the length of the ridge system.

Detailed seismic studies, coring, and the direct observation of large transform faults from submersibles have provided ocean scientists with considerable information about the process of rifting, the formation of oceanic crust, and the structure of ridges and rises. A generalized section across a ridge is shown in figure 2.28. This figure shows the four basic layers that make up the oceanic lithosphere.

The topmost layer (layer 1 on fig. 2.28) is composed of sediment; it may not be present at the ridge axis if the sea floor is too young to have acquired a covering of this loose material. Layer 2 is made up of two forms of basalt. The surface portion of this layer is the result of the rapid cooling of magma to form fine-grained, or glassy, lava. The deeper part of the layer is formed from magma that solidifies to form a series of vertically oriented basalt dikes. Layer 3 is also basaltic, but it has cooled more slowly, forming a granular, dark-colored igneous rock called **gabbro.** The Moho is found at the bottom of this layer. Layer 2 and 3 together are usually 6–7 km (≈4 mi) thick, but they are thinner at the rift zones. The lithosphere contains these layers and a thick fourth layer of rock known as peridotite that forms the rigid upper mantle and the base of the lithospheric plate. The partly molten asthenosphere portion of the mantle lies below this fourth layer.

The mechanism whereby peridotite produces basalt and gabbro was determined by subjecting laboratory samples of peridotite to the high temperature and pressures found 100 km (60 mi) below the surface of the oceanic crust. When cooled and decompressed, the partially melted peridotite produces rock similar to the basalt found in the upper layers of oceanic crust.

Transform Boundaries

The ocean ridge system is divided into segments that are offset by transform faults (fig. 2.29). As two plates move away from the ridge crest, they simply slide past one another along the transform fault. In this manner the ridge crest segments and transform faults form a continuous plate boundary that alternates between being a divergent boundary and a transform

separate from the rest of the continent and a new ocean basin will form with seafloor spreading and a central spreading ridge (see fig. 2.24*d*).

Most divergent boundaries are located along the axis of the mid-ocean ridge system (fig. 2.26). Mantle upwelling beneath the ridge heats the overlying oceanic lithosphere, causing it to expand, and creates a submarine mountain range. As the places on either side of the boundary move apart, cracks form along the crest of the ridge, allowing molten rock to seep

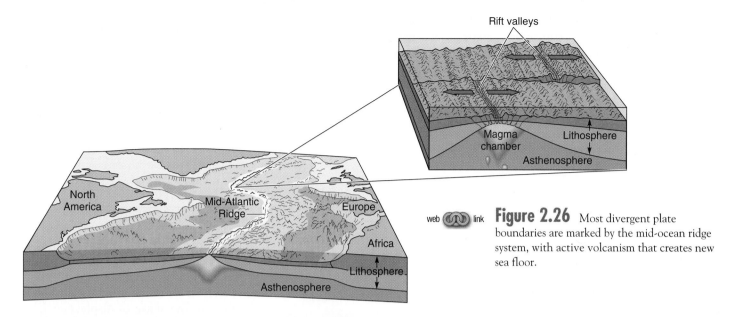

Rift valleys

Magma chamber

Lithosphere

Asthenosphere

web link **Figure 2.26** Most divergent plate boundaries are marked by the mid-ocean ridge system, with active volcanism that creates new sea floor.

North America

Mid-Atlantic Ridge

Europe

Africa

Lithosphere

Asthenosphere

web link **Figure 2.27** Pillow lavas are mounds of elongate lava "pillows" formed by repeated oozing and quenching of extruding basalt magma on the sea floor. First, a flexible glassy crust forms around the newly extruded lava, forming an expanded pillow. Next, pressure builds within the pillow until the crust breaks and new basalt extrudes like toothpaste, forming another pillow.

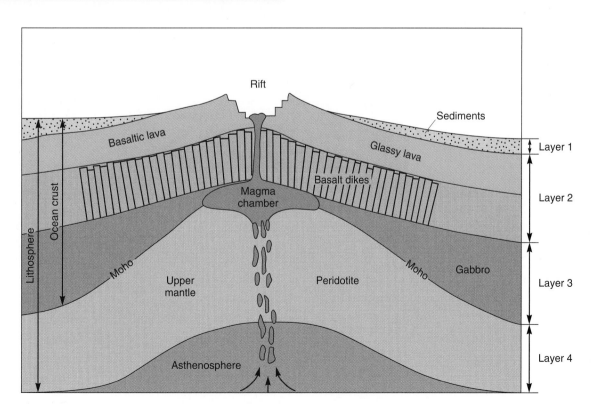

Rift

Sediments

Basaltic lava

Glassy lava

Layer 1

Basalt dikes

Layer 2

Lithosphere

Ocean crust

Magma chamber

Moho

Moho

Gabbro

Upper mantle

Peridotite

Layer 3

Asthenosphere

Layer 4

Figure 2.28 A mid-ocean ridge cross section showing the four layers of the oceanic lithosphere.

boundary. Differences in the age and temperature of the plates across a transform boundary can create significant and rapid changes in the elevation of the sea floor called **escarpments** (look at fig. 2.29 and imagine walking across the transform fault along a line that begins very near the ridge crest in *plate A* and ends far from the ridge crest in *plate B*). These changes in elevation are propagated into each plate by seafloor spreading, where they are preserved as "fossil" transform faults. The two symmetrical fossil transform faults in each plate together with the active transform fault between the adjacent segments of ridge crest form a single linear feature called a **fracture zone** that is roughly perpendicular to the ridge crest. The longest fracture zones can be up to 10,000 km (6200 mi) long, extending deep into each plate on either side of the ridge. It is important to remember that there is no movement or earthquake activity along the fracture zone where it extends into either plate. However, there is relative motion and seismicity along the transform fault between adjacent segments of ridge crest. The direction of motion of the two plates on either side of a transform fault is determined by the direction of seafloor spreading in each plate and is opposite the sense of displacement of the ridge segments.

While most transform faults join two divergent boundaries (segments of ridge crest), they may also join different combinations of other types of plate boundaries. Transform faults can join two convergent boundaries (ocean trenches), as they do between the Caribbean and North American Plates, or a convergent boundary and a diver-

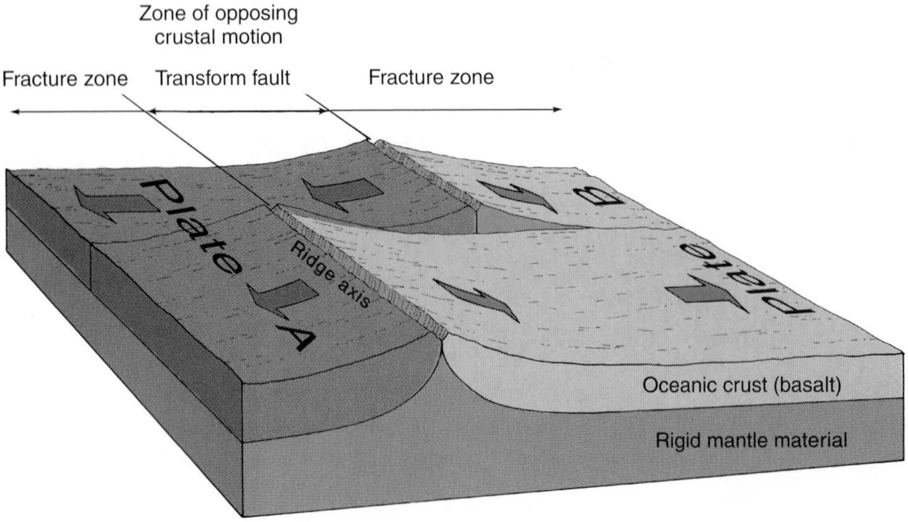

Figure 2.29 Transform boundaries are marked by transform faults. Transform faults usually offset segments of ocean ridges (divergent boundaries). Plates slide past one another along transform boundaries. The direction of motion is opposite the sense of displacement of the ridge segments.

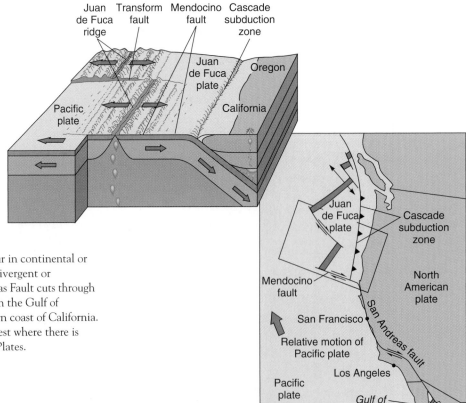

Figure 2.30 Transform boundaries can occur in continental or oceanic crust. In addition, they may join either divergent or convergent boundaries together. The San Andreas Fault cuts through continental crust, joining a divergent boundary in the Gulf of California with a subduction zone off the northern coast of California. Other transform faults offset segments of ridge crest where there is spreading between the Pacific and Juan de Fuca Plates.

gent boundary, as they do on either side of the Scotia Plate (see fig. 2.22). The diversity of settings in which transform faults play a role in marking the boundaries of plates can be seen along the west coast of the United States (fig. 2.30). Much of the boundary between the North American and Pacific Plates is marked by a long transform fault; its southern portion is known as the San Andreas Fault (see this chapter's opening photograph), cutting through continental crust from the Gulf of California to Cape Mendocino, California. Offshore of Cape Mendocino a transform fault known as the Mendocino Fault links a spreading center and a trench, forming a boundary between the Pacific and Juan de Fuca Plates. Further to the north, additional transform faults offset segments of the Juan de Fuca Ridge.

The San Andreas Fault system allows crustal sections on either side to move horizontally relative to each other. The land on the Pacific side, including the coastal area from San Francisco to the tip of Baja California, is actively moving northward with respect to the land on the east side of the fault system. The motion is not uniform; when the accumulated stress exceeds the strength of the rock there is a sudden movement and displacement along the fault, causing an earthquake. Movement along this fault system caused the famous 1906 and 1989 (Loma Prieta) earthquake in the San Francisco Bay area and the 1994 (Northridge) Los Angeles earthquake.

The reason for the many transform faults that are found associated with the ridge system is related to changes in the speed and direction of the plates as they move apart on a spherical surface. Variations in the strength or location of convection cells and collisions between sections of lithosphere may also result in transform faults.

Convergent Boundaries

Plates collide along convergent plate boundaries. The result of plate collision depends largely on whether the colliding edges of the plates are oceanic or continental lithosphere. Three possible combinations occur, and hence three different outcomes to plate convergence: ocean-continent, ocean-ocean, and continent-continent convergence (fig. 2.31). The geological features and processes that are characteristic of these different types of convergent boundaries are determined by the relative density of continental and oceanic lithosphere. Roughly 20%–40% of the thickness of continental lithosphere consists of continental crust. Because continental crust has a relatively low density, continental lithosphere is very buoyant and will not sink into the underlying mantle. The density of oceanic lithosphere is greater than continental lithosphere and changes with age. The density of oceanic lithosphere increases with increasing age as seafloor spreading carries it away from the ridge crest and it cools and thickens. Roughly 90% of old oceanic lithosphere consists of cold, dense mantle rock. Old oceanic lithosphere can actually be more dense than the hot mantle material below it, making it relatively easy for it to be subducted.

Along ocean-continent convergent boundaries the dense oceanic lithosphere sinks into the mantle and dips

under the continental lithosphere, creating an ocean trench on the sea floor marking the boundary between the two plates (fig. 2.31a). Some of the marine sediments may be scraped off the descending plate by the edge of the continental plate. The remainder of the sediments are carried into the mantle with the descending plate. As the oceanic plate descends it gradually heats up. When it reaches a depth of about 100–150 km (roughly 60–100 mi), the temperature is high enough to drive water and other volatiles out of the subducted sediments and into the surrounding mantle rock that the plate is penetrating. The addition of

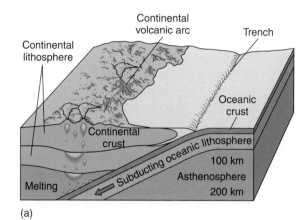

(a)

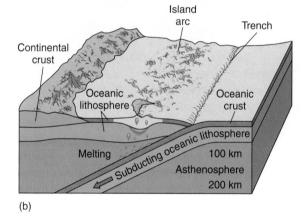

(b)

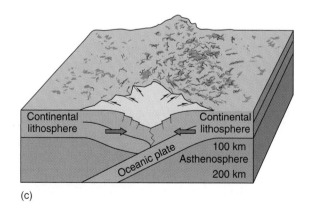

(c)

Figure 2.31 The three different types of convergent plate boundaries are (a) ocean-continent convergence, (b) ocean-ocean convergence, and (c) continent-continent convergence.

web link **Figure 2.32** Mount St. Helens erupted violently on May 18, 1980. The mountain lost nearly 4.1 km³ (1 mi³) from its once symmetrical summit, reducing its elevation from 2950 to 2550 m (a change of 1312 ft). The force of the lateral blast blew down forests over a 594 km² (229 mi²) area. Huge mud floes of glacial meltwater and ash flowed down the mountain.

water to the mantle reduces its melting temperature and it begins to partially melt, producing a basaltic magma that rises beneath the edge of the continent. This magma may partially melt the overlying continental crust as well, resulting in a magma with a mixed composition between basalt and granite called **andesite.**

Eventually, this rising magma produces active volcanoes along the edge of the continental plate. Such volcanic action is distinctly different from the volcanic action associated with spreading centers, where basaltic magma is extruded on the sea floor in nonviolent eruptions. These volcanoes often erupt explosively, due to high concentrations of volatiles (or gases) as well as higher concentrations of silicon dioxide in the melt than are found in basaltic magmas. When they erupt they eject large amounts of rock, ash, and gas that can rise high into the atmosphere. Examples of ac-

tive continental volcanic chains include the Andes along the west coast of South America and the Cascade Range of the Pacific Northwest, including Mount St. Helens (fig. 2.32). Subducting oceanic plates are also associated with intense earthquake activity as they bend and descend into the mantle. Dipping zones of earthquakes, called **Benioff zones** in honor of seismologist Hugo Benioff, mark the location of the subducted plate. Some of the deepest earthquakes in the world occur in Benioff zones extending to depths of as much as 650 km (400 mi) in the western Pacific Ocean.

Ocean-ocean convergent boundaries are also marked by ocean trenches where one of the plates is subducted beneath the other (see fig. 2.31b). The plate that is subducted generally is the one whose convergent edge is the oldest, and hence the furthest from the ridge where the sea floor was first created. Just as in the case of ocean-continent convergence, the subducted plate partially melts and produces a basaltic magma. This magma rises beneath the overriding oceanic plate and produces a line of active volcanoes on the ocean floor that may eventually break the surface to form a volcanic **island arc.** Island arcs can also produce andesitic eruptions. They are most commonly found in the Pacific Ocean. Andesitic volcanoes form the island arc chains of the Aleutians, Japan, the Philippines, and Malaysia. There are only two island arcs in the Atlantic Ocean, the Sandwich Islands in the South Atlantic and the Lesser Antilles on the eastern side of the Caribbean Sea. In some cases island arcs can form on pieces of continental crust that have broken away from the mainland. Examples of this include Japan and the Philippines.

Continent-continent collision is the end result of the closing of an ocean basin by ocean-continent convergence (see fig. 2.31c). As the ocean basin closes, the continental lithosphere of the subducting plate moves progressively closer to the trench and the continental edge of the overriding plate. When the oceanic lithosphere is completely subducted and the ocean basin has closed, the continental lithosphere of the subducting plate collides with the continental edge of the overriding plate. The subducted oceanic plate tears away and continues to descend. The subduction process ceases however, along with the associated volcanic activity, since continental lithosphere is too buoyant to descend into the mantle. Along the edges of the two colliding continents, the continental crust buckles, fractures, and thickens. What was once a well-defined convergent boundary marked by a narrow ocean trench between the two plates becomes a broad zone of intense deformation with no single discrete boundary. This produces relatively shallow earthquakes over a large area.

Continent-continent convergence created the Himalayas as India collided with Asia. India continues to collide with Asia at the rate of about 5 cm (2 in) per year, and the Himalayas are still rising in elevation. Other examples of mountain ranges created in this manner are the Appalachians, Alps, and Urals. Continent-continent collision can trap some of the marine sediment of the old ocean basin and incorporate it into the mountains that are built. Old marine fossils and sediments, including limestone remains of coral reefs, are found in the summits of peaks in the Himalayas and the Alps.

Continental Margins

When a continent rifts and moves away from a spreading center, the resultant continental margin is known as a **passive,** or **trailing, margin.** These are also frequently referred to as Atlantic-style margins since they are found on both sides of the Atlantic Ocean as well as around Antarctica, around the Arctic Ocean, and in the Indian Ocean. Continental and oceanic lithosphere are joined along passive margins so there is no plate boundary at the margin. As passive margins move away from the ridge the oceanic lithosphere cools, increases its density, thickens, and subsides. This causes the edge of the continent to slowly subside as well.

Passive margins are modified by the erosion of waves and currents; they may also be modified by shell or coral reefs. These margins accumulate sediment deposits eroded from the continents, to a depth of about 3 km (1.8 mi). Currents move the sediments downslope to the sea floor, where thick deposits of graded sediments build on top of the oceanic crust. Old passive margins are not greatly modified by tectonic processes because of their distance from the ridge. These margins are often broad and shallow and have thick sedimentary deposits, as along the eastern coast of the United States. While passive margins begin at a divergent plate boundary, they end up in a midplate position as a result of seafloor spreading and the opening of the ocean basin.

When a plate boundary is located along a continental margin, the margin is called an **active,** or **leading, margin.** Active continental margins are often marked by ocean trenches where oceanic lithosphere is subducted beneath the edge of the continent. These margins are typically narrow and steep with volcanic mountain ranges, as along the west coast of South America as well as Oregon and Washington. Because sediments moving from the land into the coastal ocean move directly downslope into deeper water, into trenches, or into adjacent ocean basins, thick sediment deposits do not routinely accumulate to form broad shallow shelves at active margins. Active margins are found primarily in the Pacific Ocean.

Motion of the Plates   2.5

Mechanisms of Motion

The mechanism that drives the plates apart is still not fully understood. The plates could be forced apart at the ridges by the formation of new crust, moved sideways, and then thrust downward at the trenches. Another possibility is that the thick, dense, oceanic crust and its sediment load sink into the mantle and the resulting tension drags the remainder of the plate with it. This latter mechanism could result in tension cracks through which the mantle material escapes upward to form the ridges and new crust. In this theory the weight of the dense, cold, descending slab of oceanic lithosphere pulls the remainder of the plate with it, and the friction between the moving plate and the asthenosphere helps to drive the convection cell rather than the reverse.

New computer models of this "slab pull" theory lead some researchers to suggest that the fracturing and sinking of old ocean plates can drive plate tectonics without the forces of convection cells driving seafloor spreading. The fact that the speed of a plate seems to increase as the amount of subducting lithosphere along the edge of the plate increases further supports theories that slab pull is an important driving force. The situation is probably more complex, combining both mechanisms. Other factors that may be involved include the shape of the Earth, changes in the weight of the plates resulting from land erosion or the accumulation of sediments, the rate at which magma wells up into the ridges, and the thickening of the older lithosphere as it cools.

Listening to Seafloor Spreading

Volcanic events on land are often predicted in advance by seismometers recording the weak, low-frequency Earth tremors associated with moving magma; in active areas, tilt meters are installed to detect the gradual swellings of the Earth's crust prior to eruptions. Volcanism on land is easily recognized as it occurs, and lava and gas samples are taken directly while aircraft and satellites document each event. Seafloor volcanism is rarely visible, and monitoring equipment has been installed in very few undersea areas. The discovery of a volcanic event in progress has depended on chance; observers must be in the right place at the right time. For example, if researchers are studying the properties of the water, changes in deep-water chemistry or higher water temperature at depth may indicate seafloor volcanism. Surveys of regions thought to be active have been made by staffed submersibles or, if the task is too risky, remote operating vehicles (ROVs) might scan these areas.

In August 1991 the U.S. Navy allowed the National Oceanic and Atmospheric Administration (NOAA) access to environmental data collected from its Sound Surveillance Systems (SOSUS). SOSUS are networks of underwater hydrophones (listening devices) installed by the U.S. Navy in the early 1950s. They were set up primarily on continental slopes and seamounts to detect sound signals from submarines and to monitor coastal vessel traffic; however, the system also detects nearby, weak, low-frequency Earth tremors (Richter magnitude 2 or less) and can locate the area where the tremors are formed. On June 22, 1993, NOAA installed a direct-line system to allow instantaneous acoustic monitoring of the SOSUS network near the Juan de Fuca Ridge, off the Washington and Oregon coasts. On June 26 scientists detected a burst of low-level seismic activity that migrated northward for the next forty hours to a site along the spreading axis of the Juan de Fuca Ridge, where most of the 676 seismic events were measured during the next three weeks. The intensity, frequency, and duration of the signals indicated that a seafloor spreading episode had occurred,

forming a magma dike about 60 km (36 mi) long as well as a possible seafloor extrusion of lava. This area of the Juan de Fuca Ridge is known as the CoAxial segment; it had been surveyed by side-scan bathymetry (see box titled "Undersea Robotic Technology" later in this chapter) in a two-year survey (1981–82) and again in 1991. These surveys provided the data against which the 1993 changes were measured.

Good fortune in 1993 made available two oceanographic research vessels, the NOAA ship *Discoverer* and the Canadian research vessel *John P. Tully*, to respond to these events. Both proceeded to the area of disturbance and began looking for changes in the sea floor and the seawater over the area. The *Discoverer* detected a recent lava flow covering 3.4×10^5 m² (0.13×10^5 mi²) and decreasing water depth by about 29 m (95 ft). The *Tully* deployed the robotic device ROPOS (Remotely Operated Platform for Ocean Science), which observed an area of fresh glassy lava 2.5 km (1.5 mi) long and 300 m (0.2 mi) wide. ROPOS also observed nearly continuous diffuse venting of warm water (<30°C). Orange flocculent material covered 7%–14% of the new lava flow. Samples of this material were collected by ROPOS in July 1993 and by the submersible *Alvin* in October of that year. The electron microscope showed the "floc" to contain disorganized aggregations of bacteria and debris.

Water samples taken by ROPOS south of the new lava flow showed elevated levels of manganese, iron, calcium, silica, and lithium but depleted levels of magnesium. Between July 1 and August 2, 1993, towed sensors detected three huge ($1.3–4 \times 10^{10}$ m³ [$0.3–1.0 \times 10^{10}$ mi³]) masses of warm water or hydrothermal event plumes called megaplumes. The production of these vent-water megaplumes is thought to be directly related to the spreading rate and frequency of magma intrusions at a spreading center ridge crest. Before the researchers left the area, they deployed current meters and temperature sensors to continue monitoring the hydrothermal event plumes.

Rates of Motion

The sea floor, acting like a conveyor belt, moves away from the ridge system where it is created by volcanic activity. The rate at which each plate moves away from the axis of the ridge is commonly known as the half-spreading rate, while the rate at which the two plates move away from each other is called the full spreading rate, or simply the **spreading rate.** Spreading rates vary between about 1 and 20 cm (0.4 and 8 in) per year but are generally between about 2 and 10 cm (0.8 and 4 in) per year. The average spreading rate is about 5 cm (2 in) per year, roughly the rate at which fingernails grow. In the course of a typical lifetime of seventy-five years, a plate moving at an average spreading rate would travel 3.75 m (12.3 ft), or roughly the length of an automobile. Although these rates are slow by everyday standards, they produce large changes over geologic time. For example, if a plate moved at the rate of just

1.6 cm (0.6 in) per year it would take 100,000 years for it to travel 1.6 (1 mi); therefore, in the 200 million years since the breakup of Pangaea it could move more than 3200 km (2000 mi), which is more than half the distance between Africa and South America. The spreading rate affects the physical structure of divergent plate boundaries. Slow spreading rates are found along ridges that have steep profiles and deep central valleys, like the Mid-Atlantic Ridge. Fast spreading rates produce ridges with gentler slopes and shallower, or nonexistent, central valleys, like the East Pacific Rise. Spreading rates are estimated at about 2.5–3 cm (1–1.2 in) per year for the Mid-Atlantic Ridge and about 8–13 cm (3–5 in) per year for the East Pacific Rise. Keep in mind that the process of spreading does not occur smoothly and continuously but goes on in fits and starts, with varying time periods between occurrences. Compare the width of the age band centered on the East

Three months after the eruption, the submersible *Alvin* descended four times to the area of volcanic eruption, making direct visual observations and measuring the magnetic properties of the new lava flow at the south end of the rift site (box fig. 1). The magnetic signal from the new lava was significantly higher than that from older lava; successive surveys indicated a decrease in magnetism at the new lava source. Future studies of this flow will add to our understanding of the rate at which magnetic properties decrease as fresh lava ages and cools. At present, the data point to a 10% decrease in magnetism in the first year after lava formation, due to changes in magnetic material exposed to seawater.

There are still unanswered questions in the interpretation of this event. The origin of the burst of seismic activity at the north end of the rift zone is still unclear. The presence of megaplumes appears to point to a preexisting hydrothermal system for which there is no crustal evidence at this time. Analysis of the 1993 data and monitoring of the SOSUS network continue as oceanographers listen for more signals of volcanic events. Detecting mid-ocean ridge events as they occur enables scientists to be in the right place at the right time to observe and investigate the dynamics of seafloor spreading.

Internet References

Visit the book's Online Learning Center at www.mhhe.com/sverdrup (click on the book's cover) to explore links to further information on related topics.

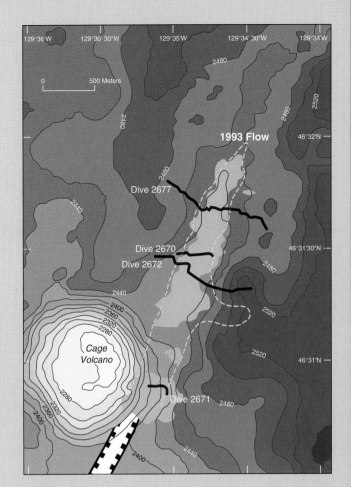

 web link **Box Figure 1** A bathymetric chart of the 1993 Juan de Fuca volcanic event. Fresh lava is shown in *pink*. The edge of the flow is shown by *dashes*. Dive tracks of the *Alvin* are shown in *black* and identified by dive numbers. *From H. Paul Johnson, Alvin Magnetic Survey of Zero-Age Crust: Coaxial Segment Eruption, Juan de Fuca Ridge, 1993 in Geophysical Research Letters, Vol. 22(2) 171–174, January 15, 1995.*

Pacific Rise with the width of the band centered on the Mid-Atlantic Ridge in figure 2.20; the wider the age band, the faster the spreading rate.

Although seafloor spreading can be observed directly only with great expense and difficulty at sea, there is one place where many of the processes can be seen on land—in Iceland, the only large island lying across a mid-ocean ridge and rift zone. Spreading in Iceland occurs at rates similar to those found at the crest of the Mid-Atlantic Ridge. Northeastern Iceland had been quiet for 100 years, until volcanic activity began in 1975; in six years this activity widened by 5 m (17 ft), an 80 km (50 mi) long stretch of the ridge's rift zone. Over 100 years this spreading rate is 5 cm (2 in) per year, which is within the typical range of spreading rates. The spreading motion between the Earth's surface plates can now be monitored by satellite. The satellite system known as GPS, for Global Positioning System (see chapter 1, section 1.5), detects the positions of land control points with great accuracy. Recent GPS measurements in the eastern Mediterranean Sea indicate that the African Plate is moving northward at 10 mm (0.4 in) per year.

Recent investigations of ancient granitic rocks, formed during the Archean eon and found exposed in West Greenland, eastern Labrador, Wyoming, western Australia, and southern Africa, indicate that 3.5 billion years ago crustal plates existed and moved granitic continental blocks at an average rate of about 1.7 cm (0.67 in) per year, again within today's average rates of plate motion.

Hot Spots

Scattered around the Earth are approximately forty fixed areas of isolated volcanic activity known as **hot spots** (fig. 2.33). They are found under continents and oceans, in the center of plates, and at the mid-ocean ridges. These hot spots periodically

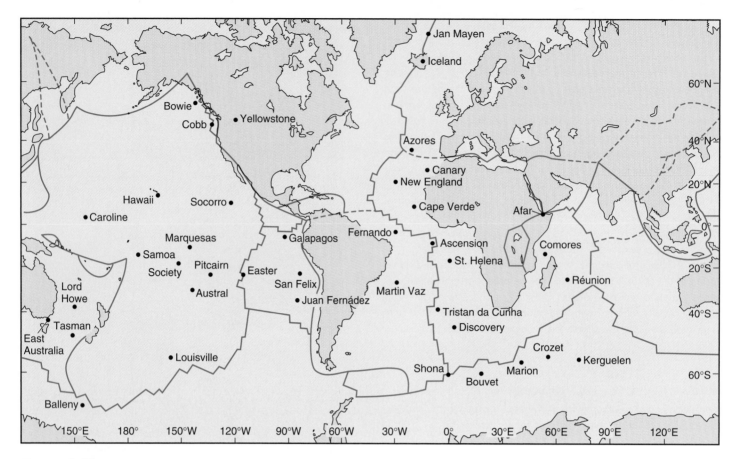

Figure 2.33 Location of major hot spots.

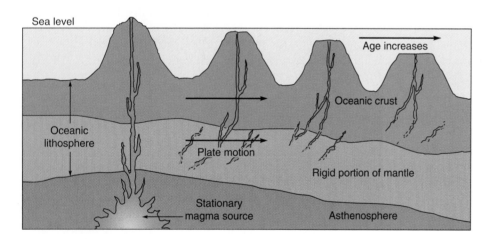

Figure 2.34 A chain of islands is produced when oceanic lithosphere moves over a stationary hot spot.

channel hot material to the surface from deep within the mantle, possibly from the core-mantle boundary. At these sites, a plume of mantle material may force its way through the lithosphere and form a volcanic peak, or seamount, directly above. If the hot spot does not break through, it may produce a broad swelling of the ocean floor or the continent that subsides as the plate moves over and away from the magma source. Hot spots may also resupply the asthenosphere, which is constantly cool-ing and becoming attached to the base of the lithosphere, thickening the plates. Some people believe that the breakup of Pangaea began when a chain of hot spots developed under the supercontinent.

Hot-spot plumes of magma are not uniform; they differ in chemistry, suggesting that they come from different mantle depths, and it has been suggested that their discharge rates may also vary. Hot spots may fade away and new ones may form. The life span of a typical hot spot appears to be about 200 million years. Although their positions may change slightly, they tend to remain relatively stationary in comparison to moving plates and they can be useful in tracing plate motions.

As the oceanic lithosphere moves over a hot spot, successive eruptions can produce a linear series of peaks, or seamounts, on the sea floor. In such a series, the youngest peak is above the hot plume, and the seamounts increase in age as their distance from the hot spot increases (fig. 2.34). For example, in the islands and seamounts of the Hawaiian Islands system, the island of Hawaii, with its active volcanoes, is presently located over the hot spot. The newest volcanic seamount in the series is Loihi, found in 1981 45 km (28 mi) east of Hawaii's southernmost tip and rising 2450 m (8000 ft) above the sea floor but still under water. At its current rate of growth it should become an island sometime between 50,000 and 100,000 years from now.

The land features of the island of Hawaii show little erosion, for it is comparatively young. To the west are the islands of Maui, Oahu, and Kauai, which have been displaced by the moving crust. Kauai's canyons and cliffs are the result of erosion over the longer period of time that it has been exposed to the winds and rains. Although these four islands are the most familiar, other islands and atolls attributed to the same hot spot stretch farther west across the Pacific. They are peaks of eroded and subsided seamounts formed by the same hot spot. Subsidence occurs when the heated and expanded plate moves slightly downhill and away from the mantle bulge over the hot spot. As the plate moves away from the hot spot it cools and contracts, and when combined with the weight of the seamount it is carrying, the result is a depression in the mantle that gradually carries the seamount below the ocean surface. When seamounts in tropical areas sank slowly, coral grew upward and coral atolls were the result. The formation of atolls is discussed further in chapter 3.

West of Midway Island, the chain of peaks changes direction and stretches to the north, indicating that the Pacific Plate moved in a different direction some 40 million years ago. This line of peaks is the Emperor Seamount Chain, volcanic peaks that once were above the sea surface as islands but have since eroded and subsided over time, resulting in many flat-topped seamounts known as **guyots** 1000 m (0.6 mi) below the surface. The northern end of the Emperor Seamount Chain is estimated to be 75 million years old, and Midway Island itself may be 25 million years old. Refer to figure 2.9 to follow this seamount chain. Notice in figure 2.34 that the peaks formed by the hot spot get older in the direction in which the plate is moving. The plate is presently moving westward; Midway Island is northwest of the main Hawaiian Islands, and it is also older.

It is possible to check the rate at which the plate is moving by using the distance between seamounts in conjunction with radiometric dating. For example, the distance between the islands of Midway and Hawaii is 2700 km (1700 mi). Midway was an active volcano 25 million years ago, when it was located above the hot spot currently occupied by Hawaii. In other words, Midway has moved 2700 km (1700 mi) in 25 million years, or 11 cm (4 in) per year.

There is another hot spot at 37° 27'S, in the center of the South Atlantic, which is marked by the active volcanic island of Tristan da Cunha (see fig. 2.33). Volcanic activity at this more slowly diverging plate boundary produces seamounts that can be carried either to the east or to the west, depending on the side of the spreading center on which the seamount was formed. The hot spot produces a continuous series of seamounts very close together, forming a **transverse,** or **aseismic, ridge:** the Walvis Ridge to the east,

between the Mid-Atlantic Ridge and Africa, and the Rio Grande Rise to the west, between the Mid-Atlantic Ridge and South America. Refer to figures 2.9 and 3.14.

If a hot spot is located at a spreading center, the flow of material to the surface is intensified. The crust may thicken and form a platform. Iceland is an extreme example of this process in which the crust has become so thick that it stands above sea level.

Seamount chains, plateaus, and swellings of the sea floor, all products of hot spots, are being used to trace the motion of the Earth's plates over known hot-spot locations. Recently, hot-spot tracks have been used to reconstruct the opening of the Atlantic and Indian Oceans. Although the exact mechanism is not known, hot spots also appear to have formed huge basaltic plateaus: the Kerguelen Plateau in the southern Indian Ocean about 110 million years ago, and the Ontong Java Plateau in the western Pacific about 122 million years ago.

History of the Continents 2.6

The Breakup of Pangaea

Figure 2.35 traces the recent plate movements that led to the configuration of the continents and oceans as we know them today. In the early Triassic, when the first mammals and dinosaurs appeared, the continents were all joined in a single landmass we call Pangaea (fig. 2.35a). Most of the rest of the globe at this time was covered by the massive Panthalassic Ocean, also known as Panthalassa. A second, much smaller ocean called the Tethys occupied an indentation in Pangaea between what would eventually become present-day Australia and Asia. About 200 million years ago, Pangaea began to break apart into Laurasia and Gondwana. By about 150 million years ago, when the dinosaurs were flourishing, Laurasia and Gondwana were separated by a narrow sea that would grow to be the central Atlantic Ocean (fig. 2.35b). At the same time, India and Antarctica were beginning to move away from South America and Africa. Water flooded the spreading rift between South America and Africa about 135 million years ago and the South Atlantic Ocean began to form. As seafloor spreading opened the Atlantic Ocean, Panthalassa (what we now call the Pacific Ocean) was growing progressively smaller. At the same time, India was moving north and the southern Indian Ocean was forming as the Tethys Ocean was closing.

By the late Cretaceous, 97 million years ago, the North Atlantic Ocean was opening and the Caribbean Sea was beginning to form. India was continuing to move toward Asia, and the Tethys Ocean was being consumed from the north as the Indian Ocean was expanding from the south (fig. 2.35c). By the end of the Cretaceous, shortly before the meteor impact that led to the extinction of the dinosaurs and many other species, the Atlantic Ocean was well developed and Madagascar had separated from India (fig. 2.35d). During the middle Eocene, 50 million years ago, Australia had separated from Antarctica, India was on the verge of colliding with Asia, and the Mediterranean Sea had formed (fig. 2.35e). About 40 million years ago India collided with Asia and initiated a continent-continent convergent boundary, destroying the last of the Tethys Ocean and beginning the formation of the Himalayas. Twenty million years ago, Arabia moved away from Africa to form the Gulf of Aden and the new and still opening Red Sea.

Before Pangaea

Because the Earth is about 4.6 billion years old, there is no reason to believe that Pangaea and its breakup into today's continental configuration was the first or only time during which the sea floor spread or the continents changed position and recombined. Scientists are now searching for new and more extensive evidence to indicate the position of the landmasses before the breakup of Pangaea. Because there is no record of pre-Pangaean relationships in the present 200-million-year-old oceanic crusts, scientists must depend on evidence from the continents, where the oldest rocks are found.

Radiometric dating and magnetic and mineral analyses indicate that ancient granitic formations are common to all landmasses; they have been present in continental crust for more than 3 billion years. Ancient mountain ranges located in the interiors of today's continents are evidence of collisions between old plates; seismic evidence has been used to mark old plate edges below the Ural and Appalachian Mountains. The magnetic and fossil evidence frozen into the land rocks during the 350 million years prior to Pangaea have been used to propose a series of pre-Pangaea plate movements during the Paleozoic era, which began about 570 million years ago.

In this proposed sequence of events, six major continents are recognized from the Paleozoic era: Gondwana (Africa, South America, India, Australia, Antarctica), Baltica (Scandinavia), Laurussia (North America), Siberia, China, and Kazakhstania. Approximately 555 million to 540 million years ago these landmasses were strung along the Earth's equator; there was no land above 60°N and below 60°S, and the polar regions were wide expanses of ocean. Gondwana and Baltica shifted to the east and south 490 million to 475 million years ago. Gondwana's southward motion carried it across the South Pole 435 million to 430 million years ago and then north again on the opposite side of the Earth, reaching a position that would eventually make it a part of Pangaea about 350 million years ago.

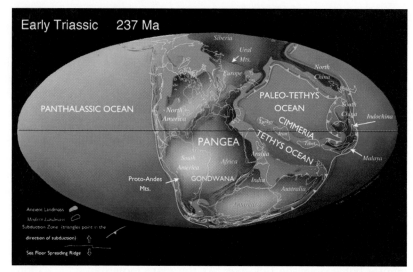

(a)

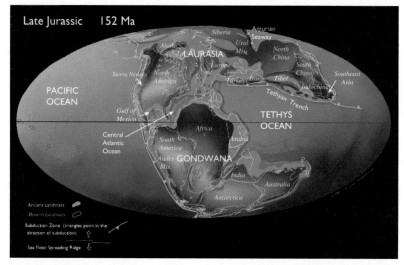

(b)

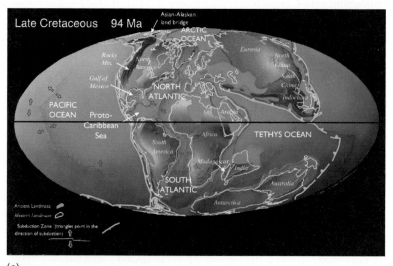

(c)

web link **Figure 2.35** The configuration of landmass and oceans from Pangaea to the present. Geologic period (or epoch) and time before the present in millions of years (Ma) are given in the upper left-hand corner of each illustration.

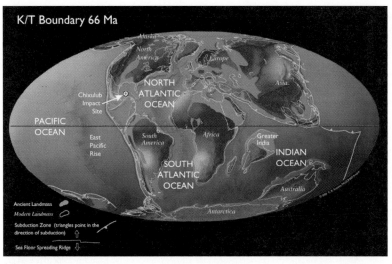

(d)

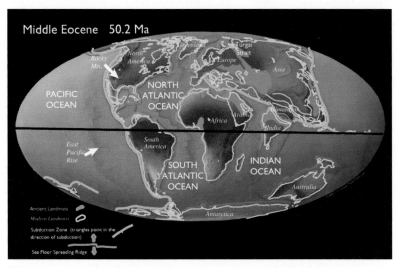

(e)

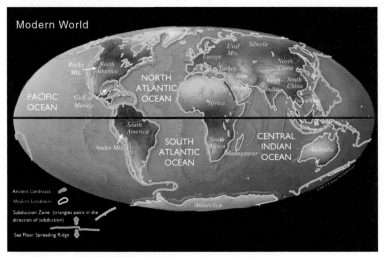

(f)

 web link **Figure 2.35** *continued*

This movement completely reversed the north-south orientation of modern Africa and South America. Their present-day southern tips pointed to the north at the beginning of the journey, but as the landmass drifted across the South Pole to the other side of the world, they pointed to the south, as we know them today. About 310 million years ago the ocean between Gondwana and the other continental fragments began to close. Siberia moved from low to high latitudes and merged with Kazakhstania, while China moved westward. The assembly of Pangaea 260 million to 250 million years ago resulted from a series of collisions between landmasses that formed great mountain belts that are still detectable. This scenario is not the only possible solution to the continent puzzle. As time passes, new evidence may point to a different fit and movement of the puzzle pieces through time.

Because the tectonic processes are continuous, there probably has never been a stable geography of the Earth. The modern configuration is no more stable than in the Paleozoic era. They are both steps in an ever-changing pattern. We have followed this changing pattern of the Earth's landmasses from wide dispersal 555 million years ago, to a single joined continental mass 260 million years ago, and then through another period of dispersal to the present time.

It has been suggested that this is an orderly, cyclic, 500-million-year pattern powered by the heat from radioactive decay in the Earth's interior. Although the production of heat is continuous, this model proposes that the heat is released in periodic bursts that are related to the movement of the continents. As the continents shift, coalesce, and compress, the sea level drops. When sufficient heat accumulates under the continental mass, the rifting process begins unstitching the continents, and as the continents move apart, they cool and subside, then the sea level rises and covers their border lands.

Using this model, we see the Atlantic Ocean opening and closing as the continents move apart and then reassemble over the 500-million-year cycle, while the Pacific Ocean boundaries, which have remained more stable, approximate the wide, ancient hemispheric ocean of that period. We can even look forward in time to see the Mediterranean Sea being closed and lost as Africa continues to move northward, and the Atlantic and Indian Oceans continuing to expand while the Pacific Ocean narrows as North and South America collide with Asia. Australia will continue to move northward, eventually colliding with Eurasia, while Los Angeles and coastal southern California will pass San Francisco on their way toward the Aleutian Trench.

History of the Continents **85**

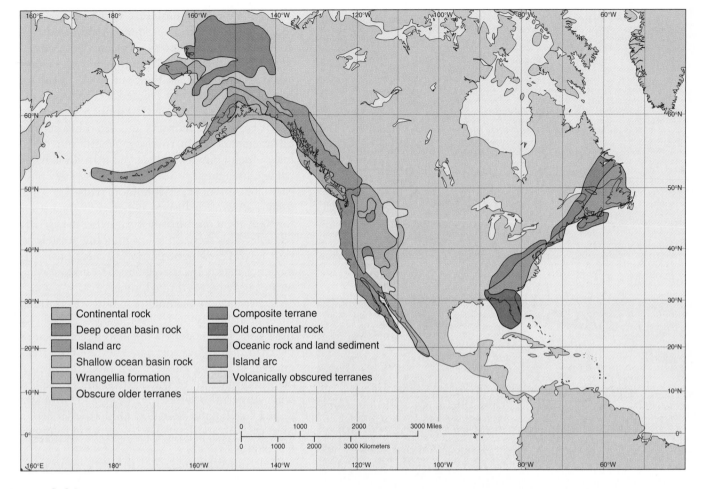

Figure 2.36 Examples of North American terranes, crustal fragments with histories distinct from adjoining fragments.

Legend:
- Continental rock
- Deep ocean basin rock
- Island arc
- Shallow ocean basin rock
- Wrangellia formation
- Obscure older terranes
- Composite terrane
- Old continental rock
- Oceanic rock and land sediment
- Island arc
- Volcanically obscured terranes

Terranes

Studies of ancient crustal rocks from the interior of North America indicate that the core of the continent was assembled about 1.8 billion years ago from several large pieces of granitic crust more than 3 billion years old. It appears that four or five large pieces of crust called **cratons** collided and joined over a 100-million-year period. Mixtures of sediments from the craton margins and the deep-sea floor, the remains of ancient island arc systems, and other small pieces of continental crust were trapped between the colliding blocks of continent. Radiometric dating has identified boundaries between North American cratons at 1.78 billion and 1.65 billion years old.

Geological studies of Alaska show that the characteristics of one area are not necessarily related to the history, age, structure, or mineral composition of an adjacent area. Alaska is made up of crustal fragments, some showing the characteristic homogeneity and higher density of oceanic crust and others showing the highly varied mineral content and lower density of continental crust. Smaller crustal fragments associated with craton margins, bounded by faults, and with a history distinct from adjoining crustal fragments are known as **terranes.** See figure 2.36 for a map of Alaskan terranes.

Terranes are often elongate, as if produced by an island arc system that has collided with a craton, or by faulting,

which has cut off a sliver of continent similar to the land west of California's San Andreas Fault. The terranes of Alaska appear to have arrived from the south, rotated clockwise, and faulted and stretched as the Pacific and North American Plates moved relative to each other. A terrane known as Wrangellia originated at or to the south of the equator, moved northward, and crashed into western North America about 70 million years ago. Faulting has spread its fragments throughout eastern Oregon, Vancouver Island, the Queen Charlotte Islands, and the Wrangell Mountains of southeast Alaska.

India is considered a single giant terrane by some; north of the Himalayas are terranes that predate the arrival of India. The area of North America west of Montana, south into New Mexico, and north into Canada and Alaska appears to be an assemblage of terranes that have been modified by faulting and vulcanism. In Oregon, some volcanic peaks appear to be basalt seamounts that have moved landward from relatively short distances offshore. Rock formations found in the San Francisco area are typical of rock formed in the South Pacific. Fossils collected between Virginia and Georgia indicate that a long section up and down the East Coast was formed somewhere adjacent to an island arc system; the fossils point to a past European connection (fig. 2.36).

web link **Figure 2.37** Oblique view of the research vessel *Atlantis* with the submersible *Alvin* hanging on an A-frame.

Research Projects and Plans 2.7

The exploration of the sea floor that began in the 1950s has continued for fifty years, providing oceanographers with the data to understand the basic features of ocean crust and plate tectonics. The present scientific effort is more specialized and seeks to understand ongoing seafloor changes as well as the processes of the past. Using more and more sophisticated technology the search is on for the details, and numerous programs and projects are underway to these ends.

Project FAMOUS

Oceanographers made their first descent into the rift valley of the mid-ocean ridge in 1973. The previous year more than twenty U.S., French, Canadian, and British oceanographic vessels had mapped a small section of the Mid-Atlantic Ridge. In the summer of 1973, in an area centered at 36° 50'N, southwest of the Azores, French scientists aboard the submersible *Archimede* made the first visual study of a rift valley, or spreading center. The following summer the French submersible *Cyana* and the U.S. submersible *Alvin* (fig. 2.37) joined the dive program. This effort was known as the French-American Mid-Ocean Undersea Study, or Project FAMOUS.

The submersibles descended nearly 3000 m (9800 ft) to the floor of the huge rift valley. In a series of fifty dives, they took more than 50,000 photographs and made 100 hours of film recording. They also recovered 150 rock samples and 70 water samples from the bottom. The cracks and fissures running parallel to the direction of the rift indicated continuous spreading with episodic, localized volcanic eruptions occurring in the central rift area.

Seafloor Spreading and Hydrothermal Vents

In 1972, measurements of water temperature and water chemistry were made in the Galápagos Rift. This spreading area lies between the East Pacific Rise and the South American mainland, 964 km (599 mi) west of Ecuador. Measurement data showed active circulation of seawater through newly formed oceanic crust. This circulation appeared to be the cause of plumes of hot water rising from **hydrothermal vents** along the rift. Water issuing from these vents had temperatures of 17°C (63°F), compared with 2°C (36°F) for the surrounding seawater. The hot water mixed with the cold

Undersea Robotic Technology

The deep-ocean floor is at or near the operating limits of staffed submersibles, and research programs for this area of the oceans are difficult and expensive. A variety of remotely operated devices and vehicles have been developed to decrease the risks and costs associated with deep-sea observations. They act as eyes, samplers, and manipulators for oceanographers in submersibles and surface vessels.

Remotely operated vehicles, or ROVs, are under the continuous control of their operators, who respond to the ROVs' observations. ROVs are tethered to a surface vessel, a staffed submersible, or an intermediate, below-the-surface-system that isolates the ROV from ship motion. The tether is the umbilical cord between the ROV and its operator; it supplies power and transmits information. ROVs use video, digital, and still cameras to explore their environment. They are equipped with mechanical hands to manipulate objects and, depending on their mission, other sensors such as sonar, lights, cameras, and temperature and salinity monitors. The operator "flies" the ROV over the sea floor while the ROV sends data to the operator and receives directions to change position, to manipulate or retrieve objects, or to use cameras and other sensors. Full-ocean-depth ROVs that are required to perform a variety of tasks over long periods weigh several tons and must be towed slowly because of the stresses on their long cables. Small, limited-use, shallow-water vehicles that may weigh less than 45 kg (100 lb) can be towed by small surface vessels. These smaller robotic devices are used to service, inspect, and aid in construction of offshore oil and gas facilities, bridge piers, and dam footings. *MKI* ROV is shown in box figure 1.

ROV *Jason* is a tethered vehicle operated by the Woods Hole Oceanographic Institution. *Jason* weighs about 1350 kg (3000 lb), can operate to a depth of 6000 m (20,000 ft), and provides simultaneous use of up to four color video channels, sonar, electronic cameras, and manipulators. A smaller ROV, *Jason Jr.* (or *JJ*), oper-

ated from the submersible *Alvin*, was used to inspect the wreck of the *Titanic* and to photograph its interior spaces (box fig. 2). In 1993, ROPOS (Remotely Operated Platform for Ocean Science) was used by Canadian and U.S. scientists to explore fresh lava and hot water venting on the Juan de Fuca Ridge (see box earlier in chapter titled "Listening to Seafloor Spreading"). Three years later ROPOS had returned to the ridge when a sudden and severe storm produced high waves that interfered with its operation. While ROPOS was being held along the side of the research vessel before being hoisted to the deck, the waves broke it from its tether. The loss cost approximately $750,000 and reinforced the oceanographer's adage that anything that goes over the side may not come back. In this case, however, insurance came to the rescue. Construction of ROPOS II began in early 1997, and it was doing its job at sea by July 1998.

The Japanese ROV *Kaiko* is the only ROV presently capable of descending below 6500 m (21,000 ft). In 1995 *Kaiko* dove into the Mariana Trench and photographed fish at 10,911 m (35,799 ft), 99 cm (39 in) short of the depth reached by the Swiss-built, U.S. Navy's staffed bathyscaphe *Trieste* in 1960.

ROVs are much less costly than any submersible designed to carry humans; they operate to deeper depths, work underwater for longer periods of time, and explore areas of higher risk (box fig. 3 shows the ROV *Ventana*). Their two-dimensional video images are a drawback, however, for they do not measure up to the three-dimensional image seen by a human in a submersible, and ROVs do require the expensive support of surface vessels.

Autonomous underwater vehicles, or AUVs, further reduce costs and extend exploration and survey time. These independent, mobile instrument platforms and their sensors have no tether, giving them much greater range of movement than ROVs. They may be linked by acoustic modem to a buoy or to an operator in a surface

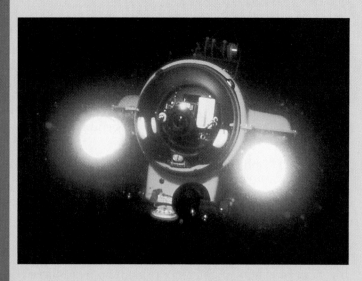

Box Figure 1 *MKI* ROV (remotely operated vehicle), a MiniRover, is an example of a small "low-cost" system.

web link **Box Figure 2** The camera-equipped ROV *Jason Jr.* peers into a cabin of the luxury liner *Titanic*, 3800 m (12,500 ft) below the surface. A tether connecting *Jason Jr.* to the submersible *Alvin* is seen at left.

Box Figure 3 The *Ventana* ROV is owned and operated by Monterey Bay Aquarium Research Institute. This ROV is used to study marine life in Monterey Canyon.

Box Figure 4 The Autonomous Benthic Explorer (ABE) is retrieved by a research ship. ABE makes repeated underwater surveys returning to its mooring platform on the sea floor.

vessel. They may be preprogrammed to complete surveys and take samples with no additional human supervision. Many of these vehicles weigh less than 25 kg (50 lb) and have specialized capabilities and restricted depth ranges. Others work effectively in deep water and inaccessible regions, for example, under the polar ice.

The U.S. Navy's Advanced Unmanned Search System (AUSS) is an AUV that locates unidentified objects on the sea floor at depths from 5 m (17 ft) to 6 km (20,000 ft); it is 23 m (77 ft) long, 78 cm (31 in) in diameter, and weighs 1270 kg (2800 lb). It follows a programmed search pattern while transmitting sonar images to the surface, or it may be ordered to suspend its search pattern to take a closer look at an object. The Autonomous Benthic Explorer (ABE) is designed to survey deep-sea hydrothermal vent areas for up to one year, making a daily survey of its area (box fig. 4). It monitors currents, takes pictures, and makes temperature and other measurements, then returns to its mooring platform on the sea floor and remains quiet to conserve energy.

A new generation of AUVs known as UUVs (unmanned undersea vehicles, or untethered undersea vehicles) is being developed by the U.S. Navy. These highly mobile vehicles are being designed for underwater surveillance, ocean data and intelligence collection, and mine countermeasures. They are launched and recovered from submarines and surface vessels. They are completely autonomous and approximately 60 cm (24 in) in diameter, 8–10 m (25–35 ft) long, and travel at 4–12 knots (10–25 mi/h).

Satellites can monitor large areas, but they track mainly surface features. To monitor the interior ocean on a global scale requires the deployment of very large numbers of small, cost-effective, autonomous devices. One type of vehicle developed to do this is gliders (see the box titled "Ocean Gliders" in chapter 7).

All AUVs obtain more data in more detail over shorter periods than any single research vessel. They are rapidly becoming the survey tool of choice for work in the oceans—the Earth's last, largest, and least known frontier.

To Learn More About Undersea Robotic Technology

Bradley, A., D. Yoerger, and B. Walden. 1995. An AB(L)E-Bodied Vehicle. *Oceanus* 38 (1): 18–20.

Coudeville, J., and H. Thomas. 1998. A Primer: Using GPS Underwater. *Sea Technology* 39 (4): 31–34. Methods for positioning AUVs.

Fornari, D., A. Bowen, and D. Foster. 1995. Visualizing the Deep Sea. *Oceanus* 38 (1): 10–13.

Fornari, D., S. Humphris, and M. Perfit. 1997. Deep Submergence Science Takes a New Approach. *EOS* 78 (30): 402, 408.

Hui, L. 1997. New Robotic Vessel Extends Deep-Ocean Exploration. *Science* 278 (5344): 1705. Chinese AUV.

Kunzig, R. 1996. A Thousand Diving Robots. *Discover* 17 (4): 60–71.

Nadis, S. 1998. Mixed Results from the Labrador Sea. *Science* 280 (5362): 375.

Sea Technology. 37 (12), 1996. Issue devoted to marine robotics.

Simonetti, P. 1998. Low-Cost, Endurance Ocean Profiler. *Sea Technology* 39 (2): 17–21.

Storkersen, N., J. Kristensen, A. Indreeide, J. Seim, and T. Glancy. 1998. Hugin-UUV for Seabed Surveying. *Sea Technology* 39 (2): 99–104.

Tivey, M., A. Bradley, D. Yoerger, R. Catanach, A. Duester, S. Liberatore, and H. Singh. 1997. Autonomous Underwater Vehicle Maps Seafloor. *EOS* 78 (22): 229–30.

Wernli, R. 1997. Trends in UUV Development. *Sea Technology* 38 (12): 17–23.

Internet References

Visit the book's Online Learning Center at www.mhhe.com/sverdrup (click on the book's cover) to explore links to further information on related topics.

bottom water to produce shimmering upward-flowing streams rich in silica, barium, lithium, manganese, hydrogen sulfide, and sulfur. Rocks in the vicinity of the vents were coated with chemical deposits rich in metals precipitated out of the vent waters.

In 1979, the Mexican-French-American Riviera Submersible Experiment (RISE) research program investigated ocean-floor hot springs near the tip of Baja California. Two thousand m (6000 ft) down, *Alvin* found mounds and chimney-shaped vents that were 20 m (65 ft) high ejecting hot (350°C, 660°F), black streams of particles containing sulfides, lead, cobalt, zinc, silver, and other minerals.

Hydrothermal vent activity is now known to occur at seafloor spreading centers worldwide. Cold, dense seawater circulates through magma chambers close to the sea floor; there, it is heated and then released through cracks in the newly created upper lithosphere. Some of these vents discharge low-temperature, clear waters up to 30°C (86°F); others, known as white smokers, give rise to milky discharges with temperatures ranging from 200°–330°C (392°–626°F); and still others, black smokers, release jets of sulfur-blackened water at temperatures between 300°–400°C (572°–752°F). This hydrothermal vent activity is thought to go on for years, perhaps for decades. Rapid, unpredictable temperature fluctuations have been recorded on a timescale of days to seconds, indicating a very unstable environment surrounding the vents.

During a dive on the Juan de Fuca Ridge an electronic camera carried on the *Alvin* detected light coming from seafloor vents (fig. 2.38). Preliminary studies of this long-wavelength light, undetectable by the human eye and known as "vent glow," indicate that thermal radiation alone is responsible for the phenomenon. On another dive, the *Alvin's* camera recorded the features of massive sulfide structures with large protruding flanges that trap beneath them layers of hot vent water (350°C, 660°F) (fig. 2.39).

Researchers investigating a section of the East Pacific Rise in the winter of 1993–94 discovered an area of extreme volcanic activity. Scientists mapping the sea floor northwest of Easter Island in the South Pacific found about

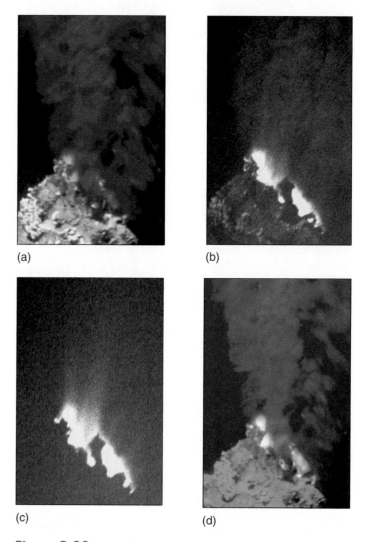

(a)   (b)

(c)   (d)

Figure 2.38 (a) A digital image of vent light taken with an electronic still camera and illuminated by a thallium iodide lamp mounted on the outside of *Alvin*. (b) A digital image of vent light taken with an electronic still camera and the light of a flashlight shining from inside *Alvin*. (c) A digital image of vent light taken with an electronic still camera and vent radiation. (d) False-color image produced from parts (a), (b), and (c) by passing different wavelengths of light through those images.

(a)

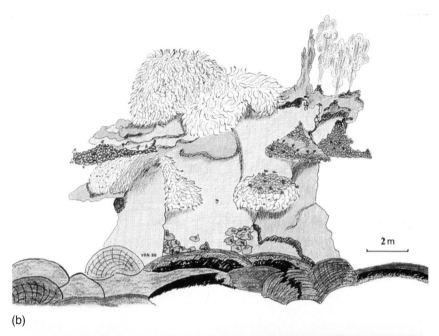

(b)

(c)

Figure 2.39 (a) A large sulfide structure in the main vent field of the Endeavour Segment of the Juan de Fuca Ridge. The structure is about 10–15 m (33–49 ft) wide and 15–20 m (49–66 ft) high. The *Alvin* carried the life-sized plywood cutout down to the sea floor and placed it beside the structure. The cutout points to a large flange on the side of the structure. The light color of the flange is caused by organisms growing on the flange's upper surface.
(b) A composite illustration of an Endeavour sulfide structure demonstrates a variety of features. Flanges protrude from the sides. Vented hot water (350°C) is often trapped under these flanges. As this water leaks around the flange edge and rises, it forms a deposit that increases the flange's outward growth. Some flanges have vent water channels within them and form smokers on upper surfaces. Organisms may also be found on flange surfaces.
(c) A closeup of a flange protruding from the side of a large sulfide structure. *(b) By Veronique Robigou-Nelson.*

1200 seamounts in a region of about 21,000 km² (8100 mi²). No other region of the Earth is known to have so many volcanoes. The lack of sediment in the volcanic area indicates that eruptions are frequent and recent. The area lies at the boundary between the Pacific and Nazca Plates and is a zone of rapid spreading. This same region was explored in 1993 with the French submersible *Nautile*. Many lavas appeared fresh—less than fifty years old—and many hydrothermal vents were noted as was the presence of warm water rising off the lava beds.

Until recently it was thought that nearly all hydrothermal vent systems shared some basic characteristics. Most of the known vent systems were located on the axis of ocean ridges on very young sea floor right along the plate boundary. The fluids emanating from them passed through basaltic rock, and the resultant minerals that precipitated from the fluids were iron and sulfide-rich, especially at the dramatic black smokers. There is a growing body of evidence, however, that hydrothermal systems with a different chemical signature located off the axis of the ridge system in older sea floor may be more common than previously thought. The recent discovery and investigation of a spectacular new type of hydrothermal vent field located just off the axis of the Mid-Atlantic Ridge at 30°N, which has been named "Lost City," is a case in point. Lost City is located nearly 15 km (9 mi) away from the axis of the ridge on sea floor that is

Recovery of Black Smokers

In 1998 an ambitious expedition was undertaken by scientists at the University of Washington and the American Museum of Natural History in cooperation with the Canadian Coast Guard to recover black smokers from an active hydrothermal vent system on the Endeavour Segment of the Juan de Fuca Ridge. The Endeavour, located about 300 km (180 mi) off Washington State and Vancouver Island, is one of the most active hydrothermal vent regions ever discovered. There are at least four major hydrothermal vent fields along the Endeavour Segment with hundreds of black smokers, some of which emit fluids at temperatures of 400°C (750°F). One giant smoker, Godzilla, stood roughly 45 m (150 ft) tall before it collapsed in 1996. The southernmost field in this area, the Mothra Hydrothermal Field, was chosen as the recovery site because it contains abundant steep-sided smokers that host diverse macrofaunal communities. Diffusely venting sulfide structures in this area are up to 24 m (79 ft) high and extend linearly for over 400 m (1300 ft). The tops of black smokers targeted for recovery were initially identified during a preliminary survey conducted in 1997 using the ROV *Jason* and the staffed submersible *Alvin*. The smokers identified as

possible targets for retrieval were relatively small; the largest was about 3 m (10 ft) tall and weighed roughly 6800 kg (15,000 lb).

The expedition used the Canadian ROPOS (Remotely Operated Platform for Ocean Science) to recover four sulfide chimneys from a depth of about 2250 m (7400 ft). ROPOS works out of a cage that acts as a garage for the vehicle while it is being lowered to the bottom or brought back to the surface (box fig. 1). The fiber-optic tether that connects the cage to the ship is also used to transmit images from cameras on ROPOS and other data, including piloting commands to the vehicle. After the cage is lowered to the target area, ROPOS "swims" out on a separate tether connected to the cage. Because ROPOS is on its own tether, it is unaffected by any motion of the cage caused by ship motion at the surface.

Once the expedition had arrived at the site, ROPOS was lowered from the University of Washington's research vessel *Thomas G. Thompson* to the bottom, where it moved to the first target, an inactive black smoker called Phang. Its initial task was to take pictures of this structure to document its characteristics for before-and-after studies. It then cinched cables around the smoker using a recovery cage that had been specially designed to fit over it. The next step was for ROPOS to approach the smoker and cut into its base using a chain saw with carbide and diamond-embedded blocks in the chain (box fig. 2). After a series of cuts had been made, ROPOS was backed away from the chimney. The recovery cage and cables were attached to a line brought over by ROPOS from a previously deployed recovery basket on the bottom holding 2400 m (8000 ft) of line. ROPOS was then brought on deck, and the recovery line was floated to the surface, where it was brought onboard the Canadian Coast Guard's vessel *John P. Tully* and attached to a winch. The *Tully* then took up slack on the line, broke the smoker off its base, and brought it to the surface (box fig. 3), where the structure was cut in half in preparation for geological and microbiological studies.

This process was repeated three times during the course of the expedition to recover a total of four sulfide chimneys (box fig. 4). The sulfide edifices were chosen in part for their diversity. Phang, the

Box Figure 1 The Canadian ROPOS (Remotely Operated Platform for Ocean Science) in its deployment cage.

Box Figure 2 With the recovery cage in place, ROPOS approaches the base of the chimney in preparation for cutting with a carbide and diamond-embedded chain saw.

Box Figure 3 A sulfide chimney breaking the surface during recovery, with the recovery cage surrounding it.

dead chimney, did not have any water flowing through it, but another chimney, Finn, was very active, with 304°C (579°F) hydrothermal fluid flowing out. Neither chimney had many large organisms attached to it. The remaining two chimneys recovered, Roane and Gwenen, were venting diffuse fluids at temperatures of 20°–200°C (68°–392°F) and had diverse communities of organisms living on them, including tube worms, limpets, and snails (box fig. 5). These two structures remained hot when brought up from a water depth of over 2000 m (nearly 1.5 mi); Roane had temperatures of 90°C (194°F) and Gwenen had temperatures of 60°C (140°F). Their temperatures were measured from within the structures when they were on deck. The largest chimney recovered is 1.5 m (5 ft) tall and weighs 1800 kg (4000 lb); it is now on display at the American Museum of Natural History in New York City.

The smokers will be studied in detail by geologists, biologists, and chemists to learn more about the extreme chemical and thermal gradients that characterize these environments, the conditions under which sulfide structures grow and evolve, and how nutrients may be delivered to the organisms that live on and within the structures. Such studies are likely to find new species of microorganisms that thrive in these high-temperature, sunlight-free environments. Preliminary studies on these samples indicate that microorganisms living within the structures at temperatures of 90°C (194°F) derive their energy from carbon-bearing species in the hydrothermal fluids and from mineral-fluid reactions within the rocks. These and similar findings are exciting in that they show that microorganisms are capable of living in the absence of sunlight in volcanically active water-saturated regions of our planet. Other hydrothermally active planets may harbor similar life forms.

Internet References

Visit the books' Online Learning Center at www.mhhe.com/sverdrup (click on the book's cover) to explore links to further information on related topics.

(a)

(b)

Box Figure 4 (a) Several segments of recovered chimneys secured on deck. (b) Section of a chimney that was cut in half for study of its interior.

Box Figure 5 Tube worm colony attached to one of the recovered chimneys.

roughly 15 million years old. The Mid-Atlantic Ridge is a slow-spreading ridge. Along slow-spreading ridges, hydrothermal fluids not only react with shallow basaltic rock but may also circulate through deeper rocks that have been chemically altered. This is apparently the case at Lost City, where the venting fluids are very cool (40°–75°C, 104°–167°F) and abnormally alkaline (pH 9.0–9.8 compared with the average pH of seawater of 7.8; see the discussion of pH in chapter 5). The low-temperature venting at Lost City has created numerous chimneys, one of which is 60 m (197 ft) tall. These chimneys are all composed of carbonate and magnesium-rich minerals, making them chemically very different from the other well-studied vent systems on young sea floor.

At present two major U.S. programs are associated with identifying and understanding processes along the mid-ocean ridges: NOAA's VENTS program and the National Science Foundation's Ridge Inter-Disciplinary Global Experiments (RIDGE). The VENTS program is exploring the source and strength of hydrothermal discharges as well as the time intervals between discharges. The program is interested in the pathways followed by the materials issuing from vents, and its researchers have set up monitors to discover the impact of vent discharges on the chemistry of the water.

The RIDGE program's mission is to integrate observations, experimental results, and theoretical studies in a decade-long program to understand the geological, geochemical, and biological processes responsible for creating new oceanic crust along the mid-ocean ridge systems. The Mantle Electromagnetic and Tomography (MELT) experiment is the largest geophysical field program ever organized. It was designed to test the validity of the many proposed models of magma generation beneath oceanic ridges. Along ocean spreading centers, new basaltic crust is created within 1–2 km (0.6–1.2 mi) of the ridge axis from upwelling magma that is generated in the mantle. The goal of the MELT experiment is to obtain better models of the geometry of the region of magma generation in the mantle and the movement of magma to the axis of the ridge by collecting seismic and electromagnetic data across the East Pacific Rise. Collection of the seismic data was completed in May 1996, and the first results from that phase of the experiment were published in May 1998.

Detailed mapping of the East Pacific Rise has revealed a significant asymmetry to the sea floor in the study area. The sea floor is subsiding more slowly and the underlying mantle is warmer on the Pacific Plate side to the west than on the Nazca Plate side to the east. In addition, the Pacific Plate is moving much faster and has more seamounts and recent lava flows away from the axis of the ridge.

Fifty-one seismometers were deployed in two linear arrays about 800 km (500 mi) long across the East Pacific Rise. The seismometers were used to record seismic waves generated by regional and distant earthquakes. The velocity of seismic waves decreases with decreasing density and increasing temperature, characteristics that would be found where magma is being generated in the mantle. The first results from the seismic experiment indicate that there is a broad region of melting in the mantle beneath the ridge. The magma generation zone is thought to be several hundred kilometers wide and extend to a depth of 100–150 km (60–90 mi). The magma zone is also asymmetric. It extends farther to the west of the ridge axis. This difference would explain why the mantle temperatures are higher west of the ridge and why the sea floor is shallower there. The seismic data also indicate that the mantle is relatively homogeneous below depths of about 300 km (190 mi), indicating that the geographic location of the ridge is not controlled by deep mantle processes. An unexpected result was the observation that the lowest seismic velocities are not directly beneath the axis of the ridge but to the west of the ridge beneath the Pacific Plate. This finding suggests that the center of magma generation and upwelling may be west of the ridge axis. More-detailed models of ridge processes will be produced after the electromagnetic data have been analyzed and the two data sets have been synthesized.

The RIDGE program has also installed sampling devices, recording cameras, and seismic event detectors in the Juan de Fuca Rift area and is making use of the U.S. Navy's sound surveillance system present in that area. (See box earlier in the chapter titled "Listening to Seafloor Spreading.")

Hydrothermal Vent Communities

The submersible *Alvin* carries two scientists and a pilot. It is equipped with cameras inside and out, as well as baskets for samples taken with its mechanical claws (fig. 2.40). It dives at a speed of 2 knots (2.3 mi/h) to a depth of 3000 m (10,000 ft) and spends four to five hours on the bottom. In 1977 and 1979 expeditions from Woods Hole Oceanographic Institution (WHOI) returned with the submersible *Alvin* to explore the vent areas along the Galápagos Rift. An underwater camera assembly called ANGUS (Acoustically Navigated Underwater Survey System) was slowly towed above the ocean floor to pinpoint the vent area, and the *Alvin* made twenty-four dives to the rift, which is 2500 m (8000 ft) deep.

The most surprising discovery of these dives was the perplexing presence of large communities of animals so far away from surface food sources. Clams, mussels, limpets, tube worms, and crabs were viewed, photographed, and collected. New species were identified; giant tube worms and clams with red blood and flesh similar to beef were collected for laboratory analysis. The presence of so many large animals at such depths immediately brought up the question of their source of food. Such dense communities could not be supported by the fall of organic matter from the sea surface. Instead of depending on plants using the Sun's energy to produce organic material, these ocean-bottom populations rely on bacteria that rely on the hydrogen sulfide and particulate sulfur in the hot vent water to provide their food. The vent animals feed on the bacteria, which were calculated to reach concentrations as high as 0.1–1 g/L vent water. No sunlight is necessary in these chemosynthetic (rather than photosynthetic) communities, and no food is needed from

Figure 2.40 *Alvin's mechanical arm collects a sample*

the surface; the populations are sustained by the vents themselves. In areas where the vents have become inactive, the animal communities have died. Since the initial discoveries along the Galápagos Rift, communities of vent organisms have been discovered in association with the Gorda Plate's Juan de Fuca Rift off the coasts of Oregon and Washington, in sites along the Mid-Atlantic Ridge, along the East Pacific Rise off Central America, and at several places in the western Pacific. Further discussion of these communities, the organisms, and their food supply is found in chapter 17.

The Ocean Drilling Program

The Ocean Drilling Program (ODP) is an international effort that began in 1983. It is presently guided by the Joint Oceanographic Institutions for Deep Earth Sampling (JOIDES) located at Texas A&M University and uses the specialized drilling ship *JOIDES Resolution* (see fig. 2.15). Since 1985 the *JOIDES Resolution* has conducted drilling in all of the world's ocean basins seeking to obtain answers to questions related to global climate change, the creation and destruction of ocean basins, volcanoes, earthquakes, sea-level fluctuations, and the evolution of life forms.

The *JOIDES Resolution* makes six scientific expeditions, or legs, each year. Each cruise is approximately two months in length and has specific scientific goals chosen

through a careful review process. Each leg is staffed with thirty shipboard scientists (including graduate students) from the ODP member institutions. Upon completion of a cruise, the recovered cores are transported to one of four repositories for curation, storage, and future research. ODP scientists are able to use these repositories much as the general public uses a library. Scientists can also access the ODP databases containing vast amounts of data gathered during each cruise.

In 1997 ODP legs were carried out in the Southern Ocean off Antarctica to study global climate change, off the tip of South Africa to study the arid southern African climate, in the North Atlantic to study seawater circulation in the ocean crust, and off the coast of New Jersey to study the effect of the ice ages on sea level. In 1998 cruises were conducted in the southwest Pacific and Indian Oceans to study unusual cool-water coral reefs and climate change. In addition, cores were obtained along large faults in a young spreading basin near Papua New Guinea to learn more about submarine earthquakes. To improve the detection of earthquakes in remote areas an ocean-bottom seismometer package, for the detection and recording of earthquakes, was placed in a drill hole in the central Indian Ocean. The 1999 legs included studies of earthquake and subduction processes in the Japan and Mariana Trenches and an investigation of the properties of a large igneous plateau roughly one-third the size of the United States in the southern Indian Ocean, called the Kerguelen Plateau; cores indicate that it was a land of lush vegetation before it sank 800 m (2600 ft) below the ocean surface.

The ODP is scheduled to end in 2003 and plans are now underway for a follow-up drilling program called the Integrated Ocean Drilling Program (IODP). As envisioned, the IODP will use two drill ships, the *JOIDES Resolution* and a new, larger vessel built by Japan. The new Japanese vessel, named the *Chikyu*, is 210 m (689 ft, or roughly the length of two and one-third football fields) long. The hull of the *Chikyu* is expected to be launched in January 2002, and the ship is scheduled to be finished in October 2003. The deepest hole drilled by the *JOIDES Resolution* to date is 2111 m (6926 ft) below the sea floor. The new ship will be designed to drill up to 7000 m (nearly 23,000 ft) beneath the sea floor. State-of-the-art drilling technology will also allow the new vessel to drill safely in areas with gas or hydrocarbon deposits along continental margins as well as into regions with thick sediment deposits or fault zones.

Future research priorities of ocean drilling are currently envisioned to investigate global climate change, assist in the emplacement of geophysical and geochemical observatories on the sea floor, and explore the deep structure of continental margins and oceanic crust.

Summary

The Earth is made up of a series of concentric layers: the crust, the mantle, the liquid outer core, and the solid inner core. The evidence for this internal structure comes indirectly from studies of the Earth's dimensions, density, rotation, gravity, and magnetic field and of the remains of meteorites. It also comes from the ways in which seismic waves change speed and direction as they move through the Earth. Seismic tomography is being used to describe the Earth's interior layers.

Continental crust is formed from granite-type rock, which is less dense than oceanic crust (formed of basalt). The top of the mantle is fused to the crust to form the rigid lithosphere. The lithosphere floats on the deformable upper mantle, or asthenosphere. The pressures beneath the elevated continents and depressed ocean basins are kept in balance by vertical adjustments of the crust and mantle, a process known as isostasy.

Alfred Wegener's theory of drifting continents was based on the geographic fit of the continents and the similarity of fossils collected on different continents. His ideas were ignored until the discovery of the mid-ocean ridge system and until the proposal of convection cells in the asthenosphere led to the concept of seafloor spreading. New lithosphere is formed at the ridges, or spreading centers. Old lithospheric material descends into trenches at subduction zones. Seafloor spreading is the mechanism of continental drift. Evidence for lithospheric motion includes the match of earthquake zones to spreading centers and subduction zones, the greater heat flow along ridges, age measurement of seafloor rocks, age and thickness measurements of sediment from deep-sea cores, and the magnetic stripes in the sea floor on either side of the ridge system. Rates of seafloor spreading are generally 1–20 cm (0.4–8 in) per year, averaging about 5 cm (2 in) per year. Plate tectonics is the unifying concept of lithospheric motion. Plates are made up of continental and oceanic lithosphere bounded by ridges, trenches, and faults. Plates move apart at divergent plate boundaries and together at convergent plate boundaries.

Rift zones separate ocean basins, and, in the past, they have separated landmasses and produced new ocean basins. Lithosphere is made up of layers of sediment, fine-grained basalt over vertical basalt dikes, igneous rock, and, under the Moho, mantle rock. Subduction produces island arcs, mountain ranges, earthquakes, and volcanic activity. The passive, or trailing, continental margin is closest to the ridge system; the active, or leading, margin borders a subduction, or collision, zone. Hot spots and polar wandering curves are used to trace plate motions. The mechanism that drives the plates is not fully known but is probably related to convection in the mantle and variations in the density and thickness of oceanic lithosphere with time.

Plate movements traced over the last 225 million years show the breakup of Pangaea. Paleozoic plate movements in the 225 million years prior to Pangaea have recently been estimated using climatological evidence. Much of North America appears to be made up of continental fragments, or terranes, around a core continent, or craton.

Submersibles have been used to explore the animal communities and hydrothermal vents found in many areas of the world's oceans. Numerous ongoing programs and projects are exploring and monitoring the formation of new oceanic crust; sampling devices and cameras have been installed at several locations. Deep-sea drilling continues in all the oceans, working to understand the formation of seafloor features and looking for evidence of global environmental change.

Key Terms

All key terms from this chapter can be viewed by term, or by definition, when studied as flashcards on this book's Online Learning Center at www.mhhe.com/sverdrup (click on this book's cover).

density, 54	hypocenter, 61
seismic wave, 54	sediment, 61
P-wave, 54	core, 61
S-wave, 54	dipole, 64
refraction, 55	Curie temperature, 66
inner core, 55	paleomagnetism, 66
outer core, 55	polar reversal, 66
mantle, 56	polar wandering curve, 70
seismic tomography, 56	plate tectonics, 71
crust, 56	divergent plate boundary, 71
Mohorovicic discontinuity, 56	inertia, 71
Moho, 56	convergent plate boundary, 71
granite, 56	transform boundary, 71
basalt, 56	transform fault, 71
lithosphere, 58	rift zone, 73
asthenosphere, 58	graben, 73
mesosphere, 58	pillow basalt, 74
isostasy, 58	gabbro, 74
Gondwanaland, 60	escarpment, 76
continental drift, 60	fracture zone, 76
Pangaea, 60	andesite, 78
Laurasia, 60	Benioff zone, 78
ridge, 60	island arc, 79
rise, 60	passive/trailing margin, 79
rift valley, 60	active/leading margin, 79
trench, 60	spreading rate, 80
convection cell, 61	hot spot, 81
seafloor spreading, 61	guyot, 83
spreading center, 61	transverse/aseismic ridge, 83
subduction zone, 61	craton, 86
epicenter, 61	terrane, 86
focus, 61	hydrothermal vent, 87

Study Questions

1. What is meant by the term *polar wandering*? Have the magnetic poles actually wandered?
2. Describe the three types of plate boundaries. What processes take place at each type of boundary? In what direction do the plates move at each boundary?
3. What mechanisms have been proposed to account for plate motion?

4. What is the difference between the leading edge and the trailing edge of a continent? Between a divergent plate boundary and a convergent plate boundary?

5. If the ability of the oceanic crust to transmit heat were uniform, the rate of heat flow through the ocean floor would depend only on the temperature change across the oceanic crust. Under such a condition, how would the heat flow measurements in figure 2.13 indicate the presence of ascending convection cells in the asthenosphere?

6. If the polar wandering curves for North America and Europe are made to coincide, how will these continents move relative to each other?

7. Using the techniques and reasoning employed to discover the properties of the interior of the Earth, explain how you would determine what is inside a sealed box (for example, measuring the box, weighing it, spinning it, balancing it on different axes, sampling its exterior). What clue would each of these measurements give you to the contents of the box?

8. What had to be learned about the Earth before Alfred Wegener's ideas could be accepted?

9. Why does a newly formed mid-ocean volcanic island gradually subside?

10. Explain the formation and symmetry of the magnetic stripes found on either side of the mid-ocean ridge system. What is their significance when the magnetic information is correlated with the age of the crust?

11. Under what conditions will a subduction zone form a mountain range? An island arc system? Why do volcanoes associated with subduction zones usually erupt more explosively than mid-ocean volcanoes associated with hot spots and spreading centers?

12. On an outline may of the world draw in (a) earthquake belts, (b) mid-ocean ridges, and (c) trenches. Relate your map to figure 2.9. What do you conclude?

13. How have recent advances in tomography modified our ideas of the Earth's internal layers as shown in figure 2.3?

14. Why do P-waves pass through the Earth's outer core?

15. What is a terrane? What role do terranes play in our understanding of today's continents?

Study Problems

1. If a plate moves away from a spreading center at the rate of 5 cm/yr, what is the displacement of a landmass carried by that plate after 180×10^6 years?

2. Magnetic stripes with the same magnetic orientation are measured on either side of a ridge crest. The stripe on the west side of the ridge is displaced 11 km from the crest; the stripe on the east side is displaced 9 km from the crest. The age of the rock in both stripes is 4×10^5 years. Calculate the average spreading rate at this ridge.

3. If the north end of the Emperor Seamount Chain is 75 million years old and Midway Island is 25 million years old, what was the rate of movement of the Pacific Plate during the period of the seamount chain's creation? (You will have to use an atlas to determine the distance between the north end of the Emperor Seamount Chain and Midway Island.) What can you deduce about the past direction of the Pacific Plate's movement compared to its present direction of motion? Base your deduction on the orientation of the islands and seamounts from Hawaii to Midway and from Midway to the north end of the Emperor Seamount Chain.

Links to Related Websites

Visit the book's Online Learning Center at www.mhhe.com/sverdrup (click on the book's cover) to find live Internet links for additional topics related to this chapter's content.

- Earth's interior
- Earthquakes and volcanoes
- Plate tectonics
- Paleomagnetism
- Ocean drilling

Visit the book's Online Learning Center at www.mhhe.com/sverdrup (click on this book's cover) to find these additional chapter tools: Suggested Readings; links to further information on boxed readings, selected figures, and related chapter topics; and additional study aids.

chapter 3

The Sea Floor and Its Sediments

The fog continued through the night, with a very light breeze, before which we ran to the eastward, literally feeling our way along. The lead was hove every two hours and the gradual change from black mud to sand showed that we were approaching Nantucket South Shoals. On Monday morning, the increased depth and deep blue color of the water, and the mixture of shells and white sand which we brought up, upon sounding, showed that we were in the channel, and nearing George's; accordingly, the ship's head was put directly to the northward, and we stood on, with perfect confidence in the soundings, though we had not taken an observation for two days, nor seen land; and the difference of an eighth of a mile out of the way might put us ashore. Throughout the day a provokingly light wind prevailed, and at eight o'clock, a small fishing schooner, which we passed, told us we were nearly abreast of Chatham lights. Just before midnight, a light land-breeze sprang up, which carried us well along; and at four o'clock, thinking ourselves to the northward of Race Point, we hauled upon the wind and stood into the bay, north-north-west, for Boston light, and commenced firing guns for a pilot.

Richard Henry Dana, Jr.
From *Two Years Before the Mast*

Eleuthera Island in the Bahamas from the space shuttle *Columbia*. The *light blue* of the shallow-water Bahama Bank contrasts with the *dark Blue* of the deep ocean.

Early mariners and scholars believed that the oceans were large basins or depressions in the Earth's crust, but they did not conceive that these basins held features that were as magnificent as the mountain chains, deep valleys, and great canyons of the land. As maps became more detailed and as ocean travel and commerce increased, measurement of water depths and recording of seafloor features in shallower regions became necessary to maintain safe travel and ocean commerce. The secrets of the deeper oceanic areas had to wait for hundreds of years until the technology of the late twentieth century made it relatively easy to map and sample the sea floor. It was only then that large numbers of survey vessels accumulated sufficient data to provide the details of this hidden terrain.

What we know about the sea floor and its covering of sediments comes almost entirely from the observations by surface ships, and, more recently, submersibles, robotic devices, and satellites have added to our knowledge. Some areas of the sea floor have been measured in great detail; charts of other areas have been made from scanty data. The demand for more measurements to describe and explain the features of the sea floor continues to the present.

In this chapter we survey the world's ocean floors and discuss their topography and geology. We examine the sources, types, and sampling of sediments and also discuss seabed mineral resources.

Measuring the Depths   3.1

In about 85 B.C. a Greek geographer named Posidonius set sail, curious about the depth of the ocean. He directed his crew to sail to the middle of the Mediterranean Sea, where they eased a large rock attached to a long rope over the side. They lowered it nearly 2 km (1.2 mi) before it hit bottom and answered Posidonius's question. Crude as this method was, it continued with minor modifications as the means of obtaining **soundings,** or depth measurements, for the next 2000 years.

An early modification was the use of hemp line or rope with a greased lead weight at its end. This line was marked in equal distances (usually **fathoms;** a fathom is the length between a person's fully outstretched hands, standardized at 6 ft). The change in line tension when the weight touched bottom indicated depth, and the particles from the bottom adhering to the grease confirmed the contact and brought a bottom sample to the surface. This method was quite satisfactory in shallow water, and the experienced captain used the properties of the bottom sample to aid in navigation, particularly at night or in heavy fog. In deep water, however, the weight of the hemp line was so great that it was difficult to sense when the lead weight touched the bottom. The sediments adhering to the grease could still confirm a touch, but there was no way of knowing how much slack line lay on the bottom. For this reason the deeper areas measured by this technique were often thought to be greater than their real depth.

Later, piano wire with a cannonball attached was used in deep water. The heavy weight of the ball compared to the weight of the wire made it easier to sense the bottom, but the time (eight to ten hours) to winch the wire out and in and the effort consumed for each measurement were so great that by 1895 only about 7000 depth measurements had been made in water greater than 2000 m (6600 ft) and only 550 measurements had been made of depths greater than 9000 m (29,500 ft) over all the world's oceans.

It was not until the 1920s, when acoustic sounding equipment was invented, that deep-sea depth measurements became routine. The **echo sounder,** or **depth recorder,** which measures the time required for a sound pulse to leave the surface vessel, reflect off the bottom, and return, allows continuous measurements to be made easily and quickly when a ship is underway. The behavior of sound in seawater and its uses as an oceanographic tool are discussed in chapter 4. A trace from a depth recorder is shown in figure 3.1.

In 1925 the German vessel *Meteor* made the first large-scale use of an echo sounder on a deep-sea oceanographic research cruise and detected the Mid-Atlantic Ridge for the first time. After this expedition, depth measurements gradually accumulated at an ever-increasing rate. As the acoustic equipment improved and was used more frequently, knowledge of the ocean floor's bathymetry expanded and improved, culminating in the 1950s with the first detailed mapping of all the mid-ocean ridge and trench systems.

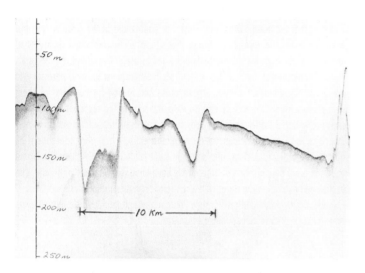

Figure 3.1 Depth recorder trace. A sound pulse reflected from the ocean floor traces a depth profile as the ship sails a steady course. The horizontal scale depends on ship speed.

Today, a wide variety of methods are used to obtain even more detailed seafloor bathymetry at scales that range from centimeters (inches) to thousands of kilometers (thousands of miles). The specific technique used often depends on the amount of time that can be spent, the scale of the feature that is being examined, and the amount of detail that is required. When necessary, direct observation of small-scale structures is possible with the use of staffed submersibles and remotely operated vehicles (ROVs) carrying video cameras. These images can be transmitted to surface ships and relayed by satellite anywhere in the world in real time (fig. 3.2). Investigations by staffed submersibles or ROVs provide great detail, but they typically cover very small areas and are both time-consuming and expensive for the amount of sea floor surveyed. On large scales of tens or hundreds of square kilometers, sophisticated multibeam sonar systems can rapidly map extensive regions at relatively low cost with great accuracy (see fig. 1.13).

A new system has been developed for making detailed bathymetric surveys in shallow coastal water using an airborne laser. The laser airborne depth sounder (LADS) system is flown in a small fixed-wing aircraft 350–550 m (1200–1800 ft) over the surface; the exact position of the aircraft is determined by the Global Positioning System (GPS). The laser is used to measure the distance between the aircraft and the sea floor. The LADS system can take up to 900 soundings per second (3.24 million soundings per hour). Individual measurements typically are taken on 5 × 5 m (16.5 × 16.5 ft) spacing in a swath 240 m (790 ft) wide along the survey line. For greater detail, the spacing can be reduced to 2 × 2 m (6.5 × 6.5 ft). Because light is rapidly attenuated in water, LADS has an operational depth range of 0.5–70 m (1.5–230 ft). The actual maximum depth in a specific area depends on water clarity. In pristine coral reef environments, soundings as great as 70 m (230 ft) can be obtained, whereas in clear to moder-

Figure 3.2 This internally recording television camera is placed on the sea floor, where it automatically photographs events until it is retrieved by the researchers. The *red box* contains the camera's power supply.

ately turbid coastal waters the effective depth of penetration decreases to 20–50 m (66–164 ft). In very turbid water, the system is restricted to 0–15 m (0–49 ft).

Very-large-scale seafloor surveys use satellite measurements of changes in sea surface elevation caused by changes in the Earth's gravity field due to seafloor bathymetry. These changes in sea surface elevation can be detected by radar altimeters that measure the distance between the satellite and the sea surface. The sea surface is not flat even when it is perfectly calm. Changes in gravity caused by seafloor topography create gently sloping hills and valleys in the sea surface. The excess mass of features such as seamounts and ridges creates a gravitational attraction that draws water toward them, resulting in an elevation of the sea surface. Conversely, the deficit of mass along deep-ocean trenches, and subsequent weaker gravitational attraction, results in a depression of the sea surface as water is drawn away toward surrounding areas with greater gravitational attraction. Sea level over large seamounts is elevated by as much as 5 m (16 ft) and over ocean ridges by about 10 m (33 ft); it is depressed over trenches by about 25–30 m (80–100 ft). These changes in elevation occur over tens to hundreds of kilometers, so the slopes are very gentle. The sea surface is always perpendicular to the local direction of gravity, so precise measurements of the slope of the surface can be used to determine the direction and magnitude of the gravitational field at any point. Because these changes in gravity are related to seafloor topography, it is possible to use them to reconstruct what the bathymetry must look like to produce the observed variations in sea surface topography (fig. 3.3). Tides, currents, and changes in atmospheric pressure can cause undulations of more than a meter (3 feet) in the ocean surface. These effects are filtered out to produce the bathymetric details. Bathymetric features with "footprints," or horizontal dimensions as small as about 10 km (6.2 mi), can be resolved with satellite altimetry data. Satellite maps are particularly valuable in the Southern Ocean, where the weather and sea conditions are frequently bad and it is difficult to conduct general bathymetric surveys to locate areas of scientific interest.

Bathymetry of the Sea Floor 3.2

The land below the sea surface is as rugged as any land above it. The Grand Canyon, the Rocky Mountains, the desert mesas in the Southwest, and the Great Plains all have their undersea counterparts. In fact, the undersea mountain ranges are longer, the valley floors are wider and flatter, and the canyons are often deeper than those found on land. Features of land topography, such as mountains and canyons, are continually and aggressively eroded by wind, water, and ice as well as changes in temperature and the chemical alteration of minerals in rocks. The erosion of ocean bathymetry is generally slow. Physical weathering is accomplished primarily by waves and currents, and chemical erosion occurs by the dissolution of minerals in the water. More rapid erosion is generally restricted to the continental margin, as discussed in section 3.3.

The most important agents of physical change on the deep sea floor are the gradual burial of features by a constant rain of sediments falling from above and volcanism associated with the mid-ocean ridge system, hot spots, island arcs and active seamounts, and abyssal hills. Movements of the Earth's crust may displace features and fracture the sea floor, and the weight of some underwater volcanoes may cause them to subside, but the appearance of the bathymetric features of the ocean basins and sea floor has remained much the same through the last 100 million years. Computer-drawn profiles of crustal elevations across the United States and the Atlantic Ocean are shown in figure 3.4. At 40°N latitude the height above zero elevation (*dotted line*) and width of mountains in the western United States are about the same as the height and width of the Mid-Atlantic Ridge system on the sea floor. Compare the topography of the Rocky Mountains and the undersea peaks in this figure.

Continental Margin

The edges of the landmasses at present below the ocean surface and the steep slopes of these landmasses that descend to the sea floor are known as the **continental margin.** There are two basic types of continental margins. These are often called passive, or Atlantic, margins and active, or Pacific, margins. Passive margins have little seismic or volcanic activity and involve a transition from continental crust to oceanic crust in the same lithospheric plate. They form when continents are rifted apart, creating a new ocean basin between them. Passive margins tend to be relatively wide. Active margins are tectonically active and associated with earthquakes and volcanism. They are most often associated with plate convergence and the subduction of oceanic lithosphere beneath a continent. Active margins are plate boundaries and are frequently relatively narrow. The continental margin is made up of the continental shelf, shelf break, slope, and rise. The **continental shelf** lies at the edge of the continent; continental shelves are the nearly flat borders of varying widths that slope very gently toward the ocean basins. Shelf widths average about 65 km (40 mi) but are typically much narrower along active margins than passive margins. The width of the continental shelf can be as much as 1500 km (930 mi). Water depth at the outer edge of the continental shelf varies from 20–500 m (65–1640 ft), with an average of about 130 m (430 ft).

The distribution of the world's continental shelves is shown in figure 3.5. The width of the shelf is often related to the slope of the adjacent land; it is wide along low-lying land and narrow along mountainous coasts. Note the narrow shelf along the western coast of South America and the great wide expanses of continental shelf along the eastern and northern coasts of North America, Siberia, and Scandinavia. The continental shelves are geologically part of the continental crust; they are the submerged seaward edges of the continents. There are several processes that contribute to the formation of continental shelves. Storm waves may erode continental shelves (fig. 3.6*i*), and, in some areas, natural dams trap sediments between the offshore dam and the coast (fig. 3.6*ii, iii*). Seamounts and island arcs (fig. 3.6*iv*) and coral reefs (fig. 3.6*vi*) also trap sediments. In the Gulf of Mexico sediments from the land are trapped behind folds and domes of the sea floor containing salt deposits (fig. 3.6*v*). Along the northeastern coast of North America sediments are trapped behind upturned rock near the outer edges of the continental shelf (fig. 3.6*ii*).

During past ages, the shelves have been covered and uncovered by fluctuations in sea level. During the glacial ages of the Pleistocene epoch there were a number of short-term changes in sea level, some of which were greater than 120 m (400 ft). When the sea level was low, erosion deepened valleys, waves eroded previously submerged land, and rivers left sediments far out on the shelf. When the glacial ice melted, these areas were flooded and sediments built up in areas closer to the new shore. At present, although submerged, these areas still show the scars of old riverbeds and glaciers they acquired when part of the landmass. Today, some continental shelves are covered with thick deposits of silt, sand, and mud sediments derived

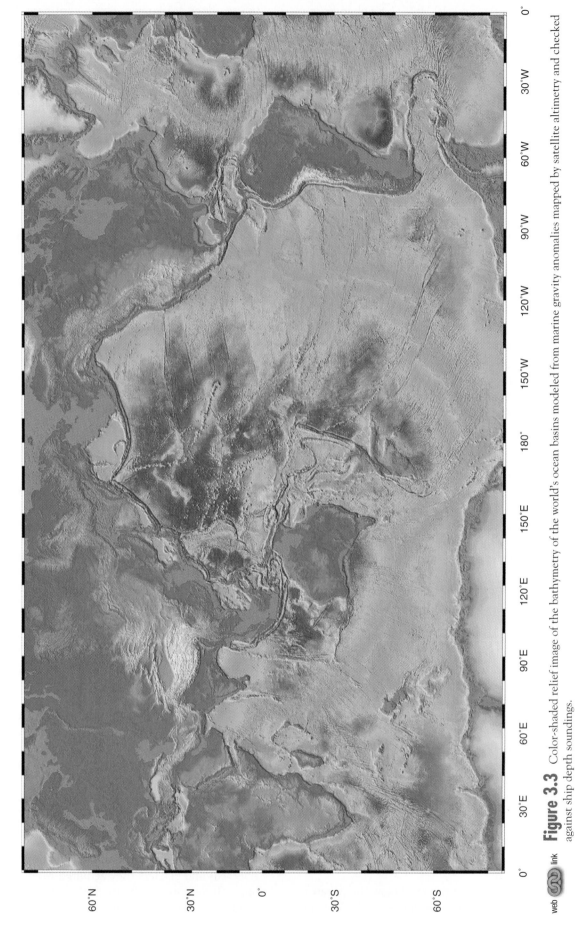

Figure 3.3 Color-shaded relief image of the bathymetry of the world's ocean basins modeled from marine gravity anomalies mapped by satellite altimetry and checked against ship depth soundings.

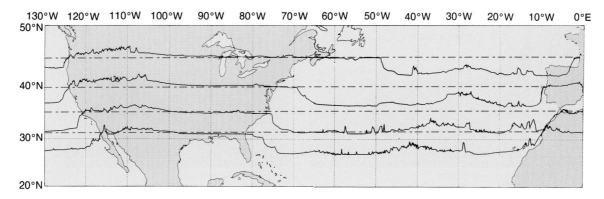

Figure 3.4 Computer-drawn topographic profiles from the western coast of Europe and Africa to the Pacific Ocean. The elevations and depths above and below 0 m are shown along a line of latitude by using the latitude line as zero elevation. For example, the ocean depth at 40°N and 60°W is 5040 m (16,531 ft). The vertical scale has been extended about 100 times the horizontal scale. If both the horizontal and vertical scales were kept the same, a vertical elevation change of 5000 m would measure only 0.05 mm (0.002 in).

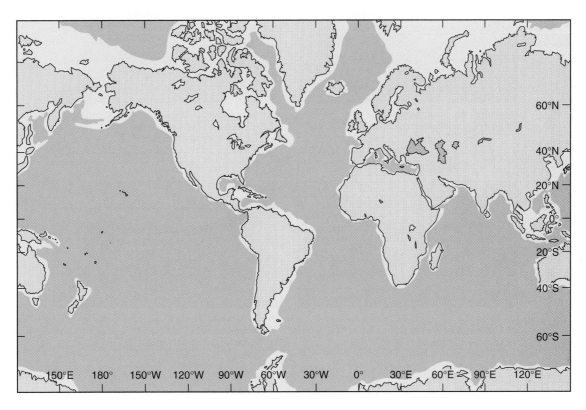

Figure 3.5 Distribution of the world's continental shelves (shown in *light blue*). The seaward edges of these shelves are at an average depth of approximately 130 m (430 ft).

from the land: for example, the mouths of the Mississippi and Amazon Rivers, where large amounts of such sediments are deposited annually. Other shelves are bare of sediments, such as where the fast-moving Florida Current sweeps the tip of Florida, carrying the sediments northward to the deeper water of the Atlantic Ocean.

The boundary of the continental shelf on the ocean side is determined by an abrupt change in slope and a rapid increase in depth. This change in slope is referred to as the **continental shelf break;** the steep slope extending to the ocean basin floor is known as the **continental slope.** These features are shown in figure 3.7. The angle and extent of the slope vary from place to place. The slope may be short and steep (for example, the depth may increase rapidly from 200 m [650 ft] to 3000 m [10,000 ft], as in fig. 3.7), or, along an active margin, it may drop as far as 8000 m (26,000 ft) into a great deep-seafloor depression or trench (for example, off the western coast of South America, where the narrow continental shelf is bordered by the Peru-Chile Trench). The continental slope may show rocky outcroppings and be relatively bare of sediments because of its steepness, tectonic activity, or a low supply of sediments from land.

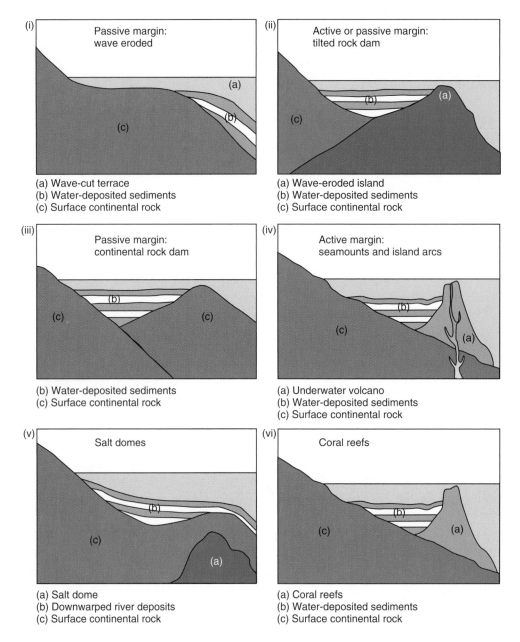

Figure 3.6 Examples of how continental shelves are formed by trapping land-derived sediments at the edge of the continental landmasses.

Figure 3.7 A typical profile of a passive continental margin. Notice both the vertical and horizontal extent of each subdivision. The average slope is indicated for the continental shelf, slope, and rise. The vertical exaggeration is 100 times greater than the horizontal scale.

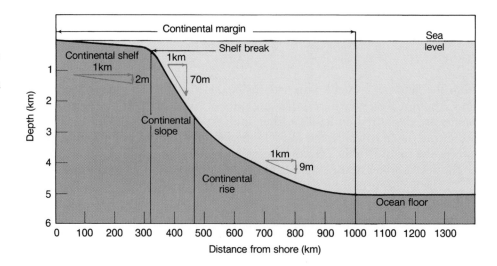

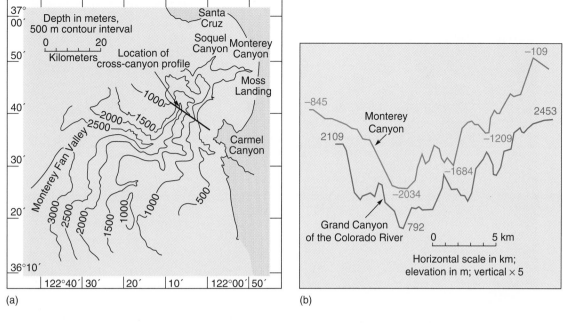

(a)   (b)

Figure 3.8 (a) Depth contours depict the Monterey and Carmel Canyons off the California coast as they cut across the continental slope and up into the continental shelf. (b) Cross-canyon profile of the Monterey Canyon shown in (a). Compare this profile to that of the Grand Canyon drawn to the same scale. *From* Submarine Geology. *Copyright © 1963 by Francis P. Shepard. Reprinted by permission of Addison-Wesley Educational Publishers.*

The most outstanding features found on the continental slopes are **submarine canyons.** These canyons sometimes extend up, into, and across the continental shelf. A submarine canyon is steep-sided and has a **V**-shaped cross section, with tributaries similar to those of river-cut canyons on land. Figure 3.8*a* shows the Monterey and Carmel Canyons off the coast of California. Figure 3.8*a* is a bathymetric chart; figure 3.8*b* compares the profile of the Monterey Canyon with the profile of the Grand Canyon of the Colorado River. A submarine canyon is also shown in figures 1.11 and 1.12.

Many of these submarine canyons are associated with existing river systems on land and were apparently cut into the shelf during periods of low sea level, when the glaciers advanced and the rivers flowed across the continental shelves. Ripple marks on the floor of the submerged canyons, and sediments fanning out at the ends of the canyons, suggest that they were formed by moving flows of sediment and water called **turbidity currents.** These sediment-laden currents can travel at speeds up to 90 km (56 mi) per hour and carry in suspension up to 300 kg of sediment per cubic meter (18.7 lb/ft³). Caused by earthquakes or the overloading of sediments on steep slopes, turbidity currents are fast-moving avalanches of mud, sand, and water that flow down the slope, eroding and picking up sediment as they gain speed. In this way the currents erode the slope and excavate the submarine canyon. As the flow reaches the bottom it slows and spreads, and the sediments settle. Because of their speed and turbulence, such currents can transport large quantities of materials of mixed sizes. The settling process produces graded deposits of coarse materials overlain by successive layers of decreasing particle size. These graded deposits are called **turbidites.** Figure 3.9 shows a turbidite preserved in compacted

Figure 3.9 This beach cliff shows a series of ancient turbidite deposits that have been uplifted and then exposed by wave erosion. Each individual turbidite is a graded deposit, with the largest particles in the deposit at the bottom of the turbidite and the smallest particles at the top.

Bathymetry of the Sea Floor **105**

Figure 3.10 Sand fall in San Lucas submarine canyon. Sand moving down the continental slope is deflected seaward to the ocean basin floor.

seafloor sediments that have been uplifted and exposed by wave erosion. These large and occasional currents have never been directly observed, although similar but smaller and more continuous flows, such as sand falls, have been observed and photographed (fig. 3.10).

Research on turbidity currents began with laboratory experiments in the 1930s. These experiments were based on earlier observations of the silty Rhone River water moving along the bottom of Lake Geneva in Switzerland. Later analysis of a 1929 earthquake that broke transatlantic telephone and telegraph cables on the continental slope and rise off the Grand Banks of Newfoundland showed a pattern of rapid and successive cable breaks high on the continental slope, followed by a sequence of downslope breaks after the earthquake. These breaks were calculated to have been caused by a turbidity current that ran for 800 km (500 mi) at speeds of 40–55 km (25–35 mi) per hour. Later samples taken from the area showed a series of graded sediments at the end of the current's path. Searches of cable company records showed similar patterns of cable breaks in other parts of the world.

A turbidity flow can be demonstrated by placing water and loose sediments of mixed particle size in a 6 ft section of 2–3 in diameter clear plastic tube. Cap both ends securely and stand the tube on end for a day or so. Then carefully tilt the tube until it is horizontal and slowly elevate the end with the sediments. Tap the tube gently as you raise it, and the sediments will move down slope as a turbidity flow. When the sediment settles at the other end, a turbidite pattern is formed.

Canyons with U-shaped cross sections and flat bottoms have been found off rivers that supply large quantities of sediments, such as the Ganges River. Continental shelves also show numerous gullies, probably formed by loose sediments moving down the slope.

At the base of the steep continental slope there may be a gentle slope formed by the accumulation of sediment. This portion of the sea floor is the **continental rise,** made up of sediment deposited by turbidity currents, underwater landslides, and any other processes that carry sands, muds, and silt down the continental slope. Continental rises may be compared to the landforms known as alluvial fans found where outwashes from steep gullies spread across a valley floor. The continental rise is a conspicuous feature at passive margins in the Atlantic and Indian Oceans and around the Antarctic continent. Few continental rises occur in the Pacific Ocean, where active margins border the great seafloor trenches located at the base of the continental slope. Refer to figure 3.7 to see the relationship of the continental rise to the continental slope.

Ocean Basin Floor

The true oceanic features of the sea floor occur seaward of the continental margin. The deep-sea floor, between 4000 and 6000 m (13,000 and 20,000 ft), covers more of the Earth's surface (30%) than do the continents (29%). In many places the ocean basin floor is a flat plain extending seaward from the base of the continental slope. It is flatter than any plain on land and is known as the **abyssal plain.**

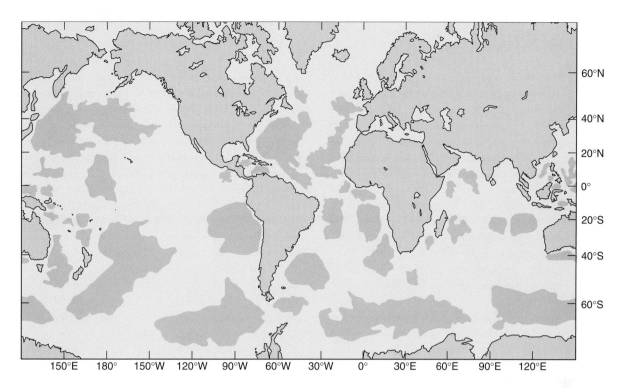

Figure 3.11 The major ocean basins of the world (*dark blue*) are separated by ridges, rises, and continents.

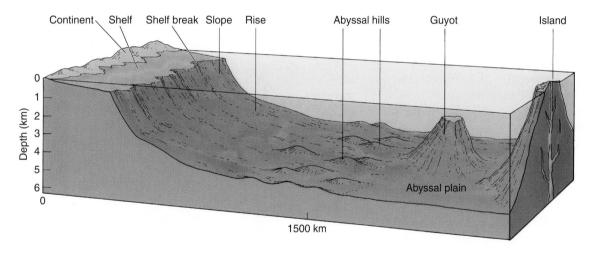

Figure 3.12 An idealized portion of ocean basin floor with abyssal hills (less than 1000 m of elevation), a guyot (a flat-topped seamount), and an island on the abyssal plain. The island was previously a seamount before it reached the surface. Seamounts and guyots are known to be volcanic in origin (vertical × 100).

The abyssal plain is formed by sediments that fall from the surface and are deposited by turbidity currents to cover the irregular topography of the oceanic crust. An area of the abyssal plain that is isolated from other areas by continental margins, ridges, and rises is known as a basin, and some basins may be subdivided into subbasins by ridge and rise subsections. The distribution of these basins and subbasins is shown in figure 3.11. Low ridges allow some exchange of water between adjacent basins, but if the ridge is high, both the water and the deep-dwelling marine organisms within the basin are effectively cut off from other basins. For exam-ple, the deep water of the Angola Basin, in the bight of the western coast of Africa, is cut off from the Brazil Basin to the west by the Mid-Atlantic Ridge and from the Cape Basin to the south by the Walvis Ridge. A physiographic chart of the sea floor (see fig. 2.9) provides another view of these basins.

Abyssal hills and **seamounts** are scattered across the sea floor in all the oceans. Abyssal hills are less than 1000 m (3300 ft) high, and seamounts are steep-sided volcanoes ris-ing abruptly and sometimes piercing the surface to become islands. These features are shown in figure 3.12. Abyssal hills are probably the Earth's most common topographic feature.

Bathymetrics

Visualizing the sea floor began with single soundings made with a lead line and continued with simple echo sounders and contour maps drawn by hand. In one hour, an individual with a lead line could take twenty measurements in water 10 m (33 ft) deep, and in four hours only one measurement in water 4000 m (13,200 ft) deep; the echo sounder allowed 36,000 measurements to be made each hour in 10 m of water and 680 in the same time in 4000 m of water. Today's multibeam sound systems can take 293,000 measurements per hour in 10 m of water and 20,000 measurements in 4000 m. Advances in multibeam sound system technology and improved computer graphics, combined with satellite navigation for precise positioning, are opening dramatic new windows to the sea floor.

A single sound beam device releases a cone of sound; as the depth of the water increases, the area of the sea floor from which the echo is reflected also increases. Depth is averaged over the "footprint" of the sound beam, and therefore seafloor features smaller than the footprint are difficult to detect and detail is reduced. Two new technologies using multiple sound devices are being used to produce detailed, high-resolution seafloor maps: (1) side-scan acoustical imaging and (2) swath bathymetry.

Side-scan measurements can be made either from a surface vessel or from a submerged system towed behind a vessel. If the ship is pitching and rolling, the path of the sound beam from a surface vessel will be displaced from its intended direction, resulting in inaccurate data. A towed system is below the depth of surface waves and winds; it is also closer to the sea floor, allowing the use of a conical sound beam that produces a smaller sound footprint. The smaller footprint increases the detail that is imaged but decreases the scanned area for each cone of sound. The area surveyed is increased by sending out multiple sound beams obliquely on either side of the sound device; no image is obtained from directly under the side scanner.

Side-scan acoustical images are the product of the reflectivity of the seafloor materials and the angle at which the sound beams strike the sea floor. Changes in the reflection of the sound come from the irregularities and the changing properties of the bottom being scanned. The sides of a seamount, a fault, and other objects with strong topographic relief act as good reflectors.

Side-scan sonar systems that are housed in torpedo-like casings and towed behind ships are known as towfish; GLORIA (geological long-range inclined asdic) is one of the most sophisticated. It is towed at 10 knots (nautical miles per hour), has a depth capability of 5000 m (16,400 ft), and scans the sea floor with two sound beams 30 km (18 mi) wide. The sound beams are composed of sound pulses that last four seconds, with forty-second intervals between pulses to allow the echo to return to the towfish for recording (box fig. 1).

Side-scan acoustical imaging also works very well to detect sunken ships, planes, or other structures, because the reflecting surfaces of these structures are at an angle to the sea floor, and their acoustical properties are very different from those of the sea floor. The object's shape is accompanied by an acoustical shadow (seen behind the plane in box fig. 2) that provides strong image definition and indicates elevation above the sea floor.

The GLORIA surveys of the 1980s produced the first accurate maps of the deep-sea floor within U.S. resource limits, but it was too slow and the sound beam was too narrow to obtain detailed information over the shallow continental shelves. In the 1990s the USGS (United States Geological Survey) began a high-resolution survey of U.S. continental shelves using a system based on multiple sound

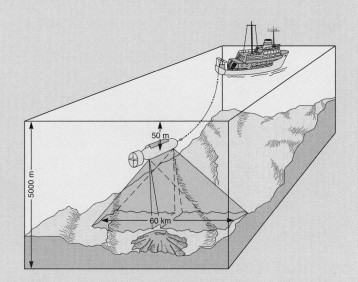

Box Figure 1 A surface vessel tows a side-scan sonar system, or towfish, to acoustically map a swath of sea floor. The *dark triangle* shows the area covered by the sound beams. Vertical scale distorted for clarity.

They are found over 50% of the Atlantic sea floor and about 80% of the Pacific floor; they are also abundant in the Indian Ocean. Most abyssal hills are probably volcanic, but some may have been formed by other movements of the sea floor. Submerged, flat-topped seamounts, known as **guyots,** are found most often in the Pacific Ocean; a guyot is also shown in figure 3.12. The Pacific guyots are 1000–1700 m (3300–5600 ft) below the surface; many are at the 1300 m (4300 ft) depth. Many guyots show the remains of shallow marine coral reefs and evidence of wave erosion at their summits. These features indicate that at one time they were warm-water surface features and that their flat tops are the result of wave erosion. They have since subsided owing to their weight, the accumulated sediment load bearing down on the oceanic crust, and the natural subsidence of the sea floor with increasing age as the crust cools and increases in

ZOOMED IMAGE

AMS 120 kHz Side Scan Sonar (150 Meter Swath)

WILLIAMSON & ASSOCIATES, INC.
Seattle, Wa.

Box Figure 2 The image of a plane on the sea floor obtained by side-scan sonar. Note the shadow generated when sound was not returned from the seabed.

sources known as swath bathymetry. Swath bathymetry determines depth and produces images by analyzing the sound interference patterns between outgoing sound beams and returning echoes. This method uses hull-mounted sonar arrays of 60 to more than 150 electronically separated sound sources mounted along the length of the hull and multiple sound receivers mounted across the hull. Both sound sources and receivers are directed vertically downward. The swath length is the length of the area returning echoes. The width of the swath is in the direction of the vessel's motion and is determined by the sound cone angle. A computer analyzes the sound interference patterns between outgoing sound pulses and their returning echoes. Once a computer image has been assembled, the image can be viewed from any angle and enhanced with color to indicate depth (see fig. 1.13). The ship is positioned by GPS, and measurements can be collected at vessel speeds in excess of 15 knots (17 mi/h).

Between 1994 and 1997 the USGS mapped five areas using swath bathymetry: Massachusetts Bay, parts of the continental shelf off New York, the Hudson River, the coastal margin of Santa Monica, and the central portion of San Francisco Bay. In 1998 the project continued, mapping portions of the slopes of the Hawaiian Islands and the shelf areas of San Diego and Newport, California. The images obtained with these multibeam systems are providing highly detailed and accurate images of the sea floor, similar to those seen in aerial photographs. These maps are providing basic information, fundamental for geological research and biological management of the coastal margin.

To Learn More About Bathymetrics

Gardner, J., P. Butman, and L. Mayer. 1998. Mapping U.S. Continental Shelves. *Sea Technology* 39 (6): 10–17.
Pratson, L. F. and W. F. Haxby. 1997. Panoramas of the Seafloor. *Scientific American* 276 (6): 82–87.

Internet References

Visit the book's Online Learning Center at www.mhhe.com/sverdrup (click on the book's cover) to explore links to further information on related topics.

density as it moves farther away from the ridge where it formed. They have also been submerged by rising sea level during periods in which land ice melted.

In the warm waters of the Atlantic, Pacific, and Indian Oceans, coral reefs and coral islands are formed in association with seamounts. Reef-building coral is a warm-water animal that requires a place of attachment and grows in intimate association with a single-cell, plantlike organism; it is confined to sunlit, shallow tropical waters. When a seamount pierces the sea surface to form an island, it provides a base on which the coral can grow. The coral grows to form a **fringing reef** around the island. If the seamount sinks or subsides slowly enough, the coral continues to grow upward at a rate that is not exceeded by the rising water, and a **barrier reef** with a lagoon between the reef and the island is formed. If the process continues, eventually the seamount

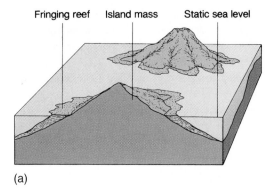

(a)

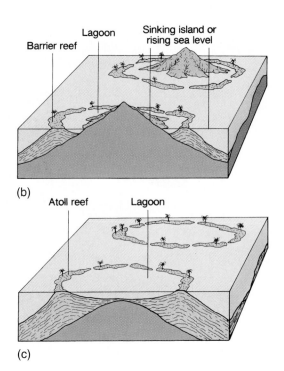

(b)

(c)

Figure 3.13 Types of coral reefs and the steps in the formation of a coral atoll shown in profile. (a) Fringing reef. (b) Barrier reef. (c) Atoll reef.

disappears below the surface and the coral reef is left as a ring, or **atoll.** This process is illustrated in figure 3.13.

On the basis of the observations he made during the voyage of the *Beagle* from 1831–36, Charles Darwin suggested that these were the steps necessary to form an atoll. Darwin's ideas have been proved to be substantially correct by more recent expeditions that drilled through the debris on a lagoon floor and found the basalt peak of a seamount that once protruded above the sea surface. The organisms that inhabit coral reefs are discussed in chapter 17.

Ridges, Rises, and Trenches

The most notable features of the ocean floor are the mid-ocean ridge and rise systems stretching for 65,000 km (40,000 mi) around the world and running through every ocean. Their origin and their role in plate tectonics were discussed in chapter 2 (see fig. 2.9). Review their distribution using figure 3.14. Recall also the roles played by their rift valleys and transform faults.

The relationship of the deep-sea trenches to plate tectonics was also discussed in chapter 2 (see fig. 2.9). Use figure 3.15 to trace the Japan-Kuril Trench, the Aleutian Trench, the Philippine Trench, and the deepest of all ocean trenches, the Mariana Trench. All these trenches are associated with **island arc systems.** The Challenger Deep, a portion of the Mariana Trench, has a depth of 11,020 m (36,150 ft), making it the deepest known spot in all the oceans. The longest of the trenches is the Peru-Chile Trench, stretching 5900 km (3700 mi) down the western side of South America. To the north, the Middle America Trench borders Central America. The Peru-Chile and Middle America Trenches are not associated with volcanic is-

land but are bordered by land volcanoes. In the Indian Ocean the great Sunda-Java Trench runs for 4500 km (2800 mi) along Indonesia. In the Atlantic there are only two comparatively short trenches: the Puerto Rico-Cayman Trench and the South Sandwich Trench, both associated with chains of volcanic islands. To view the bathymetry of the ocean floor as it is known to exist today, see figure 2.9.

In figure 3.16 the topography of the land and the bathymetry of the sea floor are summarized as percentages of the Earth's area. Compare the tectonically active areas of trenches and ridges. Compare the area of low-lying land platforms with the area of the ocean basins.

Sediments 3.3

The margins of the continents and the ocean basin floors receive a continuous supply of particles from many sources. Whether these particles have their origin in living organisms, the land, the atmosphere, or the sea itself, they are called sediment when they accumulate on the sea floor. The thickest deposits of sediment are generally found near the continental margins, where sediment is deposited relatively rapidly; in contrast, the deep-sea floor receives a constant but slow accumulation of sediment that produces a thinner layer that varies in thickness with the age of the oceanic crust.

Oceanographers study the rate at which sediments accumulate, the distribution of sediments over the sea bottom, their sources and abundance, their chemistry, and the history they record in layer after layer as they slowly but continuously accumulate on the ocean floors. In order to describe and catalog the sediments, geological

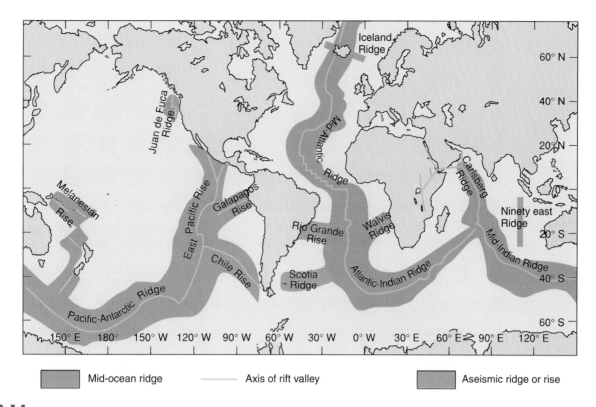

| Mid-ocean ridge | Axis of rift valley | Aseismic ridge or rise |

Figure 3.14 The mid-ocean ridge and rise system of divergent plate boundaries. Locations of major aseismic (no earthquakes) ridges and rises are added. Aseismic ridges and rises are elevated linear features thought to be created by hot-spot activity.

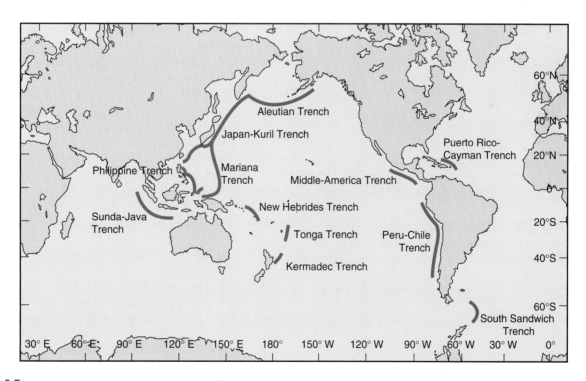

Figure 3.15 Major ocean trenches of the world. The deepest ocean depth is 11,020 m (36,150 ft), east of the Philippines in the Mariana Trench. It is known as the Challenger Deep.

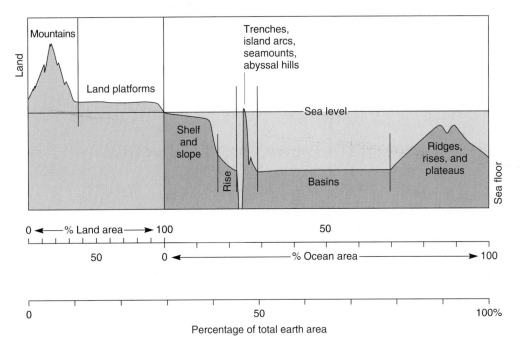

Figure 3.16 The Earth's main topographic features shown as percentages of the total Earth's surface and as percentages of the land and of the oceans.

Table 3.1 Sediment Size Classifications

Descriptive Name		Diameter (mm)
Gravel	Boulder	> 256
	Cobble	64–256
	Pebble	4–64
	Granule	2–4
Sand	Very coarse	1–2
	Coarse	0.5–1
	Medium	0.25–0.5
	Fine	0.125–0.25
	Very fine	0.0625–0.125
Mud	Silt	0.0039–0.0625
	Clay	< 0.0039

From Fundamentals of Oceanography, *4th edition, Duxbury, Duxbury, and Sverdrup. Copyright 2000 The McGraw-Hill Companies. All rights reserved.*

oceanographers classify sediments by particle size, location, origin, and chemistry.

Particle Size

Sediment particles are classified by size, as indicated in table 3.1. Familiar terms such as *gravel, sand,* and *mud* are used to identify broad size ranges of large, intermediate, and small particles, respectively. Within each of these ranges particles are further ranked to produce a more detailed scale from boulders to the very smallest clay-sized particles, which can only be seen with a microscope. Note that a boulder is a particle whose diameter is greater than 256 mm (10 in), a size

much smaller than what we commonly envision when we hear the word *boulder.*

The individual particles in a sediment sample can be sorted using a series of sieves with woven mesh of decreasing size. A sample is said to be "well sorted" if it is nearly uniform in particle size and "poorly sorted" if it is made up of many different particle sizes (fig. 3.17). Particle size influences the deposition of sediment by the horizontal distance it is transported before settling out of the water and the rate at which it sinks. In general, it takes more energy to transport large particles than it does small particles. In the coastal environment, when poorly sorted sediment is transported by wave or current action, the larger particles will settle out and be deposited first, while the smaller particles may be carried farther away from the coast and deposited elsewhere. In the open ocean, the variation in sinking rate between large and small particles has a tremendous influence on how long it takes for a particle to sink to the deep-sea floor and, hence, how far the particle may be transported by deep horizontal currents while it is settling (table 3.2). A very fine sand-sized particle may settle to the deep-sea floor in a matter of days, where it could come to rest a short horizontal distance away from the point at the surface where it began its journey. In contrast, it may take clay-sized particles over 125 years (nearly 50,000 days) to make the same journey (see Stokes Law for small-particle settling velocity in appendix C). The speed of deep horizontal currents in the oceans is generally quite slow, but even at a speed of 5 cm (2 in) per second a clay-sized particle could theoretically be transported around the world five times before it reached the deep-sea floor. Smaller soluble particles also have time to dissolve as they slowly sink in the deep ocean.

Scientists have puzzled over the fact that there is a close correlation between the particle types found in surface waters and those found almost directly below on the sea floor. This observation seems to contradict the inevitable large horizontal displacement of very slowly sinking particles due to currents in the water. Some mechanisms must be working to aggregate the tiny particles into larger particles. Scientists have observed that small particles often attract each other owing to their electrical charges. This attraction forms larger particles, which sink more rapidly. This process is important in the formation of the abundant sediment deposits in river deltas. Also, when predators eat tiny plants and animals, they process the organic material to produce energy for life and package the inorganic remains, often small shells called **tests,** and expel them as larger fecal pel-

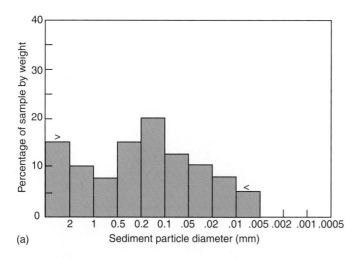

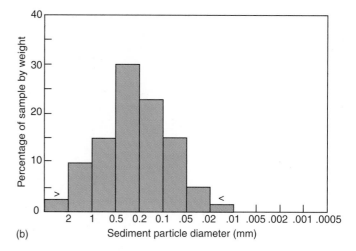

Figure 3.17 (a) A poorly sorted sample. Particles fall into a wide variety of size ranges in approximately equal amounts. (b) A well-sorted sample. One size range predominates in a limited distribution of sizes.

Table 3.2 Sediment Sinking Rate and Distance Traveled

Sediment Size	Approximate Sinking Rate (m/s)	Time for a Vertical Fall of 4 km (days)	Horizontal Distance Traveled in a 5 cm/s Current (km)
Very fine sand	9.8×10^{-3}	4.7	20.4
Silt	9.8×10^{-5}	470	2040
Clay	9.8×10^{-7}	47,000	204,000

Note: The sinking rate of a particle depends on its density, shape, and diameter. These rates are based on the assumption that the particles are spherical and have a density similar to that of quartz. Estimates of the speed of deep currents vary. A conservative estimate of 5 cm/s is chosen for purposes of illustration.

From Fundamentals of Oceanography, *4th edition, Duxbury, Duxbury, and Sverdrup. Copyright 2000 The McGraw-Hill Companies. All rights reserved.*

lets, which sink more rapidly. It is estimated that as many as 100,000 tests of small organisms can be packaged in a single fecal pellet. These processes help to decrease the time it takes for small particles to sink to the sea floor from years to just ten to fifteen days, minimizing their horizontal displacement by water movements. Once the pellets have been deposited on the bottom, the breaking down of the organic portion of the pellets liberates the small particles.

Location

Marine sediments are classified as either **neritic** (*neritos* = of the coast) or **pelagic** (*pelagios* = of the sea) based on where they are found (fig. 3.18). Neritic sediments are found near continental margins and islands and have a wide range of particle sizes. Most neritic sediments are eroded from rocks on land and transported to the coast by rivers. Once they enter the ocean they are spread across the continental shelf and down the slope by waves, currents, and turbidity currents. The largest particles are left near coastal beaches while smaller particles are transported farther from shore.

Pelagic sediment deposits are fine-grained sediments that collect slowly on the deep-sea floor. The thickness of pelagic sediments is related to the length of time they have been accumulating or the age of the sea floor they cover.

Consequently, their thickness tends to increase with increasing distance from mid-ocean ridges (see fig. 2.16).

Rates of Deposit

The rates at which marine sediments accumulate have a wide range, due to the natural variability of the processes that produce and transport sediments. Accumulation rates of neritic sediments are highly variable. In river estuaries the rate may be more than 800,000 cm (315,000 in) per 1000 years, or 8 m (over 26 ft) per year. Each year the rivers of Asia, such as the Ganges, the Yangtse, the Yellow, and the Brahmaputra, contribute more than one-quarter of the world's land-derived marine sediments. In quiet bays, the rate may be 500 cm (197 in) per 1000 years, and on the continental shelves and slopes values of 10–40 cm (3.9–15.7 in) per 1000 years are typical, with the flat continental shelves receiving the larger amounts. Many of the sediments covering the continental shelves away from river mouths are sediments laid down by processes that no longer exist at that location. Such sediments are called **relict sediments** and represent conditions that existed many thousands of years ago, when sea level was lower than it is today because of the accumulation of water in ice caps and glaciers.

Accumulation rates for pelagic sediments are much slower than those of typical neritic sediments. An average

Figure 3.18 Classification of sediments by location of deposit. The distribution pattern is partially controlled by the sediments' proximity to source and rate of supply.

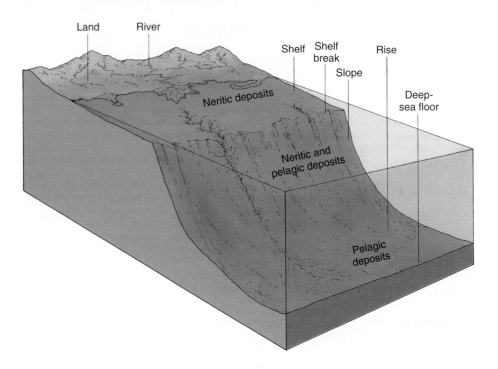

accumulation rate for deep-ocean pelagic sediment is 0.5–1.0 cm (0.2–0.4 in) per 1000 years. Although deep-sea sedimentation rates are excessively slow, there has been plenty of time during geological history to accumulate the average deep-sea sediment thickness of approximately 500–600 m (1600–2000 ft) at the continental rises and areas of older sea floor. At a rate of 0.5 cm (0.2 in) per 1000 years, it takes only 100 million years to accumulate 500 m (1600 ft) of sediment, and the oldest sea floor is known to be roughly twice that age, or about 200 million years old.

Source and Chemistry

Marine sediments are also classified by the source of the particles that make up the sediment and may be further subdivided by their chemistry. Sedimentary particles may come from one of four different sources: preexisting rocks, marine organisms, the seawater itself, or from space.

Sediments derived from preexisting rocks are classified as **lithogenous** (*lithos* = stone, *generare* = to produce) **sediments.** These are also commonly called **terrigenous** (*terri* = land, *generare* = to produce) **sediments.** While terrigenous sediment technically includes any type of material coming off the land, such as rock fragments, wood chips, and sewage sludge, the majority of terrigenous material consists of lithogenous particles. Active volcanic islands in the ocean basins are also an important source of lithogenous sediment. Rocks on land are weathered and broken down into smaller particles by wind, water, and seasonal changes in temperature that result in freezing and thawing. The resulting particles are transported to the oceans by water, wind, ice, and gravity. Windblown dust from arid areas of the continents, ash from active volcanoes, and rocks picked up by glaciers and embedded in icebergs are additional sources of lithogenous materials.

Lithogenous material can be found everywhere in the oceans. It is the dominant type of neritic sediment because the supply of lithogenous particles from land simply overwhelms all other types of material. Lithogenous deposits in pelagic sediments on the deep-sea floor are called **abyssal clay,** composed of at least 70% by weight clay-sized particles. Abyssal clay accumulates very slowly at rates that are generally less than 0.1 cm (0.04 in) per 1000 years. Because the accumulation rate is so slow, even a thin deposit represents a very long period of time. It is important to understand that where abyssal clay is the dominant pelagic sediment, it is only because of the lack of other types of material that would otherwise dilute it, not because of an increase in the supply of clay particles. This is generally the case in regions where there is little marine life in the surface waters above. This fine rock powder, blown out to sea by wind and swept out of the atmosphere by rain, may remain suspended in the water for many years. These clays are often rich in iron, which oxidizes in the water and turns a reddish brown color, hence they are frequently called **red clay** (fig. 3.19a). The distribution of red clay is illustrated in figure 3.20.

The composition of lithogenous sediments, generally various clays and quartz, is controlled by the chemistry of the rocks they came from and their response to chemical and mechanical weathering. Most lithogenous sediments have quartz because it is one of the most abundant and stable minerals in continental rocks. Quartz is very resistant to both chemical and mechanical weathering, so it can easily be transported long distances from its source. The distribution pattern of quartz grains in the sediment can provide important information concerning changes in wind patterns and intensity through time.

Clays are abundant because they are produced by chemical weathering. Four clay minerals make up the deep-sea clays: chlorite, illite, kaolinite, and montmorillonite. The distribu-

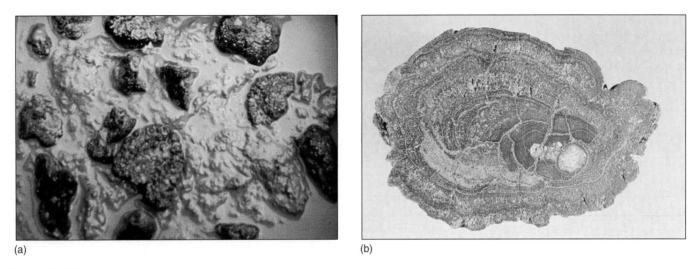

(a)   (b)

Figure 3.19 (a) Manganese nodules resting on red clay photographed on deck in natural light. Nodules are 1–10 cm in diameter. (b) A cross section of a manganese nodule showing concentric layers of formation.

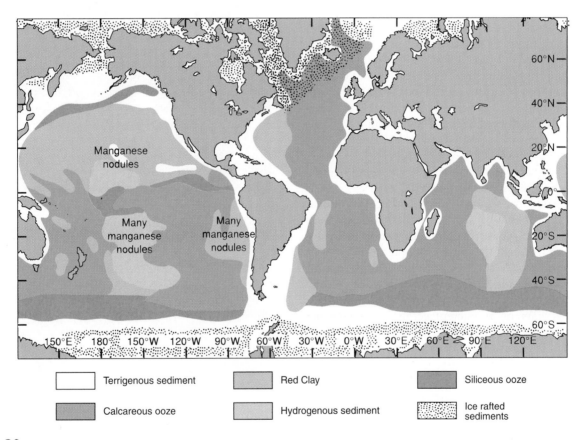

Figure 3.20 Distribution of the principal sediment types on the deep-sea floor. Sediments are usually a mixture but are named for their major component.

tion of these four clays reflects different climatic and geological conditions in the areas where, and when, they originated as well as along the paths they traveled before settling on the sea floor. These conditions often have a strong dependence on latitude. The warm, moist climate of low latitudes supports strong chemical weathering. Mechanical weathering tends to be dominant in the cold, dry climate typical of high latitudes.

Chlorite is highly susceptible to chemical weathering and can be altered to form kaolinite. Consequently, chlorite is abundant in deep-sea clays at high latitudes, where chemical weathering is less effective. Kaolinite is produced in the strong chemical weathering of minerals to form soil. It is ten times as abundant in the tropics as in polar regions, where soil-forming processes are very slow. Illite is the most widespread clay

Sediments **115**

mineral. It has a clear hemispheric rather than climatic distribution. In the Southern Hemisphere, it comprises up to 20%–50% of the clay minerals present; in the Northern Hemisphere, it usually accounts for more than 50% of the clay minerals. Montmorillonite is produced by the weathering of volcanic material on land and on the sea floor. It is common in regions of low sedimentation near sources of volcanic ash. It is more abundant in the Pacific and Indian Oceans than in the Atlantic Ocean, where there is little volcanic activity along the surrounding coastlines.

Sediments derived from organisms are classified as **biogenous** (*bio* = life, *generare* = to produce) **sediments.** These may include shell and coral fragments as well as the hard skeletal parts of single-celled plants and animals that live in the surface waters. Pelagic biogenous sediments are composed almost entirely of the shells, or tests, of single-celled organisms (fig. 3.21). The chemical composition of these tests is either calcareous (calcium carbonate: $CaCO_3$, or seashell material) or siliceous (silicon dioxide: SiO_2, clear and hard). If pelagic sediments are more than 30% biogenous material by weight, the sediment is called an **ooze;** specifically either a **calcareous ooze** or **siliceous ooze,** depending on the chemical composition of the majority of the tests found in the sediment. The distribution of calcareous and siliceous oozes on the sea floor is related to the supply of organisms in the overlying water, the rate at which the tests dissolve as they descend through the water, the depth at which they are deposited, and dilution with other sediments types (see fig. 3.20).

Calcareous tests are created by small, single-celled plants called **coccolithophores** (covered with calcareous plates called **coccoliths**), small snails called **pteropods,** and minute amoebae-like animals called **foraminifera** (fig. 3.21a, c). These deposits are often named for their principal constituent: coccolithophore ooze, pteropod ooze, or foraminiferan ooze. Calcareous oozes are the dominant pelagic sediments (see fig. 3.20). The dissolution, or destruction, rate of calcium carbonate varies with depth and temperature and is different in different ocean basins. Calcium carbonate generally dissolves more rapidly in cold, deep water, which characteristically has a higher concentration of CO_2 and is slightly more acidic (this is discussed in detail in the sections on the pH of seawater and dissolved gas in chapter 5). The depth at which calcareous skeletal material first begins to dissolve is called the **lysocline.** Below the lysocline, there is a progressive decrease in the amount of calcareous material preserved in the sediment. The depth at which the amount of calcareous material preserved falls below 20% of the total sediment is called the **carbonate compensation depth (CCD).** The CCD is also commonly defined as the depth at which the amount of calcium carbonate produced by organisms as skeletal material in the overlying water column is equal to the rate at which it is dissolved in the water. Calcareous ooze tends to accumulate on the sea floor at depths above the CCD and it is generally absent at depths below the CCD. The CCD has an average depth of about 4500 m (14,800 ft), or roughly midway be-

(a)

(b)

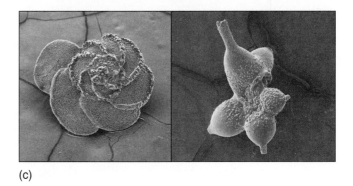
(c)

Figure 3.21 Scanning electron micrographs of biogenous sediments: (a) Diatoms and coccolithospheres. The small disks are detached coccoliths. (b) Radiolarians. (c) Foraminifera. *Fig. 3.21c images courtesy of Steve Nathan and R. Mark Leckie.*

tween the depth of the crests of ocean ridges and the deepest regions of the abyssal plains. In the Pacific, the CCD is generally at depths of about 4200–4500 m (13,800–14,800 ft). An exception to this is the deepening of the CCD to about 5000 m (16,400 ft) in the equatorial Pacific, where high rates of biological productivity result in a large supply of calcareous material. In the North Atlantic and parts of the South Atlantic, it is at or below depths of 5000 m (16,400 ft). Calcareous oozes are found at temperate and tropical latitudes in shallower areas of the sea floor such as the Caribbean Sea, on elevated ridge systems, and in coastal regions.

Siliceous tests are created by small, single-celled photosynthetic organisms called diatoms and animals called radiolaria (fig. 3.21a,b). Their skeletal remains are the dominant components of diatomaceous and radiolarian ooze, respectively. The pattern of dissolution of siliceous tests is opposite to that of calcareous tests. The oceans are undersaturated in silica everywhere, so siliceous material will dissolve at all depths, but it dissolves most rapidly in shallow, warm water. Siliceous oozes are only preserved below areas of very high biological productivity in the surface waters (see fig. 3.20). Even in these areas, an estimated 90% or more of the siliceous tests produced are dissolved, either in the water or on the sea floor, before they can be preserved.

Diatomaceous ooze is found at cold and temperate latitudes around Antarctica and in a band across the North Pacific. Because diatoms are plants, they require sunlight and fertilizers, or **nutrients,** for growth. The sunlight is available at the ocean's surface; the nutrients are produced by the decomposition of all plant and animal life in the ocean, and these nutrients are liberated in the deeper water as decomposition takes place. Only at certain locations are these nutrients returned to the surface by the large-scale upward flow of deeper water. Where this upward flow occurs, the availability of sunlight combines with the nutrients to produce the conditions needed for high levels of plant production. Large populations are found in the areas that combine suitable light, nutrients, and the correct temperature; because diatoms reproduce rapidly in cold waters, the waters of the North Pacific and Antarctic Oceans are best suited for their growth.

Radiolarian ooze is found beneath the warm waters of equatorial latitudes. Radiolaria thrive in warm water, producing siliceous outer shells that are often covered with long spines. Figure 3.21b shows radiolaria tests at high magnification.

Sediments derived from the water are classified as **hydrogenous** (hydro = water, generare = to produce) **sediments.** Hydrogenous sediments are produced in the water by chemical reactions. Most of the sediments are formed by the slow precipitation of minerals onto the sea floor, but some are created by the precipitation of minerals in the water column in plumes of recirculated water at hydrothermal vents along the ocean ridge system. Hydrogenous sediments include some **carbonates** (limestone-type deposits), **phosphorites** (phosphorus in the form of phosphate in crusts and nodules), **salts,** and **manganese nodules.** In addition, hydrothermally generated sulfides rich in iron and manganese form along the axis of spreading centers on young sea floor and carbonates and magnesium-rich minerals form off the axis of spreading centers on older sea floor, as discussed in chapter 2.

Hydrogenous carbonates are known to form by direct precipitation in some shallow, warm-water environments as a result of an increase in water temperature or a slight de-crease in the acidity of the water. In shallow, warm water with high biological productivity, plants can remove enough dissolved carbon dioxide in the water during photosynthesis to decrease the acidity of the water and trigger the precipitation of calcium carbonate (see the discussion of carbon dioxide as a buffer in section 5.1 of chapter 5). The calcium carbonate often precipitates in small pellets called **ooliths** (oon = egg) about 0.5–1.0 mm (0.02–0.04 in) in diameter. In the present oceans there are relatively few places where this is known to be occurring. The largest modern deposits of hydrogenous carbonates are currently forming on the Bahama Banks. Additional deposits are forming on Australia's Great Barrier Reef and in the Persian Gulf.

Phosphorites contain phosphorus in the form of phosphate and are most abundant on the continental shelf and upper part of the continental slope. They are occasionally found as nodules as much as 25 cm (10 in) in diameter, but more often they form thick crusts. Most phosphate deposits on continental margins do not appear to be actively accumulating. Phosphorite deposits are currently forming in regions of high biological productivity off the coasts of southwestern Africa and Peru.

Salt deposits occur when a high rate of evaporation removes most of the water and leaves a very salty brine in shallow areas. Chemical reactions occur in the brine, and salts are precipitated or separated from solution and then deposited on the bottom. In such processes carbonate salts are formed first, followed by sulfate salts, and then rock salt. Studies of precipitated material on the floor of the Mediterranean Sea have provided clues to its past isolation from the Atlantic Ocean.

Manganese nodules are composed primarily of manganese and iron oxides but also contain significant amounts of copper, cobalt, and nickel. They were first recovered from the ocean floor in 1873 during the *Challenger* expedition. They are found in a variety of marine environments, including the abyssal sea floor, on seamounts, along active ridges, and on continental margins. Their chemistry is related to the ocean basin they are found in as well as the specific marine environment where they have grown (tables 3.3 and 3.4). Nodules from the Pacific Ocean tend to have the highest concentrations of metals, with the exception of iron. Nodules in the Atlantic Ocean generally have the highest iron concentration. The average weight percent of manganese and iron in nodules is about 18% and 17%, respectively, while the average weight percent of nickel, cobalt, and copper varies from about 0.5% down to 0.2%. Nodules that form on continental margins are very distinct chemically. They have very high manganese concentrations combined with very low iron concentrations. The chemistry of nodules can also be influenced by their position on the sea floor with respect to other sediments. Manganese nodules may lay on top of the other sediment (see fig. 3.19a) or be buried at shallow depth in the sediment. Nodules lying on top of the sediment react chemically with the seawater and can become enriched in iron and cobalt. Those that are buried react with both the seawater and the sediment and can become enriched in manganese and

Table 3.3 Average Chemistry of Manganese Nodules from the Three Ocean Basins

Element	Atlantic	Pacific	Indian	Average for All Three Oceans
Mn	16.18	19.75	18.03	17.99
Fe	21.2	14.29	16.25	17.25
Ni	0.297	0.722	0.510	0.509
Co	0.309	0.381	0.279	0.323
Cu	0.109	0.366	0.223	0.233

Note: The average abundances of manganese (Mn), iron (Fe), nickel (Ni), cobalt (Co), and copper (Cu) in manganese nodules from the Atlantic, Pacific, and Indian Ocean basins. Numbers are weight % of each metal.

From Ocean Chemistry and Deep-Sea Sediments, *Pergamon Press, 1989.*

Table 3.4 Average Chemistry of Manganese Nodules from Different Environments

Element	Seamounts	Active Ridges	Continental Margins	Abyssal Depths
Mn	14.62	15.51	38.69	17.99
Fe	15.81	19.15	1.34	17.25
Ni	0.351	0.306	0.121	0.509
Co	1.15	0.400	0.011	0.323
Cu	0.058	0.081	0.082	0.233
Mn/Fe	0.92	0.81	28.9	1.04

Note: The average abundances of manganese (Mn), iron (Fe), nickel (Ni), cobalt (Co), and copper (Cu), and the manganese-to-iron ratio in manganese nodules, from different environments. Numbers are weight % of each metal.

From Ocean Chemistry and Deep-Sea Sediments, *Pergamon Press, 1989.*

Figure 3.22 These splash-form tektites were collected in Southeast Asia.

copper. The concentric layers in a nodule typically have slightly different chemistries (see fig. 3.19b). This chemical layering is the result of changes in the chemistry of the seawater through geologic time during nodule growth.

On the deep-sea floor, manganese nodules form black or brown rounded masses typically 1–10 cm (0.5–4 in) in diameter, roughly the size of a golf ball or a little larger. Continental margin manganese and iron oxide deposits can take a variety of forms, from nodules similar to those found on the deep-sea floor to extensive slabs, or crusts. Most manganese nodules grow very slowly, 1–10 mm (0.004–0.04 in) per million years for deep-sea nodules; roughly 1000 times slower than other pelagic sediments. Nodules grow layer upon layer, often around a hard skeletal piece such as a shark's tooth, rock fragment, or fish bone, which acts as a seed, much as a pearl grows around a grain of sand. They generally form in areas of very little sediment supply from other sources or where rapid bottom currents prevent them from being

deeply buried. Manganese nodules on continental margins are unique in their rapid growth, having growth rates on the order of 0.01–1 mm per year; from 1000 to 1 million times faster than their deep-sea counterparts. Manganese nodules have been mapped in all oceans except the Arctic. They are most abundant in the central Pacific north and south of the biogenous oozes along the equator (see fig. 3.20). In the Atlantic and Indian Oceans there are higher rates of lithogenous and biogenous sedimentation and consequently fewer deposits of manganese nodules.

Sediments derived from space are classified as **cosmogenous** (*cosmos* = universe, *generare* = to produce) **sediments.** Particles from space constantly bombard the Earth. Most of these particles burn up as they pass through the atmosphere, but roughly 10% of the material reaches the surface of the Earth. Cosmogenous particles are very small, and those that survive the passage through the atmosphere and fall in the ocean stay in suspension in the water long enough that they dissolve before they reach the sea floor. These iron-rich sediments are found in small amounts in all oceans, mixed in with the other sediments. The pattern of related cosmic materials can indicate the direction of the particle shower that supplied them. The particles become very hot as they pass through the Earth's atmosphere and partially melt; this melting gives the particles a characteristic rounded or teardrop shape. Cosmic bodies can disintegrate and melt surface materials as they strike the Earth. Their impact can cause a splash of melted particles that spray outward and produce splash-form **tektites** (fig. 3.22). Microtektites are found on the ocean floor and on land.

A brief summary of the major sediment types is given in table 3.5.

Patterns of Deposit on the Sea Floor

The patterns formed by the sediments on the sea floor reflect both distance from their source and processes that control the rates at which they are produced, transported, and deposited. Seventy-five percent of marine sediments are terrigenous. The majority of these terrigenous sediments are initially deposited on the continental margins but are moved seaward by the waves, currents, and turbidity flows that

Table 3.5 Sediment Summary

Type	Source	Areas of Significant Deposit	Examples
Lithogenous (terrigenous)	Eroded rock, volcanoes, airborne dust	Dominantly neritic, pelagic in areas of low productivity	Coarse beach and shelf deposits, turbidites, red clay
Biogenous	Living organisms	Regions of high surface productivity, areas of upwelling, dominantly pelagic, some beaches, shallow warm water	Calcareous ooze (above the CCD), siliceous ooze (below the CCD), coral
Hydrogenous	Chemical precipitation from seawater	Mid-ocean ridges, areas starved of other sediment types, neritic and pelagic	Metal sulfides, manganese nodules, phosphates, some carbonates
Cosmogenous	Space	Everywhere but in very low concentration	Meteorites, space dust

From Fundamentals of Oceanography, *4th edition*, Duxbury, Duxbury, and Sverdrup. Copyright 2000 The McGraw-Hill Companies. All rights reserved.

move across the continental shelves and down the continental slopes. The terrigenous sediments of coastal regions are primarily lithogenous, supplied by rivers and wave erosion along the coasts. Worldwide river sediment transport is about $12-15 \times 10^9$ metric tons per year. The majority of this transport enters the tropical and subtropical oceans.

Coarse sediments are concentrated close to their sources in high-energy environments, for example, beaches with swift currents and breaking waves. The waves and currents move quite large rock particles in the shore zone, but these larger particles settle out quickly. Finer particles are held in suspension and are carried farther away from their source. This pattern results in a gradation by particle size: coarse particles close to shore and to their source, with finer and finer particles predominating as the distance from the source increases.

Finer sediments are deposited in low-energy environments, offshore away from the currents and waves or in quiet bays and estuaries. In higher latitudes, deposits of rock and gravel carried along by glaciers are found in coastal environments, whereas in low latitudes fine sediments predominate and are considered to be products of large rivers, heavy rainfall, and loose surface soils.

At the present time, most of the land-derived sediments are accumulating off the world's river mouths and in estuaries. Estuaries and river deltas serve as sediment traps, preventing these terrigenous sediments from reaching the deep-sea floor in such places as the Chesapeake and Delaware Bay systems along the North Atlantic coast, in the Georgia Strait of British Columbia, and in California's San Francisco Bay along the North Pacific coast. If sediments are supplied to a delta faster than they can be retained, the sediments will move across the shelf into the deeper water environments. This is currently the case with the sediments of the Mississippi River. Much of the total thickness of sediments on the outer continental shelf is sediment laid down during the ice ages when the sea level was lower, for example, at Georges Bank southeast of Cape Cod. Little is currently being added to these outer regions of the continental shelf.

The accumulation of sediments on the passive shelves of continents results in unstable, steep-sided deposits that may slump, sending a flow of terrigenous sediment moving rapidly down the continental slope in a turbidity current (see section 3.2). Turbidity currents move coarse materials of land origin seaward; in doing so they distort the general deep-sea sediment pattern and reduce the abundance of pelagic deposits. Near shore, the spring flooding of the rivers alternates with periods of low river discharge in summer and fall. The floods bring large quantities of sediment to the coastal waters, and a record of the contribution of this flooding is seen in the layering of the sediments. Sudden contributions of sediment material from the collapse of a cliff or the eruption of a volcano are seen in the sediment pattern as specific additions of large quantities of sand or ash.

Oceanic sediments form visually distinct layers characterized by color, particle size, type of particle, and supply rate (fig. 3.23). Seasonal variations and the patterns of long and short growing seasons among marine life can also be determined from the properties and thicknesses of the layers of biogenous material. Over long periods of geologic time, climatic changes such as the ice ages have altered the biological populations that produce sediment and have left their record in the sediment layers.

In shallow coastal areas, cycles of climate change cause variation in rates of sediment production. Along passive continental margins, biogenous sediments may also be diluted by large amounts of lithogenous sediment washing from the land. In coastal areas where marine life is very abundant and river deposits are reduced, biogenous sediments are formed from both shell fragments and broken corals. In the more homogeneous environment of the deep sea, biogenous sediments make up the majority of the pelagic deposits. There is less dilution with terrigenous materials, and few environmental changes disturb the deep bottom deposits, allowing them to remain relatively unchanged for long periods of time. Calcareous oozes are found where the production of organisms is high, dilution by other sediments is small, and depths are less than 4000 m (13,000 ft). (See the areas including the mid-ocean ridges and the

Figure 3.23 A deep-sea sediment core obtained by the drilling ship *Glomar Challenger*. Note the layering of the sediments.

kelp, which grow attached to rocks in coastal areas, are dislodged by storm waves. The kelp may have enough buoyancy to float away, carrying with it the rock to which it is attached. When the plant dies or sinks, the rock is deposited on the ocean floor at some distance from its origin. The deposit of larger rocks by this process of rafting is infrequent and irregular.

The wind is an effective agent for the movement of lithogenous materials out to sea in some parts of the world. Winds blowing offshore from the Sahara Desert or other arid regions transfer sand particles directly from land to sea, sometimes 1000 km (600 mi) or more offshore. A similar process can occur between sand dunes and coastal waters. In the open ocean, the airborne dust probably supplies much of the deep-sea red clay material. Figure 3.24 indicates the frequency with which winds carry airborne dust, or haze, out to sea. The world's volcanoes are another source of airborne particles. Volcanic ash is present in seafloor sediments and can be found in layers of significant thickness associated with past and present volcanic events. The annual supply of airborne particles to the sediments is estimated at 100×10^6 metric tons.

Formation of Rock

Loose sediments on the sea floor are transformed into **sedimentary rock** by a series of relatively low-temperature (below 200°C) processes known as **lithification.** As one layer of sediment covers another, the weight of the sediments puts pressure on the lower sediment layers, and the sediment particles are squeezed more and more tightly together. The particles begin to stick to each other, and the seawater between the sediment particles with its dissolved minerals moves through the sediments. As it does so, it begins to precipitate dissolved minerals on the surfaces of the particles, and, in time, the minerals act to glue the sediment particles together into a mass of sedimentary rock. The sediments in these processes are also exposed to increasing temperature with increasing depth of burial. Chemicals dissolve because of combined temperature and pressure increases, and these chemicals may react with minerals in the sediment to form other minerals.

Sedimentary rock may preserve the layering of the sediments in visually distinct folds or strata. Ripple marks from the motion of waves and currents across the sediments may be seen, and fossils may also be present. Sedimentary rocks are found beneath the sediments of the deep-sea floor, along the passive margins of continents, and on land where they have been thrust upward along active margins or formed in ancient inland seas. Sedimentary rocks include sandstone, shale, and limestone.

If sediments are subjected to greater changes in temperature, pressure, and chemistry, **metamorphic rock** results. Slate is a metamorphic rock derived from shale, and marble is recrystallized metamorphosed limestone.

Sampling Methods

To analyze sediments, the geological oceanographer must have an actual bottom sample to examine. A variety of de-

warmer shallower areas of the South Pacific in fig. 3.20). Siliceous oozes cover the deep-sea floor beneath the colder surface waters of 50°–60°N and S latitude and in equatorial regions where cold deeper water is brought to the surface by vertical circulation processes. Deep basin areas of the Pacific have extensive deposits of red clay (again, see fig. 3.20).

Large rock particles of land origin are also moved out to sea by a process known as **rafting.** Glaciers carry sand, gravel, and rocks with them as they cut through the Earth's crust. When the glacier reaches the sea, parts of it break off and fall into the water as icebergs. The icebergs are carried away from land by the currents and winds, taking with them terrigenous materials far from their original sources. As the ice melts, rocks and gravel that were frozen in the ice sink to the sea floor. In addition, sea ice formed in shallow water against the shore can incorporate material from the sea floor and transport it out to sea. Figure 3.20 indicates areas of terrigenous deposits that are affected by ice rafting. It is estimated that ice-rafted material can be found over about 20% of the sea floor. Sometimes large, brown seaweeds known as

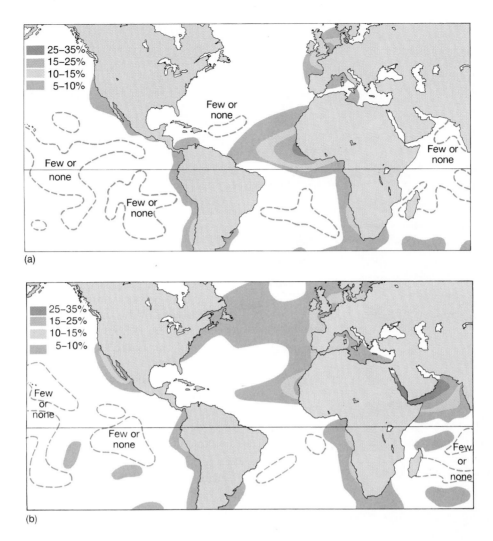

Figure 3.24 Frequency of haze as a result of airborne dust during the Northern Hemisphere's (a) winter and (b) summer. Values are given in percentages of total observations.

vices have been developed to take a sample from the sea floor and return it to the laboratory for analysis. **Dredges** are net or wire baskets that are dragged across a bottom to collect loose bulk material, surface rocks, and shells in a somewhat haphazard manner (fig. 3.25). **Grab samplers** are hinged devices that are spring- or weight-loaded to snap shut when the sampler strikes the bottom. See figure 3.26 for examples of this type of device. Grab samplers sample surface sediments from a fixed area of the sea floor at a single known location.

A **corer** is essentially a hollow pipe with a sharp cutting end. The free-falling pipe is forced down into the sediments by its weight or, for longer cores, by a piston device that enables water pressure to help drive the core barrel into the sediment; coring devices are shown in figure 3.27a–e. The result is a cylinder of mud, usually 1–20 m (3.3–65.6 ft) long, that contains undisturbed sediment layers (see fig. 3.23). Box corers (fig. 3.27e) are used when a large and nearly undisturbed sample of surface sediment is needed. These corers drive a rectangular metal box into the sediment; they have doors that close over the bottom before the sample is retrieved. Long cores required to penetrate the thick sediment overly-

ing older sea floor and to reach the older sediment layers nearer the oceanic basalt may be obtained by drilling through both loose sediments and rock. For a discussion of the highly sophisticated drilling techniques used by the research vessel *JOIDES Resolution*, refer to chapter 2.

Geological oceanographers and geophysicists also study sediment distribution and seafloor structure with high-intensity sound, a technique known as **acoustic profiling.** Bursts of sound are directed toward the sea floor, where the sound waves either reflect from or penetrate into the sediments. Sound waves that penetrate the sediments are refracted and change speed as they pass through the different layers of sediments. A surface vessel tows an array of underwater microphones, or hydrophones, to sense the returning sound waves, and a recorder plots the returning sound energy to produce a profile of the sediment structure. This technique details the structure of the continental margin, finds buried faults and filled submarine canyons, and searches for clues to oil and gas deposits. Acoustic profiling is very similar to the techniques used to study the interior of the Earth (see chapter 2, section 2.1).

(a)

(b)

Figure 3.25 (a) Rocks can be recovered from the sea floor with a dredge having a chain basket. Sediments and other fine material escape through the chains. (b) Basalt dredged from a depth of about 8 km (5 mi) near the Tonga Trench in the western Pacific Ocean. The dredge is in the *foreground*.

Figure 3.26 Grab samplers: Van Veen (*left*) and orange peel (*right*), both in open positions. Grabs take surface sediment samples.

Today's ocean scientists are searching for records of the Earth's history in the sediment and rock layers of the ocean floor. These layers hold evidence for understanding the formation of the Earth's ocean basins and continents, its changing climate, its periods of unusual volcanism, the presence and absence of various life forms, and much more. The information is there, but it requires a combination of sophisticated technical know-how at sea combined with increasingly detailed scientific research in the laboratory to first discern and then understand it.

Sediments as Historical Records

Marine sediments and the skeletal materials contained in them provide important information about processes that have shaped the planet, and its ocean basins specifically, over the past 200 million years. The study of the ocean's through an analysis of its sediments is called **paleoceanography.** Two examples of the use of marine sediments to unravel history are (1) the study of the distribution of skeletal remains of marine organisms to time the initiation of the Antarctic Circumpolar Current (ACC) and (2) the study of the relative abundance of different oxygen isotopes in skeletal parts preserved in the sediment to determine variations in climate and seawater temperature over time.

Prevailing westerly winds at high southern latitudes cause the ACC to flow continuously from west to east around Antarctica. The interaction of prevailing winds and ocean currents is discussed in detail in chapter 6, and the general pattern of surface currents is discussed in chapter 8. The ACC is a very deep current, extending to depths of 3000–4000 m (9800–13,000 ft), and it is able to flow unimpeded all the way around the globe because there are no shallow seafloor features to block its path. This situation has not always existed, however. The southern continents began to break apart significantly at different times about 135 million years ago as Africa and India first began to separate from Antarctica, South America, and Australia (see fig. 2.35). As recently as 80 million years ago, South America, Antarctica, and Australia were still effectively one landmass. Sometime around 55 million years ago, sea floor

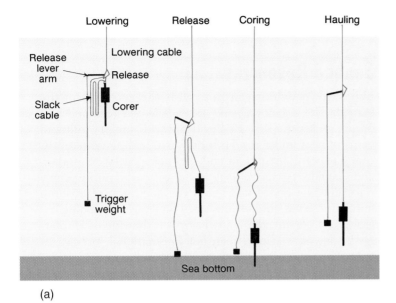

(a)

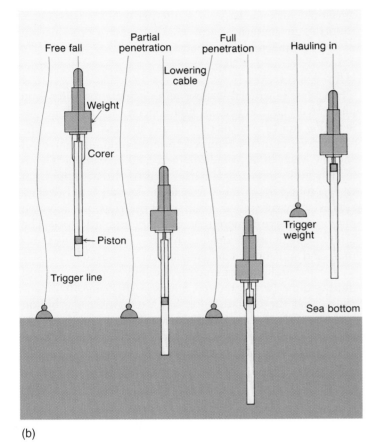

(b)

(c)

(d)

(e)

Figure 3.27 (a) The Phleger corer is a free-fall gravity corer. The weights help to drive the core barrel into the soft sediments. Inside the corer is a plastic liner. The sediment core is removed from the corer by removing the plastic tube, which is capped to form a storage container for the core. (b) A sketch of a piston corer in operation. The corer is allowed to fall freely to the sea bottom. The action of the piston moving up the core barrel owing to the tension on the cable then allows water pressure to force the core barrel into the sediments. (c) Loading a piston corer with weights to prepare it for use. (d) A gravity corer ready to be lowered. (e) A box corer is used to obtain large, undisturbed seafloor surface samples.

existed between Australia and Antarctica, and by 35 million years ago they had separated sufficiently to create a narrow expanse of water called the Austral Gulf. South America had not yet separated from Antarctica. Marine sediments deposited at this time indicate that a small, single-celled, shallow-water marine organism called *Guembelitria* lived in the restricted waters of the Austral Gulf. The absence of its skeletal remains in all other Southern Hemisphere sediments of the same age indicates that it had not been spread to other areas by ocean currents. Skeletal remains of *Guembelitria* appear quite suddenly in sediments deposited all around Antarctica about 30 million years ago. Even though the Drake Passage between South America and Antarctica did not fully open before 20 million years ago, there must have been a shallow channel a few hundred meters deep as early as 30 million years ago that allowed the ACC to first flow around the continent, carrying *Guembelitria* with it.

The chemistry of skeletal parts found in successive layers of sediment can provide information about changes in climate and seawater temperature over time through a careful analysis of the relative abundance of different oxygen isotopes in these remains. **Isotopes** are atoms of elements that have different numbers of neutrons in the nucleus; thus, they have different atomic masses but behave identically chemically. Some marine organisms remove oxygen from water molecules in the ocean to construct calcareous hard parts (see the discussion of foraminiferans in chapter 15, for example). Seawater contains two main isotopes of oxygen: the common ^{16}O and the rarer ^{18}O. These isotopes are stable and do not decay radioactively, so once they have been incorporated into an organism's skeletal material their relative proportion ($^{18}O:^{16}O$) remains constant even after the organisms dies. The $^{18}O:^{16}O$ ratio in a skeletal fragment depends in part on the relative abundance of the isotopes in the seawater at the time the organism formed it. Thus, calcareous biogenous remains record changes in the isotopic chemistry of seawater that are related to changes in global temperature.

Water molecules containing ^{16}O are lighter than molecules containing ^{18}O, so they are more easily removed from the oceans by evaporation. During glacial periods, the water evaporated from the sea surface is trapped in ice sheets; the sea level is lowered and ^{16}O is removed from the ocean system. This process increases the $^{18}O:^{16}O$ ratio in the seawater and in skeletal parts that organisms are forming at that time. When these organisms die, their skeletal parts sink to the sea floor and are incorporated into the sediment. During warmer, interglacial periods the melting of ice sheets causes a rise in sea level and returns ^{16}O-enriched fresh water to the oceans. The result is a drop in the $^{18}O:^{16}O$ ratio in the seawater and in the skeletal parts that are being formed. The isotopic composition of skeletal parts is also influenced by seawater temperature. As temperature decreases, organisms preferentially take up more ^{18}O than ^{16}O in their skeletons. The actual $^{18}O:^{16}O$ ratio preserved in a skeletal fragment is primarily due to changes in seawater composition related to the growth and decay of global ice sheets and consequent fall and rise of sea level.

Seabed Resources 3.4

Long ago, people began to exploit the materials of the seabed. The ancient Greeks extended their lead and zinc mines under the sea, medieval Scottish miners followed veins of coal under the Firth of Forth, and, more recently, coal has been mined from undersea veins off Japan, Turkey, and Canada. As technology has developed, and as people have become concerned about the depletion of onshore mineral reserves, interest in seabed minerals and mining has grown. At present, the United States is showing little interest in new seabed resources, but international interest remains strong; there is continued research in exploration, technology development, and environmental studies, especially in Japan, India, China, and South Korea. Keep in mind that each potential deep-sea source is in competition with an onshore supply. Whether the seabed source will be developed depends largely on international markets, needs for strategic materials, and whether offshore production costs can compete with onshore costs.

Sand and Gravel

The largest superficial seafloor mining operation is for sand and gravel, used in cement and concrete for buildings, for landfills, and to construct roads and artificial beaches. The technology and cost required to mine sand and gravel in shallow water differ very little from land operations. This is a high-bulk, low-cost material tied to the economics of transport and the distance to market. Annual world production is approximately 1.2 billion metric tons; the reported potential reserve is more than 800 billion metric tons. The United Kingdom and Japan each take 20% of their total annual sand and gravel requirements from the sea floor.

Sand and gravel mining is the only significant seabed mining done by the United States at this time. It is estimated that the United States has a reserve of 450 billion tons of sand off its northeastern coast; there are large deposits of gravel along Georges Bank off New England and also in the area off New York City. Along the coasts of Louisiana, Texas, and Florida, shell deposits are mined for use in the lime and cement industries, as a source of calcium oxide used to remove magnesium from seawater, and, when crushed, as a gravel substitute along roads and highways.

Sands are mined as a source of calcium carbonate throughout the Bahamas, which have an estimated reserve of 100 billion metric tons. Coral sands are mined in Fiji, in Hawaii, and along the U.S. Gulf Coast. Other coastal sands contain iron, tin, uranium, platinum, gold, and diamonds. The "tin belt" stretches for 3000 km (1800 mi) from northern Thailand and western Malaysia to Indonesia. Here, sediments rich in tin have been dredged for hundreds of years and supply more than 1% of the world's market. Iron-rich sediments are dredged in Japan, where the reserve of iron in shallow coastal waters is estimated at 36 million tons. The

United States, Australia, and South Africa recover platinum from some sands, and gold is found in river delta sediments along Alaska, Oregon, Chile, South Africa, and Australia. Diamonds, like gold, are found in sediments washed down the rivers in some areas of Africa and Australia; mining of diamonds at depths of 300 m (1000 ft) began off southwestern Africa in 1994. Muds bearing copper, zinc, lead, and silver also occur on the continental slopes, but they lie too deep for exploitation, considering the present demand and their market value. At present, Japan is exploring for gold at 1400 m (4600 ft) off its shores.

Phosphorite

Phosphorite, which can be mined to produce the phosphates needed for fertilizers, is found in shallow waters as phosphorite muds and sands containing 12%–18% phosphate and as nodules on the continental shelf and slope. The nodules contain about 30% phosphate, and large deposits are known to exist off Florida, California, Mexico, Peru, Australia, Japan, and northwestern and southern Africa. Recently a substantial source of phosphorite was located in Onslow Bay, North Carolina. Eight beds have been located, and five are thought to be economically valuable; they have been estimated to contain 3 billion metric tons of phosphate concentrates.

The world's ocean reserve of phosphorite is estimated at about 50 billion tons. Readily available land reserves are not in short supply, but most of the world's land reserves are controlled by relatively few nations. Therefore, political considerations may make these marine deposits attractive as mining ventures for some countries. No commercial phosphorite mining occurs at present in the oceans.

Sulfur

Sulfur is necessary for the production of sulfuric acid, which is needed in nearly all industrial processes. Presently, the most economical way to acquire sulfur is to recover sulfur waste from pollution-control equipment. In the past sulfur has been mined in the Gulf of Mexico by injecting high-pressure steam into wells to melt the sulfur and then pumping it ashore to processing plants. If sulfur must be mined again, millions of tons of sulfur reserves are known to exist in the Gulf of Mexico and the Mediterranean Sea.

Coal

Coal is produced by the burial and alteration of large amounts of land plant material in swampy environments with low oxygen concentration. Plant material undergoes a series of changes as more volatiles and impurities are removed at higher temperatures and pressures. Initially, the partially altered plant material forms peat. Peat can then progress through a series of stages to form coal of increasing hardness, from lignite, to bituminous, and finally to anthracite coal. Large accumulations of coal occurred during periods of time when sea level was relatively low. Changes in sea level and land geography over geologic time have caused some coal deposits to be submerged. Coal deposits under the sea floor are mined when the coal is present in sufficient quantity and

quality to make the operation worthwhile. In Japan, the undersea coal deposits are reached by shafts that stretch under the sea from the land or descend from artificial islands.

Oil and Gas

Oil and gas represent more than 95% of the mineral value presently taken from the sea. Oil and gas deposits are almost always associated with marine sedimentary rocks and are believed to be produced by the slow conversion of marine plant and animal organic matter to hydrocarbons. Conditions must be just right for marine organic material to eventually be converted to oil and gas. It must first accumulate in relatively shallow, quiet water with low oxygen content. Anaerobic bacteria can then oxidize the organic matter to produce methane and other light hydrocarbons. As these simple hydrocarbons are buried beneath deeper layers of sediment, they are subjected to higher pressure and temperature. Over a period of millions of years they will be converted to either oil or gas. Oil forms if the depth of burial is on the order of about 2 km (1.2 mi). If the organic material is buried even deeper, or cooked for a longer period of time at higher temperature, gas is produced. Oil deposits are generally found at depths less than 3 km (1.9 mi), and below 7 km (4.3 mi) only gas is found.

Because oil and gas are very light, they migrate upward over time, moving slowly out of the source rock they formed in and into porous rocks above. This upward migration continues until it reaches an impermeable layer of rock. The oil and gas then stop their ascent and fill the pore spaces of the reservoir rock below this impermeable layer.

Petroleum-rich marine sediments are more likely to accumulate during periods of geologic time when sea level is unusually high and the oceans flood extensive low-lying continental regions to create large shallow basins. Most oil and gas reserves are found in marine rocks that formed from sediments deposited during a relatively short period of time during the Jurassic and Cretaceous, between about 85 and 180 million years ago, when sea level was high.

Data for 1997 show that 24.8% of U.S. oil and 20.7% of U.S. gas production came from offshore areas. Major offshore oil fields are found in the Gulf of Mexico, the Persian Gulf, and the North Sea and off the northern coast of Australia, the southern coast of California, and the coasts of the Arctic Ocean. At the present time, many U.S. companies are finding it more profitable to drill for oil and gas in foreign waters and are moving their rigs to the waters of the North Sea, West Africa, and Brazil. The opening of trade with Vietnam and expanding markets in China are attracting U.S. companies to an estimated 850 million barrels of oil and 3.7 trillion cubic feet of natural gas in Vietnamese coastal areas.

Bringing the offshore oil fields into production has required the development of massive drilling platforms and specialized equipment to withstand heavy seas and fierce storms and to allow drilling and development of wellheads at great depth (fig. 3.28). Although the cost of drilling and equipping an offshore well is three to four times greater than

Figure 3.28 The oil-drilling platform Hibernia, located 320 km (200 mi) offshore Newfoundland, produced its first oil in November 1997.

that of a similar venture on land, the large size of the deposits allows offshore ventures to compete successfully. The gas and oil potential of the much greater depths of the ocean floor is still unknown, but the deeper the water in which the drilling must be done, the higher the cost. Exxon is studying a project that would require drilling in 1400 m (4600 ft), more than twice the depth of its current deepest well.

The new methods and equipment used and developed for deep-sea oceanographic drilling and research have provided the prototypes for new generations of deep-sea commercial drilling systems. Even though legal restraints, environmental concerns, and worldwide political uncertainties will continue to contribute to the slow development of offshore deposits, petroleum exploration and development will undoubtedly continue to be the main focus of ocean mining in the near future.

Gas Hydrates

In recent years there has been increasing interest in gas hydrates trapped in deep marine sediments. Gas hydrates are a combination of natural gas, primarily methane (CH_4), and water, which forms a solid icelike structure under pressure at low temperatures. Drill cores of marine sediment have recovered samples of gas hydrates that melt and bubble as the natural gas escapes. These melting samples burn if lit. Gas hydrates are a subject of intense interest for three reasons: they are a potential source of energy, they may contribute to slumping along continental margins, and they may play a role in climate change.

When 1 cubic foot of gas hydrate melts, it releases about 160 cubic feet of gas. A gas hydrate accumulation thus can contain a huge amount of natural gas. Scientists have mapped two relatively small areas, each the size of the State of Rhode Island, off the coasts of North and South Carolina that are thought to contain more than 1300 trillion (1.3×10^{15}) cubic feet of methane gas, an amount that is more than seventy times the 1989 gas consumption of the United States. In 1997 the U.S. Geological Survey estimated that U.S. gas hydrate reserves may be on the order of 200,000 trillion (2×10^{17}) cubic feet, an amount that is about 143 times the estimated reserves of conventional natural gas. Worldwide gas hydrate reserves are estimated at 400 million trillion (4×10^{20}) cubic feet, roughly 80,000 times the known conventional natural gas reserves. Japan, interested in reducing or eliminating its dependence on imported oil, began drilling in the Nankai Trough in 1999 east of its main island to see if hydrates could be harvested, part of a five-year, $60-million research program. India has also begun a five-year, $50-million effort. Many other countries are investigating the feasibility of mining this potentially important source of energy as well.

A second reason gas hydrates are significant is their effect on seafloor stability. Along the southeastern coast of the United States a number of submarine landslides, or slumps, have been identified that may be related to the presence of gas hydrates. The hydrates may inhibit normal sediment consolidation and cementation processes, creating a weak zone in the sediments. Alternately, the lowering of sea level during the last glacial period may have reduced the pressure on the sea floor enough to allow some of the gas to escape from the hydrates and accumulate in the sediment, decreasing its strength.

A final reason for studying gas hydrates is their potential link to climate changes. The amount of methane stored in hydrates is believed to be about 3000 times the amount currently present in the atmosphere. Since methane is a greenhouse gas, its release from hydrates could affect global climate.

Manganese Nodules

Manganese nodules are found scattered across the world's deep-ocean floors, with particular concentrations in the red clay regions of the northeastern Pacific (see figs. 3.19 and 3.20). The mineral content varies from place to place, but the nodules in some areas contain 30% manganese, 1% copper, 1.25% nickel, and 0.25% cobalt: these are much higher concentrations than are usually found in land ores. Cobalt is of particular interest since it is classified as being of "strategic" importance to the United States and hence essential to the national security. This is because it is an important component in the manufacture of strong alloys used in tools and aircraft engines. The nodules grow very slowly, but they are present in huge quantities. An estimated 16 million additional tons of nodules accumulate each year.

Since the 1960s, large multinational consortia have spent hundreds of millions of dollars to locate the highest nodule concentrations and to develop technologies for their collection. However, their expectations of rapid development have been disappointed. In the 1980s, some of these consortia had withdrawn completely and others were dormant. The primary reason for the lack of development of this industry is the presently depressed international market in metals. Another reason involves the history of ownership of the pelagic nodules (see the section in this chapter titled "Laws and Treaties").

Cobalt-enriched manganese crusts, or hard coatings on other rocks, were discovered in relatively shallow water on the slopes of seamounts and islands within U.S. territorial waters in the 1980s. The concentration of cobalt in these deposits is roughly twice that found in typical pelagic manganese nodules and about one and one-half times that found in known continental deposits. These crusts are not being actively mined because of the relatively low cost and continued availability of continental sources.

Progress in the 1990s continued to be slow and mostly outside the United States. An organization of twelve South Pacific Island nations (Cook Islands, Federated States of Micronesia, Fiji, Guam, Keribati, Marshall Islands, Papua New Guinea, Solomon Islands, Tonga, Tuvalu, Vanuatu, and Western New Guinea), with Australia and New Zealand as associate members, has supported over twenty-five deep-ocean survey cruises. The Cook Island government is accepting proposals for the mining of cobalt crusts in its coastal waters. These crusts have four to five times the cobalt content of manganese nodules. India has applied to the United Nations (UN) to develop manganese nodule deposits in the Indian Ocean and is working to advance mining systems and processing plant designs. Japan continues its research interests.

Sulfide Mineral Deposits

Expeditions to the rift valleys of the East Pacific Rise near the Gulf of California, the Galápagos Ridge off Ecuador, and the Juan de Fuca and Gorda Ridges off the northwestern United States have found deposits of minerals combined with sulfur to form sulfides of zinc, iron, copper, and possibly silver, molybdenum, lead, chromium, gold, and platinum. Molten material from beneath the Earth's crust rises along the rift valleys, fracturing and heating the rock. Seawater percolates into and through the fractured rock, forming mineral-rich hot solutions. When these solutions rise from the cracks and cool, the metallic sulfides precipitate to the sea floor. Deposits may be tens of meters thick and hundreds of meters long. Too little is presently known about these deposits for us to know whether they might be of economic importance at some future date. No practical technology ex-

ists to sample or retrieve them at this time, and, like the manganese nodules, these deposits are found outside national economic zones, so there are ownership problems (see "Laws and Treaties").

In the 1960s metallic sulfide muds were discovered in the Red Sea. Deposits of mud 100 m (330 ft) thick were found in small basins at depths of 1900–2200 m (6200–7200 ft). High amounts of iron, zinc, and copper and smaller amounts of silver and gold were found. The salty brines over these muds contained hundreds of times more of some metals than normal seawater contains.

Laws and Treaties

Because of the potential value of deep-sea minerals, specifically manganese nodules, and because the nodules are found in international waters, outside the usual 200-mile economic zones of coastal nations, the developing nations of the world feel they have as much claim to this wealth as those countries that are presently technically able to retrieve the nodules. The developing nations want access to the mining technology and a share in the profits. For nearly ten years, the United Nations' Law of the Sea Conferences worked to produce a treaty to regulate deep-ocean exploitation, including mining. The Law of the Sea Treaty was completed in April 1982. The treaty recognizes deep-sea mineral resources as the heritage of all humankind, to be regulated by a UN seabed authority that would license private companies to mine in tandem with a UN company. The quantities removed would be limited, and the profits would be shared. A UN cartel would both regulate and compete with those mining the sea.

The United States chose not to sign the treaty. In 1984 the United States, Belgium, France, West Germany, Italy, Japan, and the Netherlands signed a separate Provisional Understanding Regarding Deep Seabed Matters, and, under this provisional understanding, four international consortia have been awarded exploration licenses by the United States, West Germany, and the United Kingdom. By 1991 forty-five countries had ratified the Law of the Sea Treaty. Tempers cooled as the years passed, and the UN, realizing that it needs U.S. support to ensure international cooperation for the exploration and exploitation of ocean resources, included the United States in the working groups that met to resolve mining conflicts.

The UN Convention on the Law of the Sea, UNCLOS, came into force in 1994 without U.S. endorsement, but the ongoing negotiations to resolve the issues related to the section governing seabed mining have produced some results. The same year an agreement that changes the provisions covering deep-seabed mining beyond the 200-mile Exclusive Economic Zone (EEZ) of individual nations was completed by the UN. The United States announced that it intends to sign the agreement and to begin the process of submitting both the new agreement and UNCLOS to the U.S. Senate for advice and consent. The process of gaining U.S. Senate approval and ratification will probably be long and arduous. It will raise many questions over existing U.S. laws and policies that may not be consistent with the provisions of UNCLOS. The political climate may change several times between the submission of the treaty and the final votes on its ratification. In any case, it is unlikely that any deep-sea mining will occur until well into the twenty-first century. The high costs of sea mining, low metal prices, and still-undeveloped land sources combine to make rapid commercialization unlikely under any regulatory system.

Because there may be rich deposits of minerals and oil in Antarctica and its surrounding seas and because both national and private corporations are interested in surveying these areas for possible future mining and drilling, a multinational meeting in June 1988 produced an agreement to regulate mining exploration in and around that continent. This treaty must be ratified by sixteen of the twenty nations that signed the 1959 Antarctica Treaty, which banned all military activity and permitted scientific research. According to the 1988 treaty, no activities will be permitted if they will cause "significant changes" in atmospheric, terrestrial, or marine environments. Because no one really knows for sure how much mineral and oil wealth may be under Antarctica's ice and snow, the extent of future operations is also unknown.

In the United States, there is no offshore mining except for the sand and gravel operations in the state-owned waters of a few coastal states. Mining for hard minerals, which include manganese nodules, cobalt crusts, metallic sulfides, and phosphorites, would be carried out within U.S. waters under provisions of U.S. domestic laws. Under these laws the Department of the Interior is authorized to issue leases for mineral exploration and development on the continental shelf. No leasing or regulatory program has yet been developed, but federal and state task forces are currently working to develop programs for cobalt-rich crusts in the Hawaiian Islands, metallic sulfides on the Gorda Ridge off the coast of Oregon and northern California, and phosphorites off North Carolina.

Summary

Ocean-depth measurements were made first with a hand line, then with wire, and since the 1920s with echo sounders. Today they are made with precision depth recorders. Seafloor features can also be sensed by satellites that measure the distance between the satellite and the sea surface.

The bathymetric features of the ocean floor are as rugged as the topographic features of the land but erode more slowly. The continental margin includes the continental shelf, slope, and rise. The continental shelf break is located at the change in steepness between the continental shelf and the continental slope. Submarine canyons are major features of the continental slope and, in some cases, the continental shelf. Some canyons are associated with rivers; others are believed to have been cut by turbidity currents. Turbidity currents deposit graded sediments known as turbidites.

The ocean basin floor is a flat abyssal plain, but it is interrupted by scattered abyssal hills, volcanic seamounts, and flat-topped guyots. In warm shallow water, corals have grown up around the seamounts to form fringing reefs. A barrier reef is formed when a seamount subsides while the coral grows. An atoll results when the seamount's peak is fully submerged. The mid-ocean ridges and rises extend through all the oceans; trenches are associated with island arcs and are found mainly in the Pacific Ocean.

Sediment classifications are based on their size, location, origin, and chemistry. Sediment particles are broadly categorized in order of decreasing size as gravel, sand, and mud. Within each of these categories particles can be further subdivided by size. The sinking rate and distance traveled in the water column are related to sediment size. Small particles sink more slowly than large particles. The very smallest particle sizes, silts and clays, sink so slowly that they may be transported large distances while falling to the sea floor. The sinking rate of particles is increased by clumping and incorporation into fecal pellets of small organisms.

Sediments that accumulate on continental margins and the slopes of islands are called neritic sediments. Sediments of the deep-sea floor are pelagic sediments. In general, pelagic sediments accumulate very slowly and neritic sediments accumulate more rapidly.

Sediments formed from preexisting rocks are called lithogenous sediments. These sediments are also sometimes referred to by the more general term *terrigenous sediments*. Since lithogenous sediments are typically derived from the land, they are also known as terrigenous sediments. Pelagic lithogenous sediment is dominated by red clay. Red clay dominates marine sediments only in regions that are starved of other sources of sediment. Biogenous sediments come from living organisms. Sediments composed of at least 30% biogenous material are called oozes; this material accumulates in regions of high biological productivity. Siliceous sediments are subjected to dissolution everywhere in the oceans while calcareous sediments dissolve rapidly in deep, cold water below the CCD. Sediments that precipitate directly from the water are called hydrogenous sediments. These include manganese nodules on the deep-sea floor and metal sulfides along mid-ocean ridges. Sediments that originate in space are called cosmogenous sediments.

Patterns of sediment deposit result from the distance from their source, the abundance of living forms contributing their remains, the seasonal variations in river flow, the waves and currents including turbidity currents, the variability in land sources, the prevailing winds, and sometimes rafting.

Coarse sediments are concentrated close to shore; finer sediments are found in quiet offshore or nearshore environments. Terrigenous sediments are found mainly along coastal margins; most deep-sea sediments come from biogenous sources. Mixtures of particle sizes reveal the processes that formed the deposit. Sediment layers provide clues to ancient climate patterns.

In general, sedimentation rates are slowest in the deep sea and greatest near the continents. Relict sediments were deposited under conditions that no longer exist. Mechanisms that increase the sinking rates of particles include clumping and incorporation of sediment particles into the larger fecal pellets of small marine organisms. Loose sediments are transformed into sedimentary rock, which may preserve the layering of the sediments.

Sediments are sampled with dredges, grabs, and corers; deep-sea drilling takes samples through the sediments into the seafloor rock below.

Calcareous biogenous sediments preserve records of changes in the oxygen isotopic composition of seawater that are related directly to water temperature, and hence can be used to study changes in global climate. Consequently, variations in $^{18}O:^{16}O$ isotopic ratios in calcareous skeletal remains record fluctuations in global coverage by ice sheets and in sea level.

Seabed resources include sand and gravel used in construction and landfills. Sands and muds that are rich in mineral ores are mined. Phosphorite nodules have potential as fertilizer. Oil and gas are the most valuable of all seabed resources. Manganese nodules are rich in copper, nickel, and cobalt; they are present on the ocean floor in huge numbers. Retrieval of seafloor mineral resources is slowed by disputes over international law, high mining costs, and low market prices. Sulfide mineral deposits have been discovered along rift valleys; their economic importance is unknown.

Large deposits of gas hydrates are now being studied to determine their potential as economically important sources of methane gas. These deposits are icelike accumulations of natural gas and water that form at low temperature and high pressure on the sea floor. Scientists are also studying their possible role in the occurrence of submarine landslides and what effect they may have on global climate.

Key Terms

All key terms from this chapter can be viewed by term, or by definition, when studied as flashcards on this book's Online Learning Center at www.mhhe.com/sverdrup (click on this book's cover).

soundings, 99
fathom, 99
echo sounder, 99
depth recorder, 99
continental margin, 101
continental shelf, 101
continental shelf break, 103
continental slope, 103
submarine canyon, 105
turbidity current, 105
turbidite, 105
continental rise, 106
abyssal plain, 106
abyssal hill, 107
seamount, 107
guyot, 108
fringing reef, 109

barrier reef, 109
atoll, 110
island arc system, 110
test, 112
neritic, 113
pelagic, 113
relict sediment, 113
lithogenous sediment, 114
terrigenous sediment, 114
abyssal clay, 114
red clay, 114
biogenous sediment, 116
ooze, 116
calcareous ooze, 116
siliceous ooze, 116
coccolithophore, 116
coccolith, 116

pteropod, 116
foraminiferan, 116
lysocline, 116
carbonate compensation
  depth, 116
nutrient, 117
hydrogenous sediment, 117
carbonate, 117
phosphorite, 117
salt, 117
manganese nodule, 117
oolith, 117

cosmogenous sediment, 118
tektite, 118
rafting, 120
sedimentary rock, 120
lithification, 120
metamorphic rock, 120
dredge, 121
grab sampler, 121
corer, 121
acoustic profiling, 121
paleoceanography, 122
isotope, 124

Study Questions

1. Are calcareous oozes more common in the southern Pacific Ocean or in the northern Pacific Ocean? Why?
2. What is a turbidity current? Where would you expect a turbidity current to occur? How does the structure of a sediment deposit left by a turbidity current differ from the structure of other sediment deposits?
3. List the four basic sediment types classified by source. Where is each sediment type most likely to be found?
4. Discuss the future of commercial development and exploitation of deep-sea mineral resources.
5. Imagine that you are in a submersible on the ocean bottom. You leave New York and travel across the North Atlantic to Spain. Draw a simple ocean-bottom profile showing each major bathymetric feature you see as you move across the ocean. Name each feature. Do the same for the South Pacific between the coast of Chile and the west coast of Australia. Compare the two profiles. Did your depth scale differ from your horizontal scale? How much?
6. What processes form submarine canyons?
7. What is the continental margin? What pattern of sediment deposit would you expect to find associated with it? What processes produce these patterns of deposit?
8. What combination of factors is required to form a coral atoll?
9. What is a relict sediment? Where would you be likely to find such a deposit, and why would you find it in that place?
10. Describe several ways in which a continental shelf may be formed.
11. Describe methods used to recover sediment samples from the sea floor. Discuss the advantages and disadvantages of each method.
12. What is the average depth of the oceans in meters? In miles? In fathoms?
13. How is particle size used in understanding the pattern of ocean-floor deposits?
14. What are the implications for the marine environment as exploitation of seabed resources continues? Consider mining and drilling on continental shelves, in Antarctic waters, and in the open ocean.

15. The Grand Banks is an extensive, relatively shallow area southeast of Newfoundland, Canada, and off the U.S. northeastern coast. Why are there large boulders scattered across this area so far from shore?

Particle Type	Particle Diameter (mm)	Settling Rate (V) (cm/s)
Very fine sand	0.1	6.6×10^{-1}
Silt	0.06	2.4×10^{-1}
Clay	0.004	1.05×10^{-3}

Study Problems

1. If underwater cables are spaced 14 km apart on the sea floor, and if monitoring equipment shows that they break in sequence from shallow to deeper water at fifteen-minute intervals, what can you determine about the event causing the breaks?

2. If the average concentration of suspended sediment in the water is 1 g/m^3 and the volume of water in a harbor is 158 km^3, what is the average residence time of sediment in the water of this harbor? The daily sediment supply rate averages 1×10^7 kg.

3. In how many days will each of the particles listed in the following table reach the sea floor if the particles fall through 4000 m of seawater? All the particles are derived from land rock of the same density (2.8 g/cm^3). Settling rate is calculated from Stokes Law:

$$V \text{ cm/s} = 2.62 \times 10^4 r^2$$

where r is the radius expressed in centimeters. (2.62×10^4 contains gravity, viscosity of water, the difference between the density of the particle and the density of water, and a constant for particle shape.) How does the settling rate change if the diameter remains constant but the density of a particle changes?

4. Assuming a constant sedimentation rate of 0.4 cm per 1000 years, how thick will the sediments be in a portion of an ocean basin where the underlying crust is 130 million years old?

Links to Related Websites

Visit the book's Online Learning Center at www.mhhe.com/sverdrup (click on the book's cover) to find live Internet links for additional topics related to this chapter's content.

- Sea floor and sediments
- Gas hydrates
- Law of the sea

Visit the book's Online Learning Center at www.mhhe.com/sverdrup (click on this book's cover) to find these additional chapter tools: Suggested Readings; links to further information on boxed readings, selected figures, and related chapter topics; and additional study aids.

chapter 4

The Physical Properties of Water

Heceta Head Light, Oregon Coast.

A sudden fog-drift muffled the ocean,
  A throbbing of engines moved in it,
At length, a stone's throw out, between the rocks and the vapor,
One by one moved shadows
Out of the mystery, shadows, fishing-boats, trailing each other
Following the cliff for guidance,
Holding a difficult path between the peril of the sea-fog,
And the foam on the shore granite.
One by one, trailing their leader, six crept by me,
Out of the vapor and into it.
The throb of their engines subdued by the fog, patient and cautious,
Coasting all around the peninsula
Back to the buoys in Monterey harbor. A flight of pelicans
Is nothing lovelier to look at;
The flight of the planets is nothing nobler; all the arts lose virtue
Against the essential reality
Of creatures going about their business among the equally
Earnest elements of nature.

Robinson Jeffers
From *Boats in a Fog*

W ater is one of the commonest substances on our Earth, yet it is uncommon in many of its properties. Water is a unique liquid. It makes life possible, and its properties largely determine the characteristics of the oceans, the atmosphere, and the land. To understand the oceans, one must examine water as a substance and learn something of its physical and chemical characteristics. In this chapter we learn about the structure of the water molecule and explore the properties of water. We also review three interesting and potentially hazardous forms of water: sea ice, icebergs, and fog.

The Water Molecule   4.1

The properties of water have excited scientists for over 2000 years. The early Greek philosophers (500 B.C.) counted four basic elements from which they believed all else was made: fire, Earth, air, and water. In 1783, more than 2200 years later, the English scientist Henry Cavendish determined that water was not a simple element but a substance made up of hydrogen and oxygen. Shortly afterward another Englishman, Sir Humphrey Davey, discovered that the correct formula for water was two parts hydrogen to one part oxygen, or H_2O.

The chemical properties that make water such a special, useful, and essential substance result from its molecular structure. These properties are the subject of this chapter and are summarized in table 4.1.

The water molecule is deceptively simple, made up of three atoms: two hydrogen atoms and one oxygen atom. An atom is the smallest unit of matter that retains the properties of an element. We can envision an atom as consisting of three distinct types of particles located in two different regions. At the center of an atom is the nucleus. The nucleus contains positively charged particles called protons tightly packed with electrically neutral particles called neutrons. Each individual element has a characteristic number of protons in the nucleus of every atom of that element. For instance, every atom with one proton in the nucleus is an atom of hydrogen, and every atom with eight protons in the nucleus is an atom of oxygen. Negatively charged particles called electrons orbit the nucleus in a series of energy levels. The different energy levels can hold different numbers of electrons. Atoms are generally electrically neutral; they have the same number of electrons as protons. In the case of hydrogen, there is one electron in the first energy level; an energy level that can hold a maximum of two electrons. In the case of oxygen, there are two electrons that fill the first energy level and six additional electrons in the second energy level; a level that can hold a total of eight electrons when it is full. Thus a hydrogen atom's outermost energy level is one electron short of being full and an oxygen atom's outermost energy level is two electrons short of being full. When two hydrogen atoms and one oxygen atom combine to form a water molecule, each hydrogen atom shares its single electron with the oxygen atom and the oxygen atom shares one of its electrons with each hydrogen atom. Shared pairs of electrons form **covalent bonds.** The formation of covalent bonds in the water molecule has the effect of filling the outer energy levels of all three atoms in the molecule (fig. 4.1a).

The angle between the hydrogen atoms in the water molecule is about 105° (fig. 4.1b). A molecule of water is electrically neutral, but the negatively charged electrons within the molecule are distributed unequally. The shared electrons of the covalent bonds spend more time around the oxygen nucleus than they do around the hydrogen nuclei, giving the oxygen end of the molecule a slightly negative charge. The hydrogen end of the molecule carries a slightly positive charge. As a consequence, the opposite ends of the water molecule have opposite charges, and the molecule is an electrically unbalanced, or **polar, molecule.**

When one end of a water molecule comes close to the oppositely charged end of another water molecule, a bond forms between their positively and negatively charged ends (fig. 4.1c). These bonds are known as **hydrogen bonds,** and each water molecule can establish hydrogen bonds with four other water molecules. Any single hydrogen bond is weak (less than one-tenth the strength of the

Table 4.1 Properties of Water

Definition	Comparison	Effects
Physical States Gas, liquid, solid Addition or loss of heat breaks or forms bonds between molecules to change from one state to another.	The only substance that occurs naturally in three states on the Earth's surface.	Important for the hydrologic cycle and the transfer of heat between the oceans and atmosphere.
Heat Capacity One calorie per gram of water per °C.	Highest of all common solids and liquids.	Prevents large variations of surface temperature in the oceans and atmosphere.
Surface Tension Elastic property of water surface.	Highest of all common liquids.	Important in cell physiology, water surface processes, and drop formation.
Latent Heat of Fusion Heat required to change a unit mass from a solid to a liquid without changing temperature.	Highest of all common liquids and most solids.	Results in the release of heat during freezing and the absorption of heat during melting. Moderates temperature of polar seas.
Latent Heat of Vaporization Heat required to change a unit mass from a liquid to a gas without changing temperature.	Highest of all common substances.	Results in the release of heat during condensation and the absorption of heat during vaporization. Important in controlling sea surface temperature and the transfer of heat to air.
Compressibility Average pressure on total ocean volume 200 atmospheres; ocean depth decreased by 37 m (121 ft).	Seawater is only slightly compressible, 4–4.6×10^{-5} cm^3/g for an increase of 1 atmosphere of pressure.	Density changes only slightly with pressure. Sinking water can warm slightly due to its compressibility.
Density Mass per unit volume: grams per cubic centimeter, g/cm^3.	Density of seawater is controlled by temperature, salinity, and pressure.	Controls the ocean's vertical circulation and layering. Affects ocean temperature distribution.
Viscosity Liquid property that resists flow. Internal friction of a fluid.	Decreases with increasing temperature. Salt and pressure have little effect. Water has a low viscosity.	Some motions of water are considered friction free. Low friction dampens motion; retards sinking rate of single-celled organisms.
Dissolving Ability Dissolves solids, gases, and liquids.	Dissolves more substances than any other solvent.	Determines the physical and chemical properties of seawater and the biological processes of life forms.
Heat Transmission Heat energy transmitted by conduction, convection, and radiation.	Molecular conduction slow; convection effective. Transparency to light allows radiant energy to penetrate seawater.	Affects density; related to vertical circulation and layering.
Light Transparency Transmits light energy.	Relatively transparent for visible wavelength light.	Allows plant life of grow in the upper layer of the sea.
Sound Transmission Transmits sound waves.	Transmits sound very well compared to other fluids and gases.	Used to determine water depth and to locate targets.
Refraction The bending of light and sound waves by density changes that affect the speed of light and sound.	Refraction increases with increasing salt content and decreases with increasing temperature.	Makes objects appear displaced when viewed by light and sound.

From Fundamentals of Oceanography, 4th edition, Duxbury, Duxbury, and Sverdrup. Copyright 2000 The McGraw-Hill Companies. All rights reserved.

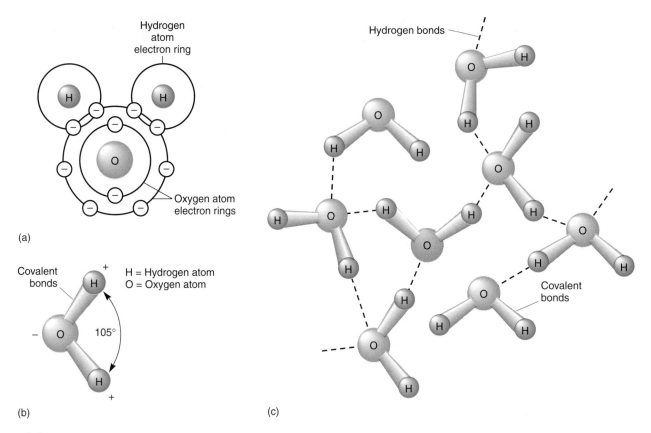

Figure 4.1 The water molecule. (a) The hydrogen atoms share electrons with the outer ring of the oxygen atoms. (b) The angle at which the hydrogen atoms form covalent bonds with the oxygen atom results in a polar molecule. (c) The positive and negative charges allow each water molecule to form hydrogen bonds with other water molecules.

covalent bonds between the hydrogen and oxygen atoms), but as one hydrogen bond is broken, another is formed. Consequently, water is characterized by an extensive, but ever-changing, three-dimensional network of hydrogen-bonded molecules; the result is an atypical liquid with the extraordinary properties that are the subject of this chapter.

Temperature and Heat   4.2

The atoms and molecules in any gas, liquid, or solid are always in motion. Thus, each individual atom or molecule has a certain amount of kinetic energy that is equal to one-half its mass times its velocity squared (kinetic energy = $\frac{1}{2}\ mv^2$). Even in solids, where atoms and molecules are tightly packed and fixed in place, unable to move from one location to another as in a gas or liquid, those atoms and molecules will vibrate around their average position; they will have kinetic energy. The **temperature** of a substance is a measure of the average kinetic energy of the atoms and molecules in the substance. For a homogeneous material consisting of identical atoms or molecules all having the same mass, temperature is simply related to the average velocity of the atoms or molecules. The colder a substance is, the slower the motion of its atoms and molecules. Conversely, the warmer a substance is, the faster the motion of its atoms and molecules. Temperature is measured in **degrees** using one of the three different

scales. The two most commonly used scales are the Fahrenheit (°F) and Celsius (°C) scales (see "Temperature," appendix B). The third scale is the Kelvin (K) scale. It is used to measure extremes of hot and cold and is constructed in such a manner that 0 K corresponds to absolute zero, the temperature at which all atomic and molecular motion ceases. Absolute zero, 0 K, is equal to –273.2°C or –459.7°F; it is a temperature that can never be reached. **Heat** is a measure of the total kinetic energy of the atoms and molecules in a substance. The amount of heat in a substance is determined by the sum of the product of one-half the mass of every atom or molecule in the substance and its velocity squared. Heat is measured in **calories.** One calorie is the amount of heat needed to raise the temperature of 1 g of water by 1°C from 14.5°–15.5°C (see "Energy," appendix B). One thousand of these calories is equivalent to 1 Calorie, kilocalorie (kcal), or food calorie.

The difference between heat and temperature can be readily understood with a simple example. Imagine comparing boiling water in a pot on your stove to the water in a swimming pool on a warm day. The boiling water in the pot clearly has a higher temperature than the water in the pool; the average kinetic energy, or velocity, of the water molecules in the pot is much faster than in the pool. However, there is far more heat in the pool water. Even though the average kinetic energy, or velocity, of the water molecules in

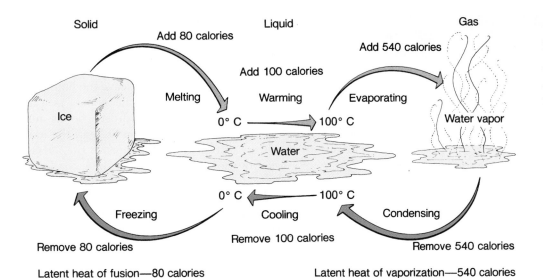

Figure 4.2 Heat energy must be added to convert a gram of ice to liquid water and to convert liquid water to water vapor. The same quantity of heat must be removed to reverse the process.

the pool is relatively small, there are so many of them that the total kinetic energy, or heat, is very large.

Changes of State 4.3

Water exists on the Earth in three physical states: solid, liquid, and gas (fig. 4.2). When it is a solid we refer to it as ice, and when it is a gas we call it water vapor. Pure water ice melts at 0°C and pure water boils at 100°C at standard atmospheric pressure. Pure water is defined as fresh water without suspended particles or dissolved substances, including gases. Standard atmospheric pressure is equal to 76 cm of mercury measured by barometer and is discussed in chapter 6. If you are unfamiliar with the Celsius temperature scale, see appendix B.

Because of the hydrogen bonding between the water molecules, it takes energy (heat) to separate them from each other: that is, for liquid water to form vapor or for ice to form liquid water. In the natural environment this heat is supplied by the Sun.

When pure water makes any of the changes among liquid, solid, and gas, it is said to change its state. Changes of state are due to the addition or loss of heat. When enough heat is added to solid water or ice, the hydrogen bonds break and the ice melts, forming water. When heat is added to liquid water, the temperature of the water rises and some of the water molecules escape from the liquid or evaporate to form water vapor. When heat is removed from water vapor and the temperature falls below the **dew point,** or temperature of water vapor saturation, the water vapor condenses to liquid. When liquid water loses heat and its temperature is lowered to its freezing point, ice is formed as hydrogen bonds link molecules into a lattice pattern.

To change pure water from its solid state (ice) to liquid water at 0°C requires the addition of 80 calories for each gram of ice. There is no change in temperature; there is a change in the physical state of the water as hydrogen bonds

break. The reverse of this process is required to change liquid water to ice. For each gram of liquid water that becomes ice, 80 calories of heat must be removed at 0°C. The heat necessary to change the state of water between solid and liquid is known as the **latent heat of fusion.** This addition or loss of heat takes time in nature. A lake does not freeze immediately, even though the surface water temperature is 0°C, nor does the ice thaw on the first warm day. Time is required to remove or add the heat needed for the change of state.

One gram of liquid water requires 1 calorie of heat to raise its temperature 1°C. Therefore, 100 calories are needed to raise the temperature of 1 g of water from 0° to 100°C. In comparison, the 80 calories of heat required per gram of ice to convert ice to and from liquid water at 0°C with no change in temperature is relatively large. Ice is a very stable form of water; it takes a large addition of heat to melt it and a large loss of heat to form it.

The change of state between liquid water and water vapor requires 540 calories of heat to convert 1 g of water to water vapor at 100°C. When 1 g of water vapor condenses and returns to the liquid state, 540 calories of heat are liberated. Again, there is no change in temperature; there is only a change in the water's physical state. The heat needed for a change between the liquid and vapor states is the **latent heat of vaporization.**

The energy required to change the state of water from solid to liquid to vapor is shown in figure 4.2. Both the latent heat of fusion and the latent heat of vaporization are presented in this diagram, as well as the calories needed for each change of state and the temperatures at which these changes occur. Water converts from the liquid to the vapor state at temperatures other than 100°C; for example, rain puddles evaporate and clothes dry on the clothesline. This type of change requires slightly more heat to convert the liquid water to a gas at these lower temperatures. Water can remain in the liquid phase at temperatures above 100°C when the pressure is greater than 1 atmosphere. One atmosphere

Table 4.2 Energy Required to Convert Water to Water Vapor (Latent Heat of Vaporization)

Temperature of Water (°C)	Calories Required (per g of water)
0°	596.0
10°	590.8
20°	585.6
30°	580.4
40°	575.2
50°	568.5
60°	563.2
70°	557.5
80°	551.7
90°	545.8
100°	539.5
110°	532.9
120°	525.7

(atm) is equal to the pressure of a column of mercury 76 cm (30 in) high; in English units, 1 atm equals 14.7 pounds per square inch (lb/in^2). The conversion from liquid to water vapor at temperatures above 100°C requires slightly less heat than at lower temperatures (table 4.2).

As water is evaporated from the world's lakes, streams, and oceans, and is then returned as precipitation, heat is being removed from the Earth's surface and liberated into the atmosphere, where condensation occurs to form clouds. This heat energy is a major source of the energy used to power the Earth's weather systems (see chapter 6).

Under certain conditions it is possible (1) to cool liquid water below 0°C and keep it as a liquid, (2) to change ice directly to a gas, a process known as **sublimation,** or (3) to boil water at temperatures below 100°C. Cooling below 0°C occurs in some clouds and in the laboratory by controlled cooling of pure water to produce supercooled water. Sublimation is seen in nature when snow or ice evaporates directly under very cold and dry conditions, as on the high desert lands of Utah and Arizona. Anyone living at high altitudes knows that potatoes must be boiled longer and that brewed coffee is cooler there than at sea level because of the decrease in atmospheric pressure and the lower boiling temperature of the water.

The behavior of water, as described, is explained by considering the processes occurring between the water molecules. At the molecular level, the addition of heat energy increases the speed with which molecules move while the loss of heat energy decreases their rate of motion. Heat energy is required to break hydrogen bonds between water molecules, and heat is released when hydrogen bonds are formed. The addition of heat to water causes a relatively small change in the temperature of the water, because much of the heat energy is used to disrupt the hydrogen bonds. These bonds must be broken before the molecules are able to move more rapidly. When heat is removed from water, the molecules slow, many additional hydrogen bonds are formed, and con-

siderable energy is released as heat; this process prevents any rapid drop in temperature.

Water molecules stay close together because of their polarity. If the molecules are moving fast enough they overcome the attractions between them and leave the liquid, entering the air as a gas. Relatively large amounts of heat are needed to evaporate water because hydrogen bonds must first be broken. The greater the addition of heat, the more hydrogen bonds are disrupted and the greater the average energy of motion of the molecules. Under these conditions more water molecules leave the liquid more quickly. In the same way, large amounts of heat must be extracted from water to form the hydrogen bonds required to freeze water. On Earth, natural temperatures required for boiling are rare and those for freezing are frequent but geographically limited; the average Earth temperature of 16°C ensures that liquid water is abundant.

The addition of salt to water changes its boiling and freezing points. The boiling temperature is raised and the freezing temperature is lowered. The amount of change is controlled by the amount of salt added. The rise in the boiling temperature is of little consequence to oceanography because seawater does not normally reach such high temperatures in nature, but the lowering of the freezing point is important in the formation of sea ice. Seawater freezes at about –2°C.

Heat Capacity   4.4

Of all the naturally occurring Earth materials, water changes its temperature the least for the addition or removal of a given amount of heat. The ability of a substance to give up or take in a given amount of heat and undergo large or small changes in temperature is a measure of the substance's **heat capacity,** or specific heat. The heat capacity of water is very high compared to that of the land and the atmosphere. For example, summer temperatures in the Libyan desert reach 50°C and temperatures in the Antarctic drop down to –50°C, for a world temperature range of 100°C. Ocean temperatures vary from a nearly constant high of approximately 28°C in the equatorial areas to a low of –2°C in Antarctic waters, for a world range of 30°C. The water in a lake changes its temperature very little between noon and midnight, but the adjacent land and air temperature changes are large during this same time. The high heat capacity of water and the ability of water to redistribute heat over depth allow the world's lakes and oceans to change temperature slowly, helping to keep the Earth's surface temperature stable.

Heat capacity of a material is the quantity of heat required to produce a unit change of temperature in a unit mass of that material. The heat capacity of water is 1.0 calorie per gram per degree Celsius (cal/g/°C). This is much higher than that of most other liquids because of water's extensive hydrogen bonding. Energy that is used to break hydrogen bonds between water molecules is used in other liquids to contribute directly to increases in molecular motion and elevation in temperature. The heat capacity of

Table 4.3 Heat Capacity of Common Materials

Material	Heat Capacity (calories/g/°C)
Acetone	0.51
Aluminum	0.22
Ammonia	1.13
Copper	0.09
Grain alcohol	0.23
Lead	0.03
Mercury	0.03
Silver	0.06
Water	1.00

Table 4.4 Densities of Common Materials

Material	Density (g/cm³)
Ice (pure) 0°C	0.917
Water (pure) 0°C	0.99987
Water (pure) 3.98°C	1.0000
Water (pure) 20°C	0.99823
White pine wood	0.35–0.50
Olive oil 15°C	0.918
Ethyl alcohol 0°C	0.791
Seawater 4°C (salt 35 g/kg)	1.0278
Steel	7.60–7.80
Lead	11.347
Mercury	13.6

some common materials is given in table 4.3. The high heat capacity of water allows water to gain or lose large quantities of heat with little change in temperature. When salt is added to pure water the changes in heat capacity, latent heat of fusion, and latent heat of vaporization are small.

Cohesion, Surface Tension, and Viscosity   4.5

In the liquid state, the bonds between water molecules are quite fragile; they form, break, and re-form with great frequency. Each bond lasts only a few trillionths of a second. However, at any instant, a substantial percentage of all water molecules are bonded to their neighbors. Therefore, water has more structure than other liquids. Collectively, the hydrogen bonds hold water together; this property is known as **cohesion.**

Cohesion is related to **surface tension,** which is a measure of how difficult it is to stretch or break the surface of a liquid. At the surface between air and water, water molecules arrange themselves in an ordered system, hydrogen-bonded to each other laterally and to the water molecules beneath. This arrangement forms a weak elastic membrane that can be demonstrated by filling a water glass carefully; the water can be made to brim above the top of the glass but not overflow. A steel needle can be floated on water; insects such as the water strider walk about on the surface of lakes and streams. These things are possible because water has a high surface tension. This property is important in the early formation of waves. A gentle breeze stretches and wrinkles the smooth water surface, enabling the wind to get a better grip on the water and to add more energy to the sea surface (see chapter 9, section 9.1).

The addition of salt to pure water increases the surface tension. Decreasing the water temperature also increases the surface tension, and increasing the water temperature decreases it.

Liquid water pours and stirs easily. It has little resistance to motion or to internal friction. This property is called **viscosity.** Water has a low viscosity when compared to motor oil, paint, or syrup. Viscosity is affected by temperature. Consider pancake syrup. When it is stored in the refrigerator it becomes thick and slow to pour. It has a high viscosity. When the syrup is returned to room temperature or heated, it becomes thin and runny; it has a low viscosity. The same is true of water, but the change in viscosity is much less, and it is not noticeable with normal temperature variations. Surface water at the equator is warmer and therefore less viscous than surface water in the Arctic. Minute microscopic organisms find it easier to float in the more viscous polar waters; their tropical-water cousins have adapted to the less-viscous water by developing spines and frilly appendages to help keep them afloat. The addition of salt to pure water increases the viscosity of the water, but the change is small.

Density   4.6

Density is defined as mass per unit volume of a substance. Water density is usually measured in grams per cubic centimeter (g/cm³). The density of pure water is often determined at 3.98°C, the temperature of maximum density, or approximately 4°C, and is 1 g/cm³. Therefore a cube of pure water that is 1 cm high, 1 cm wide, and 1 cm deep has a mass of 1 g or a density of 1 g/cm³. The density of seawater is greater than the density of pure water at the same temperature because seawater contains dissolved salts. At 4°C, the density of seawater of average salinity is 1.0278 g/cm³. The densities of other substances may be considered in the same way (table 4.4). Less-dense substances will float on denser liquids (for example, oil on water, dry pine wood on water, and alcohol on oil). Ocean water is denser than fresh water; therefore fresh water floats on salt water.

The Effect of Pressure

Pure water is nearly incompressible, and so is seawater. Pressure in the oceans increases with increasing depth. For every 10 m (33 ft) in depth, the pressure increases by about 1 atmosphere, or 14.7 lb/in². See figure 4.3 for the effect of

Figure 4.3 These oceanography students hold research cruise souvenirs—polystyrene coffee cups that were attached to a sampler and lowered 2000 m (6560 ft) into the sea. The water pressure compressed the cups to the size of thimbles.

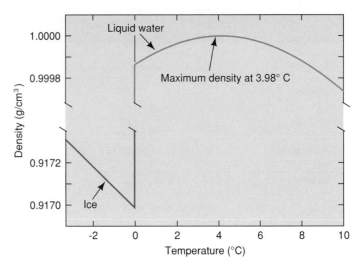

Figure 4.4 The density of pure water reaches its maximum at 3.98°C. Pure water is free of dissolved gases.

water pressure at 2000 m (6560 ft). Pressure in the deepest ocean trench, 11,000 m (36,000 ft) deep, is about 1100 atm. These great pressures have only a small effect on the volume of the oceans because of the slight compressibility of the water. A cubic centimeter of seawater at the surface will lose only 1.7% of its volume if it is lowered to 4000 m (13,000 ft) with a pressure of 400 atm. The average pressure acting on the total world's ocean volume results in the reduction of ocean depth by about 37 m (121 ft). In other words, if the ocean water were truly incompressible, sea level would stand about 37 m higher than it does at present. The pressure effect is small enough to be ignored in most instances except when very accurate determination of seawater density is required. Further information on units of pressure can be found in appendix B.

The Effect of Temperature

Water density is very sensitive to temperature changes. When water is heated, energy is added and the water molecules speed up and move apart; therefore the mass per cubic centimeter becomes less because there are fewer water molecules per cubic centimeter. For this reason, the density of warm water is less than that of cold water, and the warm water floats on the cold water. When water is cooled it loses heat energy, and the water molecules slow down and come closer together; there are then more water molecules, or a greater mass, per cubic centimeter. Because cold water is denser than warm water, it sinks below the warm water. To see these changes, try the following experiment. Chill a small quantity of water in the refrigerator; mix it with a little dye such as ink or food coloring to make it easy to see the effect. Allow the water from the tap to run hot and fill a glass half full of the hot water. Now slowly and carefully pour the colored water down the side of the glass on top of the hot water and watch what happens. You should see the cold, colored water sink below the hot water to create two distinct

layers; the higher-density cold water will be on the bottom and the lower-density hot water will be on top.

In pure fresh water, molecules move more slowly and come closer and closer together as water is cooled to 4°C. At this temperature the water molecules are so close together and moving so slowly that each molecule can form hydrogen bonds with four other molecules. Unlike other fluids that continuously increase in density as they cool, water reaches its greatest density, 1 g/cm³, at 3.98°C, or about 4°C. As the temperature of the water falls below 4°C, the molecules move slightly apart, and at 0°C they form an open lattice-work that is the most stable structure for an ice crystal. Water as a solid takes up more space than water as a liquid; there are fewer water molecules per cubic centimeter, and therefore, ice is less dense than water and floats on water (figs. 4.4 and 4.5). Frozen water's increase in volume is dramatically and often disastrously demonstrated every winter when water pipes freeze and then burst. However, if water continued to contract as it froze, the ice would sink, and lakes would accumulate ice from the bottom up, and eventually all their life forms would die.

When ice absorbs enough heat, the hydrogen bonds between molecules are broken, the lattice starts to collapse, and the molecules move closer together. Above 4°C, the molecules move apart as they increase their motion (fig. 4.5). If enough heat is added, the energy level of the molecules increases until they overcome the attractive forces between each other and the vapor pressure of the air at the surface of the water. The molecules can then escape, or evaporate, into the air as water vapor. Because water vapor is less dense than the mixture of gases that form the atmosphere, a mixture of water vapor and dry air is less dense than dry air alone at the same temperature and pressure.

The Effect of Salt

When salts are dissolved in water, the density of the water increases because the salts have a greater density than water.

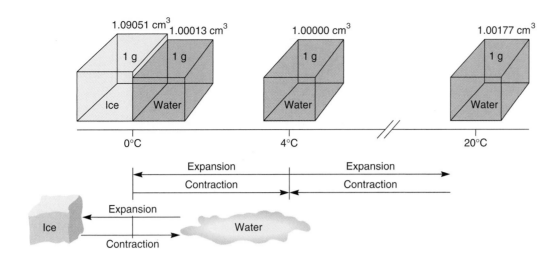

Figure 4.5 Expansion and contraction, or change in volume of 1 g of water with changes in temperature. At 0°C water expands during change of state to ice. Between 0° and 4°C water contracts as it warms and expands as it cools. Above 4°C water expands as it warms and contracts as it cools.

In other words, they have more mass per cubic centimeter. The average density of seawater at 4°C is approximately 1.0278 g/cm³, in comparison with 1 g/cm³ for fresh water. Therefore fresh water floats on salt water. To see this happen, take two small quantities of fresh water; add dye to one and some salt to the other. Carefully and slowly add the salty water to the dyed fresh water. You should see the salty water sink below the dyed fresh water to create two layers: the lower-density fresh water on top and the higher-density salty water on the bottom.

If seawater contains less than 24.7 grams of salt per kilogram of seawater, then the seawater, although it is denser, will behave much like fresh water. Seawater of low salt concentration will reach its maximum density before it freezes, at a temperature less than 0°C. Cooling surface water with a salt content of less than 24.76 g/kg causes the surface water to increase in density and sink. This process continues with surface cooling until the temperature of maximum density is reached for water with a salt content of less than 24.7 g/kg. Further cooling causes this low-salinity surface water to become less dense and remain at the surface.

At a salt content of 24.7 g/kg, the freezing point and the temperature of maximum density of seawater coincide at −1.332°C. If the salt content is greater than 24.7 g/kg, freezing occurs before a maximum density is reached. This relationship is shown graphically in figure 4.6. Open-ocean water generally has an average salt content of 36 g/kg; therefore the density of seawater increases, and the seawater sinks continuously as it is cooled to its freezing point. The effect of both temperature and salt content on the density of water is shown in table 4.5. Note that density decreases as temperature increases when salt content is constant and greater than 24.7 g/kg and that density increases with increasing salt content when temperature is constant.

Dissolving Ability 4.7

More substances—solids, liquids, and gases—dissolve in water than in any other common liquid. Water has been

called the universal solvent because of its exceptionally good dissolving ability for a wide range of materials.

It is the polar nature of the water molecule that makes it such an excellent solvent. For example, common table salt, chemically sodium chloride (NaCl), dissolves in water. Each salt molecule is made up of one atom of sodium with a positive charge and one atom of chloride with a negative charge. These charged atoms are known as **ions.** When sodium chloride is in its natural crystalline lattice, the sodium ions and the chloride ions are held together by the attraction of their opposite electrical charges. To dissolve in water, the sodium and chloride ions must overcome their attraction for each other and interact instead with the molecules of water. Because of their polarity, the water molecules can form spheres around both ions. Figure 4.7a shows water molecules clustered around a sodium ion (Na^+) with their negative (oxygen) ends pointing toward it. Around the chloride ion (Cl^-) the orientation of the water molecules is reversed (fig. 4.7b), with the positive (hydrogen) ends of the solvent molecules pointing toward the ion. The salt ions are surrounded and separated by water molecules. The many salts in seawater are discussed in chapter 5.

For millions of years underwater volcanism and rain washing over the land have been supplying the oceans with dissolved salts. Once these dissolved substances reach the ocean basins there is no waterborne mechanism to return them to the land. Water is recycled to the land by oceanic evaporation, but salts remain in the sea and the bottom sediments. Some land salt deposits resulted from geological uplift processes that bring seafloor deposits above the ocean surface. Other salt deposits on land are the remnants of ancient shallow seas that became isolated over geologic time; the water evaporated, leaving the salt deposits behind.

Transmission of Energy 4.8

Fresh water and salt water both transmit energy. The energy may be in the form of heat, light, or sound. Because the principles of transmission are the same in fresh water and salt water, our discussion will center on energy transmission in the oceans.

Figure 4.6 The initial freezing point and the temperature of maximum density of water decrease as the salt content increases.

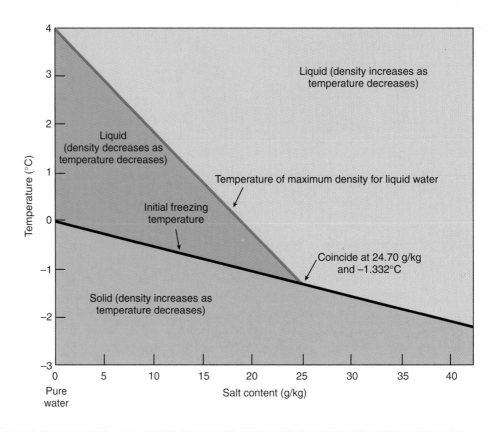

Table 4.5 Density of Water with and Without Dissolved Salt (g/cm³)

°C	No Salt	Salt 20 g/kg	25 g/kg	30 g/kg	35 g/kg
−1	Ice 0.917	1.01606	1.02010	1.02413	1.02817
0	0.99984	1.01607	1.02008	1.02410	1.02813
1	0.99990	1.01605	1.02005	1.02406	1.02807
2	0.99994	1.01603	1.02001	1.02400	1.02799
3	0.99996	1.01598	1.01995	1.02393	1.02791
4	0.99997	1.01593	1.01988	1.02384	1.02781
5	0.99996	1.01586	1.01980	1.02374	1.02770
10	0.99970	1.01532	1.01920	1.02308	1.02697
15	0.99910	1.01450	1.01832	1.02215	1.02599
20	0.99820	1.01342	1.01720	1.02098	1.02478
25	0.99704	1.01210	1.01585	1.01960	1.02336
30	0.99565	1.01057	1.01428	1.01801	1.02175

Heat

There are three ways in which heat energy may be transmitted within a material: by **conduction,** by **convection,** and by **radiation.** Conduction is a molecular process. When heat is applied at one location the molecules move faster because of the addition of energy; gradually this more rapid molecular motion passes on to the adjacent molecules and the heat spreads. For example, if the bowl of a metal spoon is placed in a hot liquid, the handle soon becomes hot. Heat has been conducted from the part of the spoon that is in contact with the heat source to the handle. Metals are excellent conductors; water is a poor conductor and transmits heat slowly in this way.

Convection is a density-driven process in which a heated fluid moves and carries its heat with it to a new lo-cation. In older home heating systems, hot air is supplied through vents at floor level; the hot air rises because it is less dense than the cooler air above it, and the air carries the heat with it. When this air is cooled it drops toward the floor because it has become denser. In these energy-conscious days, ceiling fans are used to force the less dense hot air down, to keep the room warm at all levels rather than accumulating excess heat at the ceiling. Water behaves in the same way; it rises when it is heated from below, and its density decreases; it sinks when it is cooled at its surface, and its density increases. Remember that the principle of convection cells was discussed in chapter 2.

Radiation is the direct transmission of heat from its energy source. Switch on a heat lamp and feel from a foot

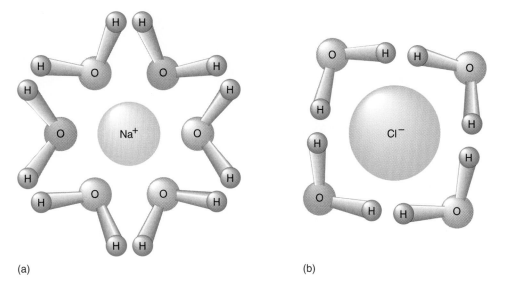

(a)   (b)

Figure 4.7 Salts dissolve in water because the polarity of the water molecule keeps positive ions separated from negative ions. (a) Sodium ions are surrounded by water molecules with their negatively charged portion attracted to the positive ion. (b) Chloride ions are surrounded by water molecules with their positively charged portion attracted to the negative ion.

away the instantaneous heat released. This is radiant energy. The Sun provides the Earth with radiant energy that penetrates and warms the surface waters of the planet. Unlike conduction and convection, which require a medium for the transfer of heat, radiated heat can be transferred through the vacuum of space and through transparent materials.

Think about trying to warm the water in each of two containers. If solar radiation from above is used to heat the surface water of the first container, some of the radiation will be reflected from the water's surface, adding no heat to the water, and some of the radiation will be absorbed by the surface water, raising the water's temperature. Because warming water makes it less dense, the warm water remains at the surface. A small amount of heat may be transmitted slowly downward by molecular conduction, but unless the container is stirred by mechanical energy from another source, the heat remains at the water's surface. If the second container of water is heated from below, the water at the bottom of the container gains heat, decreases in density, and rises, taking heat with it and distributing the heat through the water volume. As the warmed water rises, it is replaced by colder water from the surface, which is warmed in turn. Convection is a much more rapid and efficient method of distributing heat in water than is conduction.

The oceans are heated from above by solar radiation, which is absorbed in the upper surface layers of water. The heat gained at the surface of the ocean is transmitted slowly downward by the natural turbulent stirring action of the wind and the currents, as well as by the slower molecular conduction. In addition, the ocean's surface water may lose heat to the overlying atmosphere as a result of two processes: (1) direct transfer of heat from warm water to colder air by conduction and convection processes, and (2) the transfer of water vapor to the atmosphere by evaporation and atmospheric condensation. Heat transfers that warm the atmosphere from

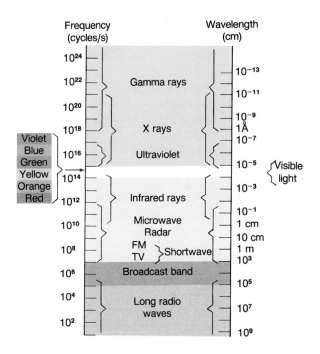

Figure 4.8 The electromagnetic spectrum.

below create atmospheric convection that enhances this heat transfer. The role of these processes in oceanic and atmospheric circulation is discussed in chapters 6 and 7.

Light

The light striking the ocean surface is one of the many forms of **electromagnetic radiation** the Earth receives from the Sun. The full range of this radiation may be seen in the **electromagnetic spectrum,** shown in figure 4.8. Note that visible light occupies a very narrow segment of the spectrum. Wavelengths shorter than visible light include

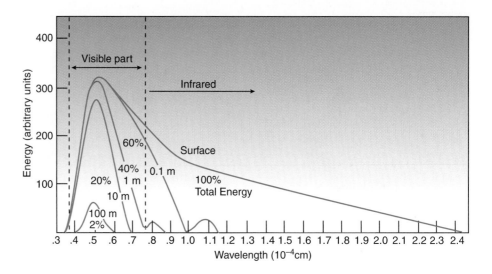

Figure 4.9 The percentage of solar energy in the sea decreases as depth increases. The long, red wavelengths are absorbed first, and the color peak shifts toward the shorter, blue wavelengths. *From Sverdrup/Johnson/Fleming, The Oceans, 1942, renewed 1970, p. 105. Adapted by permission of Prentice Hall, Inc., Upper Saddle River, NJ.*

ultraviolet light, X rays, and gamma rays, whereas infrared (heat) waves, microwaves, TV, and FM and AM radio waves are all wavelengths longer than visible light. Visible light may be broken down into the familiar spectrum of the rainbow: red, orange, yellow, green, blue, and violet. Each color represents a range of wavelengths; the longest wavelengths are at the red end of the spectrum, and the shortest are at the blue-violet end. When combined, these wavelengths produce white light.

Seawater transmits only the visible light portion of the electromagnetic spectrum. About 60% of the entering light energy is absorbed in the first meter, and about 80% is gone after 10 m (33 ft). Only 1% of the total light available at the surface is left in the clearest water below 150 m (500 ft), and no light penetrates below 1000 m (3300 ft). Wavelengths of visible light are not all transmitted equally (fig. 4.9). The long wavelengths at the red end of the spectrum are absorbed rapidly within the upper 10 m, and the shorter wavelengths of blue-green light are transmitted to the greater depths.

Color perception is due to the reflection back to our eyes of wavelengths of a particular color. Because all wavelengths, or colors, are present to illuminate objects in shallow water, objects are seen in their natural colors at the surface. Objects in deeper waters usually appear dark because they are illuminated by blue light. Ocean water usually appears blue-green because wavelengths of this color, being absorbed the least, are most available to be reflected and scattered back to an observer. Coastal waters vary in color, appearing green, yellow, brown, or red; these waters usually contain silt from rivers as well as large numbers of microscopic organisms and dissolved organic substances. The colors of these inclusions are revealed by the light entering the water. Open-ocean water beyond the influence of land is often clear and blue. The clearer the water, the less suspended matter present and the deeper the light penetration.

When light passes from air into water it is bent, or **refracted,** because the speed of light is faster in less dense air than in denser water. Because of the light refraction, objects seen through the water's surface are not where they appear

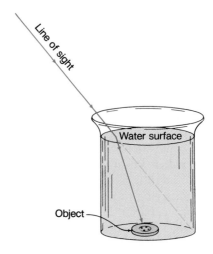

Figure 4.10 Objects that are not directly in the individual's line of sight can be seen in water because of the refraction of the light rays. The refraction is caused by the decreased speed of light in water.

to be (fig. 4.10). Refraction is affected slightly by changes in salinity, temperature, and pressure.

As light passes through the water it is **absorbed, scattered,** and reflected by suspended particles, including silt, single-celled organisms, and the water and salt molecules. It is also absorbed by plants, to be used in their life processes. This decrease in the intensity of light over distance is known as **attenuation.** The clearer the water, the greater the light penetration and the smaller the attenuation.

The simplest way of measuring light attenuation in surface water is to use a **Secchi disk** (fig. 4.11). This is a white disk, about 30 cm (12 in) in diameter, that is lowered on a line to the depth at which it just disappears from view. The measure of this depth can be used to determine the average attenuation of light. In water rich with living organisms or suspended silt, the Secchi disk may disappear from view at depths of 1–3 m (3–10 ft); in the open ocean visibility usually extends down to 20–30 m (65–100 ft), although a Secchi disk reading of 79 m (260 ft) was reported from Antarctica's Weddell Sea in 1986. Although crude, a Secchi

Figure 4.11 A Secchi disk measures the transparency of water.

web link **Figure 4.12** The Marine Optical Buoy (MOBY) is lowered into the sea. MOBY measures radiant energy entering and exiting the sea surface at three different depths. The data are used to adjust ocean color in satellite images. *Courtesy of Dennis K. Clark, MOBY Project Team Leader, NOAA.*

disk has the advantages of low cost, never needing adjustment, never leaking, and never requiring replacement of electronic components. However, this measurement technique explains little about how seawater affects light.

Changes in light attenuation are measured by devices that project a beam of light over a fixed distance to an electronic photoreceptor. Such devices are lowered meter by meter into the sea, and measurements are taken at chosen intervals. The cumulative effect of attenuation may also be calculated to any depth. Different wavelengths of light are used to separate the effects of absorption and scattering by inorganic and organic particles as well as absorption by dissolved substances and single-celled plantlike organisms found in the ocean's surface layer. The role of natural fluorescence of matter exposed to sunlight is also under investigation. Most light is produced when atoms become excited and bounce into one another until they give off radiation in the form of light and heat. In flourescence, only the atom's electrons are agitated; they give off radiation in the form of light but practically no heat. Flourescence is sometimes described as "cold light." Multichannel color sensors are used to observe both the changing properties of solar radiation as it penetrates the ocean and the radiance emitted by the fluorescence of dissolved compounds. Light transmission data are used to adjust color quality and clarity of underwater photographs and video.

The Marine Optical Buoy (MOBY; fig. 4.12) was launched in 1997; it carries instruments to measure light and color at the surface and at three different depths. Measurements are transmitted by fiber optics from the buoy to instruments onboard ship. These measurements are used to improve satellites' sensing of sea surface color. The satellites monitor sea surface color to determine the abundance of single-celled plantlike organisms at the sea surface, which are used as indicators of the biological health of the oceans. In the future, it may be possible to identify blooms of toxic organisms in this way (see the discussion of marine toxins in chapter 15).

Sound

The sea is a noisy place; waves break, fish grunt and blow bubbles, crabs snap their claws, and whales whistle and sing. Sound travels farther and faster in seawater than it does in air; the average velocity of sound in seawater is 1500 m/s (5000 ft/s) compared with 334 m/s (1100 ft/s) in dry air at 20°C.

The speed of sound in seawater is a function of the axial modulus and density of the water.

$$\text{speed of sound} = \sqrt{\frac{\text{axial modulus}}{\text{density}}}$$

The axial modulus of a material is a measure of how easy it is to compress. A material with a high axial modulus is more difficult to compress than a material with a low axial modulus; a golf ball has a higher axial modulus than a tennis ball. Both the axial modulus and the density of seawater depend on temperature, salinity, and pressure (or depth). The speed of sound in seawater increases with increasing temperature, salinity, or depth and decreases with decreasing temperature, salinity, or depth. Sound speed increases with increasing temperature because the density of the water decreases as it becomes warmer. Sound speed increases with increasing salinity and depth, despite the fact that both these changes increase seawater's density, because the axial modulus of the water increases faster than its density.

Water dissipates the energy of high-frequency sounds faster than that of low-frequency sounds. Therefore, high-frequency sounds do not travel as far as the lower-frequency sounds.

Sound is reflected back after striking an object, and therefore sound can be used to find objects, sense their shape, and determine their distance from the sound's source. If a sound signal is sent into the water and the time required for the return of the reflected sound, or echo, is measured accurately,

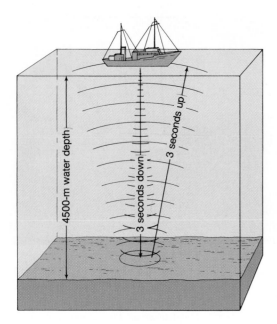

Figure 4.13 Traveling at an average speed of 1500 m per second, a sound pulse leaves the ship, travels downward, strikes the bottom, and returns. In 4500 m of water, the sound requires three seconds to reach the bottom and three seconds to return.

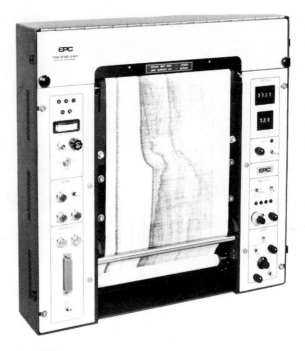

Figure 4.14 A precision depth recorder displaying bottom and subbottom profiles.

the distance to the object may be determined. For example, if six seconds elapse between the outgoing sound pulse and its return after reflection, the sound has taken three seconds to travel to the object and three seconds to return. Because sound travels at an average speed of 1500 m/s in water, the object is 4500 m away, as shown in figure 4.13.

Water depth is measured by directing a narrow sound beam vertically through the seawater to the sea floor. The sound beam passes through the nearly horizontal layers of water, in which salt content, temperature, and pressure vary. The sound speed continuously changes as it passes from layer to layer, until it reaches the sea floor and is reflected back to the ship. Little refraction, or bending, of the sound beam occurs, because its path is perpendicular to the water layers. Echo sounders, or depth recorders, are used by all modern vessels to measure the depth of water beneath the ship. Oceanographic vessels record the depths on a chart, producing a continuous reading of depth as the vessel moves along its course. The precision depth recorder (PDR) on an oceanographic research vessel uses a very narrow sound beam to give detailed and continuous traces of the bottom while the ship is in motion (fig. 4.14).

Geologists can detect the properties of the sea floor by studying echo charts, because some seafloor materials reflect back a stronger signal than other materials. If high-intensity sound pulses are transmitted, some sound energy can penetrate the sea floor and reflect back from layers in the sediments. In this case, the echo chart displays the layering of the seabed materials (fig. 4.14).

Depth recorders can also record large numbers of small organisms, including fish, that move toward the surface during the night and sink down to greater depths during the day. These organisms form a layer known as the **deep scattering layer (DSL).** This layer reflects a portion of the sound beam energy and creates the image of a false bottom on the depth recorder trace.

Other echoes are returned from mid-depths by fish swimming in schools or by large individual fish. Echo sounders are designed and marketed as "fish finders," and persons who fish learn to recognize fish school reflections. The echoes provide information on the depth to set nets, while the location and the appearance of the echo pattern give information on the fish species.

Porpoises and whales use sound in water in the same way that bats use sound in air. The animal produces a sound, which travels outward until it reaches an object, from which it is reflected. The animal is able to judge the direction from which the sound returns, the distance to the reflecting object, and the properties of that object (see chapter 16). Human technology has produced an underwater location system called **sonar** (sound navigation and ranging) that uses sound in a similar way. Sonar technicians send directional pulses through the water, searching for targets that return echoes. They are then able to determine the distance and direction of the target. An electronic screen is used to display the direction of sound pulses relative to the sending vessel. If a reflective target is found, the screen also portrays the distance to the target, using the time difference between the sending of the sound pulse and the return of the echo. After much training and practice, technicians are able to distinguish between the echoes produced by a whale, a school of fish, and a submarine. It is even possible to determine the type and class of a vessel merely by listening to the sounds produced by its propellers and engines. (See the box

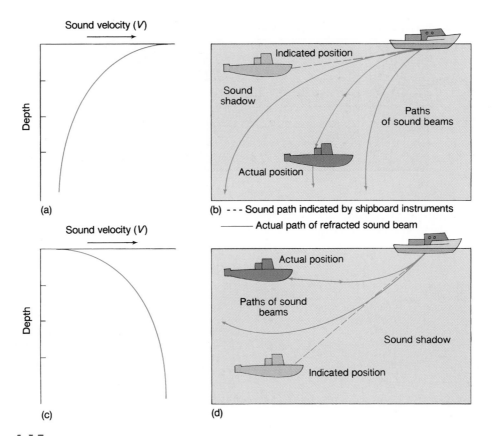

Figure 4.15 Sound waves change velocity and refract as they travel at an angle through water layers of different densities. The angle at which the sound beam leaves the ship indicates a target in the indicated, or ghost, position. The degree of refraction and change in velocity with depth must be known in order to determine the actual position of the target.

titled "Bathymetrics" in chapter 3 for a discussion of multiple sonar devices being used in seafloor mapping.)

The target, however, may not be at the depth, distance, and angle indicated, because the sound beam may change its speed as it passes obliquely through water layers of differing densities (fig. 4.15*a*, *c*). Figure 4.15 *b* and *d* illustrate the refraction of sound beams and the formation of **sound shadow zones,** areas of the ocean into which sound does not penetrate. Sound beams bend toward regions in which sound travels more slowly and away from regions in which sound waves travel more rapidly.

To interpret the returning echo and to determine distance and depth correctly, the sonar operator must have information about the properties of the water through which the sound passes. Governments and their navies have conducted intensive research in the field of underwater sound. Survival for a surface vessel depends on its accuracy in locating a submarine's position, while the submarine must remain at the correct depth and distance from the sonar detector to remain invisible in the shadow zone.

At about 1000 m (3280 ft) the combination of salt content, temperature, and pressure creates a zone of minimum velocity for sound, the **sofar** (sound fixing and ranging) **channel** (fig. 4.16*a*). Sound waves produced in the sofar channel do not escape from it unless they are directed outward at a sharp angle. Instead, the majority of the sound energy is refracted and bounces back and forth along the channel for great distances (fig. 4.16*b*). Test explosions set off in the channel near Australia have produced sound heard as far off as Bermuda. This channel is being used by a project known as ATOC (Acoustic Thermometry of Ocean Climate) to send sound pulses over long distances to look for long-term changes in the temperature of ocean waters. (See the box titled "Acoustic Thermometry of Ocean Climate.")

In the 1950s and 1960s, the U.S. Navy installed large arrays of hydrophones on the sea floor in areas that were potential sailing routes for foreign submarines. These acoustical nets, known as sound surveillance systems (SOSUS), were used to track and identify vessels at and below the sea surface. Oceanographers and other scientists were given access to this Cold War legacy for acoustic monitoring of marine mammals and seafloor seismic events. (See the box titled "Listening to Seafloor Spreading" in chapter 2.)

Recently, a French oceanographic team studied the pure low-frequency tones of long duration received by undersea seismic stations in French Polynesia; SOSUS archives provided similar recordings. The sound source was tracked to a region of seamounts at shallow depths. It appears that erupting seamounts produce clouds of bubbles between their tops and the sea surface and that the bubbles act as a resonating chamber that allows sound waves to oscillate vertically between the volcanic source and the sea surface, producing the low-frequency tones.

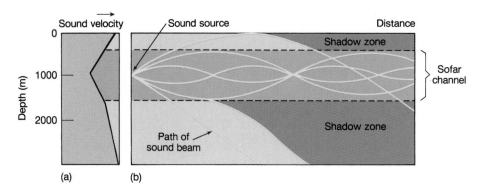

Sound velocity   Sound source   Distance

Shadow zone

Sofar channel

Path of sound beam

Shadow zone

(a)   (b)

 Figure 4.16 (a) The temperature, salinity, and pressure variation with depth combine to produce a minimum sound velocity at about 1000 m (3280 ft). (b) Sound generated at this depth is trapped in a layer known as the sofar channel.

Acoustic Thermometry of Ocean Climate

Based on the principle that sound's speed in seawater is determined primarily by the temperature of the water, oceanographers Carl Wunsch of the Massachusetts Institute of Technology and Walter Munk of the Scripps Institution of Oceanography in California envisioned a transoceanic experiment to follow ocean temperature response to global warming. If the oceans are warming, detectable worldwide decreases in the travel time of sound will occur, because sound waves travel faster in warm water than in cold water.

In 1991 the travel time of low-frequency sound pulses transmitted through the sofar channel were repeatedly measured as they moved from a site near Heard Island in the southern Indian Ocean to special listening stations around the world. The precision of sound travel time measurements from source to receiver was about 1 millisecond over a path 1000 km (660 mi) long, so temperature changes of a few thousandths of a degree could be detected over long distances. Test results were promising, and a new series of tests called ATOC, acoustic thermometry of ocean climate, began in the Pacific Ocean in 1995.

ATOC broadcasts sound from underwater sources, near San Francisco and Hawaii. The sound is picked up by arrays of sensitive hydrophones as far away as Christmas Island and New Zealand (box fig. 1). After eighteen months of testing, temperature readings of the Pacific Ocean were found to be even more precise than had been projected. Scientists can detect variations as small as 20 milliseconds in the hour-long travel time of the pulses. This precision enables researchers to calculate the average ocean temperature along the sound pulse path to within 0.006°C. Repeating the measurements will allow long-term temperature changes at mid-ocean depths to be measured before they can be deduced by any other method.

Questions concerning the effect of sound signals on marine mammals began with the Heard Island tests and delayed the start of

the ATOC experiment. Marine mammalogists have monitored the behavior of whale and elephant seal populations near the sound source off California. They report having seen no changes in the animals' swimming activity or distribution. Mammal researchers have also released deep-diving elephant seals farther out to sea and used satellite tags to track the paths of the animals as they returned to shore. The seals made no attempt to avoid the sound source. Additional experiments are being conducted to see whether whale vocalizations are affected.

A similar experiment that measured water temperatures in the Arctic Ocean was performed in the spring of 1994 by a joint U.S.-Russian-Canadian project, Transarctic Acoustic Propagation Experiment (TAP). Sound signals were sent from an ice camp north of Spitsbergen to a camp 900 km (540 mi) in the Lincoln Sea and to another camp 2600 km (1600 mi) away in the Beaufort Sea. Travel times were predicted by using water temperatures from earlier research, but the measured travel times were shorter, implying that the mid-depth Atlantic water that penetrates the Arctic Ocean had warmed by 0.2°–0.4°C since the mid-1980s. The U.S.-Canada Arctic Ocean Section cruise had also measured such a warming trend, but scientists emphasize that it is too early to say whether the change is due to global warming or whether it is a part of some other natural cycle.

To Learn More About Acoustic Thermometry of Ocean Climate

Forbes, A. 1994. Acoustic Monitoring of Global Ocean Climate. *Sea Technology* 35 (5): 65–67.

Georges, T. M. 1992. Taking the Ocean's Temperature with Sound. *The World & I* (July): 282–89.

The Heard Island Experiment. 1991. *Oceanus* 34 (1): 6–8.

Ice and Fog   4.9

Ice and fog are forms of water that are extremely hazardous to ocean navigation and transportation. At polar latitudes in the Northern Hemisphere, ice prevents the safe use of shipping routes for much of the year, and ice driven by winds and currents hampers polar research in both Arctic and Antarctic regions. Although radar helps ships underway in foggy seas, the restricted visibility is a danger to all vessels, from the largest to the smallest. Processes producing ice and fog at sea are discussed in this section.

Sea Ice

Ice is present year-round at all latitudes if the elevation is high enough to keep the average Earth surface temperature below the freezing point of water. This elevation varies between sea level in polar regions and approximately 5000 m (16,400 ft) at the equator. On land, ice and snow are the result of low temperatures and precipitation. In the sea and in lakes, precipitation is not necessary; the ice is formed when the temperature of the air drops below the freezing point of water.

Sea ice is formed at polar latitudes because of the low air and water temperatures. As the seawater begins to freeze, the surface water becomes dull, and clouds of ice crystals are produced. As the number of ice crystals increases, a layer of slush forms, covering the ocean in a thin sheet of ice. Sheets of new sea ice are broken into "pancakes" by waves and wind (fig. 4.17a). As the freezing continues, the pancakes move about, unite, and form floes. Ice floes move with the

Mikhalevsky, P., A. Braggeroer, A. Gavrilov, and M. Slavinsky. 1995. Experiment Tests Use of Acoustics to Monitor Temperature and Ice in Arctic Ocean *EOS* 76 (27): 265, 268–69.

Internet References
Visit the book's Online Learning Center at www.mhhe.com/sverdrup (click on the book's cover) to explore links to further information on related topics.

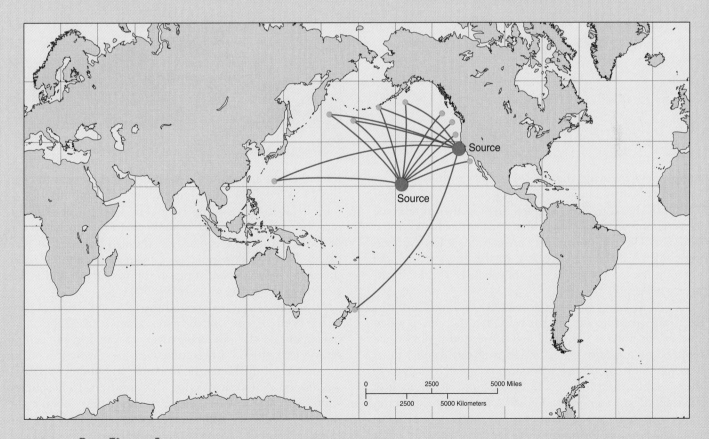

web **link**   **Box Figure 1**   Acoustic thermometry of ocean climate (ATOC) takes the temperature of the Pacific Ocean using two sound sources, California and Hawaii, and twelve receivers.

(a)

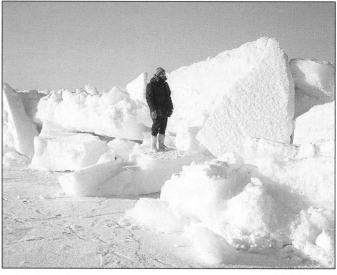

(b)

web link **Figure 4.17** Sea ice. (a) Pancake ice, an early stage of ice formation. (b) A pressure ridge formed by colliding floes. (c) Meltwater forms ponds on the Arctic sea ice in summer.

(c)

currents and the wind (fig. 4.17*b*), collide with each other, and form ridges and hummocks, or low hills. Some floes shift constantly, breaking apart and freezing together in response to winds and water motion. Others remain anchored to the landmass. These stationary floes are called **fast ice.** As freezing continues, the ice increases in thickness. Snow and water on the ice also freeze and contribute to the flat, floating ice masses.

The dissolved salts of the seawater do not fit into the ice crystal structure and are left behind in the surface water. This saltier water is very cold and dense, and it sinks to be replaced by less-salty water, which is in turn cooled to its freezing point. As the ice is formed, seawater is trapped in the voids of the forming ice. If the ice forms slowly, most of the trapped seawater drains out and escapes; if the ice forms quickly at subfreezing temperatures, more salt water is trapped. As time passes and the ice ages, the salt water slowly escapes through the ice, and eventually the sea ice becomes fresh enough to drink when melted.

In one season about 2 m (6 ft) of ice forms at the ocean's surface at polar latitudes. The thickness of the ice is limited by the necessity of extracting sufficient heat through the ice to cool the underlying water, allowing new ice to form beneath the existing layer. The latent heat of fusion must be removed from the water beneath the ice, and it can be extracted only by conduction through the ice that is in place. This is a slow process at polar temperatures, made slower by snow accumulating on the ice surface and acting as an insulator to the water below.

Sea ice is seasonal around the bays and shores of parts of the northeastern United States, Canada, Russia, Scandinavia, and Alaska. The ice exists year-round in the central Arctic and around Antarctica. Sea ice covers the entire Arctic Ocean in the northern winter and pushes far out to sea around Antarctica during the Southern Hemisphere's winter (fig. 4.18). Waves, currents, winds, and tides fracture the edges of the ice and cause floes to collide and pile into irregular masses to form pressure ridges (see fig. 4.17*b*). Ice ridges formed at the collision boundaries between ice floes may reach 10 m (30 ft) in thickness. In summer the ice melts back along its edges and at its surface (fig. 4.17*c*). In areas where the ice does not melt entirely during the year,

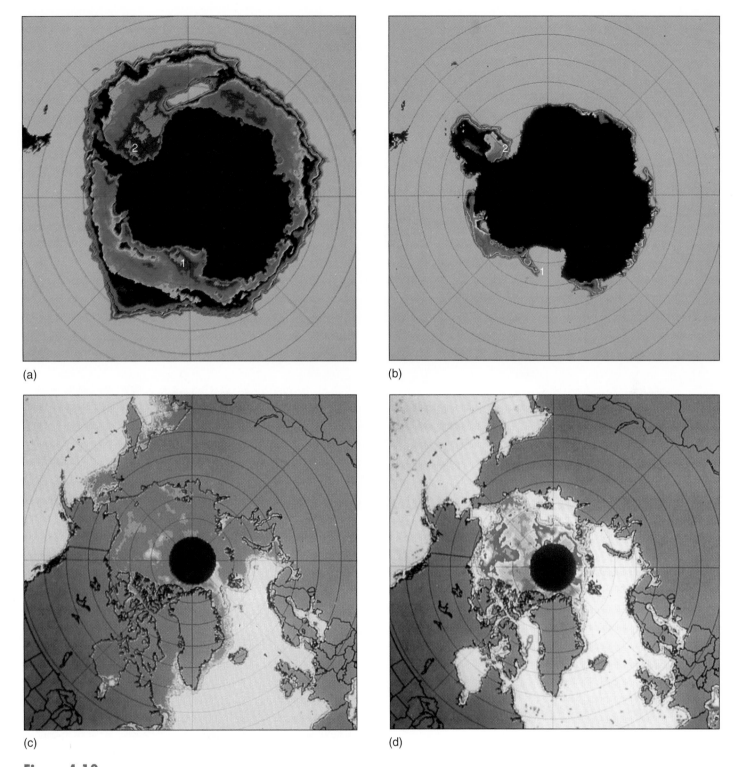

(a)

(b)

(c)

(d)

Figure 4.18 Sea ice around the Arctic and Antarctic. Data from NASA's *NIMBUS-5* satellite were used to construct seasonal maps of sea ice coverage at high latitudes. This information is used to help develop a better understanding of global climate. The Antarctic ice extends more than 100 km (60 mi) from the land into the Ross (*1*) and Weddell (*2*) Seas during the winter (a) and is much reduced during the summer (b). The Arctic Ocean is filled with ice in the winter (c), but by the next fall the ice cover is much less (d). Short-term motions of the ice boundaries can be calculated from repeated satellite images. These show that the ice floes move up to 50 km (30 mi) per day. *Purple* indicates high ice concentration, and *blue* indicates open water in (a) and (b). *Yellow* areas are open water in (c) and (d).

(a)

Figure 4.19 (a) A castle berg drifting in the Atlantic Ocean just north of the Grand Banks. (b) A tabular berg in the South Atlantic Ocean. It is approximately 500 m (1640 ft) long.

(b)

new ice is added each winter, and a thickness of 3–5 m (10–16 ft) can accumulate.

Remember that sea ice is a product of salt water; it should not be confused with freshwater glacial ice, which covers Greenland and the Antarctic continent. When pieces of land ice break off and fall into the sea, these pieces of freshwater ice are called **icebergs.**

Icebergs

Icebergs are massive and irregular in shape; they float with only about 12% of their mass above the sea surface (fig. 4.19). They are formed by glaciers, large rivers of ice that begin inland in the snows of central Greenland, Antarctica, and Alaska, and inch their way toward the sea. The forward movement, the melting at the base of the glacier where it meets the

Figure 4.20 Iceberg B-9 as observed by *Landsat 4*, November 28, 1987, about four to six weeks after detachment from the Ross Ice Shelf, Antarctica.

ocean, and waves and tidal action cause blocks of ice to break off and float out to sea. The production of icebergs by a glacier is called calving, and it occurs mainly during the summer.

Icebergs produced in the Arctic drift south with the currents as far as New England and the busy shipping lanes of the North Atlantic. It was one of these icebergs that sank the *Titanic* on its maiden voyage in 1912. This iceberg had most probably become detached from a glacier in Greenland and then floated south. After the sinking, the United States began an iceberg patrol to warn vessels of iceberg locations, and this effort, aided by satellite detection of the bergs, continues to the present.

Icebergs produced in the Antarctic usually stay close to the polar continent, caught by the circling currents, although they have been known to reach latitudes of 40°S to 50°S. Alaskan icebergs are usually released in narrow channels and semienclosed bays; they do not often escape into the open ocean. Rather than being a hazard to shipping, Alaskan icebergs are tourist attractions in Glacier Bay and Prince William Sound.

Icebergs from narrow valley glaciers are irregularly shaped, resembling towers and battlements; they are known as castle bergs. Icebergs from broad, flat, continental ice sheets are usually much larger than castle bergs. These huge, flat ice-

bergs are called tabular bergs and form the ice islands that are used as bases for polar research. Some of these icebergs have been large enough to accommodate aircraft runways, housing, and research facilities. Such an ice island may be used for years before it eventually breaks up.

In late 1987, a large tabular berg known as B-9, 155 km (96 mi) long and 230 m (755 ft) thick, with a surface area about the size of Long Island, New York, broke from Antarctica and drifted 2000 km (1250 mi) along the Antarctic coast (fig. 4.20). This berg was so large and so deep that its drift was the product of subsurface currents; conventionally sized tabular bergs drift with the wind. It drifted clockwise and to the west in the Ross Sea for two years before it grounded and broke into three pieces. B-9 was tracked by plane and satellite; its drift path helped increase our knowledge of currents in the Ross Sea.

The calving of a spectacularly large tabular berg is infrequent, for the seaward movement of glacial ice is only about 1 m (3 ft) per day, and a berg of B-9's size contains about seventy years of glacial advance. Formation of a berg of this size is also controlled by the spacing of the major crevasses that produce weak spots along which a berg separates from the glacier. The developing rift along which B-9 broke away was first identified in aerial photos taken between 1965 and 1971.

Figure 4.21 Advective fog obscures the Golden Gate Bridge at the entrance to San Francisco Bay.

Ongoing studies in the Antarctic indicate that the glaciers that reach the sea and form the large expanses of permanent shelf ice that surround Antarctica are receding. At the same time, the shelf ice is calving large bergs at an increasing rate. Whether this activity is due to global warming driving an increased rate of melting or whether the flow rate of the glaciers moving toward the sea is changing and altering the balance between the rate of ice supply and the rate of meltback is unknown.

Fog

The ability of water to remain as water vapor is a function of air temperature. Warm air can hold more moisture than cold air, and when air is cooled, its water vapor condenses around any small particles in the air, forming liquid droplets. Large accumulations of water droplets form clouds, and when clouds form close to the ground they are called **fog.** Fog at sea is a navigation hazard to vessels, but the ocean scientist is interested in fog and in predicting its occurrence because of the role fog plays in transferring water and heat between the atmosphere and the ocean surface.

There are three basic types of fog, each formed by a different process. The most common fog at sea is advective fog, forming when warm air saturated with water vapor moves over colder water. This advective fog hugs the sea surface like a blanket. For example, when the warm, moist air over the Gulf Stream flows over the cold water of the Labrador Current, the famous fogs of the Grand Banks re-

sult. In northern California, Oregon, and Washington, when offshore warm, moist air moves over the cold coastal waters, particularly in summer, a coastal fog results, giving moisture to the redwood forests and grassy cliffs and cooling the city of San Francisco (fig. 4.21).

Streamers of **sea smoke** rising from the sea surface on a cold winter day are a spectacular sight. Dry, cold air from the land or from the polar ice pack moves out over the warmer water, and the water warms the air above it. This warmed air picks up water vapor from the sea surface and begins to rise rapidly. As it rises, it is cooled below the dew point, and rising ribbons of fog are formed. When sea smoke is formed the vertical convection results in high evaporation that rapidly cools the water's surface, while advective fog's condensation returns moisture and heat to the Earth's surface.

Radiative fog is the result of warm days and cold nights. The Earth's surface warms and cools, and so does the air above it. If the air holds enough moisture during the day, it condenses as the Earth cools at night, forming the low-lying, thick white fogs that are often found in river valleys and occasionally in bays and inlets along the coast. These fogs usually disappear during the morning as the Sun gradually warms the air, changing the water droplets back to water vapor.

The Earth's water, whether found as a liquid filling our lakes, streams, and oceans; as a solid in the land glaciers, snowpacks, and sea ice; or as a gas in the atmosphere, is a remarkable substance. The properties of this water as well as its abundance give the Earth its habitable characteristics and make it the unique planet in our solar system.

Green Icebergs

Icebergs are ordinarily blue to white, although they sometimes appear dark or opaque because they carry gravel and bits of rock. They may change color with changing light conditions and cloud cover, glowing pink or gold in the morning or evening light, but this color change is generally related to the low angle of the Sun above the horizon. However, travelers to Antarctica have repeatedly reported seeing green icebergs in the Weddell Sea and more commonly close to the Amery Ice Shelf in East Antarctica (box fig. 1).

One explanation for green icebergs attributes their color to an optical illusion when blue ice is illuminated by a near-horizon red Sun; however, green icebergs stand out among white and blue icebergs under a variety of light conditions. Another suggestion is that the color might be related to ice with high levels of metallic compounds, including copper and iron. Recent expeditions have taken ice samples from green icebergs and ice cores from the glacial ice shelves along the Antarctic continent. Analyses of these cores and samples provide a different solution to the problem.

The ice shelf cores, with a total length of 215 m (700 ft), were long enough to penetrate through glacial ice formed from the compaction of snow and to continue into the clear, bubble-free ice formed from seawater that freezes onto the bottom of the glacial ice. The properties of this clear sea ice were very similar to the ice from the green iceberg. The scientists concluded that green icebergs form when a two-layer block of shelf ice breaks away and capsizes, exposing the clear shelf ice that was formed from seawater.

A green iceberg that stranded just west of the Amery Ice Shelf showed two distinct layers: bubbly blue-white ice and bubble-free green ice separated by a 1 m (3 ft) ice layer containing sediments. The green ice portion was textured by seawater erosion. Where cracks were present the color was light green because of light scattering; where no cracks were present, the color was dark green. No air bubbles were present in the green ice, suggesting that the ice was not formed from the compression of snow but instead from the freezing of seawater. Large concentrations of single-celled organisms with green photosynthetic pigments occur along the edges of the ice shelves in this region, and the seawater is rich in their decomposing organic material. The green iceberg did not contain large amounts of particles from these organisms, but the ice had accumulated dissolved organic matter from the seawater. It appears that, unlike salt, dissolved organic substances are not excluded from ice in the freezing process. Analysis shows that the dissolved organic material absorbs enough blue wavelengths from solar light to make the ice appear green.

Isotopic and chemical evidence shows that platelets of ice form in the water and then accrete or stick to the bottom of the ice shelf. Ice platelets and ice crystals are buoyant and accumulate under the ice shelf to form a slush. The slush is compacted by an unknown mechanism, and solid, bubble-free ice is formed from water high in soluble organics. When an iceberg separates from the ice shelf and capsizes, the green ice is exposed. See box figure 2 for a diagram of this process.

web link **Box Figure 1** A green iceberg from the Amery Ice Shelf in Antarctica.

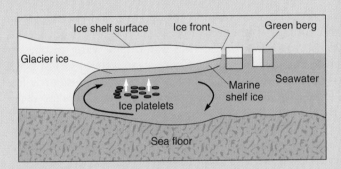

Box Figure 2 When a glacier moves over seawater, it forms an ice shelf ending in an ice front. Seawater, rich in dissolved organic matter, freezes onto the underside of the ice shelf. If bergs calved from the ice front capsize, the green shelf ice is exposed.

The Amery Ice Shelf appears to be uniquely suited to the production of green icebergs. Once detached from the ice shelf, these bergs drift in the currents and wind systems surrounding Antarctica and can be found scattered among Antarctica's less colorful icebergs.

To Learn More About Green Ice

Kipfstuhl, J., G. Dieckmann, H. Oerter, H. Hellmer, and W. Graf. 1992. The Origin of Green Icebergs in Antarctica. *Journal of Geophysical Research* 79 (C12): 20, 319–20, 324.

Thomas, D., and G. Dieckmann. 1994. Life in a Frozen Lattice. *New Scientist* 142 (1929): 33–37.

Internet References

Visit the book's Online Learning Center at www.mhhe.com/sverdrup (click on the book's cover) to explore links to further information on related topics.

Summary

A water molecule is made up of two positively charged atoms of hydrogen and one negatively charged atom of oxygen. The molecule has a specific shape with oppositely charged sides; it is a polar molecule. Because of the distribution of the charges, water molecules interact with each other by forming hydrogen bonds between molecules. As a result of its structure, the water molecule is very stable. The structure of the water molecule is responsible for the properties of water.

Water exists as a solid, a liquid, and a gas. Changes from one state to another require the addition or extraction of heat energy. Water has a high heat capacity; it is able to take in or give up large quantities of heat with a small change in temperature. The surface tension of water, which is related to the cohesion between water molecules at the surface, is high. The viscosity of water is a measure of its internal friction; it is primarily affected by temperature. Water is nearly incompressible. Pressure increases 1 atm for every 10 m of depth in the oceans.

The density, or mass per unit volume, of water increases with a decrease in temperature and an increase in salt content. The effect of pressure on density is small. Less dense water floats on denser water. Pure water reaches its maximum density at 4°C. Open-ocean water does not reach its maximum density before freezing. Water vapor is less dense than air; a mixture of water vapor and air is less dense than dry air.

The ability of water to dissolve substances is exceptionally good. River water dissolves the salts from the land and carries the salts to the sea.

Seawater transmits energy as heat, light, and sound. The sea surface layer is heated by solar radiation, and heat is transmitted downward by conduction. This is an inefficient process compared to convection.

The long red wavelengths of light are lost primarily in the first 10 m (33 ft) of seawater; only the shorter wavelengths of blue-green light penetrate to depths of 150 m (500 ft) or more. Light passing into water is refracted. Attenuation, or the decrease in light over distance, is the result of absorption and scattering by the water and by particles suspended in the water. Light attenuation is measured by a Secchi disk and photoreceptors.

Sound travels farther and faster in water than in air. Its speed is affected by the temperature, pressure, and salt content of the water. Echo sounders are used to measure the depth of water, and sonar is used to locate objects. Sound is refracted as it passes at an angle through the density layers, and sound shadows are formed. The sofar channel, in which sound travels for long distances, is the result of salt content, temperature, and pressure in the oceans. The deep scattering layer, formed by small animals moving toward the surface at night and away from it during the day, reflects portions of a sound beam and creates a false bottom on a depth recorder trace.

Sea ice is formed in the extreme cold of polar latitudes. The process is self-insulating, so that large thicknesses of ice do not form each winter. Icebergs are formed from glaciers that break off into the sea.

Fog occurs when water vapor condenses to form liquid droplets. Three types of fog occur: advective fog occurs when warm, water-saturated air passes over cold water; sea smoke occurs when dry, cold air moves over warm water; and radiative fog occurs when warm, moist air is cooled at night.

Key Terms

All key terms from this chapter can be viewed by term, or by definition, when studied as flashcards on this book's Online Learning Center at www.mhhe.com/sverdrup (click on this book's cover).

covalent bond, 132	radiation, 140
polar molecule, 132	electromagnetic radiation, 141
hydrogen bond, 132	electromagnetic spectrum, 141
temperature, 134	refraction, 142
degree, 134	absorption, 142
heat, 134	scattering, 142
calorie, 134	attenuation, 142
dew point, 135	Secchi disk, 142
latent heat of fusion, 135	deep scattering layer (DSL), 144
latent heat of vaporization, 135	
sublimation, 136	sonar, 144
heat capacity, 136	sound shadow zone, 145
cohesion, 137	sofar channel, 145
surface tension, 137	fast ice, 148
viscosity, 137	iceberg, 150
ions, 139	fog, 152
conduction, 140	sea smoke, 152
convection, 140	

Study Questions

1. Discuss the structure of the water molecule.
2. What happens to the arrangement of water molecules when water freezes?
3. What would happen to the sea level of the oceans if all the sea ice melted? Assume that the ocean area does not change and that ocean temperature stays the same.
4. Compare the heat capacity of the oceans with that of the land and the atmosphere. Compare the climate of the west and east coasts of the United States with that of the central and midwestern states. How is climate related to heat capacity, solar heating and cooling, and the direction of the prevailing wind?
5. How do the properties of the water molecule affect (a) surface tension and (b) the dissolving ability of water?
6. Explain why heating the atmosphere from below is an efficient way to distribute heat in the atmosphere, whereas heating the oceans from above is an inefficient way to distribute heat downward in the oceans.

7. If there were no scattering and only absorption of light by seawater, what color would the oceans appear? Why?

8. Both advective fog and sea smoke transfer heat between the atmosphere and the oceans. How do the rate and direction of heat transfer differ between the two? Why?

9. How are the amounts of heat energy lost and gained when water changes its state related to the bonding and motion of the water molecules?

10. If the substances in table 4.3 are exposed to the same quantity of heat energy, which liquid and which solid undergo the greatest temperature change?

11. Explain the effect of increasing and decreasing the salinity, temperature, and pressure on the density of seawater.

12. How are light and sound in seawater affected by changes in the water's density?

13. Where do icebergs come from? How are they formed?

14. Explain differences between *sonar* and *sofar*.

15. Why might one expect to find that traces of every known naturally occurring substance are dissolved in seawater?

Study Problems

1. If surface cooling and evaporation remove heat from the surface of a deep lake at the rate of 800×10^3 calories per hour, and if the surface temperature of the lake is 0°C, at what rate must deep water be pumped to the surface of the lake to prevent the formation of ice?

2. Determine the freezing point of seawater at 32 g/kg salt content. Use figure 4.6.

3. If an echo sounder measures the depth of the water as 3500 m, but the instrument is shown to have a timing error of ±0.001 second over the total time period of the measurement, what is the error in the depth measurement?

4. If water from a given area of sea surface evaporates at the rate of 1 metric ton per day at 20°C, at what rate is heat being transferred to the atmosphere above this surface area? See table 4.2.

5. If heat is removed from a kilogram of water at a constant rate that lowers the temperature of the water from 25°C to 0°C in thirty minutes, how long will it take to convert 0°C water to ice?

Links to Related Websites

Visit the book's Online Learning Center at www.mhhe.com/sverdrup (click on the book's cover) to find live Internet links for additional topics related to this chapter's content.

- Water: General properties
- Water density
- Light
- Sound
- Sea ice

Visit the book's Online Learning Center at www.mhhe.com/sverdrup (click on this book's cover) to find these additional chapter tools: Suggested Readings; links to further information on boxed readings, selected figures, and related chapter topics; and additional study aids.

chapter 5

The Chemistry of Seawater

nd truly, though we were at sea, there was much to behold and wonder at, to me, who was on my first voyage. What most amazed me was the sight of the great ocean itself, for we were out of sight of land. All round us, on both sides of the ship, ahead and astern, nothing was to be seen but water—water—water; not a single glimpse of green shore, not the smallest island, or speck of moss anywhere. Never did I realize till now what the ocean was: how grand and majestic, how solitary, and boundless, and beautiful and blue; for that day it gave no tokens of squalls or hurricanes, such as I had heard my father tell of, nor could I imagine how anything that seemed so playful and placid could be lashed into rage, and troubled into rolling avalanches of foam, and great cascades of waves, such as I saw in the end.

Herman Melville
From *Redburn*

Evaporate salt Pond—Scammon's Lagoon, Baja California, Mexico.

Seawater is salt water, and historically seawater has been valued for its salt. Until recently salt was enormously important as a food preservative, and at one time salt formed the basis for a major commercial trade. Today, although salt is still extracted from seawater, it is the water that has become increasingly valuable in many areas of the world. Seawater is also much more than salt water. Seawater is a complex mixture containing dissolved gases, nutrient substances, and organic molecules as well as salts.

In this chapter we investigate this seawater mixture, and we explore the physical, chemical, and biological processes that regulate this mixture. We also review the commercial extraction of salts from seawater and the possibilities of increasing our supply of fresh water by desalination.

The pH of Seawater 5.1

Compounds can break apart into individual atoms or groups of atoms that have opposite electrical charges. A charged atom or group of atoms is an ion. An ion with a positive charge is a **cation;** an ion with a negative charge is an **anion.** The water molecule, H_2O, can dissociate (break apart) to form a hydrogen cation, H^+, and a hydroxide anion, OH^-. Consequently, in any water solution there will always be a combination of H_2O molecules, H^+ ions, and OH^- ions. The concentration of H_2O molecules always greatly exceeds the concentrations of the two ions. In a pure water solution (one in which there is only water molecules) at 25°C, a very small fraction of the water molecules, about 10^{-7}, will spontaneously dissociate into H^+ and OH^- ions. In other words, the concentration of both H^+ and OH^- ions will be 10^{-7}, as one in every 10 million (10^7) water molecules breaks apart. Solutions in which the concentrations of these two ions are equal are called neutral solutions.

In solutions that are not pure water, chemical reactions can remove or release hydrogen ions, making the concentrations of H^+ and OH^- unequal. The concentrations of H^+ and OH^- in a water solution are inversely proportional to each other. In other words, a tenfold increase in the concentration of one of these ions results in a tenfold decrease in the concentration of the other. An imbalance in the relative concentrations of these ions results in either an acidic (if there are more H^+ cations than OH^- anions) or alkaline, also called basic, (if there are more OH^- anions than H^+ cations) solution.

The acidity or alkalinity of a solution is measured using the **pH** scale, which ranges from a low of 0 to high of

14 (fig. 5.1). The pH scale is a logarithmic scale that measures the concentration of the hydrogen ion (written $[H^+]$) in a solution. The formal definition of pH is:

$$pH = -\log_{10}[H^+]$$

In pure water, where the concentrations of H^+ and OH^- are both 10^{-7}, the pH is equal to 7,

$$pH = -\log_{10}[10^{-7}] = -(-7) = 7,$$

and the solution is neutral. If the concentration of the hydrogen ion is increased by a factor of ten to 10^{-6}, or one part in 1 million instead of one part in 10 million, then the pH would drop to 6 and the solution would be slightly acidic. Solutions with a pH less than 7 (high H^+ concentrations) are acidic and those with a pH greater than 7 (low H^+ concentrations) are alkaline or basic.

Because the hydrogen ion is very reactive with other compounds, acidic water (water with a relatively high concentration of H^+ and pH less than 7) is an effective chemical weathering agent capable of dissolving rock. The normal pH of rainwater is about 5.0–5.6 (slightly acidic), but in some heavily industrialized regions where emissions combine with water droplets to form acid rain it can be much lower. Typical rain in the eastern United States has a pH of about 4.3, roughly ten times more acidic than normal, and can in some cases drop as low as 3, or 100 times more acidic than normal.

Seawater is slightly alkaline with a pH between 7.5 and 8.5. The average pH value for the world's oceans over all depths is approximately 7.8. The pH of seawater remains relatively constant because of the buffering action of carbon dioxide in the water. A **buffer** is a substance that prevents sudden, or large, changes in the acidity or alkalinity of a

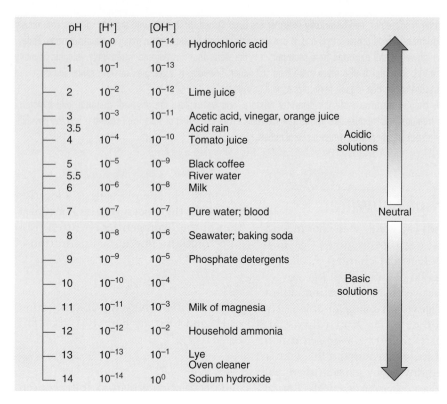

pH	[H⁺]	[OH⁻]	

pH	$[H^+]$	$[OH^-]$	
0	10^0	10^{-14}	Hydrochloric acid
1	10^{-1}	10^{-13}	
2	10^{-2}	10^{-12}	Lime juice
3	10^{-3}	10^{-11}	Acetic acid, vinegar, orange juice
3.5			Acid rain
4	10^{-4}	10^{-10}	Tomato juice
5	10^{-5}	10^{-9}	Black coffee
5.5			River water
6	10^{-6}	10^{-8}	Milk
7	10^{-7}	10^{-7}	Pure water; blood
8	10^{-8}	10^{-6}	Seawater; baking soda
9	10^{-9}	10^{-5}	Phosphate detergents
10	10^{-10}	10^{-4}	
11	10^{-11}	10^{-3}	Milk of magnesia
12	10^{-12}	10^{-2}	Household ammonia
13	10^{-13}	10^{-1}	Lye
			Oven cleaner
14	10^{-14}	10^0	Sodium hydroxide

Acidic solutions

Neutral

Basic solutions

Figure 5.1 The pH scale is a measure of the concentration of hydrogen ions in a solution. A neutral solution has a pH of 7. Acidic solutions have a pH less than 7 and alkaline, or basic, solutions have a pH greater than 7.

solution. If some process changes the concentration of hydrogen ions in seawater, causing the pH to rise above or fall below its average or mean value, the buffer becomes involved in chemical reactions that release or capture hydrogen ions, returning the pH to normal. When carbon dioxide dissolves in seawater, the CO_2 combines with the water to form carbonic acid (H_2CO_3). The carbonic acid rapidly dissociates into bicarbonate (HCO_3^-) and a hydrogen ion (H^+), or carbonate (CO_3^{2-}) and two hydrogen ions ($2\ H^+$). The CO_2, H_2CO_3, HCO_3^-, and CO_3^{2-} exist in equilibrium with each other and with H^+ as shown in the following equation. The double arrows indicate that the reactions can move in either direction, either producing or removing hydrogen ions as is necessary to maintain a relatively constant pH.

$$CO_2 + H_2O \Leftrightarrow H_2CO_3 \Leftrightarrow HCO_3^- + H^+ \text{ or } CO_3^{2-} + 2\ H^+$$

If seawater becomes too alkaline or basic, then the reactions in this equation progress to the right, releasing hydrogen ions and decreasing the pH. If seawater becomes too acidic, then the reactions progress to the left, removing free hydrogen ions from the water and increasing the pH. This buffering capacity of carbon dioxide in seawater is important to organisms requiring a relatively constant pH for their life processes and to the chemistry of seawater, which is controlled, in part, by its pH.

From the reactions described, it is clear that the pH of seawater is inversely proportional to the concentration of CO_2 in the water. The higher the concentration of CO_2 in the

water, the lower its pH and the more acidic it becomes. While the average pH of seawater tends to be fairly constant, it does vary with changes in the concentration of CO_2 in the water, as discussed further in section 5.3.

Salts 5.2

Ocean Salinities

In the major ocean basins, 3.5% of the weight of seawater is, on the average, dissolved salt and 96.5% is water, so a typical 1000 g or 1 kg sample of seawater is made up of 965 g of water and 35 g of salt. Oceanographers measure the salt content of ocean water in grams of salt per kilogram of seawater (g/kg), or parts per thousand (‰). The total quantity of dissolved salt in seawater is known as **salinity,** and the average ocean salinity is approximately 35‰. There are about 1338.5 million km³ ($\sim 1.34 \times 10^9$ km³) of seawater in the oceans, with an average density of approximately 1.03 g/cm³ (1.03×10^{12} kg/km³) (see tables 1.4 and 4.4). The total weight of this seawater is about 1378.7 million trillion kg ($\sim 1.4 \times 10^{21}$ kg); the weight of the salt in the oceans is about 3.5% of the total weight, or approximately 48 million trillion kg ($\sim 1.1 \times 10^{20}$ lb, or $\sim 5.3 \times 10^{16}$ tons). If all of the water in the oceans evaporated, this amount of salt would form a layer roughly 45.5 m (150 ft) thick over the entire surface of the Earth.

The salinity of ocean surface water is associated with latitude. Latitudinal variations in evaporation and precipitation, as well as freezing, thawing, and freshwater runoff from the land, affect the amount of salt in seawater. The relationship between evaporation, precipitation, and mid-ocean surface salinity with latitude is shown in figure 5.2. Notice the low surface salinities in the cool and rainy 40°–50°N and S latitude belts, high evaporation rates and high surface salinities in the desert belts of the world centered on 25°N and S, and low surface salinities again in the warm but rainy tropics centered at 5°N. Sea surface salinities during the Northern Hemisphere summer are shown in figure 5.3

In coastal areas of high precipitation and river inflow, surface salinities fall below the average. For example, during periods of high flow, the water of the Columbia River lowers the Pacific Ocean's surface salinity to less than 25‰ as far as 35 km (20 mi) at sea. Also, sailors have dipped up water fresh enough to drink from the ocean surface 85 km (50 mi) from the mouth of the Amazon River. In subtropic regions of high evaporation and low freshwater input, the surface salinities of nearly landlocked seas are well above the average: 40–42‰ in the Red Sea and the Persian Gulf and 38–39‰ in the Mediterranean Sea. In the open ocean at these same latitudes the surface salinity is closer to 36.5‰. Surface salinities change seasonally in polar areas, where the

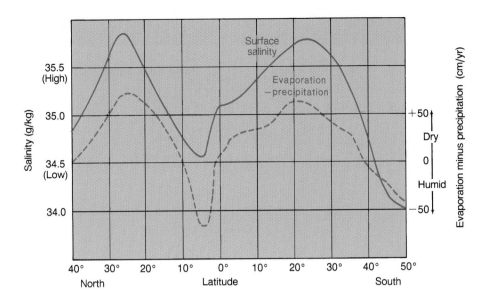

Figure 5.2 Mid-ocean average surface salinity values match the average changes in evaporation minus precipitation values that occur with latitude. *Source: From A.N. Straher, Physical Geography, 1960, John Wiley & Sons, Inc., New York, N.Y.*

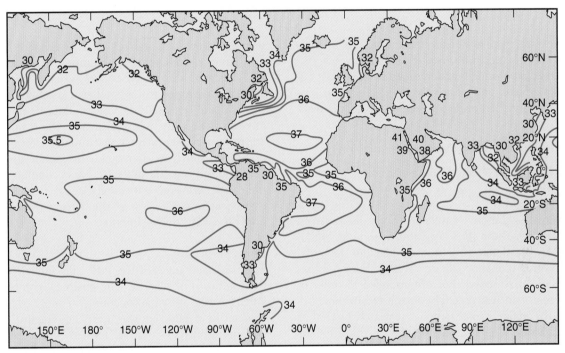

Figure 5.3 Average sea surface salinities in the Northern Hemisphere summer, given in parts per thousand (‰). (High salinities are found in areas of high evaporation; low salinities are common in coastal areas and regions of high precipitation.

surface water forms sea ice in winter, leaving behind the salt and raising the salinity of the water under the ice. In summer a freshwater surface layer forms when the sea ice melts. Deep-water samples from the mid-latitudes are usually slightly less salty than the surface waters because the deep water is formed at the surface in high latitudes with high precipitation. The formation of these deep-water types is discussed in chapter 7.

Dissolved Salts

The atoms or groups of atoms that combine to form compounds can be held together by different kinds of bonds. The two hydrogen atoms and one oxygen atom in the water molecule are held together by covalent bonds in which electrons are shared between the atoms (see section 4.1). Other compounds, such as sodium chloride (NaCl), are held together by **ionic bonds** in which electrons are transferred from a metal atom (in this case the metal sodium) to a nonmetal atom (in this case the chlorine atom), creating ions of opposite charge that attract each other. As stated earlier, an ion with a positive charge is a cation; an ion with a negative charge is an anion. Ionic bonds are easily broken in water because of the polar nature of the water molecule. Thus when salts are added to water, the salts dissolve, or dissociate (break apart),

Table 5.1 Major Constituents of Seawater[1]

Constituent		Symbol	g/kg in Seawater		Percentage by Weight	
Chloride	The six most abundant ions	Cl⁻	19.35		55.07	
Sodium		Na⁺	10.76		30.62	
Sulfate		SO₄²⁻	2.71	34.91	7.72	99.36
Magnesium		Mg²⁺	1.29		3.68	
Calcium		Ca²⁺	0.41		1.17	
Potassium		K⁺	0.39		1.10	
Bicarbonate		HCO₃⁻	0.14		0.40	
Bromide		Br⁻	0.067		0.19	
Strontium		Sr²⁺	0.008		0.02	
Boron		B³⁺	0.004		0.01	
Fluoride		F⁻	0.001		0.01	
Total			35.13		99.99	

1. Nutrients and dissolved gases are not included.

From J. P. Riley and G. Skirrow, Chemical Oceanography, *Vol. 1, 2nd edition. Copyright 1975 Elsevier Science. Reprinted by permission.*

into ions. The salts in seawater are mostly present in their ionic forms as cations and anions (see section 4.7).

When sodium chloride is added to water, the bonds between the atoms break and positively charged sodium ions and negatively charged chloride ions are formed. This reaction can be written as

sodium chloride → sodium ion + chloride ion

or it can be written with chemical abbreviations as

$$NaCl \rightarrow Na^+ + Cl^-$$

Six ions make up more than 99% of the salts dissolved in seawater. Four of these ions are cations: sodium (Na^+), magnesium (Mg^{2+}), calcium (Ca^{2+}), and potassium (K^+); two are anions: chloride (Cl^-) and sulfate (SO_4^{2-}). Table 5.1 lists these six ions and five more; they are arranged in order of the amount present in seawater. The ions listed in table 5.1 are known as the **major constituents** of seawater. Note that sodium and chloride ions account for 86% of the salt ions present in seawater.

All the other elements dissolved in seawater are present in concentrations of less than one part per million and are called **trace elements** (table 5.2). Most trace elements are present in such small concentrations that it is common to report their concentrations at the parts per billion level. The conversion of concentration units is simply a matter of powers of ten. Remember that "parts per thousand" is equivalent to "g/kg," "parts per million" is equivalent to "mg/kg," and "parts per billion" is equivalent to "μ g/kg." Thus, one part per thousand (1 g/kg) = 10^3 parts per million (1000 mg/kg) = 10^6 parts per billion (1 million μ g/kg), or one part per billion (1 μ g/kg) = 10^{-3} parts per million (0.001 mg/kg) = 10^{-6} parts per thousand (0.000001 g/kg). Some of these elements are important to organisms that are able to concentrate the ions. For example, long before the presence of iodine could be determined chemically as a trace element of seawater, it was known that shellfish and

seaweeds were rich sources for this element, and seaweed was harvested for commercial iodine extraction.

Because the major constituents of seawater do not change their ratios to each other with changes in total salt content, and because these constituents are not generally removed or added by living organisms, the major constituents are termed **conservative constituents.** Certain of the ions present in much smaller quantities, some dissolved gases, and assorted organic molecules and complexes do change their concentrations with biological and chemical processes that occur in some areas of the oceans; these are known as **nonconservative constituents.**

Sources of Salt

The original sources of the seas' salts include the crust and the interior of the Earth. The chemical composition of the Earth's rocky crust can account for most of the positively charged ions found in seawater. Large quantities of cations are present in rocks that are formed by the crystallization of molten magma from volcanic processes. The physical and chemical weathering of rock over time breaks it into small pieces and the rain dissolves out ions, which are carried to the sea by rivers. Anions are present in the Earth's interior and may have been present in the Earth's early atmosphere. Some anions may have been washed from the atmosphere by long periods of rainfall, but the more likely source of most of the anions is thought to have been the Earth's mantle. During the formation of the Earth, gases from the mantle are believed to have supplied anions to the newly forming oceans.

Gases released during volcanic eruptions (for example, hydrogen sulfide, sulfur dioxide, and chlorine) dissolve in rainwater or river water and are carried to the oceans as Cl^- (chloride) and SO_4^{2-} (sulfate). Water carrying chloride ions and sulfur-containing ions is acidic and erodes and dissolves the rock over which it flows, helping to liberate the cations.

Tests show that the most abundant ions in today's rivers (table 5.3) are the least abundant ions in ocean water,

Table 5.2 Concentrations of Trace Elements in Seawater[1]

Element	Symbol	Concentration[2]	Element	Symbol	Concentration[2]
Aluminum	Al	5.4×10^{-1}	Manganese	Mn	3×10^{-2}
Antimony	Sb	1.5×10^{-1}	Mercury	Hg	1×10^{-3}
Arsenic	As	1.7	Molybdenum	Mo	1.1×10^{1}
Barium	Ba	1.37×10^{1}	Nickel	Ni	5×10^{-1}
Bismuth	Bi	$\leq 4.2 \times 10^{-5}$	Niobium	Nb	$\leq (4.6 \times 10^{-3})$
Cadmium	Cd	8×10^{-2}	Protactinium	Pa	5×10^{-8}
Cerium	Ce	2.8×10^{-3}	Radium	Ra	7×10^{-8}
Cesium	Cs	2.9×10^{-1}	Rubidium	Rb	1.2×10^{2}
Chromium	Cr	2×10^{-1}	Scandium	Sc	6.7×10^{-4}
Cobalt	Co	1×10^{-3}	Selenium	Se	1.3×10^{-1}
Copper	Cu	2.5×10^{-1}	Silver	Ag	2.7×10^{-3}
Gallium	Ga	2×10^{-2}	Thallium	Tl	1.2×10^{-2}
Germanium	Ge	5.1×10^{-3}	Thorium	Th	(1×10^{-2})
Gold	Au	4.9×10^{-3}	Tin	Sn	5×10^{-4}
Indium	In	1×10^{-4}	Titanium	Ti	$< (9.6 \times 10^{-1})$
Iodine	I	5×10^{1}	Tungsten	W	9×10^{-2}
Iron	Fe	6×10^{-2}	Uranium	U	3.2
Lanthium	La	4.2×10^{-3}	Vanadium	V	1.58
Lead	Pb	2.1×10^{-3}	Yttrium	Y	1.3×10^{-2}
Lithium	Li	1.7×10^{2}	Zinc	Zn	4×10^{-1}
			Rare Earths		$(0.5\text{–}3.0) \times 10^{-3}$

1. Nutrients and dissolved gases are not included.
2. Parts per billion, or μ g/kg.
Note: Parentheses indicate uncertainty about concentration.
From J. P. Riley and R. Chester, Chemical Oceanography, *Vol. 8. Copyright 1983 Elsevier Science. Reprinted by permission.*

Table 5.3 Dissolved Salts in River Water

Ion	Symbol	Percentage by Weight
Bicarbonate	HCO_3^-	48.7
Calcium	Ca^{2+}	12.5
Silicon dioxide (nonionic)	SiO_2	10.9
Sulfate	SO_4^{2-}	9.3
Chloride	Cl^-	6.5
Sodium	Na^+	5.2
Magnesium	Mg^{2+}	3.4
Potassium	K^+	1.9
Oxides (nonionic)	$(Fe, Al)_2O_3$	0.8
Nitrate	NO_3^-	0.8
Total		100.00

Note: Average river salt concentration is 0.120‰.
Data from U.S. Geological Survey.

because the rivers have previously removed the most easily dissolved land salts and are now carrying the less soluble salts. Exceptions to this pattern are found in rivers used for irrigation. These rivers are flowing through arid soils that have not lost much of their salt content. The river water is frequently used several times and passes through a number of irrigation projects on its way downriver, causing the water to become increasingly salty and unfit for irrigation purposes downstream. This situation has produced years of continuous conflict between the United States and Mexico over the waters of the Colorado and Rio Grande Rivers, which irrigate much of the agricultural land of the U.S. desert Southwest before becoming available to the farmlands of Mexico.

In addition, we know that hot-water vents located on the sea floor supply chemicals to the ocean water and also remove them. Hydrothermal activity found at the mid-ocean ridges and associated with hot spots and ridge formation may play an important role in stabilizing the ocean's salt composition. When hot magma is introduced into the cold crust, it cracks and becomes permeable. Pressure from the water above the sea floor is high (1 atm for every 10 m of water depth), and it forces water into cracks and voids, where it is heated under pressure to extremely high temperatures. The salinity of seawater entering the hydrothermal system is relatively constant worldwide, but the water emerging from hydrothermal vents has a salinity that may be more than double its original value. The process controlling this salinity change and its role in the ocean's total salt balance are unclear at this time.

Regulating the Salt Balance

We know from the age of older marine sedimentary rocks that the oceans have been present on the Earth for about 3.5 billion years. Chemical and geological evidence from rocks and salt deposits leads researchers to believe that the

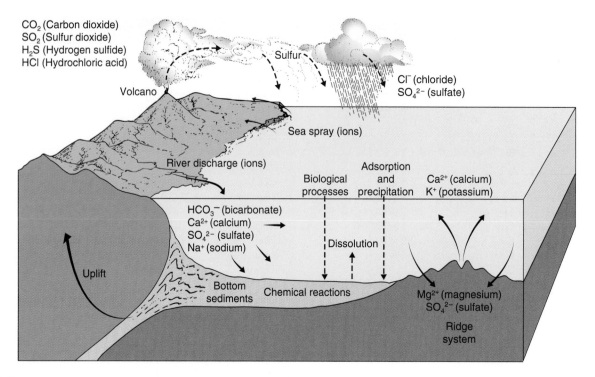

Figure 5.4 Processes that distribute and regulate the major constituents in seawater. Salt ions are added to seawater from rivers, volcanic events, ridge systems, and decay processes. Salt ions are removed from seawater by adsorption and ion exchange, spray, chemical precipitation, biological uptake, and release at ridge systems.

salt composition of the oceans has been the same for about the last 1.5 billion years. The total amount of dissolved material in the world's oceans is calculated to be 5×10^{22} g for ocean water of 36‰ salinity. Each year the runoff from the land adds another 2.5×10^{15} g, or 0.000005% of total ocean salt. If we assume that the rivers have been flowing to the sea at the same rate over the last 3.5 billion years, more dissolved material has been added by this process than is presently found in the sea. For the oceans to remain at the same salinity, the rate of addition of salt by rivers must be balanced by the removal of salt; input must balance output.

Salt ions are removed from seawater in a number of ways; some are shown in figure 5.4. Sea spray from the waves is blown ashore, depositing a film of salt on land. This salt is later returned to the oceans by runoff from the land. Over geologic time, shallow arms of the sea have become isolated, the water has evaporated, and the salts have been left behind to become land deposits called **evaporites.** Salt ions can also react with each other to form insoluble products that precipitate on the ocean floor. Biological processes concentrate salts, which are removed from the water if the organisms are harvested or if the organisms become part of the sediments. Organisms' excretion products trap ions, which are transferred to the sediments or returned to the seawater. Other biological processes remove Ca^{2+} (calcium) by incorporating it into shells, and silica is used to form the hard exterior coverings of certain plantlike organisms and animals. These hard parts are accumulated in the sediments when the organisms die. Chapter 3 discussed the sediments produced by biological processes.

A chemical process known as **adsorption,** the adherence of ions and molecules onto a particle's surface, removes other ions and molecules from seawater. In this process, tiny clay mineral particles, weathered from rock and brought to the oceans by winds and rivers, bind ions such as K^+ (potassium) and trace metals to their surfaces. The ions sink with the clay particles and are eventually incorporated into the sediments. Strongly adsorbable ions replace weakly adsorbable ions in a process known as **ion exchange.** If the clay minerals and sediments adsorb and exchange with one ion more easily than with another, there is a lower concentration of the more easily adsorbed ion in the seawater. For example, potassium is more easily adsorbed than sodium, so K^+ is less concentrated than Na^+ in seawater. Nickel, cobalt, zinc, and copper are adsorbed on nodules that form on the ocean floor. The fecal pellets and skeletal remains of small organisms also act as adsorption surfaces. It is the settling of this organic debris that transports metallic ions to the sediment, where they can be adsorbed on nodules. In all of these cases the ions are removed from the water and are transferred to the sediments.

The process of forming the new Earth's crust at the ridge system of the deep-ocean floor (see chapter 2) participates in the input and output of the ions in seawater. Where molten rock rises from the mantle into the crust, magma chambers are formed. These chambers are found mainly along plate boundaries but have also been identified under volcanic hot spots in the middle of plates. Cold water seeps down several kilometers through the fractured crust at a spreading center. The seawater becomes heated by flowing near the magma chamber. By convection, the heated water rises up through the crust and reacts chemically with the rocks of the crust. Magnesium ions (Mg^{2+}) are transferred from the water to form mineral deposits in the crust. At the same time, hydrogen ions (H^+) are released and the seawater solution becomes more acidic. Chemical reactions change sulfate (SO_4^{2-}) to sulfur and then to hydrogen sulfide (H_2S). The acidic water releases and dissolves metals from the crust, including copper, iron, manganese, zinc, potassium, and calcium. It has been estimated that a volume of water equivalent to the entire mass of the oceans circulates through the crust at oceanic ridges every 10 million years.

The most important process for the removal of most elements from seawater is still the adsorption of ions onto fine particles and their removal to the sediments. Ions deposited in the sediments are trapped there. They are not returned to or redissolved in the seawater; but geological uplift processes elevate some sediments from the sea floor to positions above sea level. Erosion works to dissolve and wash these deposits back to the sea.

Residence Time

The relative abundance of the salts in the sea is due in part to the ease with which they are introduced from the Earth's crust and in part to the rate at which they are removed from the seawater. Sodium is moderately abundant in fresh water, but, because it reacts slowly with other substances in seawater, it remains dissolved in the ocean. Calcium is removed rapidly from seawater to form limestone and the shells of marine organisms, but it is also replaced rapidly by calcium ions moving down rivers and cycling in and out of the ridge systems on the sea floor (fig. 5.4).

The average, or mean, time that a substance remains in solution in the ocean is called its **residence time** (table 5.4). Aluminum, iron, and chromium ions have short residence times, in the hundreds of years. They react with other substances quickly and form insoluble mineral solids in the sediments. Sodium, potassium, and magnesium are very soluble and have long residence times, in the millions of years. If the concentration of an ion in seawater is constant, the rates of supply and removal are equal. If the total amount of an ion present in the ocean is divided by either its rate of supply or its rate of removal, the residence time for that ion is known. For example, there are roughly 5.74×10^{20} g of Ca^{2+} in the oceans. It is estimated that Ca^{2+} is added to the oceans at a rate of about 5.55×10^{14} g per year. Therefore

$$\text{residence time (Ca}^{2+}) = \left(\frac{5.74 \times 10^{20}\,g}{5.55 \times 10^{14}\,g/yr} \right)$$
$$= 1.03 \times 10^6 \text{ years}$$

Constant Proportions

Seawater is a well-mixed solution; currents and eddies in both surface and deep water, vertical mixing processes, and wave and tidal action have all helped to stir the oceans through geologic time. Because of this thorough mixing, the ionic composition of open-ocean seawater is the same from

Table 5.4 Approximate Residence Time of Ions in the Oceans

Ion	Time in Years
Chloride	100 million–infinite
Sodium	210 million
Magnesium	22 million
Potassium	11 million
Sulfate	11 million
Calcium	1 million
Manganese	0.0014 million (1400)
Iron	0.00014 million (140)
Aluminum	0.00010 million (100)

From Fundamentals of Oceanography, 4th edition, Duxbury, Duxbury, and Sverdrup. Copyright 2000 The McGraw-Hill Companies. All rights reserved.

place to place and from depth to depth. That is, the ratio of one major ion or seawater constituent to another remains the same. Whether the salinity is 40‰ or 28‰, the major ions exist in the same proportions. This concept was first suggested by the chemist Alexander Marcet in 1819. In 1865 another chemist, Georg Forchhammer, analyzed several hundred seawater samples and found that these constant proportions did hold true. During the world cruise of the *Challenger* expedition (1872–76), seventy-seven water samples were collected from different depths and locations in the world's oceans. When chemist William Dittmar analyzed these samples, he verified Forchhammer's findings and Marcet's suggestion. These analyses led to the **principle of constant proportion** (or **constant composition**) of seawater, which states that regardless of variations in salinity, the ratios between the amounts of major ions in open-ocean water are constant. Note that the principle applies to major conservative ions in the open ocean; it does not apply along the shores, where rivers may bring in large quantities of dissolved substances or may reduce the salinity to very low values. The ratios of abundance of minor nonconservative constituents vary, because some of these ions are closely related to the life cycles of living organisms. Populations of organisms remove ions during periods of growth and reproduction, reducing the amounts of these ions in solution. Later, the population declines, and decay processes return these ions to the seawater.

Determining Salinity

Seawater transmits, or conducts, electricity because it contains dissolved salts; the more ions there are in solution, the greater the conductance. Therefore, the salt content, or salinity, can be determined by using an instrument, called a **salinometer,** that measures electrical conductivity. Salinity readings in parts per thousand (S‰) are made quickly and directly on the water sample with an electrical probe (fig. 5.5). Because conductivity of seawater is affected by both salinity and temperature, a conductivity instrument must correct for the temperature of the sample if the instrument reads directly

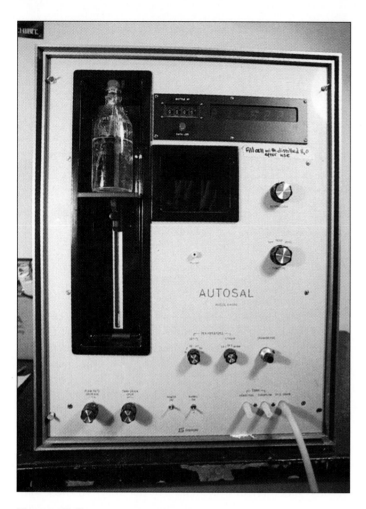

Figure 5.5 The salinometer determines salinity by measuring the electrical conductivity of a seawater sample.

in salinity units. This correction is not necessary if the instrument reads conductance only. If conductance and temperature are measured separately, a computer program calculates the salinity. Other direct measurement techniques are discussed in chapter 7.

To be sure that all salinity determinations are equalized in the same way, the world's oceanographic laboratories use a standard method of analysis and a standard seawater reference. At present, the Institute of Oceanographic Services in Wormly, England, is responsible for the production of standard seawater adjusted to both constant chloride content and electrical conductance to ensure standard calibration of laboratory instruments.

Historically, the quantity of chloride ions present in a water sample was measured to establish the salinity of the sample. To do so, standardized silver nitrate was added because the silver combines with the sample's chloride ions. If the amount of silver required to react with all the chloride ions in a sample is known, the amount of chloride is known. The chloride concentration measured in this way is termed **chlorinity (Cl‰)** and is measured in parts per thousand or grams of chloride per kilogram of seawater. When the chlo-

rinity of a sample is known, the concentration of any other major constituent can be calculated by using the principle of constant proportion.

Chlorinity and salinity are related by the equation

$$\text{salinity (‰)} = 1.80655 \times \text{chlorinity (‰)}$$

or

$$S‰ = 1.80655 \times Cl‰$$

Gases 5.3

Gases move between the sea and the atmosphere at the sea surface. Atmospheric gases dissolve in the seawater and are distributed to all depths by mixing processes and currents. The most abundant gases in the atmosphere and in the oceans are nitrogen (N_2), oxygen (O_2), and carbon dioxide (CO_2). The percentages of each of these gases in the atmosphere and in seawater are given in table 5.5. Oxygen and carbon dioxide play important roles in the ocean because they are necessary to life, and biological activities modify their concentrations at various depths. Nitrogen is not used directly by living organisms except for certain bacteria. Gases such as argon, helium, and neon are present in small amounts, but they are inert and do not interact with the ocean water, nor are they used by its inhabitants.

The amount of any gas that can be held in solution without causing the solution either to gain or to lose gas is the **saturation value.** The saturation value changes because it depends on the temperature, salinity, and pressure of the water. If the temperature or salinity decreases, the saturation value for the gas increases. If the pressure decreases, the saturation value decreases. In other words, colder water holds more dissolved gas than warmer water, less-salty water holds more gas than more-salty water, and water under more pressure holds more gas than water under less pressure.

Distribution with Depth

In the process known as **photosynthesis,** plants and plant-like organisms produce oxygen and use carbon dioxide to form organic molecules. Because plants require sunlight for photosynthesis, they are confined to the sea surface, on average the top 100 m (330 ft) of the water column, where sufficient light is available to energize the photosynthesis process. Therefore, oxygen is produced in surface water and carbon dioxide is consumed there. By contrast, **respiration,** which breaks down organic substances to provide energy, requires oxygen and produces carbon dioxide. All living organisms respire in order to produce energy in living cells, and therefore respiration occurs at all depths of the ocean. Photosynthesis and respiration are discussed in more detail in chapter 14. Decomposition, the bacterial breakdown of nonliving organic material, also requires oxygen and releases carbon dioxide. Decomposition becomes the more important factor in the removal of oxygen from seawater at greater depths because usually the densities of living populations diminish with depth.

Table 5.5 Abundance of Gases in Air and Seawater

Gas	Symbol	Percentage by Volume in Atmosphere	Percentage by Volume in Surface Seawater[1]	Percentage by Volume in Total Oceans
Nitrogen	N_2	78.03	48	11
Oxygen	O_2	20.99	36	6
Carbon dioxide	CO_2	0.03	15	83
Argon, helium, neon, etc.	Ar, He, Ne	0.95	1	
Totals		100.00	100	100

1. Salinity = 36‰, temperature = 20°C.

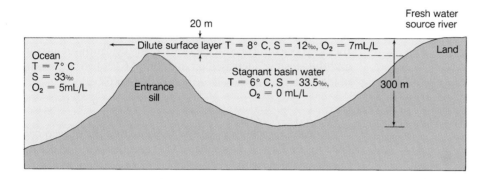

Figure 5.6 High-density ocean water is trapped behind the sill. The trapped water becomes anoxic at depth owing to continual respiration and decomposition. At the surface, sunlight enhances photosynthesis and the production of oxygen.

The depth at which the rate of photosynthesis balances the rate of respiration by plants and plantlike organisms is called the **compensation depth.** Above the compensation depth, plants and plantlike organisms produce oxygen at the expense of carbon dioxide; below it, carbon dioxide is produced at the expense of oxygen. Oxygen can be added to the oceans only at the surface, from exchange with the atmosphere or as a waste product of photosynthesis. Carbon dioxide also enters from the atmosphere at the surface, but it is available at all depths from respiration and decomposition.

Dissolved oxygen concentrations vary from 0–10 mL/L of seawater. Very low or zero concentrations occur in the bottom waters of isolated deep basins, which have little or no exchange or replacement of water. Such an area can occur at the bottom of a trench, in a deep basin behind a shallow entrance sill (as in the Black Sea), or at the bottom of a deep fjord (300–400 m; 900–1300 ft). If the deep water is stagnant, respiration and decomposition use up the oxygen faster than the slow circulation to this depth is able to replace it. The bottom water becomes **anoxic,** or stripped of dissolved oxygen; **anaerobic** (or nonoxygen-using) bacteria live in such water. This condition is shown in figure 5.6. Because oxygen is more soluble in cold water than in warm water, more oxygen is found in surface waters at high latitudes than at lower latitudes. If the water is quiet, the nutrients and sunlight are abundant, and a large population of plants is present, oxygen values at the surface can rise above the equilibrium (or saturation) value to 150% or more. This water is **supersaturated.** Wave action tends to liberate oxy-

gen to the atmosphere and return the water to its 100% saturation state.

Carbon dioxide levels range between 45 and 54 mL/L of seawater throughout the oceans. The average concentration is not substantially affected by biological processes but tends to remain almost constant, controlled by temperature, salinity, and pressure.

Figure 5.7 shows typical oxygen and carbon dioxide concentrations with depth. At the surface, the concentration of oxygen is high because of the available atmospheric oxygen and photosynthesis. Below the surface layer, oxygen decreases as animal respiration and the decomposition of organic material remove the oxygen; the **oxygen minimum** occurs at about 800 m (2600 ft). Below this depth, the rate of removal of oxygen decreases because the population density of animals and the abundance of decaying organic matter have decreased.

The slow supply of oxygen to greater depths by water sinking from the surface gradually increases the oxygen concentration above that found at the oxygen minimum. The carbon dioxide graph in figure 5.7 shows almost the direct opposite effect. The CO_2 concentration at the surface is low, because it is used in photosynthesis. Below the surface layer the concentration increases with depth as animal respiration and decomposition continually produce CO_2 and add it to the water. The deep water is able to hold high concentrations of CO_2 because the saturation value is high at low temperatures and high pressures. This makes it easy to understand why calcareous oozes are preserved in the warmer, shallower water

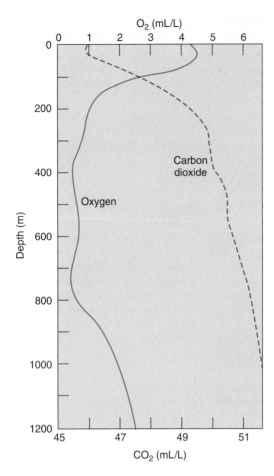

Figure 5.7 The distribution of O_2 and CO_2 with depth. Changes in O_2 concentration may exceed 400% from the surface to depth, whereas CO_2 concentrations change by less than 15%.

above the CCD and dissolved below it (review the discussion of biogenous sediments in chapter 3). At shallow depths the concentration of CO_2 in the oceans is relatively low, in part because of high rates of photosynthesis and in part because the warm, shallow water has a low saturation value. Remember that the pH of seawater is inversely proportional to the concentration of CO_2, so the pH of the surface water tends to be slightly higher, or more alkaline, than average. The pH at the ocean surface may be as high as 8.5 if the water is warm and if the rate of primary production, or photosynthesis, is high. Raising the pH of the water releases carbonate ions, CO_3^{2-}, that bond with the abundant Ca^{2+} in solution to form calcium carbonate, $CaCO_3$. In cold, deep water, where the concentration of CO_2 is high, the pH drops, making the water more acidic and dissolving calcium carbonate shells.

The Carbon Dioxide Cycle

At present, the net annual ocean uptake of carbon as carbon dioxide (CO_2) from the atmosphere is estimated to be 2–3 billion metric tons per year. The rate at which the oceans absorb CO_2 is controlled by water temperature, pH, salinity, the chemistry of the ions (presence of calcium and carbonate ions), and biological processes, as well as mixing and circulation patterns of seawater (fig. 5.8).

The transfer of carbon from CO_2 to organic molecules by photosynthesis results in the addition of CO_2 to the intermediate and deep-ocean water when the organic material sinks and decays. This process is often called the **biological pump.** Small marine plants called phytoplankton (see chapter 14) are responsible for about 40% of the Earth's total production of organic material by photosynthesis. These organisms inhabit the shallow surface water, where sufficient sunlight is available for photosynthesis (see sections 4.8 and 14.2). About 90% of the phytoplankton organic matter is recycled near the surface as a result of consumption by predators and decomposition by bacteria. Most of the remaining 10% (between 70% and 90% of it) is recycled before it reaches the sea floor, where the remainder is consumed by bottom-dwelling animals, decomposed by bacteria, or preserved in the sediment. This biochemical pump works along with the chemical solubility pump discussed previously to concentrate carbon in the deep ocean. The storage of CO_2 in the Earth's reservoirs and the effect of increased production of CO_2 by the burning of fossil fuels are discussed further in chapter 6.

The Oxygen Balance

The oceans also play a large role in regulating the oxygen balance in the Earth's atmosphere. Photosynthesis in the oceans releases oxygen, which is consumed by respiration and decay processes in the same way as on land. However, some organic matter is incorporated into the seafloor sediments, preventing the organic matter's decay and decomposition. Therefore, oxygen is not consumed to balance the oxygen produced in photosynthesis, and 300 million metric tons of excess oxygen should be released into the atmosphere each year, resulting in an increase of oxygen over time. This amount is not released because the marine sediments are formed into rocks by the Earth's geological processes. Some of these rocks are uplifted onto land, and oxygen is eventually consumed in the weathering and oxidation of the materials in the rocks. This process balances the atmosphere's oxygen budget. The mechanisms that link and control the process are not well understood.

Measuring the Gases

The amount of dissolved oxygen present in seawater samples can be measured by traditional chemical techniques in the laboratory. It is also possible to directly measure oxygen in the oceans, using specialized probes that send an electronic signal back to the ship or store the information in the testing unit. Carbon dioxide is rarely measured directly; rather, it is determined from the pH of the water, which can be measured directly.

Other Substances   5.4

Nutrients

Ions required for plant growth are known as nutrients; these are the fertilizers of the oceans. As on land, plants require nitrogen and phosphorus in the form of nitrate (NO_3^-) and phosphate (PO_4^{3-}) ions. A third nutrient required in the

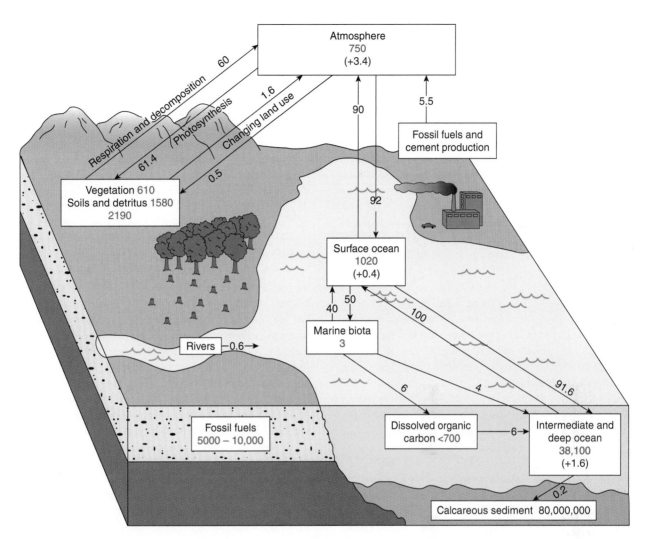

Figure 5.8 Major carbon dioxide pathways through the Earth's environment. All numbers are given in billions of metric tons (1 metric ton = 10^6 g, 1 billion metric tons = 10^{15} g) of carbon. Numbers in *black* are rates of exchange per year between reservoirs. Numbers in *green* are total amounts stored in reservoirs. Numbers in *parentheses* are net annual changes in the total amount stored. *Modified from Schimel, et al. 1995. CO_2 and the Carbon Cycle. In: Climate Change 1994, Cambridge University Press, Cambridge, UK.*

oceans is the silicate ion (SiO_4^-), which is needed to form silica (SiO_2), the hard outer wall of the single-celled plantlike organisms called diatoms and the skeletal parts of some protozoans. These three nutrients are among the dissolved substances brought to the sea by the rivers and land runoff. Despite their importance they are present in very low concentrations (table 5.6).

The concentrations of nutrient ions vary because some of these ions are closely related to the life cycles of living organisms. Nutrients are removed from the water as the plant populations grow and reproduce, temporarily reducing the amounts in solution. Later, when the populations decline, death and decay return the ions to the seawater. Nutrients are cycled into the animal populations as the animals feed on the plants. These animals may be eaten in their turn, and eventually the nutrients are returned to the oceans by way of death and bacterial decomposition. Excretory products from these animals are also added to the seawater, broken down, and used again by a new generation of plants and animals.

Table 5.6 Nutrients in Seawater

Element	Concentration µg/kg[1]
Nitrogen (N)	500
Phosphorus (P)	70
Silicon (Si)	3000

1. Parts per billion.

Nutrients are nonconservative; they do not maintain constant ratios in the way most major salt ions do. The cyclic nature of the nutrient pathways for nitrogen and phosphorus is discussed in chapter 14.

Organics

A wide variety of organic substances are present in seawater. Proteins, carbohydrates, lipids (or fats), vitamins, hormones,

Messages in Polar Ice

The chemical history of the Earth's climate for the last 200,000 years is trapped in the Earth's polar ice sheets. Scientists are going back in time by drilling ice cores in Greenland and Antarctica to discover connections between climate and atmospheric chemistry and the impact of human activities on climate and atmospheric composition (box fig. 1). The chemical composition of the ice preserves, layer by layer, a record of soluble substances (sodium, chloride, sulfate, and nitrate) and heavy metals (lead, cadmium, and zinc). Dust tells how stormy the atmosphere was and when volcanic eruptions occurred, and bubbles in the ice contain trapped air, tracing the natural changes of carbon dioxide and other atmospheric gases through time.

A collaborative European program known as the Greenland Ice Core Project (GRIP) began in 1989 and ended in 1995. In 1992 GRIP reached 3,029 m (9930 ft), where the ice is 200,000 years old or more. A new hole, NGRIP, is being drilled by Denmark north of the GRIP position. Thirty-five kilometers (20 mi) away, the U.S. Greenland Ice Sheet Project (GISP2) spent five years drilling through the ice sheet and into bedrock to recover an ice core of 3053 m (10,014 ft) in July 1993. The drilling operation has ceased, but there is ongoing research at the site.

Scientists analyzing these ice cores are looking back in history. The GISP2 ice core has identified some of the world's major volcanic eruptions by deposits of sulfur as sulfate: Italy's Vesuvius in 6955 B.C.; Oregon's Mt. Mazama's creation of Crater Lake in 5676 B.C.; Iceland's Hekla in 2310 B.C.; and Alaska's Aniakchak in 1623 and 1629 B.C.

Sections of the GRIP ice core were analyzed to follow the history of human lead production. Samples taken from ice that formed 7760 years ago (depth of 1286 m [4000 ft]) in the pre-lead-production period were compared to samples from the times of the rise of Greek civilization, the Roman Republic and Empire, and the Medieval and Renaissance periods. Large-scale pollution from lead smelters during those periods contaminated even the Arctic, leaving the oldest documented record of human pollution.

Comparison of ice layers deposited before and after the Industrial Revolution shows that the concentration of sulfate from sulfuric acid has tripled and that the concentration of nitrate from nitric acid has doubled during the last century. Lead continued to be a significant contaminant and increased dramatically as a by-product of leaded-fuel combustion tied to the world's increased dependence on gasoline.

These ice cores tell us that climate changes have occurred more rapidly and more frequently than scientists had thought. Climate changes are recorded in the ice every few thousand years until the end of the last ice age, about 15,000 years ago. The climate warmed for the next 2000 years until a short period when Greenland temperatures decreased by 7°C, followed by very rapid warming, possibly over a period as short as three years, to the present interglacial stage that the Earth enjoys today. These rapid, recent shifts in climate are thought to have been coupled with large shifts in the circulation of both the oceans and the atmosphere.

GISP2 core samples show massive changes in the quantities of dust and sea salt from aerosols that occur with the onset of glacial periods. When the ocean ice cover expanded, the dusts and salts originated in more southern latitudes, requiring parallel, large-scale changes in atmospheric circulation to deliver the particles to the Arctic. Such events may have contributed to blocking incoming solar radiation and may partly explain the duration of the glacial periods.

In Antarctica, Russian scientists set up at the Vostok Antarctic Research Station in 1957 and began drilling ice cores in 1972. They were joined by a French team in 1984, and together they drilled more than 2000 m (6,600 ft) into the ice, capturing an ice record going back 160,000 years. In February 1998 the Russians shut down their station, having drilled the world's deepest ice core, 3623 m (12,000 ft). The bottom ice of the core dates back about 450,000 years. The Russians stopped their drilling just before reaching the surface of Lake Vostok, which is twice the size of Lake Ontario and lies between the ice sheet and bedrock.

Their findings show that carbon dioxide levels increased at the end of the next to the last ice age, 140,000 years ago, and decreased rapidly again about 100,000 years ago, the start of the last ice age. Ten thousand years ago carbon dioxide levels rose again (by almost 40%); the rising levels at that time may have played a role in ending the last ice age. The Vostok core shows a rise in carbon dioxide over the last 2000 years. Methods that compare Greenland and Antarctic ice cores against a common timeline suggest that the Northern Hemisphere has a more changeable climate than the Southern Hemisphere and that the beginning and ending of ice ages in the north trigger changes in the south.

New ice core projects have begun in West Antarctica, conducted by the United States at Siple Dome (using the drill from the Greenland GISP2 hole) and by the European Union at Dome C. The snowfall at these sites is much heavier than at the Vostok site and produces a more detailed climate record of the last 100,000 years. Data from these new cores will help researchers interpret the Vostok data.

To Learn More About Messages in Polar Ice

Fiedel, S., J. Southon, and T. Brown. 1995. The GISP Ice Core Record of Volcanism Since 7000 B.C. *Science* 267 (5195): 256–58.

Hong, S., P. Candelone, C. Patterson, and C. Boutron. 1994. Greenland Ice Evidence of Hemispheric Lead Pollution Two Millennia Ago by Greek and Roman Civilizations. *Science* 265 (5180): 1841–43.

Mayewski, P.A., et al. 1994. Changes in Atmospheric Circulation and Ocean Ice Cover over the North Atlantic During the Last 41,000 Years. *Science* 263 (5154): 1747–51.

Mayewski, P.A., et al. 1994. Record Drilling Depth Struck in Greenland. *EOS* 75 (10): 113, 119, 124.

Stauffer, B. 1993. The Greenland Ice Core Project. *Science* 260 (5115): 1766–67.

Stone, R. 1998. Russian Outpost Readies for Otherworldly Quest. *Science* 279 (5351): 658–61.

Weiner, J. 1989. Glacier Bubbles Are Telling Us What Was in Ice Age Air. *Smithsonian* 20 (2): 78–87.

Internet References

Visit the book's Online Learning Center at www.mhhe.com/sverdrup (click on the book's cover) to explore links to further information on related topics.

web ⚭ link **Box Figure 1** (a) Scott-Amundsen Base, 90°S, November–December 1982. Cores 10 cm (4 in) in diameter were drilled to 227 m (745 ft) with a mechanical drill. The ice at the bottom of the core is approximately 2200 years old. (b) Greenland Ice Sheet Project 2 (GISP2) drilling station, summer 1990. The tunnel houses laboratories and leads to the drilling site. The boxes at the tunnel entrance hold core sections ready for shipment to cold storage at the scientists' home laboratories. (c) A lengthwise slab is removed from a core section for analysis in the tunnel lab. (d) GISP2, July 1989. The layering of this ice core was inspected and photographed before it was analyzed for gas and particle content, chemistry, and conductivity. The core was 13 cm (5 in) in diameter.

and their breakdown products are all present. Some are eventually reduced to their inorganic components; others are used directly by organisms and are incorporated into their systems. Another portion of the organic matter accumulates in the sediments, where over geologic time it slowly forms deposits of oil and gas. In the areas of the ocean that are high in plant and animal life, the surface layer may take on a green-yellow color owing to the presence of organic decay products. The incorporation of soluble organics into glacial ice at the Antarctic ice shelves is related to the formation of green ice (see the box on this subject in chapter 4).

Practical Considerations: Salt and Water 5.5

Chemical Resources

About 30% of the world's table salt is extracted from seawater. The industrially produced energy required to remove the water is kept to a minimum to keep extraction costs low. In warm, dry climates seawater is allowed to flow into shallow ponds and evaporate down to a concentrated brine solution. More seawater is added, and the process is repeated several times, until a dense brine is produced. Evaporation continues until a thick, white salt deposit is left on the bottom of the pond. The sea salt is collected and refined to produce table salt. This technique is used in southern France, Puerto Rico, and California (fig. 5.9).

In cold climate areas, salt has been recovered by freezing the seawater in similar ponds. The ice that forms is nearly fresh; the salts are concentrated in the brine beneath the ice. The brine is removed and heated to remove the last of the water.

Of the world's supply of magnesium, 60% comes from the sea, and so does 70% of the bromine. There are vast amounts of dissolved minerals in the world's seawater, including 10 million tons of gold and 4 billion tons of uranium, but the concentration is very low (one part per billion or less). In the case of gold, no method has been devised in which the cost of extraction does not exceed the value of the recovered element. At one time or another Japan, Germany, and the United States have expressed an interest in the extraction of uranium. Japan set up a land-based test plant in 1986 to produce 10 kg (26 lb) of uranium from seawater each year, but the operation proved to be too expensive.

Desalination

Desalination is the process of obtaining fresh water from salt water. There are three main desalination methods:

1. processes involving a change of state of the water— liquid to solid or liquid to vapor;
2. processes requiring ion exchange columns;
3. processes using a semipermeable membrane— electrodialysis and reverse osmosis.

The simplest process involving a change of state is a solar still (fig. 5.10). In this process a pond of seawater is capped by a low plastic dome. Solar radiation penetrates the

Figure 5.9 The southern end of San Francisco Bay was diked into shallow ponds, where seawater is evaporated to obtain salt.

dome and evaporates the seawater. The evaporated water condenses on the undersurface of the dome and trickles down to be caught in a trough, where it accumulates and flows to a freshwater reservoir. The rate of production is slow, and a very large system is needed to supply the water requirements of even a small community.

When water is distilled by heating water to boiling, evaporation proceeds at a rapid rate and large quantities of fresh water are produced, but the energy requirement is high. If water is introduced to a chamber with a reduced air pressure, the boiling occurs at a much lower temperature and therefore uses less energy. Change of state by freezing can also be used to recover fresh water from seawater. The energy requirement is approximately one-sixth of that needed for evaporation, but the mechanical separation of the freshwater ice from the salt brine remains difficult.

Columns containing ion exchange resins that extract ions from salt water work well with water of low salinity, but the resins need to be replaced periodically (fig. 5.11b). Small ion exchange units are manufactured for household use to improve drinking water quality.

Electrodialysis uses an electrical field to transport ions out of solution and through semipermeable membranes;

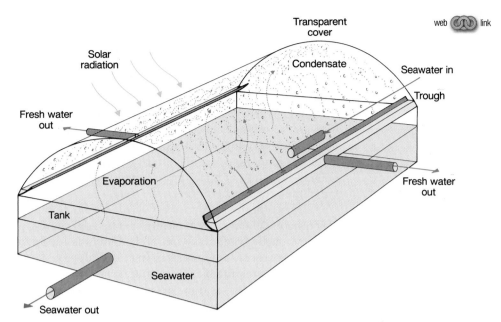

web (link) **Figure 5.10** Solar energy is used to evaporate fresh water from seawater. Solar radiation penetrates the transparent cover over the seawater contained in the tank.

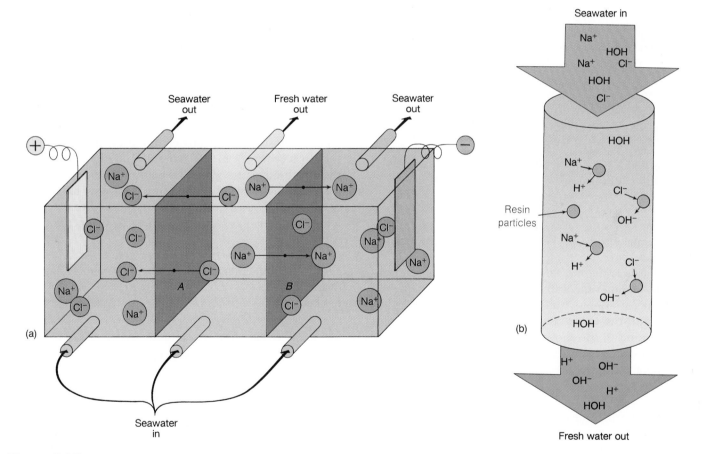

Figure 5.11 (a) Electrodialysis. A tank is separated into three compartments by semipermeable membranes *A* and *B*. Membrane *A* passes only Cl^-. Membrane *B* passes only Na^+. Electrodes that are placed in the end compartments and are supplied with a direct-current voltage will cause the salt ions to migrate out of the center compartment. Fresh water is recovered from the center compartment; excess salt water is removed from each side compartment. (b) Ion-exchange column. Seawater passes through a column of resin particles that exchange H^+ for Na^+ and Cl^- for OH^- to produce HOH, or fresh water (H_2O).

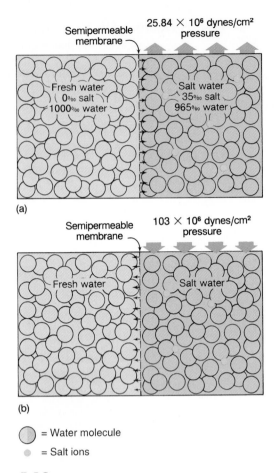

(a)

Semipermeable membrane

25.84 × 10⁶ dynes/cm² pressure

Fresh water
0‰ salt
1000‰ water

Salt water
35‰ salt
965‰ water

(b)

Semipermeable membrane

103 × 10⁶ dynes/cm² pressure

Fresh water

Salt water

⬤ = Water molecule

• = Salt ions

Figure 5.12 (a) Osmosis. Water molecules move from the freshwater side to the saltwater side through a semipermeable membrane. (b) Reverse osmosis. When pressure on the salt water exceeds 25.84×10^6 dynes/cm², water molecules move from the saltwater side to the freshwater side of a semipermeable membrane.

this technique also works best in low-salinity (or brackish) water (fig. 5.11a).

Osmosis is the movement of water across a **semipermeable membrane;** the water moves from the side with the higher concentration of water molecules (or low salinity) to the side with the lower concentration of water molecules (or high salinity); this movement creates a higher pressure on the low-water concentration (or high-salinity) side of the membrane (fig. 5.12a). **Reverse osmosis** produces fresh water from seawater by applying pressure to seawater and forcing the water molecules through a semipermeable membrane, leaving behind the salt ions and other impurities (fig. 5.12b). The pressure applied to the seawater must exceed 24.5 atm, or 25.84×10^6 dynes/cm². A pressure of about 101.5 atm, or 103×10^6 dynes/cm², is required to achieve a reasonable rate of freshwater production. The energy requirement is about one-half that needed for the evaporative process.

Reverse osmosis is the most popular and rapidly growing form of desalination technology. As older evaporative desalination plants wear out, reverse osmosis plants are re-

placing them. The advantages of reverse osmosis plants include no energy requirement for heating the water entering the plant, no thermal pollution from the discharge, and removal of unwanted contaminants—including pesticides, bacteria, and some chemical compounds. High-salinity wastewater returning to the coastal environment can be a disadvantage. The cost of the energy required to pump the water under pressure makes the cost of this type of desalinated water very high. In Southern California, desalinated water is about fifteen times more expensive than local groundwater and five times more expensive than water imported from the northern part of the state.

Because of the high price involved, desalination plants in Southern California are typically operated only during periods of drought, when reservoirs and groundwater levels are low. During years when normal or high rainfall provides sufficient fresh water, the plants are shut down and only maintenance-level work is done to ensure that they will be operational when needed.

In areas such as Kuwait, Saudi Arabia, Morocco, Malta, Israel, the West Indies, California, and the Florida Keys, water is a limiting factor for population and industrial growth. The greatest drawback to the production of fresh water from seawater is the high cost, which is linked to the energy required. In the Arab countries, water costs are low because fuel costs are low.

Large floating masses of naturally desalinated ice have attracted interest as sources of fresh water. The first of a series of meetings on this possibility was hosted in 1978 by Arab interests. Discussions centered on the feasibility of towing icebergs from the Antarctic to the Red Sea. Although the expense would be enormous and much of the ice would melt as it was towed into equatorial latitudes, the need for water is so great that it is believed that sufficient ice would remain after the journey to make the project worthwhile.

Harvesting ice from Alaskan glaciers is done under permits from the Alaskan Department of Natural Resources; quantity and places of harvesting are limited. Most of the ice has been sold to Asia for processing into gourmet ice cubes.

Summary

Seawater is slightly alkaline with a pH between 7.5 and 8.5. The average pH value for all of the oceans over all depths is about 7.8. Seawater pH remains fairly constant because of the buffering action of carbon dioxide in the water.

The average salinity of ocean water is 35‰. The salinity of the surface water changes with latitude and is affected by evaporation, precipitation, and the freezing and thawing of sea ice. Soluble salts are present as ions in seawater. Positive ions are cations; negative ions are anions. Six major constituent ions make up 99% of the salt in seawater. Trace elements that are present in very small quantities are particularly important to living organisms.

Most of the positively charged ions come from the weathering and erosion of the Earth's crust and are added to the sea by rivers. Gases from volcanic eruptions are dissolved in river water as anions. Because the average salinity of the oceans remains constant, the salt gain must be balanced by the removal of salt; input must equal output. Salts are removed as sea spray, evaporites, and insoluble precipitates, as well as by biological reactions, adsorption, chemical reactions, and uplift processes. Seawater circulating through magma chambers in the Earth's crust deposits dissolved metals and releases other chemicals in solution. The time that salts remain in solution, known as residence time, depends on their reactivity.

The proportion of one major ion to another remains the same for all open-ocean salinities. Ratios may vary in coastal areas and in association with biological processes.

Salinity is determined by measuring a sample's electrical conductivity. Historically, salinity has been determined chemically by measuring the quantity of chloride ions in a sample.

The saturation value of gases dissolved in seawater varies with salinity, temperature, and pressure. Carbon dioxide is added to seawater at the sea surface from the atmosphere and by respiration and decay processes at all depths; it is removed at the surface by photosynthesis. Oxygen is added only at the surface from the atmosphere and the photosynthetic process; it is depleted at all depths by respiration and decay. Seawater may become supersaturated with oxygen or it may become anoxic. Carbon dioxide levels tend to change little over depth. Carbon dioxide has the additional role of buffer in keeping the pH range of ocean water between 7.5 and 8.5. Large quantities of CO_2 are absorbed by the oceans. Biological processes pump carbon as carbon dioxide into deep water, where it is fixed in the marine sediments as calcium carbonate. Atmospheric oxygen is regulated by oceanic processes. The amount of oxygen present in seawater is measured chemically and electronically. Carbon dioxide content is determined from the pH of the water.

Nutrients include the nitrates, phosphates, and silicates required for plant growth. A wide variety of organic products are also present.

Salt, magnesium, and bromine are currently being commercially extracted from seawater. Direct extraction of other chemicals is neither economic nor practical at present. Fresh water is an important product of seawater. Desalination methods include change-of-state processes, movement of ions across semipermeable membranes, and ion exchange. The practicality of desalination is determined by cost and need. Reverse osmosis has become the most popular option; it is nonpolluting but costly because of its energy requirements.

Key Terms

All key terms from this chapter can be viewed by term, or by definition, when studied as flashcards on this book's Online Learning Center at www.mhhe.com/sverdrup (click on this book's cover).

Study Questions

1. How is the ocean's salt balance regulated? Make a diagram showing inputs and outputs.
2. Explain the concept of the constant proportions of the major ions in seawater.
3. What is the least expensive method of desalination? Where is it most likely to be used?
4. How does the salinity of mid-ocean surface water change with latitude? What atmospheric processes produce these changes?
5. List the sources of the salts found in seawater. How is their input regulated?
6. Silicate is a nonconservative constituent of seawater and does not obey the principle of constant proportions. Explain why.
7. Compare the distributions of oxygen and carbon dioxide to depth in seawater. What processes are responsible for the distribution of each gas?
8. Explain the relationship between carbon dioxide and the pH of the oceans. Why is the buffering of seawater important?
9. Suggest several pathways a carbon dioxide molecule might follow as it is transferred from the atmosphere to the seafloor sediments.
10. Why does the addition of rain and river water to the oceans not decrease the ocean's overall salinity?
11. Why is the residence time in seawater different for different salts?
12. Explain reverse osmosis and how it produces fresh water from seawater.
13. What is the significance of the compensation depth to plant life?

14. Why does the concentration of dissolved substances in river water differ from the concentration of the same substances in seawater?

15. Why are nutrients considered to be nonconservative materials?

Study Problems

1. If there is 1.4×10^{21} kg of water in the oceans, what is the potential mass of NaCl, sodium chloride, in the oceans? Use table 5.1.

2. If the chloride ion (Cl⁻) content of a seawater sample is 18.5‰, what is the concentration of magnesium in the same sample? Express your answer in g/kg.

3. Determine the residence time of calcium using the following information:
   Calcium ion present in seawater = 0.41 g/kg
   Seawater in the oceans = 1.4×10^{21} kg
   Calcium ion = 12.5% by weight of the average dissolved substances in river water
   Average salt content of river water = 0.12‰
   Annual river runoff = 3.7×10^{16} kg/yr

4. How many kilograms of seawater would have to be processed to obtain 1 kg of gold? Use table 5.2.

Links to Related Websites

Visit the book's Online Learning Center at www.mhhe.com/sverdrup (click on the book's cover) to find live Internet links for additional topics related to this chapter's content.

- The pH scale
- Seawater chemistry
- Carbon cycle
- Desalination

Visit the book's Online Learning Center at www.mhhe.com/sverdrup (click on this book's cover) to find these additional chapter tools: Suggested Readings; links to further information on boxed readings, selected figures, and related chapter topics; and additional study aids.

chapter 6

The Structure and Motion of the Atmosphere

Clouds cover the ocean at Kihei, Maui, Hawaii

Aeolus entertained me for a whole month asking me questions all the time about Troy, the Argive fleet, and the return of the Achaeans. I told him exactly how everything had happened, and when I said I must go, and asked him to further me on my way, he made no sort of difficulty, but set about doing so at once. Moreover, he flayed me a prime oxhide to hold the ways of the roaring winds, which he shut up in the hide as in a sack—for Zeus had made him captain over the winds, and he could stir or still each one of them according to his own pleasure. He put the sack in the ship, and bound the mouth so tightly with a silver thread that not even a breath of a sidewind could blow from any quarter. The west wind which was fair for us did he alone let blow as it chose; but it all came to nothing. . . . They [the crew] loosed the sack, whereupon the winds flew howling forth and raised a storm that carried us weeping out to sea and away from our own country.

Homer
From *The Odyssey, Book X*

The Sun's energy reaches the Earth's surface through the atmosphere, a thin shell of mixed gases we call air. The atmosphere and the ocean are in contact over 71% of the Earth's surface; their interaction is continuous and dynamic. Processes that occur in the atmosphere are closely related to processes that occur in the oceans, and together they form much of what we call weather and climate. Clouds, winds, storms, rain, and fog are all the result of interplay between the Sun's energy, the atmosphere, and the oceans. This complex of interactions provides the Earth's average climate and its daily weather, sometimes pleasant and stable, at other times severe and turbulent. Some of these interactions and processes are more predictable than others; some are better understood than others. Understanding the oceans requires an understanding of the atmosphere's influence on them. This chapter presents an overview of these self-adjusting relationships as well as specific examples of their combined effects.

Heating and Cooling the Earth's Surface 6.1

The heating and cooling of the Earth's surface are accomplished by energy exchanges that act to alter the densities of two fluid envelopes surrounding the Earth, the atmosphere and the oceans. The initial source of the energy is solar radiation, which varies in both time and location on the Earth's surface. This energy is absorbed, reflected, reradiated, converted into other forms of energy, and redistributed over the Earth to create not only the structure but also the dynamics of the atmosphere and oceans.

Distribution of Solar Radiation

Instantaneous solar radiation per unit area of Earth surface has its greatest intensity at the equator, moderate intensity in the middle latitudes, and least intensity at the poles. If the Earth had no atmosphere, the intensity of solar radiation available on a surface at right angles to the Sun's rays would be 2 calories per square centimeter per minute ($cal/cm^2/min$); this value is called the **solar constant.** The solar constant is approached only at latitudes between 23½°N (the Tropic of Cancer) and 23½°S (the Tropic of Capricorn), because only between those latitudes does sunlight strike the Earth at a right angle. Because of the Earth's spherical shape, at all other latitudes the Earth's surface is inclined to the Sun's rays at an angle other than 90°. Compare the angles at which the Sun's rays strike the Earth in figure 6.1. Because the Sun is so far away from the Earth, its rays are parallel when they reach the Earth. Where the rays strike the Earth at right angles, the same amount of radiant energy strikes each unit area. But as the angle at which the Sun's rays strike the Earth increases, the unit areas receive less energy. This difference is also shown in figure 6.1.

When the Sun stands directly above the equator at noon, during either the vernal or the autumnal equinox, the radiation value is about 1.6 $cal/cm^2/min$. This value is less than the solar constant because the Earth's atmosphere stands between the Earth's surface and the incoming solar radiation, and the atmosphere both absorbs and reflects portions of the Sun's energy. Atmospheric interference causes the solar radiation per unit of surface area to decrease with increasing latitude in each hemisphere, because the greater the latitude, the longer the distance through the atmosphere the Sun's rays must travel (fig. 6.1). The combined effects of atmospheric path length and inclination of the Earth's axis cause the Earth to receive more solar heat in the tropics, less at the temperate latitudes, and the least at the poles. The intensity of solar radiation also varies as points on the turning Earth move from darkness to light and return to darkness and as the distance between the Sun and the Earth changes seasonally.

Heat Budget

To maintain its long-term mean surface temperature of 16°C, the Earth must lose heat as well as gain it. On a world average, the Earth and its atmosphere must reradiate as much heat back to space as they receive from the Sun. These gains and losses in heat are represented by a **heat budget.** If less heat was returned to space than is gained, the Earth would become hotter, and if more heat was returned than is gained, the Earth would become colder. In both cases, the planet would change dramatically.

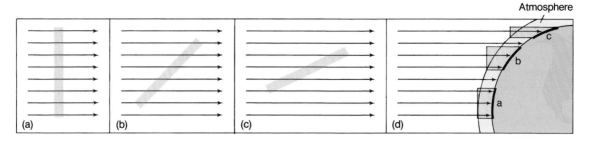

Figure 6.1 Areas of the Earth's surface that are equal in size receive different levels of solar radiation as they become more oblique to the Sun's rays (see a, b, and c). As latitude increases, the angle between the Sun's rays and the Earth increases, and the solar radiation received on equal surface areas decreases (see d). Note also that the Sun's rays must travel an increasing distance through the atmosphere as latitude increases.

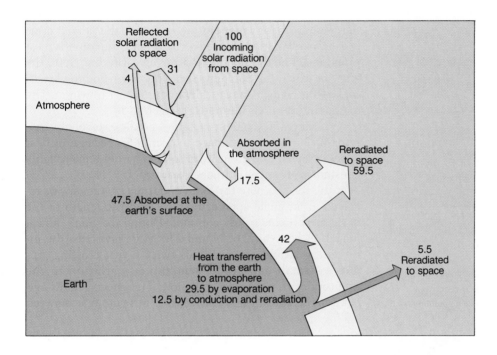

Figure 6.2 The Earth's heat budget. Incoming solar energy is balanced by reflected and reradiated energy. The atmosphere's loss of heat is balanced by heat transferred from the Earth to the atmosphere by evaporation, conduction, and reradiation. Based on 100 units of incoming radiation, or 100% of available radiation.

To follow the long-term average deposits and withdrawals in the total Earth heat budget, assume that 100 units, or 100%, of solar energy is incoming to the Earth's atmosphere. Refer to figure 6.2 as you read this discussion. Thirty-one of these units are reflected back to space by the Earth's atmosphere and 4 units are reflected by the Earth's surface through the atmosphere to outer space; they take no part in heating the Earth or its atmosphere. Of the remaining 65 units, 47.5 are absorbed by the Earth's surface, and 17.5 units are absorbed by the Earth's atmosphere. Balancing the budget requires the following: these 65 units must be lost by **reradiation** of long-wave radiation back to outer space; 5.5 units are reradiated from the Earth's surface, and 59.5 units are reradiated from the atmosphere. The budget is balanced; incoming radiation balances outgoing radiation.

Closer inspection shows that, although the Earth absorbs 47.5 incoming units, it loses only 5.5 units to space, for a gain of 42 heat units. The atmosphere, in contrast, absorbs 17.5 incoming units and loses 59.5 units to space, for a loss of 42 units. In these circumstances it appears that the atmosphere is cooled and the Earth is heated. For the Earth's surface and atmosphere to maintain their near-constant average temperatures, the 42 units gained by the surface must be transferred to the atmosphere. First, 29.5 of the units are transferred by evaporation processes, which cool the Earth's surface and liberate heat when water vapor condensation occurs in the atmosphere. The remaining 12.5 units are transferred to the atmosphere by conduction and reradiation and are absorbed as heat, raising the temperature of the air. On the world average, the atmosphere is primarily heated from below by heat given off from the Earth. This means the atmosphere can be set in convective motion.

The Earth's surface, including the oceans, is heated from above. If we consider the short-term heat budget of only a small portion of the oceans rather than the total, the following must be taken into consideration: total energy absorbed at the sea surface in that area, loss of energy due to evaporation, transfer of heat into and out of the area by currents, vertical redistribution, warming or cooling of the overlying atmosphere by heat from the sea surface, and heat reradiated to space from the sea surface. These processes vary with time, showing daily and seasonal variations.

Measurements made over the Earth's surface show that on the average more heat over the annual cycle is gained than lost at the equatorial latitudes, while more heat is lost than gained at the higher latitudes (fig. 6.3). Winds and ocean currents remove the excess heat accumulated in the tropics and release it at high latitudes to maintain the Earth's present surface temperature patterns. Figure 6.4 shows Northern Hemisphere summer ocean surface temperatures. North-south deflections in the lines of constant temperature indicate the displacement of surface water by currents carrying warm water from low to high latitudes and cold water from high to low latitudes, redistributing heat energy over the Earth's surface.

Annual Cycles of Solar Radiation

The total Earth long-term heat budget and the average distribution of incoming and outgoing radiant energy with latitude do not include the annual cycle of radiant energy changes related to the seasonal north-south migration of the Sun. When these variations are included, the annual cycle of seasonal variation in average daily solar radiation is most pronounced at the middle and higher latitudes. Here, the angle at which the Sun's rays strike the Earth and the length of daylight change dramatically from summer to winter. The seasonal variation is illustrated in figure 6.5.

Changes in incident radiation produce seasonal variations in land and sea surface temperatures because of heat losses or gains. The intensity of solar radiation remains fairly constant at tropical latitudes (between 23½°N and 23½°S) over the year, because the Sun's noontime rays are always received at an angle approaching 90° and the length of the daylight period is nearly constant. During the Sun's annual migration between 23½°N and 23½°S, its rays cross the intervening latitudes twice. This twice-a-year crossing produces a small-amplitude, semiannual variation in the intensity of tropical solar radiation. The effect is most evident at the equator and can be seen in figure 6.5. Between

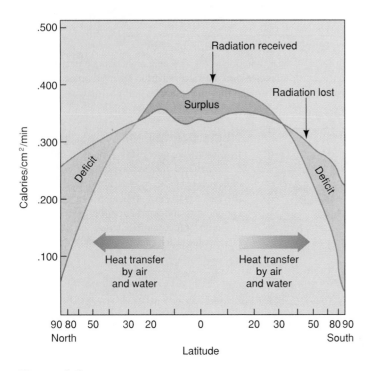

Figure 6.3 Comparison of incoming solar radiation and outgoing long-wave radiation with latitude. A transfer of energy is required to maintain a balance. *Source: NOAA Meteorological Satellite Laboratory.*

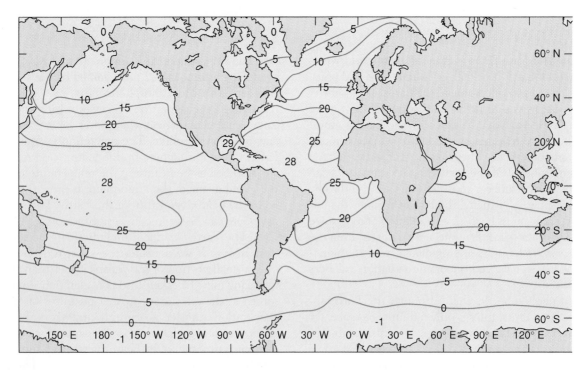

Figure 6.4 Sea surface temperatures during summer in the Northern Hemisphere, given in degrees Celsius.

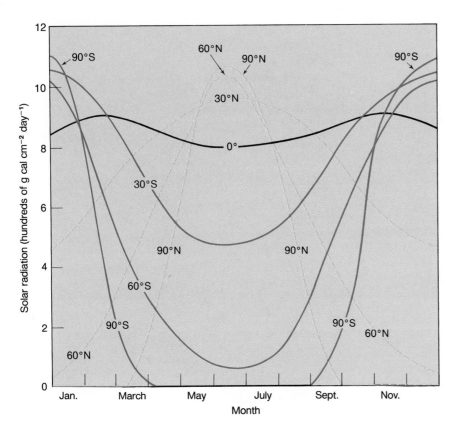

Figure 6.5 Average daily solar radiation values at different latitudes during the year. When it is summer in the Northern Hemisphere, it is winter in the Southern Hemisphere; therefore, the higher values in the Northern Hemisphere coincide with the lower values in the Southern Hemisphere. Peak values of solar radiation at the higher latitudes occur during summer in each hemisphere as a result of more hours of sunlight.

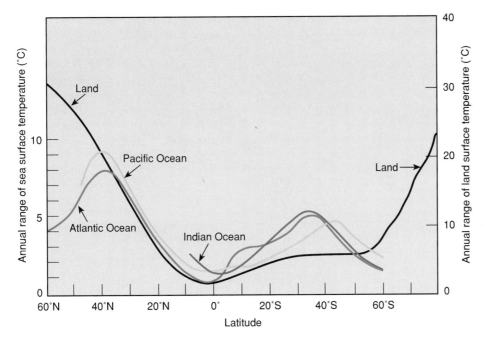

Figure 6.6 The annual range of mid-ocean sea surface temperatures is considerably less than the annual range of land surface temperatures at the higher latitudes. The maximum annual range of sea surface temperatures occurs at the middle latitudes.

23½°N and 90°N and between 23½°S and 90°S, the Sun's noontime rays always strike the Earth at an oblique angle. The long duration of daylight hours in summer at polar latitudes produces a high level of incident solar radiation averaged over the twenty-four-hour day. However, the intensity of radiation per unit surface area per minute of daylight and the annual average radiation level are much lower than those found at lower latitudes (see fig. 6.1). Refer back to chapter 1 to review the relationship of the Earth, the Sun, and the Earth's seasons.

Heat Capacity

Land and ocean areas respond differently to solar radiation. Land has a low heat capacity; it changes temperature rapidly as it gains and loses heat between day and night or summer and winter. The oceans have a high heat capacity; they absorb and release large amounts of heat at the surface with very little change in temperature. For a discussion of heat capacity, see chapter 4.

The average annual range of surface temperatures for land and ocean are given in figure 6.6. Note that the seasonal

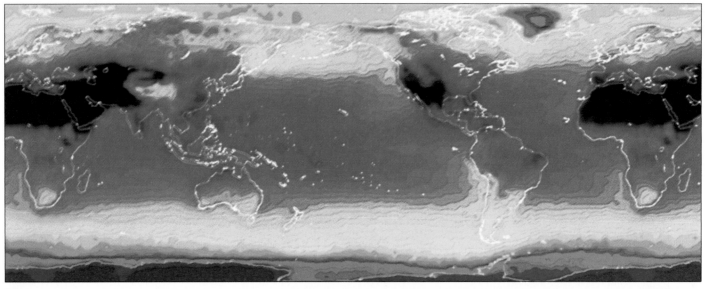

(a)

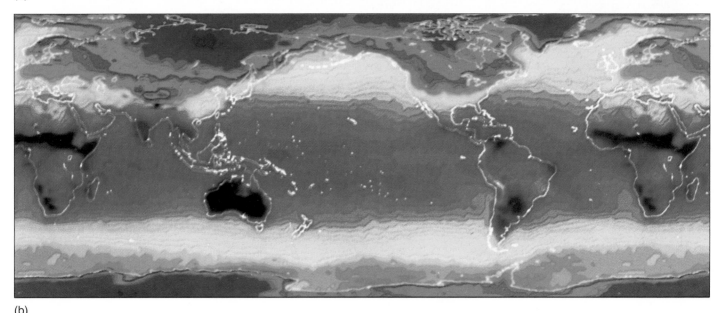

(b)

Figure 6.7 Meteorological satellites, such as NOAA's *TIROS*, carry high-resolution infrared sensors that measure long-wave radiation emitted from the Earth's surface and atmosphere. This radiation is related to the Earth's surface temperatures. *Green* and *blue* indicate temperatures below 0°C; warmer temperatures are shown in *red* and *brown*. (a) July data show the Northern Hemisphere landmasses with considerably warmer temperatures, but the ocean waters do not change dramatically from winter to summer. The oceans' surface temperature distribution moves north and south with the change in seasons. North-south currents along the coasts of continents are also visible. (b) In January, Siberia and Canada show surface temperatures near −30°C; at the same time, latitudes between 30°S and 50°S show warm summer temperatures.

temperature ranges at high latitudes are greater for land than for the oceans and that the annual range of ocean surface temperatures is greatest at the middle latitudes. Because of the unequal distribution of land between the two hemispheres, the summer-to-winter variation in temperature for land at the middle latitudes is much greater in the Northern Hemisphere than it is in the Southern Hemisphere. Because of the lack of land in the Southern Hemisphere, the oceans, with their high heat capacity, control the annual temperature range in southern middle latitudes. These differences be-

tween land and ocean annual surface temperatures can be seen in figure 6.7. Compare the summer (fig. 6.7*a*) and winter (fig. 6.7*b*) temperatures of landmasses and water areas at about 60°N.

Heat that is absorbed at the ocean surface in summer is turbulently mixed downward by winds, waves, and currents. In winter, heat is transferred upward toward the cooling surface. Heat is also transferred to and from the atmosphere at the sea surface. The net effect of these processes is the small annual change in mid-ocean surface temperatures: 0°–2°C

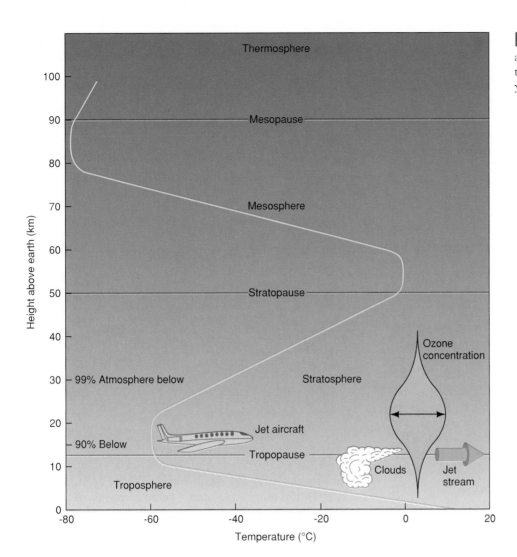

Figure 6.8 The structure of the atmosphere. The distribution of air temperature with height is shown by the *yellow line*.

in the tropics and 5°–8°C at the middle latitudes. The 2°–4°C change in temperature at polar latitudes results from the heat used in the formation and melting of sea ice rather than changing the temperature of seawater.

The Atmosphere 6.2

The absorption, reflection, and transmission of solar energy in the atmosphere depend on the gaseous composition of the atmosphere, suspended particles, and the abundance and types of clouds. The transfer of heat energy from the Earth's surface acts to heat the atmosphere from below and set this gaseous fluid in convective motion. This convective motion produces the winds, which redistribute heat over the Earth's surface and produce waves and currents in the oceans as they blow over the sea surface.

Structure of the Atmosphere

The atmosphere is a nearly homogeneous mixture of gases extending 90 km (54 mi) above the Earth. Ninety-nine percent of the mass of atmospheric gases is contained in a layer extending upward 30 km (18 mi), and 90% is within a layer

extending only 15 km (9 mi) above the Earth's surface. The first layer of atmosphere above the Earth is the **troposphere;** here, the temperature decreases with altitude, changing from a mean Earth surface value of 16°C to –60°C within 12 km (7 mi). The tropopause marks the minimum temperature zone between the troposphere and the layer above it, the **stratosphere.** In the stratosphere the temperature increases with increasing elevation until the stratopause is reached at 50 km (30 mi) (fig. 6.8).

The troposphere is warmed from below by heat energy reradiated and conducted from the Earth's surface and by evaporation of water vapor and its condensation in the atmosphere (see fig. 6.2). Precipitation, evaporation, convective circulation, wind systems, and clouds are all found within the troposphere. **Ozone,** an unstable form of oxygen, occurs principally in the stratosphere. Each ozone molecule, O_3, is made up of three molecules of oxygen instead of two, as found in oxygen, O_2. Ozone absorbs ultraviolet radiation from sunlight and therefore raises the temperature of the stratosphere. By absorbing ultraviolet radiation, the ozone lowers the incidence of ultraviolet light at the Earth's surface, protecting it from the high-intensity ultraviolet radiation that is harmful to living organisms.

At elevations extending higher than 50 km (30 mi) there is little absorption of solar radiation, so the temperature again decreases with height in the layer known as the **mesosphere.** Here the number of molecules of gas per cubic centimeter is reduced by 1000, and the pressure from the remaining atmosphere is only 1/1000 of the pressure at the Earth's surface. The mesosphere extends upward to 90 km (54 mi), and above the mesosphere, the **thermosphere** extends out into space. (See again figure 6.8.)

The emphasis here is on the troposphere, for within this layer heat and water move between the Earth's surface and the atmosphere, causing the motions that produce the winds, the weather, the ocean's waves, and ocean surface currents. The tropopause is also of interest, because this is the region of the high-altitude winds called jet streams that play a role in wind systems and storm tracks.

Composition of Air

The atmosphere is composed of a mixture of gases, suspended microscopic particles, and water droplets. This mixture is commonly simply called air. Atmospheric gases are typically categorized as being permanent or variable. The permanent gases are present in a constant relative percentage of the atmosphere's total volume while the concentration of the variable gases changes in both time and location (table 6.1).

The density of air is controlled by three variables: temperature, the amount of water vapor in the air, and elevation above sea level. Air's density is inversely proportional to all three of these variables. The density of air decreases with increasing temperature and increases with decreasing temperature. In other words, warm air is generally lighter than cold air. Air density decreases if its humidity increases, or the concentration of water vapor increases, and it increases if the humidity decreases. In general, moist air is lighter than dry air. When water vapor is added to the atmosphere, the relatively low molecular weight water molecules replace higher molecular weight permanent gases (compare the molecular weight of water to the molecular weights of the permanent gases listed in table 6.1). Thus cold, dry air is more dense than warm, moist air. Finally, the density of air decreases with increasing elevation above sea level. The air at any given elevation is compressed by the weight of the column of air above it. The greater the compression, the greater the density. Changes in the density of air allow air to move vertically and cause atmospheric convective motion.

Atmospheric Pressure

Atmospheric pressure is the force with which a column of overlying air presses on an area of the Earth's surface. The average atmospheric pressure at sea level is 1013.25 millibars (1 bar = 1×10^6 dynes/cm^2), or 14.7 lb/in^2. This standard atmospheric pressure is also equal to the pressure produced by a barometer's column of mercury standing 760 mm (29.92 in) high. Barometers may measure atmospheric pressure in millibars or in millimeters or inches of mercury. Pressure can also be recorded in torrs, where 1 torr equals the pressure of a col-

Table 6.1 Composition of the Atmosphere

Permanent Gases

Gas	Formula	Percent by Volume	Molecular Weight
Nitrogen	N$_2$	78.08	28.01
Oxygen	O$_2$	20.95	32.00
Argon	Ar	0.93	39.95
Neon	Ne	1.8×10^{-3}	20.18
Helium	He	5.0×10^{-4}	4.00
Hydrogen	H$_2$	5.0×10^{-5}	2.02
Xenon	Xe	9.0×10^{-6}	131.30

Variable Gases

Gas	Formula	Percent by Volume	Molecular Weight
Water vapor	H$_2$O	0 to 4	18.02
Carbon dioxide	CO$_2$	3.5×10^{-2}	44.01
Methane	CH$_4$	1.7×10^{-4}	16.04
Nitrous oxide	N$_2$O	3.0×10^{-5}	44.01
Ozone	O$_3$	4.0×10^{-6}	48.00

umn of mercury 1 mm high. Atmospheric pressure distribution is shown on weather charts by lines of constant pressure, or **isobars** (fig. 6.9). Where the density of the air is less than average, atmospheric pressure is below average, and a **low-pressure zone** of rising air is formed. Regions of air with a density greater than average are known as **high-pressure zones** of descending air.

Greenhouse Gases 6.3

The atmosphere exerts primary control over the Earth's climate, and during the last fifty years it has become increasingly evident that the balance of the gases present in the Earth's atmosphere is changing. Human activities appear to play a significant role in these changes, and there is particular concern about the changes in concentration of greenhouse gases, which include most importantly water vapor and carbon dioxide as well as ozone, nitrous oxide, and others.

Carbon Dioxide and the Greenhouse Effect

There are three active reservoirs for carbon dioxide (CO$_2$): the atmosphere, the oceans, and the terrestrial system; in addition there is the inactive geological reservoir of the Earth's crust. The oceans store the largest amount of CO$_2$, and the atmosphere has the smallest amount (fig. 6.10). The atmosphere is the link with the other reservoirs, and the ocean plays a major part in determining the atmosphere's concentration of CO$_2$ through physical (mixing and circulation), chemical, and biological processes (see section 5.3, chapter 5).

Atmospheric CO$_2$ is transparent to incoming short-wave solar radiation, but at the same time it reduces by absorption

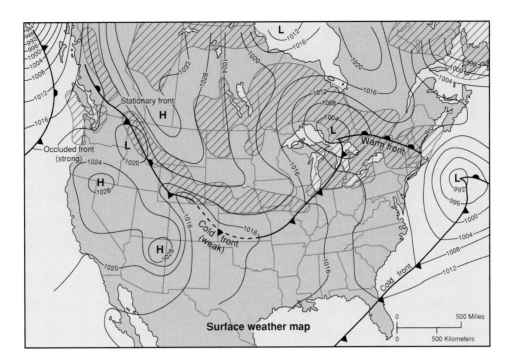

web link **Figure 6.9** Isobars are in millibars. The closer the isobars are to each other, the stronger the winds. *H* is high pressure; *L* is low pressure; *cross-hatching* is rain. Cold fronts occur when cold dense air wedges itself under less dense, warmer air. Warm fronts are produced when less dense, warm air moves over denser, colder air. *Source: NOAA; Marine Climate of Washington.*

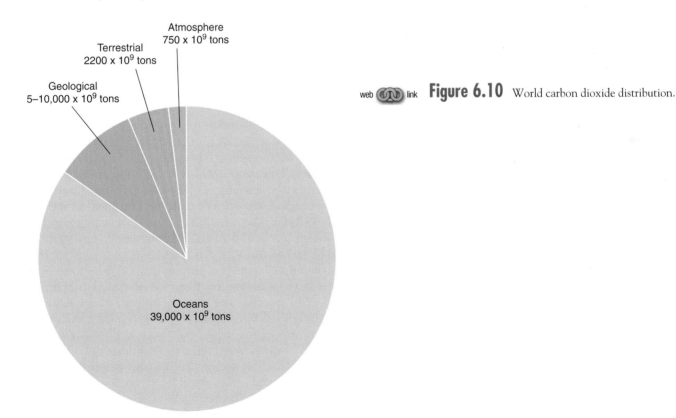

web link **Figure 6.10** World carbon dioxide distribution.

the rate of outgoing long-wave radiation from the Earth. In this way the Earth and its atmosphere are warmed by what is commonly known as the **greenhouse effect.** In the Northern Hemisphere the natural cycle of the world's carbon dioxide shows decreasing atmospheric CO_2 in the spring and summer. At this time plants increase active photosynthesis and remove CO_2 in greater amounts than that contributed by respiration and decay. In the fall and winter, the photosynthetic activity is

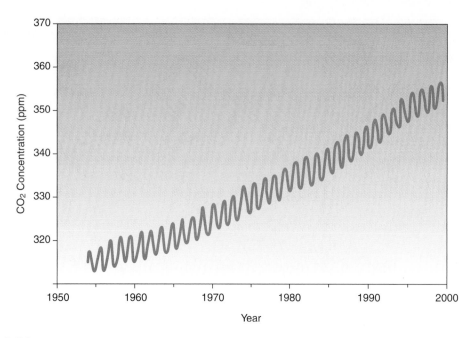

web ⊗ link **Figure 6.11** Concentration of atmospheric carbon dioxide in parts per million (ppm) of dry air versus time in years observed with a continuously recording nondispersive infrared gas analyzer at Mauna Loa Observatory, Hawaii. *After C. D. Keeling, Scripps Institution of Oceanography.*

reduced, plants lose their leaves, and decay processes release CO_2; atmospheric CO_2 increases. The effects of deforestation and conversion of forest land to agriculture, the burning of fossil fuels, and the growth of human populations have been superimposed on this natural cycle. In preindustrial times the human impact on this seasonal cycle was small and the cycle was in balance.

Since 1850 and the Industrial Revolution, however, the concentration of CO_2 in the atmosphere has increased from 280 parts per million (ppm) to 360 ppm. Recently, the average rate of increase has been 1.5–2 ppm per year. For more than forty-five years scientists have been recording a steady increase in the CO_2 concentration in the Earth's atmosphere due to the burning of coal, oil, and other fossil fuels (fig. 6.11). If the increasing trend continues, the concentration of CO_2 will double its 1850 value sometime before the end of this century. This increase will warm the Earth and alter its average heat budget by reducing the surface heat loss to space by long-wave radiation. Instead, more long-wave radiation will be absorbed into the atmosphere, increasing its temperature and forcing it to lose more long-wave radiation to space in order to maintain the heat budget of the Earth and atmosphere systems.

Based on this trend of increasing CO_2, climate researchers have predicted global warming of 2°–4°C in the next hundred years. Accurate measurements of the Earth's warming periods are relatively recent. A twenty-year warming period began in 1920; temperatures increased again in 1977 and remained high through the 1980s. Neither of these periods achieved the predicted warming rate of approximately 0.25°C per decade required by the present CO_2 concentrations.

Because different researchers assign different levels of importance to the many factors that affect the Earth's global temperature, results of research studies vary, and there is spirited debate over the causes and the rate of global warming. Some researchers point to changes in Sun spot cycles; others consider atmospheric aerosols of dust from human and natural sources significant because they could govern cloud formation. Natural events promote global cooling as well as global warming. For example the 1991 discharge of large amounts of debris and gas by the eruption of the Philippine volcano Mount Pinatubo caused Earth cooling of about 0.5°C and made 1992 the coolest year since 1986. This cool period has been followed by continued warming, each succeeding year warmer than the last. In time, reliable global models will be achieved to show us how much and by what means changes in Earth and atmospheric temperatures are related to increasing levels of CO_2.

The annual world production of CO_2 from fossil fuels and the burning of tropical forests is estimated at approximately 7 billion tons. The increase in atmospheric CO_2 accounts for about 3 billion tons annually; however, this rate is not constant but varies with events such as El Niño (see section 6.8). The most recent research on storage of CO_2 in the oceans estimates that about 2 billion tons enters the oceans and ocean sediments annually; refer to the discussion of carbonate sediments in chapter 3 and of dissolved CO_2 in seawater in chapter 5. At least 1 billion tons is thought to be taken up by the temperate forests growing back from eighteenth- and nineteenth-century logging and the intense growth of the plants of tropical regions. The exact rates at which this CO_2 is transferred from one reservoir to another are difficult to measure, and the pathways followed by the

CO$_2$ are difficult to trace. During the 1990s much was learned about the global CO$_2$ cycle, but predicting specific Earth reactions to increases of greenhouse gases requires further study.

In 1997 the Third Conference of Parties to the Framework Convention on Climate Change met in Kyoto. The Kyoto pact calls for thirty-eight industrial nations to cut their emissions of greenhouse gases by an average of 5.2% from 1990 levels by the year 2012. The United States, the world's leader in greenhouse gas emissions at 25% of the total, agreed to a 7% cut, and the nations of the European Union committed to 8%; Japan agreed to a reduction of 6%. If achieved, these levels will reduce projected 2012 emissions by two-thirds, but they will not prevent total greenhouse emissions from rising because developing nations that are not bound by the treaty, such as China, India, and Russia, are expected to account for about 58% of the world's total greenhouse emissions by the year 2012. Controversy over the details of the treaty has prevented its implementation, and the United States has not yet signed it. The United States supports a plan that would allow developed nations to count CO$_2$ absorbed by forests, so-called carbon sinks, against emissions reduction targets. The European Union is concerned that countries may claim that all their greenhouse gases are being absorbed by sinks and that they therefore do not need to make any actual reductions in the pollution they emit from vehicle exhausts and other sources. Representatives of the countries involved in the treaty failed to reach agreement on this issue once again at a meeting at The Hague, Netherlands, in the fall of 2000. The United States chose not to attend a summer 2001 meeting in Bonn, Germany, to resume talks, which was attended by representatives of 178 countries. The United States felt that limits on carbon dioxide emissions could seriously harm the economy. This meeting also ended without a final agreement on limiting emissions. Only time will tell if an agreement on this complex issue will eventually be reached.

With global warming, several sea surface scenarios are possible. It is expected that such warming would affect the higher latitudes, causing melting of polar land ice and raising the world sea level about 1 m (3 ft) by the year 2100. A decrease in sea ice could provide more open water for marine plant populations with a corresponding increase of photosynthesis, leading to increased carbon storage in the ocean reservoir. Other possibilities are related to the interaction between clouds, the oceans, and the Earth's surface (see the box titled "Clouds and Climate"). Changes in sea surface temperature could also affect the oceanic circulation (see chapters 7 and 8) and the surface wind systems (covered later in this chapter) that drive the ocean currents. Changes in the ocean current patterns would modify the transfer of heat and water vapor from low to high latitudes and alter the Earth's climate patterns. Also, increases in available atmospheric CO$_2$ have the potential to stimulate photosynthesis both on the land and in the oceans with unknown effects.

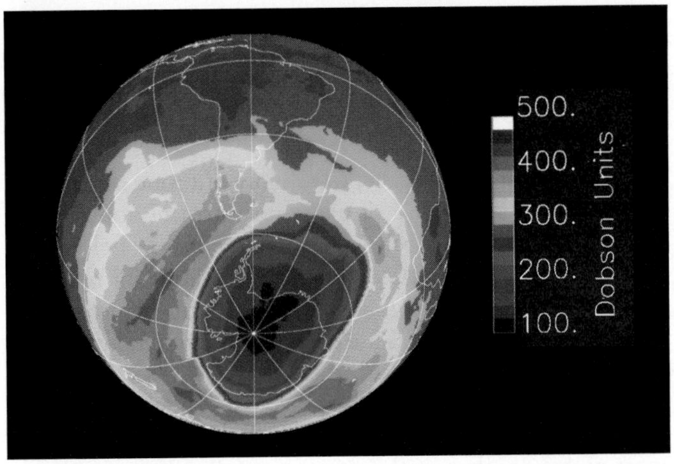

Figure 6.12 Map of ozone over Antarctica measured on September 24, 1997. A TOMS (total ozone mapping spectrometer) is carried by the NASA *Earth Probe* satellite launched in 1996. It replaces previous TOMS that have gathered data since 1978. Measurements are in Dobson Units (defined in glossary). *Courtesy of Richard McPeters, Head, Ozone Processing Team, GSFC, NASA.*

The Ozone Problem

Depletion of the stratospheric ozone layer that screens the Earth from much of the Sun's ultraviolet radiation was first reported in 1985 by members of the British Antarctic Survey, who discovered that significant ozone loss had been occurring over Antarctica since the late 1970s. The Antarctic "ozone hole" is the result of a large-scale destruction of the ozone layer over Antarctica that occurs each Antarctic spring. The word *hole* is a misnomer; the hole is really an area over the pole that experiences a significant reduction in ozone concentrations, which results in the destruction of up to 70% of the ozone normally found over Antarctica. The Antarctic ozone hole covered 18×10^6 km^2 in 1990, 20.9×10^6 km^2 in 1997, and grew to its largest recorded size of 28.3×10^6 km^2 (an area three times the size of the United States) in 2000. Satellites carrying total ozone mapping spectrometers (TOMS) have monitored the ozone situation since 1978 with increasing accuracy. Currently NASA's *Earth Probe (EP)* satellite is monitoring both the Antarctic and Arctic ozone holes (fig. 6.12).

Arctic winters are not as cold as those of the Antarctic; therefore, ozone loss over the Arctic is less and the Arctic ozone hole is smaller. Nevertheless, the Arctic ozone hole covered approximately 6×10^6 km^2 in March of 1997. The Arctic winter of 1996–97 was the fifth winter in a row in which the stratosphere over the Arctic had been particularly cold. Measurements taken early in 2000 found cumulative ozone losses of more than 60% over the Arctic. There are three possible explanations for the colder conditions in the Arctic stratosphere between 1992 and 1997. Under conditions of decreasing ozone, the stratosphere is less able to absorb solar radiation and warm itself; the result is a destructive system of increasing cold that leads to lower ozone levels.

Also, the increasing greenhouse gases trap heat in the lower atmosphere, and, because this heat does not reach the upper atmosphere, the stratosphere gets cooler. Last, some other variations in the Earth's climate or changes in the weather may be at work.

The most widely accepted theory of ozone destruction is related to the release of chlorine into the atmosphere. Chlorine is commonly released as a component of chlorofluorocarbons (CFCs). CFCs are used as coolants for refrigeration and air conditioning, as solvents, and in the production of insulating foams. CFCs are distributed throughout the troposphere by the winds and gradually leak into the stratosphere, where the ultraviolet light breaks them apart. Gases in the atmosphere react with the chlorine and trap it in inert molecules, but in the presence of stratospheric clouds, which are common during the polar winter, the chlorine gas is freed to react with the ozone. In the presence of sunlight, chlorine and oxygen are formed, and the liberated chlorine is free to attack other ozone molecules.

CFCs appear to have reached their maximum level in the troposphere and are expected to decline gradually as efforts to reduce their production continue. Because it takes many years for CFCs in the troposphere to work their way into the stratosphere, they will continue to destroy ozone for many years to come.

Recently, methyl bromide has been identified as an even more efficient agent for breaking down ozone. It is thought to be responsible for as much as 20% of the Antarctic ozone depletion and 5%–10% of the Earth's global ozone loss. Estimates place the input of methyl bromide to the atmosphere at 137 gigagrams (Gg) per year, of which 86 Gg/yr act directly to destroy ozone. Sources include microscopic single-celled organisms at the sea surface, pesticides, industry, and slow-smoldering burning of vegetation.

At ground level, ozone is a pollutant and a health hazard; in the stratosphere it absorbs most of the ultraviolet radiation from the Sun, protecting life forms on land and at the sea surface. The loss of ozone is a significant concern because a 50% decrease in ozone is estimated to cause a 350% increase in ultraviolet radiation reaching the Earth's surface. Increased ultraviolet radiation is responsible for increases in skin cancers and has been shown to affect the growth and reproduction of some organisms. The effects of increased ultraviolet radiation on the plantlike organisms of the surface layers of the Southern Ocean are discussed in chapter 14.

The Role of Sulfur Compounds   6.4

It is currently estimated that 20–40 million tons of sulfur from dimethyl sulfide (DMS) is added to the atmosphere each year. Dimethyl sulfide is a gas produced by single-celled, plantlike organisms at the ocean's surface. It is in part responsible for the characteristic odor of the sea that one notices when approaching the coast from the land. Once in the atmosphere, DMS changes rapidly. One of its main products is sulfur in the form of sulfate, which combines with at-

mospheric water to form sulfuric acid, which is returned to the Earth's surface with the rain (fig. 6.13). The quantities of sulfuric acid returned to the Earth's surface by this process are far below the values associated with the problem of "acid rain." Because DMS forms airborne particles, it plays a role in controlling the density of the clouds that appear over the ocean. Dimethyl sulfide changes the clouds' reflective properties, reduces incoming radiation, and decreases the heating of the ocean's surface. If a DMS buildup results in an excess of clouds and sulfur, less light strikes the sea surface, surface temperatures drop, and the plant production of DMS decreases. If less DMS is produced, there is less cloud cover and less reflection of short-wave radiation. In this way DMS may act as a feedback mechanism, or self-regulating thermostat, to help control ocean surface temperatures, another link in the complexity of interactions between the atmosphere and the oceans.

In industrialized areas of the Northern Hemisphere, the burning of fossil fuels creates another source of sulfates. Like the sulfates from DMS, these industrially produced sulfates contribute to acid rain; they also form clouds that block solar radiation and cool the Earth's surface.

Unlike its effect on land, acid rain does not have a great impact on the ocean. When acid rain falls on the sea surface the buffering capacity of seawater reduces its impact (review section 5.1 in chapter 5).

The Atmosphere in Motion   6.5

The air moves because at one place less dense air rises away from the Earth, while in another place denser air sinks toward the Earth. Between these areas the air that flows horizontally along the Earth's surface is the wind. This process is shown in figure 6.14; note that there are really two horizontal airflows, or wind levels, moving in opposite directions: one at the Earth's surface and one aloft. Air circulating in this manner forms a convection cell based on vertical air movements due to changes in the air's density. Less dense air rises, and denser air sinks.

Winds on a Nonrotating Earth

The heating and cooling of air and the gains and losses of water vapor in air are related to the unequal distribution of the Sun's energy over the Earth's surface, the presence or absence of water, and the variation in temperature of the Earth's surface materials in response to heating. These act to affect the air's density. Imagine a model Earth with no continents and with no rotation but heated like the real Earth. On this stationary model covered with uniform layers of atmosphere and water, the wind pattern is very simple. Around the equator the air, warmed from below, rises. Once aloft, the air flows toward the poles, where it is cooled and sinks to flow back toward the equator. Because of the unequal distribution of the Sun's heat over the model's surface, large amounts of heat and water vapor are transferred to the atmosphere around the equator. This less dense air rises, and as it

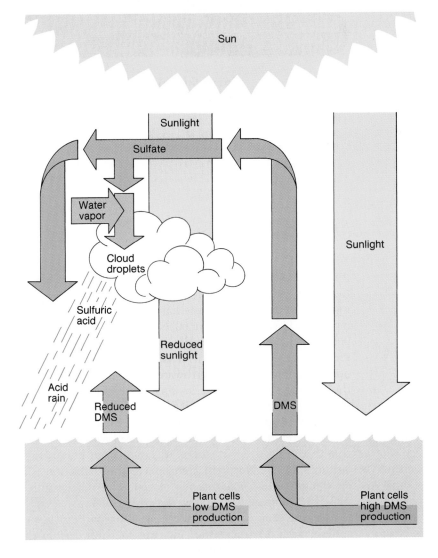

Figure 6.13 A proposed self-supporting thermal control system linking dimethyl sulfide (DMS) to the sea surface and the atmosphere, a cycle among plants, the Sun, and clouds. Acid rain, pH ~ 3.5, has little impact on seawater because of seawater's carbon dioxide buffer.

Sun

Sunlight

Sulfate

Water vapor

Cloud droplets

Sunlight

Sulfuric acid

Reduced sunlight

DMS

Acid rain

Reduced DMS

Plant cells low DMS production

Plant cells high DMS production

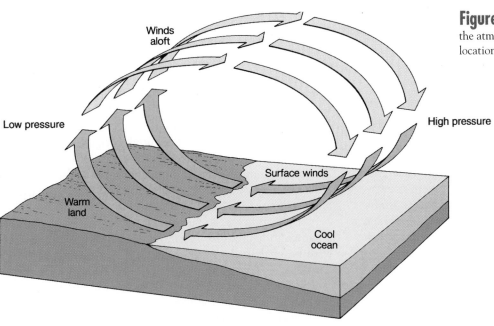

Figure 6.14 A convection cell is formed in the atmosphere when air is warmed at one location and cooled at another.

Winds aloft

Low pressure

High pressure

Surface winds

Warm land

Cool ocean

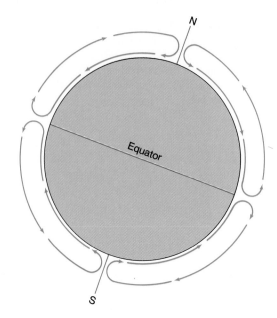

Figure 6.15 Heating at the equator and cooling at the poles produce a single large convection cell in each hemisphere on a nonrotating water-covered model Earth.

Table 6.2 Eastward Speed of Earth's Surface with Latitude

Latitude	Speed (km/h)	Speed (mi/h)
90°N and S	0	0
75°N and S	433	269
60°N and S	837	520
45°N and S	1184	735
30°N and S	1450	900
15°N and S	1617	1004
0°	1674	1040

rises it cools; condensation exceeds evaporation, clouds form, and it rains. Equatorial regions are known for their warm, wet climate. The cool dry air remains aloft and flows toward the poles, where it sinks, producing high evaporation, dense air, and a zone of high atmospheric pressure. Such air movement is shown in figure 6.15; note each hemisphere's two large convective circulation cells, each extending from a pole to the equator. In this model, the Northern Hemisphere surface winds blow from north to south, and the upper winds blow from south to north; in the Southern Hemisphere the reverse is true, with the surface winds blowing from south to north and the upper winds blowing from north to south.

It is important to remember that winds are named for the direction *from which they blow*. A north wind blows from north to south; a south wind blows from south to north. In this model, the Northern Hemisphere surface winds are north winds, and the Southern Hemisphere surface winds are south winds.

The Effects of Rotation

Consider next the same model of a water-covered Earth but with rotation added. In this case, observations of moving parcels are judged relative to coordinates, latitude and longitude, that are fixed to and rotate with the Earth. Referencing motion to this rotating coordinate system creates apparent forces that appear to deflect moving bodies. Gravity holds the Earth's atmosphere captive, but the atmosphere is not rigidly attached to the Earth's surface. Little or low friction exists between the Earth's surface and the atmosphere, and the atmosphere moves somewhat independently of the Earth's surface. For example, a parcel of air that appears to be stationary above a point on the equator is actually turning with the Earth and moving eastward at a speed

of about 1674 km (1040 mi) per hour. If a south wind blows this parcel of air due north, it is moved across circles of latitude with progressively smaller circumferences. At these higher latitudes, points on the Earth's surface move eastward more slowly (table 6.2). At 60°N, the eastward speed of a point on the Earth's surface due to rotation is only half the speed at the equator. For this reason, a parcel of air that was originally moving only northward relative to the Earth at the equator carries with it its equatorial eastward speed so that it is now moving eastward at a speed that is greater than the eastward speed of the Earth's surface at this higher latitude. Therefore, the air parcel is displaced to the east, relative to the Earth's surface, as it moves from low to high latitudes; in the Northern Hemisphere this deflection is *to the right of the direction of the air motion*. This relationship is illustrated in figure 6.16; follow the arrows at A.

If the same situation occurs in the Southern Hemisphere, with a parcel of air moving southward from the equator due to a north wind, the deflection is still to the east relative to the Earth's surface. However, this deflection is now *to the left of the direction of the air motion*. See again figure 6.16 and follow the arrows at B.

If an air parcel moves southward toward the equator in the Northern Hemisphere, then it moves from a latitude where a position on the Earth has a low eastward speed to a latitude where a position on the Earth has a higher eastward speed. These positions move to the east, relative to the air, when the air moves from a higher to a lower latitude. This relationship causes the air to fall behind the position on the Earth, so that the air appears to move westward relative to the Earth. This pattern is illustrated in figure 6.17. The deflection is still to the right of the direction of motion in the Northern Hemisphere. A similar pattern in the Southern Hemisphere shows that the deflection is to the left of the direction of motion.

Deflection due to air moving to the east or to the west is shown in figure 6.18. Air moving eastward is moving eastward faster than the Earth beneath it, with reference to the Earth's axis. The air is affected by a centrifugal force acting outward from the Earth's axis of rotation that is stronger than the centrifugal force that is acting at the Earth's surface. This small excess force acting on the air is in part acting against

Ship Emissions

As land dwellers, we all know that automobiles, power plants, industries, burning of grasslands and forests, and even cooking fires contribute to the carbon dioxide content of the atmosphere. Those of us who live near any metropolitan area have seen the brown smog and haze that accumulate when the air becomes static under a thermal inversion. When we think of the oceans, however, we think of cool, fresh breezes with a tang of salt or wild, swirling winds sweeping the clean air in from the sea. But a study by the International Maritime Organization (IMO) shows that the sea air is not as clean as we might have thought or hoped (box fig. 1).

Data on marine fuel consumption, fuel type, engine characteristics, exhaust emissions, and vessel traffic density have been combined to estimate the pollution contribution of vessels at sea. The geographic distribution of emission levels for sulfur follows the patterns of density and frequency of vessel traffic. Sulfur emissions are concentrated in the Northern Hemisphere. In the Atlantic Ocean,

they are highest in coastal regions and in the triangle stretching from the St. Lawrence Seaway to Central America and to western and northern Europe and back to Canada. In the Pacific, the area around Japan, south through Indonesia, and eastward to the west coast of the United States shows the highest concentrations. Sulfur emissions are much lower in the Southern Hemisphere, where the emissions clearly define discrete shipping routes.

World sulfur-type emissions are estimated at 8.48 million tons per year and contribute 4.24 million tons of sulfur. Ship emissions account for 16% of the sulfur from all petroleum uses and 5% when all combustible fuels are considered. This is the same as 43% of the total U.S. emissions of sulfur.

The global estimate of nitrogenous ship emissions is 10.12 million tons per year, or 3.08 million tons of nitrogen per year. This is equivalent to 42% of the total nitrogen emissions from North America, 74% of the emissions from the European Union, or nearly 50% of total U.S. emissions and is also equal to 100% of the nitrogenous emissions from automobiles and trucks.

The high nitrogen and sulfur outputs from vessels rank them as the world's highest-polluting combustion sources per ton of fuel used. Remembering that the oceans cover 71% of the Earth's surface, and that dimethyl sulfide is produced at the sea surface, the emission of sulfur from vessels approximates 20% of the worldwide marine production of sulfur by natural processes at the sea surface. The world estimate for total carbon released from the burning of fossil fuels each year is 6 billion tons. Ships contribute 2% of this amount.

The IMO approved global emission limits in 1997, but these limits apply only to new ships or vessels undergoing major conversions after January 1, 2000. Reducing vessel emissions and enforcing reductions will be difficult because vessels may register in countries other than their own and may change the country in which they register if they so desire.

Next time you are at the beach, take a deep breath of fresh ocean air!

Box Figure 1 The cruise ship M. S. *Oceanic* leaving port, New Providence/Nassau, Bahamas.

Data from J.J. Corbett and P. Fischbeck. 1998. Emissions from Ships. *Science* 278 (5339): 823–24.

the Earth's gravity and in part acting parallel to the Earth's surface directed toward lower latitudes. That part of the centrifugal force acting against the Earth's gravity is so small that it has very little effect. That part acting parallel to the Earth's surface is unopposed and causes a deflection of the moving air to lower latitudes, or to the right of its motion in the Northern Hemisphere. Westward-moving air is moving eastward more slowly, relative to the Earth's surface. This air is affected by an outward-acting centrifugal force that is weaker than the centrifugal force that is acting at the Earth's surface. Because the centrifugal force acting at the Earth's surface is greater than the centrifugal force acting on the air, there is a weak force acting toward the Earth's axis of rotation. This

force is acting in part in the direction of gravity and in part toward higher latitudes. Note that the deflection continues to be to the right of the initial wind motion in the Northern Hemisphere. In the Southern Hemisphere, the deflection of eastward-moving air toward lower latitudes and of westward-moving air toward higher latitudes results in a deflection of the air to the left of its original direction of motion.

If we were to move from the Earth's surface out into space and were able to watch the motions of Earth and atmosphere, we would see the atmosphere's independent motion and we would also see the Earth turning out from under the moving parcels of air. But we live on a moving Earth's surface and we make our measurements of air motion relative to a

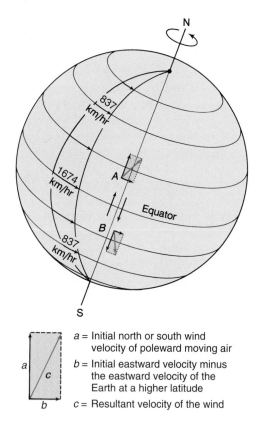

a = Initial north or south wind velocity of poleward moving air

b = Initial eastward velocity minus the eastward velocity of the Earth at a higher latitude

c = Resultant velocity of the wind

Figure 6.16 Because air moving northward from the equator to point A carries with it its higher equatorial eastward velocity, the air is deflected to the right of the initial northward wind direction in the Northern Hemisphere. In the Southern Hemisphere, air moving southward to point B is deflected to its left.

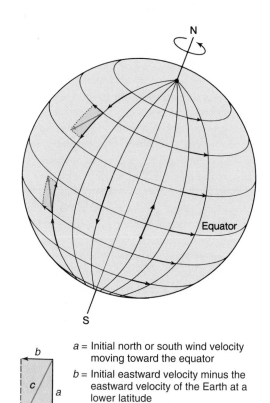

a = Initial north or south wind velocity moving toward the equator

b = Initial eastward velocity minus the eastward velocity of the Earth at a lower latitude

c = Resultant velocity of the wind

Figure 6.17 Air moving toward the equator passes from a latitude of lower eastward speed to a latitude of higher eastward speed. The result is it falls behind the eastward displacement of the Earth's surface causing a deflection to the right of the initial wind direction in the Northern Hemisphere and deflection to the left in the Southern Hemisphere. There is no deflection at the equator.

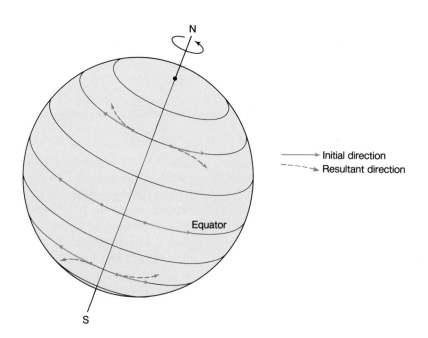

→ Initial direction
⇢ Resultant direction

Figure 6.18 The deflection of eastward-moving and westward-moving air. There is no deflection at the equator.

www.mhhe.com/sverdrup

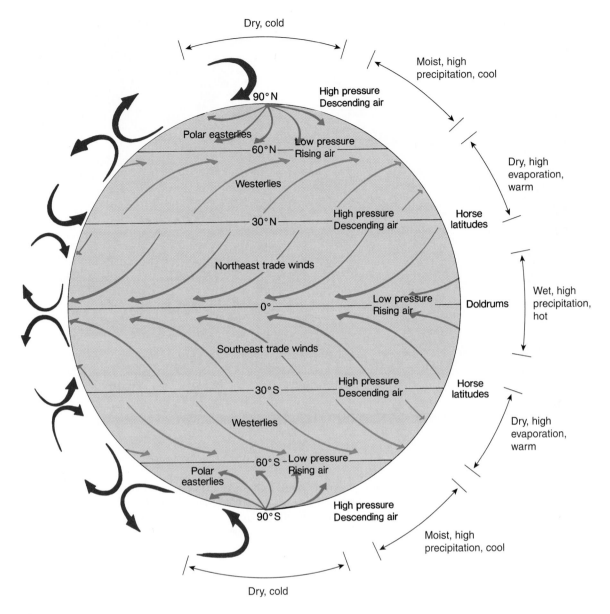

Figure 6.19 The circulation of the Earth's atmosphere results in a six-band surface wind system.

coordinate system fixed to this surface. Therefore, from the Earth's surface, we see the moving air parcels deflected from their paths—deflected to the right in the Northern Hemisphere and to the left in the Southern Hemisphere.

The apparent deflection of the moving air relative to the Earth's surface is called the **Coriolis effect,** after Gaspard Gustave de Coriolis (1792–1843), who mathematically solved the problem of deflection in frictionless motion when the motion is referred to a rotating body and its coordinate system. Because one of the laws of motion in physics tells us that a body set in motion along a straight line continues to move along that straight line unless it is acted upon by a force, such as a push or a pull, the Coriolis effect is often called the Coriolis force. If the term *force* is used, remember that it is an *apparent* force, and that a deflection appears only when the motion of an object that is subject to little or

no friction is judged against a rotating frame of reference. The magnitude of the Coriolis force increases with increasing latitude, increases with the speed of the moving air, and is dependent on the rotation rate of the Earth on its axis.

Wind Bands

Applying rotation, and therefore the Coriolis effect, to the nonrotating model Earth modifies the winds considerably. Refer to figure 6.19 as you read the following description. The air rises at the equator and flows aloft to the north and the south, but it cannot continue to move northward and southward without being deflected to the right in the Northern Hemisphere and to the left in the Southern Hemisphere. This deflection short-circuits the large, hemispheric, atmospheric convection cells of the stationary model Earth. The deflected air aloft sinks at 30°N and 30°S; it moves

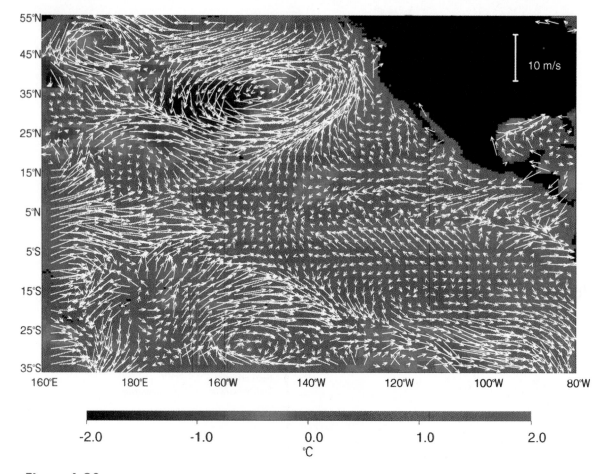

web link **Figure 6.20** Surface wind anomalies (*white arrows*) measured by NASA's high-resolution scatterometer (NSCAT) are superimposed on the map of sea surface temperature anomalies (*colors*) derived from advanced very high-resolution radiometer (AVHRR) measurements. The *white arrows* and *colors* indicate the deviation of winds and sea surface temperatures from the average conditions present for the period May 22–31, 1997. Notice the elevated temperature (*red*) in the eastern Pacific due to El Niño conditions and the weak and reversed trade winds in the same area. *Courtesy of W. Timothy Liu, JPL, California Institute of Technology.*

along the water-covered surface, either back toward the equator or toward 60°N and 60°S. The upper-level air that reaches the poles cools, sinks, and moves to lower latitudes, warming and picking up water vapor; at 60°N and 60°S it rises again. The result is three convection cells in each hemisphere wrapped around the rotating Earth.

Consider the flow of surface air in the three-cell system. Between 0° and 30°N and 30°S the surface winds are deflected relative to the Earth, blowing from the north and east in the Northern Hemisphere and from the south and east in the Southern Hemisphere. This deflection creates bands of moving air that are known as the **trade winds:** the northeast trade winds, north of the equator, and the southeast trade winds, south of the equator. Between 30°N and 60°N, the deflected surface flow produces winds that blow from the south and west, while between 30°S and 60°S they blow from the north and west. In both hemispheres these winds are called the **westerlies.** Between 60°N and the North Pole, the winds blow from the north and east, while between 60°S and the South Pole, they blow from the south and east. In

both cases they are called the **polar easterlies.** The six surface wind bands are shown in figures 6.19 and 6.20.

At 0° and 60°N and 60°S, moist, low-density air rises in areas of low atmospheric pressure; these are zones of clouds and rain. Zones of high-density descending air at 30° and 90°N and 90°S are areas of high atmospheric pressure, dry air with low precipitation, and clear skies.

The salinity of mid-ocean surface water is controlled by the distribution of the world's evaporation and precipitation zones (see figs. 5.2 and 5.3). Salinity of ocean water is measured in grams of salt per kilogram of seawater and is expressed in parts per thousand (‰). The average ocean salinity is considered to be 35‰. In the tropics, the precipitation is heavy on land and at sea. On land, the tropical rain forest is the result; at sea, the surface water has a low salinity, around 34.5‰. At approximately 30°N and 30°S, the evaporation rates are high. These are the latitudes of the world's land deserts and of increased salinity in the surface waters, around 36.7‰. Farther north and south, from 50°–60°N and from 50°–60°S, precipitation is again heavy, producing cool but less salty surface water, around 34.0‰, and the heavily forested areas of the Northern Hemisphere. At the polar latitudes, sea ice is formed in the winter. During the freezing process, the salinity of the water beneath the ice increases, and in the summer the thawing of the ice reduces the surface salinity once more (see fig. 6.19).

Wind belts are formed when air flows over the Earth's surface from regions of high atmospheric pressure to areas of low atmospheric pressure. In the zones of vertical motion, between the wind belts, the surface winds are unsteady. Such areas were troublesome to the early sailors, who depended on steady winds for propulsion. The area of rising air at the equator is known as the **doldrums,** and the high-pressure areas at 30°N and 30°S are known as the **horse latitudes.** In all these areas sailing ships could find themselves becalmed for days. The origin of the word *doldrum* is obscure, but it is likely related to its meaning of a period of low spirits, listlessness, or despondency, feelings often felt by sailors becalmed in this region. The horse latitudes are said to have gotten their name from the stories of ships' carrying horses that were thrown overboard when the ships were becalmed and the freshwater supply became too low to support both the sailors and the animals.

Every day for decades, thousands of measurements of air pressure, wind speed, and wind direction have been taken by meteorological stations on land and by ships and buoys at sea. These data are converted into atmospheric wind and pressure charts that are used to make the world's weather forecasts. Satellites can provide another way to report wind velocities at the Earth's surface. Microwave radar signals sent from the satellite are scattered by the waves at the Earth's surface and returned to the satellite. These returned signals are then interpreted as measurements related to the speed and direction of the surface winds.

The radar signals are sent and received by an instrument called a scatterometer, which is carried by the satellite. The first scatterometer was sent into space in the late 1970s aboard the satellite *SEASAT.* The most recent model, *NASA Scatterometer (NSCAT),* went into space on Japan's *Advanced Environmental Orbiting Satellite (ADEOS)* in August 1996, but *ADEOS* remained active only until June 1997. For eight-and-a-half months, *NSCAT* measured surface wind data that were verified by data from ships and buoys (see fig. 6.20). Such data are not only valuable for weather forecasting but also yield data that can be used in computer models that couple the ocean and atmosphere, enabling meteorologists and oceanographers to investigate and better understand the couplings of the ocean-atmosphere system.

Modifying the Wind Bands   6.6

Moving from the rotating, water-covered model of the Earth to the real Earth requires the consideration of three factors: (1) seasonal changes in the Earth's surface temperature due to solar heating, (2) the addition of the large continental land blocks, and (3) the difference in heat capacity of land and water. Both ocean and land surfaces remain warm at the equatorial latitudes and cold at the polar latitudes all through the year, but the middle latitudes have seasonal temperature changes, ranging from warm in summer to cold in winter. Keep in mind that land surface temperatures have a greater seasonal fluctuation than ocean surface temperatures, because the heat capacity of the ocean water is greater than the heat capacity of land and because ocean water has the ability to transfer heat from the surface to depth in summer and from depth to the surface in winter. Land does not transfer heat in this way.

Seasonal Changes

The presence of both land and water in near-equal amounts at the middle latitudes in the Northern Hemisphere produces average seasonal patterns of atmospheric pressure. During the warm summer months, the land is warmer than the ocean (see fig. 6.7). The air over the land is heated from below and rises, creating a low-pressure area, while the air over the sea cools and sinks, producing a high-pressure area over the water. In the summer, low-pressure zones at 60°N and 0° tend to combine over the land, cutting through the high-pressure belt along 30°N latitude. This breaks the high-pressure belt into several high-pressure cells over the oceans rather than maintaining the latitudinal zones of pressure stretching continuously around the Earth. In the winter, the reverse is true at the middle latitudes (see again fig. 6.7). The land becomes colder than the water. The air over the land is cooled from below, increasing its density and producing a region of descending air and high atmospheric pressure over the land. The air over the relatively warm water is heated from below, decreasing its density and producing a region of rising air and low atmospheric pressure over the water.

Over the land, the polar high-pressure zone spreads toward the high-pressure zone at 30°N, breaking the low-pressure belt centered about 60°N into discrete low-pressure cells that are centered over the warmer ocean water. This middle-latitude, seasonal alternation of high- and low-pressure cells breaks up the latitudinal pressure zones and wind belts that are seen in the water-covered model to produce the air-pressure distribution shown in figure 6.21.

During the Northern Hemisphere's summer, the air in the high-pressure cells over the central portion of the North Atlantic and North Pacific Oceans descends and flows outward toward the continental low-pressure areas. As the descending air moves outward it is deflected to its right, producing winds that spiral in a clockwise direction about the high-pressure cells. The wind circulation about a high-pressure cell in the Northern Hemisphere is always clockwise. On the northern side of these high-pressure cells are the westerlies and on the southern side are the northeasterlies; the eastern side of the cell has northerly winds, and the western side has southerly ones. In the winter the air circulates counterclockwise about the low-pressure cell over the northern oceans, and the prevailing wind directions reverse. The wind circulation pattern about a low-pressure cell in

Clouds and Climate

Clouds are beautiful, always changing, ever moving; we see them white and fluffy, high and wispy, black and threatening. Clouds form from liquid water droplets with temperatures above freezing, supercooled water droplets with temperatures below freezing, and solid particles. In the lower troposphere, condensation nuclei, small particles of dust, salt, or other matter, serve as the cores for condensing water vapor droplets. Clouds of the upper troposphere are composed of crystallized water vapor or ice crystals and snow.

In 1803 Luke Howard, an English pharmacist, proposed the first useful classification of clouds: stratus (layered), cumulus (puffy), cirrus (wispy), and nimbus (clouds releasing snow or rain that travels to the ground) (box figure 1). This system is the basis for the cloud types recognized by today's World Meteorological Organization (WMO) (box figure 2). Different cloud types absorb radiant energy at different rates and affect the heat budget and climate of the Earth.

Clouds have a dual role in the atmosphere: they simultaneously heat and cool the Earth. The tops and sides of clouds appear white because they reflect and scatter short-wave radiation. If clouds are dense enough, they absorb, reflect, and scatter sufficient incoming radiation to produce dark shadows on their undersides and on the Earth's surface. These dense clouds help to cool the Earth. Clouds absorb long-wave radiation from the Earth's surface because of their water content, and, because they are generally cooler than the Earth, they reradiate only part of this absorbed long-wave radiation and absorbed solar radiation to space. The net effect is that clouds play a significant part in warming the Earth and its atmosphere.

Recent research indicates that on a global scale the Earth's present cloud cover provides more shielding from incoming solar radiation than trapping of long-wave radiation. At present the result is negative, a total net reduction in radiation to the Earth of about 14–21 watts per square meter per month. It is estimated that if there were no clouds, average Earth surface temperature would warm by about 10°C.

Not only do the clouds affect the Earth's climate; they are also affected by it. This feedback mechanism could impose a new cloud-controlled radiation balance on the Earth's climate, if the Earth's climate changed. If the present negative radiation balance became less negative, the Earth's surface would warm; if cloud feedback produced a more negative radiation balance, the Earth's surface would become cooler than it is at present. If the Earth warmed because of the greenhouse effect, there could be an increase in evaporation and therefore an increase in low-level water-droplet clouds over the ocean; these clouds would absorb and reflect incoming short-wave radiation and help to cool the Earth. However, an increase in high-altitude ice clouds would warm the Earth because ice clouds pass incoming short-wave radiation but reduce the Earth's long-wave radiation loss.

Box Figure 1 Subtropical cumulus clouds over the Florida Keys.

the Northern Hemisphere is always counterclockwise. The wind directions related to these pressure cells are shown in figure 6.22. Along the Pacific coast of the United States, the northerly winds cool the coastal areas in the summer and the southerly winds warm them in the winter. The eastern United States receives warm, moist air from the low latitudes in the summer, and cold air moves down from the high latitudes in the winter. Although these seasonal changes modify the wind and pressure belts that were developed on the water-covered model, the generalized wind and pressure

The formation of clouds is related to the availability of condensation nuclei. If nuclei are sparse, the droplets are fewer and larger; the chance of precipitation is increased and the amount of nuclei is further decreased. Many small condensation nuclei produce abundant small droplets, increasing the cloud reflectivity and decreasing incoming radiation. In oceanic areas where precipitation is frequent, condensation particles are removed by precipitation and their low supply may hinder cloud formation. In these areas satellite photos show passing ships leaving cloud trails as condensation occurs on particles from the ships' exhausts. Dimethyl sulfide is a source of condensation nuclei over ocean areas, and its availability is controlled by its own feedback system, as discussed in this chapter. Volcanic activity contributes particles that act as condensation nuclei, as do severe dust storms and major land fires such as those that occurred in Brazil in 1997 and in Indonesia in 1998. The carbon particles in smoke increase the cloud absorption of solar energy while, at the same time, they reduce the size of water droplets.

Although clouds are among the most common of atmospheric phenomena, our understanding of cloud formation and distribution and of climate change and feedback mechanisms is presently incomplete. Scientists use large computer models known as general circulation models (GCMs) to test their ideas, modifying them as new information is acquired. There is an ongoing concerted effort to directly measure radiation levels above and below clouds to better understand the complexities of cloud-climate systems. Early results indicate that the solar radiation shielding of clouds has been underestimated. By using improved GCMs to more clearly define the role of clouds in climate changes, scientists hope to better understand the consequences of modifying our atmosphere.

To Learn More About Clouds and Climate

Glanz, J. 1993. Climate Researchers Look to Clouds. *R and D Magazine*, February, 56–62.

Kaufman, Y.J., and R.S. Fraser. 1997. The Effect of Smoke Particles on Clouds. *Science* 277 (5332): 1636–39.

Kiehl, J.T. 1994. Clouds and Their Effects on the Climate System. *Physics Today* 47 (11): 36–42.

Levi, B. 1995. Clouds Cast a Shadow of Doubt on Models of Earth's Climate. *Physics Today* 48 (5): 21–23.

Internet References

Visit the book's Online Learning Center at www.mhhe.com/sverdrup (click on the book's cover) to explore links to further information on related topics.

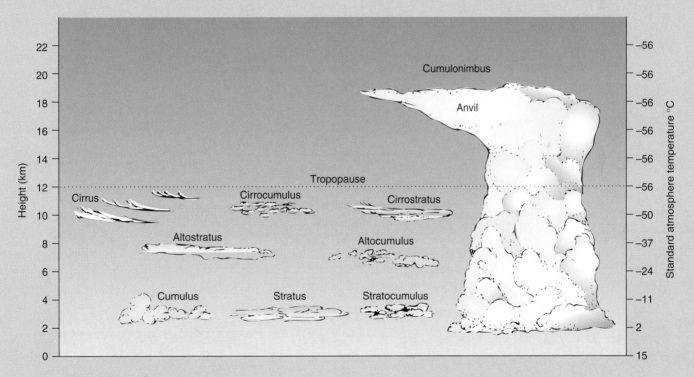

web link **Box Figure 2** Cloud types used by the World Meteorological Organization.

belts are still identifiable over the Earth in the northern latitudes when the atmospheric pressures are averaged over the annual cycle.

Rotation of the airflow around high- and low-pressure air cells is reversed in the Southern Hemisphere because the Coriolis effect is opposite to that in the Northern Hemisphere. At the middle latitudes in the Southern Hemisphere, there is little land, and the water temperature predominates. There is little seasonal effect; the atmospheric pressure and wind patterns created by surface temperatures

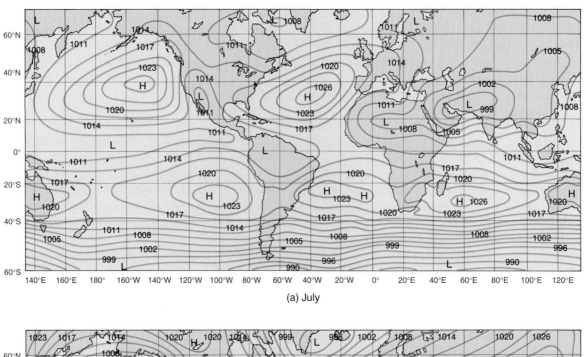

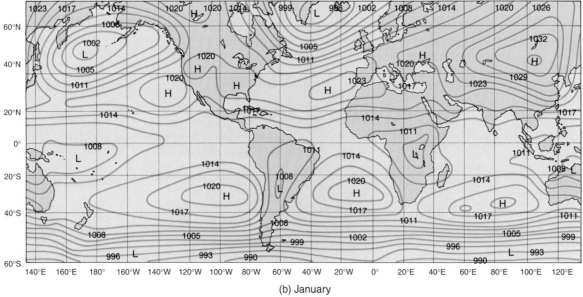

Figure 6.21 Average sea-level atmospheric pressures expressed in millibars for (a) July and (b) January. Large landmasses at mid-latitudes in the Northern Hemisphere cause high and low atmospheric pressure cells to change their positions with the seasons. In the Southern Hemisphere, large landmasses do not exist at mid-latitudes, and average air-pressure distribution changes little with the seasons.

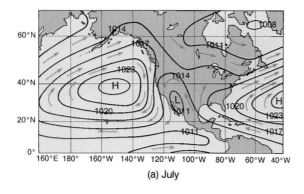

(a) July

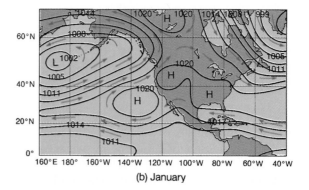

(b) January

Figure 6.22 Atmospheric pressure cells control the direction of the prevailing winds. Atmospheric pressures are expressed in millibars. In the Northern Hemisphere, air flows outward and clockwise around a region of high pressure and inward and counterclockwise around a low-pressure area. (a) Average wind conditions for summer. (b) Average wind conditions for winter.

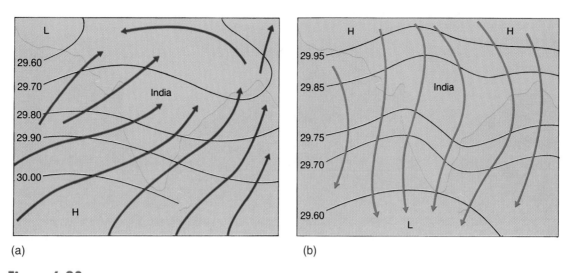

(a)   (b)

 Figure 6.23 The seasonal reversal in wind patterns associated with (a) the summer (wet) monsoon and (b) the winter (dry) monsoon. The isobars of pressure are given in inches of mercury. Land topography increases friction between winds and land, thereby decreasing the Coriolis effect. The winds blow more directly from high- to low-pressure areas across the isobars.

change little over the annual cycle and are very similar to those developed for the water-covered model (see figs. 6.19 and 6.21).

The Monsoon Effect

The differences in temperature between land and water produce large-scale and small-scale effects in coastal areas. In the summer along the west coast of India and in Southeast Asia, the air rises over the hot land, creating a low-pressure air system (fig. 6.23a). The rising air is replaced by warm, moist air carried on the southwest winds from the Indian Ocean. As this onshore airflow rises over the land it cools and causes evaporation to decrease and condensation to increase. This produces clouds and a steady heavy rainfall. This is the wet, or summer, **monsoon.** In the winter, a high-pressure cell forms over the land, and northeast winds carry the dry, cool air southward from the Asian mainland and out over the Indian Ocean (fig. 6.23b). This movement produces cool, dry weather over the land, known as the dry, or winter, monsoon. For years, coastal traders in the Indian Ocean who were dependent on sailing craft planned their voyages so that they sailed with the wind to their destination on one phase of the monsoon and returned on the next.

The monsoon effect is seen on a smaller local scale along a coastal area or along the shore of a large lake. During daylight, the land is warmed faster than the water, and the air rises over the land. The air over the water moves in to replace it, creating an **onshore** breeze. At night, the land cools rapidly, and the water becomes warmer than the land. Air rises over the water, and the air from the land replaces it, this time creating an **offshore** breeze. Such a local diurnal (once-a-day) wind shift is referred to as a land-sea breeze (fig. 6.24). The onshore breeze reaches its peak in the afternoon, when the temperature difference between the land and the water is at its maximum. The offshore breeze is

strongest in the late night and early morning hours; sometimes one can smell the land 30 km (20 mi) at sea, its odors carried by the offshore winds. This daily wind cycle helps fishing boats that depend only on their sails to leave the harbor early in the morning and return late in the afternoon or early evening. This effect brings the summer fogs to San Francisco, California. The area east of San Francisco Bay heats up during the day, and the warm air over the land rises. When the warm marine air reaches the cold coastal waters, a fog is formed that pours in through the Golden Gate, obscuring first the Golden Gate Bridge and then the city (see fig. 4.21). When the air reaches the east side of the bay, it warms, and the fog dissipates. At night the flow is reversed, and the city is swept free of the fog.

The Topographic Effect

Because the continents rise high above the sea surface, they affect the winds in another way. As the winds sweep across the ocean, they reach the land and are forced to rise to continue moving across the land, as shown in figure 6.25. The upward deflection cools the air, causing rain on the windward side of the islands and mountains; on the leeward (or sheltered) side, there is a low-precipitation zone, sometimes called a **rain shadow.** For example, the windward sides of the Hawaiian mountains have high precipitation and lush vegetation, whereas the leeward sides are much drier and require irrigation. On the west coast of Washington State, the westerlies moving across the North Pacific produce the Olympic Rain Forest as the air rises to clear the Olympic Mountains. On the western side of the Olympic Mountains the rainfall is as much as 5 m (200 in) per year; 100 km (60 mi) away, in the rain shadow on the leeward side of the mountains, it is 40–50 cm (16–20 in) per year. The west side of the mountains of Vancouver Island in British Columbia has a high rainfall; the eastern side of the island is known for its sunshine and scenic

Figure 6.24 Differences between day and night land-sea temperatures produce an onshore breeze during the day and an offshore breeze at night.

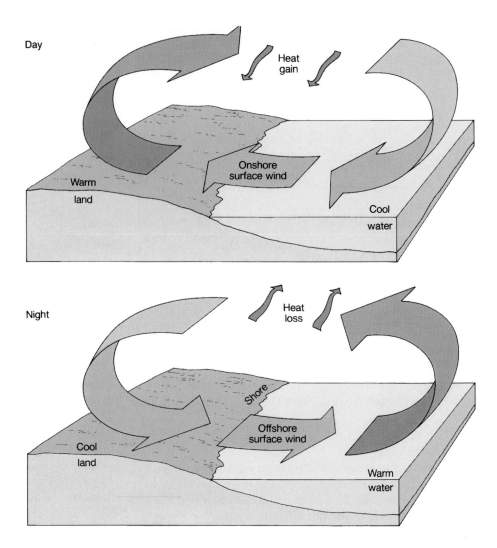

Figure 6.25 Moist air rising over the land expands, cools, and loses its moisture on the mountains' windward side. Descending air compresses, warms, and becomes drier, creating a rain shadow on the leeward side.

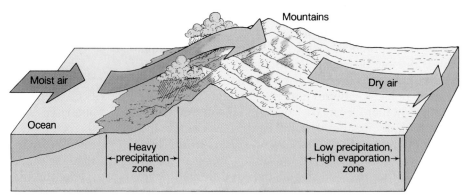

cruising. The southeast trade winds on the east coast of South America sweep up and across the lowlands and then rise to cross the Andes Mountains, producing rainfall, large river systems, and lush vegetation on the eastern side of the Andes and a desert on their western slope. Over the Indian subcontinent in summer the air approaching the Himalayas rises, intensifying the wet monsoon; air flows down from these mountains during the winter, the time of the dry monsoon. The control of precipitation patterns due to elevation changes is called the **orographic effect.**

Jet Streams

Centered over zones of sinking and rising air near 50°–60°N and S and 30°N and 30°S are the **jet streams,** high-speed winds of the upper troposphere. The polar jet streams (50°–60°N and S) are westerlies, lying above the boundary between the polar easterlies and the westerlies; the subtropical jet streams (30°N and 30°S) are easterlies, found above the zone between the trades and the westerlies (fig. 6.26). The Northern Hemisphere polar jet stream is used as an example in the following discussion.

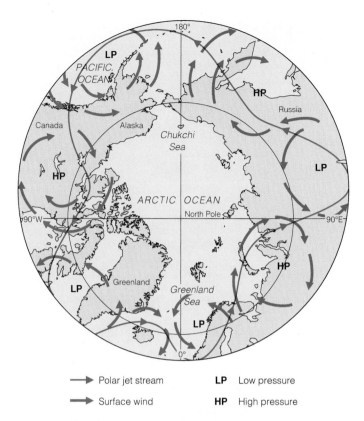

| Polar jet stream | **LP** | Low pressure |
| Surface wind | **HP** | High pressure |

web link **Figure 6.26** The polar jet stream circles the Earth in the Northern Hemisphere above the boundary between the polar easterly winds and the westerly winds. It is deflected north and south by the alternating air-pressure cells of the northern temperate zone.

In the winter the polar jet stream flows rapidly, reaching 250 km (150 m) per hour, and oscillates more than 2000 km (1250 mi) north to south. Wind speeds and displacement are reduced in the summer. The westerlies move the wave form of the stream and its associated high- and low-pressure systems eastward around the Earth. Oscillations in jet stream flow are caused by strong high-pressure systems to the south and intense low-pressure systems to the north. The location and movement of the polar jet stream are controlled by the location and strength of the boundary between subtropical and polar air and the shape of the temperate zone's alternating high- and low-pressure systems.

The position of the polar jet stream and the boundary between subtropical and polar air are important in determining the weather of the temperate zone. Circulation of air around these pressure systems transports warm subtropical air to higher latitudes and cold polar air to lower latitudes; refer to the discussion of the heat budget in this chapter and to figure 6.2. The most important source of heat in this system is the water vapor gained by excess evaporation in the subtropics and excess condensation as the air cools at higher latitudes. In the equatorial regions, great cumulus cloud systems form in the rising air of the doldrums and are moved westward by the subtropical jets.

Hurricanes 6.7

Although the trade winds of the tropics blow steadily, they may develop variations in speed and direction when the winds move over waters of changing temperature. These variations cause moving air to converge and then, as the air oscillates, to diverge; the resulting pressure disturbance is known as an easterly wave. It is visible as a sharp wrinkle in the isobars over the tropical oceans.

If the sea surface temperature is above 27°C, the atmospheric pressure decreases and the evaporation rate increases; the easterly wave develops into an intense, isolated, low-pressure cell, and a tropical depression is formed. The winds circle this depression counterclockwise in the Northern Hemisphere and clockwise in the Southern Hemisphere, build in strength, and the weather disturbance becomes a tropical storm and then a **hurricane** (fig. 6.27). The strong winds, in excess of 130 km (70 knots) per hour, that rotate about the low-pressure system associated with hurricane formation extract water vapor, and therefore heat, from the sea surface. The large amounts of heat energy liberated from the condensing water vapor fuel the storm winds, raising them to destructive levels, up to 300 km (160 knots) per hour. A major hurricane contains energy exceeding that of a large nuclear explosion; fortunately, the energy is released much more slowly. The energy generated by a hurricane is about 3×10^{12} watt-hours per day, the equivalent of the energy in 1 million tons of TNT. When these storms move over colder water or over land, the hurricane is robbed of its energy source and begins to dissipate. These storms bring not only strong winds but also high precipitation because of the very high rate of condensation associated with the rising warm, moist tropical air.

Hurricanes can form on either side of the equator but not at the equator because the Coriolis effect is zero. When a storm of this type is formed in the western Pacific Ocean it is called a **typhoon,** or **cyclone,** instead of a hurricane. Areas that give rise to hurricanes and typhoons and their typical storm tracks are shown in figure 6.28. The surge of excessive high water that accompanies these storms is discussed in section 6.9.

Forecasting the path, the strength, and the destination of hurricanes is of enormous importance to the lives and livelihoods of people around the globe. Models that incorporate satellite data, weather observations, and atmospheric data are used to make these forecasts. The relatively new geostationary operational environmental satellites (GOES) continuously monitor the position, speed, and path of these intense storms. Land-based Doppler radar can see into a hurricane as the storm approaches land and can measure its wind speed and wind distribution in great detail.

Combining these data, scientists are attempting to create other models to predict the role that global warming may play in the frequency or severity of hurricanes. Early results indicate that increasing sea surface temperature in the Pacific Ocean could increase storm intensity. However, researchers studying the records of past storms in the Atlantic find that the frequency of intense storms is decreasing and

web link **Figure 6.27** Hurricane Allen in the western Caribbean, photographed by a NOAA satellite in 1980. The center, or eye, of the storm is the clear, calm hole in the ascending air. Winds circulate counterclockwise about the eye.

that there are no data pointing to worsening conditions with a warmer climate. What is evident is that hurricane damage is growing because of greater population concentrations and property development in hurricane-prone areas.

El Niño 6.8

Under normal conditions a low-pressure area in the atmosphere is located north of Australia over Indonesia in the western equatorial Pacific Ocean and a corresponding high-pressure area is located over Easter Island in the eastern equatorial Pacific Ocean off the coast of Central and South America. This difference in atmospheric pressure helps to drive the steady trade winds, which blow from east to west along the equator. The trade winds drive water away from the west coast of Central and South America. This causes a deep accumulation of warm, high sea surface temperatures, and an elevation of the sea surface in the western equatorial Pacific Ocean (fig. 6.29a). Sea surface temperatures are about 8°C greater and sea surface elevation is about 50 cm (20 in) higher in the western Pacific compared to the eastern Pacific. As water is driven away from the west coast of Central and South America upwelling brings cold, deep water to the surface along the west coast of Peru to replace it. This results in relatively low sea surface temperatures in the eastern equatorial Pacific Ocean.

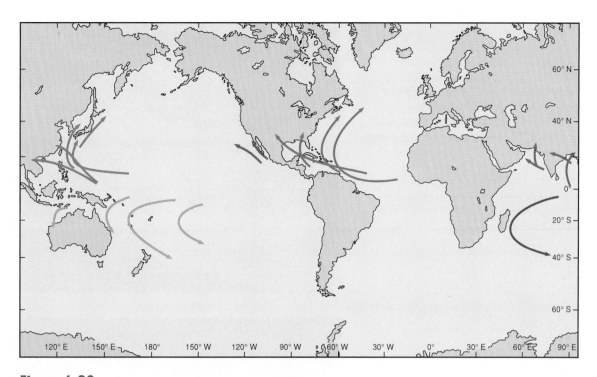

Figure 6.28 Hurricanes, typhoons, and cyclones form on either side of the equator in tropical seas. These storms follow preferred paths in different areas of the oceans.

Roughly every three to eight years some still unknown process causes the low- and high-pressure areas on either side of the Pacific Ocean basin to reverse their position; high pressure dominates the western equatorial Pacific and low pressure dominates the eastern equatorial Pacific. When this happens the trade winds often have an initial surge in strength and then falter and reverse themselves to blow from west to east. This periodic reversal in atmospheric pressure, and resulting reversal in the trade winds, is called the **Southern Oscillation.** As a result of the Southern Oscillation the deep, warm pool of water that normally occupies the western equatorial Pacific surges back toward the eastern equatorial Pacific over a period of two to three months to accumulate along the coast of the Americas (fig. 6.29b). This effectively shuts down the upwelling of deep water off the Peruvian coast and raises the sea surface temperature by several degrees centigrade. The return flow of water from the west also raises the sea surface elevation by 30 cm (12 in) or more (the change in elevation in the eastern equatorial Pacific between January 1997 and November 1997 shown in figure 6.29 was approximately 34 cm). This sequence of events is known as **El Niño,** or Christ Child, named for its frequent occurrence around the Christmas season. Eventually the normal atmospheric pressure distribution is reestablished, the trade winds again blow to the west, and the equatorial Pacific Ocean again has warm water in the west and cool water in the east, signaling the end of the El Niño. Figure 6.29c illustrates the beginning of a resumption in normal conditions following the end of the 1997–98 El Niño. Because of the relationship between El Niño and the South-

ern Oscillation the two events together are known as ENSO (an acronym for El Niño/Southern Oscillation).

Years in which more severe El Niño events have occurred are shown in figure 6.30. The climatic effects of El Niño are highly variable and appear to depend on the size of the warm pool of water and its temperature. The same region may experience higher-than-normal rainfall and flooding during one event and drought conditions during the next event. However, there do appear to be some very regular consequences of El Niño events. The northern United States and Canada generally experience warmer-than-normal winters. The eastern United States and normally dry regions of Peru and Ecuador typically have high rainfall while Indonesia, Australia, and the Philippines experience drought. In addition, El Niño years are associated with less intense hurricane seasons in the Atlantic Ocean.

During the severe El Niño of 1982–83, ocean surface temperatures off Peru rose 7°C above normal and tropical species were displaced as far north as the Gulf of Alaska. In 1982–83 the polar jet stream was displaced far southward over the Pacific Ocean, bringing unusually dry conditions to Hawaii but high winds and high precipitation to areas along the west coast of the United States while the eastern United States had its mildest winter in twenty-five years. Heavy rains occurred in Ecuador, Peru, and Polynesia, but drought conditions were experienced in Central America, Africa, Indonesia, Australia, India, and China. At the same time, lower sea surface temperatures in the North Atlantic made the hurricane season the quietest in over fifty years. There was an estimated $8 billion in damages worldwide ($2 billion

Figure 6.29 These images show sea surface topography from NASA's *TOPEX* satellite, sea surface temperatures from NOAA's *AVHRR* satellite sensor, and sea temperature below the surface is measured by NOAA's network of TAO moored buoys in the equatorial Pacific Ocean. *Red* is 30°C and *blue* is 8°C. (a) Normal conditions as seen in January 1997, (b) El Niño conditions during November 1997, (c) the end of El Niño conditions and the beginning of a return to normal conditions in March 1998. Images from NASA Goddard Space Flight Center.

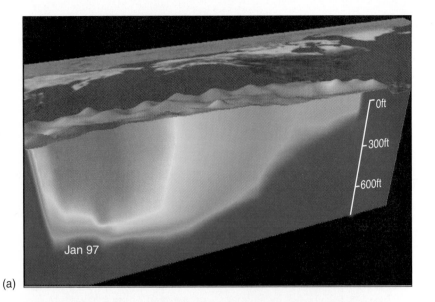

(a)

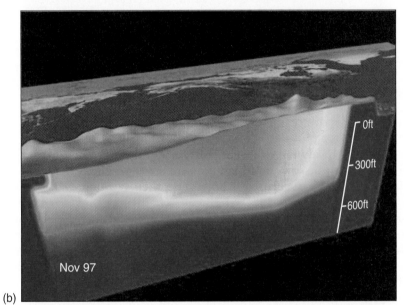

(b)

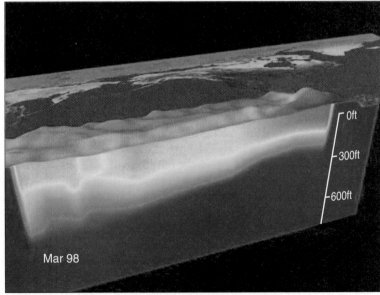

(c)

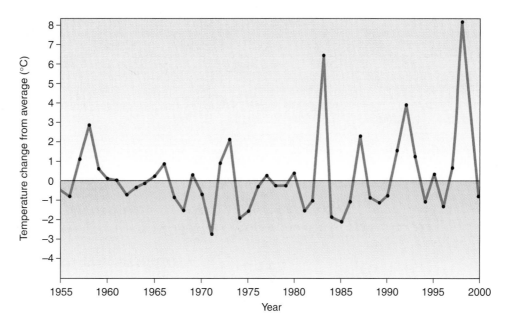

web link **Figure 6.30** Winter temperature changes above the 0 line indicate warmer than average El Niño events; temperature changes below the 0 line depict cooler-than-average La Niña events. Note the period of weak La Niña events during 1975–81 and the series of strong El Niño events since 1982. The 1955–94 sea surface temperatures at Punto Chicama, Peru, were averaged for the months of January, February, and March; the differences between this forty-year, three-month average (18.3°C) and the yearly three-month averages are plotted in this diagram. *Data courtesy of Francisco Chaves, Monterey Bay Aquarium Research Institute.*

in the United States) from this event. The severity of this El Niño and its worldwide effects led to the initiation in 1985 of a decade-long study called the Tropical Ocean Global Atmosphere (TOGA) program. TOGA was designed to monitor ocean and atmospheric conditions in the equatorial South Pacific Ocean for the purpose of predicting future El Niño events. The ability to forecast these El Niño events and so regulate fisheries, predict agricultural droughts, and prepare for severe weather conditions have great economic and commercial importance around the world. Continued monitoring is now being done using seventy stationary buoys deployed in the international Tropical Atmosphere and Ocean (TAO) project. Data collected during the TOGA and TAO projects has led to the creation of improved ocean-atmosphere models that have been used with some success to predict the onset of El Niño conditions, but further improvement is needed to accurately predict the severity and length of future El Niños.

During 1991–92 the southern displacement of the polar jet stream over the Pacific brought heavy winter rains to Southern California and the U.S. Gulf Coast, and a mild, low-precipitation winter to the coastal regions of Oregon and Washington. The jet stream's extreme oscillation, north over the central United States and south over the eastern United States, gave New England and the maritime provinces of Canada extreme cold and heavy snow.

The strongest El Niño on record occurred in 1997–98. The greatest warming occurred in the eastern topical Pacific, where surface water temperatures were as much as 8°C above normal near the Galápagos Islands and off the coast of Peru. Normally dry regions in Ecuador and Peru that usually receive

only 10–13 cm (4–5 in) of rain annually had as much as 350 cm (138 in, or 11.5 ft). Severe drought struck areas of the west Pacific Ocean in the Philippines, Indonesia, and Australia. Because El Niño tends to reduce precipitation in the wet monsoon areas of Asia, a weak rainy season was predicted, but India's rains were 2% above normal. A series of strong storms caused severe beach erosion, flooding, and landslides along the California coast. At the same time the Pacific Northwest and Midwest winters were mild and the southern portion of the United States experienced wetter-than-normal conditions. The jet stream was diverted far to the south over North America, inhibiting the growth of hurricanes in the Atlantic Ocean. By the spring of 1998, Pacific Ocean sea surface temperatures were returning to normal and signaling the end of this El Niño event.

Computerized ENSO models attempt to forecast El Niño events by means of the Southern Oscillation, sea surface temperatures, and the surge in the strength of the trade winds that appears to precede their decline and reversal. Computer models successfully predicted the onset of the 1991–92 El Niño event, but the polar jet stream split over the eastern Pacific and not all the predictions came to pass. The 1997–98 episode was predicted six months in advance, although there was difficulty distinguishing the signs of the approaching El Niño from the unusually warm conditions that persisted throughout the 1990s. The advanced warnings of the 1997–98 El Niño are estimated to have saved $1 billion to $2 billion in property damage in the United States.

Between El Niño events surface temperatures off Peru may drop below normal; an event of this type is known as **La Niña,** "the girl" (fig. 6.30). These colder-than-normal years also produce wide-scale meteorological effects. The trade

web link **Figure 6.31** Storm tide damage caused by Hurricane Hugo, Pawley's Island, South Carolina, September 1989.

winds strengthen and surface water temperatures of the eastern tropical Pacific are colder, whereas those to the west are warmer than normal. These changes help establish dry conditions over the coastal areas of Peru and Chile while rainfall and flooding increase in India, Myanmar, and Thailand.

The cyclic alternation between these two events has been quite regular for the last hundred years, except for the periods between 1880 and 1900, when La Niña conditions prevailed, and the 1975–97 period of El Niños. In 2000–2001 a La Niña prevailed. This led to decreased rainfall in the Pacific Northwest and increased rain in California. The reduced rain and snowpack in Washington State caused problems with irrigation and power production. Look carefully at figure 6.30 and notice that the 0°C line represents the long-term average sea surface temperature over the last forty or so years. If this line were redrawn between 0°C on the left and 0.5°C on the right, allowing a half-degree rise in ocean surface temperature, the cyclic alternation between El Niño and La Niña events would appear as a more regular pattern. Recent analysis of surface water temperatures off the Southern California coast during this period (1950–90) indicate an average water temperature increase of 0.8°C. There are at least two possible mechanisms that could lead to such a warming of ocean surface water: (1) the ocean's response to global warming or (2) a reduction in upwelling in the Pacific Ocean associated with a reduction in production of North Atlantic deep water (see chapter 7).

Practical Considerations: Storm Tides and Storm Surges 6.9

Periods of excessive high water along a coast associated with changes in atmospheric pressure and the wind's action on the sea surface are known as **storm surges** or **storm tides.** These storm surges, combined with normal high-tide conditions, can spell disaster for low-lying coastal areas.

Intense storms at sea, such as hurricanes or typhoons, are centered about intense low-pressure systems in the atmosphere. Under the low-pressure area in the center of the storm, the sea surface rises up into a dome, or hill, while the surface is depressed farther away from the center, where the atmospheric pressure is greater. The atmospheric pressure change between the outside of a hurricane and its center can be as large as 7.5 cm (3 in) of mercury. Because mercury is 13.6 times denser than water, this pressure change produces a 97 cm (38 in) change in elevation of the water between regions outside the storm and those at the storm's center. In addition, the surface winds spiraling toward the center of the storm add to the sea's elevation under the storm center.

The elevated dome of water travels across the sea under the storm center and raises the water level at the shore when the storm reaches the coast. During the storm, the drag of the wind on the sea surface pushes surface water in the approximate direction of the wind. On the side of the storm where the winds are toward the shore, the water moving toward the shore is piled up against the coast, increasing the height of the sea surface and producing the storm surge. The water continues to pile against the shore, steepening the slope of the water until the tendency of the water to flow downhill and back to sea is equal to the force of the wind driving the water ashore. If the water along the shore is deep, some of the water moving landward downwells and returns seaward; this process decreases the height of the water being driven ashore. Along a shallow shore the landward-moving layer of water extends to the sea floor; there is little downwelling or seaward return flow, and the height of the

web ⟨ᴡ⟩ link **Figure 6.32** A satellite image of Hurricane Andrew making its landfall south of Miami, August 24, 1992. Maximum sustained winds were 220 km per hour (138 mi per hour).

water along the shore is much greater. A storm surge may sustain a high water level for many hours until the storm winds diminish in the coastal area. Storm surges are often confused with tsunamis; these great sea waves are discussed in chapter 9.

Along shallow areas of the East and Gulf Coasts of the United States, storm surges have caused considerable damage. In 1900 a storm surge caused "the great Galveston flood." It produced water depths of 4 m (13 ft) over the island of Galveston, Texas, destroying the city and killing 5000 people. Hurricane Camille, in 1969, was one of the strongest storms ever to hit the Gulf Coast. The high water levels caused severe property damage, and several hundred people were killed. In 1989 a storm surge of 5 m (16.5 ft) came ashore with Hurricane Hugo in South Carolina, destroying much property along the coast and in the city of

Charleston. Great loss of life was averted by early warnings and mass evacuation of low-lying coastal areas (fig. 6.31). In August 1992 Hurricane Andrew (fig. 6.32) struck Florida with the same intensity that Hurricane Hugo brought to the Carolina coast, but the coastal water damage in Florida was much less. Although the winds were high (220 km/h, or 138 mi/h), the storm surge was only 2.4 m (8 ft). The storm traveled a short distance between the Bahamas and Florida at a fast rate of speed; it did not have enough time to build up a large mass of onshore moving water. In addition, some of Andrew's wave energy was absorbed by offshore reefs and coastal mangrove swamps. On shore, however, the wind damage was very severe, and Hurricane Andrew became the most costly storm in U.S. history.

Severe storms over the shallow Bay of Bengal spawned storm surges that took an estimated 300,000 lives in 1970.

Another 10,000 people were killed in 1985, and in 1991 yet another storm surge struck this same area of Bangladesh, killing an estimated 139,000 people. The expanding populations of this area need land for homes and farms, and as fast as new land appears in the delta of the Ganges River, the people move seaward. Although our ability to predict the severity and the route over which a hurricane will pass is improving, other factors work against the potential life-saving results of this ability. When a storm comes to a heavily populated, low-lying coast, the water rises rapidly and covers hundreds of square miles; such conditions make the evacuation of thousands of people extremely difficult.

The Netherlands has constructed barrier dams to protect their coast from the storm tides of the North Sea. England has installed gates across the River Thames to bar a storm tide moving up the river. The great cost of engineering projects such as these may be justified in heavily populated industrial areas, but storm tides cannot be prevented and along many stretches of shore there is no defense. The costs to taxpayers and their governments for emergency services and cleanup of storm tide damage are enormous. Would it be wiser to vacate areas historically prone to storm surges and high-water damage and use them only in nonpermanent ways?

Summary

The intensity of solar radiation over the Earth's surface varies with latitude. Incoming radiation and outgoing radiation are equal when averaged with time over the whole Earth. Reflection, reradiation, evaporation, conduction, and absorption keep the heat budget in balance with the incoming solar radiation. In any local area of the oceans there is daily and seasonal variation. Winds and ocean currents move heat from one ocean area to another to maintain the surface temperature patterns.

The oceans gain and lose large quantities of heat, but their temperature changes very little. The heat capacity of the oceans is high compared to that of the land and the atmosphere. Heat absorbed at the surface mixes downward to reduce the temperature change at the surface.

Most clouds and weather occur in the troposphere, where temperature decreases with altitude. In the stratosphere temperature increases with altitude because of the ozone and its ability to absorb ultraviolet radiation. The temperature of the higher mesosphere and thermosphere decreases with height.

The atmosphere is a mixture of gases, including water vapor. Atmospheric pressure is the force with which air presses on the Earth's surface; high-density air creates high-pressure zones, and low-density air forms low-pressure zones.

Concern that the Earth's climate may be changing is related to changes in the balance of the gases in the Earth's atmosphere. Increases in CO_2 concentration due to burning and use of fossil fuels are leading to the prediction of global warming because carbon dioxide traps outgoing long-wave radiation by the greenhouse effect. There has been a significant depletion of the ozone layer, most probably due to the release of chlorine into the atmosphere.

Dimethyl sulfide produced by plant cells at the sea surface is self-regulated by its role in cloud formation. Sulfur gases from industry and volcanic eruptions are also related to cloud formation and climate change.

The density of air is controlled by air temperature, pressure, and water vapor content. The winds are the horizontal air motion in convection cells produced by heating the atmosphere from below. Winds are named for the direction from which they blow.

Because of the Coriolis effect, winds are deflected to their right in the Northern Hemisphere and to their left in the Southern Hemisphere. This action produces a three-celled wind system in each hemisphere that results in the surface wind bands of the trade winds, the westerlies, and the polar easterlies. Zones of rising air occur at 0° and at 60°N and S; these are low-pressure areas of clouds and rain. Zones of descending air at 30° and 90°N and S are high-pressure areas of clear skies and low precipitation. Surface winds are unsteady and unreliable at the zones of rising and sinking air, producing the doldrums and the horse latitudes. The doldrum belt is displaced north of the Earth's geographic equator.

Seasonal atmospheric pressure changes modify these wind bands and cause coastal winds to change direction seasonally in the Northern Hemisphere. Differences in temperature between land and water produce the monsoon effect. The seasonal reversal in wind pattern causes the wet and dry monsoons of the Indian Ocean; a similar daily reversal causes the onshore and offshore winds of any coastal area. Winds from the ocean rising to cross the land also produce heavy rainfall. The jet streams are high-altitude, fast-moving winds found in both hemispheres; when greatly displaced, abnormal weather patterns result. Hurricanes develop from tropical low-pressure systems with high-wind systems; these storms carry enormous amounts of energy.

In some years warm tropical surface water moves eastward across the Pacific and accumulates along the west coast of the Americas, blocking the normal upwelling. This phenomenon is El Niño; it is associated with changes in atmospheric pressure and wind direction. Colder-than-normal surface water temperatures off coastal Peru and associated weather phenomena are known as La Niña; La Niña episodes appear to alternate with El Niño events.

A storm tide or a storm surge is produced by a storm-elevated sea surface when the storm reaches the shore; it is associated with severe coastal flooding and destruction.

Key Terms

All key terms from this chapter can be viewed by term, or by definition, when studied as flashcards on this book's Online Learning Center at www.mhhe.com/sverdrup (click on this book's cover).

atmospheric pressure, 182
isobar, 182
low-pressure zone, 182
high-pressure zone, 182
greenhouse effect, 183
Coriolis effect, 191
trade winds, 192
westerlies, 192
polar easterlies, 192
doldrums, 193
horse latitudes, 193
monsoon, 197
onshore, 197

offshore, 197
rain shadow, 197
orographic effect, 198
jet stream, 199
hurricane, 199
typhoon, 199
cyclone, 199
Southern Oscillation, 201
El Niño, 201
La Niña, 203
storm surge, 204
storm tide, 204

Study Questions

1. Write an equation for the whole Earth's heat budget. Include all the factors for incoming and outgoing energy.
2. How is the troposphere different from the stratosphere?
3. What controls the density of air? Why is moist air less dense than dry air?
4. Why are the carbon dioxide and ozone concentrations of the atmosphere changing? Why is there concern about these changes?
5. How do airborne sulfur compounds affect cloud cover and the acid rain problem? What are natural and industrial sources of these compounds?
6. Explain why, on the world average, the Earth's atmosphere is in motion and unstable. Take into consideration the distribution of the Earth's surface area with latitude and the variation in heat loss and heat gain with latitude.
7. Why are regions that are noted for their low barometric pressure and rising air famous for their excessive precipitation?
8. A frictionless projectile is fired from the North Pole and is aimed along the prime meridian. It takes three hours to reach its landing point, halfway to the equator. Where does it land? (Give latitude and longitude.) If the same projectile is fired from the South Pole under the same circumstances, where does it land? (Give latitude and longitude.)
9. Why does the circulation of the atmosphere depend on its transparency to solar radiation? What would happen to atmospheric circulation if the upper atmosphere absorbed most of the solar radiation?
10. Plot the six major wind belts on a map of the world. Add bands of rising air and descending air, regions of the doldrums, and the horse latitudes.
11. Explain why the northeast and southeast trade winds are steady in strength and direction year-round but the Northern Hemisphere westerlies alternate from northwest to southwest with summer and winter along the West Coast of the United States.
12. Why are the westerlies of the Southern Hemisphere more consistent than the westerlies of the Northern Hemisphere?
13. What are the early signs that alert forecasters to the onset of an El Niño event?
14. In what way does the polar jet stream influence the transfer of heat from low to high latitudes?
15. How do hurricanes produce storm tides? Why is a storm tide more severe along a coast with a wide, shallow continental shelf than along a coast with a narrow continental shelf?
16. Why do the windward sides of the Hawaiian Islands receive more rain than their leeward sides?

Links to Related Websites

Visit the book's Online Learning Center at www.mhhe.com/sverdrup (click on the book's cover) to find live Internet links for additional topics related to this chapter's content.

- Weather
- Carbon dioxide
- Climate change
- Winds and storms
- El Niño

Visit the book's Online Learning Center at www.mhhe.com/sverdrup (click on this book's cover) to find these additional chapter tools: Suggested Reading; links to further information on boxed readings, selected figures, and related chapter topics; and additional study aids.

chapter 7

Circulation and Ocean Structure

T he weeks passed. We saw no sign either of a ship or of drifting remains to show that there were other people in the world. The whole sea was ours, and, with all the gates of the horizon open, real peace and freedom were wafted down from the firmament itself.

It was as though the fresh salt tang in the air, and all the blue purity that surrounded us, had washed and cleansed both body and soul. To us on the raft the great problems of civilized man appeared false and illusory—like perverted products of the human mind. Only the elements mattered. And the elements seemed to ignore the little raft. Or perhaps they accepted it as a natural object, which did not break the harmony of the sea but adapted itself to current and sea like bird and fish. Instead of being a fearsome enemy, flinging itself at us, the elements had become a reliable friend which steadily and surely helped us onward. While wind and waves pushed and propelled, the ocean current lay under us and pulled, straight toward our goal.

Thor Heyerdahl
From *Kon-Tiki*

Waves at Cape Kiwanda, Oregon coast.

H idden below the ocean's surface is its structure. If we could remove a slice of ocean water in the same way we might cut a slice of cake, we would find that, like a cake, the ocean is a layered system. The layers are invisible to us, but they can be detected by measuring the changing salt content and temperature and by calculating the density of the water from the surface to the ocean floor. This layered structure is a dynamic response to processes that occur at the surface: the gain and loss of heat, the evaporation and addition of water, the freezing and thawing of ice, and the movement of water in response to wind. These surface processes produce a series of horizontally moving layers of water, as well as local areas of vertical motion. In this chapter, we will study both the surface processes and their below-the-surface results in order to understand why the ocean is structured in this way and how the structure is maintained. We will also explore the ways in which oceanographers gather data about this layered system, and we will survey the possibilities of extracting useful energy from it.

Density Structure   7.1

Surface Processes

Absorption of solar energy and energy exchanges between the sea surface and the atmosphere determine sea surface temperatures and the zones of precipitation and evaporation that govern surface salinities. Variations in temperature and salinity combine to control the density of ocean surface water. Many combinations of salinities and temperatures produce the same density; the result is shown in lines of constant density (fig. 7.1). Figure 7.1 is related to the density values in table 4.5 in chapter 4.

As the salinity increases, the density increases; as the temperature increases, the density decreases. Salinity may be increased by evaporation or by the formation of sea ice; it may be decreased by precipitation, by the inflow of river water, by the melting of ice, or by a combination of these factors. Changes in pressure also affect density. As the pressure increases, the density increases. However, because pressure plays a minor role in determining the density of surface water, its effects are not considered here.

Notice that in figure 7.1 the lines of constant density are curved. Therefore, two types of water having the same density but different values of salinity and temperature when mixed form a mixture that lies on their mixing line. Such a mixture has a density greater than either of the original water types, and it will sink. This mixing and sinking process is known as **caballing** and occurs in all oceans when surface waters converge.

Less dense water remains at the surface (for example, the warm, low-salinity surface water of the equatorial

latitudes). Although the surface water at 30°N and 30°S latitudes is warm, it has a higher salinity than surface waters at the equator. Therefore, it is denser than the warm, low-salinity equatorial water. This 30° latitude surface water sinks below the equatorial surface water; it extends from the surface at 30°N to below the less dense equatorial layer and back to the surface at 30°S. The combined salinity and temperature in surface waters at 50°–60°N and 50°–60°S produces a water that is denser than either the equatorial or the 30° latitude surface water. The 50°–60° latitude water therefore sinks below the equatorial and 30° surface water and extends from the surface in one hemisphere below the other water types to the surface in the other hemisphere. Winter conditions in the polar regions lower the surface water temperature and, if sea ice forms, increase the surface salinity. The result is a dense surface water that sinks at the polar latitudes. These variations in surface water properties and the resulting density changes produce a density-layered ocean. This layered system is shown in figure 7.2. The thickness and horizontal extent of each layer are related to the rate at which the water of each layer is formed and the size of the surface region over which it is formed.

Changes with Depth

The oceans have a well-mixed surface layer of approximately 100 m (330 ft) and layers of increasing density to a depth of about 1000 m (3300 ft). Below 1000 m the waters of the deep ocean are relatively homogeneous. A region between 100 m and about 1000 m, where density changes rapidly with depth, is known as a **pycnocline** (fig. 7.3).

Below the 100 m surface layer the temperature decreases rapidly with depth to the 1000 m level. A zone with a rapid change in temperature with depth is called a **thermocline.** Below the thermocline, the temperature is relatively uniform over depth, showing a small decrease to the ocean bottom. A similar situation occurs with salinity. Below the surface water at the middle latitudes, the salinity increases rapidly to about 1000 m; this zone of relatively large change in salinity with depth is called the **halocline.** Beneath the halocline, relatively uniform conditions extend to the ocean bottom. Both a thermocline and a halocline are shown in figure 7.4.

If the density of the water increases with depth, the water column from surface to depth is **stable.** If there is more dense water on top of less dense water, the water column is **unstable.** An unstable water column cannot persist; the denser surface water sinks and the less dense water at depth rises to replace the surface water. Vertical **overturn** of the water takes place. If the water column has the same density over depth, it has neutral stability and is termed **isopycnal.** A neutrally stable water column is easily vertically mixed by wind, wave action, and currents. If the water temperature is unchanging over depth, the water column is **isothermal;** if the salinity is constant over depth, it is **isohaline.**

Density-Driven Circulation

Processes that increase the water's density at the surface cause convective water movement, or **vertical circulation.** This overturn may reach only to shallow depths or extend to the deep-sea floor, ensuring an eventual top-to-bottom exchange of water. Because the density is normally controlled by surface changes in temperature and salinity, this vertical circulation is known as **thermohaline circulation.** An excellent example of thermohaline circulation occurs in the Weddell Sea of Antarctica, where the winter cooling and freezing produce dense, cold, more saline surface water that sinks to the sea floor. This water descending along the coast of Antarctica is the densest water found in the open oceans (see fig. 7.2).

At the temperate latitudes in the open ocean the surface water's temperature changes with the seasons. This effect is illustrated in figure 7.5. During the summer the surface water warms and the water column is stable, but in the fall the surface water cools, its density increases, and overturn begins. Winter storms and winter cooling continue the mixing process. The shallow thermocline formed during the previous summer is lost, and the upper portion of the water column is vertically mixed and becomes isothermal to greater depths. Spring brings warming, and the shallow thermocline begins to reestablish itself; the water column becomes stable and remains so through the summer.

Seasonal temperature changes are more important than salinity changes in altering density in the open ocean below polar latitudes. For example, in the Atlantic Ocean the surface water at 30°N and 30°S has a high salinity but is warm year-round, so it stays at the surface. In contrast, water from

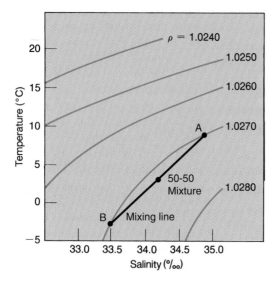

Figure 7.1 The density of seawater, measured in grams per cubic centimeter, is abbreviated as ρ (rho) and varies with temperature and salinity. Many combinations of salinity and temperature produce the same density. Low densities are at the *upper left* and high densities at the *lower right.* The *straight line* is the mixing line for waters A and B, both with the same density. A mixture of A and B lies on the mixing line and is more dense than either A or B.

Figure 7.2 Water of different densities forms a layered ocean. The density of each layer is determined at the surface by the climate at the latitude at which it is formed. Density values are given in grams per cubic centimeter.

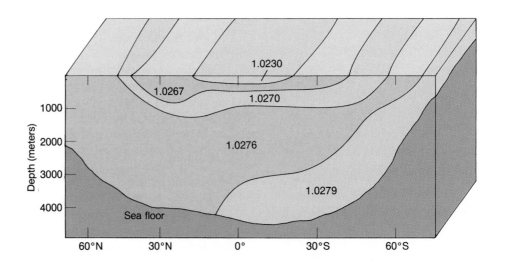

50°–60°N in the North Atlantic has a lower salinity, but it is cold, especially during the winter. This cold water sinks and flows below the saltier but warmer surface water.

Close to shore, the salinity of seawater can become more important than the temperature in controlling density.

Salinity is particularly important in semienclosed bays, sounds, and fjords that receive large amounts of freshwater runoff. Here, extremely cold (0°–1°C) freshwater ice melt is added to the rivers. When this water meets the salt water its salinity is so low that the water does not sink but remains at

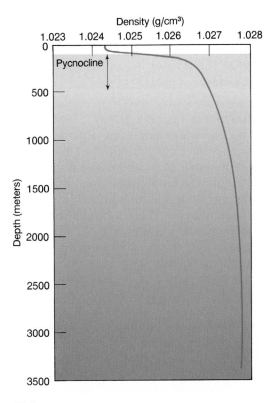

Figure 7.3 Density increases with depth in seawater. The pycnocline is the region in which density changes rapidly with depth. (Based on data from the northeastern Pacific Ocean.)

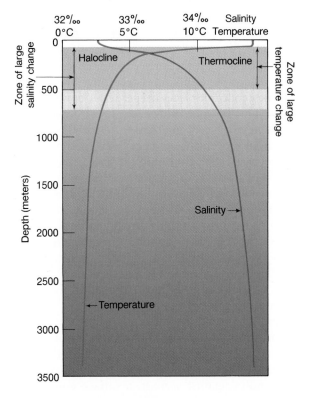

Figure 7.4 Temperature and salinity values change with depth in seawater. Rapid changes in temperature and salinity with depth produce a thermocline and a halocline, respectively. (Based on data from the northeastern Pacific Ocean.)

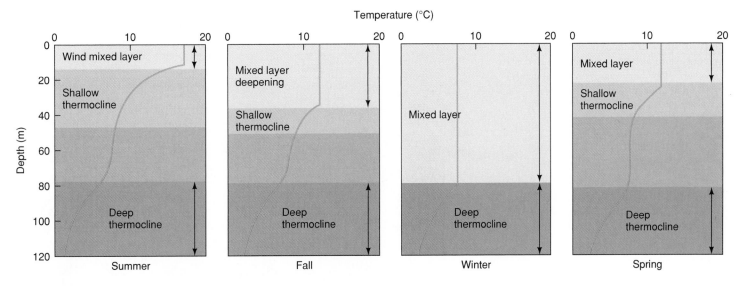

Figure 7.5 The surface-layer temperature structure varies over the year. In the absence of strong winds and wave action in the summer, solar heating produces a shallow thermocline. During the fall and winter, the surface cooling and storm conditions cause mixing and vertical overturn, which eliminate the shallow thermocline and produce a deep wind-mixed layer. In spring the thermocline re-forms. The deep thermocline is below the influence of seasonal surface changes and remains in place year-round.

the surface as a seaward-moving **freshwater lid.** In polar regions, when the sea ice melts, a layer of fresh water forms at the surface and slowly dilutes the underlying surface water.

Upwelling and Downwelling 7.2

When dense water from the surface sinks and reaches a level at which it is denser than the water above but less dense than the water below, it spreads horizontally as more water descends behind it. At the surface, water moves horizontally into the region where sinking is occurring. The dense water that has descended displaces deeper water upward, completing the cycle. Because water is a fixed quantity in the oceans, it cannot be accumulated or removed at given locations without movement of water between those locations. This concept is called **continuity of flow.** Areas of thermohaline circulation where water converges and sinks are called **downwelling zones;** areas of diverging rising waters are **upwelling zones.** Downwelling is a mechanism that transports oxygen-rich surface water to depth, where it is needed for the deep-living animals. Upwelling returns water with dissolved, decay-produced nutrients that have accumulated at depth to the surface, where they act as fertilizers to promote the production of more oxygen by photosynthesis in the sunlit surface waters.

Upwelling and downwelling refer to vertical motion of water upward or downward. They are present in thermohaline circulation but can also be caused by wind-driven surface currents. When the surface waters are driven together by the wind or against a coast, a surface **convergence** is formed. Water at a surface convergence sinks, or downwells. When the wind blows surface waters away from each other or away from a coast, a surface **divergence** occurs and water

from below is upwelled (fig. 7.6). During the slow movement of water from surface to depth and back, it continually mixes with adjacent layers of water, gradually sharing chemical and physical properties.

The speed of upwelling and downwelling water is about 0.1–1.5 m (5 ft) per day. Compare this speed to that of oceanic surface currents, which reach speeds of 1.5 m/s. Horizontal movement at mid-ocean depth due to the thermohaline flow is about 0.01 cm (0.004 in) per second. Water caught in this slow but relentlessly moving cycle may spend 1000 years at the greater ocean depths before it again reaches the surface.

Chemical Tracers

Chemical tracers are substances that occur in small amounts in seawater; they have no effect on water motion but can be used to identify kinds of water moving through the ocean system. Transient tracers are substances that change their distributions with time and have usually been added to the oceans by human means. When carefully sampled and analyzed, these transient tracers provide us with the quantitative data used to form computer and mathematical models of water pathways in the oceans.

Atmospheric testing of nuclear weapons in the late 1950s and early 1960s produced radioactive materials that still persist in traces in the oceans. Research carried out by the Geochemical Ocean Sections (GEOSECS) program in 1972–73 showed that water containing radioactive substances from the sea surface had reached the bottom of the North Atlantic at 5000 m (16,400 ft) depth north of 40°N. South of this latitude no radioactive substances were present below 3000 m (10,000 ft), indicating a much faster rate of downwelling transport than was previously thought to occur.

Figure 7.6 The anatomy of the Atlantic Ocean. Surface and subsurface circulation are related here. The surface currents converge to produce downwelling; water sinks to its density level and flows horizontally. Water at depth rises under zones of surface divergence.

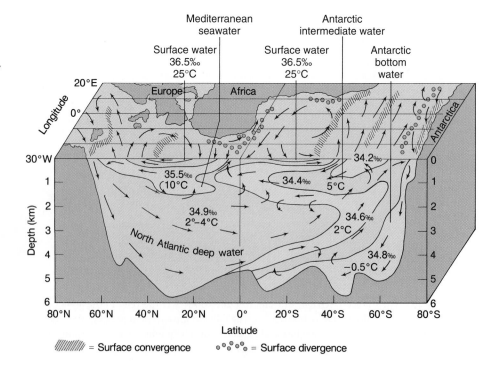

The 1981 Transit Tracers in the Ocean (TTO) cruises found that the radioactive material had moved southward since GEOSECS and was found distributed from ocean surface to ocean floor as far south as 30°N. South of this latitude the North Atlantic was free of radioactive substances at depths below 2000 m (6500 ft).

Current work on radioactive substances focuses on tritium, a heavy isotope of hydrogen left from these same tests. Because tritium is chemically hydrogen, it is present in the oceans as part of the water molecules and is therefore the perfect tracer for water movement. Tritium is radioactive, making it easily traceable, and has a half-life of 12.45 years; it decays to an inert isotope of helium. It is possible to compute the time since a water parcel was last at the surface by measuring the tritium-helium isotope concentrations in that water. These age calculations are most accurate for time periods of less than ten years and for water movements in shallow areas, where there is the least background of naturally occurring helium. Layers of constant density along which water tends to move and mix have been identified by this method in the North Atlantic. Studies of water samples taken from greater depths show that deep water may take 500–1000 years to sink, move across a deep-ocean basin, and rise again to the surface. Determination of deep-water residence times are made using the isotope C^{14}, which has a half-life of 5570 years.

The Layered Oceans   7.3

Oceanographers have taken salinity and temperature measurements with depth from many surface positions and for many years. Gradually they accumulated sufficient data to identify the layers of water that make up each ocean and the surface source of the water forming each layer. The structure of an ocean is determined by the properties of the layers of water present under the sea surface. Each layer received its characteristic salinity, temperature, and density at the surface. The water's density controls the depth to which the water sinks; the thickness and horizontal extent of each layer are related to the rate of its formation and the size of the surface region from which it came. Water that sinks from the surface to spread out at depth and slowly mixes with adjacent layers eventually rises at another location. In all cases, water that sinks to depth displaces an equivalent volume of water upward toward the surface at some other location so that the oceans' vertical circulation may continue.

The layers of water just described are associated with depth zones: surface, intermediate, deep, and bottom. The surface zone extends to 200 m (650 ft) and the intermediate zone lies between 300 and 2000 m (1000 and 6500 ft). Deep water is found between 2000 and 4000 m (6500 and 13,000 ft), and bottom water is the water below 4000 m (13,000 ft).

The Atlantic Ocean

The properties of the layers of water making up the Atlantic Ocean are shown in figure 7.6. At the surface in the North Atlantic, water from high northern latitudes moves southward, while water from low latitudes moves northward along the coast of North America and then east across the Atlantic. These waters converge in areas of cool temperatures and high precipitation at approximately 50°–60°N in the Norwegian Sea and at the boundaries of the Gulf Stream and the Greenland and Labrador Currents. The resulting mixed water has a salinity of about 34.9‰ and a temperature of 2°–4°C. This water, known as **North Atlantic deep water,** sinks and moves southward. North Atlantic deep water from the Norwegian Sea moves down the east side of the Atlantic, while water formed at the boundary of the Labrador Current and the Gulf Stream flows along the western side. Above this water at 30°N, a low-density lens of very salty (36.5‰) but very warm (25°C) surface water remains trapped by the circular movement of the major oceanic surface currents. Between this surface water and the North Atlantic deep water lies water of intermediate temperature (10°C) and salinity (35.5‰). This water is a mixture of surface water and the upwelled colder, saltier water from the subtropical regions. It moves northward to reappear at the surface south of the convergence in the North Atlantic.

Near the equator, the upper boundary of the North Atlantic deep water is formed by water produced at the convergence centered about 40°S. This is **Antarctic intermediate water.** Because it is warmer (5°C) and less salty (34.4‰) than the North Atlantic deep water, it is less dense and remains above the denser and saltier water below. Along the edge of Antarctica very cold (–0.5°C), salty (34.8‰), and dense water is produced at the surface by sea ice formation during the Southern Hemisphere's winter. This is **Antarctic bottom water,** the densest water produced in the world's oceans. After it descends to the ocean floor, it begins to move northward. When it meets North Atlantic deep water it creeps beneath it and continues to move northward through the deep South Atlantic ocean basins west of the Mid-Atlantic Ridge. Antarctic bottom water does not accumulate enough thickness to be able to flow over the mid-ocean ridge system into the basins on the African side of the ridge; it is confined to the deep basins on the west side of the South Atlantic and has been found as far north as the equator.

At the same time, the North Atlantic deep water trapped between the Antarctic bottom and intermediate waters rises to the ocean's surface in the area of the 60°S divergence. As it reaches the surface it splits; part moves northward as **South Atlantic surface water** and Antarctic intermediate water; part moves southward toward Antarctica, to be cooled and modified to form Antarctic bottom water. A mixture of North Atlantic deep water and Antarctic bottom water becomes the circumpolar water for the Southern Ocean as it flows around Antarctica.

Warm (25°C), salty (36.5‰) surface water in the South Atlantic is also caught by the circular current pattern at the surface and is centered about 30°S. Below the southern tips of South America and Africa the water flows eastward, driven by the prevailing westerly winds, which move the water around and around Antarctica.

Figure 7.7 Mid-ocean salinity and temperature profiles of the Atlantic, Pacific, and Indian Oceans. Temperature and salinity are shown as functions of depth and latitude.

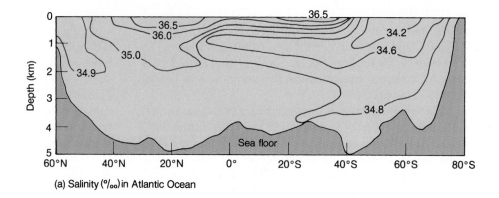

(a) Salinity (⁰/₀₀) in Atlantic Ocean

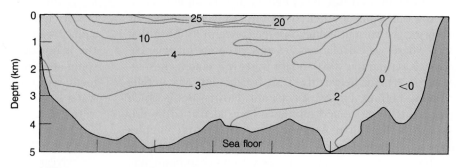

(b) Temperature (°C) in Atlantic Ocean

Because the Atlantic Ocean is a narrow, confined ocean of relatively small volume but great north-south extent, the water types are readily identifiable and the movement of the layers can be followed quite easily. In addition, the bordering nations of the Atlantic have had a long-standing interest in oceanography, so the vertical circulation and layering of the Atlantic are both the most studied and the best understood of all the oceans.

The Pacific Ocean

In the vast Pacific Ocean waters that sink from relatively small areas of surface convergences lose their identity rapidly, making the layers difficult to distinguish. Antarctic bottom water forms in small amounts along the Pacific rim of Antarctica, but it is quickly lost in the great volume of the Pacific Ocean. The deeper water of the South Pacific Ocean is the common mixture of the Antarctic circumpolar flow. Because the North Pacific is isolated from the Arctic Ocean, only a small amount of water comparable to North Atlantic deep water can be formed. In the extreme western North Pacific, convergence of the southward-flowing cold water from the Bering Sea and the Sea of Okhotsk and the northward-moving water from the lower latitudes produces only a small volume of water that sinks to mid-depths. There is no large source of deep water similar to that found in the North Atlantic. Warm, salty surface water occurs at subtropical latitudes (30°N and 30°S) in each hemisphere, and Antarctic intermediate water is produced in small quantities, but its influence is small. Deep-water flows in the Pacific are sluggish, and conditions are very uniform below 2000 m (6600 ft). The slow circulation of the Pacific means that it has the oldest water at depths where age is measured as time from the water's last contact with the surface. Residence time for deep water in the Pacific is about twice that of deep water in the Atlantic.

The Indian Ocean

Because the Indian Ocean is principally an ocean of the Southern Hemisphere, there is no counterpart of the North Atlantic deep water. Small amounts of Antarctic bottom water are soon mixed with the deeper waters to form a fairly uniform mixture of Antarctic circumpolar water brought into the Indian Ocean by the Antarctic circumpolar current. There is a small amount of Antarctic intermediate water, and in the subtropics a lens of warm, salty water occurs at the surface.

Comparing the Major Oceans

Temperature and salinity distributions as functions of depth are shown in figure 7.7 for each of the oceans. The distributions of Atlantic Ocean salinity and temperature values form patterns with depth (fig. 7.7a, b) that are clearly identifiable and can be related to figure 7.6. Salinity and temperature values in the Pacific (fig. 7.7c, d) identify the high-salinity surface water lenses and the mixture of Antarctic bottom water and circumpolar deep water, with a 34.7‰ salinity and a 1°C temperature. A minor intrusion of low-salinity water in the north is the result of the surface convergence in the far northwestern Pacific. Note the large volume of deep water that shows a uniform salinity in the Pacific Ocean. Temperature values also follow a less definite pattern than in the Atlantic, emphasizing the uniformity of most of this deep water. The Indian Ocean values in figure 7.7e, f resemble those for the South Atlantic, without the presence of

Figure 7.7 *Continued*

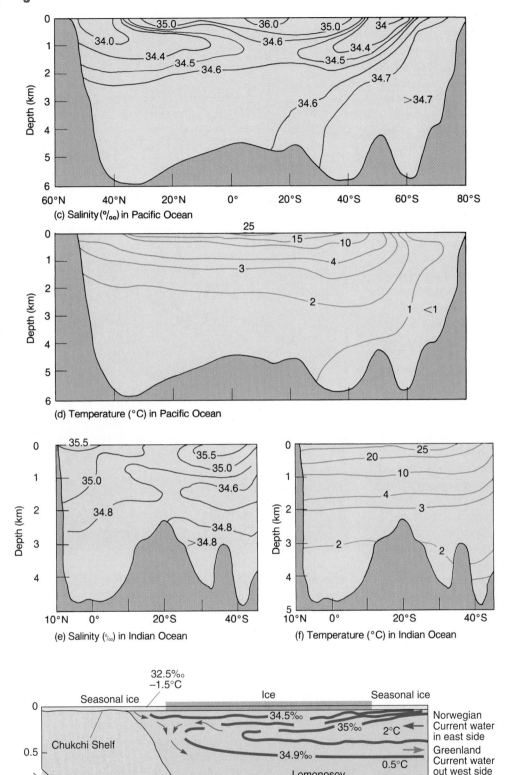

(c) Salinity (‰) in Pacific Ocean

(d) Temperature (°C) in Pacific Ocean

(e) Salinity (‰) in Indian Ocean

(f) Temperature (°C) in Indian Ocean

32.5‰
−1.5°C

Seasonal ice   Ice   Seasonal ice

Chukchi Shelf

34.5‰

35‰   2°C

Norwegian Current water in east side

34.9‰

0.5°C

Greenland Current water out west side

Lomonosov Ridge

Canadian Basin

Eurasian Basin

water that is comparable to North Atlantic deep water. The Arctic Ocean is an extension of the North Atlantic; see figure 7.8 and the following discussion.

The Arctic Ocean

Some oceanographers consider the Arctic Ocean to be an extension of the North Atlantic, but it differs from the Atlantic Ocean in many significant ways. About one-third of its area, 8×10^6 km^2 (3.1×10^6 mi^2), is covered by extensive continental shelves, the widest of any ocean. Two basins, the Eurasian to the east and the Canadian to the west, occupy the central portion of the ocean; they are separated by the Lomonosov Ridge extending due north from Greenland (fig. 7.8). The Eurasian basin is the deeper basin, 5000 m (16,400 ft); it is connected to the North Atlantic through a gap in the continental shelf between Spitsbergen and Greenland. The larger, shallower Canadian basin is about 3800 m (12,450 ft) deep. Both basins contain spreading centers that are extensions of the North Atlantic Ridge system.

The density of the Arctic Ocean water is controlled more by salinity than by temperature. Its surface layer is formed from low-salinity water entering from the Bering Sea, fresh water from Siberian and Canadian rivers, and seasonal melting of sea ice. The surface layer from these combined sources is about 80 m (250 ft) deep

Figure 7.8 The Arctic Ocean. Water is supplied to a shallow surface layer by the Atlantic Ocean's Norwegian Current. Water exits the Arctic Ocean into the Atlantic Ocean by the Greenland Current. The circulation is confined almost entirely to the upper 500 m (1650 ft) of the Arctic Ocean.

Arctic Ocean Studies

Not all oceanographic research involves ships or is conducted in warm, tropical regions. More and more interest is being shown in the Arctic Ocean, the world's northernmost but least studied ocean. Understanding Northern Hemisphere climate changes and making long-range weather forecasts for northern Europe and North America require that we understand the effects of periodic increases in ice and the large volumes of meltwater that enter the North Atlantic from the Arctic. Changes in sea ice coverage directly affect the Earth's heat budget. When the Arctic Ocean is ice-covered, reflectance of solar energy is high and absorption is low, but when there is little ice exposed, reflectance is low and absorption is high. There is new concern that circulation within the Arctic Ocean may distribute persistent organic pollutants (POPs) from adjacent land sources, contaminating marine life and placing the native peoples dependent on this marine life at risk; again we need more information on the circulation of the Arctic Ocean to evaluate this problem.

In 1994 a joint program between Canada and the United States used icebreakers to traverse 1600 nautical miles of the Arctic Ocean from the Bering Sea to the pole. The main goal was to establish the Arctic Ocean's role in global climate change. Studies were made of the water, sea ice, sea floor, marine organisms, distribution of pollutants, and polar bear populations (box figs. 1 and 2). The overlying theme was to establish the Arctic Ocean's role in global climate change by studying the area from the atmosphere through the water and ice down to the sea floor.

On January 1, 1994, the Arctic Climate System Study (ACSYS) was organized. The observational phase of this program continues for ten years. Its primary goal is to understand the role of the Arctic in producing global climate. ACSYS seeks to encourage and coordinate both national and international research activities concerning ocean circulation, ice cover, exchanges of water in the Arctic Ocean, long-term climatic research, and monitoring programs. SHEBA (Surface Heat Budget of the Arctic) is a subprogram of ACSYS and also a United States–Canada cooperative research program. This program froze a Canadian Coast Guard icebreaker into the Arctic ice in October 1997, 248 km (400 mi) north of Prudhoe Bay, Alaska. For the next year the ship acted as hotel and supply base for the SHEBA program (box fig. 3). The data from studies of air-sea exchange processes affecting Arctic Ocean heat budgets helped scientists better understand the role of Arctic processes in global climate models.

In the spring of 2000 researchers began a new five-year program known as the National Science Foundation's **North Pole Environmental Observatory.** To learn more about how the northernmost ocean regulates global climate, buoys were placed on the polar ice to drift with the ice pack. These buoys are equipped with sensors that extend through the ice and gather data on changes in the thickness of the ice and the properties of the upper ocean. Information is relayed via satellites. Another installation of drifting buoys was completed in the spring of 2001. One buoy was developed by the Japanese to measure ocean temperature and

Box Figure 1 Scientists and their gear are offloaded to the ice from the U.S. Coast Guard icebreaker *Polar Sea* during its 1994 Arctic Ocean Section expedition. *Courtesy of Capt. L. W. Brigham Ret., Commanding Officer of the* Polar Sea.

salinity, current profiles, and atmospheric temperature and pressure, as well as wind velocity. Other buoys include a meteorological buoy that records wind speed, temperature, and atmospheric pressure; two radiometer buoys that measure radiation from the sun and reflected radiation from the ice; and two ice-mass balance buoys that record ice temperature and snow thickness.

Another installation in 2001 was a 4250 m (14,000 ft) long instrumented cable with 4000 kg (9000 lb) of gear that was moored to the sea floor under the ice (box fig. 4). The mooring, supported by submerged floats, carries conductivity-temperature recorders to monitor salinity and temperature changes, current meters to record speed and direction of flow, a current profiler to detail

web link **Box Figure 2** A polar bear is tranquilized and tested for contaminants during the 1994 Arctic Ocean Section expedition. *Courtesy of Capt. L. W. Brigham Ret., Commanding Officer of the Polar Sea.*

web link **Box Figure 4** Deploying a 4250 m mooring at the North Pole. The mooring is outfitted with a suite of oceanographic instruments. An acoustic Doppler current profiler is located near the top of the mooring.

web link **Box Figure 3** The Canadian Coast Guard icebreaker *Des Groseilliers* serves as the support base for SHEBA research huts on the Arctic ice. *Courtesy of Sandra Hines, University of Washington.*

web link **Box Figure 5** The building on the ice combines living quarters and a scientific laboratory. Equipment includes an A-frame and buoys ready for deployment through the ice.

ice drift and vertical structure of currents, and sonar to measure ice thickness. Each instrument must record its data internally because a satellite cannot retrieve data from under the ice. The data will be recovered in the following year when the mooring is retrieved. Another part of the 2001 program includes using ski-equipped aircraft to establish five camps over a 485 km (300 mi) route from the pole to Alaska. At each camp (box fig. 5) members of the research team

make measurements through the ice and also collect samples for chemical tests. These data continue the effort of 2000, when similar data was obtained over a 565 km (350 mi) route between the pole and Canada. The data that are being collected will be combined with satellite and meteorological data to study changes in the Arctic Ocean and to determine how fast the changes are occurring.

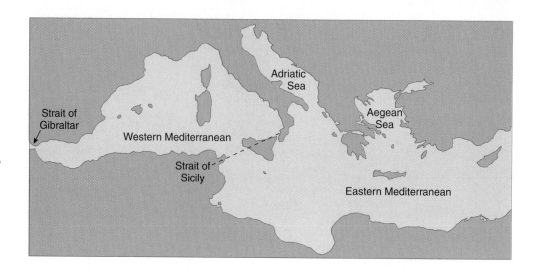

Figure 7.9 A *line* through the Strait of Sicily separates the Mediterranean Sea into the Western Mediterranean basin and the Eastern Mediterranean basin. Deep and intermediate waters are formed in the eastern basin, move through the Strait of Sicily into the western basin, continue westward, and exit through the Strait of Gibraltar into the Atlantic Ocean. Before 1987 cool, high-salinity, deep water was formed principally in the Adriatic Sea; by the winter of 1995, the source of this water had changed to the Aegean Sea.

and has a low salinity (32.5‰) and a low temperature (−1.5°C). Below the surface layer, salinity increases with depth in the halocline layer, 200 m (650 ft) thick, to reach 34.5‰ at its base. The cold, salty water of the halocline layer is produced by the annual freezing and formation of sea ice over the continental shelves. This water sinks and moves across the shelves to spread out in the central ocean basins. West of Spitsbergen, North Atlantic water (2°C and 35‰) enters the Arctic Ocean and is cooled as it flows under the halocline and fills the Arctic Ocean basins. It upwells along the edge of the continental shelves, mixing with the water formed during freezing, and exits the Arctic as water of 0.5°C temperature and 34.9‰ salinity along the edge of the shelf adjacent to Greenland. This exiting water moves down the coast of Greenland and enters the North Atlantic south of Greenland and Iceland, where it combines with Gulf Stream water to form North Atlantic deep water.

Many aspects of the structure of the Arctic Ocean are still a puzzle. Approximately 70% of the total Arctic Ocean observations by all nations to date were made by the former Soviet Union. Since the cessation of the Cold War, detailed bathymetric and water column data have been declassified and made available to the Arctic research community. One data set contained 1.3 million salinity and temperature measurements taken over a period of many years from icebreakers, drifting ice camps, and buoys. The data are being distributed in a four-volume U.S.-Russian Arctic Atlas. The first volume was released in January 1997.

Bordering Seas

Two small but specific water types from bordering seas are readily identifiable, one in the North Atlantic and one in the Indian Ocean. The water from the Mediterranean Sea has a temperature of about 13°C and a salinity of 37.3‰ as it leaves the Strait of Gibraltar. This water, mixing with Atlantic Ocean water, creates a water mixture that sinks and finds its own density in the North Atlantic at approximately 1000 m (3300 ft) in depth. The influence of Mediterranean water can be traced 2500 km (1500 mi) from the Strait of Gibraltar be-

fore it is lost through modification and mixing. In the Indian Ocean, the initially very salty (40–41‰) water from the Red Sea has been found in a spreading layer at 3000 m (10,000 ft) depth more than 200 km (124 mi) south of its source.

Studies to determine where and how the deep water of the Mediterranean is formed were made in the 1980s and again in 1995. The Mediterranean Sea is divided into east and west basins at the Strait of Sicily (fig. 7.9). Deep and intermediate Mediterranean waters formed in the east basin move through the Strait of Sicily into the west basin and then through the Strait of Gibraltar into the Atlantic Ocean. The 1980 studies showed that high evaporation rates and winter cooling in the southern Adriatic Sea caused surface water to sink, and this water then moved into the east basin and from there into the west basin on its way to the Atlantic. The 1995 studies indicate that the source of the east basin water is no longer the Adriatic Sea; instead the water is being formed in the Aegean Sea. The Aegean Sea water is saltier and denser than that produced in the Adriatic Sea, and the Aegean water is now displacing older Adriatic water upward and westward into the Atlantic Ocean. It appears that small climate changes in the eastern Mediterranean Sea affect winter cooling and evaporation rates and so play a significant role in the location and strength of the processes that produce the deep waters of the Mediterranean. Oceanographers are interested in what happens in the Mediterranean Sea because the forces that drive the Mediterranean system are comparable to the larger-scale processes governing the open ocean; thus, the Mediterranean Sea can be used as a model for the larger ocean system.

Internal Mixing

Mixing between waters in the ocean is most active when turbulence and energy of motion are available to stir the waters and blend their properties. At the sea surface wind-driven waves and currents supply energy for mixing. When surface currents converge, mixing at current boundaries may produce caballing of the mixed water. When currents and their associated turbulence are weak, mixing is reduced.

Mixing by diffusion occurs continually at the molecular level, but diffusion is much weaker than mixing by turbulent processes.

If a parcel of water is displaced vertically by turbulence, buoyancy forces tend to return the parcel to its original density level. Therefore vertical mixing between the water types that form the oceans' internal layers is weak. Horizontal mixing is more efficient because it requires less energy than vertical mixing. A parcel of water displaced horizontally along a surface of constant density remains at its new position and shares its properties with the surrounding water.

In areas under warm, high-salinity surface water with an appreciable salinity and temperature decrease with depth, internal vertical mixing processes occur despite the stability of the water column. Vertical columnar flows, approximately 3 cm (5 in) in diameter, are called salt fingers; they develop and mix the water vertically, causing a stair-step salinity and temperature change with depth. This phenomenon is caused by the ability of seawater to gain or lose heat faster by conduction than it gains or loses salt by diffusion. This causes the density of the vertically moving water to change relative to that of the surrounding water and it is propelled either up or down. Salt fingers mix water over limited depths, creating homogeneous layers 30 m (100 ft) thick. These layers exist from about 150–700 m (500–2300 ft) deep and are estimated to occur over large areas of the oceans when the required conditions are present.

Measurement Techniques   7.4

Measuring the salinity, temperature, dissolved gases, nutrients, suspended matter, and other characteristics of seawater in situ requires that the oceanographer devise specialized sampling and measuring equipment and a platform from which the equipment can be used or deployed. The research vessel is the traditional platform from which samples are taken and measurements are made (see the photo essay, "Going to Sea," pages 273-276). These vessels are equipped with winches and cables that launch, lower, and retrieve instruments; laboratories for specialized onboard research; and a large array of measuring devices, including depth recorders, sonar, speed and direction sensors, atmospheric and solar radiation sensors, and many more. However, research time at sea is expensive, for vessels require fuel, living amenities, and professional merchant crews as well as oceanographers. Total vessel and crew costs are high: $25,000 and more per operating day, not including the scientific party and equipment. Therefore, researchers need to be able to gather accurate information in the least amount of time.

The oceanography of the first fifty years of the twentieth century depended on robust mechanical devices that were lowered into the sea, took their samples at preset depths, and returned those samples to a ship for later analysis. Today's oceanographers still use bottom corers and grabs, dredges, and nets, but the standard use of long cables with their attached water bottles and thermometers (fig. 7.10) is decreasing rapidly.

Figure 7.10 Hanging a water bottle. A mechanical water bottle with three deep-sea thermometers attached is fastened to a cable. When the water bottle reaches the desired depth it is signaled to close by a weight that slides down the cable. At the time the water sample is taken, the temperatures registered by the thermometers are fixed. Sample depth is set by measuring the cable as it passes over the meter wheel at the top of the photograph.

The conductivity-temperature-depth sensor, or CTD (fig. 7.11), is today's workhorse, providing a depth profile of salinity by measuring electrical conductivity of the seawater. A temperature profile is obtained with an electrical resistance thermometer, and depth readings are made with a pressure sensor. The CTD is lowered in a protective cage that contains a series of water bottles and may also carry other sensors such as pH probes, chlorophyll sensors, optical scanners, and dissolved oxygen sensors. Conductivity and temperature are monitored continuously as the CTD is lowered and the data are returned to the ship as electronic signals through the suspending cable. Onboard ship, the CTD data may be fed directly into a computer, recorded on a chart, or made available as numerical data. Continuous profiles of data to several thousand meters can be acquired in this way, whereas the old water bottles collected only one sample at each depth for later analysis.

CTD systems still require a research vessel to stop and remain on station while the instrument is lowered,

Figure 7.11 Retrieving a conductivity-temperature-depth sensor, or CTD. The CTD is attached below a rosette of water bottles. Data are relayed to the ship's electronic processing and data center. Water samples are taken when an interesting water structure is found or when water samples are required to calibrate the CTD.

measurements are made, and the CTD is recovered. However, the CTD can also be left in place, suspended from a surface vessel or buoy; the collected data can be stored on magnetic tape for retrieval at another time.

Seasoar (fig. 7.12) is a modified CTD device that allows sampling while the vessel is underway. The Seasoar is a winged vehicle that is towed behind a ship. Its depth is controlled by the device's diving planes and can be compared to flying a kite upside down underwater. Its vertical range extends from the surface to about 350 m (1100 ft) and its dive cycle takes about ten minutes to complete, compared with the standard CTD, which, with stop and start time for the vessel, takes about forty minutes. The Seasoar's sensors record continuously, both horizontally and vertically, while it dives.

To sample larger ocean areas than can be served by a research ship and for continuous sampling over long periods of time, large, instrumented buoys are moored at sea and their data transferred by satellite or radio link to a research ship or laboratory for analysis. Smaller surface buoy systems are released to drift with the currents, monitoring water properties at changing locations and reporting back to a ship via satellite. Single instruments unattached to buoys also drift with the currents and then sink to predetermined depths. These instruments monitor water properties, rise to the surface, and send this information back to a research vessel or to a satellite.

Continual improvements in electronics, computers, and sensor systems have resulted in oceanographers devising independent and relatively inexpensive devices that are placed in the sea in large numbers. These independent, untethered instruments sink to predetermined depths, making measurements as they rise through the water column on their return to the surface. These data-gathering devices are known as profilers and carry a wide variety of sensors.

One type of profiler functions as a winged glider (fig. 7.13; see the box titled "Ocean Gliders"). When this profiler sinks and rises, it flies down and up at an angle measuring properties along a diagonal over its operating depth. This flying ability allows the profiler to wander partially independent of the currents. The Argo profiler (fig. 7.14; see prologue fig. XXII) does not fly but sinks and rises vertically only. Lateral displacement is caused by the currents encountered over its vertical path. Once at the surface both types of profilers broadcast their recorded data and positions to a satellite, then sink and repeat the process.

Specialized satellites measure a large variety of sea surface and atmospheric conditions, sea topography, wind speeds, plankton abundance, air and water temperatures, waves, and more (see box titled "Satellite Oceanography" in the Prologue). Satellite are very expensive monitoring devices, but their ability to survey the global oceans repeatedly and the huge amounts of data they collect make them cost effective as they collect data that cannot be obtained by any other means.

Not all marine studies require a large research vessel with its sophisticated and expensive equipment. Inshore, shallow-water studies usually use smaller vessels with less sophisticated equipment. Small winches driven by power or even by hand are used to lower mechanical water bottles with their thermometers to the required depth. Lightweight electronic instruments are also available and include devices to measure conductivity, temperature, dissolved oxygen, optical properties, pH, and fluorescence of chlorophyll.

Practical Considerations: Ocean Thermal Energy Conversion 7.5

An indirect form of solar energy and an alternative to fossil fuel production of energy is found in the sea. The transparency and heat capacity of water allow large amounts of

web link **Figure 7.12** Seasoar is a conductivity-temperature-depth sensor that is towed by a ship. The wings allow Seasoar to move vertically while it is towed horizontally.

web link **Figure 7.13** The SLOCUM glider has multiple sensors designed to monitor water properties independently for up to five years. While at the surface it is able to send data and update its navigation by satellite.

web link **Figure 7.14** This Argo subsurface drifter is equipped with conductivity-temperature-depth sensors. The instrument sinks to a preprogrammed depth, where it drifts for a set time and then surfaces. The position and recorded data are then transmitted to the research vessel or to a satellite from which the data may be retrieved.

Practical Considerations: Ocean Thermal Energy Conversion **221**

Ocean Gliders

An oceanographic research vessel can monitor only a small portion of an ocean during any one cruise, and the cost is very high (see the photo essay, "Going to Sea"). To increase the size of the ocean area that can be surveyed, as well as lower the cost, oceanographers have been experimenting with independent, unstaffed vehicles (see figs. 7.13 and 7.14). A new type of survey vehicle is an ocean glider, which functions much like a conventional sailplane in the atmosphere. Unlike a sailplane, which has to be towed aloft and ultimately falls back to the ground, an ocean glider has the ability to change its buoyancy so that it can glide up as well as down.

An oceanographic glider, like an airborne glider, has no propeller but is powered by its buoyancy. Buoyancy is changed by adjusting glider volume in order to sink or rise. The presence of wings translates this vertical force into motion along a slanting glide path. The limited onboard battery energy is devoted to buoyancy control, sampling sensors, navigation, and communication systems. Through energy efficiency, the glider remains operational for long periods, up to a year, and collects and reports large quantities of data.

At present, research groups at the Scripps Institution of Oceanography, the Woods Hole Oceanographic Institution, the Webb Research Corporation, and the University of Washington are developing gliders with the support of the U.S. Navy, the U.S. National Science Foundation, and the U.S. National Oceanic and Atmospheric Administration. Different glider designs are being used to address different ocean sampling needs, such as performing long open-ocean transects, maintaining geographic position while profiling vertically, or operating in shallow regions. One under development by Webb Research, known as the Slocum Glider, extracts energy from ocean thermal stratification to change buoyancy (see fig. 7.13). Others use battery energy for propulsion.

At the University of Washington, Seagliders are being developed and evaluated (box fig. 1). Seagliders are small (1.8 m

web link **Box Figure 1** A Seaglider being launched for a trial ocean run.

[7 ft] long), light (52 kg [115 lb]) examples of battery-powered gliders that can be easily launched and retrieved from small vessels rather than relying on research ships. They alternatively dive and climb through the water column as oil is transferred between an internal reservoir and an external bladder. When the glider is launched at the surface its density is just slightly less than the density of seawater. A small amount of oil is bled from the external

solar energy to be stored in the ocean, and this heat can be extracted independent of daily and seasonal changes in the available solar radiation.

Ocean thermal energy conversion (OTEC) depends on the difference in temperature between ocean surface water and water at 600–1000 m (2000–3300 ft) depth. There are two types of OTEC systems: (1) closed cycle (fig. 7.15a), which uses a contained working fluid with a low boiling point, such as ammonia or Freon, and (2) open cycle (fig. 7.15b), which directly converts seawater to steam.

In a closed system, the warm surface water is passed over the evaporator chamber containing the ammonia or Freon, and the ammonia or Freon is vaporized by the heat derived from the warm seawater. The vapor builds up pressure in a closed system, and this gas under pressure is used to spin a turbine, which generates power. After the pressure has been released, the ammonia or Freon is passed to a con-

denser, where it is cooled by cold water pumped up from depth. Cooling the ammonia or Freon returns it to its liquid state, and it is pumped as a liquid back to the evaporator to repeat the cycle.

In an open system, large quantities of warm seawater are converted to steam in a low-pressure vacuum chamber, and the steam is used as the working fluid. Because less than 0.5% of the incoming water is turned into steam, large quantities of warm water must be used. The steam passes through a turbine, is condensed in a condenser cooled by cold water from depth, and turns into desalinated water.

OTEC plants using either system can be located onshore, offshore, or on a ship that moves from place to place. Figure 7.15c is an engineering concept design for an independent, free-floating OTEC plant. The Lockheed Spar has not been built, and there are no current plans to do so. OTEC requires at least a 20°C difference in temperature between surface and depth to generate useful amounts of

bladder to the inside of the pressure case, decreasing the bladder volume and increasing the density of the Seaglider, causing it to sink. When sinking motion starts, the wings on the Seaglider provide lift, and the glider moves forward as it sinks. When the vehicle reaches its programmed depth, an electric pump transfers oil from inside the pressure case to the external bladder, decreasing the vehicle density. As the less dense glider rises toward the surface, the wings act to hold it down so that it moves forward. Its forward speed is only about 0.25 m/s (0.5 knots) but because drag increases quadratically with speed it has a range and endurance much greater than conventional propeller-driven Autonomous Underwater Vehicles (AUV) (see box fig. 4, page 00, in chapter 2).

Seagliders carry sensors and recorders for temperature, salinity, dissolved oxygen, fluorescence, and optical backscatter. GPS navigation and telemetry equipment are also onboard; antennae are housed in the trailing wand (box fig. 2). At the surface the glider determines its position using GPS and sends data via a global satellite phone link. Researchers are able to communicate with the glider and guide it under remote control.

To measure surface currents the drifting glider can be tracked by GPS. When the glider surfaces from a dive, the difference between the glider's actual position and its intended position is the result of displacement by the average of the currents over both its sinking and rising paths.

Seagliders are designed to dive as deep as 1000 m (3000 ft) and cover as much as 10 km (5.3 nautical mi) horizontally in one dive cycle. In a six-month mission they can travel as much as 5000 km (2700 nautical mi) through the ocean.

While building a glider costs roughly as much as buying an expensive automobile, its operational cost is quite low. A single one-year mission costs about the same as one day of research vessel operation. The production and annual operational cost of gliders will decrease when they are produced in quantity. Even in their de-

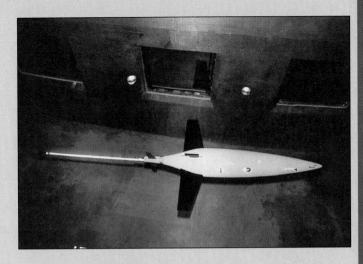

web link **Box Figure 2** A Seaglider moving upward in a testing tank.

velopment stage, gliders have been able to collect observations that are prohibitively costly using ship-based means.

To Learn More About Ocean Gliders

Eriksen, C. C., T. J. Osse, R. D. Light, T. Wen, T. W. Lehman, P.L. Sabin, J. W. Ballard, and A. M. Chiodi. 2001. Seaglider: A Long Range Autonomous Underwater Vehicle for Oceanographic Research. *I.E.E.E. Journal of Oceanic Engineering,* in press.

Internet References

Visit the book's Online Learning Center at www.mhhe.com/sverdrup (click on the book's cover) to explore links to further information on related topics.

energy. This means that these power plants would be located at latitudes between approximately 25°N and 25°S with both warm surface water and cold deep water available; seawater temperature difference over depth must average 22°C. W. H. Avery and C. Wu (*Renewable Energy from the Ocean, A Guide to OTEC,* Oxford University Press, New York, 1994.) estimated that the area of the world's oceans meeting OTEC requirements is 60×10^6 km^2 (23×10^6 mi^2). The total power generated within this area would exceed 10×10^6 MW (megawatts); the total U.S. electricity generating capacity is about 1.7×10^6 MW.

Engineers studying the potential and feasibility of OTEC have estimated that about 0.2 MW of usable power can be extracted per square kilometer of tropical ocean surface; that is about 0.07% of the average absorbed solar energy. The process is considered safe and environmentally benign; however, the pumping of cold water from depth to the sea surface brings nutrients from depth and liberates

them into the nutrient-poor tropical surface waters, changing the productivity of the surface water.

The Natural Energy Laboratory of Hawaii near Keahole Point operated a land-based, open-system OTEC plant for six years (1993–98). The project produced a net of 100 kW of power per day using the temperature difference between the water at the ocean surface and the water about 800 m (2500 ft) below the surface. The plant also produced 26,000 L (7000 gal) of desalinated water per day. Aquaculture projects using the nutrient-rich water from depth included production of oysters, shrimp fish, black pearls, and various seaweeds.

The State of Hawaii is currently installing a new seawater system in the plant; it is scheduled for completion in December 2001. Pipelines with a diameter of 140 cm (50 in) will bring ashore 4°C water from 1000 m (3000 ft) and 28°C water from 8 m (25 ft) depths. The desalinated water will be used primarily for aquaculture.

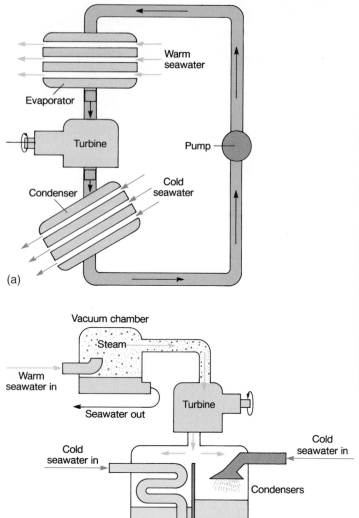

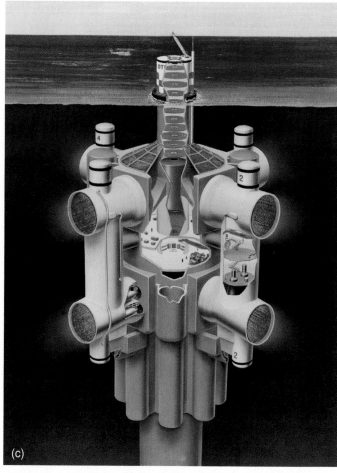

web link **Figure 7.15** (a) The simplified working system of an OTEC (ocean thermal energy conversion) closed-system electrical generator. (b) The simplified working system of an OTEC open-system electrical generator. (c) The concept drawing for the Lockheed OTEC Spar.

Other ideas for using cold deep-ocean water, or DOW, pumped from 630 m (2000 ft) include air conditioning and industrial cooling systems. The condensate that forms on pipes carrying DOW is fresh water, formed at a rate of about 5% of the flow of the cold water. A flow of 76,000 L (20,000 gal) per minute generates an estimated 3800 L (1000 gal) per minute of fresh water. Lobsters, flat fish, shrimp, abalone, oyster, and other organisms have been raised in the tropics using DOW. Tropical agricultural experiments have also used DOW, pumping it through pipes placed at root depth in the soil. This system chills the soil to 10°C and produces freshwater condensate on the pipes and on the soil.

Summary

The absorption and exchange of energy at the sea surface governs the properties of surface seawater. Sea surface exchanges of heat, radiant energy, and water alter the temperature and salinity of the surface water and also affect the density of the water. Many combinations of salinity and temperature can produce seawater of the same density. When waters with different properties but the same density are mixed, the resulting water has a greater density than either of its components. The density-driven vertical circulation that results is known as caballing. The waters of different densities produced at the sea surface and the resulting vertical circulation produce a layered ocean that is primarily stably stratified.

The geographical distribution of surface salinities reflects the Earth's latitudinal and seasonal patterns of evaporation, precipitation, and sea ice formation. Cooling, evaporation, and freezing increase the density of the sea surface water. Heating, precipitation, and ice melt decrease its density. The surface water changes its density with changes in salinity and temperature that are keyed to latitude. The densities at depth are more homogeneous. Thermoclines and haloclines form where the temperature and salt concentrations change rapidly with depth. If the density increases with depth, the water column is stable; unstable water columns overturn and return to a stable distribution. Neutrally stable water columns are mixed vertically by winds and waves.

Vertical circulation that is driven by changes in surface density is known as thermohaline circulation. In the open ocean, temperature is more important than salinity in determining the surface density. Salinity is the more important factor in areas that are close to shore and to land runoff.

Water sinks at downwellings and rises at upwellings. A downwelling occurs at the convergence of surface currents and transfers oxygen to depth. Upwellings bring nutrients to the surface and occur at zones of surface current divergence. Chemical tracers are used to determine residence time of deep water.

The oceans are layered systems. The layers (or water types) are identified by specific ranges of temperature and salinity. The water types of the Atlantic Ocean are formed at the surface at different latitudes. They sink and flow northward or southward. The water types of the Pacific Ocean lose their identity in the large volume of this ocean; their movements are sluggish. The water types of the Indian Ocean are less distinct than those of the Atlantic, with no water types that correspond to those formed in the northern latitudes. Mediterranean Sea and Red Sea waters enter the Atlantic and Indian Oceans at depth as discrete water types that can be tracked for long distances. Water enters and exits the Arctic Ocean from the North Atlantic. The density of Arctic water is controlled more by salinity than by temperature.

The CTD (conductivity-temperature-depth) sensor takes profiles of salinity by measuring electrical conductance of seawater. Temperature is obtained with an electrical resistance thermometer, and depth is obtained with pressure sensors. Seasoar is a towed CTD modified to take samples when a ship is underway. To sample large areas moored buoys, buoys that drift with the currents, and drifting instruments are used. Profilers sink to predetermined depths making measurements as they rise. All are able to send their positions and data via satellite. Satellites also monitor the oceans directly. Shallow-water measurements can be made with mechanical water bottles and thermometers.

OTEC (ocean thermal energy conversion) generates energy by using the difference in temperature between ocean surface water and deeper water. A land-based, open-system OTEC plant is operating in Hawaii.

Key Terms

All key terms from this chapter can be viewed by term, or by definition, when studied as flashcards on this book's Online Learning Center at www.mhhe.com/sverdrup (click on this book's cover).

caballing, 209
pycnocline, 209
thermocline, 210
halocline, 210

stable water column, 210
unstable water column, 210
overturn, 210
isopycnal, 210

isothermal, 210
isohaline, 210
vertical circulation, 210
thermohaline circulation, 210
freshwater lid, 212
continuity of flow, 212
downwelling zones, 212
upwelling zones, 212
convergence, 212

divergence, 212
North Atlantic deep water, 213
Antarctic intermediate water, 213
Antarctic bottom water, 213
South Atlantic surface water, 213
ocean thermal energy conversion (OTEC), 222

Study Questions

1. Why does the mixing of two water types at the sea surface result in a mixed water type that sinks to a deeper depth?
2. What natural processes alter the surface salinity of the oceans? How do these processes vary with latitude?
3. Describe the changes in water density in the upper ocean layer over the annual cycles at tropical, polar, and temperate latitudes. Indicate when periods of stable and unstable conditions exist in the upper water column.
4. Why are water layers more prominent in the Atlantic Ocean than in the Pacific Ocean?
5. How do salt fingers form, and how do they contribute to vertical mixing?
6. How can heat be transferred from place to place in the oceans? Why is this heat transfer ignored when considering the world's total heat budget?
7. If the upper layers of an ocean area are homogeneous in salinity, explain why the thermocline coincides with the pycnocline.
8. How does Mediterranean Sea water alter the salinity and temperature of the North Atlantic? Use figures 7.6 and 7.7a, b. At what depth does this change occur?
9. A water sample taken at 4000 m in the Atlantic Ocean has a salinity of 34.8‰ and a temperature of 3°C. At approximately what latitude was this water last at the surface?
10. The following data were taken from a sampling station located at 79°N, 145°W:

Depth (meters)	Temp. (°C)	Salinity (‰)	Density g/cm³
0	1.28	33.29	1.02659
50	1.29	33.30	1.02659
100	1.36	33.35	1.02669
150	1.39	33.55	1.02694
200	2.73	33.76	1.02701
300	3.07	33.87	1.02708
400	3.12	34.03	1.02713
500	3.14	34.13	1.02721

a. In what ocean region is this station?
b. Is the water column stable or unstable?
c. Does the temperature or the salt content control the density?
d. How deep is the wind-mixed layer?
e. At what time of year were these data obtained?
Explain your answers.

11. Why is the surface of the Earth not heated equally all over by the Sun's radiation?
12. What does a CTD measure, and why is the CTD the equipment of choice for routine oceanographic measurements?
13. Where do the deeper waters in the bottom of the Atlantic Ocean have their origins?
14. Distinguish between the terms in each pair:
    a. halocline—thermocline
    b. upwelling—downwelling
    c. water mass—water type
15. Why is convective overturn in the sea's upper layers more likely to occur at temperate and high latitudes than at low latitudes?

Links to Related Websites

Visit the book's Online Learning Center at www.mhhe.com/sverdrup (click on the book's cover) to find live Internet links for additional topics related to this chapter's content.

- Ocean conditions

Visit the book's Online Learning Center at www.mhhe.com/sverdrup (click on this book's cover) to find these additional chapter tools: Suggested Readings; links to further information on boxed readings, selected figures, and related chapter topics; and additional study aids.

chapter 8

The Currents

There is a river in the ocean. In the severest droughts it never fails, and in the mightiest floods it never overflows. Its banks and its bottom are of cold water, while its current is of warm. The Gulf of Mexico is its fountain, and its mouth is in the Arctic Seas. It is the Gulf Stream. There is in the world no other such majestic flow of waters. Its current is more rapid than the Mississippi or the Amazon.

Its waters, as far out from the Gulf as the Carolina coasts, are of an indigo blue. They are so distinctly marked, that their line of junction with the common sea-water may be traced by the eye. Often one half of the vessel may be perceived floating in the Gulf Stream water, while the other half is in common water of the sea, so sharp is the line, and such the want of affinity between those waters, and the reluctance, on the part of those of the Gulf Stream to mingle with the common water of the sea.

Matthew Fontaine Maury
From *The Physical Geography of the Sea,* 1855

A surging wave at the Galápagos Islands.

The Earth is surrounded by two great oceans: an ocean of air and an ocean of water. Both are in constant motion, driven by the energy of the Sun and the gravity of the Earth. Their motions are linked; the winds give energy to the sea surface and the currents are the result. The currents carry heat from one location to another, altering the Earth's surface temperature patterns and modifying the air above. The interaction between the atmosphere and the ocean is dynamic; as one system drives the other, the driven system acts to alter the properties of the driving system.

In this chapter we explore the formation of the ocean's surface currents. We follow these currents as they flow, merge, and move away from each other. We examine both horizontal and vertical circulation, inspect the coupling of these water motions, and consider the ways in which they are linked to the overall interaction between the atmosphere and the ocean.

Surface Currents 8.1

When the winds blow over the oceans they set the surface water in motion, driving the large-scale surface currents in nearly constant patterns. The density of water is about 1000 times greater than the density of air, and once in motion, the mass of the moving water is so great that its inertia keeps it flowing. The currents flow more in response to the average atmospheric circulation than to the daily weather and its short-term changes; however, the major currents do shift slightly in response to long-term seasonal changes in the winds. The currents are further modified by interactions that occur among the currents, zones of converging and diverging water, and landmasses. The major surface currents have been called the rivers of the sea; they have no banks to contain them, but they maintain their average course.

Because the friction coupling between the ocean water and the Earth's surface is small, the moving water is deflected by the Coriolis effect in the same way that moving air is deflected (see chapter 6). But because water moves more slowly than air, it takes longer to move water the same distance as air. During this longer time period, the Earth rotates farther out from under the water than from under the wind. Therefore, the slower-moving water appears to be deflected to a greater degree than the overlying air. The surface water layer acted upon by the Coriolis effect is deflected to the right of the driving wind direction in the Northern Hemisphere and to the left in the Southern Hemisphere. In the open sea, the surface flow is deflected at a 45° angle from the wind direction, as shown in figure 8.1.

The Ekman Spiral and Ekman Transport

Wind-driven surface water sets the water immediately below it in motion. But because of low-friction coupling in the water, this next deeper layer moves more slowly than the surface layer and is deflected to the right (Northern Hemisphere) or left (Southern Hemisphere) of the surface layer direction. The same is true for the next layer down and the next. The result is a spiral in which each deeper layer moves more slowly and with a greater angle of deflection than the layer above. This current spiral is called the **Ekman spiral,** after the physicist V. Walfrid Ekman, who developed its mathematical relationship. The spiral extends to a depth of approximately 100–150 m (330–500 ft), where the much reduced current will be moving exactly opposite to the surface. Over the depth of the spiral, the average flow of all the water set in motion by the wind, or the net flow **(Ekman transport),** moves 90° to the right or left of the surface wind, depending on the hemisphere in which the motion takes place (fig. 8.2). This relationship is in contrast to the surface water, which moves at an angle of 45° to the wind direction.

Ocean Gyres

Refer to figure 8.3 as you read the description of surface currents in the world's major oceans. In the Northern Hemisphere the wind-driven transport under the westerlies moves 90° to the right of the westerlies, away from an ocean's western shore or boundary, and along the 40°–50°N latitudes until it reaches its eastern shore or boundary. The Ekman transport under the trade winds moves 90° to the right of the trades, away from an ocean's eastern boundary, and along

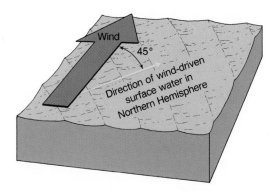

Figure 8.1 A wind-driven surface current moves at an angle of 45° to the direction of the wind; this angle is to the right in the Northern Hemisphere and to the left in the Southern Hemisphere.

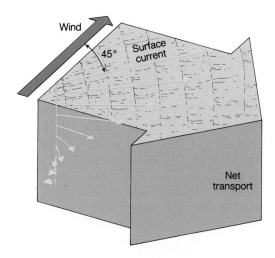

Figure 8.2 Water is set in motion by the wind. The direction and speed of flow change with depth to form the Ekman spiral. This change with depth is a result of the Earth's rotation and the inability of water, due to low friction, to transmit a driving force downward with 100% efficiency. The net transport over the wind-driven column is 90° to the right of the wind in the Northern Hemisphere and 90° to the left in the Southern Hemisphere.

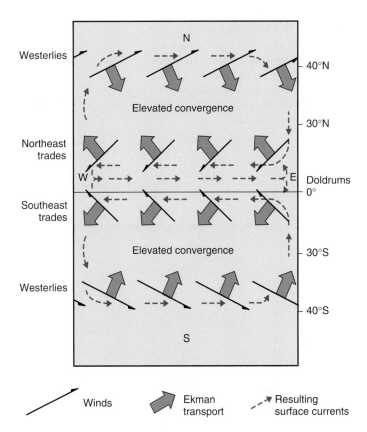

Winds — Ekman transport — Resulting surface currents

Figure 8.3 Wind-driven transport and resulting surface currents in an ocean bounded by land to the east and to the west. The currents form large oceanic gyres that rotate clockwise in the Northern Hemisphere and counterclockwise in the Southern Hemisphere.

the 10°–20° latitudes until it reaches its western boundary. When Northern Hemisphere water moving with the trade winds accumulates at the land boundary on the west side of an ocean, the water flows north to the latitude of the westerlies and then eastward across the ocean. Water accumulating at the land boundary on the east side of a northern ocean flows south, toward the region from which the water moves westward under the trade winds. In Southern Hemisphere oceans the east-west wind-driven transport is deflected 90° to the left of the trade winds and the westerlies. Water accumulating on the eastern side of a Southern Hemisphere ocean moves north, and water on the western side of a southern ocean moves south.

The Ekman transport causes an accumulation of water forming an elevated convergence. As the elevation builds, the wind-driven surface flow moves more closely in line with the driving winds and the surface current circulates around the convergence. In each hemisphere this pattern produces continuous flow and a series of interconnecting surface currents moving in a circular path centered on 30° latitude. The rotation is clockwise in the Northern Hemisphere and counterclockwise in the Southern Hemisphere. These circular-motion, wind-driven current systems are known as **gyres.** In the more southern latitudes there is no land between the Atlantic, Pacific, and Indian Oceans; here the surface currents, driven by the westerlies, continue around the Earth in a circumpolar flow around Antarctica.

Geostrophic Flow

If Ekman transport is applied to oceans with east-west land boundaries, a portion of the wind-driven surface water is deflected toward the center of each of the large, circular current gyres just described (fig. 8.3). A convergent lens of surface water is elevated more than 1 m (3 ft) above the equilibrium sea level, and this lens depresses the underlying denser water.

The denser water's vertical displacement is about 1000 times greater than the elevation of the surface mound. This is because the difference in density between the surface water and the deeper water is only about 1/1000 of the density difference between air and water at the sea surface. The

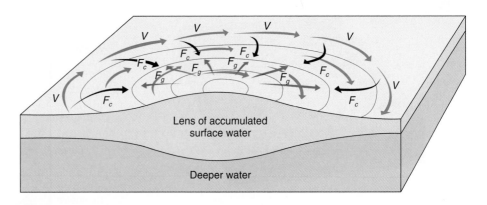

Figure 8.4 Geostrophic flow (V) exists around a gyre when F_c, the inward deflection force due to the Coriolis effect, is balanced by F_g, the outward-acting pressure force created by the elevated water and gravity. This example is of a clockwise gyre in the Northern Hemisphere.

Lens of accumulated surface water

Deeper water

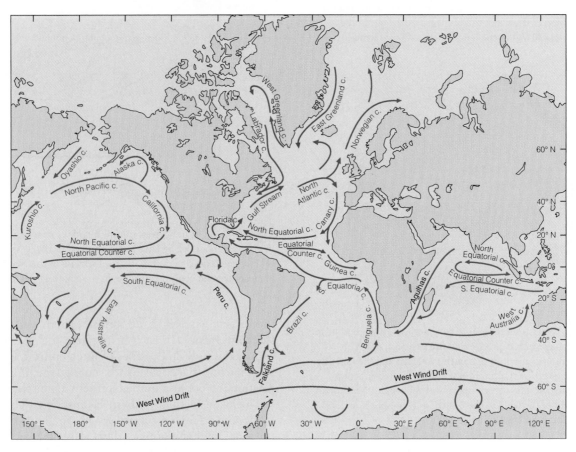

web link **Figure 8.5** The major surface currents of the world's oceans.

surface slope of the mound increases as water moves inward until the outward pressure driving the water away from the gyre center equals the Coriolis effect, acting to deflect the moving water into the raised central mound. At this balance point, **geostrophic flow** is said to exist, and no further deflection of the moving water occurs. Instead, the currents flow smoothly around the gyre parallel to its elevation contours. See figure 8.4 for a diagram of this process. Using the subsurface water-density distribution to describe the extent of the depression of the deeper water, oceanographers are able to calculate the elevation and slope of the sea surface and so calculate the velocity, volume transport, and depth of the currents present in the geostrophic flow around the mound. It is also possible to measure the topography of the

sea surface using satellites and to calculate the geostrophic flow that maintains the topography. The Sargasso Sea in the North Atlantic Ocean is the classic example of gyre in geostrophic balance; it is discussed following the Atlantic currents in section 8.2.

Wind-Driven Ocean Currents 8.2

The currents that make up the large oceanic gyre systems and other major currents have been given names and descriptions based on their average positions. These are presented here ocean by ocean and can be followed on figure 8.5. As you follow these current paths, review their associations with the large gyre systems and their overlying wind belts.

Pacific Ocean Currents

In the North Pacific Ocean, the northeast trade winds push the water toward the west and northwest; this is the **North Equatorial Current.** The westerlies create the **North Pacific Current,** or **North Pacific Drift,** moving from west to east. Note that the trade winds move the water away from Central and South America and pile it up against Asia, while the westerlies move the water away from Asia and push it against the west coast of North America. The water that accumulates in one area must flow toward areas from which the water has been removed. This movement forms two currents: the **California Current,** moving from north to south along the western coast of North America, and the **Kuroshio Current,** moving from south to north along the east coast of Japan. The Kuroshio and California Currents are not completely wind-driven currents; they provide continuity of flow and complete a circular motion centered around 30°N latitude. This circular, clockwise flow of water is called the North Pacific gyre. Other major North Pacific currents include the **Oyashio Current,** driven by the polar easterlies, and the **Alaska Current,** fed by water from the North Pacific Current and moving in a counterclockwise gyre in the Gulf of Alaska. There is little exchange of water through the Bering Strait between the North Pacific and the Arctic Ocean; no current exists that is comparable to the Atlantic Ocean's Norwegian Current, which moves warm water to the Arctic Ocean.

In the South Pacific Ocean, the southeast trade winds move the water to the left of the wind and westward, forming the **South Equatorial Current.** The westerly winds push the water to the east; at these southern latitudes the surface current so formed can move almost continuously around the Earth. This current is the **West Wind Drift.** The tips of South America and Africa act to deflect a portion of this flow northward on the east side of both the South Pacific and South Atlantic Oceans. As in the North Pacific, continuity currents form between the South Equatorial Current and the West Wind Drift. The **Peru Current,** or **Humbolt Current,** flows from south to north along the coast of South America, while the **East Australia Current** can be seen moving weakly from north to south on the west side of the ocean. These four currents form the counterclockwise South Pacific gyre.

The North Pacific and South Pacific gyres are formed not on either side of the equator (0°) but on either side of 5°N, because the doldrum belt is displaced northward owing to the unequal heating of the Northern and Southern Hemispheres. Also between the North and South Equatorial Currents and below the doldrums there is a current moving in the opposite direction, from west to east. This is a continuity current known as the **Equatorial Countercurrent,** which helps to return accumulated surface water eastward across the Pacific. Under the South Equatorial Current there is a subsurface current flowing from west to east called the **Cromwell Current.** This cold-water continuity current also returns water accumulated in the western Pacific.

Atlantic Ocean Currents

The North Atlantic westerly winds move the water eastward as the **North Atlantic Current,** or **North Atlantic Drift.** The northeast trade winds push the water to the west, forming the **North Equatorial Current.** The north-south continuity currents are the **Gulf Stream,** flowing northward along the coast of North America, and the **Canary Current,** moving to the south on the eastern side of the North Atlantic. The Gulf Stream is fed by the **Florida Current** and the North Equatorial Current. The North Atlantic gyre rotates clockwise. The polar easterlies provide the driving force for the **Labrador** and **East Greenland Currents,** which balance water flowing into the Arctic Ocean from the **Norwegian Current.**

In the South Atlantic, the westerlies continue the West Wind Drift. The southeast trade winds move the water to the west, but the bulge of Brazil splits the **South Equatorial Current.** Much of this flow is deflected northward over the top of South America, into the Caribbean Sea, and eventually into the Gulf of Mexico, where it exits as the Florida Current, which joins the Gulf Stream. A portion of the South Equatorial Current slides south of the Brazilian bulge along the western side of the South Atlantic to form the **Brazil Current.** The **Benguela Current** moves northward up the African coast. The South Atlantic gyre is complete, and it rotates counterclockwise.

Because much of the South Equatorial Current is deflected across the equator, the Equatorial Countercurrent appears only weakly in the eastern portion of the mid-Atlantic. The northward movement of South Atlantic water across the equator results in a net flow of surface water from the Southern Hemisphere to the Northern Hemisphere. This flow is balanced by a flow of water at depth from the Northern Hemisphere to the Southern Hemisphere. This deep-water return flow is the North Atlantic deep water, discussed in chapter 7. Again, the equatorial currents are displaced north of the equator, although not as markedly as in the Pacific Ocean.

The Sargasso Sea is the classic example of an ocean gyre. It is located in the central North Atlantic Ocean, and its boundaries are the Gulf Stream on the west, the North Atlantic Current to the north, the Mid-Atlantic Ridge on the east, and the North Equatorial Current to the south. The circular motion of the current gyre isolates a lens of clear, warm water 1000 m (3000 ft) deep; the water is trapped by the geostrophic flow. The region is famous for the floating mats of *Sargassum*, a brown seaweed, stretching across its surface. The extent of the floating seaweed frightened early sailors, who told stories of ships imprisoned by the weed and sea monsters lurking below the surface. Except for the floating *Sargassum*, with its rich and specialized community of small plants and animals, the clear water is nearly a biological desert.

Indian Ocean Currents

The Indian Ocean is mainly a Southern Hemisphere ocean. The southeast trade winds push the water to the west, creating the **South Equatorial Current.** The Southern Hemisphere westerlies still move the water eastward in the West

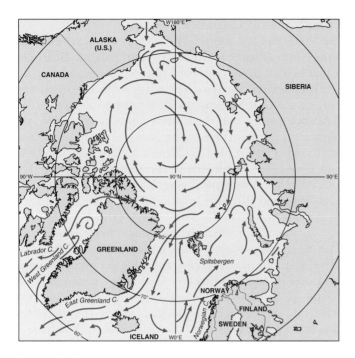

Figure 8.6 The circulation in the Arctic Ocean is driven by the polar easterlies, which produce a large, clockwise gyre. Water enters the Arctic Ocean from the North Atlantic by way of the Norwegian Current and exits to the Atlantic by the East Greenland Current and the Labrador Current.

Wind Drift. The gyre is completed by the **West Australia Current** moving northward and the **Agulhas Current** moving southward along the east coast of Africa. Because this is a Southern Hemisphere ocean, the currents move to the left of the wind direction, and the gyre rotates counterclockwise. The northeast trade winds in winter drive the **North Equatorial Current** to the west, and the **Equatorial Countercurrent** returns water eastward toward Australia. Again, these equatorial currents are displaced approximately 5°N. With the coming of the wet monsoon season and its west winds, these currents are reduced. The strong seasonal monsoon effect controls the surface flow of the Northern Hemisphere portion of the Indian Ocean. In the summer the winds blow the surface water eastward, and in the winter they blow it westward. This strong seasonal shift is unlike anything found in either the Atlantic or the Pacific Ocean.

Arctic Ocean Currents

The relentless drift of water and ice in the Arctic Ocean moves in a large clockwise gyre driven by the polar easterly winds. This gyre is centered not on the North Pole, as early explorers expected, but is offset over the Canadian basin at 150°W and 80°N (fig. 8.6). Although the currents and the winds move the ice slowly at 0.1 knot (2 mi/day), Arctic explorers trying to reach the North Pole found that they traveled south with the drifting ice and water at speeds almost equal to their difficult progress north.

The Arctic Ocean is supplied from the North Atlantic by the Norwegian Current; some of this flow enters west of Spitsbergen, but most flows over the top of Norway and Finland and moves eastward along the Siberian coast into the Chukchi Sea. A small inflow of water entering the Arctic through the Bering Strait brings water from the Bering Sea to join the eastward flow along Siberia and the large Arctic gyre. The western side of the gyre crosses the center of the Arctic Ocean to split north of Greenland. Here the larger flow forms the East Greenland Current flowing south and taking Arctic Ocean water into the North Atlantic. The lesser flow moves along the west side of Greenland to join the Labrador Current and move south along the Canadian coast.

Outflow from Siberian rivers is caught in the eastward flow of water and ice along Siberia. Eventually this discharge joins the gyre, distributing sediments and pollutants throughout the Arctic (see the box in chapter 7 titled "Arctic Ocean Studies").

Current Flow   8.3

Current Speed

Wind-driven open-ocean surface currents move at speeds that are about 1/100 of the wind speed measured 10 m (30 ft) above the sea surface. The water moves between 0.25 and 1.0 knot, or 0.1–0.5 m (0.3–1.5 ft) per second. Currents flow faster when a large volume of water is forced to flow through a narrow gap. For example, both the North and South Atlantic Equatorial Currents flow into the Caribbean Sea, then into the Gulf of Mexico, and finally exit to the North Atlantic as the Florida Current through the narrow gap between Florida and Cuba. The Florida Current's speed may exceed 3 knots, or 1.5 m (5 ft) per second. Once into the Atlantic Ocean this current turns north and becomes known as the Gulf Stream.

Major ocean currents transport very large volumes of water. For example, the Gulf Stream's transport rate to the northeast is about 55×10^6 m^3 (1.9×10^9 ft^3) per second; this rate is more than 500 times the flow of the Amazon River. The flow is distributed over the width and depth of the current. When the cross-sectional area of the current expands, the current slows down; when the cross-sectional area decreases, the current speeds up. Speed of flow then is not always directly related to surface wind speed but can be affected by the depth and width of the current as determined by land barriers, by the presence of another current, or by the rotation of the Earth, as explained in the next section, "Western Intensification."

Western Intensification

In the North Atlantic and North Pacific the currents flowing on the western side of each ocean tend to be much stronger, deeper, and narrower in cross section than the currents on the eastern side. This phenomenon is known as the **western intensification** of currents. The Gulf Stream and Kuroshio Currents are faster and narrower than the Canary and California Currents, although both the eastern and western currents transport about the same amount of water in order to preserve continuity of flow around the gyres. Western intensification of currents traveling from low to high latitudes is related to (1) increases in the Coriolis effect with latitude, (2) the changing strength and direction of the east-west wind field (trade winds and westerlies) with latitude, and (3) the friction between landmasses and ocean water currents. These factors cause a compression of the currents on the western side of the oceans, where water is moving from lower to higher latitudes. This compression requires that the current speed increase in order to transport the amount of water that circulates about the gyre. On the eastern side of the gyre, where currents are moving from higher to lower latitudes, the currents are stretched in the east-west direction. Here the current speed is reduced, but it still transports the required volume of water.

Fast-flowing, western-boundary currents move warm equatorial surface water to higher latitudes. Both the Gulf Stream and the Kuroshio Current bring heat from equatorial latitudes to moderate the climates of Japan and northern Asia (in the case of the Kuroshio) and the British Isles and northern Europe (in the case of the Gulf Stream, via the North Atlantic and Norwegian Currents). Western intensification is obscured in the South Pacific and South Atlantic, because both Africa and South America deflect portions of the West Wind Drift and create strong currents on the eastern side of these oceans. The deflection of water from the Atlantic's South Equatorial Current to the Northern Hemisphere removes water from the South Atlantic gyre and strengthens the Gulf Stream. The flow of surface water from the Pacific to the Indian Ocean through the islands of Indonesia also helps to prevent the development of strongly flowing currents on the west side of Southern Hemisphere oceans.

Eddies 8.4

When a narrow, fast-moving current moves into or through slower-moving water, the force of its flow displaces the quieter water and captures additional water as it does so. The current oscillates and develops waves along its boundary that are known as meanders. These meanders break off to form **eddies,** or packets of water moving with a circular motion; eddies take with them energy of motion from the main flow and gradually dissipate this energy through friction.

As the Gulf Stream moves away from the North American coast it is likely to develop a meandering path. The western edge of the Gulf Stream develops oscillations, and the indentations are filled by cold water from the Labrador Current side. When these indentations pinch off they become counterclockwise rotating cold-water eddies that are displaced through the Gulf Stream and into the warm-water core of the gyre. Bulges at the western edge of the Gulf Stream are filled with warm Gulf Stream water. When these bulges are cut off they become warm-water clockwise rotating eddies drifting into cold water, to the west and north of the Gulf Stream. Follow these processes in figures 8.7 and 8.8. These eddies may maintain their physical identity for weeks as they wander about the oceans; they are especially numerous in the area north of the Sargasso Sea. Current meanders and eddy formation produce surface flow patterns that differ markedly from the uniform current flows shown on current charts. These charts show average current flow, not daily or weekly variations.

Large and small eddies generated by horizontal flows or currents exist in all parts of the oceans; these eddies are of varying sizes, ranging from 10 to several hundred kilometers in diameter. Each eddy contains water with specific chemical and physical properties and maintains its identity and rotational inertia as it wanders through the oceans. Eddies may appear at the sea surface or be embedded in waters at any depth.

Eddies rotate in a clockwise or counterclockwise direction. They stir the ocean until they gradually dissipate because of fluid friction, losing their chemical and thermal identity and their energy of motion. By testing the water properties of an eddy, oceanographers are able to determine the eddy's place of origin. Small surface eddies found 800 km (500 mi) southeast of Cape Hatteras in the North Atlantic have been found with water properties of the eastern Atlantic near Gibraltar more than 4000 km (2500 mi) away. Eddies from the Strait of Gibraltar are formed in the salty water of the Mediterranean as it sinks and spreads out into the Atlantic 500–1000 m (1600–3300 ft) down; these eddies have been nicknamed "Meddies." Deep-water eddies near Cape Hatteras may come from the eastern and western Atlantic, the Caribbean, or Iceland. Researchers estimate that some of these eddies are several years old; age determination is based on drift rates, distance from source, and biological consumption of oxygen.

The rotational water speed in the large eddies that form at the western boundary of the Gulf Stream is about 0.51 m/s (1 knot), but because of the water's density, the force of the flow is similar to that generated by a 35-knot wind. The diameter of the eddies may be as much as 325 km (200 mi), and their effect may reach to the sea floor. At the sea floor the rotation rate is zero; therefore a few meters above the bottom the speed of rotation diminishes very rapidly, and considerable turbulence is generated as the energy of the eddy is dissipated. These eddies are similar in many ways to the winds rotating about atmospheric pressure cells, and they are sometimes called abyssal storms. As the eddies wander through the oceans, they stir up bottom sediments, producing ripples and sand waves in their wakes; they also mix the water, creating homogeneous water properties over large areas. Eventually the eddies lose their energy to turbulence and blend into the surrounding water.

Eddies constantly form, migrate, and dissipate at all depths. Eddy motion is superimposed on the mean flow of the oceans. To understand the role of eddies in mixing the oceans we need more data and better tracking of eddy size, position, and rate of dissipation. Satellites are important tools for detecting surface eddies because they can precisely measure temperature, increased elevation, and light reflection of the sea surface (fig. 8.9). Deep-water eddies are

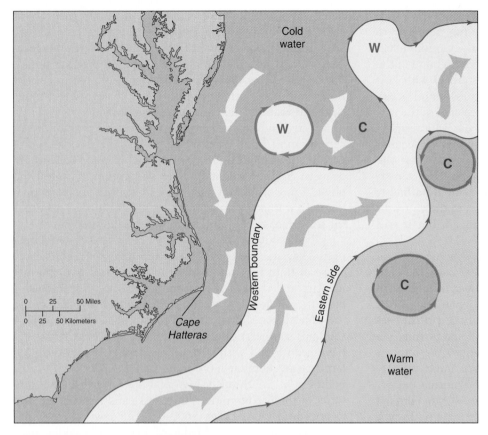

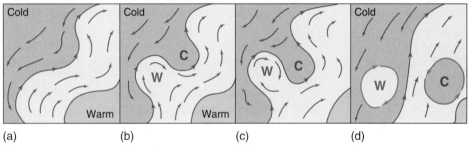

web ⬤ link **Figure 8.7** The western boundary of the Gulf Stream is defined by sharp changes in current velocity and direction. Meanders form at this boundary after the Gulf Stream leaves the U.S. coast at Cape Hatteras. The amplitude of the meanders increases as they move downstream (a and b). In time, the current flow pinches off the meander (c). The current boundary re-forms, and isolated rotating cells of warm water (W) wander into the cold water, while cells of cold water (C) drift through the Gulf Stream into the warm water (d).

ity of flow were introduced in section 7.2 of chapter 7. The convergence and divergence zones discussed in this chapter are the product of wind-driven surface currents that produce large-scale areas of convergence and divergence at the sea surface. For example, convergence zones are at the centers of the large oceanic gyres, and when wind-driven surface currents collide or are forced against land-masses they produce convergences. When surface currents move away from each other or away from a land-mass they produce a surface divergence. Upwellings and downwellings of this type are nearly permanent, reacting to seasonal changes in the Earth's surface winds.

Langmuir Cells

A strong wind blowing across the sea surface often causes streaks of foam and surface debris that are seen trailing off in the direction the wind is blowing. These streaks are called windrows and may be 100 m (330 ft) in length. Windrows mark the convergence zones of shallow circulation cells known as **Langmuir cells.** These cells are composed of paired right- and left-handed helixes (fig. 8.10). The spacing between the windrows varies between 5 and 50 m (16 and 160 ft), and the rows are closer together if the thermocline depth of the area is shallow. The distance between rows becomes larger at higher wind speeds. Upwelling between the lines of convergence sweeps surface debris into windrows. This shallow circulation process helps to mix the surface water; it also organizes food particles into concentrated bands.

monitored by using instruments designed to float at a mid-depth density layer. The instruments are caught up in the eddies, moving with them and sending out acoustic signals that are monitored through the sofar channel. In this way the rate of deep-water eddy formation, the numbers of major eddies, their movements, and their life spans can be observed.

Convergence and Divergence   8.5

Changes in density and the accompanying concepts of up-welling, downwelling, convergence, divergence, and continu-

Permanent Zones

Convergence and divergence zones on an ocean scale are shown in figures 7.6 and 8.11. There are five major zones of convergence: the **tropical convergence** at the equator and the two **subtropical convergences** at approximately 30°–40°N and S. These convergences mark the centers of the large ocean gyres. The **Arctic** and **Antarctic convergences** are found at about 50°N and S. Surface convergence zones are regions of downwelling. These areas are low in nutrients and biological productivity. There are three major divergence zones: the two **tropical divergences** and the **Antarctic divergence.** Upwelling associated with

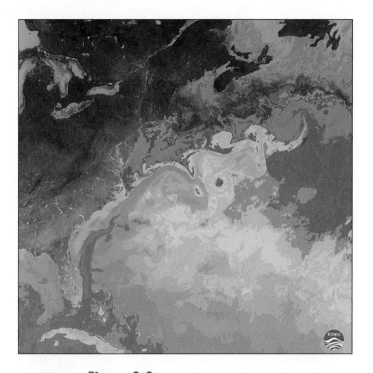

web ⦿ link **Figure 8.8** A satellite image of the sea surface reveals the warm (*orange* and *yellow*) and cold (*green* and *blue*) eddies that spin off the Gulf Stream. (*Reddish blue* areas at the top are the coldest waters.) These eddies may stir the water column right down to the ocean floor, kicking up blizzards of sediment.

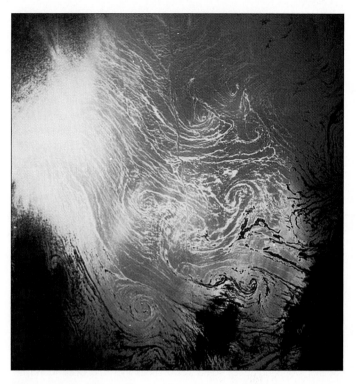

Figure 8.9 Space shuttle view. Sunlight reflected off the Mediterranean reveals spiral eddies; their effects on climate are being monitored.

Figure 8.10 Langmuir cells are helixes of near surface, wind-driven water motion. The cells extend downwind to form windrows of debris along convergences.

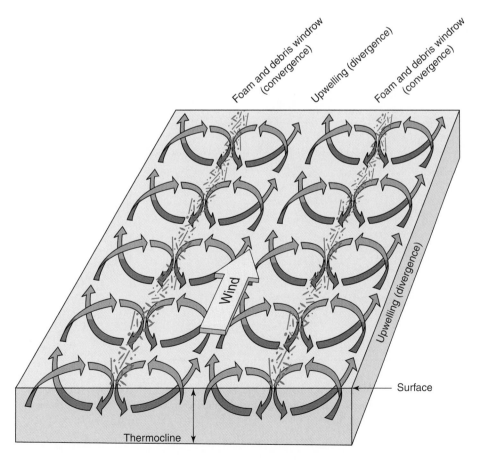

Ocean Drifters

In May 1990, a severe storm in the North Pacific caused the container ship *Hansa Carrier*, en route from Korea to the United States, to lose overboard twenty-one deck-cargo containers, each approximately 40 ft long (box fig. 1). Among the items lost were 39,466 pairs of Nike brand athletic shoes, and these began washing up along the beaches of Washington, Oregon, and British Columbia six months to a year later. Although the shoes had been drifting in the ocean for almost a year, they were wearable after washing and having the barnacles and the oil removed. However, the shoes of a pair had not been tied together for shipping, and pairs did not come ashore together. As beach residents recovered the shoes (some with a retail value of $100 a pair), swap meets were held in coastal communities to match the pairs.

In May of 1991, I was having lunch with my parents, Gene and Paul, and Gene pointed out the news of the beached Nikes. I was intrigued and realized that 78,932 shoes was a very large number of drifting objects compared with the 33,869 drift bottles used in a 1956–59 study of North Pacific currents. I contacted Steve McLeod, an Oregon artist and shoe collector who had information on locations and dates for some 1600 shoes that had been found between northern California and the Queen Charlotte Islands in British Columbia. I contacted additional beachcombers and constructed a map showing the times and locations where batches of 100 or more shoes had been found (box fig. 2). Next I visited Jim Ingraham at NOAA's National Marine Fisheries Service's offices in Seattle to study his computer model of Pacific Ocean currents and wind systems north of 30°N latitude. Using the spill date (27 May 1990), the spill

Box Figure 1 After the storm, the *Hansa Carrier* docked in Seattle.

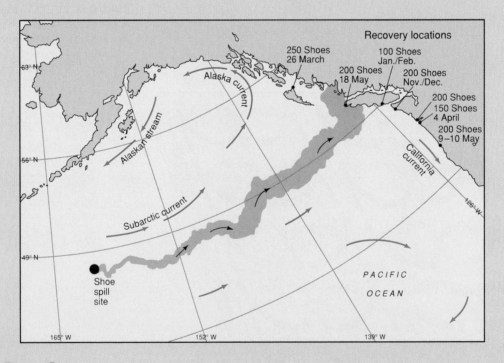

web link **Box Figure 2** Site where 80,000 Nike shoes washed overboard on May 27, 1990, and dates and locations where 1300 shoes were discovered by beachcombers (*dots at upper right*). Drift of the shoes was simulated with a computer model (*colored plume*). *From Curtis E. Ebbesmeyer and W. James Ingraham, Jr., Shoe Spill in the North Pacific in American Geophysical Union EOS, Transactions, Vol. 73, No. 34, August 25, 1992, pp. 361, 365.*

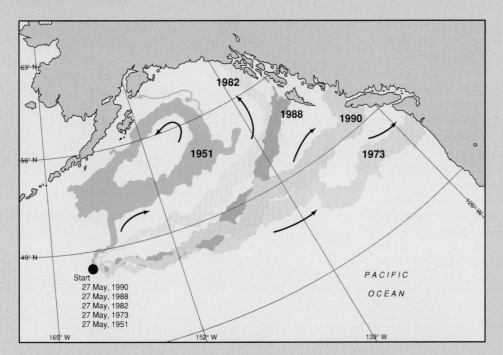

Box Figure 3 Projected drift tracks for the sneakers for the years 1951, 1973, 1982, 1988, and 1990 based on computer modeling of ocean currents and weather. *From Curtis E. Ebbesmeyer and W. James Ingraham, Jr., Shoe Spill in the North Pacific in American Geophysical Union EOS, Transactions, Vol. 73, No. 34, August 25, 1992, pp. 361, 365.*

Box Figure 4 Shoes and toys after their rescues from Pacific cargo spills.

location (161°W; 48°N), and the dates of the first shoe landings on Vancouver Island and Washington beaches between Thanksgiving and Christmas 1990, I found that the shoe drift rates agreed with the computer model's predicted currents.

News of Jim's and my interest in the shoe spill reached an Oregon news reporter and was then picked up by the Associated Press, resulting in a quick dissemination of the news nationwide. Readers sent letters describing their own shoe finds, and even reports

of single shoes were valuable because we found that each shoe had within it a Nike purchase order number that could be traced to a specific cargo container. We were able to determine from these numbers that only four of the five containers broke open so that only 61,820 shoes were left afloat.

The computer model and previous experiments with satellite-tracked drifters showed that there would have been little scattering of the shoes as the ocean currents carried them eastward and approximately 1500 miles from the spill site to shore, but the shoes were found scattered from California to northern British Columbia. The north-south scattering is related to coastal currents that flowed northward in winter, carrying the shoes to the Queen Charlotte Islands, and southward in spring and summer, bringing the shoes to Oregon and California.

I was interested to see where the shoes might have gone if they had been lost on the same date but under different conditions in other years. The computer allowed simulations for May 27 of each year from 1946–91. Box figure 3 shows the wide variation in model predicted drift routes. If the shoes had been lost in 1951 they would have traveled in the loop of the Alaska Current. If they had been lost in 1982 they would have been carried far to the north during the very strong El Niño of 1982–83, and if lost in 1973 they would have come ashore at the Columbia River.

Another spill event occurred in January 1992, when twelve cargo containers were lost from another vessel in the North Pacific at 180°W, 45°N. One of these containers held 29,000 small, floatable, plastic, bathtub toys. Blue turtles, yellow ducks, red beavers, and green frogs began arriving on beaches near Sitka, Alaska, in November 1992. Advertisements asking for news of toy strandings were placed in local newspapers and the Canadian lighthouse keepers' newsletter. Beachcombers reported a total of about 400 of these toys (box fig. 4).

None of the toys were found south of 55°N latitude, suggesting that once the toys reached the vicinity of Sitka, they drifted to the north. This time, the computer model showed that if the toys had continued to float with the Alaska Current, they would have moved with the Alaska stream through the Aleutian Islands into the Bering Sea. Some may have continued across the Bering Sea into the Arctic Ocean to the vicinity of Point Barrow, and from there would have drifted north of Siberia with the Arctic pack ice. Eventually, some plastic turtles, ducks, beavers, or frogs may have come to rest on the coast of western Europe. If the toys had turned south they might have merged with the Kuroshio Current and been carried past the location where they were spilled.

Two thousand five hundred cases of hockey gloves (34,300 gloves) were lost from a container ship in the North Pacific after a December 1994 fire. In August 1995 a fishing vessel found seven gloves 800 miles west of the Oregon coast, and by January 1996 the barnacle-covered gloves began to arrive on Washington beaches. The most northerly glove sighting came from Prince William Sound, Alaska, in August 1996. The gloves are expected to follow the tub toys along the coast of Alaska and into the Arctic. More than 4 million lego pieces lost off the English coast in 1997 are expected to be distributed along northern hemisphere shores by 2020.

Curtis C. Ebbesmeyer, Ph.D.
Seattle, Washington

To Learn More About Ocean Drifters

Krajick, K. 2001. Message in a Bottle. *Smithsonian* 32 (4): 36–47.

Internet References

Visit the book's Online Learning Center at www.mhhe.com/sverdrup (click on the book's cover) to explore links to further information on related topics.

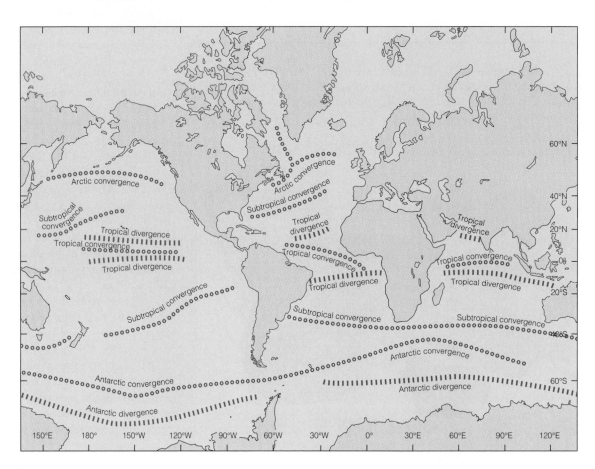

Figure 8.11 The principal zones of open-ocean surface convergence and divergence associated with wind-driven and thermohaline circulation.

divergences delivers nutrients to the surface waters to supply the food chains that support the tropical tuna fisheries and the richly productive waters of Antarctica.

In coastal areas where trade winds move the surface waters away from the western side of continents, upwelling occurs nearly continuously throughout the year. For example, the trade winds drive upwellings off the west coasts of

Africa and South America that are very productive and yield large fish catches.

Seasonal Zones

Off the west coast of North America, downwelling and upwelling occur on a seasonal basis as the Northern Hemisphere temperate wind pattern changes from southerly in

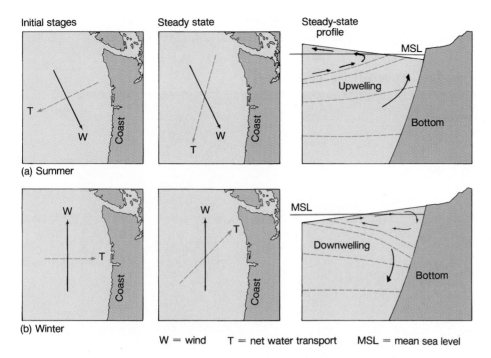

Figure 8.12 Initially, the Ekman transport is 90° to the right of the wind along the northwest coast of North America. As water is transported (a) away from the shore in summer or (b) toward the shore in winter, a sea surface slope is produced. This slope creates a gravitational force that alters the direction of the Ekman transport to produce a geostrophic balance and a new steady-state direction. *From Sverdrup/Johnson/Fleming, The Oceans, © 1942, renewed 1970, p. 501. Adapted by permission of Prentice Hall, Inc., Upper Saddle River, New Jersey.*

winter to northerly in summer. The downwelling and up-welling occur because of the change in direction of Ekman transport (fig. 8.12). Remember that the wind-driven Ekman transport moves at an angle of 90° to the right or left of the wind direction, depending on the hemisphere.

Along this coast, the average wind blows from the north in the summer, and the net movement of the water is to the west, or 90° to the right of the wind. This action results in the offshore transport of the surface water and the upwelling of deeper water along the coast to replace it. The upwelling zone is evident from central California to Vancouver Island. In winter, the winds blow from the south; the wind-driven surface waters move to the east, onshore against the coast; and downwelling occurs (fig. 8.13; see also fig. 6.22). The summer upwelling pattern is what produces the band of cold coastal water at San Francisco that helps cause the frequent summer fogs (see chapter 4). Because of the lack of land at the middle latitudes in the Southern Hemisphere, this type of seasonal upwelling is less common.

Convergence and divergence of surface currents result not only in upwellings and downwellings but also in the mixing of water from different geographic areas. Waters carried by the surface currents converge, share properties due to mixing, and form new water mixtures with specific ranges of temperature and salinity, which then sink to their appropriate density levels. After sinking, the water moves horizontally, blending and sharing its properties with adjacent water, and eventually it rises to the surface at a new location. This process is discussed as thermohaline circulation in the section on density-driven circulation in chapter 7. Thermohaline circulation and wind-driven surface currents are closely related. The connections are so close among formation of water mixtures, upwelling and downwelling, and

convergence and divergence that it is difficult to assign a priority of importance to one process over another. Changes in these flows and events that might cause such changes are discussed in the next section.

Changing Circulation Patterns 8.6

Global Currents

The interchange of water between oceans redistributes heat, salt, and dissolved gases; use figure 8.14 to trace the surface and deep-water flows of the world's oceans. Is it possible that some phenomenon could alter this water motion and trigger changes that would have major climatic consequences?

Studies of fossil marine organisms found in cores of marine sediments show that the temperature of the North Atlantic Ocean during glacial and interglacial periods can be related to Milankovitch cycles. These cycles are associated with regular long-term changes in the Earth's orbit and the relation of the Earth's axis to the Sun, producing variations in the amount and distribution of solar radiation over the Earth's surface. Twenty-three thousand years ago, the Earth had its period of minimum solar radiation. About 6000 years later, the Earth endured its coldest ocean temperatures and its maximum land ice volumes. Notice the time lag between decreasing solar energy and Earth cooling. However, sediment-core analysis shows that the warming of the oceans to today's temperatures occurred rapidly with high solar radiation values 12,000 years ago. The warming, like the cooling, is related to the Milankovitch cycles of Earth-Sun placement.

Water is a good absorber of solar radiation but ice is a poor absorber and a good reflector, so the land ice melted slowly as the oceans warmed rapidly. Eleven thousand years

Changing Circulation Patterns **239**

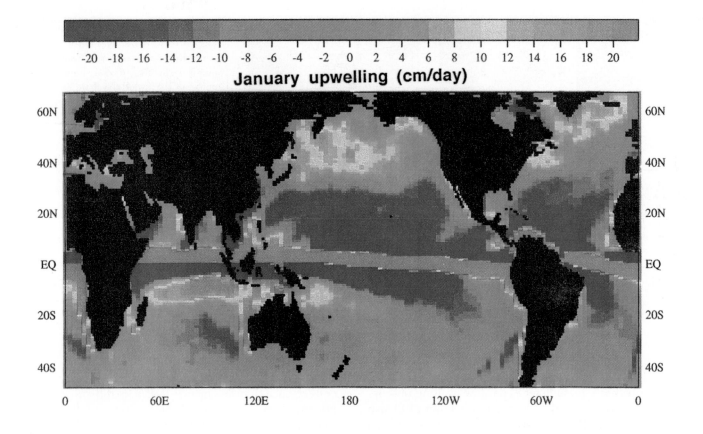

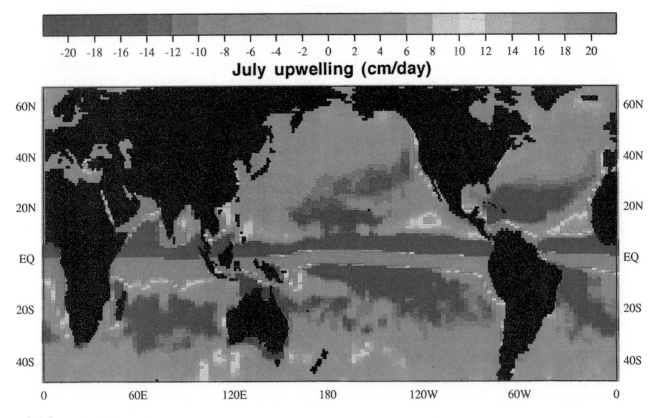

Figure 8.13 Surface winds produce surface divergences and convergences that create areas of upwelling and downwelling. Average surface wind stress values were used to produce this model of global upwelling and downwelling for January (a) and July (b). Negative upwelling values indicate downwelling. Speed of vertical motion is given in centimeters per day.

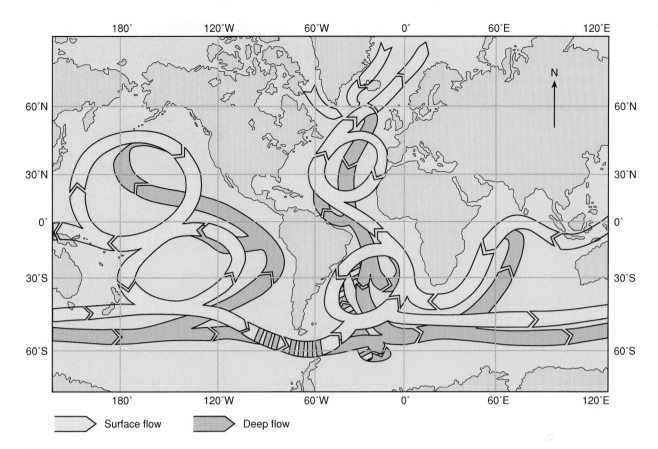

Figure 8.14 The present flow of ocean water within and between oceans circulates at the surface (*yellow arrows*) and at depth (*green arrows*). Cold surface water sinks to the sea floor in the North Atlantic Ocean, then flows south to be further cooled by the Antarctic Bottom water formed in the Weddell Sea. This deep water moves eastward around Antarctica, feeding into the surface layer of the Indian Ocean and also into the deep basins of the Pacific Ocean. A return flow of surface water from the Pacific and Indian Oceans flows north to replace the surface water in the North Atlantic Ocean.

Surface flow   Deep flow

ago, the surface temperatures of the water in the North Atlantic and the air temperatures in northern Europe decreased suddenly; the sudden decrease was unrelated to the long-term solar cycles. The temperatures remained low for about 700 years in a mini-ice age; large changes (20%) in atmospheric carbon dioxide gas trapped in ice cores of Greenland are associated with this period. These changes are interpreted to mean that large global atmospheric changes occurred and at the same time there was localized cooling of Europe and the North Atlantic. This could not have happened unless major changes occurred in the ocean circulation system that transports heat and carbon dioxide.

The required connections between global atmospheric changes and ocean circulation during the time of this mini-ice age have led Wallace Broecker of Lamont Doherty Geophysical Observatory to think that something triggered a sudden shutdown of the circulation of the North Atlantic. Broecker suggests that some event prevented the transport of warm surface water northward and interfered with the formation of North Atlantic deep water and in so doing greatly reduced the deep water flowing south. Under such conditions worldwide oceanic circulation would change and the oceans' ability to transport and store carbon dioxide would be altered. A change in the carbon dioxide storage capacity

of the oceans requires a corresponding change in atmospheric carbon dioxide levels.

Broecker has proposed that the trigger for such an episode was a sudden change in the location where meltwater from the receding North American ice sheet entered the Atlantic Ocean. Large quantities of cold, low-density, low-salinity water entering the oceans would significantly affect the salinity and density of the North Atlantic's surface waters. Geological studies support Broecker's ideas. We know that the majority of the meltwater from the early stages of the receding glaciers in North America converged on the Mississippi River system and drained into the Gulf of Mexico, not directly into the Atlantic. But gradually, as the ice margin receded, meltwater was diverted through the Hudson River drainage system, into the Great Lakes and the St. Lawrence River valley, and then into the North Atlantic. The result was a diluting and cooling of the Atlantic surface water; it no longer supplied heat to the atmosphere that had warmed northern Europe.

At the same time, the diluted surface water did not sink, and, therefore, the production of North Atlantic deep water was slowed, preventing the northward migration of warm equatorial water and affecting the circulation in all oceans. This event ceased as rapidly as it had started. As the

Figure 8.15 North-south current shifts in the Northeast Pacific are associated with changes in atmospheric pressure, winds, precipitation, and water temperature. This phenomenon is known as the Pacific Decadal Oscillation, or PDO.

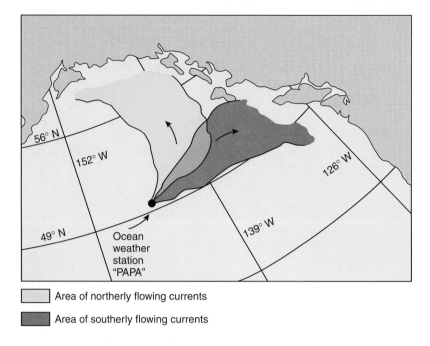

☐ Area of northerly flowing currents

■ Area of southerly flowing currents

melting of the glaciers slowed, the influx of meltwater was reduced. Surface warming and evaporation increased surface water density, and the formation of North Atlantic deep water resumed. The circulation of the North Atlantic began a rapid recovery.

North Pacific Oscillations

W. James Ingraham Jr., an oceanographer at NOAA's Seattle laboratory, has developed a North Pacific Current model, or North Pacific Ocean Surface Current Simulation (OSCURS). When the model was used to map current patterns for North Pacific surface water between 1902 and 1997, it showed a north-south current oscillation associated with changes in atmospheric pressure and climate shifts. Cold and wet conditions are associated with a southerly current flow, and warm and dry conditions predominate with a northerly flow (fig. 8.15). This evidence was compared to climate-sensitive tree-ring data from the western juniper tree that grows in eastern Oregon. These trees show wide rings during periods when currents moved to the north and narrow rings when currents moved to the south. Tree-ring data cover many more years than oceanographic and meteorological data, and from the tree rings it is calculated that there have been thirty-four north-south oscillations since the time of Columbus. The most common time period between fast and slow growth is seventeen years, with twenty-three- and twenty-six-year periods common. Tree-ring data and current oscillations agree.

The climate pattern is presently warm and dry with a northerly current flow that has not changed since 1967. This is one of the longest periods without a reversal that has been found during the last 500 years. The wet conditions of winter 1998–99 and the 1990–2000 La Niña may indicate a climate shift that will switch the currents to a more southerly flow. Yet the winter of 2000–2001 in the Pacific

Northwest was exceptionally warm and dry, and its effect has not yet been evaluated. These current shifts are associated with changes in atmospheric pressure, winds, precipitation, and water temperature; together they are known as the Pacific Decadal Oscillation, or PDO. Because the PDO affects coastal surface temperatures from California to Alaska, it is thought to affect the survival of fish stocks of the area, especially the salmon. Commercial fishing is discussed further in chapter 16.

North Atlantic Oscillations

A large pool of cold, low-salinity surface water appeared in the waters off Greenland, north of Iceland, in 1968. It was about 0.5‰ less salty and 1° and 2°C colder than usual. Within two years this cool pool had moved west into the Labrador Sea off eastern Canada; then it crossed the Atlantic and, in the mid-1970s, moved north into the Norwegian Sea. It had returned to its place of origin by the early 1980s. Follow the path of this pool of water in figure 8.16. During this period harsh winters plagued Europe, and the entire Northern Hemisphere had cooler-than-average temperatures for more than ten years.

Recent studies have further investigated the North Atlantic, seeking to improve and lengthen our ability to predict climate. The new analysis shows pools of warm and cool surface water that circle the North Atlantic; the pools seem to have life spans of four to ten years. Because of the alternation of climatic conditions this is referred to as the North Atlantic Oscillation (NAO). Investigators are finding pieces of the puzzle that are associated with NAO but have not yet found the trigger that controls the system. A counterclockwise wind circulation centered over Iceland and a high-pressure clockwise circulation residing near the Azores are the usual situation. If the air-pressure differences between these two locations are large, then strong westerly

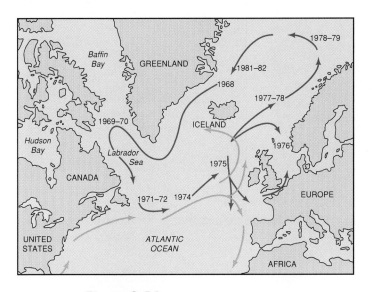

 Figure 8.16 *Purple arrows* follow the path of the North Atlantic cool pool. *Orange arrows* show warm-water flow from the Gulf Stream.

Figure 8.17 Retrieving an oceanographic research buoy in the Arabian Sea (northwest Indian Ocean). This buoy is equipped with weather, radiation and water property sensors.

winds supply Europe with heat from the North Atlantic Current. If the air-pressure difference decreases, then weaker-than-normal westerlies drive less warm water into the Norwegian Current and less heat is delivered to Europe. The periods of 1950–71 and 1976–80 were recognized as prolonged cool periods for Europe. The winds may also drive cold, low-salinity water from the Arctic into the area where North Atlantic deep water is formed. The influx of this water may cause a small-scale reduction in the formation of North Atlantic deep water and an accompanying reduction of warm surface water moving northward.

Natural cycles operate at a variety of timescales and magnitudes, and they interact and react in ways we do not yet fully comprehend. In the enormously complex series of interactions that oceanographers and meteorologists are observing they are beginning to identify the strands that unite the ocean-atmosphere and the world current systems. Eventually, longer-term predictions concerning the oceans and the atmosphere will become possible; how we use such information and whether it will benefit ocean resources are not yet clear.

Measuring the Currents 8.7

Direct measurements of currents fall into two groups: (1) those that follow a parcel of the moving water and (2) those that measure the speed and direction of the water as it passes a fixed point. Moving waters may be followed with buoys designed to float at predetermined depths. These buoys signal their positions acoustically to a research vessel or shore station; their paths are followed and their speed and displacement due to the current are calculated. Autonomous profilers such as the Argo drifter (see fig. 7.14) also measure currents. Surface water may be labeled with buoys or with dye that can be photographed from the air. Buoy positions

may also be tracked by satellites using GPS. A series of pictures or position fixes may be used to calculate the speed and direction of a current from the buoys' drift rates. Buoys can be instrumented to measure other water properties such as temperature and salinity (fig. 8.17).

A variation of this technique uses **drift bottles.** Thousands of sealed bottles, each containing a postcard, are released at a known position. When the bottles are washed ashore the finders are requested to record the time and location of the find and return the card. In this case only the release point, the recovery points, and the elapsed time are known; the actual path of motion is assumed. See the box titled "Ocean Drifters."

Sensors used to measure current speed and direction at fixed locations are called **current meters.** The current meters used by oceanographers over the last twenty years include a rotor to measure speed and a vane to measure direction of flow (fig. 8.18*a*). If the current meter is lowered from a stationary vessel, the measurements can be returned to the ship by a cable or they can be stored by the meter for reading upon retrieval. If the current meter is attached to an independent, bottom-moored buoy system, the signals can be transmitted to the ship as radio signals or they can be stored on tape in the current meter, to be removed when the buoy and current meter are retrieved.

To measure a current at a location, a current meter must not move. Although a vessel can be moored in shallow water so that it does not move, it is very difficult, if not impossible, to moor a ship or surface platform in the open sea so that it will not move and by its motion also move the current meter. The solution is to attach the current meter to a buoy system that is entirely submerged and not affected by winds or waves. See figure 8.19 for a diagram of this taut-wire moorage system. The string of anchors, current meters, wire, and floats is preassembled on deck and is launched, surface float first, over the stern of

(a)

(b)

web ⊕ link **Figure 8.18** (a) An internally recording Aanderaa current meter. The vane orients the meter to the current while the rotor determines current speed. (b) A Doppler current meter sends out sound pulses in four directions. The frequency shift of the returning echoes allows the detection of the current. These meters are also equipped with salinity-temperature-depth sensors as well as instruments for measuring water turbidity and oxygen. The data may be stored internally and collected at another time.

Figure 8.19 Taut-wire moorage. (a) Recovery is accomplished by retrieving the surface buoy and hauling in the wire. If the surface buoy is lost, it is possible to grapple for the ground wire. (b) In this system, a sound signal disconnects the anchor, and the equipment floats to the surface.

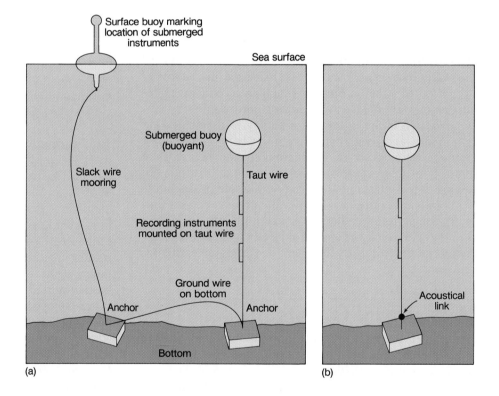

the slowly moving ship. As the ship moves away the float, meters, and cable are stretched out on the surface. When the vessel reaches the sampling site the anchor is pushed overboard to pull the entire string down into the water. The floats and meters are retrieved by grappling from the surface for the ground wire or by sending a sound signal to a special acoustical link (fig. 8.19b), which detaches the wire from the anchor. The anchor is discarded, and the buoyed equipment returns to the surface. The technique is straightforward, but many problems can occur in launching, finding, and retrieving instruments from the heaving deck of a ship at sea. Whenever oceanographers send their

Figure 8.20 Retrieving a Doppler current meter from the sea floor. The sound signals from the meter are analyzed to determine the speed and direction of the current. *Courtesy of Mark Holmes, School of Oceanography, University of Washington.*

increasingly sophisticated equipment over the side, they must cross their fingers and hope to see it again.

A new technique for measuring currents does not need the energy of the moving water to run the rotor of a current meter. This technique uses sound pulses and takes advantage of the change in the pitch, or frequency, of sound as it is reflected from particles suspended in the moving water. When sound is reflected off particles moving toward the meter, the pitch increases; when the sound is reflected off particles moving away from the meter, the pitch decreases. This is the **Doppler effect;** the same effect increases the pitch of the horn or siren of an approaching vehicle and decreases the pitch as the vehicle passes and moves away. To use this effect, the sound source is mounted on a vessel or on a buoy system, or it is placed on the sea floor. A seafloor-mounted Doppler current meter is shown in figure 8.20. Four beams of sound pulses, set at a precisely known frequency, are sent out at right angles to each other. The change in pitch of the returning echoes provides the speed and direction of the water moving along each

beam of sound pulses, and the direction of the resulting current is computed by comparison to an internal compass.

Using satellite altimeter data, scientists are able to map the topography of the sea surface on a global scale. Surface elevations and depressions are analyzed to determine the roles of gravity, periodic tidal motion, air pressure, and geostrophic flow conditions in producing sea surface topography. Oceanwide measurements of this type were not possible until satellite coverage of the oceans became available.

Practical Considerations: Energy from the Currents 8.8

The massive oceanic surface currents of the world are untapped reservoirs of energy. Their total energy flux has been estimated at 2.8×10^{14} (280 trillion) watt-hours. Because of their link to winds and surface heating processes, the ocean currents are considered as indirect sources of solar energy. If the total energy of a current was removed by conversion to electrical power, that current would cease to exist; but only a small portion of any ocean current's energy can be harnessed, owing to the current's size. Harnessing the energy from these open-ocean currents requires the use of turbine-driven generators anchored in place in the current stream. Large turbine blades would be driven by the moving water, just as windmill blades are moved by the wind; these blades could be used to turn generators and to harness the energy of the water flow. (See also the discussion on energy from tidal currents in chapter 10.)

The Florida Current and the Gulf Stream are reasonably swift and continuous currents moving close to shore in areas where there is a demand for power. If ocean currents are developed as energy sources, these currents are among the most likely. But most of the wind-driven oceanic currents generally move too slowly and are found too far from where the power is needed. In addition, the impact on other uses of the sea—transport, fishing, recreation—needs to be considered. The cost of constructing, mooring, and maintaining current-driven power-generating devices in the open sea makes them noncompetitive with other sources of power at this time.

Summary

Winds push the surface water 45° to the right of their direction in the Northern Hemisphere and 45° to the left in the Southern Hemisphere. Wind moves the water in layers that are deflected by the Coriolis effect to form the Ekman spiral; net flow over the depth of the spiral is deflected 90°. Geostrophic flow is produced when the force of gravity balances the Coriolis effect. The large surface gyres are observed in each ocean. Northern Hemisphere gyres rotate clockwise, and Southern Hemisphere gyres rotate counterclockwise. The currents of the northern Indian Ocean change with the seasonal monsoons.

Large oceanic current systems have names and descriptions based on their average locations. The water transport and speed of a current are affected by the current's cross-sectional area, by other currents, by westward intensification, and by wind speed. Eddies are formed at the surface when a fast-moving current develops waves along its boundary that break off from the parent current. Eddies occur at all depths, wander long distances, and gradually lose their identity.

Downwelling is produced by converging surface currents, and upwelling is produced by diverging surface currents. Upwelling and downwelling may be shallow and short-lived, as in Langmuir cells, or these processes may involve large volumes and large areas of the oceans. Upwelling occurs nearly continuously along the western sides of the continents in the trade-wind belts, where water diverges from the coast. Seasonal upwellings and downwellings occur in coastal areas that have changing wind patterns and an alternating coastal flow of water onshore and offshore due to the Ekman transport.

Cyclic global circulation changes are part of the Earth's normal dynamic system. Sudden changes in global ocean circulation that may lead to major climate changes are thought to be triggered by localized events in the North Atlantic. Both the North Pacific and the North Atlantic show decadal oscillations in current flow and climate.

A variety of techniques are available to measure currents: by following the water, by measuring the water's speed and direction as it moves past a fixed point, or by using changes in the pitch of sound.

Deriving energy from the oceanic current flows is not practical at the present time.

Key Terms

All key terms from this chapter can be viewed by term, or by definition, when studied as flashcards on this book's Online Learning Center at www.mhhe.com/sverdrup (click on this book's cover).

Ekman spiral, 228	subtropical convergence, 234
Ekman transport, 228	Arctic convergence, 234
gyre, 229	Antarctic convergence, 234
geostrophic flow, 230	tropical divergence, 234
western intensification, 232	Antarctic divergence, 234
eddy, 233	drift bottle, 243
Langmuir cell, 234	current meter, 243
tropical convergence, 234	

Study Questions

1. According to the net transport of the Ekman spiral, wind-driven water is directed toward the center of a large oceanic current gyre. Why does the current not flow to the gyre's center but instead flows in a clockwise circular path about a gyre in the Northern Hemisphere?

2. How is wind-driven Ekman transport related to coastal upwelling and downwelling?
3. On a map of the world, plot the oceanographic equator, the six major wind belts, the current system of each ocean, and the main areas of surface convergence and divergence.
4. Why does a flow of water that is constant in volume transport per time increase its speed when it passes through a narrow opening?
5. Explain the way in which winds and currents are named for direction.
6. What are eddies? How are they formed? Where can they be found?
7. Explain why the large Northern Hemisphere mid-ocean gyres tend to flow in a circular path, although the driving force of the winds is to the east on the higher-latitude side of the gyre and to the west on the lower-latitude side.
8. Explain why surface convergences are located at the centers of the large subtropical gyres in both the northern and southern Atlantic and Pacific Oceans.
9. Explain why upwellings on the lee side of continents at tropical latitudes function continuously while upwelling on the west side of North America is seasonal.
10. Why is there a net northward flow of surface water across the equator in the Atlantic Ocean but not in the Pacific Ocean?
11. How have tree-ring data helped describe climate and ocean current fluctuations in the North Pacific?
12. How can the cross section of a current and its average speed be used to calculate the volume of water transport?
13. Why is there a net flow of surface water from the Pacific through the Indian to the Atlantic Ocean?
14. If the production of North Atlantic deep water were severely reduced, what changes would you expect to find in (a) Gulf Stream flow, (b) upwelling in the Pacific Ocean, (c) world sea surface temperatures?
15. How are sound and the Doppler effect used to measure ocean currents?

Links to Related Websites

Visit the book's Online Learning Center at www.mhhe.com/sverdrup (click on the book's cover) to find live Internet links for additional topics related to this chapter's content.

• Ocean currents

Visit the book's Online Learning Center at www.mhhe.com/sverdrup (click on this book's cover) to find these additional chapter tools: Suggested Readings; links to further information on boxed readings, selected figures, and related chapter topics; and additional study aids.

chapter 9

The Waves

A seat in this boat was not unlike a seat upon a bucking bronco, and, by the same token, a bronco is not much smaller. The craft pranced and reared, and plunged like an animal. As each wave came, and she rose for it, she seemed like a horse making at a fence outrageously high. The manner of her scramble over these walls of water is a mystic thing, and, moreover, at the top of them were ordinarily these problems in white water, the foam racing down from the summit of each wave, requiring a new leap, and a leap from the air. Then, after scornfully bumping a crest, she would slide, and race, and splash down a long incline and arrive bobbing and nodding in front of the next menace.

A singular disadvantage of the sea lies in the fact that after successfully surmounting one wave you discover that there is another behind it just as important and just as nervously anxious to do something effective in the way of swamping boats. In a ten-foot dingey one can get an idea of the resources of the sea in the line of waves that is not probable to the average experience, which is never at sea in a dingey. As each salty wall of water approached, it shut all else from the view of the men in the boat, and it was the final outburst of the ocean, the last effort of the grim water. There was a terrible grace in the move of the waves, and they came in silence, save for the snarling of the crests.

Stephen Crane
From *The Open Boat*

Waves coming ashore.

W e have all seen water waves. This up-and-down motion occurs on the surface of oceans, seas, lakes, and ponds, and we speak of waves, ripples, swells, breakers, whitecaps, and surf. Sometimes the waves are related to the wind, or to a passing ship, or even to a stone thrown into the water. Waves may swamp a small boat, or, when large, smash and twist the bow of a supertanker. The storm waves of winter crash against our coasts, eating away the shore and damaging structures, such as docks and breakwaters. Heavy wave action may force commercial vessels to slow their speed and lengthen their sailing time between ports, to the inconvenience and increased cost of the shippers. Special waves associated with seismic disturbances have killed thousands of coastal inhabitants and have severely damaged their cities and towns. Surfers search for the perfect wave, and ancient peoples navigated by the patterns that waves form.

In this chapter we describe different kinds of waves, investigate their origin and behavior, and study their major characteristics. Our study is somewhat superficial and simplified, because waves are among the most complex of the ocean's phenomena and because no two waves are exactly the same. Books are written about waves; mathematical models have been devised to explain waves; wave tanks are built to study waves. However, there is much we can learn if we follow the sequence of events beginning at sea where a wave is born and ending at last in the spray and surf of a faraway shore.

How a Wave Begins   9.1

Imagine that you are standing on the beach looking out across a perfectly smooth surface of water. The production or generation of waves requires a disturbing force. You throw a stone into the water, or a large, naturally occurring landslide comes down into the water. A pulse of energy is introduced, and waves are produced; this disturbing force is called a **generating force.** The waves produced by the generating force move away from the point of disturbance. Of course, the magnitude of the disturbance and the resulting wave size are very different in these two examples.

In the case of the thrown stone, the stone strikes the surface of the water and displaces, or pushes aside, the water surface. As the stone sinks, the displaced water flows back from all sides into the space left behind, and as this water rushes back, the water at the center is forced upward. The elevated water falls back, causing a depression below the surface that is refilled, starting another cycle. This process sets up a series of waves, or oscillations, at the air-sea boundary that move outward and away from the point of disturbance until they are dissipated through friction among the water molecules.

In this example the wave-generating force is the stone as it strikes the surface of the water. The force that causes the water to return to its undisturbed surface level is the **restoring force.** If the stone thrown into the water is small, the waves are small, and the restoring force is the surface tension of the water surface. (Surface tension, or the elastic quality of the surface due to the cohesive behavior of the water molecules, is discussed in chapter 4.) All very small water waves are affected by surface tension.

In the example in which the landslide is the generating force, the much larger waves are pulled back to the undisturbed surface level of the water by the restoring force of gravity. When waves are of sufficiently large size so that the restoring force of the Earth's gravity is more important than surface tension, the waves are called **gravity waves.**

The most common generating force for water waves is the moving air, or wind. As the wind blows across a smooth water surface, the friction, or drag, between the air and the water tends to stretch the surface, resulting in wrinkles; surface tension acts on these wrinkles to restore a smooth surface. The wind and the surface tension create small waves, called **ripples,** or **capillary waves** (fig. 9.1). Patches of these very small waves are seen forming, moving, and disappearing as they are driven by pulses of wind. These patches darken the surface of the water and move quickly, keeping pace with the gusts of wind; sailors call these fast-moving patches **cat's-paws.** These waves die out rapidly because of friction, while new ripples form constantly in front of each moving wind gust.

When the wind blows, energy is transferred to the water over large areas, for varying lengths of time, and at different pulse rates. As waves form, the sur-

Figure 9.1 Wind-generated capillary waves appear as wrinkles atop larger waves. Capillary wavelengths are measured in centimeters.

Figure 9.2 Wind-generated gravity waves at sea.

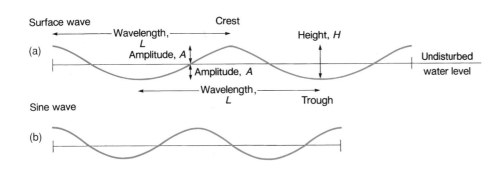

Figure 9.3 The profile of an ideal sea surface wave (a) differs from the shape of a sine wave (b).

face becomes rougher, and it is easier for the wind to grip the roughened water surface and add energy. As the wind increases in speed, the waves become larger, and the restoring force changes from surface tension to gravity (fig. 9.2).

Anatomy of a Wave 9.2

In any discussion of waves, certain terms are used; refer to figure 9.3 during the explanation of these terms. The part of the wave that is elevated the highest above the undisturbed sea surface is called the **crest;** the part that is depressed the lowest below the surface is called the **trough.** The distance between two successive crests or two successive troughs is the length of the wave, or its **wavelength.** The **wave height** is the vertical distance from the top of the crest to the bottom of the trough. Sometimes the term **amplitude** is used. The amplitude is equal to one-half the wave height, or the distance from either the crest or the trough to the undisturbed water level, or **equilibrium surface.** The oceanographer characterizes a wave not only by its length and height (or amplitude) but also by its **period.** The period is the time required for two successive crests or two successive troughs to pass a point in space. If you are standing on a piling and start a stopwatch as the crest of a wave passes and then stop the stopwatch as the crest of the next wave passes, the time you have measured is the period of the wave.

The dimensions and characteristics of waves vary greatly, but the regularity in the rise and fall of the water's surface and the relationship between wavelength and wave period allow mathematical approximations to be made, which give us insight into the behavior and properties of waves. These calculations are done by relating real waves to simple model waves. Figure 9.3*a* has been drawn to show that real waves tend to have a trough that is flatter than the crest; by contrast, figure 9.3*b* is that of a symmetrical sine wave. This regular wave, which only approximates a water wave in nature, is one of the wave forms used by physical oceanographers and mathematicians to explore and explain wave motion. The relationships presented in sections 9.3 through 9.7 are based on this regular sine-wave form.

Wave Motion 9.3

As a wave form moves across the water surface, particles of water are set in motion. Seaward, beyond the surf and breaker zone, where the surface undulates quietly, the water is not moving toward the shore. Such an ocean wave does not represent a flow of water but instead represents a flow of motion or energy from its origin to its eventual dissipation at sea or loss against the land. To understand what is happening during the passing of a wave, let us follow the motion of the water particles as the wave moves through them.

Figure 9.4 The moving wave form sets the water particles in motion (note the *arrows*). The diameter of a water particle's orbit at the surface is determined by wave height. Below the surface, the diameter decreases and orbital motion ceases at a depth (*D*) equal to one-half the wavelength.

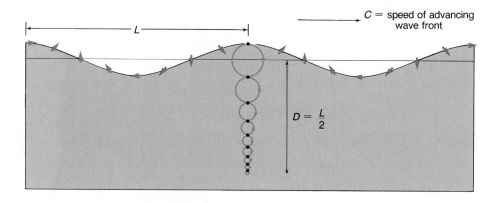

As the wave crest approaches, the surface water particles rise and move forward. Immediately under the crest the particles have stopped rising and are moving forward at the speed of the crest. When the crest passes, the particles begin to fall and to slow in their forward motion, reaching a maximum falling speed and a zero forward speed when the midpoint between crest and trough passes. As the trough advances, the particles slow their falling rate and start to move backward, until at the bottom of the trough they have attained their maximum backward speed and neither rise nor fall. As the remainder of the trough passes, the water particles begin to slow their backward speed and start to rise again, until the midpoint between trough and crest passes. At this point the water particles start their forward motion again as they continue to rise with the advancing crest. This motion (rising, moving forward, falling, reversing direction, and rising again) creates a circular path, or **orbit,** for the water particles as the wave passes. Follow this motion in figure 9.4.

It is the orbital motion of the water particles that causes a floating object to bob, or move up and down, forward and backward, as the waves pass under it. This motion affects a fishing boat, swimmer, seagull, or any other floating object on the surface seaward of the surf zone. The surface water particles trace an orbit with a diameter equal to the height of the wave. This same type of motion is transferred to the water particles below the surface, but less energy of motion is found at each succeeding depth. The diameter of the orbits becomes smaller and smaller as depth increases. At a depth equal to one-half the wavelength, the orbital motion has decreased to almost zero; notice how the orbit decreases in diameter in figure 9.4. Submarines dive during rough weather for a quiet ride because the wave motion does not extend far below the surface.

The orbits just described are based on the sine wave (see fig. 9.3*b*), not on real sea waves (see fig. 9.3*a*). The water particles of actual sea waves move in orbits whose forward speed at the top of the orbit is slightly greater than the reverse speed at the bottom of the orbit. Therefore, each orbit made by a water particle does move the water slightly forward in the direction the waves travel. This movement is due to the shape of the real waves, whose crests are sharper than their troughs; again see figure 9.3*a*. This difference means that a very slow transport of water in the direction of

the waves occurs in nature, but this motion is ignored in calculations based on simple wave models.

Wave Speed   9.4

A wave's speed across the sea surface is related to its wavelength and wave period. The speed of any surface wave (*C*) is equal to the length of the wave (*L*) divided by the period (*T*):

$$\text{speed} = \frac{\text{length of wave}}{\text{wave period}}$$

or

$$C = \frac{L}{T}$$

Once a wave is created, the speed at which the wave moves may change, but *its period remains the same* (period is determined by generating force).

In wave studies *C* stands for celerity, a term traditionally used to identify wave speed and to distinguish wave speed from the group speed of waves. Group speed is usually identified by *V* (see section 9.5).

Deep-Water Waves   9.5

"Deep water" has a precise meaning for the oceanographer studying waves. To be a **deep-water wave,** the wave must occur in water that is deeper than one-half the wave's length. Under this condition the orbits of the wave do not reach the sea floor. For example, a wave 15 m (50 ft) long must occur in water that is deeper than 7.5 m (26 ft) to be considered a deep-water wave and to behave as the waves described next.

The wavelength (*L*) of deep-water waves is derived from the wave period (*T*), and, because wavelength (*L*) is a function of wave period (*T*), wave speed (*C*) is also derived from wave period (*T*). The oceanographer at sea determines the wave period (*T*) by direct measurement and calculates the wavelength (*L*) from its relationship with gravity and wave period. In deep water, the wavelength is equal to the Earth's acceleration due to gravity (*g*) divided by 2π times

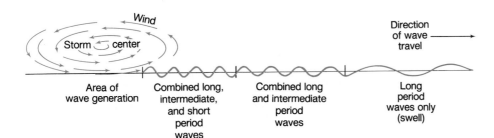

Figure 9.5 Dispersion. The longer waves travel faster than the shorter waves. Waves are shown here moving in only one direction.

the square of the wave period (T). The value of Earth's gravity, g, is 9.81 m/s².

$$L = \frac{g}{2\pi} T^2 \quad \text{or} \quad L = 1.56 \ m/s^2 \ T^2$$

This deep-water wave equation, when combined with the general wave speed equation $C = L/T$, is used to determine wave speed (C) from either wavelength (L) only or wave period (T) only:

$$C = \frac{L}{T} = \frac{g}{2\pi} T \quad \text{or} \quad C^2 = \frac{L^2}{T^2} = \frac{g}{2\pi} L$$
$$C = 1.56T \quad \text{or} \quad C^2 = 1.56L$$

See appendix C for a more complete wave equation and the method used to determine the equations for deep-water wavelength and celerity.

Storm Centers

Most waves observed at sea are **progressive wind waves.** They are generated by the wind, are restored by gravity, and progress in a particular direction. These waves are formed in local storm centers or by the steady winds of the trade and westerly wind belts. In an active storm area covering thousands of square kilometers, the winds are not steady but turbulent, varying in strength and direction. Storm-area winds flow in a circular pattern about the low-pressure **storm center,** creating waves that move outward and away from the storm in all directions. When a storm center moves across the sea surface in a direction following the waves, wave heights are increased because the winds supply energy for a longer time and over a longer distance. You may gain some idea of the size of storms and their direction of travel by viewing the cloud patterns in the satellite photographs that are presented in televised weather forecasts.

In a storm area, the sea surface appears as a jumble and confusion of waves of all heights, lengths, and periods; there are no regular patterns. Capillary waves ride the backs of small gravity waves, which in turn are superimposed on still higher and longer gravity waves. This turmoil of mixed waves is called a "sea," and sailors use the expression "There is a sea building" to refer to the growth of these waves under storm conditions. When waves are being generated they are forced to increase in size and speed by the continuing input of energy; these are known as **forced waves.** Because of variations in the winds of the storm area, energy at different intensities is transferred to the sea surface at different pulse

rates, resulting in waves with a variety of periods and heights. Remember, wave periods are a function of the generating force; the speed at which a wave moves away from a storm may change, but its period remains the same.

Dispersion

When the waves move away from the storm's influence, they are no longer wind-driven forced waves but become **free waves** moving at speeds due to their periods and wavelengths. Waves with long periods and long wavelengths have a greater speed than waves with short periods and short wavelengths. The faster, longer waves gradually move through and ahead of the shorter, slower waves; this process is called **sorting,** or **dispersion.** Groups of these faster waves move as **wave trains,** or packets of similar waves with approximately the same period and speed. Because of dispersion, the distribution of observed waves from any single storm changes with time. Near the storm center the waves are not yet sorted, while farther away the faster, longer-period waves are out ahead of the slower, shorter-period waves. This process is shown in figure 9.5.

Away from the storm, the faster, longer-period waves appear as a regular pattern of crests and troughs moving across the sea surface. These uniform, free waves are called **swell;** they carry considerable energy, which they lose very slowly (fig. 9.6). The distribution of the waves from a given storm and the energy associated with particular wave periods change predictably with time, allowing the oceanographer to follow wave trains from a single storm over long distances. Groups of large, long-period waves created by storms between 40°S and 50°S in the Pacific Ocean have been traced across the entire length of that ocean, until they die on the shores and beaches of Alaska.

Group Speed

Consider again the waves formed by a stone thrown into the water; the wave group, or train, is seen as a ring of waves moving outward from the point of disturbance. Careful observation shows that waves constantly form on the inside of the train as it moves across the water. As each wave joins the train on the inside of the ring, a wave is lost from the leading edge, or outside, of the ring and the number of waves remains the same. The outside wave's energy is lost in advancing the wave form into undisturbed water. Therefore the speed of each individual wave (C) in the group is greater than the speed of the leading edge of the wave train, and the

Figure 9.6 Waves of uniform length and period, known as swell, approaching the coast at the entrance to Grays Harbor, Washington.

wave train moves outward at a speed one-half that of the individual waves. This speed is known as the **group speed,** the speed at which wave energy is transported away from its source under deep-water conditions:

group speed = ½ wave speed = speed of energy transport

or

$$V = \frac{C}{2}$$

Wave Interaction

Waves that escape a storm and are no longer receiving energy from the storm winds tend to flatten out slightly, and their crests become more rounded. These waves moving across the ocean surface as swell are likely to meet other trains of swell moving out and away from other storm centers. When the two wave trains meet, they pass through each other and continue on. Wave trains may intersect at any angle, and many possible interference patterns may result. If the two wave trains intersect each other sharply, as at a right angle, then a checkerboard pattern is formed (fig. 9.7).

If the waves have similar lengths and heights and approach from opposite directions (fig. 9.8a), and if the crests of one wave train coincide with the crests of another train, the wave trains reinforce each other by constructive interference (fig. 9.8b). If the crests of a wave train coincide with the troughs of another wave train, the waves are canceled through destructive interference (fig. 9.8c). In this way two or more similar wave trains, traveling in the same or opposite direction and passing through each other, can phase together constructively and suddenly develop large-amplitude waves unrelated to local storms. If these waves become too

high, they may break, lose some of their energy, and create new, smaller waves. If such high waves overtake a vessel, they can cause severe damage.

Wave Height   9.6

The height of wind waves is controlled by the interaction of several factors. The three most important factors are (1) wind speed (how fast the wind is blowing), (2) wind duration (how long the wind blows), and (3) **fetch** (the distance over water that the wind blows in the same direction). The wave height may be limited by any one of these factors. If the wind speed is very low, large waves are not produced, no matter how long the wind blows over an unlimited fetch. If the wind speed is great, but it blows for only a few minutes, no high waves are produced despite unlimited wind strength and fetch. Also, if very strong winds blow for a long period over a very short fetch, no high waves form. When no single one of these three factors is limiting, spectacular wind waves are formed at sea.

Table 9.1 lists the maximum and significant wave heights possible for certain average wind speeds when fetch and wind duration are not limiting. The significant wave height is defined as the average wave height of the highest one-third of the waves in a long record of measured wave heights. For example, if a storm wave record of 1200 successive waves is made and the individual wave heights are arranged in order of height, the average height of the 400 highest waves defines the significant wave height. Significant wave heights are forecast from wind data, and the maximum wave heights are related to the significant wave heights by calculation.

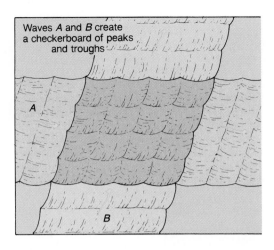

Figure 9.7 Waves meeting at right angles create a checkerboard pattern.

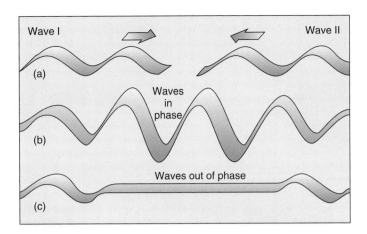

Figure 9.8 Constructive and destructive interference between similar wave series. (a) Waves approach each other. (b) If the crests of the approaching waves coincide, the height of the combined waves increases, and the waves are in phase. (c) If the crests of one wave and the troughs of the other wave coincide, the waves cancel, and the waves are out of phase.

Table 9.1 The Relationship between Wind Speed and Wave Height

Average Wind Speed		Significant Wave Height	Significant Wave Period	Significant Wave Speed	Maximum Wave Height	Minimum Fetch[1]	Minimum Wind Duration[1]
(knots)	(m/s)	(m)	(s)	(m/s)	(m)	(km)	(h)
10	5.1	1.22	5.5	8.58	2.19	16	2.4
20	10.2	2.44	7.3	11.39	4.39	110	10
30	15.3	5.79	12.5	19.50	10.43	450	23
40	20.4	14.33	18.0	28.00	25.79	1136	42
50	25.5	16.77	21.0	32.76	30.19	2272	69

1. Minimum fetch and minimum wind duration are distances and times required when wind speed is the only limiting factor in wave development.

The area of the ocean in the vicinity of 40°–50°S latitude is ideal for the production of high waves. Here, in an area noted for high-intensity storms and strong winds of long duration (what sailors have called the "roaring forties and furious fifties"), there are no landmasses to interfere and limit the fetch length. The westerly winds blow almost continuously around the Earth, adding energy to the sea surface for long periods of time and over great distances, resulting in waves that move in the same direction as the wind. Although this area is ideal for the production of high waves, such waves can occur anywhere in the open sea, given the proper storm conditions.

A typical maximum fetch for a local storm over the ocean is approximately 920 km (500 nautical mi). Because storm winds circulate around a low-pressure disturbance, the winds continue to follow the waves on the side of the storm, along which wave direction is the same as the storm direction. This increases both the fetch and the duration of time over which the wind adds energy to the waves. If the waves move fast enough, their speeds exceed the speed of the moving storm

center; the waves escape the generating wind, become free waves, and do not grow larger. Waves 10–15 m (33–49 ft) high are not uncommon under severe storm conditions; such waves are typically between 100 and 200 m (330 and 660 ft) long. This length is about the same as the length of some modern ships, and a vessel of this length encounters hazardous sailing conditions, because the ship may become suspended between the crests of two waves and break its back.

Giant waves over 30.5 m (100 ft) high are rare. In 1933 the USS *Ramapo*, a Navy tanker en route from Manila to San Diego, encountered a severe storm, or typhoon. As the ship was running downwind to ease the ride, it was overtaken by waves that, as measured against the ship's superstructure by the officer on watch, were 112 ft (34.2 m) high. The period of the waves was measured at 14.8 seconds; the wave speed was calculated at 27 m (90 ft) per second and the wavelength at 329 m (1100 ft). Other storm waves in this size category have been reported, but none have been as well documented. It is also probable that ships confronted with such waves do not always survive to report the incidents.

Wave Height **253**

Figure 9.9 A giant wave breaking over the bow of the *ESSO Nederland* southbound in the Agulhas Current. The bow of this supertanker is about 25 m (80 ft) above the water.

Episodic Waves

Large waves, or **episodic waves,** can suddenly appear unrelated to local sea conditions. An episodic wave is an abnormally high wave that occurs because of a combination of intersecting wave trains, changing depths, and currents. We do not know a great deal about these waves, as they do not exist for long and they can and do swamp ships, often eliminating any witnesses. They occur most frequently near the edge of the continental shelf, in water about 200 m (660 ft) deep, and in certain geographic areas with particular prevailing wind, wave, and current patterns.

The area where the Agulhas Current sweeps down the east coast of South Africa and meets the storm waves arising in the Southern Ocean is noted for such waves. Storm waves from more than one storm may combine constructively and run into the current and against the continental shelf, producing occasional episodic waves. This area is also one of the world's busiest sea routes, as supertankers carrying oil from the Middle East ride the Agulhas Current on their trip southward to round the Cape of Good Hope en route to Europe and America. The situation is an invitation to trouble, and

tankers have been damaged and lost in this area (fig. 9.9). In the North Atlantic, strong northeasterly gales send large storm waves into the edge of the northward-moving Gulf Stream near the border of the continental shelf, resulting in the formation of large waves. The shallow North Sea also seems to provide suitable conditions for extremely high episodic waves during its severe winter storms.

Researchers studying these waves describe them as having a height equal to a seven- or eight-story building (20–30 m, or 70–100 ft) and moving at a speed of 50 knots, with a wavelength approaching a half mile (0.9 km). If such a wave topples onto a vessel that has dropped its bow into the preceding trough, there is no escape from the thousands of tons of water crashing on the deck. In the North Sea, maximum potential wave height for an episodic wave is calculated as 33.8 m (111 ft), but 22.9 m (75 ft) is the highest that has been observed. In the Agulhas Current area, researchers searched the storm records for a twenty-year period and calculated a possible maximum wave height of about 57.9 m (190 ft).

There are many disappearances of vessels for which episodic waves are now suspected of being the chief cause. It

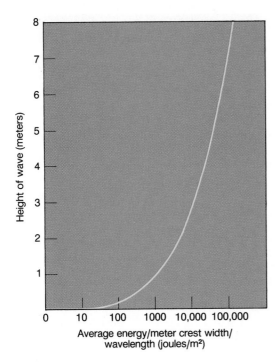

Figure 9.10 Wave energy increases rapidly with the square of the wave height. Average wave energy is calculated per unit width of crest and averaged over the wavelength (*L*) and the depth (*L/2*).

seems that many of the casualties are tankers or bulk carriers of ore, grain, and the like. These vessels are susceptible not only because so many are at sea at any one time but also because of their design. Bulk carriers comprise a bow section and an aft (or rear) section that includes the engine and the crew accommodations. These sections are separated by a series of flat-bottomed boxes, or storage tanks, which make up the majority of the vessel's length. Because these vessels are about 300 m (1000 ft) long, they are subject to great wrenching forces if a large portion of the hull is left unsupported or is only partially supported while suspended between wave crests. More-traditional and smaller hull forms are stronger, ride more easily, and are less likely to be destroyed by these severe waves.

Wave Energy

A wave's energy is present as **potential energy,** due to the change in elevation of the water surface, and as **kinetic energy,** due to the motion of the water particles in their orbits. The higher the wave, the larger the diameter of the water particle orbit and the greater the speed of the orbiting particle, therefore the greater the kinetic and potential energy. The energy in a deep-water wave is nearly equally divided between kinetic and potential energy. Wave energy is averaged over one wavelength per unit width (1 m) of crest from sea surface to a depth of *L/2* and is related to the square of the wave height. This relationship is demonstrated in figure 9.10.

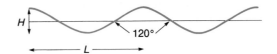

Figure 9.11 Wave steepness. When *H/L* approaches 1:7, the wave's crest angle approaches 120° and the wave breaks.

Wave Steepness

There is a maximum possible height for any given wavelength. This maximum value is determined by the ratio of the wave's height to the wave's length, and it is the measure of the **steepness** of the wave:

$$\text{steepness} = \frac{\text{height}}{\text{length}} \quad \text{or} \quad S = \frac{H}{L}$$

If the ratio of the height to the length exceeds 1:7, the wave becomes too steep and the wave breaks. For example, if the wavelength is 70 m (230 ft), the wave will break when the wave height reaches 10 m (33 ft). The angle formed at the wave crest approaches 120°, and the wave becomes unstable. Under this condition the wave cannot maintain its shape; it collapses and breaks (fig. 9.11).

Small, unstable breaking waves are quite common. When wind speeds reach 8–9 m/s (16–18 knots), waves known as whitecaps can be observed. These waves have short wavelengths (about 2 m), and the wind increases their height rapidly. As each wave reaches the critical steepness and crest angle, it breaks and is replaced by another wave produced by the rising wind.

Long waves at sea usually have a height well below their maximum value. Sufficient wind energy to force them to their maximum height rarely occurs. If a long wave does attain maximum height and breaks in deep water, tons of water are sent crashing to the surface. The energy of the wave is lost in turbulence and in the production of smaller waves. Rather than breaking, under such conditions it is more likely that the top of a large wave will be torn off by the wind and cascade down the wave face. This action does not completely destroy the wave and is not considered to be the true collapse, or breaking, of such a wave.

In addition, waves break and dissipate if intersecting wave trains pass through each other in proper phase to form a combined wave with sufficient height to exceed the critical steepness. Waves sometimes run into a strong opposing current, forcing the waves to slow down. Remember that the speed of all waves equals the wavelength divided by the period (*C = L/T*) and that a wave's period does not change. If the speed of a wave is reduced by an opposing current, its wavelength must shorten. In such a case, the wave's energy is confined to a shorter length, so the wave increases in height to satisfy the height-energy relationship. If the increase in height exceeds the maximum allowable height-to-length ratio for the shorter wavelength, the wave breaks. Crossing a sandbar into a harbor or river mouth during an outgoing or falling tide is dangerous because the waves

Table 9.2 Universal Sea State Code

Sea State Code	Description	Average Wave Height
SS0	Sea like a mirror; wind less than one knot	0
SS1	A smooth sea; ripples, no foam; very light winds, 1–3 knots, not felt on face	0–0.3 m 0–1 ft
SS2	A slight sea; small wavelets; winds light to gentle, 4–6 knots, felt on face; light flags wave	0.3–0.6 m 1–2 ft
SS3	A moderate sea; large wavelets, crests begin to break; winds gentle to moderate, 7–10 knots; light flags fully extend	0.6–1.2 m 2–4 ft
SS4	A rough sea; moderate waves, many crests break, whitecaps, some wind-blown spray; winds moderate to strong breeze, 11–27 knots; wind whistles in the rigging	1.2–2.4 m 4–8 ft
SS5	A very rough sea; waves heap up, forming foam streaks and spindrift; winds moderate to fresh gale, 28–40 knots; wind affects walking	2.4–4.0 m 8–13 ft
SS6	A high sea; sea begins to roll, forming very definite foam streaks and considerable spray; winds a strong gale, 41–47 knots; loose gear and light canvas may be blown about or ripped	4.0–6.1 m 13–20 ft
SS7	A very high sea; very high, steep waves with wind-driven overhanging crests; sea surface whitens due to dense coverage with foam; visibility reduced due to wind-blown spray; winds at whole gale force, 48–55 knots	6.1–9.1 m 20–30 ft
SS8	Mountainous seas; very high-rolling breaking waves; sea surface foam-covered; very poor visibility; winds at storm level, 56–63 knots	9.1–13.7 m 30–45 ft
SS9	Air filled with foam; sea surface white with spray; winds 64 knots and above	13.7 m and above 45 ft and above

moving over the bar and against the tidal current steepen and break. Entering a harbor or river should be done at the change of the tide or on the rising tide, when the tidal current moves with the waves, stretching their wavelengths and decreasing their heights.

Universal Sea State Code

In 1806, Admiral Sir Francis Beaufort of the British navy adapted a wind-estimation system from land to sea use. On land, the clues to wind speed included smoke drift, the rustle of leaves, the flapping of flags, slates blowing from roofs, and uprooting of trees. Admiral Beaufort related observations of the sea surface state to wind speed and designed a 0–12 (calm to hurricane) wind scale with typical wave descriptions for each level of wind speed. This Beaufort Scale was adopted by the U.S. Navy in 1838, and the scale was extended from 0–17. At present, a Universal Sea State Code of 0–9, based on the Beaufort Scale, is in international use for wind speeds and related sea surface conditions (table 9.2).

Shallow-Water Waves   9.7

As a deep-water wave approaches the shore and moves into shallow water, the reduced depth begins to affect the shape of the orbits made by the water particles. The orbits gradually become flattened circles, or ellipses (fig. 9.12). The wave begins to "feel" the bottom, and the resulting friction and compression of the orbits reduce the forward speed of the wave.

Remember that (1) the speed of all waves is equal to the wavelength divided by period, and (2) the period of a wave does not change. Therefore, when the wave "feels bottom," it slows, and the accompanying reduction in the wavelength and speed results in increased height and steepness as the wave's energy is condensed in a smaller water volume.

When the wave finally enters water with a depth of less than one-twentieth the wavelength ($D < L/20$), the wave becomes a **shallow-water wave** (fig. 9.13). While the length and speed of a deep-water wave are determined by the wave period, the shallow-water wave's length and speed are controlled only by the water depth. Here the wave and speed length are determined by the square root of the product of the acceleration due to gravity (g) and depth (D):

$$C = \sqrt{gD} \quad \text{or} \quad C = 3.13\sqrt{D}$$
$$\text{or } L = 3.13\sqrt{D} \cdot T$$

When the condition for a shallow-water wave is met, the orbits of the water particles are elliptical; they become flatter with depth until, at the sea floor, only a back-and-forth oscillatory motion remains (fig. 9.13). Note that the horizontal dimension of the orbit remains unchanged in shallow water. See appendix C for a more complete equation and the method used to calculate approximations of shallow-water wavelength and celerity.

When the water depth is between $L/2$ and $L/20$, the speed of the wave is also slowed. Waves in this depth range are called intermediate waves. There is no simple algebraic equation to determine the speed of these intermediate

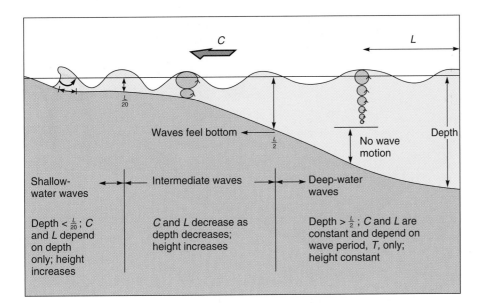

Figure 9.12 Deep-water waves become intermediate waves and then shallow-water waves as depth decreases and wave motion interacts with the sea floor.

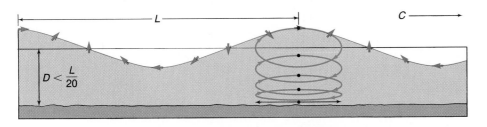

Figure 9.13 Shallow-water wave particles move in elliptical orbits. The orbits flatten with depth due to interference from the sea floor.

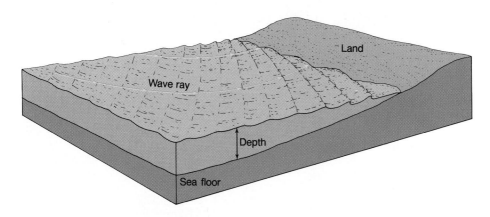

Figure 9.14 Waves moving inshore at an oblique angle to the depth contours are refracted. One end of the wave reaches a depth of *L/2* or less and slows while the other end of the wave maintains its speed in deeper water. Wave rays drawn perpendicular to the crests show the direction of wave travel and the bending of the wave crests.

waves. Methods for working with intermediate waves using the full wave-speed equation are found in appendix C.

Refraction

Waves are refracted, or bent, as they move from deep to shallow water, begin to feel the bottom, and change wavelength and wave speed. When waves from a distant storm center approach the shore, they are likely to approach the beach at an angle. One end of the wave crest comes into shallow water and begins to feel bottom, while the other end is in deeper water. The shallow-water end moves more slowly than that portion of the wave in deeper water. The result is that the wave crests bend, or refract, and tend to become oriented parallel to the shore. This pattern is shown in figure 9.14. The **wave rays** drawn perpen-

dicular to the crests show the direction of motion of the wave crests. Note that the refraction of water waves is similar to the refraction of light and sound waves, described in chapter 4.

When waves approach an irregular coastline, with headlands jutting into the ocean and bays set back into the land, a submerged ridge seaward from a headland and a depression in front of a bay often form. When shallow-water waves approach this coastline, the portion of the wave crest over the ridge slows down more than the wave crests on either side. The crests wrap around the headland in the pattern shown in figure 9.15. This refraction pattern results from the shortening of the wavelengths and focuses wave energy on the point of the headland. Since wave energy is proportional to the wave height, the waves gain in height as their wavelength is shortened over the

Figure 9.15 The energy of waves refracted over a shallow, submerged ridge is focused on the headland. The converging wave rays show the wave energy being crowded into a small volume of water, increasing the energy per unit length of wave crest as the height of the wave increases.

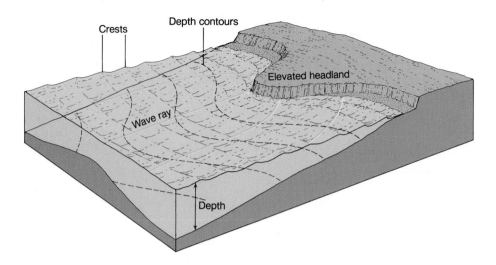

Figure 9.16 Waves refracted by the shallow depths on each side of the bay deliver lower levels of energy inside the bay. The diverging wave rays show the spreading of energy over a larger volume of water, decreasing the energy per unit length of wave crest as the wave height decreases.

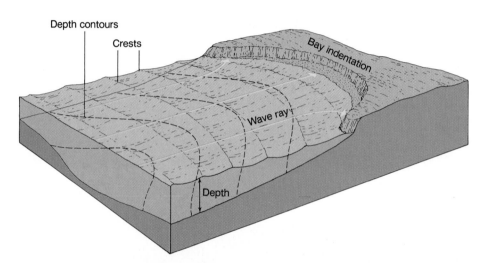

submerged ridge, and more energy is expended on a unit length of shore at the point of the headland than on a unit length of shore elsewhere.

The central area at the mouth of the bay is usually deeper than the areas to each side, so the advancing waves slow down more on the sides than in the center. Because the wavelengths remain long in the center and shorten on each side, the wave crests bulge toward the center of the bay. The waves in the center of the bay do not shorten as much; therefore, they have less height and less energy to expend per length of shoreline. This pattern is shown in figure 9.16. The result is an environment of low wave energy, providing sheltered water in the bay. Overall, this unequal distribution of wave energy along the coast results in a wearing down of the headlands and a filling in of the bays, as the sand and mud settle out in the quieter water. If all coastal materials had the same resistance to wave erosion, this process would lead in time to a straightening of the coastline; but the rocky structure of many headlands resists wave erosion, and therefore the cliffs remain.

Reflection

A straight, smooth, vertical barrier in water deep enough to prevent waves from breaking reflects the waves (fig. 9.17). The barrier may be a cliff, steep beach, breakwater, bulk-

Figure 9.17 The wave from a ship's wake is reflected from a concrete wall. The principal wave is moving from *lower left* to *upper right* of the photograph. The reflected wave is moving from *lower right* to *upper left*. A checkerboard pattern is formed as the principal wave and reflected wave move through each other at approximately right angles.

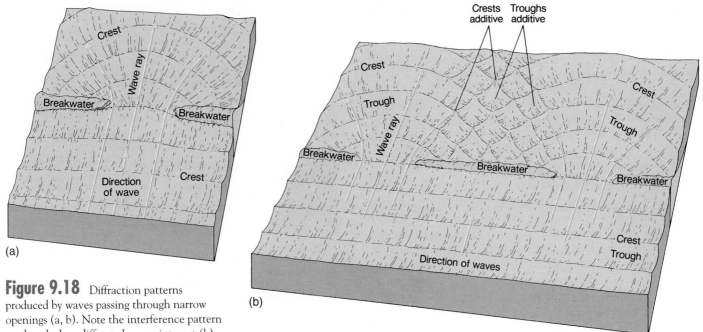

Figure 9.18 Diffraction patterns produced by waves passing through narrow openings (a, b). Note the interference pattern produced when diffracted waves interact (b).

head, or other structure. The reflected waves pass through the incoming waves to produce an interference pattern, and steep, choppy seas often result. If the waves reflect directly back on themselves, the resulting waves appear to stand still, rising and falling in place. If waves are reflected from a curved vertical surface, the curve may spread the reflected wave rays, dispersing the wave energy, but if the surface curves so that the reflected wave rays converge, the energy is focused. This situation is similar to the reflection of light from curved mirrors. Great care must be taken in designing walls and barriers to protect an area from waves to be sure that the energy of reflected waves is not focused in such a way that another area will be damaged.

Diffraction

Another phenomenon that is associated with waves as they approach the shore or other obstacles is **diffraction.** Diffraction is caused by the spread of wave energy sideways to the direction of wave travel. If waves move toward a barrier with a small opening (fig. 9.18a), some wave energy passes through the small opening to the other side. Once the waves have passed through the opening, their crests decrease in height, radiating out and away from the gap. A portion of the wave energy is diffracted, transported sideways from its original direction. If more than one gap is open to the waves, the patterns produced by the spreading waves from the openings may intersect and form interference patterns as the waves move through each other (fig. 9.18b).

If the waves approach a barrier without an opening, diffraction can still occur. Energy will be transported at right angles to the wave crests as the waves pass the end of the barrier (fig. 9.19). Note that the pattern produced is one-half of the pattern observed when the waves pass through a

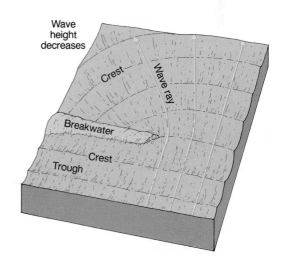

Figure 9.19 Diffraction occurs behind the breakwater.

narrow opening, as in figure 9.18a. Energy is transported behind the sheltered (or lee) side of the barrier. This effect is an important consideration in planning the construction of breakwaters and other coastal barriers intended to protect vessels in harbors from wave action and possible damage.

Navigation from Wave Direction

In areas of the world where the winds blow steadily and from one direction, as in the trade-wind belts, waves at sea are very regular in their direction of motion, and this regularity allows a vessel to maintain a constant course relative to the waves. Because waves change speed, shape, and height with water depth, and because waves change direction and pattern because of refraction, diffraction, and reflection, it is possible to deduce the presence of shoals, bars, islands, and coasts from the changes in wave patterns.

Careful direct observation of wave patterns, or the sensing of wave patterns through the motions of a small craft, even at night, allows the detection of a change in the angle between waves and wind direction or a change in the wave pattern. Although these changing patterns are subtle, they can be detected many miles downwind from islands and can be used to bring a small vessel to shore when no landmarks are visible.

The Polynesians of the past lived with the sea; it was their home, and they were acutely sensitive to its changes. They made long voyages in their small canoes using their knowledge of wave patterns. They had no theory to explain these patterns, but they understood their association with winds, shores, and islands. Combining their knowledge of star positions, cloud forms over land and sea, and bird flights with their knowledge of waves, they sailed many hundreds, even thousands, of miles across the open ocean to reach their destinations. No navigational tools were required other than charts constructed from twigs and shells, showing island positions relative to stars, wind direction, and swell. An example of such a chart is found in prologue figure III. These peoples learned by living with and observing nature in an area where nature cooperated by behaving in regular and predictable ways.

The Surf Zone 9.8

The surf zone is the shallow area along the coast in which the waves slow rapidly, steepen, break, and disappear in the turbulence and spray of expended energy. The width of this zone is variable and is related to both the length and height of the arriving waves and the changing depth pattern. Longer, higher waves, which feel the bottom before shorter waves, become unstable and break farther offshore, in deeper water. If shallow depths extend offshore for some distance, the surf zone is wider than it is over a sharply sloping shore.

Breakers

Breakers form in the surf zone because the water particle motion at depth is affected by the bottom. Orbital motion is slowed and compressed vertically, but the orbit speed of water particles near the crest of the wave is not slowed as much. The particles at the wave crest move faster toward the shore than the rest of the wave form, resulting in the curling of the crest and the eventual breaking of the wave. The two most common types of breakers are **plungers** and **spillers** (fig. 9.20).

Plunging breakers form on narrow, steep beach slopes. The curling crest outruns the rest of the wave, curves over the air below it, and breaks with a sudden loss of energy and a splash. The more common spilling breaker is found over wider, flatter beaches, where the energy is extracted more gradually as the wave moves over the shallow bottom. This action results in the less dramatic wave form, consisting of

(a)

(b)

Figure 9.20 Breaking waves. A plunger (a) loses energy more quickly than a spiller (b).

turbulent water and bubbles flowing down the collapsing wave face. The spilling breakers last longer than the plungers because they lose energy more gradually. Therefore, spillers give surfers a longer ride, but plungers give them a more exciting one.

The slow curling over of the crest observed on some breakers begins at a point on the crest and then moves lengthwise along the wave crest as the wave approaches shore. This movement of the curl along the crest occurs because waves are seldom exactly parallel to a beach. The curl begins at the point on the crest that is in the shallowest water or the point at which the crest height is slightly greater, and it moves along the crest as the rest of the wave approaches the beach. The result is the "tube" so sought after by surfers (fig. 9.21).

If the waves approaching the beach are uniform in length, period, and height, they are the swells from some far-distant storm, which have had time and distance to sort into uniform groups. For example, the long surfing waves of the

Figure 9.21 A surfer rides the tube of a large curling wave in Hawaii.

California beaches in summer begin their lives in the winter storms of the South Pacific and Antarctic Oceans. If the waves are of different heights, lengths, and periods and break at varying distances from the beach, then unsorted waves have arrived and are probably the product of a nearby local storm superimposed on the swell.

Water Transport

The small net drift of water in the direction the waves are traveling (refer to the previous discussion on the water particle orbits of real waves) is intensified in the surf zone, as the shoreward motion of the water particles at the crest becomes greater than the return particle motion at the trough. Because the crests usually approach the beach at an angle, the surf zone transport of water flows both toward the beach and along the beach. The result is that water accumulates against the beach and flows along the beach until it can flow seaward again and return to the area beyond the surf zone. This return flow generally occurs in quieter water with smaller wave heights, for example, in areas with troughs or depressions in the sea floor.

Because regions of seaward return flow may be narrow and some distance apart, the flow in these areas must be swift in order to carry enough water beyond the surf zone to balance the slower but more extensive flow toward the beach. These regions of rapid seaward flow are called **rip currents.**

Figure 9.22 Small rip currents carry turbid water seaward through the surf zone.

Rip currents can be a major hazard to surf swimmers. In the spring of 1994, five swimmers were drowned and six others were hurt in a single rip current system at American Beach, Florida, and in 1995 three more swimmers drowned in severe rip currents caused by storm waves from Hurricane Felix.

Swimmers who unknowingly venture into a rip current will find themselves carried seaward and unable to swim back to shore against the flow; they must swim parallel to the beach, or across the rip current, and then return to shore. Because of the danger associated with rip currents, swimmers should be on the lookout for indicators of these currents' presence, including (1) turbid water and floating debris moving seaward through the surf zone, (2) areas of reduced wave heights in the surf zone, and (3) depressions in the beach running perpendicular to the shore.

Wave action on the beach stirs up the sand particles and keeps them suspended in the water. The sand is carried along the beach parallel to the shore until the rip current is reached, and the sand is transported seaward. Viewed from a height, such as a high cliff or a low-flying airplane, rip currents are seen as streaks of discolored turbid water extending seaward through the clearer water of the outer surf zone (fig. 9.22).

The Surf Zone **261**

Energy Release

Watching the heavy surf pounding a beach from a safe vantage point is an exciting and exhilarating experience; the trick is to determine at what point one is safe. In a narrow surf zone during a period of very large waves, the wave energy must be expended rapidly over a short distance. Under these conditions, the height of the waves and the forward motion of the water particles combine to send the water high up on the beach. The accompanying release of energy is explosive and can result in rocks and debris from the water's edge being hurled high up on the beach by the force of the water. Minot's Lighthouse on the south side of Massachusetts Bay is 30 m (100 ft) high, but it is regularly engulfed in spray. Lonely Tillamook Light off the Oregon coast had to have steel gratings installed to protect the glass that shields the light 40 m (130 ft) above sea level, after the glass had been broken several times by wave-thrown rocks. In every winter storm, waves displace boulders weighing many tons from breakwaters along the world's coasts.

Waves do not always expend their energy on the shore. Some break farther seaward on sandbars such as those associated with river mouths and estuaries. The famous bar at the mouth of the Columbia River is responsible for extremely hazardous conditions for both fishing and commercial vessels, and in winter it often produces casualties. On an ebbing (or falling) tide, waves approaching the bar against the current rise up and break at heights up to 20 m (66 ft). The Coast Guard uses the Columbia River entrance to train its crews to operate roll-over, or self-righting, rescue boats, which are used during storms and heavy surf conditions.

The great driftwood logs found stranded on some beaches may be set afloat again at high tide during severe winter storms and can become lethal battering rams when hurled shoreward by the surf. Even with less severe waves, beach logs lying near the water's edge have a dangerous potential. The unsuspecting vacationer sits or plays on such a log, and when the occasional higher wave rolls it over, the person may become trapped under the log and crushed or drowned by the succeeding waves. Logs in or near the surf represent a great danger for the unwary. Where are you safe? You must be the judge.

Tsunamis 9.9

Sudden movements of the Earth's crust may produce **seismic sea waves,** or **tsunamis.** These waves are often incorrectly called tidal waves. Because a seismic sea wave has nothing to do with tides, oceanographers have adopted the Japanese word tsunami, meaning harbor wave, to replace the misleading term tidal wave. Tsunami is a synonym for seismic sea wave.

If a large area, maybe several hundred square kilometers, of the Earth's crust below the sea surface is suddenly displaced, it may cause a sudden rise or fall in the overlying sea surface. In the case of a rise, gravity causes the suddenly elevated water to return to the equilibrium surface level; if a depression is produced, gravity causes the surrounding water to flow into it. Both cases result in the production of waves with extremely long wavelengths (100–200 km, or 60–120 mi) and long periods as well (ten–twenty minutes). The average depth of the oceans is about 4000 m (4 km, or 13,000 ft); this depth is less than one-twentieth the wavelength of these waves, so tsunamis are shallow-water waves. These seismic waves radiate from the point of the seismic disturbance at a speed determined by the ocean's depth ($C = \sqrt{gD}$ and move across the oceans at about 200 m/s (400 mph). Because they are shallow-water waves, tsunamis may be refracted, diffracted, or reflected in mid-ocean by changes in seafloor topography and by the presence of mid-ocean islands.

When a tsunami leaves its point of origin, it may have a height of 1–2 m (3–6 ft), but this height is distributed over its many-kilometer wavelength. It is not easily seen or felt when superimposed on the other distortions of the sea's surface, and a vessel in the open ocean is in little or no danger if a tsunami passes. The danger occurs only if the vessel has the misfortune to be directly above the area of the original seismic disturbance.

The energy of a tsunami is distributed from the ocean surface to the ocean floor and over the length of the wave. When the path of the wave is blocked by a coast or an island, the wave behaves like any other shallow-water wave; it slows (to approximately 80 km [50 mi] per hour), its wavelength decreases, and its energy is compressed into a smaller water volume as the depth rapidly decreases. This sudden confinement of energy causes the wave height to build rapidly, and the loss of energy is equally rapid when the wave breaks. A tremendous surge of moving water races up over the land, flooding the coast for a period that lasts five to ten minutes before the water flows seaward, exposing the nearshore sea floor. The surge destroys buildings and docks; large ships may be left far up on the shore. The receding water carries the debris from the surge, battering and buffeting everything in its path. In bays and inlets the rising and falling of the water may not be as destructive as the extreme currents that form at the entrance of a bay or an inlet where large volumes of water oscillate in and out, producing dangerous and destructive current surges. If these oscillating flows occur at the natural oscillation period of the bay, large changes in water level are created (see section 9.11).

The leading edge of the tsunami wave group may be either a crest or a trough. If the initial crustal disturbance is an upward motion, a crest arrives first; if the crustal motion is downward, a trough precedes the crest. If a trough arrives at the shore first, the water level drops by as much as 3–4 m (10–13 ft) within two to three minutes. People who follow the receding water to inspect exposed sea life may drown, for in another four to five minutes the water rises 6–8 m (20–26 ft) and they will not be able to outrun the wall of advancing water. In still another four to five minutes a trough with a water-level drop of 6–8 m (20–26 ft) arrives, and the water flowing away from the beach will carry debris out to sea.

Tsunamis are most likely to occur in ocean basins that are tectonically active. The Pacific Ocean, ringed by crustal faults and volcanic activity, is the birthplace of most tsunamis.

They also appear in the Caribbean Sea, which is bounded by an active island arc system, and in the Mediterranean Sea. The spectacular havoc and destruction caused by these waves is well recorded. In August 1883 the Indonesian Island of Krakatoa erupted and was nearly destroyed in a gigantic steam explosion that hurled several cubic miles of material into the air. A series of tsunamis followed; these waves had unusually long periods of one to two hours. The town of Merak, 33 mi (53 km) away and on another island, was inundated by waves over 30.5 m (or 100 ft) high, and a large ship was carried nearly 3 km (2 mi) inland and left stranded 9.2 m (30 ft) above sea level. More than 35,000 people died as a result of these enormous waves, and as the waves moved across the oceans, water-level recorders were affected as far away as Cape Horn (12,500 km, or 7800 mi) and Panama (18,200 km, or 11,400 mi). The waves were traveling with a speed calculated at about 200 m/s, or 400 mi/h.

On April 1, 1946, a crustal disturbance in the area of the Aleutian Trench off Alaska produced a series of tsunamis. These waves heavily damaged Hilo, Hawaii, killing more than 150 people. Not only the portion of the island facing the oncoming waves was damaged; the waves bent around the island by refraction and were diffracted when passing between islands, producing high waves that struck the sheltered side of the island. The tsunami waves from the 1946 earthquake were highest near their source in the Aleutian Islands, where they destroyed a concrete lighthouse 10 m (33 ft) above sea level at Scotch Cap and killed the crew. A radio mast mounted 33 m (108 ft) above sea level in the same area was also destroyed. In 1957, Hawaii was hit again, with waves higher than the 1946 series; but no lives were lost because early warnings alerted people to evacuate (fig. 9.23). In 1964, the Alaska earthquake that severely damaged Anchorage and Alaskan coastal towns produced tsunamis that selectively hit areas on the west coast of Vancouver Island and the northern coast of California.

On September 1, 1992, a magnitude 7.0 earthquake occurred 100 km (60 mi) off the Nicaraguan coast. The tsunami produced by this earthquake killed 170 people, injured 500, and destroyed 1500 homes. Eyewitnesses recall a single large wave, preceded by a trough that lowered the harbor depth by 7 m (23 ft). In this case the maximum run-up was 10 m (33 ft); run-ups of 2–6 m (7–19 ft) were more common. This event was triggered by an earthquake in a previously stationary section of a fault region associated with the Cocos Plate where it is being subducted under the Caribbean Plate.

On December 12, 1992, an earthquake of magnitude 7.5 struck Flores Island in the Sunda and Banda island arc systems. The quake epicenter was only 50 km (30 mi) northwest of the town of Maumere, and five minutes after the quake, the first of a series of tsunamis struck the town. The earthquake and the tsunamis caused 2080 reported deaths and 2144 injuries. Average water run-up on the shore was 5 m (16 ft), with a maximum of 19.8 m (65 ft) at the northeast corner of the island.

After ten o'clock at night on July 12, 1993, a magnitude 7.8 earthquake occurred in the Sea of Japan at the

(a)

(b)

(c)

 web link **Figure 9.23** Sequential photos of the major wave of a tsunami at Laie Point, Oahu, Hawaii, from an earthquake in the Aleutian Islands, March 9, 1957. The highest wave in Hawaii was 3.6 m (11.8 ft) above sea level.

Tsunami Warning Systems

Seismic stations throughout the world locate earthquakes by using the arrival times of P- and S-waves to establish the earthquake epicenters (see chapter 2). If an epicenter is oceanic and the earthquake is strong, shallow, and of long duration, and if it also progresses slowly along a fault, then seafloor displacement may occur. It is the vertical displacement of the sea floor that produces the sea surface disturbance known as a tsunami. The hourly positions of sea waves that radiate out from such an undersea epicenter are then plotted using a computer model. Tsunami Prediction and Warning Centers were located in Hawaii and Alaska in 1946 after a tsunami struck Hawaii. This warning network determines arrival times of possible tsunamis along the model's wave paths and sends a tsunami warning to populated areas, where water-level gauges are watched to see if any significant rise or fall in sea level occurs.

Computer models that predict tsunamis from seismic data are not yet fully accurate. A model may predict a tsunami that does not exist, or it may fail to predict a true tsunami. Primarily on the basis of seismic data, twenty tsunami warnings have been issued since 1946. Fifteen of these warnings were false alarms. An accurate warning system must be able to sense the actual presence of a tsunami close to its source.

At the present time, the Pacific-wide system can issue a warning within about one hour of the seismic event. Because of this time lag, such warnings are useful only to populations located more than 750 km (465 mi) from the source. Regional systems (two in the United States: Alaska and Hawaii; one each in Japan, Russia, and French Polynesia) warn in about ten minutes and are useful 100–750 km (62–465 mi) from the source. Local systems (Japan and Chile) that warn in about five minutes are useful less than 100 km (62 mi) from the source. All three systems use location and earthquake data to trigger warnings and then rely on coastal tide stations for verification of the waves. A tsunami does not produce uniform water-level responses at coastal tide stations, for no two harbors or estuaries are exactly the same. Therefore, the data transmitted from one tide station do not forecast what will happen only a short distance away. Until accurate deep-ocean tsunami measurements are available, it will not be possible to forecast site-specific information in hazardous areas.

In 1996, fifty years after the establishment of the Tsunami Prediction and Warning Centers, the states of Alaska, Washington, Oregon, California, and Hawaii joined with three federal agencies—the National Oceanic and Atmospheric Administration (NOAA), the U.S. Geological Service (USGS), and the Federal Emergency Management Administration (FEMA)—to fund twelve new seismic stations and two ocean sensing stations in the North Pacific. This is the National Tsunami Hazard and Mitigation Program (box fig. 1). At present two

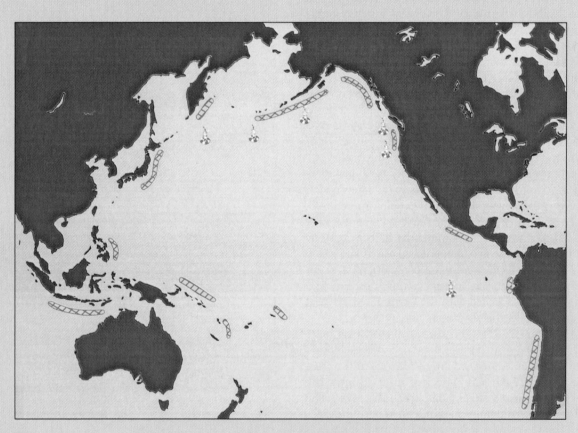

web link **Box Figure 1** NOAA Deep Ocean Tsunami Measurement Program. Tsunami reporting stations are shown as small buoys. Three ocean stations south of the Aleutian Islands and two west of the Oregon coast are presently in place; ten more stations are planned for the future. Undersea earthquake zones are associated with the subduction zones (shown in *yellow* and *red*) surrounding the Pacific Ocean. *Courtesy of National Tsunami Hazard and Mitigation Program, PMEL, NOAA.*

ocean stations are in place, with bottom-mounted pressure sensors (long-wave detectors) that are linked to surface buoys (box fig. 2) that transmit data from tsunamis to satellites. The satellites then relay the information to scientists working in the hazard and mitigation program. Three buoys are located south of Kodiak Island; these monitor any waves generated in the Aleutian Trench or the Gulf of Alaska. Two more buoys are located west of Washington and southern Oregon; these buoys monitor the Cascadia subduction zone, which lies between the Gorda-Juan de Fuca Ridge and the west coast of the United States. More ocean sensors and seismometers will be installed in future years.

Although more information is now available, the warning problem is not completely solved; distant tsunamis detected by the present warning system provide residents of a threatened area with time to evacuate, but tsunamis that are formed locally give residents only a few minutes to seek safety. The Cascadia subduction zone is very close to the west coast of the United States and Canada, and more than half a million people live in potential tsunami flooding areas. Residents of Oregon and California would have less than five minutes' warning after a large earthquake in this zone; and those living along the more northerly coast might have only twenty to forty minutes. Residents of tsunami-prone areas need to remember that any local earthquake is the signal to move away from the immediate coastline. The mitigation portion of the tsunami warning program is working to assess the local hazards and to educate the public to the situation.

To Learn More About Tsunami Warning Systems

Bernard, E. 1998. Program Aims to Reduce Impact of Tsunamis on Pacific States. *EOS* 79 (22): 258, 262–63.

Internet References

Visit the book's Online Learning Center at www.mhhe.com/sverdrup (click on the book's cover) to explore links to further information on related topics.

Box Figure 2 A tsunami detection buoy in place off the western coast of the United States. The buoy is linked to a sensor resting on the sea floor. Data are transmitted from the buoy to a satellite and relayed from the satellite to tsunami warning centers.

web link **Figure 9.24** Aonae at the southern tip of Okushiri Island in the Sea of Japan after the tsunami and fire of 1993. The area on the *left* extending to the tip of the island was leveled despite the protecting seawall. Debris left in the streets by the surging water prevented fire control equipment from reaching the burning buildings.

boundary of the Eurasian and North American Plates and very close to Okushiri Island off the southwest coast of Hokkaido. Six minutes after the earthquake began, a tsunami swept over a 5 m (16 ft) seawall, destroyed buildings, and ruptured propane tanks that set fire to the town of Aonae at the southern end of Okushiri Island. In this case the longest wave run-up was 30 m (100 ft), with more typical values at 10–15 m (33–50 ft). One hundred eighty-five people died, and the property loss was estimated at $600 million (fig. 9.24).

A massive tsunami struck the northwest coast of Papua, New Guinea on July 17, 1998. The 7–15 m (22–50 ft) high tsunami swept across the barrier beach parallel to Sissano Lagoon, destroying four fishing villages (fig. 9.25). Three waves swept debris and bodies into the lagoon and the mangrove trees beyond. More than 2000 persons were killed or reported missing. The tsunami was triggered by a close-to-shore underwater earthquake (magnitude 7 on the Richter scale) that caused a 2 m (6 ft) vertical drop along a 40 km (24 mi) fault in the sea floor followed by an underwater landslide. Although the island experienced tremors thirty minutes before the arrival of the first wave, there was no recognition of their significance in the shore area. See box titled "Tsunami Warning Systems."

Internal Waves 9.10

The waves discussed to this point have all formed at the interface of the atmosphere and the ocean. This interface marks the common boundary between two fluids of different densities, air and water. Another interface between two flu-

Figure 9.25 The tsunami that struck the northwest coast of Papua, New Guinea, July 17, 1998, destroyed a chain of villages along 30 km (18 mi) of beach bordering Sissano Lagoon. Buildings, trees, and people were swept away by three waves 7–15 m (22–50 ft) high. The eroded beach and the shattered palms were all that was left of many communities.

ids lies below the ocean surface at the pycnocline that separates the shallow mixed layer from the denser underlying water. In this case the boundary is less abrupt and the density difference is not as great as it is at the air-water boundary. The waves that form along this boundary are known as **internal waves.**

Internal waves are slower than surface waves; their periods are longer than most wind waves, and their height is

water depth, the density (rho = ρ) of both layers must be included in the equation. Wave speed squared (C^2) is equal to the Earth's acceleration due to gravity (g) divided by 2π times the wavelength (L) times a ratio of densities, where ρ (rho) is the density of the lower layer and ρ' is the density of the upper layer:

$$C^2 = \frac{g}{2\pi} L \left[\frac{r - r'}{r + r'} \right]$$

When the internal waves are long relative to the water depth, it is necessary to include other relationships. For this equation see appendix C.

Standing Waves   9.11

Deep-water waves, shallow-water waves, and internal waves are all progressive waves; they have a speed and move in a direction. **Standing waves** do not progress; they are progressive waves reflected back on themselves and appear as an alternation between a trough and a crest at a fixed position. They occur in ocean basins, partly enclosed bays and seas, and estuaries. A standing wave can be demonstrated by slowly lifting one end of a container partially filled with water and then rapidly but gently returning it to a level position. If this is done, the surface alternately rises at one end and falls at the other end of the container. The surface oscillates about a point at the center of the container, the **node**; the alternations of low and high water at each end are the **antinodes** (fig. 9.27a). A standing wave is a progressive wave reflected back on itself; the reflection cancels out the forward motions of the initial and reflected waves. If different-sized containers are treated the same way, the period of oscillation increases as the length of the container increases or its depth decreases.

Notice that the single-node standing wave contains one-half of a wave form (fig. 9.27a). The crest is at one end of the container and the trough is at the other end. As the wave oscillates, a trough replaces the crest, and a crest replaces the trough. One wavelength, the distance from crest to crest or from trough to trough, is twice the length of the container. By rapidly tilting the basin back and forth at the correct rate, one can produce a wave with more than one node (fig. 9.27b). In the case of two nodes, there is a crest at either end of the container and a trough in the center; this configuration alternates with a trough at each end and the crest in the center. The two nodes are one-quarter of the basin's length from each end. In this case, note that the wavelength is equal to the basin's length. The oscillation period of the wave with two nodes is one-half that of the wave with a single node.

Standing waves in bays or inlets with an open end behave somewhat differently than standing waves in closed basins. A node is usually located at the entrance to the open-ended bay, so only one-quarter of the wavelength is inside the bay. There is little or no rise and fall of the water surface at the entrance, but a large rise and fall occurs at the

Figure 9.26 Internal waves forming as seawater moves through the Strait of Gibraltar into the Mediterranean Sea. This satellite image shows wave diffraction as the waves move away from the strait.

limited by the thickness of the surface layer. When wave heights are large, internal waves may show at the sea surface as banding across the sea surface. The water over the crest of the internal wave shows ripples, but the water over the trough of the internal wave is smooth, and the surface bands move along the sea surface as the internal waves pass below. Sometimes, instead of bands, elevations and depressions of the sea surface occur as the internal waves pass (fig. 9.26).

Many processes are responsible for internal waves. A low-pressure storm system may elevate the sea surface and depress the pycnocline. When the storm moves away, the displaced pycnocline will oscillate as it returns to its equilibrium level. If the speed of a surface current changes abruptly at the pycnocline, internal waves may be generated. Currents moving over rough bottom topography may also produce internal waves. If a thin layer of low-density surface water allows a ship's propeller to reach the pycnocline, the energy from the propeller creates internal waves; under this condition the ship's propeller becomes inefficient because internal waves created at the pycnocline carry energy away from the vessel instead of driving the vessel forward. This results in a loss of speed that mariners call the "dead water effect."

The relationship of wave length and depth to wave speed is similar for both internal and surface waves (see section 9.5). When the internal waves are short relative to the

closed end of the bay (fig. 9.28). Multiple nodes may also be present in open-ended basins.

Standing waves that occur in natural basins are called **seiches,** and the oscillation of the surface is called seiching. In natural basins, the length dimension usually greatly exceeds the depth. Therefore, a standing wave of one node in such a basin behaves as a reflecting shallow-water wave, with the wavelength determined by the length or width of the basin. In water with distinct layers having sharp density boundaries, standing waves may occur along the fluid boundaries as well as at the air-sea boundary. The oscillation of the internal standing waves is slower than the oscillation of the sea surface.

Standing waves may be triggered by tectonic movements that suddenly shake a basin, causing the water to oscillate at a period defined by the dimensions of the basin. This phenomenon occurs during an earthquake when water sloshes back and forth in swimming pools. If storm winds create a change in surface level to produce storm surges, the surface may oscillate as a standing wave in the act of returning to its normal level when the wind ceases. The movement of an air-pressure disturbance over a lake may also cause periodic water-level changes, reaching a meter or more in height. Tidal currents moving through an area with a sharp pycnocline and an irregular bottom topography may create internal waves that sometimes produce seiches.

If the period of the disturbing force is a multiple of the natural period of oscillation of the basin (an ocean basin or a smaller coastal basin), the height of the standing wave is greatly increased. For example, if a child is riding on a swing, a gentle push timed with each swing period forces the swing higher and higher. The push may be delivered each time the swing passes, every other time, or every third time; all are multiples of the natural period of the swing. In chapter 10,

Figure 9.27a A standing wave oscillating about a single node in a basin. The time for one oscillation is the period of the wave, T.

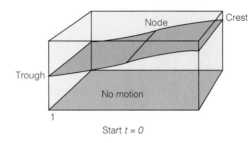
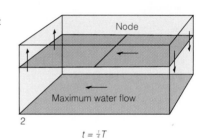
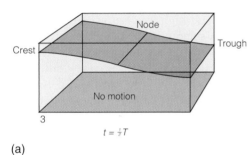
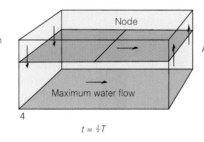
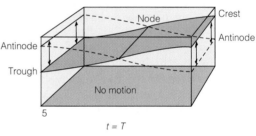

(a)

Figure 9.27b A standing wave oscillating about two nodes. The time for one oscillation is the period of the wave.

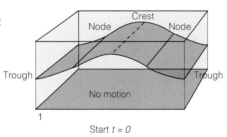
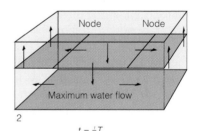
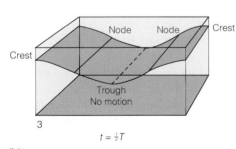
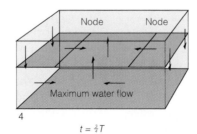
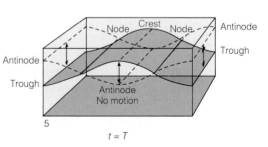

(b)

we will learn that repeating tidal forces at the entrance to a bay can produce standing waves in those basins that have natural periods of oscillation approximating the tidal period.

A standing wave in a basin is like a water pendulum. The wave's natural period of oscillation is

$$T = \left(\frac{1}{n}\right)\left(\frac{L}{\sqrt{gD}}\right)$$

where n is the number of nodes present, D is the depth of water in the basin, g is the Earth's acceleration due to gravity, and L is the length of the wave. L equals twice the basin length, l, in a closed basin and four times the basin length in a basin with an open end. This equation is related to the shallow-water wave equation when the number of nodes is equal to 1:

$$\frac{L}{T} = n\sqrt{gD}$$

A progressive wave directly reflected back on itself produces a standing wave, because the two waves—original and reflected—are moving at the same speed but in opposite directions. The checkerboard interference pattern produced by two matched wave systems approaching each other at an angle also creates standing waves, with crests and troughs alternating with each other in fixed positions (see fig. 9.7).

Figure 9.29 shows the relationship between the distribution of total oceanic wave energy and wave period. The energy of ordinary wind waves is high because these waves are always present and well distributed through all the oceans. Storm waves are larger and carry more energy, but they do not occur as frequently and are present over much less of the ocean area. Therefore storm waves have less total energy than ordinary wind waves. Tsunami-type waves contain a large amount of energy, but they are infrequent and confined to fewer areas of the oceans. The tides, when considered as waves, concentrate their energy in two narrow bands centered on the twice-daily and once-daily tidal periods. Tide wave forms are discussed in chapter 10.

Practical Considerations: Energy from Waves 9.12

A tremendous amount of energy exists in ocean waves. The power of all waves is estimated at 2.7×10^{12} watts, which is about equal to 3000 times the power-generating capacity of Hoover Dam. Unfortunately for human needs, this energy is widely dispersed and is not constant at any given location or time. It is, therefore, difficult to tap this supply to produce power, except in small quantities.

Wave energy can be harnessed in three basic ways: (1) using the changing level of the water to lift an object, which can then do useful work because of its potential energy; (2) using the orbital motion of the water particles or the changing tilt of the sea surface to rock an object to and fro; and (3) using rising water to compress air or water in a chamber. A combination of these may also be used. If the wave motion is used directly or indirectly to turn a generator, electrical energy may be produced.

Consider a large buoyant surface float with a hollow cylinder extending down into the sea (fig. 9.30). Inside the cylinder is a piston, and the up-and-down motion of the surface float causes the cylinder to move up and down over the piston, while the large drag plate restricts the motion of the piston. The system takes in water as the surface buoy rises on the crest of the waves and squirts water out as the surface

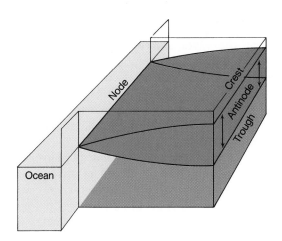

Figure 9.28 A standing wave oscillates about the node located at the opening to a basin. The antinodes produce the rise and fall of water at the closed end of the basin. This type of oscillation is produced by alternating water inflow and outflow at a period equal to the natural period of the basin.

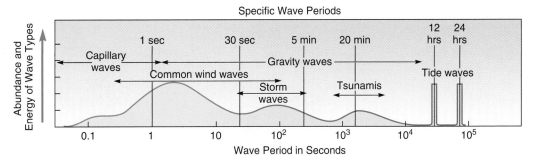

Figure 9.29 The distribution of wave energy with wave period.

buoy drops with the passing of the trough. The pumped water can be used to turn a turbine, but because wave energy is distributed over a volume of water, this mechanism does not withdraw much of the passing wave's energy. This system could be adapted to pump air rather than water.

Another system constructs a tapered channel perpendicular to the shore. Incoming waves force the water up 2–3 m (6–10 ft) in the narrow end of the channel, where it spills into

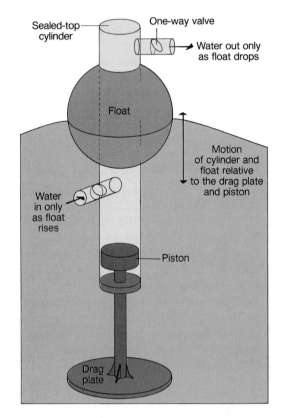

Figure 9.30 The vertical rise and fall of the waves can be used to power a pump.

 Figure 9.31 Each rise and fall of the waves pumps pulses of compressed air into a storage tank. A smooth flow of compressed air from the storage tank turns a turbine that generates electricity.

an elevated storage tank or down through a turbine. This system is used to generate power by a 75-kilowatt plant on Scotland's isle of Islay, a 350-kilowatt plant at Toftestalen, Norway, and two 1500-kilowatt plants, one in Java and the other in Australia. Both the surface float and the tapered channel are examples of changing the level of an object or the water itself to create potential energy.

Wave power systems that use the orbital or rocking motion of the waves are under development in Great Britain. Long strings of mechanical power units are moored in water where waves are abundant. Each passing wave causes the power units to move relative to each other, causing pumps to move oil that passes through a turbine in a closed system.

In Western Australia, the Azores, and Japan other systems using wave energy to compress air are being developed (fig. 9.31). Air traps can be installed along a wave-exposed coast so that the crest of a wave moving into the trap compresses air forcing it through a one-way valve, the air traps can also be constructed to pass air in either direction. This compressed air powers a turbine. The trough of the wave allows more air to enter the trap, readying it for compression by the following wave crest.

Japan launched a prototype unit, "The Mighty Whale," in 1998 (fig. 9.32). The unit is 50 m (164 ft) long, 30 m (98 ft) wide, and 12 m (39 ft) in depth. The Mighty Whale faces the waves, allowing the water level to rise and fall in three internal air chambers. The oscillating water forces air past turbines that function in two directions. The power capacity of each unit is set at 110 kilowatts. A series of these units linked side by side is expected to act as a breakwater and furnish energy, aeration, and purification of seawater at fish farms.

Shores that are continually pounded by large-amplitude waves are most likely to be developed for wave power. Great Britain has a coastline with frequent high-energy waves and an average wave power of about 5.5×10^4 watts (or 55 kilowatts) per meter of coastline. If the wave energy could be

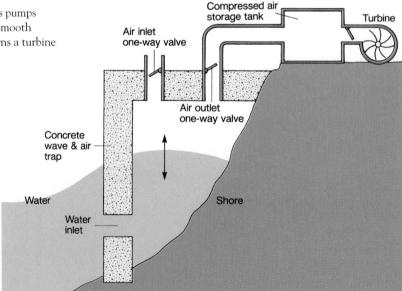

Figure 9.32 Japan's "Mighty Whale" was launched in March 1998 and has been undergoing trials as a wave absorber and an energy source.

completely harnessed along 1000 km (620 mi) of coast, it would generate enough power to supply 50% of Great Britain's present power needs. Along the northern California coast, waves are estimated to expend 23×10^6 kilowatts of power annually; it is thought that 4.6×10^6 kilowatts, or 20%, could be harvested to generate electrical power. The Pacific Gas & Electric Company, a northern California utility, is considering installing a generating device in a breakwater planned for Fort Bragg, California.

When we think about wave energy systems, thoughtful consideration needs to be given to items other than cost. If all the wave energy were extracted from the waves in a coastal area, what effect would this action have on the shore area? If the near-shore areas are covered with wave energy absorbers 5–10 m (15–33 ft) apart, what will the effect be on other ocean uses? Since the individual units collect energy at a slow rate, can they collect enough energy over their projected life span to exceed the energy used to fabricate and maintain them? Answers to these questions will help us understand that the harvesting of wave energy is not without an effect on the environment, that it may not be either cost- or energy-effective, and that its location may present enormous problems for installation, maintenance, and transport of energy to sites of energy use.

Summary

When the water's surface is disturbed, a wave is formed by the interaction between generating and restoring forces. The wind produces capillary waves, which grow to form gravity waves. The elevated portion of a wave is the crest; the depressed portion is the trough. The wavelength is the distance between two successive crests or troughs. The wave height is the distance between the crest and the trough. Wave period measures the time required for two successive crests or troughs to pass a location. The moving wave form causes water particles to move in orbits. The wave's speed is related to wavelength and period.

Deep-water waves occur in water deeper than one-half the wavelength. Wind waves generated in storm centers are deep-water waves. The period of a wave is a function of its generating force and does not change. Long-period waves move out from the storm center, forming long, regular waves, or swell. The faster waves move through the slower waves and form groups, or trains, of waves. The longer waves are followed by the shorter waves. This process is known as sorting, or dispersion. The speed of a group of waves is half the speed of the individual waves in deep water. Swells from different storms cross, cancel, and combine with each other as they move out across the ocean.

Wave height depends on wind speed, wind duration, and fetch. Single large waves unrelated to local conditions are called episodic waves. The energy of a wave is related to its height. When the ratio of the height to the length of a wave, or its steepness, exceeds 1:7, the wave breaks. The Universal Sea State Code relates wind speeds and sea surface conditions.

Shallow-water waves occur when the depth is less than one-twentieth the wavelength. The speed of a shallow-water wave depends on the depth of the water. As the wave moves toward shore and decreasing depth, it slows, shortens, and increases in height. Waves coming into shore are refracted, reflected, and diffracted. The patterns produced by these processes helped people in ancient times to navigate from island to island.

In the surf zone, breaking waves produce a water movement toward the shore. Breaking waves are classified as plungers or spillers. Water moves along the beach as well as toward it; it is returned seaward through the surf zone by rip currents.

Tsunamis are seismic sea waves. They behave as shallow-water waves, producing severe coastal destruction and flooding.

Internal waves occur between water layers of different densities. Standing waves, or seiches, occur in basins as the sea surface oscillates about a node. Alternate troughs and crests occur at the antinodes.

It is possible to harness the energy of the waves by using either the water-level changes or the changing surface angle associated with them. Difficulties include cost, location, environmental effects, and lack of wave regularity.

Key Terms

All key terms from this chapter can be viewed by term, or by definition, when studied as flashcards on this book's Online Learning Center at www.mhhe.com/sverdrup (click on this book's cover).

generating force, 248	swell, 251
restoring force, 248	group speed, 252
gravity wave, 248	fetch, 252
ripple, 248	episodic wave, 254
capillary wave, 248	potential energy, 255
cat's-paws, 248	kinetic energy, 255
crest, 249	wave steepness, 255
trough, 249	shallow-water wave, 256
wavelength, 249	wave ray, 257
wave height, 249	diffraction, 259
amplitude, 249	breaker, 260
equilibrium surface, 249	plunger, 260
wave period, 249	spiller, 260
water particle orbit, 250	rip current, 261
deep-water wave, 250	seismic sea wave, 262
progressive wind wave, 251	tsunami, 262
storm center, 251	internal wave, 266
forced wave, 251	standing wave, 267
free wave, 251	node, 267
dispersion/sorting, 251	antinode, 267
wave train, 251	seiche, 268

Study Questions

1. A surfboard slides downward on the face of a wave. The steepness of the wave face is governed by the decrease of L (wavelength) and the increase of H (wave height) as the wave slows in shallow water. How must the surfer adjust the board in order to stay on the face of the wave as the wave approaches shallow water?
2. Locate a small pond or pool and drop a stone into it. Describe what happens (1) to an individual wave and (2) to the group of waves. Try to determine the group speed and the individual wave speed.
3. Drop two stones into the pond at a short distance from each other. Describe what happens when the wave rings produced by the two stones pass through each other. Do the heights of the waves change when they intersect? Do the wave trains pass through each other and continue on?
4. List the forces that act on a smooth-water surface to create deep-water wind waves.
5. If you were ocean-sailing at night in the trade-wind belt, how could you use the waves to keep you on a course of constant direction?
6. Make a sketch of an ideal progressive wind wave in deep water. Label the parts.
7. What happens to a deep-water progressive wave when it moves into shallow water and up a sloping beach?
8. Compare a tsunami and a storm surge (see chapter 6). How are they the same? How are they different?
9. Distinguish between (a) sea and swell, (b) wave height and wave steepness, (c) wave height and wave amplitude, (d) plunger and spiller, and (e) node and antinode.
10. What is the effect of sorting (dispersion) on waves moving away from a storm center?
11. How do refraction, reflection, and diffraction affect a wave?
12. How is a standing wave related to a progressive wave?
13. Explain two ways in which wave energy could be harnessed to provide useful power. What are the advantages and disadvantages of each method?
14. If a group of mixed waves is generated in a sudden storm, why does it take more time for the group to pass an island far from the storm center than to pass an island near it?
15. A depression in the sea floor at right angles to a straight coastline may be the site of a rip current. Why?

Study Problems

1. Using the equations $C = L/T$ and $L = (g/2\pi)T^2$, show that wave speed can be determined from (a) wave period only and (b) wavelength only.
2. What is the period of a wave moving in deep water at 10 m/s, if its wavelength is 64 m? When it enters shallow water what will happen to the wave's speed, length, period, and height? How high will the wave have to be to break in deep water?
3. A submarine earthquake produces a tsunami in the Gulf of Alaska. How long will it take the tsunami to reach Hawaii, if the average depth of the ocean over which the waves travel is 3.8 km and the distance is 4600 km?
4. Explain wave dispersion. How far from a storm center will waves with periods of twelve seconds, nine seconds, and six seconds have traveled after twelve hours? If the six-second waves arrive at your beach ten hours after the twelve-second waves, how far away is the storm?
5. Fill a rectangular aquarium or dishpan approximately one-third full of water. Measure the water depth (D). Carefully lift one end of the container and set it down rapidly and smoothly. Time the period between successive high waters at one end (T); this is the wave period. The wavelength (L) is twice the length of the container. Show that C (the wave speed) determined from L/T is equal to C determined from $n\sqrt{gD}$.

Links to Related Websites

Visit the book's Online Learning Center at www.mhhe.com/sverdrup (click on the book's cover) to find live Internet links for additional topics related to this chapter's content.

- Waves
- Tsunami
- Energy

Visit the book's Online Learning Center at www.mhhe.com/sverdrup (click on this book's cover) to find these additional chapter tools: Suggested Readings; links to further information on boxed readings, selected figures, and related chapter topics; and additional study aids.

Oceanographic research vessels are operated by the U.S. Navy, NOAA, universities, private corporations, and foundations. They are the platforms staffed by professional crews that provide accommodations, laboratory space, and the equipment required to handle deep-sea sampling gear. Each vessel is scheduled for long periods at sea, so scientists and equipment are flown to and from the nearest port as the scientific mission of the vessel changes. Research vessels are often shared by scientific programs from different institutions to create the most cost-effective use. Though ships operate approximately 250 days at sea each year, a specific research cruise may last only a month.

The University of Washington's research vessel, Thomas G. Thompson, *began its research duties in 1992.*

The stern of a research vessel getting ready for sea is a bewildering sight. All these scientific supplies and equipment must be properly stowed before the ship can leave the dock.

Discussing the day's work in a quiet corner. Analysis of data at sea enables researchers to modify sampling routines and change procedures while their research is in process.

The deck of a research vessel is the working platform from which all over-the-side sampling gear is launched and retrieved. Winches, cranes, and A-frames are used to move heavy gear between the deck and the ocean surface. Controlling this equipment and successfully launching and retrieving expensive scientific instruments while the deck rolls and heaves require close cooperation between the ship's crew, who operate the vessel and the winches, and the scientists, who prepare and use the sampling gear.

This inflatable power-craft is carried on board and is used in deployment and retrieval of samplers and other equipment.

Water sampling bottles surround conductivity-temperature-depth (CTD) sensors that are being lowered into the sea.

Scientists from the Woods Hole Oceanographic Institution retrieve an Autonomous Benthic Explorer after a survey run on the Juan de Fuca Ridge in the North Pacific. Oceanographers from other institutions often share vessel facilities and ship time.

Getting a plankton net ready for its tow. This type of net samples plankton from the sea's surface waters.

A scientist takes a first look at a seafloor sediment sample obtained with a Vanveen grab.

Sometimes gear is retrieved in a less-than-desirable condition. Patience, a sharp knife, and cable cutters may be required before order is restored. The important thing is to save the instrument and its data.

A tangle of gear comes on board.

Do you have the end or do I?

The ship's laboratory spaces are divided into wet labs for handling bulk water samples, geological labs, analytical labs for chemical and biological work, and dry labs with electronic and computer facilities. Plywood, two-by-fours, and steel-angle iron are nailed and bolted together, then attached to deck and overheads to make the required arrangements of tables and benches. Each research group is able to set up its labs to meet its own special requirements. Instruments must be bolted down; storage boxes are tied in place in the lab and on deck. Nothing can be left free to break loose when the ship rolls (except the researchers).

Consulting a seafloor chart in the scientific command center. Here, charts are drawn by computer from data as they are received.

Much time is spent in the ship's laboratories analyzing water samples.

It costs over $25,000 per sea-day to operate a major research vessel, including maintenance, crew costs, fuel, and food but excluding scientific costs. Because the costs are high and maximum efficiency is required, the vessel works twenty-four hours each day, taking samples whenever the vessel reaches the required location. Standard shifts or watches are four hours on duty and eight hours off duty over the twenty-four-hour day, and sometimes a schedule of six hours on duty and six hours off duty is used. Scientists may be on the same schedule; they may work twelve hours on duty and twelve hours off duty when research schedules require continuous work, or they may get only short naps during a busy period.

Day and night, the work goes on. Here at sunset a sediment sampling device returns from the sea.

The work continues twenty-four hours a day and this oceanography student catches up on his sleep.

Days at sea are repetitive, and the routine leaves little to distinguish one day from another. Holidays and special occasions are enjoyed and celebrated with gusto.

Scientists and students cool off in a research tank.

Everyone on board enjoys dragging a line when the ship is under way. Sometimes they are lucky and fresh fish is on the menu.

A first-timer across the equator kneels before King Neptune.

Geological samples clutter the stern at the end of a cruise. Friends, relatives, and the media are all there to welcome the ship home.

Links to Related Websites

Visit the book's Online Learning Center at www.mhhe.com/sverdrup (click on the book's cover) to find live Internet links for additional topics related to this section's content.

- NOAA ships
- Oceanographic research vessels

chapter 10

The Tides

S pring tides are not the tides of spring as many landsmen suppose. They are the very high and very low tides which occur twice a month, with the new and the full Moon, when solar and lunar magnetism pull together to make the circumterrestrial tide wave higher than at other times. The opposite to the spring tide is the neap tide, halfway between these phases of the Moon; down East, in Maine, there may be as much as five feet difference in range between springs and neaps.

Spring tides are beloved by all who live by or from the sea. At a spring low, rocky ledges and sandbars which you never see ordinarily are bared; the kinds of seaweed that require air but twice a month appear; sand dollars like tarnished pieces of eight are visible on the bottom. Clam specialists can pick up the big "hen" clams or the quahaugs, and with a stiff wire hook deftly flip out of his long burrow the elusive razor clam. Shore birds—sandpiper, plover and curlew—skitter over the sea-vacated flats, piping softly and gorging themselves on the minor forms of life that cling to this seldom-bared shelf.

Samuel Eliot Morison
From *Spring Tides*

Old wharf pilings at low tide.

The Tides

B est known as the rise and fall of the sea around the edge of the land, the tides are caused by the gravitational attraction between the Earth and the Sun and between the Earth and the Moon. Far out at sea tidal changes go unnoticed, but along the shores and beaches the tides govern many of our water-related activities, both commercial and recreational. Early sailors from the Mediterranean Sea, where the daily tidal range is less than 1 m (3 ft), ventured out into the Atlantic and sailed northward to the British Isles; to their amazement, they found a tidal range in excess of 10 m (30 ft). The movement of the tide into and out of bays and harbors has been helpful to sailors beaching their boats and to food-gatherers searching the shore for edible plants and animals, but it is also recognized as a hazard by navigators and can produce some spectacular effects when rushing through narrow channels.

In this chapter we survey tide patterns around the world and explore the tides in two ways: one is a theoretic consideration of the tides on an Earth with no land; the other is a study of the natural situation. We also show how to use available tide data to predict water-level changes and coastal tidal currents.

Tide Patterns 10.1

Measurements of tidal movements around the world show us that the tides behave differently in different places. In some coastal areas a regular pattern occurs of one high tide and one low tide each day; this is a **diurnal tide.** In other areas a cyclic high water–low water sequence is repeated twice in one day; this is a **semidiurnal tide.** In a semidiurnal tidal pattern, the two high tides reach about the same height and the two low tides drop to about the same level. A tide in which the high tides regularly reach different heights and the low tides drop regularly to different levels is called a **semidiurnal mixed tide.** This type of tide has what is called diurnal (or daily) inequality, created by combining diurnal and semidiurnal tide patterns. The tide curves in figure 10.1 show each type of tide. Curves for typical tides at some U.S. coastal cities are shown in figure 10.2.

Tide Levels 10.2

In a uniform diurnal or semidiurnal tidal system, the greatest height to which the tide rises on any day is known as **high water,** and the lowest point to which it drops is called **low water.** In a mixed-tide system, it is necessary to refer to **higher high water** and **lower high water,** as well as **higher low water** and **lower low water** (see fig. 10.1).

Tide measurements taken over many years are used to calculate the **average** (or **mean**) **tide** levels. Averaging all water levels over many years gives the local mean tide level. Averages are also calculated for the high-water and low-water levels, as mean high water and

mean low water. For mixed tides, mean higher high water, mean lower high water, mean higher low water, and mean lower low water are calculated.

Because the depth of coastal water is important to safe navigation, an average low-water reference level is established; depths are measured from this level for navigational charts. The tide level is added to this charted depth to find the true depth of water under a vessel at any particular time. In areas of uniform diurnal or semidiurnal tide patterns the zero depth reference, or **tidal datum,** is usually equal to mean low water. The use of mean low water assures the sailor that the actual depth of the water is, in general, greater than that on the chart. In regions with mixed tides, mean lower low water is used as the tidal datum, for the same reason. When the low-tide level falls below the mean value used as the tidal datum, a **minus tide** results. A minus tide can be a hazard to boaters, but it is cherished by clam diggers and students of marine biology, because it exposes shoreline usually covered by the sea.

As the water level along the shore increases in height, the tide is said to be rising or flooding; a rising tide is a **flood tide.** When the water level drops, the tide is falling or ebbing; a falling tide is an **ebb tide.**

Tidal Currents 10.3

Currents are associated with the rising and falling of the tide in coastal waters. These **tidal currents** may be extremely swift and dangerous as they move the water into a region on the flood tide and out of the region on the ebb tide. When

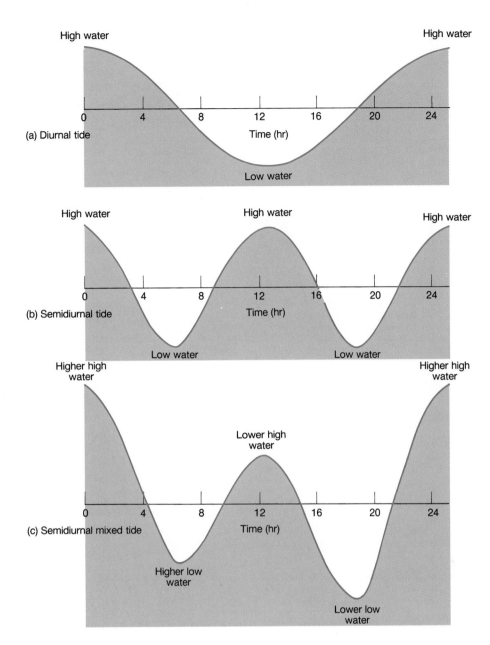

Figure 10.1 The three basic types of tides: (a) a once-daily, diurnal, tide; (b) a twice-daily, semidiurnal, tide; and (c) a mixed semidiurnal tide with diurnal inequality.

the tide turns, or changes from an ebb to a flood or vice versa, there is a period of **slack water,** during which the tidal currents slow and then reverse. Slack water may be the only time that a vessel can safely navigate a narrow channel with swiftly moving tidal currents, sometimes in excess of 5 m/s (10 knots). The relationships of tidal currents to standing wave tides, to progressive tides, and to tidal current prediction are discussed in later sections of this chapter.

Equilibrium Tidal Theory   10.4

Oceanographers analyze tides in two ways. The tides are studied as mathematically ideal wave forms behaving uni-

formly in response to the laws of physics. This method of study is called **equilibrium tidal theory.** It is based on an Earth covered with a uniform layer of water, in order to simplify the study of relationships between the oceans and the tide-rising bodies, the Moon and the Sun. The tides are also studied as they occur naturally; this method is called **dynamic tidal analysis.** It studies the oceans' tides as they occur, modified by the landmasses, the geometry of the ocean basins, and the Earth's rotation.

The effects of the Sun's and the Moon's gravity and of the rotation of the Earth on tides are most easily explained by studying equilibrium tides. In this discussion the Earth and the Moon act as a single unit, the Earth-Moon system that orbits the Sun (fig. 10.3). The Moon

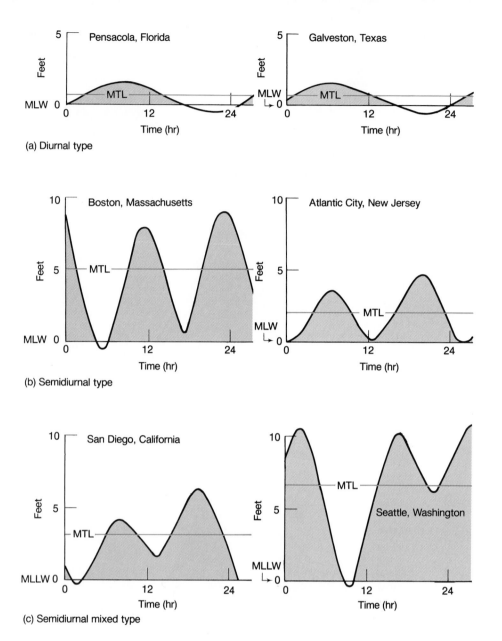

Figure 10.2 Tide types and tidal ranges vary from one coastal area to another. The zero tide level equals mean low water (*MLW*) or mean lower low water (*MLLW*), as appropriate. *MTL* equals mean tide level. All tide curves are for the same date.

orbits the Earth, held by the Earth's gravitational force acting on the Moon (B in fig. 10.3*a*). There is also a force acting to pull the Moon away from the Earth and send it out into space (B′ in fig. 10.3*b*). In the first part of this discussion B′ is considered a **centrifugal force.** A centrifugal force is an apparent force present when one judges motion against a rotating frame of reference. Forces B and B′ must be equal and opposite to keep the Moon in its orbit. Likewise the Moon's gravitational force acting on the Earth (C in fig. 10.3*a*) must be balanced by the apparent, or centrifugal, force (C′ in fig. 10.3*b*). B′ and C′ are present because the Earth-Moon system is rotating about a common axis at the system's center of mass, which is located 4640 km (2880 mi) from the Earth's center along a

line between the Earth and the Moon. The Earth-Moon system is held in orbit about the Sun by the Sun's gravitational attraction (A in fig. 10.3*c*), while a centrifugal force again acts to pull the Earth-Moon system away from the Sun (A′ in fig. 10.3*c*). For the Earth-Moon system to remain in solar orbit, the gravitational forces must equal the centrifugal forces (fig. 10.3*c*).

Sir Isaac Newton's universal law of gravitation tells us that the force of attraction between any two bodies is proportional to the product of the two masses divided by the square of the distance between the centers of the masses:

$$F = G\left(\frac{m_1 m_2}{R^2}\right)$$

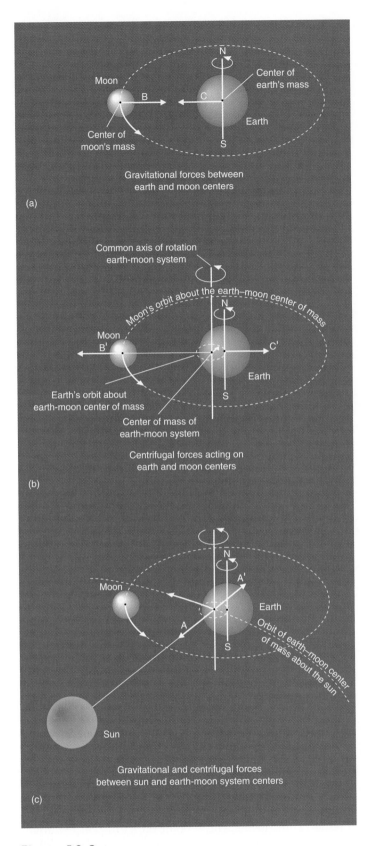

Figure 10.3 Gravitational and centrifugal forces act to keep the Earth–Moon system in balance.

where

G = universal gravitional constant,
6.67×10^8 cm^3/g/s^2
m_1 = mass of body 1 in grams
m_2 = mass of body 2 in grams
R = distance between centers of masses
in centimeters

If the gravitational forces are calculated for each unit mass of material at the Earth's surface and at its center, the gravitational forces at the Earth's surface change with location because of the changing distance between a unit mass on the Earth's surface and the center of the Sun or Moon. The centrifugal force acting on each unit mass on the Earth is constant and is equal and opposite to the gravitational force acting on a unit mass at the Earth's center.

The difference in force per unit mass between the gravitational force of the Moon or the Sun and the centrifugal force at an Earth surface point is proportional to $G(M/R^3)$, where G is the gravitational constant, M is the mass of the Moon or the Sun, and R is the distance between the centers of the Earth and the Moon or between the centers of the Earth and the Sun.

The gravitational force exerted by the Moon on a unit mass is larger at the point on the Earth's surface that is closest to the Moon than on a unit mass at the Earth's center, because the distance between the Earth's surface and the Moon is $R - r$. The quantity r is the radius of the Earth. The gravitational force difference or excess acting toward the Moon pulls a unit mass away from the Earth's center and produces a tide-raising force. At a point on the side of the Earth opposite the Moon, the gravitational force is less than the constant centrifugal force because the distance is $R + r$. The centrifugal force difference or excess that pulls a unit mass away from the Earth's center is also a tide-raising force. The calculation of these tide-raising forces is shown in appendix C.

A second way to understand tide-raising forces is to think of the Sun or the Moon as exerting a gravitational force that continuously accelerates the Earth toward the Sun or toward the Moon. In this case the gravitational force, or **centripetal force,** of the Sun or the Moon holds the Earth in orbit. The centripetal force is constant and is equal to the average gravitational force of the Sun or the Moon acting at the Earth's center. In this case the difference between the average gravitational attraction and the gravitational attraction at individual points on the Earth produces the tide-raising force. The force difference per unit mass between a surface point and the Earth's center is proportional to $G(MR^3)$ in this second case as well. The two approaches yield the same results. The derivation of $G(MR^3)$ is found in appendix C.

In the Earth-Moon-Sun system, the mass of the Sun is very great, but the Sun is very far away. By contrast, the Moon is small, but it is close to the Earth. Calculating the distribution of these unbalanced forces for each water particle at the Earth's surface shows that the Moon has a greater

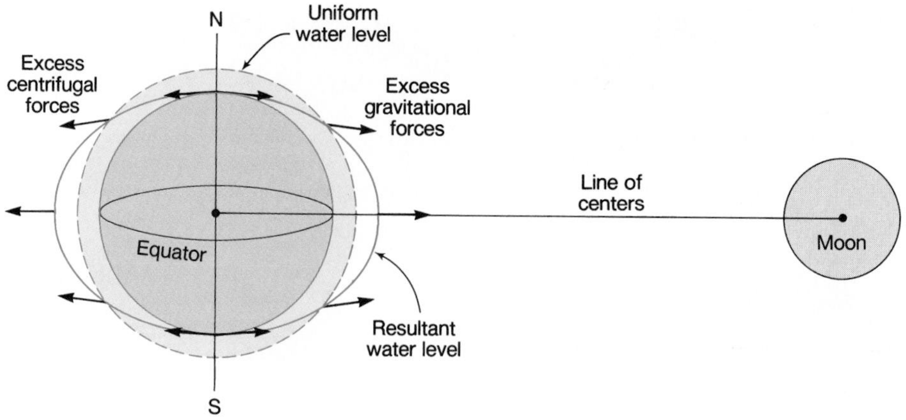

Figure 10.4 Distribution of tide-raising forces on the Earth. Excess lunar gravitational and centrifugal forces distort the Earth model's water envelope to produce bulges and depressions.

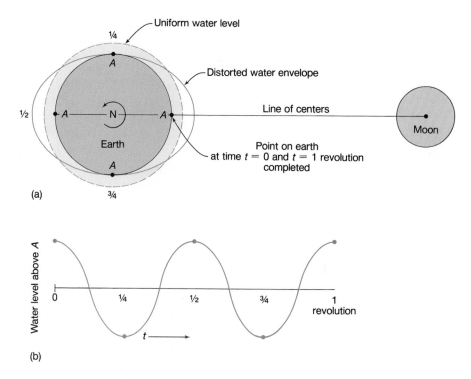

Figure 10.5 The change in water level at point A during one Earth rotation through the distorted water envelope (see fig. 10.4). Fractions indicate portions of a revolution.

attractive effect on the water particles than the Sun. In the following discussion the effects of each tide-raising body are considered separately.

The Moon Tide

The water particles on the side of the Earth facing the Moon are closest to the Moon and are acted on by the excess Moon gravitational force. Because the water covering is liquid and deformable, this force moves the water particles of the Earth's surface toward a point directly under the Moon. This movement produces a bulge in the water covering. At the same time, the centrifugal force of the Earth-Moon system acting on the water particles at the Earth's surface opposite the Moon creates an opposing bulge. This centrifugal force is equal to the magnitude of the excess gravity tide-raising force and is proportional to $-G(M/R^3)$. The minus sign indicates that this excess centrifugal force is acting opposite to the Moon's tide-producing gravity force. If we place the Moon opposite the Earth's equator and then stop the

Earth model and its Moon in space and time, we see the bulges in the water covering, as shown in figure 10.4. Remember that our model Earth initially had a water covering of uniform depth; therefore, as the two bulges are created, an area of low-water level is formed between the bulges. We now have a water covering with two bulges and two intervening depressions (or two crests and two troughs, or two high-tide levels and two low-tide levels) distributed around the equator.

The Earth makes one rotation in about twenty-four hours, and the bulges (or crests) in the water covering tend to stay under the tide-producing body as the Earth turns. As a result of this movement, a point on Earth that is initially at a crest (or high tide) passes to a trough (or low tide), to another high tide, to another low tide, and back to the original high tide as the Earth completes one revolution. You can follow this process in figure 10.5. The effect created by the motion of the Earth as it turns to the east when observed from space can also be interpreted as a wave form, for the

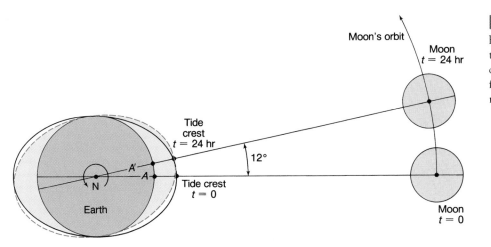

Figure 10.6 Point A requires twenty-four hours to complete one Earth rotation. During this time the Moon moves 12° east along its orbit, carrying with it the tide crest. To move from A to A′ requires an additional fifty minutes to complete a tidal day.

bulges in the water covering move westward when observed relative to a fixed position on the Earth.

The Tidal Day

While the Earth turns eastward upon its axis, the Moon is moving in the same direction along its orbit about the Earth. After twenty-four hours the Earth point that began directly under the Moon is no longer directly under the Moon. The Earth must turn for an additional fifty minutes, about 12°, to bring the starting point on the Earth back in line with the Moon. Therefore, a **tidal day** is not twenty-four hours long but twenty-four hours and fifty minutes long. This difference also explains why corresponding tides arrive at any location about one hour later each day. This relationship is shown in figure 10.6.

The Tide Wave

The tides produced in this example are semidiurnal, with two highs and two lows each day. The tidal distortion of the model's water covering produces a wave form known as the **tide wave.** The crest of the tide wave is the high-water level, and the trough of the tide wave is the low-water level. The wavelength in this example is half the circumference of the Earth, and the tide wave's period is about twelve hours and twenty-five minutes.

The Sun Tide

Although the Moon plays the greater role in the tide-producing process, the Sun produces its own tide wave. Despite the Sun's large mass, it is so far away from the Earth that its tide-raising force is only 46% that of the Moon. The time required for the Earth to revolve on its axis with respect to the Sun is on the average twenty-four hours, not twenty-four hours and fifty minutes, as in the case of the Moon tide. For this reason, the tide wave produced by the Moon is not only of greater magnitude than that produced by the Sun, but it also continually moves eastward relative to the tide wave produced by the Sun. Because the tidal forces of the Moon are greater than those of the Sun, the

tidal period of the Moon is more important; and the tidal day is considered to be twenty-four hours and fifty minutes.

Spring Tides and Neap Tides

The Moon's orbit requires 29½ days relative to a point on the Earth. During this period the Sun, the Earth, and the Moon move in and out of phase with each other. At the new Moon, the Moon and Sun are on the same side of the Earth, so the high tides, or bulges, produced independently by the Moon and the Sun coincide (fig. 10.7). Because the water level is the result of adding the two wave forms together, tides of maximum height and depression, or tides with the greatest **range** between high water and low water, are produced. These tides are known as **spring tides.** The vertical displacement, or amplitude, of the tide is one-half the range—the distance above or below mean tide level.

In a week's time, the Moon is in its first quarter; it has moved eastward along its orbit (about 12° per day) and is located approximately at right angles (or 90°) to the line of centers of the Earth and Sun. The crest, or bulge, of the Moon tide is at right angles to the tide wave created by the Sun; the crests of the Moon tide will coincide with the troughs of the Sun tide, and the same will be true of the Sun's tide crests and the Moon's tide troughs (fig. 10.7). The crests and troughs tend to cancel each other out, and the range between high water and low water is small, producing low-amplitude **neap tides.**

At the end of another week the Moon is full, and the Sun, Moon, and Earth are again lined up, producing crests that coincide and tides with the greatest range between high and low waters, or spring tides. These spring tides are followed by another period of neap tides, produced by the Moon in its last quarter when it again stands at right angles to the Sun (fig. 10.7). The tides follow a four-week cycle of changing tidal amplitude, with spring tides occurring every two weeks, and a period of neap tides occurring in between. This progression can be seen in the portions of the tide records reproduced in figure 10.8. The effect occurs each lunar month and is the result of the Moon's tide wave moving around the Earth relative to the Sun's tide wave.

Figure 10.7 Spring tides result from the alignment of the Earth, Sun, and Moon during the full Moon and the new Moon. During the Moon's first and last quarters, neap tides are produced.

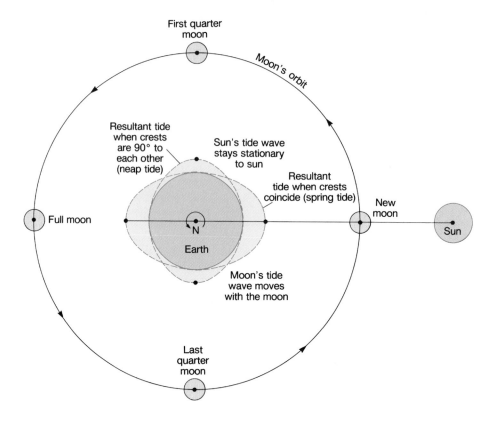

Figure 10.8 Spring and neap tides alternate during the tides' monthly cycle. *MHWS* is the mean high water spring tides; *MLWS* is mean low water spring tides. (a) A semidiurnal tide from Port Adelaide, Australia. (b) A diurnal tide from Pakhoi, China.

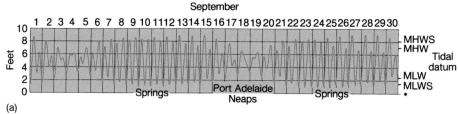

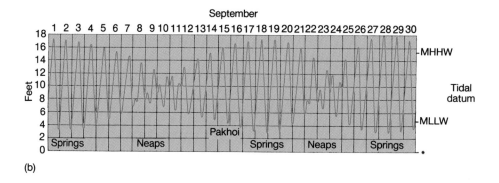

Declinational Tides

If the Moon or the Sun stands north or south of the Earth's equator, one bulge, or high water, is in the Northern Hemisphere and the other is in the Southern Hemisphere (fig. 10.9). Under these conditions, a point at the middle latitudes on the Earth's surface passes through only one crest, or high tide, and one trough, or low tide, each tidal day. A diurnal tide, often called a **declinational tide,** is formed, because the Moon or the Sun is said to have declination when it stands above or below the equator.

Declinational (or diurnal) tides are influenced by both the Moon and the Sun. The Sun stands above 23½°N at the summer solstice and above 23½°S at the winter solstice. This variation causes the bulge created by the Sun to oscillate north and south of the equator in a regular fashion each year and tends to create more diurnal Sun tides during the winter and summer than during the spring and fall. The Moon's declination varies between 28½°N and 28½°S with reference to the Earth. The Moon's orbit is inclined 5° to the Earth-Sun orbit, and it takes 18.6 years for the Moon to

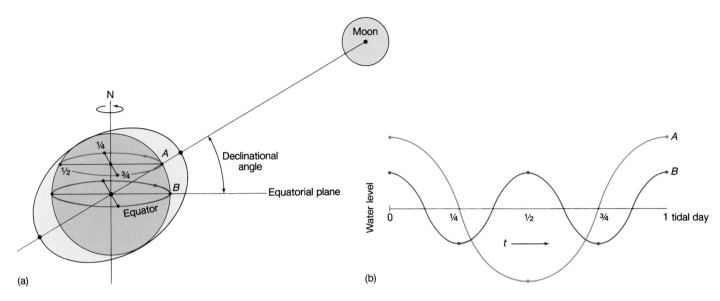

Figure 10.9 (a) The declination of the Moon produces a diurnal tide at latitude A and a semidiurnal tide at latitude B. (b) Fractions indicate portions of the tidal day.

complete its cycle of maximum declination. When the Sun's and the Moon's declinations coincide, both tide waves become more diurnal. Each lunar month the Moon travels from a declination of 5° above the Earth-Sun orbit plane to a declination of 5° below the plane and back to the high declination.

Elliptical Orbits

The Moon does not move about the Earth in a perfectly circular orbit, nor does the Earth orbit the Sun at a constant distance. These orbits are elliptical, and therefore the Earth is closer to a certain tide-raising body at some times during an orbit than at other times. During the Northern Hemisphere's winter, the Earth is closest to the Sun; therefore the Sun plays a greater role as a tide producer in winter than in summer.

Dynamic Tidal Analysis 10.5

Equilibrium tidal theory helps us understand the distribution of wave-level changes and tide-raising forces, but it does not explain the tides as observed on the real Earth. Return to figure 10.2 and notice the variety of tidal ranges and tidal periods that appear at different locations on the same date. Refer also to figure 10.8, which shows different tides at different places during the same time period. The Sun-Moon-Earth system for all locations in the two figures is the same, but the equilibrium tidal theory does not explain natural tides at any particular location. Investigating the actual tides requires the dynamic approach, a mathematical study of tide waves as they occur.

The Tide Wave

The behavior of the natural tide wave varies considerably from the tide wave of the water-covered model. Because the continents separate the oceans, the tide wave is discontinuous; the wave starts at the shore, moves across the ocean, and stops at the next shore. Only in the Southern Ocean around Antarctica do the tide waves move continuously around the Earth. A tide wave has a long wavelength compared to the depth of the oceans; therefore it behaves like a shallow-water wave, with its speed controlled by the depth of the water. Because the wave is contained within the ocean basins, it can oscillate in the basin as a standing wave, and it is also reflected from the edge of the continents, refracted by the change in water depth, and diffracted as it passes through gaps between continents. In addition, the persistence of the tidal motion and the scale on which it occurs are so great that the Coriolis effect plays a role in the water's movement. All these factors together produce the Earth's real tides, and because their interactions are complex, it is not possible to understand them together until each is first considered separately.

The tide wave's speed as a **free wave** moving across the water's surface is determined by the depth of the water. The tide wave moves as a free wave at about 200 m/s (400 mi/h) in a water depth of 4000 m (13,200 ft), but at the equator the Earth moves eastward under the tide wave at 463 m/s (1044 mi/h). This is more than twice the speed at which the tide can travel freely as a shallow-water wave. Under these conditions, the tide moves as a **forced wave** that is the result of the Moon's attractive force and the Earth's rotation. Because the Earth turns eastward faster than the tide wave moves freely westward, friction displaces the tide crest to the east of its expected position under the Moon. This eastward displacement continues until the friction force is balanced by a portion of the Moon's attractive force. These two forces, when balanced, hold the tide crest in a position to the east of the Moon rather than directly under it. This process is illustrated in figure 10.10.

Figure 10.10 The crest of the tide wave is displaced eastward until the *B* component of the gravitational force balances the friction (*F*) between the Earth and the tide wave. Component *A* of the gravitational force is the tide-raising force. Component *B* causes the tide wave to move as a forced wave.

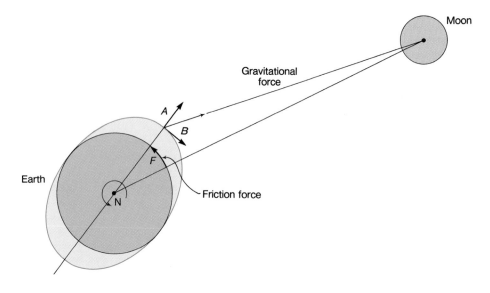

Figure 10.11 Cotidal lines for the world's oceans. The high-tide crest occurs at the same time along each cotidal line. Positions of the high tide are indicated for each hour over a twelve-hour period for semidiurnal tides.

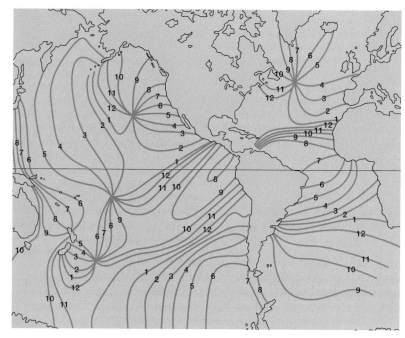

Above 60°N or 60°S, the distance around a latitude circle is less than one-half the distance around the equator. Here, the free propagation speed of the tide wave equals the speed at which the sea floor moves under the wave form. Under these conditions, the crest of the tide stays aligned with the Moon, and less friction is generated between the rotating Earth and the moving wave form. Friction between the moving tide wave and the turning Earth also acts to slow the rotation rate of the Earth, adding about 1½ milliseconds per 100 years to the length of a day.

Progressive Wave Tides

In a large ocean basin the tide wave moving across the sea surface like a shallow-water wave is a **progressive tide.** Examples of progressive tides are found in the western North Pacific, the eastern South Pacific, and the South Atlantic

Oceans. **Cotidal lines** are drawn on charts to mark the location of the tide crest at set time intervals, generally one hour apart. The cotidal lines for the world's ocean tides are shown in figure 10.11 (see also box figure 1 in box titled "Measuring Tides from Space," found later in this chapter).

Because the tide wave is a shallow-water wave, the water particles move in elliptical orbits, and their motion extends to the sea floor. The horizontal component of the motion greatly exceeds the vertical motion. Because the time in which the water particles move in one direction is so long (one-half the tide period), the Coriolis effect becomes important. In the Northern Hemisphere the water particles are deflected to the right, and in the Southern Hemisphere they are deflected to the left. This deflection causes a clockwise rotation of the water in the Northern Hemisphere and a counterclockwise rotation in the Southern Hemisphere.

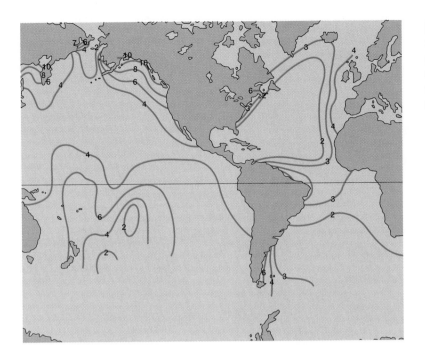

Figure 10.12 Corange lines for the world's oceans. These corange lines connect positions with the same spring tidal range. Open-ocean tidal ranges are less well known than nearshore ranges, where tide-level recorders are commonly used. Satellites now measure water elevation changes in mid-ocean; see the box in this chapter.

This circular (or rotary) movement is the oceanic tidal current described at the end of the next section.

Standing Wave Tides

In some ocean basins or parts of ocean basins, the tide wave is reflected from the edge of the continents, and a **standing wave tide** is produced (see chapter 9). Remember that if a container of water is tipped so that the water level is high at one end and low at the other, the water flows to the low end, raising the water level at that end as the water level at the high end drops. This movement produces a wave having a wavelength that is twice the length of the container, with antinodes at the ends of the basin and a node at the basin's center. This same process occurs in ocean basins, but with some important modifications.

For the water to flow from the high-water side to the low-water side of an ocean basin requires a long time, and the Coriolis effect must be included. The moving water, deflected to the right in the Northern Hemisphere, does not reach the low-tide end but instead is deflected to a position to the right of the initial high-tide position. This movement causes the tide crest to rotate counterclockwise around the basin in which it oscillates, but the tidal current rotates clockwise, because the current is deflected to the right in the Northern Hemisphere (see fig. 10.13). In the Southern Hemisphere these directions are reversed.

In the **rotary standing tide wave,** the node becomes reduced to a central point, while the tide crests (shown as cotidal lines) progress around the edges of the basin. (See figure 10.11 for a demonstration of this pattern in the northeastern and southwestern Pacific and in the north Atlantic.) The central point, or node, for a rotary tide is called the **amphidromic point.** The distance between low- and high-water levels, or the tidal range, for a rotary tide is shown on a chart

by a series of lines decreasing in value as they approach the amphidromic point. The lines of equal tidal range are called **corange lines** (fig. 10.12). Near the amphidromic point the tidal range is small; the farther from the amphidromic point, the greater the range. Because the amphidromic point is located near the center of an ocean basin, many mid-ocean areas have small tidal ranges, while the shores of the landmasses forming the sides of the basins have larger tidal ranges. The value and position of corange lines in mid-ocean are not as well known as they are near shore.

The flow of water from the high-water side to the low-water side of a standing tide wave produces a rotating tidal current, as shown in figure 10.13. A rotating tidal current is also produced by the orbital motion of water particles in a progressive tide wave. If the tide wave is diurnal, the water particles travel in one complete circle in a tidal day. A semidiurnal tide causes two circles, and a semidiurnal mixed tide produces two circles of unequal size. An example of a rotating semidiurnal mixed tidal current recorded at the Columbia River lightship in the North Pacific is presented in figure 10.14.

Rotary standing tides occur in basins in which the natural period of the basin approximates the tidal period. If the tidal period and the oscillation period of the basin coincide, the tides increase in amplitude. Table 10.1 relates basin depths and lengths or widths that produce natural oscillations equal to tidal periods. Remember from chapter 9 that the natural period of oscillation of a standing wave in a basin is

$$T = \left(\frac{1}{n}\right)\left(\frac{L}{\sqrt{gD}}\right)$$

where L, the wavelength, is twice the basin length, l, in closed basins and four times l in open-ended basins. A

Figure 10.13 Rotary standing tide waves. At the instant high tide occurs at one side of the basin (a), the water begins to flow toward the low-tide side, creating a tidal current (b) that is deflected to its right in the Northern Hemisphere. The current displaces the high tide counterclockwise from (a) by way of (b) to (c). The process continues from (c) by way of (d) to (e), from (e) by way of (f) to (g), and from (g) by way of (h) to (a). This process results in a tide wave that rotates counterclockwise about the amphidromic point (i) and a tidal current that flows clockwise (j).

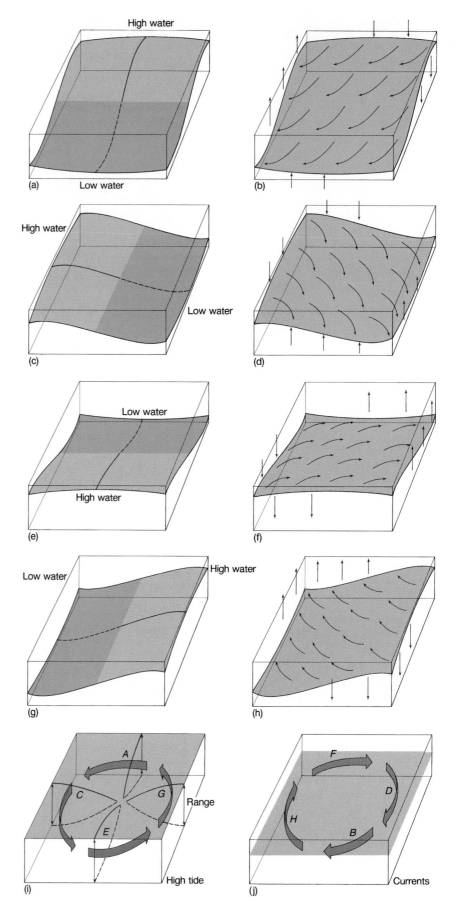

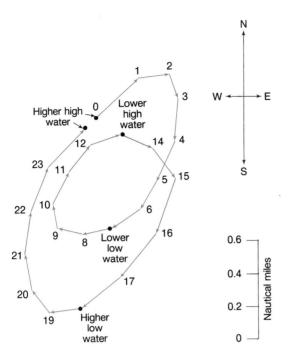

Figure 10.14 Ocean tidal currents are rotary currents. The *arrows* trace the path followed by water particles in a tide wave during a mixed tidal cycle. Two unequal tidal cycles are shown. *Numbers* indicate consecutive hours in each tidal stage. The Coriolis effect deflects the horizontal component of the water particles' orbital motion, causing them to move in a circular path.

Table 10.1 Dimensions of Closed Ocean Basins with Natural Periods Equaling Tidal Periods

Tidal Period	Depth (m)	Length or Width (km)
Semidiurnal	4000	4428
12.42 h	3000	3835
44,712 s	2000	3131
	1000	2214
	500	1566
	100	700
	50	495
Diurnal	4000	8853
24.83 h	3000	7667
89,388 s	2000	6260
	1000	4427
	500	3130
	100	1399
	50	989

comparison of the values given in table 10.1 shows that deep-ocean basins must have great length to accommodate standing waves with tidal periods, whereas shallow basins may be much shorter. Most tides are semidiurnal, but dimensions of some basins cause the basin to resonate with a diurnal tidal period rather than a semidiurnal period. See the tide curves in figure 10.2 for Pensacola, Florida, and Galveston, Texas. In an open-ended tidal basin with a mixed tide, the semidiurnal portion of the tide may cause the basin to resonate with a node at the entrance to the basin and another node within the basin. The diurnal portion of the tide may have only a single node at the basin entrance. The result is a diurnal water-level pattern at the second node of the semidiurnal tide and a semidiurnal mixed pattern in all other parts of the basin. The harbor of Victoria, British Columbia, Canada is located near a semidiurnal node so that it registers a diurnal tidal pattern in an inlet system of mixed tides.

Ocean tides are the result of combining progressive and standing wave tides with diurnal and semidiurnal characteristics. Different kinds of tides interact with each other along their boundaries, and the results are exceedingly complex.

Tide Waves in Narrow Basins

Unlike ocean basins, coastal bays and channels are often long and narrow with a length that is considerably greater than their width. These narrow basins have an open end toward the sea, so that the reflection of the tide wave occurs only at the head of the basin. To resonate with a tidal period, the length dimension need be only one-half the length cited for the closed basins in table 10.1. In this type of open basin, the node is at the entrance, and only one-fourth of the tide wave's form is present; an antinode is at the head of the bay. This situation is illustrated in chapter 9, figure 9.28. If the open basin is very narrow, oscillation occurs only along the length of the basin; there is no rotary motion because the basin is too narrow. For example, the Bay of Fundy in northeastern Canada has a tidal range near the entrance node of about 2 m (6.6 ft), whereas the range at the head of the bay is 10.7 m (35 ft). This particular bay has a natural oscillation period that is so well matched to the tidal period that every tidal impulse at its entrance creates a large oscillation at the head of the bay (fig. 10.15). In another bay along another coast, the shape of the basin may be such that it decreases rather than increases the tide's range. Every naturally occurring basin is unique in this regard.

Tidal Bores 10.6

In some areas of the world, large-amplitude tides cause large and rapid changes in water volume along shallow bays or river mouths. Under these conditions, the rising tide is forced to move toward the land at a speed greater than that of the shallow-water wave, whose speed is determined by the depth of the water or the speed of the opposing river flow. When the forced tide wave breaks, it forms a spilling wave front that moves into the shallow water or up into the river. This wave front appears as a wall of turbulent water called a **tidal bore** and produces an abrupt change in water levels as it passes. A single bore may be formed, or a series of bores may be produced. The bores are usually less than a meter in height but can be as much as 8 m (26 ft) high, as in the case of spring tides on the Qiantang River of China.

(a)

Figure 10.16 The fast-rising tide in the Bay of Fundy produces a tidal bore that sweeps across the shallows.

(b)

Figure 10.15 Low tide (a) and high tide (b) at The Rocks Provincial Park, Hopewell Cape, New Brunswick, Canada. The tidal range at the head of the bay exceeds 10 m (28 ft).

The Amazon, Trent, and Seine Rivers have bores. Fast-rising tides also send bores across the sand flats surrounding Mont-Saint-Michel in France and into Turnagain Arm of Cook's Inlet in Alaska. The bore in the Bay of Fundy in Canada (fig. 10.16) has been reduced by the construction of a causeway. Towns in areas having tidal bores often post warnings; their turbulence can be a severe hazard because they suddenly flood areas that were open stretches of beach only minutes before.

Predicting Tides and Tidal Currents 10.7

Because of all the natural combinations of progressive and standing tides and the factors that affect them, it is not possible to predict the Earth's tides from knowledge of the tide-raising bodies alone; equilibrium tidal theory is not adequate for the task. Accurate, dependable daily tidal predictions are made by combining actual local measurements with astronomical data.

The rise and fall of the tides are measured over a period of years at selected locations. Primary tide stations make these water-level measurements for at least 19 years, to allow for the 18.6-year declinational period of the Moon. From these data, mean tide levels are calculated. Oceanographers use a technique called **harmonic analysis** to separate the tide record into components with magnitudes and periods that match the tide-raising forces of the Sun and the Moon. They are then able to isolate the effect of the local geography, known as the **local effect.** Tides for any location are predicted by combining the local effect with the predicted astronomical data. Complex and cumbersome mechanical computers or tide machines were once used to predict the tides, but today computers quickly and easily recombine the data and predict the time, date, and elevation of each high-water and low-water level.

Tide Tables

Tide and tidal current tables for North America were published by the National Ocean Survey (NOS) of the National Oceanic and Atmospheric Administration (NOAA) until 1996. These tables are now available on the Internet and published by private companies using NOAA data. Tide tables give the dates, times, and water levels for high and

Measuring Tides from Space

Until recently monitoring and recording of open-ocean tides had been done only by inference from measurements at coastal sites, some mid-ocean islands, and a few seafloor-mounted pressure gauges. Today's satellites, first *SEASAT*, then *GEOSAT*, and now the *ERS* series and *TOPEX/Poseidon*, can measure the absolute elevation of the sea surface by radar altimetry. The satellite uses a radar beam to measure the distance from the satellite to the sea surface, and this distance is compared to the distance between the satellite and the Earth's center to obtain an elevation of the sea surface above the Earth's center. *TOPEX/Poseidon* provides more accurate sea-level measurements than previous satellites, collecting ten measurements per second and requiring ten days to repeat its ground track between approximately 65°N and 65°S.

The changes in sea surface elevation recorded by the satellite are caused by climate variations, water-density shifts due to temperature change, wind and atmospheric pressure fluctuations, and ocean current meanders, as well as by the passing tide wave. Tidal elevations can be separated from other height changes, because of their definite and recognizable periods, allowing oceanic tidal maps based on satellite data to be drawn with high accuracy.

Oceanic tidal maps for specific tidal components have been derived from the *TOPEX/Poseidon* data. The astronomical tidal component with the largest effect on the tides is the M_2 or principal lunar semidiurnal component. Box figure 1 is a cotidal map of global M_2 tides derived from one year of *TOPEX/Poseidon* sea surface elevation data that have been fitted to a computer model. As longer-term records are collected it will be possible to clearly separate more of the tidal components, improve computer models, and predict the total ocean tide.

Internet References

Visit the book's Online Learning Center at www.mhhe.com/sverdrup (click on the book's cover) to explore links to further information on related topics.

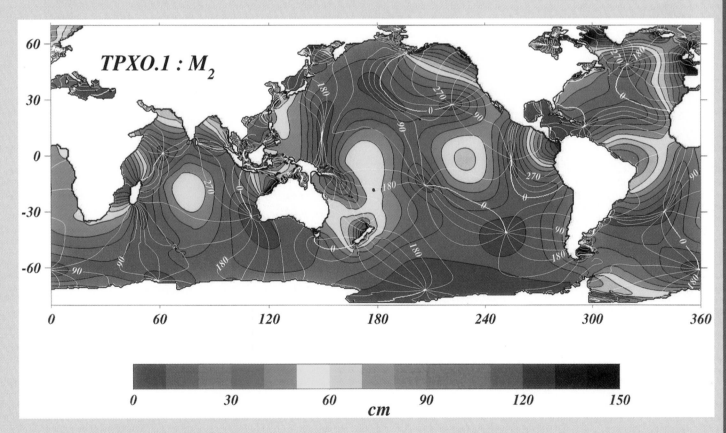

web link **Box Figure 1** *TOPEX/Poseidon* data produced this cotidal chart of the principal lunar semidiurnal tidal component (M_2). The rotation of the tides about amphidromic points and areas of progressive tides are shown. Cotidal lines mark the position of the tidal wave crest, hour by hour; each starting point is a *thick white line* representing 0000 hour GMT. The *numbers* represent the degrees of rotation of the tidal crest about the amphidromic points. Cotidal amplitude (one-half the cotidal range) increases with distance from the amphidromic points. The color scale is in centimeters.

Day	Time (h min)[1]	Height (ft)	Day	Time (h min)[1]	Height (ft)	Day	Time (h min)[1]	Height (ft)
1	0131	1.91	12	0419	9.84	22	0106	9.58
	0729	6.48		1037	−1.28		0743	−1.10
	1306	2.62		1707	9.80		1949	2.02
	1940	8.54		2309	0.41			
2	0233	1.52	13	0514	9.10	23	0148	9.50
	0841	6.21		1121	−0.34		0820	−0.91
	1406	3.21		1748	9.89		1454	8.78
	2031	8.66					2031	1.77
3	0334	0.93	14	0005	0.30	24	0228	9.33
	0956	6.30		0614	8.25		0855	−0.57
	1512	3.55		1209	0.73		1525	8.83
	2127	8.88		1833	9.86		2110	1.57
4	0432	0.22	15	0108	0.25	25	0306	9.07
	1105	6.68		0721	7.47		0928	−0.10
	1617	3.61		1303	1.77		1554	8.85
	2224	9.21		1924	9.72		2148	1.42
5	0526	−0.55	16	0215	0.13	26	0345	8.71
	1204	7.20		0836	6.98		0959	0.47
	1716	3.42		1406	2.61		1622	8.85
	2318	9.62		2023	9.55		2226	1.33
6	0615	−1.28	17	0323	−0.11	27	0426	8.25
	1256	7.73		0956	6.93		1029	1.11
	1811	3.05		1513	3.09		1649	8.85
				2127	9.44		2307	1.29
7	0009	10.03	18	0428	−0.45	28	0509	7.72
	0702	−1.89		1110	7.26		1058	1.80
	1343	8.21		1619	3.17		1718	8.82
	1902	2.58		2231	9.44		2351	1.31
8	0058	10.39	19	0526	−0.79	29	0559	7.16
	0747	−2.32		1211	7.72		1129	2.50
	1427	8.63		1721	2.98		1751	8.74
	1951	2.05		2329	9.52			
9	0147	10.60	20	0617	−1.04	30	0044	1.33
	0830	−2.51		1300	8.15		0659	6.67
	1509	9.00		1816	2.66		1208	3.20
							1833	8.63
10	0236	10.61	21	0020	9.58	31	0145	1.25
	0913	−2.41		0703	−1.15		0810	6.39
	1549	9.32		1342	8.47		1313	3.81
	2127	1.02		1905	2.32		1930	8.53
11	0326	10.36						
	0955	−2.00						
	1628	9.60						
	2216	0.65						

1. hours, minutes. Time meridian 120°W. 0000 is midnight; 1200 is noon—Pacific Daylight Time. Heights are referred to mean lower low water, which is the chart datum of soundings.

Data from WWW Tide Predictor, *http://tbone.biol.sc.edu/tide/sitesel.html*

low water at primary tide stations (table 10.2). There are 196 primary tide stations in the United States, but many more locations require accurate tide predictions. The data for these auxiliary stations are determined by correcting nearby primary station data for time and tide height.

Tidal Current Tables

Tidal currents in the open ocean have been explained as rotary currents formed by the passing tide wave form and the deflection of water particles due to the Coriolis effect. Tidal currents in the deep sea are of scientific interest to oceanographers

Day	Slack Water Time (h min)[1]	Maximum Current Time (h min)[1]	Velocity (knots)	Day	Slack Water Time (h min)[1]	Maximum Current Time (h min)[1]	Velocity (knots)	Day	Slack Water Time (h min)[1]	Maximum Current Time (h min)[1]	Velocity (knots)
1		0235	7.56E	12	0242	0532	10.35F	23	0038	0320	8.56F
	0559	0904	6.41F		0834	1145	11.31E		0614	0930	11.22E
	1232	1522	4.48E		1443	1748	11.95F		1224	1536	12.50F
	1816	2102	4.33F		2101				1859	2207	11.89E
	2344			13		0016	12.30E				
2		0327	7.10E		0326	0622	10.46F	24	0114	0359	8.92F
	0650	1004	6.89F		0932	1236	10.21E		0657	1009	10.87E
	1342	1636	4.47E		1535	1836	10.27F		1304	1612	11.77F
	1934	2205	3.60F		2142				1932	2241	11.57E
3	0033	0425	7.03E	14		0101	11.64E	25	0148	0435	8.98F
	0742	1102	7.80F		0414	0717	10.32F		0737	1047	10.20E
	1447	1747	5.18E		1034	1334	8.92E		1342	1646	10.73F
	2049	2308	3.45F		1634	1930	8.42F		2002	2315	10.95E
4	0130	0525	7.41E		2227			26	0221	0511	8.80F
	0833	1157	9.05F	15		0152	10.76E		0818	1125	9.26E
	1543	1846	6.39E		0507	0818	10.10F		1420	1720	9.45F
	2151				1143	1440	7.77E		2032	2348	10.13E
5		0007	3.85F		1743	2031	6.70F	27	0255	0548	8.43F
	0229	0620	8.20E		2318				0900	1204	8.14E
	0923	1246	10.48F	16		0250	9.89E		1459	1755	8.02F
	1631	1935	7.80E		0606	0925	10.05F		2101		
	2242				1258	1556	7.17E	28		0021	9.20E
6		0100	4.67F		1902	2140	5.47F		0331	0628	7.94F
	0326	0711	9.23E	17	0017	0356	9.28E		0945	1246	6.90E
	1010	1332	11.89F		0710	1033	10.35F		1543	1834	6.50F
	1713	2018	9.21E		1412	1714	7.41E		2131		
	2325				2023	2252	5.01F	29		0057	8.23E
7		0148	5.76F	18	0124	0506	9.17E		0410	0714	7.44F
	0420	0758	10.32E		0813	1139	10.99F		1039	1335	5.77E
	1055	1415	13.10F		1519	1823	8.34E		1635	1919	4.99F
	1753	2058	10.50E		2133				2206		
8	0005	0233	6.96F	19		0001	5.31F	30		0139	7.31E
	0511	0843	11.29E		0235	0612	9.57E		0456	0809	7.07F
	1139	1457	13.93F		0913	1238	11.78F		1142	1437	4.87E
	1831	2137	11.55E		1615	1920	9.53E		1742	2016	3.68F
9	0043	0317	8.13F		2231				2248		
	0601	0927	11.19E	20		0101	6.10F	31		0231	6.57E
	1224	1539	14.27F		0340	0711	10.22E		0551	0913	7.06F
	1908	2215	12.29E		1008	1330	12.46F		1255	1555	4.60E
10	0122	0401	9.15F		1704	2009	10.60E		1906	2126	2.90F
	0650	1012	12.23E		2318				2346		
	1309	1621	14.05F	21		0153	7.05F				
	1945	2254	12.67E		0438	0802	11.19E				
11	0201	0445	9.91F		1057	1416	12.86F				
	0741	1057	12.01E		1746	2052	11.43E				
	1355	1704	13.26F	22	0000	0239	7.92F				
	2023	2334	12.68E		0529	0847	10.83E				
					1142	1457	12.88F				
					1824	2130	11.85E				

1. hours minutes. Time meridian 120°W. 0000 is midnight; 1200 is noon—Pacific Daylight Time. F, flood dir. 180° true. E, ebb dir. 000° true.
Data from WWW Tide Predictor, http://tbone.biol.sc.edu/tide/sitesel.html

concerned with removing this circular motion from their data to obtain the net flows of the major ocean currents. Tidal currents in harbors and coastal waters are of major interest to commercial vessels and pleasure boaters, because these currents can be very strong and must be taken into account by anyone who wants to navigate in such waters.

Like the tides, the tidal currents are first measured at selected primary locations in important inland waterways and channels. These current data are studied to determine how the speed and direction of the tidal current are related to the predicted tide-level changes. As before, the local effect is determined and is used to predict tidal currents on the basis of the tide tables.

Once tidal currents have been predicted for a location, they can be graphed with time and compared to the tidal height curves for that area. If maximum current times coincide with the times for either low or high water, the tide has a progressive wave form. If maximum current times coincide with mid-tide stages, the tide is a standing wave–type tide.

Tidal current data (table 10.3) are published in a format similar to that of the tide tables. The times of slack water, maximum flood currents, and maximum ebb currents, as well as the speed of the currents in knots, and the direction of flow for ebb and flood currents are given for primary channel stations. Auxiliary tidal current stations are keyed to the primary current stations with correction factors to determine current speed, time, and direction at the secondary stations. This information allows the master of a vessel to decide at what time to arrive at a particular channel in order to find the current flowing in the right direction or how long to wait for slack water before choosing to proceed through a particularly swift and turbulent passage.

Practical Considerations: Energy from Tides   10.8

Tidal energy was used to turn mill wheels in the coastal towns of northern Europe during the nineteenth century. The possibility of obtaining large amounts of energy from the tides still exists where there are large tidal ranges or narrow channels with swift tidal currents. There are two systems for extracting energy from the rise and fall of the tides. Both require building a dam across a bay or an estuary so that seawater can be held in the bay at high tide. When the tide ebbs, a difference in water-level height is produced between the water behind the dam and the ebbing tide. When the elevation difference becomes sufficient, the seawater behind the dam is released through turbines to produce electrical power. The reservoir behind the dam is refilled on the next rising tide by opening gates in the dam. This single-action system produces power only during a portion of each ebb tide (fig. 10.17a). A tidal range of about 7 m (23 ft) is required for this system to produce power.

This same arrangement can be used as a double-action system. In this system, power is produced on the ebbing tide. At the end of this power cycle, when the level behind the dam is not sufficient to produce power, the gates are opened and the last of the water in the reservoir is spilled out to sea. The gates are then closed and some of the remaining water in the bay is pumped seaward to further decrease the water level. Although the pumps consume power, this expenditure of energy is worthwhile if the pumping period occurs when power demands are low and if the additional increase in height between water levels allows more power to be produced on the next cycle.

The gates are kept closed and the tide rises on the seaward side of the dam while the reservoir level remains low. When the difference in height is great enough, the incoming water is released through the turbines into the reservoir to produce power. At the end of the rising tide the dam is sealed, and water is pumped into the reservoir to raise the reservoir to its maximum level; when the tide drops on the seaward side, the cycle is repeated. This system requires very specialized turbine systems because the water moves through the turbines in two directions, and the turbines must be able to generate power on both the flood and ebb tides. A sketch of the power cycle of such a double-action system is presented in figure 10.17b.

This system appears to be a simple and cost-effective method for producing electrical power, but there are few places in the world where the tidal range is sufficient and where natural bays or estuaries can be dammed at their entrances at reasonable cost and effort. Moreover, the appropriate tides and bays are not necessarily located near population centers that need the power. Installation and power-distribution costs in addition to periodic low-power production because of the changing tidal amplitude over the tide's monthly cycle make this type of power more expensive than other sources.

There is a more than thirty-five-year-old commercial tidal power installation on the La Rance River Estuary in France that produces 5.4×10^{10} watt-hours per year. Present global energy demands could be satisfied by 250,000 plants of this capacity, but there are only about 255 sites that have been identified around the world with the potential for tidal energy development.

Tidal power has been under consideration for the Bay of Fundy since the 1930s. Canadian and American interest was casual because of the expense of the project, until the rising cost of fossil fuels focused interest on alternative energy sources. Nova Scotia is currently actively pursuing the development of sites in the Minas and Cumberland Basins at the head of the Bay of Fundy. The Minas Basin site would have a capacity of 5300 megawatts and the Cumberland Basin site would produce 1400 megawatts. Ultimately, the development of one or both of these sites will depend upon obtaining financing for the multimillion-dollar projects, securing markets in the northeastern United States for excess power, coordinating intermittent tidal energy with other, more conventional methods of producing electrical power, and mitigating potential environmental effects of the project. On the U.S. side of the bay, Cobscook Bay and Half-Moon Cove in Maine are being considered as potential

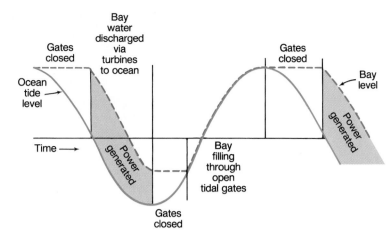

Figure 10.17 Periods of power generation related to ocean tidal heights and water storage levels. (a) A single-action tidal power system. Power is generated on the ebb tide. (b) A double-action tidal power system. Power is generated on the ebb and flood tides.

(a) Single-action power cycle; ebb only

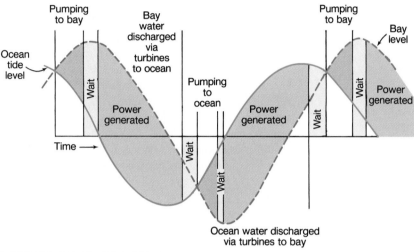

(b) Double-action power cycle; ebb and flood

power-generating sites. The U.S. Army Corps of Engineers is interested in the Cobscook Bay site, and the Passamaquoddy Indian Tribe has proposed a 12-megawatt development at Half-Moon Cove.

The province of Nova Scotia commissioned a power station in the tidal estuary of the Annapolis River in 1984 (fig. 10.18). The world's largest straight-flow–rim type turbine generator has been used to produce power in a single-basin–single-effect scheme. Tidal ranges at the Annapolis site vary from 8.7 m (29 ft) during spring tides to 4.4 m (14 ft) during neap tides. The unit generates up to 20 megawatts of power, from the head of water developed between the upstream basin and the sea level downstream at low tide. Initially, this project was intended as a pilot project to demonstrate the feasibility of a large-scale straight-flow turbine in a tidal setting. The station has now been added to the province's principal electrical utility's hydro-generating system. Annual production from the unit is 3–4 × 10^{10} watt-hours. Power availability has been in excess of 95%.

Although tidal power does not release pollutants, it is not without environmental consequences. The dams isolate the bay from the rivers and estuaries with which it was previously connected. At present, the natural period of os-

Figure 10.18 The Annapolis River tidal power project is the first tidal power plant in North America.

cillation of the Bay of Fundy is about thirty minutes longer than the tidal period; these periods are sufficiently alike to resonate with the tides. Damming the bay will shorten the period of oscillation and increase the resonance in the bay.

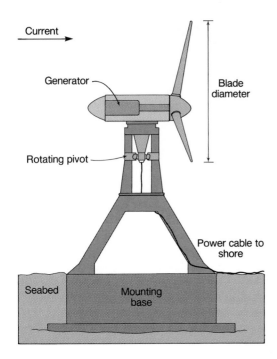

Figure 10.19 A water mill designed to use tidal current energy. Power production is related to blade size and current speed.

It is estimated that this may increase tidal ranges by 0.5 m (20 in) and tidal currents by 5% along the coast of Maine. Coastal residents fear increased erosion and changes affecting the shellfish populations.

The damming of a bay or estuary interferes with ship travel and with port facilities. The dams are barriers to migratory species and alter the circulation patterns of the isolated basin.

Swift tidal currents in inshore channels represent another possible energy source. Flowing water has been used for several centuries to turn the equivalent of windmills or water wheels for limited power. Because the tidal currents reverse with the tide, these "water mills" must be installed to operate with the current flowing in either direction (fig. 10.19).

Power generated by windmills is dependent on the density of the air, the blade diameter, and the cube of the wind speed. Water mills with a similar design are dependent on the density of the water, the blade diameter, and the cube of the current speed (table 10.4). A windmill in a 20-knot wind produces about the same power as a water mill with blades of the same diameter in a 2-knot current, because the density of water is about 1000 times the density of air:

$$0.001 \text{ g/cm}^3 \ (20 \text{ knots})^3 = 1 \text{ g/cm}^3 \times (2 \text{ knots})^3$$

$$\text{air density} \times (\text{speed})^3 = \text{water density} \times (\text{speed})^3$$

A 2×10^6 watt wind-generating unit installed on land has a blade nearly 210 ft in diameter mounted on a 120 ft tower. Installation costs are around $1.6 million. The cost of installing and maintaining a similar-sized unit in a tidal channel submerged in seawater would be much greater, and such a unit would still suffer from periodic loss of power production during slack water. These enormous units could be navigation

Table 10.4 Estimated Power Generation (in kW) for Water Mills at Various Current Speeds

| Blade Diameter (m) | Current Speed | | |
	5 Knots (2.5 m/s)	4.5 Knots (2.25 m/s)	3.5 Knots (1.75 m/s)
2	8.5	5.5	3.7
5	53	35	23
10	210	140	90
15	480	310	205
20	850	550	370
30	1910	1250	820

hazards, and channels of sufficient depth with currents of the required speed are few and not easily accessible.

A submerged current turbine has been tested in England. The turbine has a 4 m (13 ft) rotor and is designed to drive a generator producing a maximum of 10 kilowatts of power. The power directly generated by the currents will be stored in batteries to be used at a lower rate over a longer period of time. Perhaps this device could be useful in remote areas that are difficult to supply with electrical power from land sources.

Summary

Diurnal tides have one high tide and one low tide each tidal day; semidiurnal tides have two high tides and two low tides. A semidiurnal mixed tide has two high tides and two low tides, but the high tides reach different heights and the low tides drop to different levels. For diurnal and semidiurnal tides, the greatest height reached by the water is high water, and the lowest point is low water. Mixed tides have higher high water, lower high water, higher low water, and lower low water. The zero depth on charts is referenced to mean low water or mean lower low water; low tides falling below these levels are minus tides. Rising tides are flood tides; falling tides are ebb tides.

Equilibrium tidal theory is used to explain the tides as a balance between gravity and an apparent centrifugal force. An equatorial tide is a semidiurnal tide, or a tide wave with two crests and two troughs. Because the Moon moves along its orbit as the Earth rotates on its axis, a tidal day is twenty-four hours and fifty minutes long. The period of the semidiurnal tide is therefore twelve hours and twenty-five minutes.

The Sun's effect is less than half that of the Moon's, and the tidal day with respect to the Sun is twenty-four hours. Because the tidal force of the Moon is greater than that of the Sun, the tidal day is still considered to be twenty-four hours and fifty minutes.

Spring tides have the greatest range between high and low water; they occur at the new and full Moons, when the Earth, Sun, and Moon are in line. Neap tides have the least tidal range; they occur at the Moon's first and last quarters, when the Moon is at right angles to the line of centers of the Earth and Sun.

When the Moon or Sun stands above or below the equator, equatorial tides become more diurnal. Diurnal tides are often called declinational tides. The elliptic orbits of the Earth and Moon also influence the tide.

The dynamic approach to tides investigates the actual tides as they occur in the ocean basins. The tide wave is discontinuous, except in the Southern Ocean. It is a shallow-water wave that oscillates in some ocean basins as a standing wave, and its motions persist long enough to be acted on by the Coriolis effect. The tide wave is reflected, refracted, and diffracted along its route. Because a point on the Earth moves eastward faster at the equator than the tide wave progresses westward as a free wave, the tide wave moves as a forced wave, and its crest is displaced to the east of the tide-raising body. Above 60°N and 60°S latitudes, the crest is more nearly in line with the Moon, because the speeds of Earth and the tide wave match more closely.

In large ocean basins, the tide wave can move as a progressive wave. The Coriolis effect causes a rotary tidal current. Standing wave tides can also form in ocean basins; they rotate around an amphidromic point as they oscillate. Cotidal lines mark the progression of the tide crest, and regions of equal tidal range are identified by corange lines. The tidal range increases with the distance from an amphidromic point. Standing wave tides in narrow open-ended basins oscillate about a node at the entrance to the bay or basin; the antinode is at the head of the basin, as in the Bay of Fundy. There is no rotary motion when a bay is very narrow.

A rapidly moving tidal bore is caused when a large-amplitude tide wave moves into a shallow bay or river.

Tidal heights and currents are predicted from astronomical data and actual local measurements. NOAA provides data for annual tide and tidal current tables.

Single- and double-action dam and turbine systems extract energy from the tides. Tidal power plants are in use in France, the former Soviet Union, and Canada. Few places have large enough tidal ranges and suitable locations for tidal dams. Tidal power has environmental drawbacks as well as high developmental costs. Tidal currents are another energy-producing possibility, but installation and service costs are considered high.

Key Terms

All key terms from this chapter can be viewed by term, or by definition, when studied as flashcards on this book's Online Learning Center at www.mhhe.com/sverdrup (click on this book's cover).

diurnal tide, 278
semidiurnal tide, 278
semidiurnal mixed tide, 278
high water, 278
low water, 278

higher high water, 278
lower high water, 278
higher low water, 278
lower low water, 278
mean tide/average tide, 278

tidal datum, 278
minus tide, 278
flood tide, 278
ebb tide, 278
tidal current, 278
slack water, 279
equilibrium tidal theory, 279
dynamic tidal analysis, 279
centrifugal force, 280
centripetal force, 281
tidal day, 283
tide wave, 283
range, 283
spring tide, 283

neap tide, 283
declinational tide, 284
free wave, 285
forced wave, 285
progressive tide, 286
cotidal line, 286
standing wave tide, 287
rotary standing
   tide wave, 287
amphidromic point, 287
corange line, 287
tidal bore, 289
harmonic analysis, 290
local effect, 290

Study Questions

1. Distinguish between the terms in each pair:
   a. Diurnal tide—semidiurnal tide
   b. Tidal day—tidal period
   c. Spring tide—neap tide
   d. Flood tide—ebb tide
   e. Cotidal lines—corange lines
2. What is the path of a water particle in the tidal current shown in figure 10.14 if a 1-knot current flowing south is also present?
3. Why is it more efficient to generate power by means of a tidal dam than to erect water mills in tidal currents?
4. Why are standing wave tides produced in small coastal basins as well as in large ocean basins? Use table 10.1.
5. Explain why it is necessary to have both a large tidal range and a relatively large volume of water behind a tidal dam to generate electrical power from the rise and fall of the tides.
6. Sketch each of the three different tidal patterns during a spring tide and a neap tide. Label the tide levels of the spring tide sequences.
7. Explain why a tide is a wave.
8. Explain the relationship between the tides and tidal currents.
9. Why does the tide act as a shallow-water forced wave in the ocean basins?
10. How do progressive tides and standing wave tides differ?
11. How is a tidal bore produced?
12. What would happen if there were no force counteracting the Sun's gravitational force on the center of mass of the Earth-Moon system?
13. Why are high tides higher during the winter at mid-latitudes in the Northern Hemisphere than they are in the summer?
14. Why must the tides be measured for approximately nineteen years at a location before the data can be used in tide forecasting?
15. Why are accurate satellite sea surface elevation measurements over the entire Earth so important to oceanography?

Study Problems

1. How many days will pass before a high tide reoccurs at the same clock time?
2. Using the information in table 10.2 plot the Aberdeen, Washington, tide curves for August 3 and 10, 2002. Which is a spring tide? Which is a neap tide? What type of tide is the Aberdeen tide? Label the water levels on each curve.
3. Choose the best date and time for clam digging during early morning in August at Aberdeen, Washington. Refer to table 10.2.
4. Using table 10.2, find the highest Aberdeen tides for August 2002.
5. At what time on August 13, 2002, should you arrive at Seymour Narrows to navigate the narrows at slack water between breakfast and dinner? See table 10.3.

Links to Related Websites

Visit the book's Online Learning Center at www.mhhe.com/sverdrup (click on the book's cover) to find live Internet links for additional topics related to this chapter's content.

• Tides and tide currents

Visit the book's Online Learning Center at www.mhhe.com/sverdrup (click on this book's cover) to find these additional chapter tools: Suggested Readings; links to further information on boxed readings, selected figures, and related chapter topics and additional study aids.

chapter 11

Coasts, Beaches, and Estuaries

The seashore is a sort of neutral ground, a most advantageous point from which to contemplate this world. It is even a trivial place. The waves forever rolling to the land are too far-traveled and untamable to be familiar. Creeping along the endless beach amid the sun-squawl and the foam, it occurs to us that we, too, are the product of sea-slime.

It is a wild, rank place, and there is no flattery in it. Strewn with crabs, horseshoes, and razor-clams, and whatever the sea casts up—a vast morgue, where famished dogs may range in packs, and crows come daily to glean the pittance which the tide leaves them. The carcasses of men and beasts together lie stately up upon its shelf, rotting and bleaching in the sun and waves, and each tide turns them in their beds, and tucks fresh sand under them. There is naked Nature—inhumanly sincere, wasting no thought on man, nibbling at the cliffy shore where gulls wheel amid the spray.

Henry David Thoreau
From *Cape Cod*

Land's End, Cornwall, England.

Shores and beaches are the most familiar areas of the ocean for most people, but even the casual visitor senses changes in these areas. On any visit the tide may be high or low, the logs and drift may have changed position since the last visit, and the dunes and sandbars may have shifted since the previous summer. A visit to a beach farther along the coast or along a different ocean presents a different picture. The sand is a different color or there is no sand at all; the waves break higher or lower across the beach; the slope of the beach is steeper or flatter; and so on. No beach is static, and no two beaches are exactly the same.

On our visits to the coast we may stop to admire an inlet, photograph a quiet harbor, or visit a river estuary. These places are where the salt water from the oceans and the fresh water from the land meet. In some ways all these areas behave like small oceans, but the frequent addition of fresh water gives them characteristics of their own. All of these areas share certain features, but, just as beaches differ, one estuary differs from another.

In this chapter we study the types of coasts and beaches and the natural processes that create and maintain them. We also investigate the estuaries and partially enclosed seas of coastal areas and the circulation patterns that are unique to them. These are complex and sensitive parts of the ocean system; to preserve them we must understand them.

Major Zones   11.1

The **coasts** of the world's continents are the areas where the land meets the sea. The terms *coast, coastal area,* and *coastal zone* are used to describe these land edges that border the sea, including areas of cliffs, dunes, beaches, bays, coves, and river mouths. Some examples of the various kinds of coasts are presented in figures 11.1 and 11.2. The width of the coast, or the distance to which the coast extends inland, varies and is determined by local geography, climate, and vegetation, as well as by the perception of the limits of marine influence in social customs and culture. However, the coast is most generally described as the land area that is or has been affected by marine processes such as tides, winds, and waves, even though the direct effect of these processes may be felt only under extreme storm conditions. The seaward limit of the coast usually coincides with the beginning of the beach or shore but sometimes includes nearby offshore islands.

The term *coastal zone* includes the open coast as well as the semi-isolated and sheltered bays and estuaries that interrupt it. The coastal zone incorporates land and water areas and has become the standard term used in legal and legislative documents affecting U.S. coastal areas. In these circumstances the coastal zone's landward boundary is defined by a distance (often 200 ft or about 60 m) from some chosen reference (usually high water); the seaward boundary's limit is defined by state and federal laws.

Coastal areas are regions of change in which the sea acts to alter the shape and configuration of the land. Sometimes these changes are extreme and occur rapidly (for instance, the damage caused by a hurricane). Sometimes the changes are subtle and so slow that they are not perceived by people during their lifetimes but are very impressive when considered over long periods (for example, the formation of the Mississippi River delta or the gradual erosion of Cape Hatteras, North Carolina). In general, coasts that are composed of soft, unconsolidated materials, such as sand, change more rapidly than coasts composed of rock. Over the life of the Earth coasts have changed dramatically as they have emerged and submerged with tectonic processes, climatic variations, and changing sea level; some remnants of these changes can still be seen in present coastlines.

The **shore** is a part of the coast; it is that region from the outer limit of wave action on the bottom (seaward of the lowest tide level) to the limit of the waves' direct influence on the land. This land limit may be marked by a cliff or an elevation of the land above which the sea waves cannot break. Such features act as barriers to the wave-tossed drift of logs, seaweeds, and human debris. The **beach** is an accumulation of sediment (sand or gravel) that occupies a portion of the shore. The beach is not static but moving and dynamic, because the beach sediments are constantly being moved seaward, landward, and along the shore by nearshore wave and current action. Between the high-tide mark and the upper limit of the shore there may be dunes or

web link **Figure 11.1** (a) The Oregon coast is famed for its bold rocky headlands and its intervening pocket beaches. (b) Bayou La Loutre at Ysclosky, Louisiana. The bayou is an old stream course of the Mississippi River. The salt content of the water changes as the wind drives fresh water from the river or salt water from the coast through the delta. (c) Glacial bays and fjords produce a rugged topography along the coast of Glacier Bay in southeastern Alaska.

Figure 11.2 (a) Cape of Good Hope, South Africa, where the Indian and Atlantic Oceans meet. (b) Sea stacks are common features along the southern coast of Australia, occurring in a wide variety of shapes and sizes. These sea stacks are part of a group called the Twelve Apostles. (c) These chalky cliffs along the Dorset coast of England were formed from the remains of microscopic, single-celled foraminifera, principally *Globigerina*.

Figure 11.3 Driftwood accumulates at the high-tide line in areas where timber is plentiful.

grass flats dotted with drift logs left behind by an exceptionally high tide or a severe storm (fig. 11.3).

Types of Coasts 11.2

Any land form is the product of the processes that have changed it through time, and the study of land forms and the processes that have fashioned them is known as **geomorphology.** Coastal geomorphology considers tectonic processes; wave, wind, and current exposure; tidal range and tidal currents; sediment supply and coastal transport; and climate and climate changes.

Climate change can result in a worldwide, or **eustatic, change** in sea level that may submerge previous coasts or expose previous sea floor. During the last period of global glaciation sea level dropped by an estimated 100 m (328 ft) or more due to the storage of water on land as glacial ice and the cooling and subsequent decrease in volume of the remaining water in the ocean basins. With the global warming that signaled the end of the last ice age sea level rose to its present height as land ice melted and the increase in average temperature of seawater caused a corresponding increase in volume. Sea level is continuing to rise even now as a result of global warming (see the discussion of greenhouse gases in chapter 6).

There are different ways to classify coasts for the purpose of describing and studying them. One way is to classify them as erosional or depositional, depending on whether they predominantly lose or gain sediments. This is often influenced by whether the coast is located at the edge of the plate as part of an active, or leading, continental margin or away from the edge of the plate as part of a passive, or trailing, continental margin (the formation and characteristics of active and passive continental margins are discussed in chapter 2, section 2.4). In the United States eastern seaboard coasts are part of a passive continental margin and are generally considered depositional. Western seaboard coasts are located along active continental margins and are

generally considered erosional. These categories should not be confused with U.S. coastal erosion problems that are found mainly along the U.S. eastern coast.

Coasts of both types may be modified by eustatic sea-level changes. Some coasts (both erosional and depositional) have been eroded and their sediments swept away to be deposited elsewhere, while other coasts of both types have received sediments from rivers and shore currents. In some parts of the world, coasts are modified by seasonal storms; other areas enjoy more benign weather. Coasts of polar regions are modified by their interaction with sea ice; tropical coasts are altered by reef-building corals. Each coast has its own identity, but different types of coasts can be recognized as the results of certain processes.

A system devised by the late Francis P. Shepard of the Scripps Institution of Oceanography organizes coasts into two process categories: (1) coasts that owe their character and appearance to processes that occur at the land-air boundary and (2) coasts that owe their character and appearance to processes that are primarily of marine origin. Further classification depends on whether their large-scale features are the product of tectonic, depositional, erosional, volcanic, or biological processes.

In the first category are coasts that have been formed by (1) erosion of the land by running surface water, wind, or land ice, followed by a sinking of the land or a rise in sea level; (2) deposits of sediments carried by rivers, glaciers, or the wind; (3) volcanic activity, including lava flows; and (4) uplift and subsidence of the land by earthquakes and associated crustal movements. Coasts formed by these processes are **primary coasts,** since there has not yet been time for the sea to substantially alter or modify the appearance given to them by nonmarine processes. The second category includes coasts formed by (1) erosion due to waves, currents, or the dissolving action of the seawater; (2) deposition of sediments by waves, tides, and currents; and (3) alteration by marine plants and animals. These coasts are **secondary coasts;** their character, even though it may have been originally land-derived, is now distinctly a result of the sea and its processes.

A relatively young coast may be rapidly modified by the sea, while a coast that is old may retain its land-derived characteristics. Keep in mind that absolute age is not really important because the classification is based only on whether the characteristics are derived from the land or from the sea, and it is possible to find both types of features along the same coast.

Primary Coasts

Coasts formed by erosion at the land-air boundary followed by a sinking of the land or a rise in sea level include those that were covered by glaciers during the ice ages. During these glacial periods, sea level was lower than it is at present because much of the water was held as ice on the continents. Glaciers moved slowly across the land, scouring out valleys as they inched along to the sea. The weight of the ice caused the land to subside, and in some cases, when the ice began to melt, sea level rose faster than land could rebound

Figure 11.4 The narrow channel of a fjord, Milford Sound, New Zealand.

upward from its depressed state. In other cases, the glacial troughs were scoured below sea level and filled with seawater as the ice receded. The **fjords** of Norway, Greenland, New Zealand, Chile, and southeastern Alaska are the results of these processes (fig. 11.4). Fjords are long, deep, narrow channels with a U-shaped cross section. Where the glacier met the sea, there was often a collection of debris that formed a lip, creating a shallow entrance, or **sill.**

When a glacier or ice sheet ceases its forward motion and retreats, it leaves a mound of rubble, called a **moraine,** along the border of its greatest extension. If the glacier has reached the edge of a continent, this material becomes a

part of the coastal area. Long Island, off the New York and Connecticut coasts, and Cape Cod, Massachusetts, are moraines. Moraines act as protective barriers to the continental coast. In some areas, land that was heavily covered by ice during the last ice age is still slowly rising; in Scandinavia the rate of rise is about 1–5 cm (0.4–2 in) per year. Along other coasts, tectonic forces rather than loss of ice have caused land uplift. Along the western coasts of North and South America old wave-cut terraces can be identified now well above sea level.

When sea level was lowered during the ice ages, rivers flowed over the exposed shore to the sea. The river flow cut

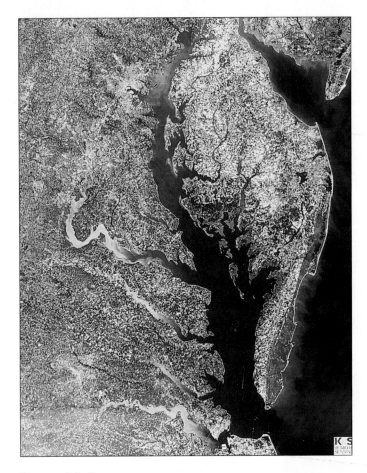

Figure 11.5 Delaware Bay (*upper right*) and Chesapeake Bay (*center*) are examples of drowned river valleys.

V-shaped river and stream channels into these areas, and in many cases the channels had numerous side branches formed by feeder streams. As the sea level rose, these channels were filled with seawater, producing areas such as Chesapeake Bay and Delaware Bay on the eastern coast of the United States. A coast of this type is called a **drowned river valley,** or **ria coast,** shown in figure 11.5.

Rivers carrying extremely heavy sediment loads build deltas on top of the continental shelves. The **delta,** or deposit of river-borne sediment left at a river mouth, produces a flat, fertile coastal area. Examples of these deposits are found at the mouths of the Mississippi, Ganges, Nile, and Amazon Rivers. A similar type of coast is produced when the eroded materials carried down from the hills by surface runoff and many small rivers join together to form an **alluvial plain.** The eastern seacoast of the United States south of Cape Hatteras was produced in this way.

It is estimated that every second, the rivers of the world carry 530 tons of sediment to the sea. This rate of removal is equal to the erosion of a layer 6 cm (2.4 in) thick from all land above sea level every 1000 years. This sediment helps to form and maintain the world's beaches, and much of it finally finds its way to the deep-ocean floor. All of it passes through the coastal zone and takes part in coastal processes. Refer to chapter 3 for sources of terrigenous material.

A **dune coast** is a wind-modified depositional coast. In Africa, the Western Sahara is gradually growing westward toward the Atlantic Ocean as the prevailing winds move sand from the inland desert to the coast. Along other dune coasts, for example, along central and northern Oregon (fig. 11.6),

Figure 11.6 Coastal sand dunes near Florence, Oregon.

the winds move the sands inland to form dunes; elsewhere dunes driven by the wind migrate along the shore.

The Hawaiian Islands are the tops of large seamounts and have excellent examples of coasts formed by volcanic activity. Lava flows extending to the sea form black sand beaches or **lava coasts.** Volcanic explosions just beneath the sea surface at the water's edge formed craters that became concave bays when they lost their rims on the seaward side, forming **cratered coasts** (fig. 11.7).

When tectonic activity results in faulting and displacement of the Earth's crust, the coast is changed in characteristic ways. The California San Andreas Fault system lies along a boundary where crustal plates are moving parallel to each other along a transform fault. (Plate movements and transform faults are described in chapter 2.) Over long periods of time, faults formed in this area filled with seawater. The Gulf of California, also known as the Sea of Cortez, located between Baja California and the mainland of Mexico, is found at the southern end of the fault system. At the northern end lie San Francisco and Tomales Bays, where the fault runs out into the Pacific Ocean. Tomales Bay is a particularly good example of a **fault bay.** See the satellite image of this fault system in figure 11.8. In other parts of the world, the Red Sea and the Scottish coast provide examples of **fault coasts.**

Secondary Coasts

Secondary coasts owe their present-day appearance to marine processes. As waves batter against the coast, they constantly erode and grind away the shore; rocks and cliffs are undercut by the wave action and fall into the sea, where they are ground into sand. A coastal area formed of uniform material may have been irregular in shape and composed of headlands and bays at one time, but the concentrated energy of the waves wears down the headlands more rapidly than the straighter shore and cove areas. With time, a more regular coastline results; examples are found in southern California, southern Australia, New Zealand (fig. 11.9), and England's cliffs of Dover. If the original coastline is made up of materials that vary greatly in composition and resistance to wave erosion, the result will be an increasingly irregular coastline of hard rock headlands separated by sand coves. In some places, such as northern California, Oregon, Washington, Australia, and New Zealand, the headlands erode irregularly and small rock islands and tall slender pinnacles of resistant rock are formed; these are known as **sea stacks** (figs. 11.2b and 11.10). Headlands still connected to the mainland are undercut to produce sea caves or are cut through to produce arches or small windows.

In some cases, eroded materials are carried seaward by the waves and currents to areas just off the coast. If sufficient

Figure 11.7 A volcanic crater coast, Hanauma Bay, Hawaii, is a crater that has lost the seaward portion of its rim. Today it is a park and marine preserve.

Figure 11.8 The San Andreas Fault runs along the California coast to the west of San Francisco, separating Point Reyes from the mainland and producing Bolinas and Tomales Bays. Tomales Bay is seen as a *dark line* at the *upper left*.

sand material is deposited in offshore shallows paralleling the beach, **bars** are produced. If still greater amounts of material are added, the bars may grow until they break the surface and form **barrier islands** (fig. 11.11). The barrier islands of the southeastern coast of the United States were formed during a period of rising sea level; seawater flooded low coastal areas, isolating the high dunes at the sea's edge and converting a primary coast to a secondary coast. Once a barrier island has formed, plants begin to grow and the vegetation helps to stabilize the sand and increase the island's elevation by trapping sediments and accumulating organic matter.

Figure 11.9 A cliffed coast made regular by erosion, New Zealand.

Figure 11.10 Sea stacks are common features along the coasts of northern California, Oregon, and Washington. They occur in a wide variety of shapes and sizes.

A line of barrier islands along a coast protects the continental coastline from storm waves and erosion, but the islands sustain the damage that the continental coast is spared. During 1989's Hurricane Hugo, Folly Island, along the coast of South Carolina, suffered extensive storm damage. Eighty-six of 290 oceanfront structures were more than 50% destroyed, and 50 were damaged beyond repair. The U.S. National Flood Insurance Program issued nearly $3 million in damage payments to Folly Island property owners. In 1997 Tropical Storm Josephine moved eastward across the Gulf of Mexico toward Florida. This storm did not develop hurricane-force winds, but it did cause extensive damage to coastal areas of Texas in the vicinity of Galveston, which is located on a barrier island. Josephine remained 482 km (300 mi) off the coast, producing steady, onshore winds of approximately 12 m/s (27 mi/h) for over a hundred hours. The winds raised the sea level at Galveston 80 cm (32 in) above normal high-tide levels and caused average wave heights of 3.5 m (11 ft) along the coast. The sustained period of high water and wave action brought by this minor storm produced severe, hurricane-like beach erosion, loss of homes, and loss of the dunes that stood between the remaining homes and the water. Attempts are made to halt storm-caused erosion by building seawalls and beach-holding devices, but these do not always prove successful. Although well intentioned, they often result in aggravating the loss of material from barrier islands rather than halting it.

Sand spits and **hooks** are bars that remain connected to the shore at one end (figs. 11.12 and 11.30). Spits and hooks may grow, shift position, wash away in a storm, or rebuild under more moderate conditions. The area between the mainland and these spits and hooks is protected from turbulence and is therefore often the site of beach flats formed of sand or mud. If a spit grows sufficiently to close off the mouth of an inlet, a shallow lagoon is formed. Water percolates through the gravel and sand of the spit, and inside the lagoon the water rises and falls with the tides.

In the tropical oceans, **reef coasts** are the result of the activities of sea organisms. Corals grow in the shallow, warm waters surrounding a landmass, and the small animals gradually build a fringing reef, which is attached directly to the landmass. In other places, a lagoon of quiet water may lie between the barrier reef and the land, and, in a few cases, the coral encircles a submerged island to form an atoll. The formation of

web link **Figure 11.11** Sea Island, Georgia, is a barrier island that has been extensively developed. The shallow water between the island and the coast is visible on the *left*.

Figure 11.12 A spit has formed across the entrance to Sequim Bay, Washington.

Figure 11.13 Mangrove trees growing along the shore of Florida Bay.

these reef types is discussed in chapter 3, and the life forms that make up the reef are discussed in chapter 17.

The Great Barrier Reef of Australia, stretching along its northeastern coast toward New Guinea, is the largest and most famous of the world's coral reefs. Coral atolls include the Pacific islands of Tarawa, Kwajalein, Eniwetok, and Bikini. The reef-encircled islands of Iwo Jima and Okinawa became familiar as the sites of major battles in the southwest Pacific during World War II.

Other marine animals form reeflike structures as their shells are deposited layer on layer, gradually building up a mass of hard material. There are large reef deposits of oyster shells in the Gulf of Mexico off the coasts of Louisiana and Texas. These reefs are so large that the shells are harvested commercially for lime production. Along the eastern coast of Florida, large populations of shell-bearing animals have contributed their shells directly to the shore; small shell fragments form the sand on the beaches.

Plants as well as animals may modify a coastal area. Along low-lying coasts in warm climates, mangrove trees grow in shallow water. Their great roots form a nearly impenetrable tangle, providing shelter for a unique community of other plants and animals (fig. 11.13). Coastal mangrove swamps are found along the Florida coast, northern Australia, and the Bay of Bengal and in the West Indies. In more temperate climates, low-lying protected coasts with areas of sand and mud are often thickly covered with grasses, forming another type of plant-maintained environment (fig. 11.14). These **salt marshes** may extend inland a considerable distance if the land is flat enough to permit periodic tidal flooding. Such marshes also form around protected bays and coves that have large

Figure 11.14 A coastal saltwater marsh, Gaspé, Québec, Canada.

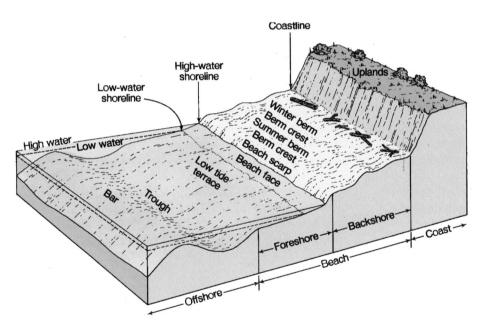

Figure 11.15 A typical beach profile with associated features.

changes in tide level. Salt marshes are extremely productive in terms of organic material and extend the shoreline seaward by trapping sediments. The effect of human activities on mangrove swamps and salt marshes is discussed in chapter 12.

Anatomy of a Beach   11.3

The beach includes the sand and sediment lying along the shore and the sediments carried along the shore by the nearshore waves and currents. Let us consider first the beach where we walk, sunbathe, and relax. A specific terminology for beaches and beach features has developed from the many studies of beaches around the world. These terms allow each area to be described and prevent confusion when one beach is compared to another. A beach profile cuts through a beach perpendicular to the coast; the profile in figure 11.15 shows the major features that might appear on any beach. Use this figure while reading this section, but remember that

(a)　　　　　　　　　　　　　　　　　　　　　　　　　　(b)

Figure 11.16 (a) Berms on a gravel beach. The winter storm berm with drift logs is located high on the beach. The summer berm is between the winter berm and the beach face in the *foreground*. (b) A wave-cut scarp on a gravel beach.

a given beach may not have all these features, for each beach is the product of its own singular environment.

The **backshore** is the dry region of a beach that is submerged only during the highest tides and the severest storms. The **foreshore** extends out past the low-tide level, and the **offshore** includes the shallow-water areas seaward of the low-tide level to the limit of wave action on the sea floor.

Terraces called **berms** appear on beaches in the area where foreshore and backshore meet. Berms are formed by wave-deposited material and have flat tops like the top of a terrace. Berms are recognizable by their slope, or rise in elevation. Some berms have a slight ridge crest that runs parallel to the beach; if there is a ridge, it is called the **berm crest.** When two berms are found on a beach (figs. 11.15 and 11.16*a*) the berm higher up the beach, or closer to the coastline, is the **winter,** or **storm, berm.** It is formed during severe winter storms when the waves reach high up the beach and pile up material along the backshore. The seaward berm, the berm closer to the water, is the **summer berm** formed by the gentle waves of spring and summer that do not reach as far up the beach. Once the winter berm is formed, it is not disturbed by the less intense storms during the year. The waves that produce a winter berm erase the summer berm, but after the storm season is over, a new summer berm is formed by the reduced wave action.

Between the berm and the water level there may be a wave-cut **scarp** at the high-water level. The scarp is an abrupt change in the beach slope that is caused by the cutting action of waves at normal high tide (fig. 11.16*b*). Berms and scarps do not form in the foreshore, between the normal high- and low-tide levels, because of the wave action and the continual rise and fall of the water. The foreshore below

the low-tide level is often flat, forming a **low-tide terrace;** the upper portion has a steeper slope and is known as the **beach face.** The slope of the beach face is related to the particle size of the loose beach material and to the wave energy.

Seaward of the low-tide level, in the offshore region, there may be **troughs** and bars that run parallel to the beach. These structures change seasonally as beach sediments move seaward, enlarging the bars during winter storms, and then shoreward in summer, diminishing the bars. When a bar accumulates enough sediment to break the surface and is stabilized by vegetation, it becomes an island of the barrier coast type.

Beaches do not occur along all shores. Along rocky cliffs, there may be no exposed beach area between low and high tides. However, there may be small pocket beaches, each separated from the next by rocky headlands or cliffs (fig. 11.17).

Beach Dynamics 11.4

A sand beach exists because there is a balance between the supply and the removal of the material that forms it. A sand beach that does not appear to change with time is not necessarily static (no new material supplied, no old material removed) but is more likely to represent a **dynamic equilibrium,** the supply of material equaling the removal of material. The beach is continually changing, but it appears unchanged and remains in balance.

Natural Processes

The gentler waves of summer move sand shoreward and deposit it, where it remains until the winter. The large storm

Figure 11.17 A series of pocket beaches cuts an elevated marine terrace along the California coast. The elevated terrace was at one time sea floor. Tectonic processes raised it to its present position.

waves of winter remove the sand from a beach and transport it back offshore, to a sandbar. These processes occur alternately, winter and summer, leaving the beach rocky and bare each winter and covered with sand during the summer (fig. 11.18). These seasonal changes are fluctuations about the equilibrium state of the beach over a yearly cycle. Other changes also appear in a cyclic fashion, such as the arrival of more sand during periods of late winter and spring river flooding. If a beach annually receives as much material as it loses, the beach is in equilibrium and does not appear to change from year to year.

Single violent events may alter a beach. A great storm may arrive from such a direction that the usual beach current is reversed. A landslide may pile rubble across the beach and some distance into the water, interfering with the flow of sand along the beach. In either case, the beach changes because of changes in rates of sediment supply, transport, and removal.

Waves moving toward a beach produce a current in the surf zone that moves water onshore and along the beach. This landward motion of water is called the **onshore current,** and this current transports suspended sediment toward the shore in what is called **onshore transport.**

Waves usually do not approach a shore with their crests completely parallel to the beach but strike the shore at a slight angle. This pattern sets up a surf-zone **longshore current** that moves down the beach. Both onshore and longshore flow in the surf zone are illustrated in figure 11.19. The turbulence that occurs as the waves break in the surf zone tumbles the beach material into suspension in the water. The wave-produced longshore current displaces this sediment down the shore in the surf zone, producing a **longshore transport.** Along both coasts of the North American continent, the predominant longshore transport steadily moves the sediments in a southerly direction.

The up-rush, or **swash,** of water from each breaking wave moves the sand particles diagonally up and along the beach in the direction of the longshore current. The backwash from the receding wave moves downslope toward the surf zone, but the backwash is weaker than the swash because much of the receding water percolates down into the sand. The combined action of the swash and backwash moves particles in a zigzag, or sawtoothed, path along the swash zone of the beach as part of the longshore transport.

Evenly spaced crescent-shaped depressions called **cusps** are sometimes found along the sand and cobble beaches of both quiet coves and exposed shores (figs. 11.16b and 11.20). How cusps are formed is still somewhat of a mystery. They tend to form during neap tide periods, when the tidal range is at a minimum, and are often destroyed during periods of spring tide, when the tidal range is at a maximum. The size of cusps appears to be directly related to the amount

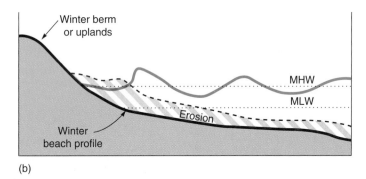

(a)

(b)

Figure 11.18 Seasonal beach changes. (a) The small waves and average tides of summer move sand from offshore to the beach and form a summer berm. (b) The high waves and storm tides of winter erode sand from the beach and store it in offshore bars. Winter conditions remove the summer berm, leaving only the winter berm on the beach.

Figure 11.19 Waves in the surf zone produce a longshore current that transports sediments down the beach. *Arrows* indicate onshore and longshore water movement.

of wave energy on the beach. Large cusps are associated with high wave energy and smaller cusps seem to form where the wave energy is low. Cusp formation may be related to some form of wave interference as waves approach the beach.

Studies of the movement of sediments in the surf zone as they travel along coasts, into navigational channels, and around coastal structures provide important information to property owners along the coast, to engineers involved in the design and construction of shoreline installations—for example, piers, breakwaters, and seawalls—and also to be military when planning troop and equipment movements in coastal areas.

The path traveled by beach sediments as the longshore current transports them from their source to their area of deposition is called a **drift sector.** Beaches within a drift sector usually remain in dynamic equilibrium and change little in appearance. Although there may be a large flow of material along the central portion of a drift sector, the amounts of sediment supplied and lost from the area are usually equal. At either end of the drift sector the beaches change with time. Areas at the sediment source are erosional; they are losing sediment. Areas at the depositional end of the drift sector are growing, or **accreting.** These processes are shown in figure 11.21. Although the longshore current is the usual transport mechanism within a drift sector, in some places tidal currents and coastal currents associated with large-scale oceanic circulation affect sediment transport.

Coastal Circulation

Onshore transport accumulates water as well as sand along a beach. This water must flow back out to sea and will do so one way or another. A headland may deflect the longshore current seaward, or the water flowing along the beach can return to the area beyond the surf zone in quieter water, such as areas over troughs or depressions in the sea floor. Regions of seaward return are frequently narrow and fast-moving.

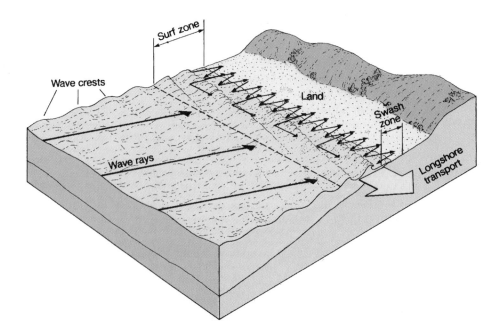

Figure 11.20 At Point Reyes in California, wave action forms a series of cusps along a shore that has been made regular by wave erosion. How cusps form is not clearly understood.

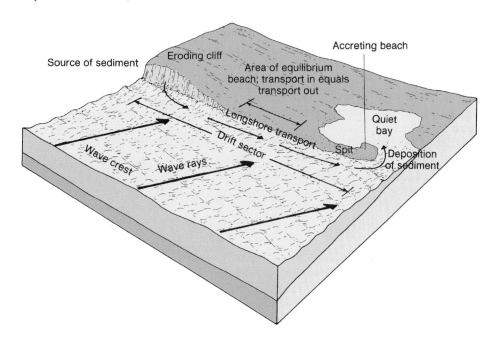

Figure 11.21 A drift sector extends down the coast from the sediment source to the area of deposit.

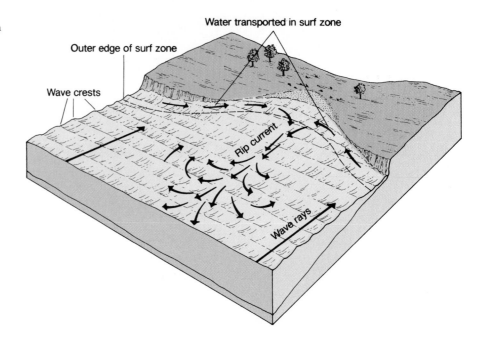

Figure 11.22 Rip currents form along a beach in areas of low surf and reduced onshore flow.

Labels in figure: Water transported in surf zone; Outer edge of surf zone; Wave crests; Rip current; Wave rays

These are the rip currents that carry sediment through the surf zone (fig. 11.22); rip currents are also discussed in chapter 9 (see fig. 9.22).

Just seaward of the surf, the rip current dissipates into an eddy. Some of the sediment is deposited here in quieter deeper water and is lost from the downbeach flow. However, some is returned to shore with the onshore transport on either side of the rip current. This cycle of partial return of sediments to the beach, transport along the beach, transport back to sea in another rip current, and return again to the beach is a part of a drift sector.

A series of these drift sectors, when linked together along a stretch of coast, forms a **coastal circulation cell.** In a major coastal circulation cell, the seaward transport of sediment results in deposits far enough offshore to prevent any recycling. At the end of a coastal circulation cell the longshore transport is deflected away from the beach and the sand moves across the continental shelf and down a submarine canyon to the ocean basin floor. This sand has been removed from the coastal system, and sand beaches disappear beyond this point until a new source contributes sand to form the next coastal circulation cell.

South of Point Conception in Southern California, oceanographers recognize four distinct coastal circulation cells (fig. 11.23). Each cell begins and ends in a region of rocky headlands, where beaches are sparse and submarine canyons are found offshore. Beaches become wider as river sources contribute their sand to the longshore current, and at the end of each cell, the current deposits the sediment in a submarine canyon, where it cascades to the ocean basin floor. Beaches just south of the canyon are sparse, and a new cell begins. Refer to chapter 3 for a discussion of submarine canyons.

Estimates of the rate at which sediments are moved along a beach are made by observing the rate at which sand

is deposited, or accreted, on the upstream side of an obstruction or by observing the rate at which sand spits migrate. Transport of sand along a section of coast varies between zero transport and several million cubic meters per year. Average values fall between 150,000 and 1,500,000 m^3 per year; 150,000 m^3 is more than 30,000 dump-truck loads. Once the enormous volume of the naturally moving sand is recognized, it becomes apparent why poorly designed harbors fill very quickly, beaches disappear, and spits migrate a considerable distance during a year. It also becomes obvious that the effort required to keep pace with the supply by dredging or pumping operations is enormously expensive and in many cases impossible.

The forces that supply the energy for the movement of beach materials come from the waves and the wave-produced currents. Surface winds supply about 10^{14} watts of power to the ocean surface to produce the waves and currents. There are about 440,000 km (264,000 mi) of coastal zone in the world, and approximately half of this coast is exposed directly to the ocean waves. Ocean waves average 1 m (3.3 ft) in height and produce power that is equivalent to 10^4 watts for each meter of exposed coastline. Under storm conditions, the wave height averages about 3 m (10 ft), and a meter of coastline receives 10^5 watts. It is this supply of energy that causes the beach erosion, the suspension of beach material, and the migration of beach sand along the narrow coastal zone.

Beach Types 11.5

Beaches are described in terms of (1) shape and structure, (2) composition of beach materials, (3) size of beach materials, and (4) color. In the first category, beaches are described as wide or narrow, steep or flat, and long or discontinuous (pocket beaches). A beach area that extends outward from

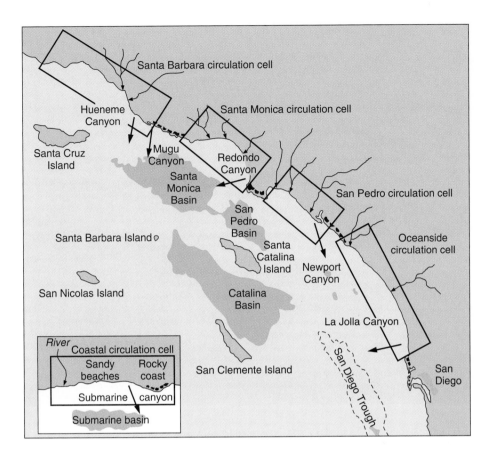

Figure 11.23 Major coastal sediment circulation cells along the California coast. Each cell starts with a sediment source and ends where beach material is transported into a submarine canyon. *D. L. Inman and J. D. Frautschy*, Littoral Processes and the Development of Shorelines, *1965, pp. 511–536. Coastal Engineering, ASCE, New York. Reprinted by permission of the authors.*

the main beach and turns and parallels the shore is called a spit (see fig. 11.12). Spits frequently change their shape in response to waves, currents, and storms. In a wave-protected environment, a spit may extend out at an angle from the shore to an offshore island. If this spit builds until it connects the rock or island to the shore, it forms what is called a **tombolo.** A spit extending offshore in a wide sweeping arc bending in the direction of the prevailing current is called a hook. The sediment moves around the end of a hook and is deposited in the quiet water behind the point, often producing a broad point on its end. See figure 11.24 for an example of a tombolo and figure 11.30 for an example of a hook.

Materials that form beaches include shell, coral, rock particles, and lava. Shingles are flat, circular, smooth stones formed from layered rock when the beach slope, wave action, and stone size combine to slide the stones back and forth with the water movement. When stones roll rather than slide, round stones or cobbles are formed. The terms *sand, mud, pebble, cobble,* and *boulder* describe the size of beach particles (see chapter 3).

Both the composition of the beach material and its size are related to the source of the material and the forces acting on the beach. Land materials are brought to the coast by the rivers or are derived from the local cliffs by wave erosion. The sands of many beaches are the ground-up, eroded products of land materials rich in minerals such as quartz and feldspar. Small particles such as sand, mud, and clay are easily transported and redistributed by waves and currents; larger rocks are usually found in the area close to their source, because

they are too large to be moved any distance. The finer particles are carried away by the currents, and the larger rocks are left scattered on the beach. The Moeraki boulders (fig. 11.25), found along the eastern coast of New Zealand's South Island, are the world's largest examples of calcite-type mud balls. Some are 2 m (6.5 ft) in diameter and are eroded from the beach cliffs. In general, a beach littered with large rocks is an eroded beach, and the remaining large rocks are called **lag deposits.** If the lag deposits accumulate in sufficient quantity to protect a beach from further wave and water erosion, the beach is known as an **armored beach** (fig. 11.26).

Some beach materials come to the beach from offshore areas. Coral and shell particles broken by the pounding action of the waves are carried to the beaches by the moving water. A beach covered with uniform small particles of sand or mud is or has been a depositional beach.

Some of the world's beaches have distinctive colors. In Hawaii, there are white sand beaches derived from coral and black sand beaches derived from lava. Green sands can be found in areas where a specific mineral (such as olivine or glauconite) is available in large enough quantities, and pink sands occur in regions with sufficient shell material.

Modifying Beaches 11.6

People like beaches and have great desires for property along these sensitive fringes of land. Sixty percent of the world population (3.8 billion people) live within 100 km (60 mi)

Figure 11.24 A spit connected to an offshore island forms a tombolo. Tombolos are often formed by converging currents from nearshore eddies; the currents transport sediment toward the tombolo from both sides.

Figure 11.25 Mud balls are washed out of the cliffs of many beaches. These "Moeraki boulders," hardened by the deposition of calcite minerals, are among the world's largest, approximately 2 m (6.5 ft) in diameter. The shape of the boulders is not due to wave action. They are found on the east coast of New Zealand's South Island.

of a coastline. Population experts expect more than 6.3 billion people to live in coastal areas worldwide within the next thirty years. In the United States today, over one-half of the population lives within 80 km (50 mi) of the coasts (including the Great Lakes). By the year 2010, coastal population in the United States is estimated to grow to more than 127 million people. The shore is affected directly and indirectly to a greater and greater extent as increasing numbers of people make their homes in these vulnerable areas.

Coastal Structures

People change beaches by damming rivers to control floods and generate power. When a dam is built, a lake is formed behind the dam, and sand, gravel, and rock that once moved down the river to join the sediments at the coast are deposited in the lake behind the dam; an important source of material to the continued balance of beaches has been removed. Although the sediment supply has been reduced, the longshore current continues to flow and carry away suspended beach material. The net result is a loss of sediment and sand from the beaches, and the beaches erode.

Coastal zone engineering projects such as **breakwaters** and **jetties** are built to protect harbors and coastal areas from the force of the waves. Breakwaters are usually built parallel to the shore; jetties extend seaward to protect or partially enclose a water area. The protected areas behind jetties and breakwaters are quiet, and the sediments suspended by the wave action and carried by the longshore current settle out in these quiet basins. The result is less material available for beaches farther down the coast, and, as the longshore current continues to flow, the next beach begins to disappear. Small-scale examples of this process are seen when **groins,** rock or timber structures, are placed perpendicular to the beach to

Figure 11.26 An armored beach. Lag deposits of large rocks and cobbles are left on an eroded beach. The rocks on this beach are eroded from glacial deposit cliffs that contain a large assortment of particle sizes, from clays to boulders.

trap sand carried in the longshore transport (fig. 11.27); sand is deposited on the up-current side of the groins and sand is lost on the down-current side.

Individual property owners, trying to stop the erosion of their property, build protective bulkheads and seawalls along the backshores of their beaches. These expensive structures are made of timbers, poured concrete, or large boulders. The land is armored in hopes of preventing erosion by storm waves, high tides, and boat wakes. Many times these structures make the erosion problem worse because the waves expend their energy over a very narrow portion of the beach. If only a portion of the coastline is protected, wave energy may concentrate at the ends of the seawall and move behind it, or severe erosion of the beach may occur at the base of the seawall, and the wall may collapse. Seawalls can reflect waves to combine with incoming waves and cause higher, more damaging waves at the seawall or at some other place along the shore.

The natural shore allows waves to expend their energy over a wide area, eroding the land but maintaining the beaches, all a part of nature's processes of give and take. Time and time again, coastal facilities have been built in response to the demands of people, only to trigger a chain of events that results in newly created problems.

Despite our efforts, our beaches are disappearing into the sea. According to a U.S. Army Corps of Engineers

Figure 11.27 These groins have been placed along the beach at Hastings in southern England to trap sand moving along the coast. Beach height difference at this groin is 4 m (12 ft).

National Marine Sanctuaries

One hundred years after the establishment of Yellowstone as our first national park, the U.S. Congress passed the Marine Protection, Research and Sanctuaries Act of 1972. The first sanctuary was Monitor National Marine Sanctuary, created in 1975 around the wreck of the Civil War ironclad *Monitor,* which lies in 70 m (230 ft) of water off Cape Hatteras, North Carolina. Today there are twelve marine sanctuaries: five off the U.S. west coast, four off the U.S. east coast, including the Florida Keys, and one each in the Gulf of Mexico, Hawaii, and American Samoa (box fig. 1). Together, these twelve sites cover 46,000 km² (18,000 mi²), less than one-half of 1% of the ocean area within U.S. national boundaries. Two more sites are proposed: one in Puget Sound, Washington, and one in Lake Huron; still others are under consideration.

Monterey Bay National Marine Sanctuary is the largest U.S. marine sanctuary. It stretches along the Pacific coast for 580 km (350 mi) and reaches seaward as much as 90 km (53 mi). This area supports a great diversity of marine life in various habitats, including an ocean canyon more than 3300 m (2 mi) deep. This sanctuary is closely associated with the research and education pro-

grams of the Monterey Bay Aquarium and Monterey Bay Research Institute (MBARI). To the south, the Channel Islands National Marine Sanctuary is home to myriad seabirds, more than twenty kinds of sharks, thousands of sea lions, and northern fur seals, elephant seals, and the rare Guadalupe fur seal, which was hunted nearly to extinction in the nineteenth century. The edges of the islands are ringed with forests of giant kelp, seaweeds that provide food and shelter for a rich diversity of marine species.

North of Monterey Bay, the Gulf of the Farallones National Marine Sanctuary lies 50 km (30 mi) west of the Golden Gate. These islands provide a dwelling place for twenty-six species of marine mammals and the largest concentrations of breeding seabirds in the continental United States. More than a quarter of a million petrels, puffins, terns, auklets, murres, and other species visit this area every year. Still farther north, the beaches of the Olympic Coast National Marine Sanctuary remain almost as they have always been, rich in marine life among rocky shores and tide pools.

Hawaiian Islands Humpback Whale National Marine Sanctuary is the breeding, calving, and nursing home for two-thirds of the

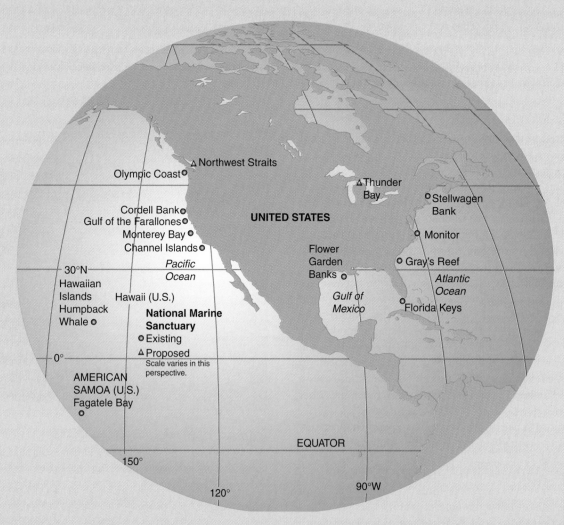

Box Figure 1 U.S. National marine sanctuaries.

North Pacific humpback whale population (box fig. 2). The very rare Hawaiian Monk seal is also found here. On the Atlantic coast of the U.S. mainland, Stellwagen Bank National Marine Sanctuary, a part of Massachusetts Bay east of Boston and north of Cape Cod, lies over an area of gravel and coarse sand left by receding glaciers. This is a very productive area and a feeding ground for humpback, fin, and northern right whales.

Sixty kilometers (100 mi) off the Texas-Louisiana coast in the Gulf of Mexico, a pair of salt domes support the northernmost coral reefs in North America. This is Flower Garden Banks National Marine Sanctuary, home to manta rays, hammerhead sharks, and loggerhead turtles. Off the tip of Florida, a 370 km (220 mi) arc of reefs, seagrass beds, and mangroves make up the Florida Keys National Marine Sanctuary (box fig. 3).

These sanctuaries are National Marine Sanctuary multiple-use areas managed by the National Oceanic and Atmospheric Administration (NOAA), a branch of the Commerce Department. Commercial fishing, sport fishing, and spear fishing are permitted in the sanctuaries; so are boating, snorkeling, diving, and other kinds of marine recreation. Shipping traffic is not prohibited, but it is restricted to certain lanes and areas. In general, drilling, mining, dredging, dumping waste, and removing artifacts are forbidden.

Sanctuary designations have not been welcomed by all local populations. The bay surrounding Stellwagen Bank is heavily fished, and those who fish are opposed to further regulation even though catches continue to decline. Also, ships bound for Boston cross sanctuary waters, producing concern that they may interfere with whale migrations. Although the Florida Keys have a permanent population of more than 80,000 people who are economically dependent on the surrounding waters and 2.5 million tourists who come to snorkel, dive, and sport fish each year (bringing in more than a $1 billion), there has been significant resistance to the presence of NOAA. In the Channel Islands a squid fishery is developing, and there is concern over the quantity of the catch, its effect on the food chains, and the numbers of other species being caught in the squid nets. How do we reconcile commercial fishing with its tradition of taking maximum sustainable harvests with the establishment of havens that seek to protect an entire sanctuary system?

Box Figure 3 A diver watches a school of grunts in a coral garden in the Florida Keys.

The annual budget for the sanctuary system is less than $12 million. Staffs are small, and sanctuaries depend heavily on volunteers. Educating the public has become a top priority, with some monitoring of water and populations. The sanctuaries are works in progress and will require continued support if they are to protect and restore these coastal environments.

To Learn More About National Marine Sanctuaries

Chadwick, D. 1998. Blue Refuges. *National Geographic* 193 (3): 2–13.
Monterey Bay Aquarium
National Marine Sanctuaries—NOAA

Internet References

Visit the book's Online Learning Center at www.mhhe.com/sverdrup (click on the book's cover) to explore links to further information on related topics.

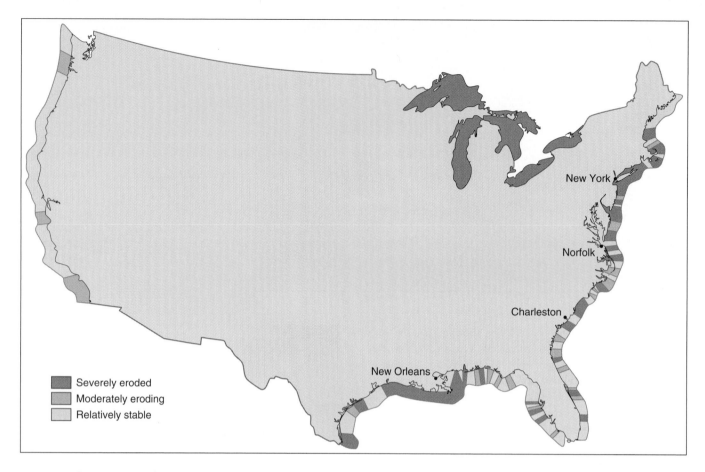

web ⬡ link **Figure 11.28** Shoreline erosion of the thirty coastal and Great Lakes states. *Source: From MMS Today 3(4) 1993, Minerals Management Service, U.S. Department of Interior.*

study, more than 40% of the U.S. continental shoreline is losing more sediment than it receives (fig. 11.28). Of the 24,000 km (15,000 mi) of shoreline, 4300 km (2700 mi) are considered critical areas meriting public protection. Most of the areas designated critical are along the Atlantic and Gulf coasts, and many are on barrier islands. Ocean City, Maryland, has lost about 25% of its beachfront in recent years. Federal officials have a plan to pump more than 1.3 million cubic yards of sand onto nearly 13 km (8 mi) of coast; the cost is estimated at $8.9 million. Louisiana's Isles Dernieres are barrier islands that are losing up to 20 m (66 ft) of shore a year; as a result, the mainland suffers increasing coastal flooding and storm damage.

Designing coastal structures that correct or solve one problem without creating more problems is as much an art as it is a science. Measurements and calculations are made, but more is required, for there must be a thorough understanding of the natural processes in an area before the consequences of the new structure can be foreseen. A scale working model is a powerful aid in understanding the present state of a shore area and the results of future changes. Natural processes such as tides, waves, and currents are reproduced on a scale model and then the proposed structure is introduced into the system. Its effect is observed, and design modifications can be made before construction begins. Models of this type are expensive, but they are much less expensive than the costs of continually correcting the results of poorly designed structures that disturb the natural environment in an undesired manner.

The Santa Barbara Story

The Santa Barbara, California, harbor project is a classic example of interference with coastal zone processes. At Santa Barbara, a jetty and a breakwater were constructed to form a boat harbor. The jetty at the north side of the harbor juts out into the sea before it turns southward, forming a breakwater that runs parallel to the coast (fig. 11.29). The longshore current and sediment transport move down the coast of Southern California from north to south (see fig. 11.23). This jetty-breakwater system creates a wave-sheltered area on the jetty's north side and also blocks the longshore current. Sand was deposited on this north side, where the longshore current was blocked, and the beach began to grow. The beaches to the south began to disappear because they were starved of the sand that had been deposited to the north.

When the beach to the north had grown until it reached the seaward limit of the jetty, the longshore current could again move beach sediments southward along the ocean side of the breakwater. When the longshore current

Figure 11.29 Santa Barbara harbor as seen from the south. In the *foreground*, a dredge removes the sediment that has built up in the protected, quiet water behind the jetty.

and the suspended sediment reached the south end of the breakwater and the entrance to the harbor, the current formed an eddy that spiraled into the quiet water of the harbor. The sediment settled out and began to fill the harbor, forming a spit connected to the end of the breakwater. The sediment settling in the harbor deprived the beaches farther south of their supply but did not alter the forces acting to remove the sediment from these same beaches.

Today, a dredge pumps the sediments deposited in the harbor through a pipe and back into the longshore current. In this way, the harbor remains open and the beaches to the south receive their needed supply of sand. Interference with a natural process requires the expenditure of much time, effort, and money to do the work nature did for nothing.

The History of Ediz Hook

Twenty-four hundred kilometers (1500 mi) north of Santa Barbara, another harbor has been altered through interference with the supply of beach material. The harbor of Port Angeles, on the Strait of Juan de Fuca in Washington State, is protected from the storm waves in the strait by a naturally occurring 5.6 km (3.5 mi) long curving spit known as Ediz Hook (fig. 11.30). The hook is composed mainly of sand and gravel and protects an area large enough and deep enough to accommodate the largest and most modern commercial vessels. In recent years the hook has undergone considerable erosion, and the danger of waves breaking through the hook near its base was severe.

To understand how the hook came to exist at all and how its present condition has come about one must go back about 124,000 years, to the time when the glaciers of the last ice age retreated from this area. At that time sea level was lower than it is now, and the Elwha River west of Port Angeles carried loose glacial deposits to the sea, forming a delta that curved to the east under the influence of local currents and waves. As sea level began to rise, the river, currents, and waves continued to carry sediment eastward, and the waves began to erode the cliffs, adding still more sediment to the longshore current. The result was a hook-type spit that built up where the shoreline makes an abrupt angle with the strait. Ediz Hook was gradually produced and was kept constantly supplied with sand and gravel from cliffs and the river.

In 1911, a dam was constructed across the Elwha River to provide power and a freshwater reservoir. The sediments that had been flowing toward the hook (estimated at 30,000 m^3 per year) were now being deposited behind the dam. In 1930 a pipeline to deliver fresh water from the Elwha Reservoir to Port Angeles was constructed around the cliffs at beach level. Bulkheads built along the face of the cliffs to protect the pipeline cut off the hook's supply of cliff-eroded sediments, estimated at 380,000 m^3 per year.

Ediz Hook is more than protection for the harbor; it is also the site of a large paper mill, a U.S. Coast Guard Station, and several small harbor facilities. All are connected by a road running the length of the hook. In the 1950s the city acted to protect the seaward side of the

Figure 11.30 Ediz Hook is a long spit forming a natural breakwater at Port Angeles, Washington. The protected harbor behind the spit is deep enough for mooring the largest supertankers (infrared false-color photo).

hook's base by applying large boulder armoring as well as steel bulkheads. Cliff material was blasted into the water in hopes of supplying the required sediments. A constant battle between the people and the sea began. Nearly as fast as the armor was applied to the base of the spit, the waves tore away the protective barriers and removed the sediments. Only one-seventh of the total sediments once available to feed the spit remained in the longshore transport, and studies by the Army Corps of Engineers showed a possible loss of 270,000 m³ of sand each year from the outside of the hook. The hook was now sediment starved.

In the winter of 1973–74 severe storms again damaged the hook, and a Corps of Engineers' program to strengthen the hook began. Over a fifty-year period, annual maintenance costs have been projected at about $30 million compared with revenue from the harbor of $425 million. Since the benefit-to-cost ratio is about 14:1, the project seems to be economically viable.

Estuaries 11.7

River mouths, fjords, fault bays, and other semienclosed bodies of salt water are all part of the coastal zone. If one of these has free connection with the ocean and is diluted with fresh water so that its average salinity is less than that of the adjacent sea, it is an **estuary.** Estuaries can be described by the geological processes that formed them, and they can also be described by the dynamics of their freshwater and saltwater circulation. Geomorphology may be used to distinguish between estuary types in the same way that it is used to classify coasts. However, such a system does not consider the complex and dynamic combination of factors that come together when fresh and salt water meet. In this chapter, estuaries are described mainly in terms of water exchange and salinity variation. Tides, river flow, and the geometry of an inlet are additional factors to be considered. Four basic estuary types are distinguishable.

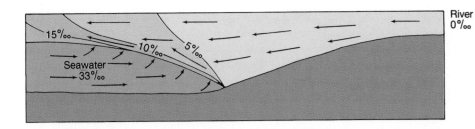

Figure 11.31 The salt wedge estuary. The high flow rate of the river holds back the salt water, which is drawn upward into the fast-moving river flow. Salinity is given in parts per thousand (‰).

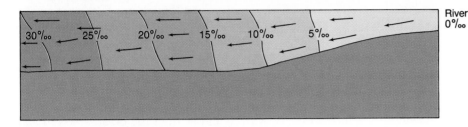

Figure 11.32 The well-mixed estuary. Strong tidal currents distribute and mix the seawater throughout the shallow estuary. The net flow is weak and seaward at all depths. Salinity is given in parts per thousand (‰).

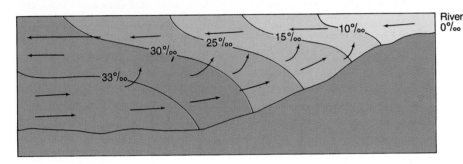

Figure 11.33 The partially mixed estuary. Seawater enters below the mixed water that is flowing seaward at the surface. Seaward surface net flow is larger than river flow alone. Salinity is given in parts per thousand (‰).

Types of Estuaries

The simplest type of estuary is the **salt wedge estuary.** Salt wedge estuaries occur within the mouth of a river flowing directly into salt water. The fresh water flows rapidly out to sea at the surface, while the denser seawater attempts to flow upstream along the river bottom. The seawater is held back by the flow of the river that forms a sharply tilted density boundary between the intruding wedge of seawater and the fresh water moving downstream. A salt wedge estuary is shown in figure 11.31. The net seaward surface flow is almost entirely fast-moving river water, because the large discharge of fresh water is forced into a thin layer above the salt wedge. The salt wedge moves upstream on the rising tide or when the river is at a low flow stage; the salt wedge moves downstream on the falling tide or when the river flow is high. The boundary between the salt water and the overriding river water is kept sharp by the rapidly moving river water. The moving fresh water erodes seawater from the face of the wedge, mixing it upward into the turbulent river water and raising the salinity of the seaward-moving fresh water. This one-way mixing process is called **entrainment;** very little river water is mixed downward into the salt wedge. Seawater from the ocean is continually added to the salt wedge to replace the salt water entrained into the seaward-flowing river water. In a salt wedge estuary, the circulation and mixing are controlled by the rate of river discharge; the influence of tidal currents is generally small compared to the influence of the river flow. Examples of salt wedge estuaries are found in the mouths of the Columbia, Hudson, and Mississippi Rivers. In the Columbia River, the salt wedge moves as far as 25 km (15 mi) upstream at times of high tide and low river flow, maintaining its identity with a sharp boundary between the two water types. Salt wedges also occur in the mouths of rivers where the rivers enter other types of estuaries: for example, the mouth of the Sacramento River in San Francisco Bay.

Estuaries that are not of the salt wedge type are divided into three additional categories on the basis of their net circulation and the vertical distribution of salinity. These categories are the well-mixed estuary, the partially mixed estuary, and the fjord-type estuary.

Well-mixed estuaries (fig. 11.32) have strong tidal mixing and low river flow, creating a slow net seaward flow of water at all depths. Note that the mixing due to strong tidal turbulence is so complete that the salinity of the water is uniform over depth and decreases from the ocean to the river. There is little or no transport of seawater inward at depth; instead, salt is transferred inward by turbulent diffusion. The nearly vertical lines of constant salinity move seaward on the falling tide or when river flow increases, and they move landward on the rising tide or when river flow decreases. Shallow estuaries, including the Chesapeake and Delaware Bays, are well-mixed estuaries.

Partially mixed estuaries (fig. 11.33) have a strong net seaward surface flow of fresh water and a strong inflow of seawater at depth. The seawater is mixed upward and combined

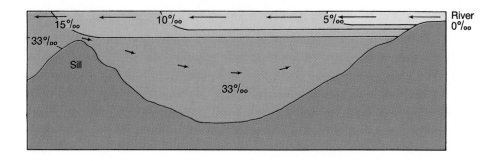

Figure 11.34 The fjord-type estuary. River water flows seaward over the surface of the deeper seawater and gains salt slowly. The deeper layers may become stagnant owing to the slow rate of inflow. Salinity is given in parts per thousand (‰).

with the river water by tidal current turbulence and entrainment to produce a seaward surface flow that is larger than that of the river water alone. This two-layered circulation acts to rapidly exchange water between the estuary and the ocean. Salt moves into a partially mixed estuary by turbulent diffusion, but also, and more important, by **advection,** the inflow of seawater at depth. Examples of partially mixed estuaries include deeper estuaries, for example, Puget Sound, San Francisco Bay, and British Columbia's Strait of Georgia.

Fjord-type estuaries (fig. 11.34) are deep, small-surface-area estuaries with moderately high river input and little tidal mixing. This pattern occurs in the deep and narrow fjords of British Columbia, Alaska, the Scandinavian countries, Greenland, New Zealand, Chile, and other glaciated coasts. In these estuaries, the river water tends to remain at the surface and move seaward with little mixing of the underlying salt water. Most of the net flow is in the surface layer, and there is little influx of seawater at depth. Salt is supplied at a slow rate by advective transport. The deeper water may stagnate because the entrance sill isolates it (see fig 5.6). There is little inflow below the surface layer.

The salt wedge estuary exists only as a strongly stratified estuary. These estuaries are never very well mixed vertically, except above the boundary between the layer of seaward-moving river water and the salt wedge. The other estuary types have various degrees of vertical stratification between the highly stratified, or poorly mixed, and the weakly stratified, or well mixed. The processes that govern the degree of vertical mixing and stratification are the strength of the oscillatory tidal currents, the rate of freshwater addition, the roughness of the topography of the estuary over which these currents flow, and the average depth of the estuary. Tidal flow that oscillates with the rise and fall of the tide produces the energy for mixing; it does not usually result in a preferred directional flow and should not be confused with the net flow in each estuary.

Not all estuaries fit neatly into one of these categories. There are estuaries that fit between the types, and some estuaries change from one type to another seasonally with changes in river flow or weekly as the tides change from springs to neaps. Studying how an individual estuary compares to examples of the different types helps the oceanographer understand the processes that govern an estuary and its water exchange with the ocean.

Circulation Patterns

This discussion of estuary circulation uses the partially mixed estuary, with its inflow from the ocean at depth and its surface outflow of mixed river and seawater. Tidal currents are produced as the water volume of the estuary increases on the rising tide and decreases on the falling tide. Water in these currents moves back and forth in the estuary, but it does not immediately leave the system and enter the open sea. However, the surface water moves farther seaward on the falling tide than it moves inward on the rising tide, and a parcel of surface water is moved progressively farther seaward on each tidal cycle. In the same way, the water from the sea moves into the estuary on the rising tide farther than it drops back seaward on the falling tide. Averaged over many tidal cycles, this pattern produces a net movement seaward at the surface and a net movement landward at depth.

The **net circulation** of an estuary out (or seaward) at the top and in (or landward) at depth is of great importance, for it is this circulation that carries wastes and accumulated debris seaward and disperses them in the larger oceanic system. It is also through this circulation that organic materials and juvenile organisms produced in the estuaries and their marsh borderlands are moved seaward, while nutrient-rich water is brought inward at depth and replenishes the estuary's water and salt.

Understanding the net circulation of an estuary and evaluating the flows of surface water seaward and the underlying ocean water landward can be a long process. To determine this pattern directly the oceanographer must install recording current meters at various depths and at several cross-channel locations. Currents are measured over many tidal cycles for both spring and neap tides and at times of low, intermediate, and high freshwater discharge. The current records from each depth and location are averaged to discover the net current distribution for a given cross section in the estuary and the net flow of the estuary. Data are averaged over one-week periods to determine the importance of spring and neap tides, over monthly cycles to find changes in patterns due to seasonal river and climate fluctuations, and over several years to produce an annual mean pattern. The cost of installing and maintaining equipment, as well as the costs of processing the collected data, make this direct approach a very expensive one.

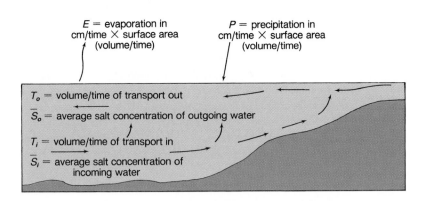

Figure 11.35 Water and salt enter and exit a partially mixed estuary.

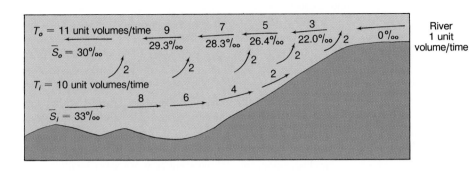

Figure 11.36 The seaward-moving surface water (*odd-numbered arrows*) increases in flow volume as the inflowing seawater (*even-numbered arrows*) moves upward. Upward mixing due to tidal turbulence is indicated by the vertical arrows. For every 2 units of seawater mixed upward, the inflow decreases and the outflow increases by the same 2 units. Salinity is given in parts per thousand (‰).

A less expensive, indirect approach is to assume a **water budget** of inflow equal to outflow and a constant estuary volume averaged over time; all processes adding water to the estuary equal all processes removing water. A **salt budget** is also assumed; salt added is equal to salt removed. At the entrance to the estuary, measurements of salinity are made and averaged over space and time, establishing the average salinity of the outflowing surface water, $\bar{S}_o$, and the average salinity of the deeper inflowing seawater, $\bar{S}_i$. The ratio

$$\frac{\bar{S}_i - \bar{S}_o}{\bar{S}_i}$$

is used to determine the fraction of river water in the seaward-moving surface layer at the estuary entrance.

If the rate of total river water inflow (R) is known for this same period, the volume rate of the surface layer's seaward flow (T_o) is found by using the following formulas:

$$\frac{\bar{S}_i - \bar{S}_o}{\bar{S}_i}(T_o) = R, \quad \text{or} \quad T_o = \frac{\bar{S}_i}{\bar{S}_i - \bar{S}_o}(R)$$

In a partially mixed estuary the net seaward flow (T_o) is always larger than the river flow, (R), and the river flow and the net saltwater inflow (T_i) combine to produce the seaward flow (T_o) and maintain the water budget. Inflow equals outflow.

$$T_o = T_i + R$$

Both evaporation, E, and precipitation, P, also remove and add water at the surface, as shown in figure 11.35. If both E and P are large, then R in the previous equations becomes ($R - E + P$).

This method assumes that, over the time period for which the calculations are used, the average water volume and the total salt content of the estuary remain constant. For more information on these equations see appendix C.

Temperate-Zone Estuaries

In the middle latitudes of North America, most estuaries gain their fresh water from rivers, and the evaporation of water from the estuary surface is minor or is nearly balanced by direct precipitation. The salt concentration of the entering seawater, $\bar{S}_i$, is about 33‰ and $\bar{S}_o$ is about 30‰. When we use the preceding equation and the example in figure 11.36, T_o is shown to be about eleven times the rate of river inflow, or R. T_i in this case is ten times R. The influx of seawater at depth decreases as it moves into the estuary, and the surface flow increases because of mixing and the incorporation of seawater from below as it moves seaward. While this mixing is occurring, the average salt concentration of the seaward-moving surface flow is also increasing, and T_o is formed when 10 unit volumes of seawater have combined with the 1 unit volume of river water. These relatively large values for T_o and T_i compared to R make these net flows the key to the exchange of estuary water with ocean water.

In some fjordlike estuaries, the surface layer is only as deep as the shallow sill, and the average salt content of the surface layer is very low, owing to weak tidal mixing. In this case, T_o is approximately equal to R; T_i is negligible, and the deep water of the fjord can stagnate.

High Evaporation Rates 11.8

Bays located near 30°N and 30°S latitudes have low precipitation and high evaporation rates. Although rivers may flow into these bays, these bays are not estuaries if they do not

Rising Sea Level

Sea level is once again on the rise around the coasts of the world. Over the past century, sea level has risen about 15 cm (6 in), but recently the rate of rise has increased. Satellite altimetry measurements indicate that sea level worldwide is presently rising at a rate of 3.9 ± 0.8 mm (0.15 ± 0.032 in) per year. It is thought that sea level may gain another 20 cm (8 in) in the next fifty years, a rate that is about three times greater than that experienced over the last century. By the year 2100, sea level is expected to be 31–110 cm (1–3.5 ft) higher than at present; the most knowledgeable estimate is for an increase of 66 cm (2 ft).

Rising sea level is a global phenomenon, but the problems that it causes are not distributed evenly around the world. Areas in mid-ocean such as small tropical atolls with elevations of only 1–2 m (3–7 ft) may be severely affected. Low coastal marshlands will undergo greater changes than coasts with steep rocky shores. Each year 100–130 km² (40–50 mi²) of Louisiana's Mississippi River wetland marsh disappears under the Gulf of Mexico (box fig. 1). Twelve thousand years ago, Louisiana's shoreline extended 200 km (120 mi) farther into the Gulf of Mexico. In the mid-1880s Bailize was a busy river town at the tip of the delta; today it is under 4.5 m (15 ft) of water.

Significant sea-level changes over hundreds to thousands of years are governed by variations in the amount of water stored in the Earth's reservoirs. Water is transferred from ocean to ice during glacial periods and back to the ocean during interglacial periods; this transference can account for a change in sea level of 100–200 m (300–650 ft).

Sediment accumulation in ocean basins also displaces seawater upward. If the oceans of the world warmed by just 1°C, world sea levels would rise by about 60 cm (24 in) owing to the thermal expansion of seawater. Over much longer periods of time, tectonic forces rearrange the continents, the shapes of the ocean basins change, and sea level tends to fall as landmasses combine and to rise as landmasses disperse. Coasts may also rise or sink relative to sea level in response to isostatic forces.

The global average change in sea level may not be reflected in the gradual changes observed at any particular location. In the tropics, sea temperatures are nearly constant over the annual cycles, so changes attributable to thermal expansion are small, but the seasonal oscillation of the meteorological equator (the intertropical convergence zone) between approximately 0° and 10°N causes sea level to vary in response to atmospheric pressure changes. Changes in sea surface temperature associated with El Niño (see chapter 6) affect sea level, and in the North Atlantic and North Pacific the annual seasonal shift from high to low pressure over the oceans causes a related change in sea level (rising with low atmospheric pressure and falling with high pressure). Onshore winds elevate and offshore winds depress the coastal water level. Ocean currents under the influence of the Coriolis effect create a sea surface topography that varies by as much as a meter. Any climate-generated changes in speed or location of winds or currents alter sea level. Climate changes associated with atmospheric warming can be expected to alter the winds, the currents, and atmospheric pressure distributions as well as the thermal expansion of seawater. These changes may be transferred to the sea surface, resulting in sea-level changes of as much as 1–2 m (3–7 ft).

Whatever the predictions, it is unlikely that most people will give up their dreams of a summer cabin on the shore or a home on the beach. There may be a greater awareness of sea-level changes as coastal erosion problems increase in the future, but human efforts are likely to be too little and too late in confronting the forces that produce these changes.

To Learn More About Rising Sea Level

ATOC Consortium. 1998. Ocean Climate Change: Comparison of Acoustic Tomography, Satellite Altimetry and Modeling. *Science* 281 (5381): 1327–32.

Schneider, D. 1998. The Rising Seas. *Oceans, Scientific American Quarterly* 9 (3): 28–35.

Internet References

Visit the book's Online Learning Center at www.mhhe.com/sverdrup (click on the book's cover) to explore links to further information on related topics.

---

experience net dilution. The contribution from the rivers is usually minor when compared to the loss of water due to evaporation. The Red Sea and the Mediterranean Sea are examples of such areas. In these seas, evaporation increases the surface water salinity and density directly, causing surface water to sink and accumulate at depth, where it flows seaward. Ocean water flows into the sea at the surface, because it is less dense than the high-salinity outflow of deep water. The T_o and T_i flows have reversed, with T_o at depth and T_i at the surface (fig. 11.37). Because of this reversal, these bays are sometimes called **inverse estuaries.**

Compare figures 11.36 and 11.37. The evaporative bay in figure 11.37 shows a typical $\bar{S}_i$ value (36‰) for ocean surface conditions at latitudes of 30°N and 30°S. The evaporative loss from the surface of this bay is 1 unit volume per time, resulting in a T_i flow of 20 units and a T_o flow of 19 units. The resulting $\bar{S}_o$ value is 37.89‰. The estuary in figure 11.36 has a river input of 1 unit volume per time and T_o and T_i flows of 11 and 10 units, respectively. In general, if the evaporation removal rate of water is similar to the river input of a temperate estuary, the evaporative bay has a better rate of exchange with the ocean. Not only is

Box Figure 1 The Mississippi River delta. Twelve thousand years ago, Louisiana's shoreline extended 200 km (120 mi) farther into the Gulf of Mexico. Today, the river is kept in its channel by levees, and the sediments continue to the river's mouth. Sediments no longer replenish the delta and its marshes; instead they are carried out into deep water.

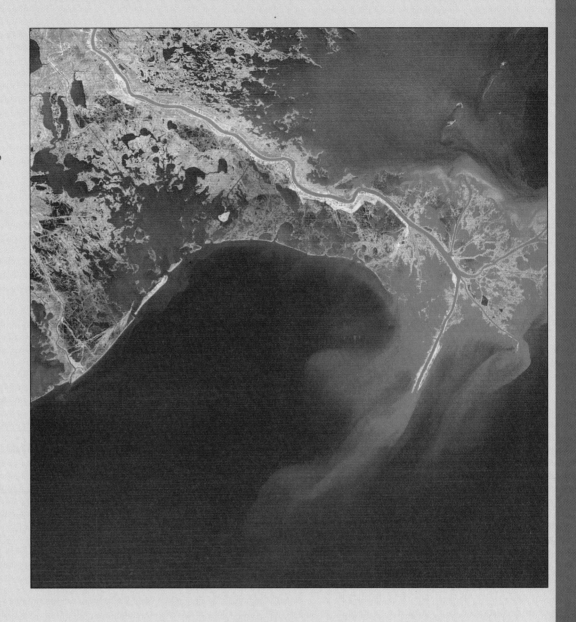

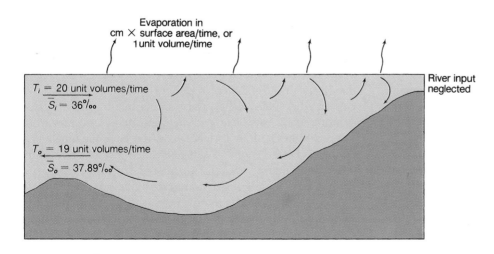

Evaporation in cm × surface area/time, or 1 unit volume/time

$T_i = 20$ unit volumes/time
$\overline{S_i} = 36‰$

$T_o = 19$ unit volumes/time
$\overline{S_o} = 37.89‰$

River input neglected

Figure 11.37 The evaporative sea or inverse estuary. Evaporation at the surface removes water (1 unit), and river inflow is negligible. Seawater flows inward at the surface ($T_i = 20$ units), and the seaward flow ($T_o = 19$ units) is at depth. Salinity is given in parts per thousand (‰).

T_o larger, but the exchange of water is more complete because the evaporative bay has a continuous overturn as surface water sinks and moves seaward.

Flushing Time   11.9

If the mean volume of an estuary is divided by T_o, an estimate of the length of time required for the estuary to exchange its water is obtained; this is known as **flushing time.** If the net circulation is rapid and the total volume of an estuary is small, flushing is rapid, and the flushing time is short. A rapidly flushing estuary has a high carrying capacity for wastes, because the dissolved or suspended wastes are moved rapidly out to sea and diluted. A slowly flushing estuary risks accumulating wastes and building up high concentrations of land-derived pollutants. Population pressures on estuaries are heavy, for these areas are used as seaports, recreational areas, and industrial terminals as well as for their fishing resources. Understanding the circulation of our estuary systems is essential to maintaining them as healthy, productive, and useful bodies of water.

If the waste products are associated only with the fresh water entering the estuary, it is possible to determine the ability of the fresh water to carry these products through the estuary without considering the action of the entire estuary. To do so the amount of fresh water in the estuary at any one time must be known. The freshwater volume is estimated from the average salinity of the estuary. For example, an estuary with an average salt concentration of 20‰ is one-third fresh water if the adjacent ocean salinity is 30‰. Dividing the freshwater volume by the rate of river water addition yields the flushing time for fresh water and its pollutants.

Some bays and harbors do not have a freshwater source for dilution; these areas are not estuaries but are flushed by tidal action. On each change of the tide, a volume of water equal to the area of the bay multiplied by the water-level change between high and low water exits and enters the bay. When this volume, known as the **intertidal volume,** leaves the bay and enters the ocean, it may be displaced down the coast by a prevailing coastal current. On the next rising tide, an equivalent volume of different ocean water enters the bay. Under these conditions, the flushing time can be estimated by the number of tidal cycles required to remove and replace the bay volume. This number is found by dividing the mean bay volume by the intertidal volume exchanged per tidal cycle. However, the flushing is often not complete, because currents along the coast do not always displace the exiting intertidal volume a sufficient distance to prevent some portion of it from recycling into the bay. Therefore, the number of tidal cycles required for flushing may be greater than the calculated estimate.

The same conditions occur in estuaries that have a two-layered flow if mixing due to turbulence incorporates some of the exiting surface water into the deeper water flowing in from the sea. This mixing results in partial recycling of the T_o flow and leads to modification of the T_i flow and aeration of the deeper parts of the estuary.

Occasionally, the net circulation in an estuary is altered by ocean conditions. For example, if coastal upwelling occurs, the inflow of denser water to the estuary increases, and this increased inflow accelerates the estuary's circulation and causes the outflow to increase. When upwelling ceases and downwelling occurs, less dense seawater is present at the estuary entrance, and the inflow is reduced. In these circumstances, the estuary's circulation may temporarily reverse as the denser, deeper water flows back to sea and the less dense seawater moves inward at the surface.

Although each estuary has its own distinctive characteristics, the knowledge gained from studying one estuary can be used to understand other similar estuary systems. Understanding the circulation patterns and the processes that form these patterns allows us to judge the degree to which an estuary can be modified and used while still preserving its environmental and economic value. Because estuaries lie at the contact zone between the land and the sea, and because humans are inhabitants of the land, alterations have been made in these areas and probably cannot be entirely avoided in the future. But we owe it to ourselves and to those who come after us to make the most intelligent and knowledgeable uses possible of these areas, for they represent a great natural resource that can renew itself if treated with care.

Practical Considerations: Case Histories   11.10

The Development of San Francisco Bay
The San Francisco Bay estuary (fig. 11.38) has a surface area of 1240 km^2 (480 mi^2). Its river systems drain 40% of the surface area of California. When visited by the early Spanish missionaries in 1769, the bay was surrounded by wetlands, where the 10,000–15,000 native people gathered much of their food. Development came slowly until 1848, when gold was discovered, and the population of San Francisco increased from approximately 400 to 25,000 in 1850. Within fifty years of the initial gold boom, the marshes were nearly gone, the bay was shallowed, fresh water had been siphoned off for irrigation, and nonnative species of animals had been introduced to the bay's waters, displacing native species.

Early in the bay's development, the fisheries' resources (salmon, sturgeon, sardines, flatfish, crabs, and shrimp) were heavily exploited to feed the rapidly expanding population. By 1900, many of the fish, shellfish, and wildfowl stocks had been overharvested and depleted. Crabs were fished out in the 1880s, and the fishery was moved offshore, until its collapse in the 1960s. When the transcontinental railroad was completed in 1869, carload after carload of oysters were shipped from the East to mature on the bay's mudflats. The oysters did not become a self-reproducing stock, but many other small marine animals introduced with the oysters did naturalize, including the eastern soft-shell clam, the Japanese littleneck clam, and the marine pests: the oyster drill and the shipworm. Even the striped bass, the bay's best-known sport fish, was introduced from the East Coast in 1879. The only

web link **Figure 11.38** The San Francisco Bay system. The city of San Francisco lies at the *bottom center*. The large, light-colored northern portion of the San Francisco Bay system is known as San Pablo Bay. The Sacramento River is joined by the San Joaquin River at the Y in the *upper center* of the photograph.

Practical Considerations: Case Histories **329**

commercially harvested fish left in the bay are herring and anchovy. For the story of more recent destructive invaders, see chapter 12.

Until 1884, hydraulic mining for gold in the surrounding watershed washed huge amounts of silt and mud directly into the streams and rivers. The sands and muds in the rivers destroyed the salmon spawning areas. Much of the sand and mud reached the bay, where it shallowed some areas, expanded the marshes, and altered the tidal flow. The rivers were flooded with the runoff from the barren land, and the resulting winter and spring floods produced changes in the estuaries, where the rivers enter the bay.

The marshlands of the bay and its river deltas, particularly the deltas of the Sacramento and the San Joaquin Rivers, were diked, first to increase agricultural land and then for homes and industries. The conversion of wetland to dry land has reduced the tidal marshes from an area of about 2200 km² (860 mi²) to 125 km² (49 mi²). The increased cropland required increased irrigation, and the state built a series of dams, reservoirs, and canals with a water-storage capacity of 20 km³ (0.5 mi³). Today, 40% of the Sacramento and San Joaquin Rivers' flow is removed for irrigation, and another 24% is sent by aqueduct to central and southern California.

The diversion of water to the fields so reduces the amount of river water during periods of low flow in the summer that the irrigation pumps cause the water to flow upstream from the bay, carrying with it hundreds of juvenile salmon and striped bass, which are drawn into the pumps and die on the fields or in the ditches. It is thought that this has been a major cause of a drop in the population of striped bass since the 1960s. The loss of freshwater flow also changes the distribution and abundance of small, floating species on which the juvenile fish feed.

Intensive agricultural practices using fertilizers and pesticides, plus the runoff of water high in salts leached from the irrigated fields, have changed the quality of the fresh water entering the bay. An effort to lessen this effect involved directing runoff water from agriculture into holding reservoirs, some of which were also wildlife refuges. In 1982, dead vegetation, deformed wildlife, and depressed reproduction in bird species were discovered at the Kesterson Reservoir. Levels of the element selenium, 130 times that of normal, were found in plant and animal tissues. The selenium was being leached from the irrigated soils and concentrated in the reservoir. Kesterson Reservoir was closed in 1986, but the water from the fields continues to flow into the San Joaquin and on into the bay. Recently, high concentrations of selenium have been found in south bay ducks.

Domestic and industrial wastes from urban areas also enter the bay. Residence time of water in north San Francisco Bay fluctuates between one day during peak river flow in the winter to two months during low summer flow. Winter flows carry fresh water into the south bay and increase its exchange with the central bay, decreasing south bay residence time from months to weeks at this period of the year. In the south bay, the city of San Jose has increased its fresh-water sewage-treatment discharge as its population has grown. In addition to the pollution problems of the increased effluent, since the 1980s this freshwater increase has converted hundreds of acres of salt marsh to freshwater and brackish marsh, producing loss of habitat for native birds and small animals. San Jose has been ordered to create new habitat to offset this loss and to reclaim water from its sewage effluent in order to decrease the freshwater flow. If these improvements fail, a limit will be imposed on the volume of discharge, requiring the city to place limits on its population growth.

San Francisco Bay is considered the most modified major estuary in the United States. The city of San Francisco's population is roughly 775,000 in a metropolitan area of about 7 million. Despite all the factors involved in this increasingly crowded area, the bay appears to have suffered less total water-quality degradation than other major estuaries. Water quality has been maintained, in part, because the greatest urbanization has occurred near the mouth of the bay, allowing wastes to exit quickly through the Golden Gate. Enhanced sewage treatment since the 1960s has improved conditions. There are patches of contamination from PCBs, oil, and chemical spills, but at this time the bay does not seem to be suffering from an overproduction of marine plants or a depletion of oxygen. Shellfish collection has recently been permitted for the first time in decades. The pressure for increased growth, the recent years of drought, and the loss of fresh water to irrigation continue to affect bay plant and animal populations; brackish-water marshes disappear or convert to saltwater systems, displacing the native organisms. The delta smelt has declined by 90% from past levels, while adult striped bass have decreased from over 3 million in the past to 500,000 in 1991. In 1969, 118,000 winter chinook salmon made their way up the Sacramento River; in 1992 only 191 were counted. Since 1993 rainfall and river flow have improved in Northern California, and in the summer of 1995 a large coastal run of chinook salmon appeared in San Francisco Bay.

Local concern and five years of work resulted in the 1965 creation of the Bay Conservation and Development Commission (BCDC). Under the BCDC public access to the bay's 276 miles of shoreline increased from 4 miles to 110 miles; the surface area of the bay increased by 800 acres, and filling was reduced from 2300 acres per year to nearly zero. The BCDC has permanently protected 99,000 acres of bay wetlands and has purchased an additional 10,000 acres in the north bay for wildlife habitat.

The Situation in Chesapeake Bay

The Chesapeake Bay estuary (fig. 11.39) is a shallow, drowned river valley with a surface area of 11,500 km² (7000 mi²) and an average depth of 6.5 m (21 ft). Chesapeake Bay flushes slowly; the residence time of the water is 1.16 years. The Chesapeake estuary has a long history as a major provider of oysters, blue crab, wildfowl, rockfish, and shad. It has also been under continuously increasing population pressure since colonial times and has been absorbing more and more wastes

web link **Figure 11.39** Chesapeake Bay is a shallow estuary located on the heavily populated eastern coast of the United States. The entrance to the estuary is at the lower right.

from the expanding urban centers of Washington, D.C., Baltimore, Norfolk, and dozens of towns in Maryland, Virginia, and Delaware. After heavy harvesting in the 1950s the oyster catch dropped by two-thirds between 1960 and 1983. The oyster population at one time was large enough to filter a volume of water equal to the bay's volume of floating microscopic algae in about two weeks; the present low population takes about one year to filter an equivalent volume. Also between 1960 and 1983, the annual rockfish catch decreased from 6 million to 600,000 pounds, commercial fishing for shad has nearly disappeared in the upper Chesapeake, and the wildfowl population drop has been equally dramatic.

Although present harvests are low compared to those of the early years, the blue crab harvest for 1992 increased, and some areas report increased oyster catches. The bay's present total harvest of seafood is approximately 45 million kilograms per year (99×10^6 lb) with a dockside value of nearly $1 billion. The total value of the bay to Maryland and Virginia is estimated at $678 billion.

Partially treated and untreated sewage was recognized as the cause for outbreaks of typhoid fever among people who drank water and ate shellfish taken from the system in the early 1900s. Then, decaying untreated sewage robbed the water of oxygen; now plant growth, stimulated by nutrients from the effluent of sewage treatment plants and the runoff from fertilized fields, decays and consumes the dissolved oxygen. Increased plant growth also reduces the clarity of the water and shades the bottom. The low oxygen levels and the reduction in bottom plant life due to decreased light alter the bottom habitat. In addition, wastes from some 5000 factories, from military bases, and from sewage plants located between Virginia and New York find their way into the Chesapeake, adding conventional pollutants and toxic compounds that are trapped in the bottom sediments.

Efforts to improve water quality in Chesapeake Bay have produced nearly 4000 studies since the 1970s. In 1972, the Federal Clean Water Act provided enforcement standards and a system of permits governing the amount of pollutants that individual dischargers could dump into any body of water. Gradually, industrial and municipal dischargers were regulated, and efforts have reduced the input of phosphorus from sewage by 75% since 1981. At the same time populations in all the areas of the Chesapeake have increased, and industries have expanded, causing the rate of effluent discharge to increase as well.

At the present time, it is estimated that industries and sewage plants along the shores of Maryland and Virginia discharge about 4 trillion gallons of waste water annually, or about 20% of the water in the bay at any one time. An EPA study costing $27 million over seven years launched the "Save the Bay" campaign in 1983. Scientists working for the campaign reported that industries in Maryland dump more than 2700 tons of heavy metals into the bay each year and that Virginia industries dump more than 400 tons during the same period. Much of the money has been spent to manage soil and fertilizer runoff from surrounding fields. About 41 million acres of the Chesapeake watershed are farmed by less than 3% of the watershed's population. Farming uses almost 700 million pounds of commercial fertilizer each year. Use of fertilizers has led to an increase in the ratio of nitrogen to phosphorus in bay waters, and many observers feel that this is the major cause of outbreaks of the toxic organism *Pfiesteria piscicida*, also known as "the cell from hell" (see chapter 15).

Critics charge that the Save the Bay program ignores industrial and municipal discharges in excess of the levels allowed under the current permit system. The federal permit system is a self-policing effort, with dischargers setting their own limits depending on available technology and cost. The permit system has not responded to new technologies and enforcement and, until recently, has not been independently monitored. Many industries trying to avoid the discharge permit process direct their discharges to municipal sewage systems that are principally equipped to treat domestic and organic wastes. Chesapeake Bay is shallow and has a large surface area; it has a very large ratio of watershed drainage area to volume of water. River systems that carry pollutants to the bay drain parts of New York, Pennsylvania, Delaware, Maryland, Washington, D.C., Virginia, and West Virginia. An unusual degree of cooperation must be achieved to solve the bay's problems.

While trying to understand the effects of human degradation of the Chesapeake, researchers have discovered that the natural changes in rainfall and temperature have large effects as well. At irregular intervals, tropical storms and hurricanes produce flood conditions that have catastrophic effects on the water quality and the organisms. Cyclic variation in survival of eastern oysters and striped bass have been related to such weather changes. In 1972, 1975, and 1979, the annual freshwater inflow was double the average. In 1972 in particular, Tropical Storm Agnes dropped 25.5 cm (10 in) of rainfall in a short period of time. This sudden surge of fresh water reduced the bay's salinity to about one-fourth its normal minimum level and rinsed out the small floating organisms on which the larger animals feed. Severe winters, occurring at six- to eight-year intervals, promote heavy ice formation, which depresses the oxygen levels beneath the ice. These irregular but significant events superimposed on the increasingly degrading human influences make it difficult to determine exactly which factors are responsible for which changes in the Chesapeake estuary system.

Summary

The coast is the land area that is affected by the ocean. The shore extends from the low-tide level to the top of the wave zone. The beach is the accumulation of sediment along the shore.

Primary coasts are formed by nonocean processes (for example, land erosion; river, glacier, or wind deposition; volcanic activity; and faulting). Primary coasts include fjords, drowned river valleys, deltas, alluvial plains, dune coasts, lava and cratered coasts, and fault coasts. Secondary coasts are modified by ocean processes (for example, ocean erosion; deposition by waves, tides, or currents; and modification by marine plants and animals). Erosion produces regular and irregular coastlines. Deposition of eroded materials creates sandbars, sand spits, and barrier islands. Barrier islands protect the continental coastline from storms, but the islands receive each storm's energy and destructive force.

Beaches exist in a dynamic equilibrium; supply balances removal of beach material. Gentle summer waves move the sand toward shore during onshore transport. In winter, high-energy storm waves scoop the sand off the beach and deposit it in a sandbar during offshore transport. In the surf zone, the breaking waves produce a longshore current. The longshore current moves the sediment down the shore as longshore transport. A beach accumulating as much material as it loses is in equilibrium. Rip currents move water and sediment seaward through the surf zone.

A series of drift sectors forms a coastal circulation cell that defines the paths followed by beach sediments from source to deposition. The energy for the movement comes from the wind-driven waves and the wave-produced currents.

A typical beach has features that include an offshore trough and bar and a beach area comprising a low-tide terrace, beach face, beach scarp, and berms. Winter, or storm, and summer berms are produced by seasonal changes in wave action. Beaches are described by their shape; the size, color, and composition of beach material; and their status as eroded or depositional beaches.

When people build dams, breakwaters, groins, and jetties, they interfere with the dynamic processes that create and maintain the oceans' beaches. When beaches are destabilized they erode, and today 40% of the U.S. continental shoreline loses more sediment than it receives. Santa Barbara harbor and Ediz Hook are examples of human interference with natural processes.

An estuary is a semi-isolated portion of the ocean that is diluted by fresh water. In the salt wedge estuary seawater becomes a sharply defined wedge moving under the fresh water with the tide; circulation and mixing are controlled by the rate of river discharge. In a shallow, well-mixed estuary, there is a net seaward flow at all depths due to strong tidal mixing and low river flow. Salinity is uniform over depth but varies along the estuary. A partially mixed estuary has strong tidal turbulence, seaward surface flow of mixed fresh water and seawater, and an inflow of seawater at depth. Salt is transported in this system by advection and mixing. Fjord-type estuaries are deep estuaries in which the fresh water moves out at the surface and there is little tidal mixing or inflow at depth.

In a partially mixed estuary, progressive movement of surface water seaward and deep water inward occurs during each tidal cycle. The circulation of the estuary can be measured directly with current meters, which is a long and expensive process, or it can be measured indirectly by determining the water and salt budgets of the estuary.

In temperate latitudes, the volume transport of water between the estuary and the sea is much greater than the addition of fresh water. In fjords, the inward flow is small and the deep water tends to stagnate. In semienclosed seas with high evaporation rates, a net removal of fresh water occurs; the seaward flow is at depth and the ocean water enters at the surface.

Estuaries that flush rapidly have a higher capacity for dissipating wastes than those that flush slowly. In bays and harbors that are flushed only by tidal action, resident water and wastes can be recycled. Partial recycling can also happen in estuaries with two-layered flow as exiting surface water is mixed with incoming water.

Case histories of two estuaries, San Francisco Bay and Chesapeake Bay are presented.

Key Terms

All key terms from this chapter can be viewed by term, or by definition, when studied as flashcards on this book's Online Learning Center at www.mhhe.com/sverdrup (click on this book's cover).

coast, 300	scarp, 310
shore, 300	low-tide terrace, 310
beach, 300	beach face, 310
geomorphology, 302	trough, 310
eustatic change, 302	dynamic equilibrium, 310
primary coast, 302	onshore current, 311
secondary coast, 302	onshore transport, 311
fjord, 303	longshore current, 311
sill, 303	longshore transport, 311
moraine, 303	swash, 311
drowned river valley, 304	cusp, 311
ria coast, 304	drift sector, 312
delta, 304	accretion, 312
alluvial plain, 304	coastal circulation
dune coast, 304	cell, 314
lava coast, 305	tombolo, 315
cratered coast, 305	lag deposits, 315
fault bay, 305	armored beach, 315
fault coast, 305	breakwater, 316
sea stack, 305	jetty, 316
bar, 306	groin, 316
barrier island, 306	estuary, 322
sand spit, 307	salt wedge estuary, 323
hook, 307	entrainment, 323
reef coast, 307	well-mixed estuary, 323
salt marsh, 308	partially mixed estuary, 323
backshore, 310	advection, 324
foreshore, 310	fjord-type estuary, 324
offshore, 310	net circulation, 324
berm, 310	water budget, 325
berm crest, 310	salt budget, 325
winter berm, 310	inverse estuary, 326
storm berm, 310	flushing time, 328
summer berm, 310	intertidal volume, 328

Study Questions

1. The processes that form a stretch of flat, uniform coast depend on whether the land is rising or sinking. Discuss the processes that form a flat coastal area under these two conditions.

2. Why are multiple berms more likely to be seen on a beach between March and August than between September and February?

3. Fjord coasts and drowned river valleys, or ria coasts, are primary coasts. Explain why their appearances are distinctly different. Give examples of each.

4. Describe the processes required to create a tombolo. Consider and discuss (a) the distribution of wave energy and (b) the longshore transport.

5. What conditions are required to maintain a beach with a constant profile and composition? Consider both a static and a dynamic environment.

6. Why does an eroding beach that is supplied with material from the cliffs of an old glacier deposit become an armored beach, while an eroding beach that is supplied by river sediments does not?

7. How does the profile of a sand beach change during alternating seasonal periods of storm waves and more gentle small waves?

8. Sketch a beach section that is stable with respect to the supply and removal of beach sediments. Indicate the source of the sediments and the final deposition of sediments in the system. If the longshore transport or wave energy distribution is altered by a barrier perpendicular to the beach, what changes will occur?

9. What is causing the apparent rise of sea level around the world? Is the rise similar in all places?

10. What processes move beach sediments in a coastal circulation cell?

11. Compare the circulation of a semienclosed basin at 30°N with that of an estuary located at 60°N. Which is less likely to accumulate waste products at depth? Why?

12. Estuaries are classified by their net circulation and salt distribution in this chapter. What other features could be used to describe and classify them?

13. Why are estuaries with short flushing times less apt to degrade when used as the receiving water for urban runoff than estuaries with long flushing times?

14. Compare the circulations and histories of San Francisco Bay and Chesapeake Bay. How do these estuaries differ; how are they similar? What do you see in the future for each?

15. In order to prevent erosion of his beach, your neighbor wants to build a groin. What effect might the groin have on your neighbor's beach? On your beach if you live on the upstream side of the longshore current? On your beach if you live on the downstream side of the longshore current?

Study Problems

1. Determine the flushing time of an estuary in which $T_o = 9 \times 10^7$ m^3/day and the volume of water is 30×10^8 m^3.

2. An estuary has a volume of 50×10^9 m^3. This water is 5% fresh water, and the fresh water is added at the rate of 6×10^7 m^3/day. Why does the rate of addition of fresh water from the estuary to the ocean equal the input of fresh water from the land to the estuary? Consider the average salinity of the estuary as constant. What is the residence time of the fresh water?

3. Water entering an estuary at depth has a salinity of 34.5‰. The water leaving the estuary has a salinity of 29‰. The river inflow is 20×10^5 m^3/day. Calculate T_o, the seaward transport.

4. A bay with no freshwater input can flush only by tidal exchange. If on each tidal cycle (ebb and flood) 10% of the bay's water volume is exchanged with ocean water, how much of the original water will still be in the bay after four tidal cycles? Solve for (1) no mixing between bay water and ocean water; (2) complete mixing between bay water and ocean water.

5. What is the flushing time of the bay in problem 4 for each case? Which case best duplicates natural conditions?

Links to Related Websites

Visit the book's Online Learning Center at www.mhhe.com/sverdrup (click on the book's cover) to find live Internet links for additional topics related to this chapter's content.

- Coastal and marine geology
- Coastal management

Visit the book's Online Learning Center at www.mhhe.com/sverdrup (click on this book's cover) to find these additional chapter tools: Suggested Readings; links to further information on boxed readings, selected figures, and related chapter topics; and additional study aids.

chapter 12

Environmental Issues and Concerns

The history of life on Earth has been a history of interaction between living things and their surroundings. To a large extent, the physical form and the habits of the Earth's vegetation and its animal life have been molded by the environment. Considering the whole span of earthly time, the opposite effect, in which life actually modifies its surroundings, has been relatively slight. Only within the moment of time represented by the present century has one species—man—acquired significant power to alter the nature of his world.

Rachel Carson
From *Silent Spring,* 1962

The tanker *Exxon Valdez* on Bligh Reef, Prince William Sound, Alaska.

The oceans have always had a profound influence on the people who settled along their shores. They have provided food, clothing, and waste disposal. They have fascinated and inspired generations of artists, writers, poets, and musicians. They have provoked the imaginations of adventurers and explorers and stimulated the curiosity of scientists. For centuries, the oceans were safe from interference by humans because human populations and technology continued at low levels. But that is no longer so. In this chapter we review ways in which our species affects the ocean: water pollution, uncontrolled harvesting, and the introduction of nonnative species.

Science documents changes in the marine environment and identifies causes, but science alone is not able to make the required corrections. Problems, once identified, may become issues that require policy changes, and changing policy may require making choices that are expensive and unpopular. Gathering a consensus in this arena is difficult, and it is often complicated by political pressures. However, whether you live by the ocean or not, the ocean is a part of your world, and you need to consider the situations that are discussed in this chapter.

Water and Sediment Quality 12.1

Our growing populations, with their necessary industries, energy-generating facilities, and waste treatment plants, place a great burden on coastal zones. In the past, the Earth's natural waters were assumed to be infinite in their ability to absorb and remove the by-products of human populations. However, too much use and too many wastes discharged into too small an area at too rapid a rate have produced problems that cannot be ignored. These discharges can and do exceed the ability of the natural systems to flush themselves and disperse their wastes into the open ocean.

Solid Waste Dumping

Using the sea as a dump for trash and garbage was and is a common practice around the world. Probably more than 25% of the mass of all material dumped at sea is dredged material from ports and waterways, and one of the major methods of industrial waste disposal is dumping at sea. After each modern war, obsolete military hardware and munitions have been disposed of at sea. The North Atlantic was the dumping ground for toxic gases confiscated from Germany at the end of World War II, and the bays and lagoons of many South Pacific islands were used for the disposal of Jeeps, tanks, bombs, and other items. After the United States left Vietnam, there was a similar unloading of vehicles and explosives in the waters around Southeast Asia. Today, Russia is living with the health and environmental problems caused by the disposal of nuclear wastes and power plants from obsolete nuclear submarines along the Arctic coast of northwest Russia.

The United States has stopped its ocean dumping except for uncontaminated dredge material with a permit from the U.S. Corps of Engineers, but the well-documented history of one U.S. area allows us to see what is still taking place in other countries. Off the mouth of the Hudson River, between New York and New Jersey, lies the New York Bight. Street sweepings, garbage, dredge spoils, cellar dirt, and waste chemicals were dumped into this area beginning in the 1890s. Increasing quantities of floating debris found their way back to the beaches until 1934, when laws were passed to prohibit the dumping of "floatables." Refuse continued to be dumped, including building and subway construction debris, toxic wastes from industry, acids, and sewage sludge. Sludge is a product of today's required secondary sewage treatment; it is the result of biological degradation of suspended solids after large solids and about half of the suspended organic material are removed during primary treatment.

It has been estimated that between 1890 and 1971 the volume of solid waste dumped into the water of the New York Bight was 1.4×10^6 m^3 or the equivalent of a layer covering all of Manhattan Island to the height of a six-story building, or 20 m (66 ft). The dumping of toxic and oxygen-demanding wastes degraded the quality of the water and the sediments. Occasionally, this degraded water upwelled along the coast and killed the marine organisms in shallow water.

In 1972 the U.S. Marine Protection, Research, and Sanctuaries Act (commonly called the Ocean Dumping

Act) prohibited dumping of all types of materials into ocean waters without a permit from the Environmental Protection Agency (EPA). In 1987 the EPA closed this offshore dump but agreed to let New York City and several New Jersey communities continue to dump sewage sludge at a new site 171 km (106 mi) from shore, at the edge of the continental shelf. Over 8 million tons of sewage sludge were dumped annually until all dumping was halted in 1992. Despite the dumping ban, not all the problem is solved because now the sludge must be recycled or dumped on land, where it has the potential to leach its nutrients and contaminants into freshwater drainage systems that flow to the sea.

Sewage Effluent

Most urban sewer systems in the United States were built in the late 1800s and early 1900s; these systems carried raw sewage to the closest body of water: river, lake, or ocean. Since then the systems have grown by adding to or combining old systems, building new systems, and adding treatment facilities. Until 1972 the law allowed the discharge of anything into a body of water until the water was polluted; "polluted" was defined by the individual states. In 1972 the Federal Clean Water Act, in an effort to make U.S. water "fishable and swimmable," mandated the upgrade of sewage treatment to the secondary treatment level by 1977.

For years Boston Harbor has received sewage effluent in various stages of treatment from an enormous urban area, and the harbor is seriously polluted. Under a provision of the Clean Water Act, Boston was able to apply for a series of treatment waivers because it discharged into marine waters rather than fresh water, and Boston continued to apply for waivers until 1985, when the EPA denied its last waiver application and the state ordered compliance with the Clean Water Act.

Construction of new sewage collecton systems and a sewage treatment plant began in 1989 and is now nearly finished, at a cost of $3.8 billion. Important components of the new Deer Island Sewage Treatment Plant have been placed in operation in a series of steps since January 1995. The plant was constructed to remove human, household, business, and industrial pollutants from the wastewater that originates in homes and businesses in forty-three different communities with a population of roughly 2.5 million in the greater Boston area. It is the second-largest sewage treatment plant in the United States. The plant's peak capacity is 1270 million gallons of wastewater per day (mgd). Average daily flow is about 380 mgd. After the sewage is treated in the plant it is discharged through an outfall tunnel 7 m (24 ft) in diameter beneath the sea floor that extends roughly 15.3 km (9.5 mi) offshore. The actual discharge of treated effluent is through a diffuser, which consists of more than fifty pipes that rise to the seabed over the last 2000 m (6600 ft) of the tunnel's length. Each pipe connects to a diffuser cap, which splits the flow into several streams. The purpose of the diffuser is to ensure that the maximum practicable dispersion and dilution is achieved for the wastewater flow. Extensive monitoring plans are in place to evaluate

what impact the discharge from the plant may have both near the outfall site and further south along the coast where wind-driven currents may carry it.

Another approach was taken by seven Southern California municipal discharge systems that began in 1971 to monitor the coastal discharge of their treated sewage. Although the volume of treated sewage effluent discharge has increased over twenty years, the discharge of oxygen-consuming constituents, nutrients, and toxicants such as heavy metals and chlorinated hydrocarbons has decreased significantly. A comparison of 1975, 1985, and 1995 discharges are given in table 12.1. The decreases are due to (1) increased efforts that prevent harmful materials from entering waste streams at their sources and (2) increases in the level of wastewater treatment.

Toxicants

Surface runoff from coastal urban areas feeds directly into the marine environment via storm sewers. Storm sewers carry a wide mix of materials, including silt, hydrocarbons from oil, lead from gasoline, residues from industry, pesticides and fertilizers from residential areas, and coliform bacteria from animal wastes. Even the chlorine added to drinking water and used to treat sewage effluent as a bactericide may form a complex with organic compounds in the water to produce chlorinated hydrocarbons that are toxic in the marine environment.

From rural and agricultural lands, runoff finds its way through lakes, streams, and rivers to the estuaries. This runoff supplies pesticides and nutrients, which can poison or overfertilize the waters. Excess nutrients can be as destructive as pesticides because the nutrients stimulate plant growth that eventually dies and decays, removing large quantities of oxygen. Lack of oxygen then kills other organisms, which in turn decay and continue to remove oxygen from the water. Animal wastes and failed septic tank systems also contribute contamination to rural runoff.

Many toxicants reaching the estuaries do not remain in the water but become adsorbed onto the small particles of matter suspended in the water column. These particles clump together and settle out because of their increased size, and toxicants become concentrated in the sediments. Analyses of core samples of estuary sediments show changes in the concentration of toxicants in bottom sediments over time. Figure 12.1a shows the concentrations of heavy-metal ions in sediments that were laid down in Southern California between 1845 and 1990. Concentration levels for sediments deposited before 1845 are assumed to be at natural levels; those since 1845 are considered to be the products of human activities. Between 1970 and 1990 the discharge to the sediments of these heavy metals decreased, except for copper, which is still commonly used for water pipes.

Figure 12.1b shows concentrations of the pesticide DDT (dichloro-diphenyl-trichloro-ethane) and of the compound PCB (polychlorinated biphenyl) in Puget Sound between 1940 and 1980. Widespread use of DDT did not begin until the 1940s, and it was followed by PCBs in the 1950s.

Table 12.1 A Twenty-Year Comparison of Combined Discharges from Seven Southern California Municipal Wastewater Treatment Facilities

	1975	1985	1995
Flow			
mgd (million gallons/day)	975	1143	1106
Liters ($\times 10^9$/yr)	1346	1579	1529
General constituents (metric tons/yr)			
Total suspended solids	284,900	204,500	73,000
5-day BOD (biological oxygen demand)	233,500	253,500	138,000
Oil and grease	56,500	34,300	19,000
NH3-N	36,300	44,500	41,000
Trace metals (metric tons/yr)			
Silver	25	26	5.4
Arsenic	6	16	5.0
Cadmium	51	16	1.0
Chromium	579	110	7.0
Copper	510	239	5.3
Mercury	2.2	0.9	0.02
Nickel	282	118	30.0
Lead	198	118	2.4
Selenium	11	5.6	7.8
Zinc	1087	375	86.0
Chlorinated hydrocarbons (kg/yr)			
Total DDT	1158	48	3.1
Total PCB	3065	46	not reported

Data from Southern California Coastal Water Research Project Annual Report, *1996, Southern California Coastal Water Research, Long Beach, CA.*

Figure 12.1 Toxicant concentrations in the sediments. (a) Heavy-metal ion concentrations increased between 1845 and 1960; environmental awareness and legislation have helped to lower concentrations in the last forty-plus years. *Pb*, lead; *Zn*, zinc; *Cr*, chromium; *Cu*, copper. (b) DDT and PCB concentrations show increases until the control of DDT in the 1960s and the banning of PCBs in the 1970s.

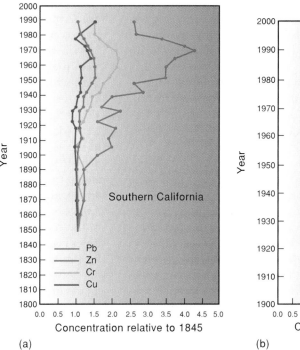

(a)

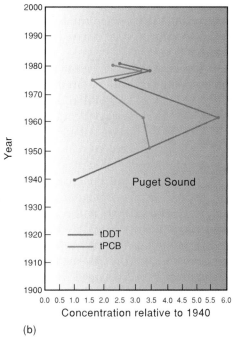

(b)

Sediment values indicate that the greatest use and discharge of PCBs occurred in the early 1960s. In 1998, a global pact (Protocol on Persistent Organic Pollutants) was negotiated by the United States and twenty-eight other countries; the pact requires most countries in western and central Europe, North America, and the former Soviet Union to ban production of compounds such as DDT and PCBs. Russia was granted a special exemption to continue production of PCBs

Table 12.2 Food Chain Concentration of DDT

	DDT Residues in Parts per Million (ppm)
Water	0.00005
Plankton	0.04
Silverside minnow	0.23
Sheepshead minnow	0.94
Pickerel (predator)	1.33
Needlefish (predator)	2.07
Heron (small animal predator)	3.57
Tern (small animal predator)	3.91
Herring gull (scavenger)	6.00
Osprey egg	13.8
Merganser (fish-eater)	22.8
Cormorant (fish-eater)	26.4

Reprinted with permission from George M. Woodwell, et al., DDT Residues in an East Coast Estuary in Science, Vol. 156, pp. 821–24, 12 May 1967. Copyright 1967 American Association for the Advancement of Science.

until 2005 and to postpone destruction of stocks until 2020 while it converts its electrical power grid transformers to other insulating fluids. Whether these goals can be met remains to be seen.

Some of the particulate matter that adsorbs toxicants has a high organic content and forms a food source for marine creatures. In this way, heavy metals and organic toxicants associated with the particles find their way into the body tissues of organisms, where they may accumulate and be passed on to predators. Throughout the United States, toxic residues are found in estuarine bottom fish. Shellfish have been found to have concentrations of heavy metals at levels that are many thousands of times over the levels found in the surrounding waters. Scallops may have levels of cadmium about 2 million times over the water concentration, and oysters may have DDT levels 90,000 times over the water concentration. A classic 1967 study conducted after DDT had been sprayed on Long Island marshes to control mosquitoes documented the effect of this long-lived toxin as it was concentrated by the food chain (table 12.2).

A tragic example of what can happen when humans ingest organisms that had accumulated a toxin occurred between 1952 and 1968 in Minamata, Japan. An estimated 200–600 tons of waste mercury from an industrial source was released into the coastal bay from which villagers gathered their shellfish harvest. The mercury formed a complex that was readily taken up by the marine life; the accumulation of mercury in the shellfish led to severe mercury poisoning and death among those eating the shellfish. The physical and mental degenerative effects were especially severe on children whose mothers had eaten large amounts of shellfish during pregnancy. The condition produced is named Minamata disease.

This ability of adult mussels and oysters to absorb and concentrate contaminants has been put to use in the NOAA Mussel Watch Project. Since 1986 divers have collected samples from these shellfish to monitor the concentrations of some chemical contaminants in U.S. estuaries and coastal waters. Ten years after the first samples were taken, samples showed decreasing national trends for contaminants such as DDT, PCB, tin, and cadmium. More than 200 coast and estuary sites are sampled each year; almost half of the sites are located in waters near urban areas, within 20 km (12 mi) of population centers with more than 100,000 people. Three sites are located in Hawaii and two in Alaska. National Mussel Watch Programs have been established in the United Kingdom and France, and there are multinational efforts in northern Europe. Mussel watches are now being developed in the Asian Pacific Rim area.

Considering all the pathways, sources, and types of materials that can be classed as toxicants or pollutants, managing and eventually excluding them from the marine environment are extremely difficult and complicated. As seen in this discussion, efforts to reduce the discharge of toxic materials are showing signs of success, but it will take continued, long-term effort on the part of scientists, citizens, and government to achieve a cleaner environment for the ocean world and the land world.

Gulf of Mexico Dead Zone   12.2

Each year off the mouth of the Mississippi River a low-oxygen, or **hypoxic,** area forms in the northern Gulf of Mexico. It begins to appear in February, peaks in the summer, and dissipates in the fall, when storms stir up the gulf waters and increase mixing. It has been called the *dead zone* because the oxygen level in the water is too low to support most marine life. As it grows in size highly mobile organisms such as fish and shrimp flee the area. More slowly moving bottom-dwellers such as crabs, snails, clams, and worms can be overtaken by the hypoxic waters and suffocate. The low-oxygen water does not just hug the bottom; it can affect 80% of the water column in shallow areas, extending upward to within a few meters of the surface.

The dead zone is caused by the introduction of large amounts of nitrogen as nitrate, much of it from fertilizers, into the gulf by the Mississippi and Atchafalaya Rivers. The Mississippi River alone delivers an average 580 km^3 (140 mi^3) of fresh water to the Gulf of Mexico each year. Thirty-one states (41% of the area of the continental United States) and half of the nation's farmlands have rivers that drain into the Mississippi. Records show that from the 1950s onward the use of fertilizers in these areas increased dramatically. The amount of nitrate from fertilizer that flowed into the gulf tripled between 1960 and the late 1990s, and the amount of another fertilizer, phosphate, doubled. It is estimated that about 30% of the nitrogen entering the gulf comes from fertilizers used in agriculture, another 30% comes from natural soil decomposition, and the remainder comes from a variety of sources including animal manure, sewage treatment plants, airborne nitrous oxides from the burning of fossil fuels, and industrial emissions.

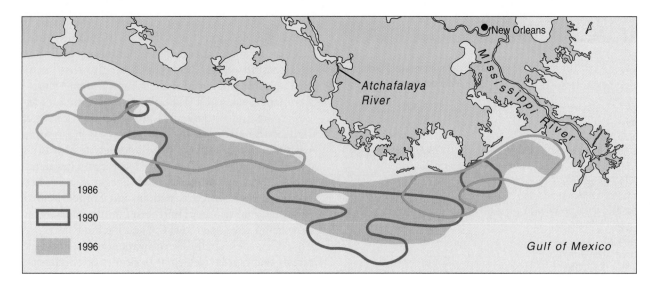

Figure 12.2 The location and size of the dead zone in the northern Gulf of Mexico vary from year to year. Research continues in an effort to better understand the factors controlling the pattern of hypoxic water. *From Ferber, D., Keeping the Stygian Waters at Bay in Science, 291:968–73, 9 Feb. 2001.*

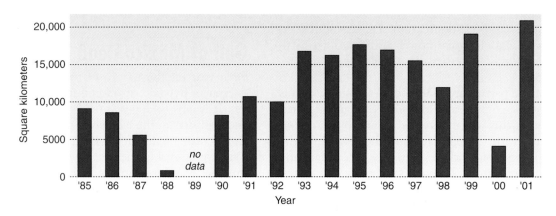

Figure 12.3 The size of the Gulf of Mexico dead zone. The dead zone grew in size during the particularly wet year of 1993, remained high through the rest of the decade, and then shrank significantly during the dry year of 2000. *Data Source: N. Rabalais, R. Turner & W. Wiseman.*

When nitrates are added to the water they trigger the rapid growth of tiny marine algae. This phenomenon is commonly referred to as a bloom (see chapter 15). Massive blooms can produce a biological chain reaction in which there is a corresponding increase in small animals that feed on the algae, producing more plant and animal material than can be consumed by fish and other predators. Billions of these small organisms die and sink to the sea floor, where they decompose and are consumed by bacteria, using up oxygen in the water in the process.

The seasonal appearance of the dead zone was first detected in the 1970s, but it was not until twenty years later that scientists began to map its location. The dead zone occupies different regions of the northern gulf from one year to the next, making it difficult to predict exactly what areas will be affected (fig. 12.2). The size of the dead zone seems to cor-

relate both with the amount of agricultural fertilizer used in the Mississippi River drainage basin and with annual changes in rainfall, and hence water runoff into the gulf. Measurements of the size of the dead zone over the last sixteen years indicate that it covered 15,000–20,000 km^2 (5800–7700 mi^2) throughout much of the 1990s (fig. 12.3). In 1993 the Mississippi River flooded and the dead zone grew to an area of 17,500 km^2 (6750 mi^2), an area twice the size of Chesapeake Bay. It reached its greatest size in 1999, when it covered nearly 20,000 km^2 (7700 mi^2), roughly the size of New Jersey. In 2000, when dry conditions prevailed and the runoff from the Mississippi River was very low, the zone shrank to its smallest recorded size. Then it rebounded in 2001.

The dead zone in the Gulf of Mexico is not an isolated situation, but it is the largest zone of this type in the Western Hemisphere. Similar problems occur in Chesapeake Bay,

Japan, Australia, New Zealand, the Kattegat Strait between Denmark and Sweden, and the Black Sea. Recent efforts to limit the use of fertilizers and restore wetlands in Sweden and Denmark have reduced the introduction of nutrients to the Kattegat Strait and led to increased levels of oxygen in the water. The dead zone in the Black Sea first began to form in the 1960s with the introduction of large amounts of nutrients from fertilizer runoff. At its peak it was larger than the dead zone in the Gulf of Mexico. When the Soviet Union collapsed in 1991 the broad government support for agriculture did also, and fertilizer usage dropped dramatically. By 1996 the dead zone in the Black Sea had disappeared for the first time in thirty years.

Plastic Trash 12.3

Walk any beach in any estuary or along any open coast and see the tide of plastics being washed ashore. It is estimated that every year more than 135,000 tons of plastic trash have been routinely dumped by naval, merchant, and fishing vessels. The National Academy of Sciences estimates that the commercial fishing industry yearly loses or discards about 149,000 tons of fishing gear (nets, ropes, traps, and buoys) made mainly of plastic and dumps another 26,000 tons of plastic packaging materials. Recreational and commercial vessels and oil and gas drilling platforms add their share. A total of 7 million tons of cargo and crew wastes were added during each year of the 1980s. Plastics are a worldwide problem; they come from many sources and are distributed by the currents to even the remotest ocean areas (fig. 12.4a).

In 1960 the U.S. production of plastics was 3 million tons; forty years later in 2000 the production was more than 50 million tons. Plastic is inexpensive, strong, and durable; these characteristics make it the most widely used manufacturing material in the world today and a major environmental problem. Nobody knows how long plastic stays in the marine environment, but an ordinary plastic six-pack ring could last 450 years.

Thousands of marine animals are crippled and killed each year by these materials. As many as 30,000 fur seals a year are estimated to become entangled in lost or discarded plastic fishing nets or to choke to death in plastic cargo straps. Lost lobster and crab traps made entirely or partially of plastic continue to trap animals; one year 25% of the 96,000 traps that had been set off of Florida's western coast were lost. Seabirds die entangled in six-pack rings and plastic fishing line; both seabirds and marine mammals swallow plastics. Porpoises and whales have been suffocated by plastic bags and sheeting, and the sheeting clings to coral and to rocky beaches, smothering plant and animal life. Fish are trapped in discarded netting; sea turtles eat plastic bags and die. Plastics are as great a source of mortality to marine organisms as oil spills, toxic wastes, and heavy metals (fig. 12.4b, c, d).

The Marine Plastic Pollution and Control Act of 1987 is the U.S. law that implements an international convention for the prevention of pollution from ships. This law prohibits the dumping of plastic debris everywhere in the oceans; other types of trash may be dumped at specific distances from shore, and ports are required to provide waste facilities for debris. The U.S. Coast Guard is the agency that must enforce this law in U.S. waters; there is no international enforcement.

Some manufacturers make biodegradable plastic by adding light-absorbing molecules that break down the plastic after a few months' exposure to sunlight. There is no evidence that this will solve the problem at sea; even at the surface, the water keeps the plastic cool, and it becomes coated with a thin film of organisms that shade it from the light. It may be that only people educated to act in a responsible manner will be able to reduce the tide of plastics pollution rising around the world.

Ocean Waste Management Proposals 12.4

Expanding populations, industrialization, and economic growth have increased the amounts of our garbage and changed its characteristics. Landfill areas are being overwhelmed, and much of their contents is increasingly recognized as a threat to the Earth's soil and its freshwater reservoirs. Aluminum and plastics displace traditional materials such as paper, steel, and glass, and household cleaners, solvents, and pesticides are often environmentally harmful. Estimates of world wastes produced each year are shown in table 12.3.

At the Woods Hole Oceanographic Institution in 1991 a group of scientists held a workshop to assess the potential of the deep ocean as an option for future waste management. They argued that controlled disposal in some ocean environments presents less risk to life than current projected land disposal, with its associated groundwater contamination, or incineration accompanied by atmospheric pollution. They pointed out that the abyssal plain of selected mid-ocean areas supports only sparse populations of marine organisms, and the slowly moving water above it stays at the bottom for thousands of years before it returns to the surface. They suggested that the deep-ocean floor far from shore may be a more appropriate site than coastal landfills that pollute inshore water and degrade beaches, and they proposed a large-scale experiment to assess the environmental effects of depositing 1 million tons of sewage sludge each year for ten years on the floor of the Hatteras Abyssal Plain in the Atlantic Ocean. Implementing such a project would require changes in the Ocean Dumping Act; this is considered unlikely at the present time. Some people feel that any experiments with ocean waste disposal will have to proceed by affiliation with other countries and should fall under UN regulations.

Another suggestion is to dispose of highly radioactive and long-lived nuclear wastes away from the continents in the muds of the deep-sea floor or in subduction zones, where the waste products will gradually recycle back into the mantle. Examining the wreckage of a Soviet nuclear submarine that sank in 1986 in the deep Atlantic with two nuclear reactors and thirty-four nuclear warheads has been suggested as a way to determine whether deep-sea muds can trap radioactive substances. Such proposals seek to avoid the possible contamination of land areas and freshwater supplies while eliminating the problem of long-term hazardous waste storage on land. No experiments involving any of these wastes are planned at this time.

Should we consider dumping to be an ocean resource that has a human and economic value just like any other resource? If so, do these suggestions deserve discussion? As world populations and their refuse products continue to in-

web link **Figure 12.4** The impact of plastics on the marine environment. (a) Persistent litter includes plastic nets, floats, and containers. (b) A common murre entangled in a six-pack yoke. (c) A young gray seal caught in a trawl net. (d) A sea turtle with a partially ingested plastic bag. Photos (a), (b), and (c) are from Sable Island in the North Atlantic, approximately 240 km (150 mi) east of Nova Scotia, Canada.

crease and available storage sites decrease, consideration of these questions may become inevitable.

Oil Spills   12.5

Humans' activities in the twenty-first century depend heavily on oil, and this dependence requires the bulk transport of crude oil by sea to the land-based refineries and centers of use. This transport creates the potential for accidents that release large volumes of oil and expose the world's coasts and estuaries to spills associated with vessel casualties and transfer procedures. Drilling offshore wells exposes marine areas to the risks of blowouts, spills, and leaks. Because industry, agriculture, and private and commercial transportation require petroleum and petroleum products, oil is constantly being released into the environment, to find its way directly or indirectly down to the sea. The average discharge of oil into the world's oceans and navigable waters during each of the last twenty years has been 135,000 metric tons (42.7×10^6 gal), not including the occasional megaspill (1983 and 1991). Figure 12.5 shows the annual spillage of oil in the world between 1978 and 1999, and table 12.4 lists marine oil spills of more than 20 million gallons that occurred between 1960 and 1997.

The damage caused by spills far out at sea is difficult to assess, because there is no direct visual or economic impact on coastal areas and damage to marine life cannot be accurately evaluated. Spills due to the grounding of vessels or accidents during transfer or storage happen in coastal areas, where the environmental degradation and loss of marine life can be observed. Spills that occur in estuaries and along coasts affect regions that are oceanographically complex, biologically sensitive, and economically important. Three oil spills between 1978 and 1991 have become ecological landmarks; these are the sinking of the *Amoco Cadiz*, the grounding of the *Exxon Valdez*, and the wartime discharge of crude oil into the Persian Gulf.

In March 1978 the *Amoco Cadiz* lost its steering in the English Channel and broke up on the rocks of the Brittany coast of France (fig. 12.6a, b). Gale-force winds and high tides spread the oil over more than 300 km (180 mi) of the French coast; more than 3000 birds died; oyster farms and fishing suffered severely. Of the approximately 210,000 metric tons

Table 12.3 Estimate of World Wastes Produced Each Year

Type	Millions of Metric Tons	Percentage of Total Waste
Municipal solid waste	1500	36.2
Dry sewage sludge	65	1.6
Dredged material	1075	26.0
Industrial wastes	1500	36.2
Total	4140	100.0

From D. Spencer, An Abyssal Ocean Option for Waste Management, 1991. © Woods Hole Oceanographic Institution.

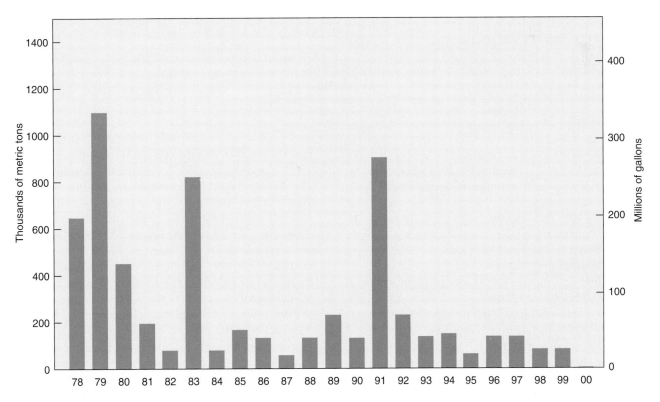

web link **Figure 12.5** World annual oil spillage.   *Source: Data from* Oil Spill Intelligence Report, *Arlington, MA.*

Table 12.4 The World's Largest Marine Oil Spills (greater than 20 million gallons spilled) 1960–97, Ranked by Volume

Rank	Incident	Millions of Gallons
1	January 26, 1991; terminals, tankers; eight sources total Sea Island installations; Kuwait; off coast in Persian Gulf and in Saudi Arabia	240.0
2	June 3, 1979; exploratory well Ixtoc I well; Mexico; Gulf of Mexico, Bahia Del Campeche, 80 km NW of Ciudad del Carmen, Campeche	140.0
3	February 4, 1983; platform No. 3 well (Nowruz); Iran; Persian Gulf, Nowruz Field	80.0
4	August 6, 1983; tanker Castillo de Bellver; South Africa; Atlantic Ocean, 64 km off Table Bay	78.5
5	March 16, 1978; tanker Amoco Cadiz; France; Atlantic Ocean, off Portsall, Brittany	68.7
6	November 10, 1988; tanker Odyssey; Canada; North Atlantic Ocean, 1175 km NE of Saint John's, Newfoundland	43.1
7	July 19, 1979; tanker Atlantic Empress; Trinidad and Tobago; Caribbean Sea, 32 km NE of Trinidad-Tobago	42.7[1]
8	April 11, 1991; tanker Haven; Italy; Mediterranean Sea, port of Genoa	42.0
9	August 2, 1979; tanker Atlantic Empress; 450 km east of Barbados	41.5[1]
10	March 18, 1967; tanker Torrey Canyon; United Kingdom; Lands End	38.2
11	December 19, 1972; tanker Sea Star; Oman; Gulf of Oman	37.9
12	February 23, 1980; tanker Irenes Serenade; Greece; Mediterranean Sea, Navarino Bay, off Pylos port	36.6
13	December 7, 1971; tanker Texaco Denmark; Belgium; North Sea	31.5
14	February 23, 1977; tanker Hawaiian Patriot; United States; Pacific Ocean 593 km west of Kauai Island, Hawaii	31.2[2]
15	November 15, 1979; tanker Independentza; Turkey, Bosporus Strait near Istanbul, 0.8 km from Hydarpasa port	28.9
16	February 11, 1969; tanker Julius Schindler; Portugal; Ponta Delgada, Azores Islands	28.4
17	May 12, 1976; tanker Urquiola; Spain; La Coruña Harbor port	28.1
18	January 5, 1993; tanker Braer; United Kingdom; Garth Ness, Shetland Islands	25.0
19	January 29, 1975; tanker Jacob Maersk; Portugal; Porto de Leisoes, Oporto	24.3
20	December 3, 1992; double-bottom tanker Aegean Sea; Spain; La Coruña Harbor port	21.9
21	December 6, 1985; tanker Nova; Iran; Persian Gulf, 140 km south of Kharg Island	21.4
22	February 15, 1996; tanker Sea Empress; United Kingdom; Mill Bay near entrance to Milford Haven Harbor port	21.3
23	February 27, 1971; tanker Wafra; South Africa; Atlantic Ocean	20.2
24	December 19, 1989; tanker Khark 5; Morocco; Atlantic Ocean, 185 km from Moroccan coast	20.0

1. On July 19, 1979, the *Atlantic Empress* spilled 42.7 million gallons of oil as the result of a collision with the *Aegean Captain* in the Caribbean Sea near Trinidad-Tobago. In this incident, the *Aegean Captain* spilled 4.31 million gallons of oil. On August 2, 1979, while under tow from the original spill site, the *Atlantic Empress* spilled an additional 41.5 million gallons.

2. In this incident the vessel sank with some of its cargo still onboard after spilling an undetermined amount of oil. The spill size reported reflects the total amount of oil spilled and oil left onboard the sinking vessel. Estimates of the two amounts are not available.

Note: There have been fifty-four marine oil spills exceeding 10 million gallons since 1960.

Source: Data from Oil Spill Intelligence Report, *Arlington, MA http://cutter.com/osir/biglist.htm.*

$(66.4 \times 10^6$ gal) of oil spilled, roughly 30% evaporated and was carried over the French countryside; 20% was cleaned up by the army and volunteers; another 20% penetrated down into the sand of the beaches to stay until winter storms washed it away; 10% (the lighter, more toxic fraction) dissolved in the seawater, and 20% sank to the sea floor in deeper water, where it continued to contaminate the area for an unknown period of time.

The United States experienced a large-volume oil spill in March 1989 when, 40 km (25 mi) out of Valdez, Alaska, the *Exxon Valdez*, loaded with 170,000 metric tons $(53.7 \times 10^6$ gal) of crude oil, ran onto a reef, tearing huge gashes in its hull and spilling 35,000 tons (almost 10.8×10^6 gal) of oil into Prince William Sound. Local contingency plans for cleaning up oil spills were not prepared to handle a situation of this magnitude; delays, caused in part by out-of-service equipment, as well as a lack of equipment and personnel, the rugged coast-line, the weather, and the tidal currents of the enclosed area in which the spill occurred, combined to intensify the problem. The oil spread quickly and was distributed unevenly over more than 2300 km² (900 mi²) of water, taking a predictable and severe toll of seabirds, marine mammals, fish, and other marine organisms. The oil moving out of Prince William Sound was not washed out to the open sea but was captured by the nearshore currents and moved westward parallel with the coast, repeatedly oiling the rocky wilderness beaches in the weeks that followed (fig. 12.6 c, d). In the subarctic conditions of Prince William Sound, the photochemical and microbial degradation of the oil proceeded more slowly than in warmer temperate regions. Plant and animal populations also recover more slowly in the cold water, where organisms tend to live longer and reproduce more slowly. Researchers inspecting the area in 1992, three years after the disaster, reported that the area was recovering and repopulating with organisms as the oil

web ⬭ link **Figure 12.6** (a) The tanker *Amoco Cadiz* aground and broken in two off the coast of France, March 1978. (b) A bay along the French coast was fouled with oil. (c) Hot water under high pressure was used to clean Prince William Sound beaches after the 1989 *Exxon Valdez* oil spill. (d) Booms collect oil washed from a beach and hold it for pickup. (e) Oil remained along the Persian Gulf coast long after the 1991 Gulf War. (f) Cleaning up a 1993 oil spill along the beaches of Tampa Bay one more time in 1994.

aged and degraded. They also noted that the areas most intensively cleaned to remove oil are recovering more slowly and with less balanced populations than the areas in which the oil has been allowed to degrade naturally. In the 1992 inspection, estimates were made of the dispersal of the spilled oil: 13% in the sediments, 2% in the beaches, 50% degraded naturally in the water, 20% degraded in the atmosphere, and 14% recovered and removed.

Much of the world's crude oil comes from the oil wells of the Persian Gulf, and it must be transported the length of the gulf on its way to refineries around the world. Each year a quarter of a million barrels of oil are routinely spilled into the gulf's shallow waters, making it one of the world's most polluted bodies of water. However, it is also an area of vigorous plant growth and supports fisheries of shrimp, mackerel, mullet, snapper, and grouper.

During the eight-year Iran-Iraq war, the bombing of oil facilities produced major spills, including one from an oil rig that poured out 172 metric tons (54,350 gal) per day for nearly three months, but the greatest catastrophe came in the 1991 Gulf War, when an estimated 800,000 metric tons $(252.8 \times 10^6$ gal) of crude oil gushed into gulf waters, the world's greatest oil spill to-date. Some of this oil was deliberately released, some came from a refinery at a gulf battle site, and bombing contributed additional quantities. This spill was much larger than the estimated output of Ixtoc 1, an offshore Mexican well that blew out in 1979 and was the previous largest spill on record.

The average depth of the Persian Gulf is 40 m (131 ft), and the circulation of its high-salinity water is sluggish. Changes in the wind in the weeks after the spill stopped the oil from drifting the entire length of the gulf; still, some 570 km (350 mi) of Saudi Arabia's shoreline were oiled. Oil spill experts from the United States, the United Kingdom, the Netherlands, Germany, Australia, and Japan rushed to help. They were able to protect the water-intake pipes of desalination plants and refineries, but other cleanup efforts were less successful (fig. 12.6e). About half of the oil evaporated and about 300,000 metric tons $(94.8 \times 10^6$ gal) were recovered. Much sank to the bottom of the gulf, where oil may still be seeping from sunken tankers.

The tragedies of these spills continue to demonstrate that no adequate technology exists to cope with large oil spills, particularly under difficult weather and sea conditions, along irregular coastlines, or far from land-based supplies. On the average only between 8% and 15% of oil spilled is recovered, and these may be inflated estimates because the recovered oil has a high water content. The technology for oil cleanup at sea includes oil booms and oil skimmers (fig. 12.7). These devices are useful in confining and recovering small spills in protected waters, but they are not very effective in open and exposed ocean areas or in rough or stormy seas.

The onshore cleanup is often as destructive as the spill, especially in wilderness situations, where the numbers of people, their equipment, and their wastes further burden the environment. The toxicity of weathered crude oil is low, and

Figure 12.7 A catamaran oil skimmer. The vessel cruises at about 3 knots, guiding surface oil between the twin hulls. The oil adheres to a moving belt and is lifted into onboard storage tanks.

many oil spill experts consider that cleanup efforts should be concerned not with removing oil from the beaches but with moving oil seaward to prevent it from reaching the beaches. Follow-up studies indicate that many of the cleanup methods used along shorelines cause more immediate and long-term ecological damage than leaving the oil to degrade naturally. Beach habitats treated with high-pressure, hot water take longer to recover than those left untreated; backhoeing and high-pressure washing (100 lb/in²) destabilize gravel and sand beaches, killing animals and working the oil into the sediments (see fig. 12.6c). To prevent oil from reaching the beaches, some support newly developed, low-toxicity dispersants; the dispersed oil moves into deeper water, where it is diluted and its toxicity is lessened.

More data are needed to assess the effectiveness of attempts to speed up natural degradation by adding nutrients to the beaches to increase populations of biodegrading microorganisms. Releasing other strains of oil-degrading bacteria has been tested with limited success; the new organisms may not compete effectively with natural populations.

Once oil is removed from a beach, the first cleaning operation may not be the last. On the east coast of Florida, the beaches of Tampa Bay are lined with resorts and condominiums, and the economy depends heavily on these beaches for tourism. A freighter and two oil barges collided in August 1993. One barge carried diesel fuel and gasoline that caught fire and burned. The second barge carried heavy fuel oil; this barge sank and released the fuel oil to mix with the sediments on the bottom of the bay. Each time strong winds stir the bay, contaminated sediments are moved onshore to create a new oil contamination problem. Figure 12.6f shows the end of a costly January 1994 cleanup episode along St. Petersburg Beach.

Figure 12.8 The industrialized estuary of the Duwamish River at Seattle, Washington. The commercially developed flat areas adjacent to the river were once tidelands and salt marshes.

The immediate damage from a large spill is obvious and dramatic; by contrast, the effect of the small but continuous additions of oil that occur in every port and harbor are much more difficult to assess, because they produce a chronic condition from which the environment has no chance to recover. Refined products such as gasoline and diesel fuel are more toxic to marine life than crude oil, but they evaporate rapidly and disperse quickly. Crude oil is slowly broken down by the action of water, sunlight, and bacteria, but the portion that settles on the sea floor moves down into the sediments, which it continues to contaminate for years.

Marine Wetlands   12.6

The value of shore and estuary areas as centers of productivity and nursery areas for the coastal marine environment is well known to oceanographers and biologists. Saltwater and brackish marshes and swamps, known as marine **wetlands,** border estuaries and provide nutrients, food, shelter, and spawning areas for marine species, including such commercially important organisms as crabs, shrimp, oysters, clams, and many species of fish. Nevertheless coastal countries have long histories of filling wetlands and modifying coasts to provide croplands, port facilities, and industrial space for their growing populations (fig. 12.8). For examples of these trends see the history of San Francisco Bay in chapter 11 and consider Holland, where the Dutch have been reclaiming wetlands for thousands of years, until at present about one-third of their land has been reclaimed from the North Sea. Although the Dutch will continue to develop parts of their coast, they have decided to return 15% of the total reclaimed area to rivers and estuaries over the next twenty-five years because of the high costs of dike repair and pumping water, as well as their concern over wetland habitat loss and its impact on the Dutch national symbol, the stork.

Spartina: Valuable and Productive or Invasive and Destructive?

Spartina alterniflora, known as smooth cordgrass, many-spiked cordgrass, and saltmarsh cordgrass, is a deciduous, perennial flowering plant native to the Atlantic and Gulf Coasts of the United States. It is the dominant native species of the lower salt marshes along the Atlantic seaboard from Newfoundland to Florida and on the Gulf Coast from Florida to east Texas. It grows in the intertidal zone from mean higher high water to 1.8 m (6 ft) below mean higher high water.

These natural salt marshes are among the most productive habitats in the marine environment. Nutrient-rich water is brought to the wetlands during each high tide, making a high rate of food production possible. As the seaweed and marsh grass leaves die, bacteria break down the plant material, and insects, small shrimplike organisms, fiddler crabs, and marsh snails eat the decaying plant tissue, digest it, and excrete wastes high in nutrients. Numerous insects occupy the marsh, feeding on living or dead plant tissue, and red-winged blackbirds, sparrows, rodents, rabbits, and deer feed directly on the cordgrass. Each tidal cycle carries plant material into the offshore water to be used by the subtidal organisms.

Spartina is an exceedingly competitive plant. It spreads primarily by underground stems; colonies form when pieces of the root system or whole plants float into an area and take root or when seeds float into a suitable area and germinate. *Spartina* establishes itself on substrates ranging from sand and silt to gravel and cobble and is tolerant of salinities ranging from near fresh to salt water (35‰). *Spartina* is able to tolerate high salinities because salt glands on the surface of the leaves remove the salt from the plant sap, leaving visible white salt crystals. Because of the lack of oxygen in marsh sediments, they are high in sulfides that are toxic to most plants. *Spartina* has the ability to take up sulfides and convert them to sulfate, a form of sulfur that the plant can use; this ability makes it easier for the grass to colonize marsh environments. Another adaptive advantage is its biochemical photosynthetic pathway, which uses carbon dioxide more efficiently than most other plants.

These characteristics make *Spartina* a valuable component of the estuaries where it occurs naturally. The plant functions as a stabilizer and sediment trap and as a nursery area for estuarine fishes and shellfishes. Once established, a stand of *Spartina* begins to trap sediment, changing the substrate elevation, and eventually the stand evolves into a high marsh system where *Spartina* is gradually displaced by higher-elevation, brackish-water species. As elevation increases, narrow, deep channels of water form throughout the marsh (box fig. 1). Along the East Coast *Spartina* is considered valuable for its ability to prevent erosion and marshland deterioration; it is also used for coastal restoration projects and the creation of new wetland sites.

Spartina has been introduced to and naturalized in Washington, Oregon, California, England, France, New Zealand, and China. The plant was carried to Washington State in packing material for oysters transplanted from the East Coast in 1894. Leaving its

Box Figure 1 A naturally occurring *Spartina* marsh.

insect predators behind, the cordgrass has been spreading slowly and steadily along Washington's tidal estuaries, crowding out the native plants and drastically altering the landscape by trapping sediment. *Spartina* modifies tidal mudflats, turning them into high marshes inhospitable to the many fish and waterfowl that depend on the mudflats. In 1945 *Spartina* covered 4.5 acres of Washington's Willapa Bay; it had spread to 432 acres in 1982, 2400 acres in 1992, and more than 3400 acres in 1995. State officials predict that by 2010 it will cover 30,000 of the bay's 80,000 acres if left unchecked (box fig. 2). It is already hampering the oyster harvest and the Dungeness crab fishery, and it interferes with the recreational use of beaches and waterfronts.

Spartina has been transplanted to England and New Zealand for land reclamation and shoreline stabilization. In New Zealand the plant has spread rapidly, changing mudflats with marshy fringes to extensive salt meadows and reducing the number and kinds of birds and animals that use the marsh. Another species of *Spartina* (*S. maritima*) is native to marshes along the coasts of Europe and Africa. *S. alterniflora* was introduced into Great Britain from eastern North America in about 1800 and spread to form large colonies. It coexisted with the native species, but by 1870 a sterile hybrid that reproduced by underground stems appeared. In about 1890, a vigorous seed-producing form was derived naturally from this hybrid and spread rapidly along the coasts of Great Britain and northwestern France.

Efforts to control *Spartina* outside its natural environment have included burning, flooding, shading plants with black canvas or plastic, smothering the plants with dredged materials or clay, applying herbicide, and mowing repeatedly. Little success has been reported in New Zealand and England; Washington State's management program has tried many of these methods and is presently using the herbicide glyphosphate to control its spread. Work has begun to determine the feasibility of using insects as biological controls, but effective biological controls are considered ten years away. Even with a massive effort it is doubtful that complete eradication of *Spartina* from nonnative habitats is possible, for it has become an integral part of these shorelines and estuaries during the last 100 to 200 years.

Internet References

Visit the book's Online Learning Center at www.mhhe.com/sverdrup (click on the book's cover) to explore links to further information on related topics.

Box Figure 2 Circular patches of *Spartina* spread along the mudflats of Willapa Bay, Washington. The *large circle* at *lower left* is thought to be the original colony. This is a false-color image on infrared film.

Figure 12.9 The wetlands of Barnegat Bay, New Jersey, were replaced by a housing and recreational complex.

Another type of wetland is being destroyed along the muddy shores of tropical and subtropical lagoons and estuaries where several species of mangrove trees grow. These salt-tolerant trees protect the shore from wave erosion and storm damage and provide specialized habitats for fish, crustaceans, and shellfish on and among their tangled roots. In recent years mangroves have been logged for timber, wood chips, and fuel; mangrove swamps have been cleared and filled to provide land for crops, shrimp ponds, and resorts. Eighty percent of the mangroves in the Philippines, 73% of Bangladesh's trees, and more than 50% of coastal stands in Africa have been removed. Over 50% of the world's mangroves are gone.

In the United States the population density of coastal counties has risen dramatically since the 1960s. In 1960 it was 248 persons per square mile, four times the U.S. average, and by 1988 population density reached 342 persons per square mile. By 2010 this coastal population is expected to be approaching 400 persons per square mile. Areas of recreation and retirement replace wetlands with waterfront homes, each with its own individual pleasure-craft moorage (fig. 12.9). Between the mid-1950s and the mid-1980s approximately 20,000 acres of coastal wetlands were lost each year in the contiguous United States. Estuarine wetland losses were greatest in six states: Louisiana, Florida, Texas, New Jersey, New York, and California. Much of the Louisiana loss is due to accelerated erosion and subsidence of Louisiana's coastal marshes; see the box titled "Rising Sea Level" in chapter 11.

In 1972 the federal government enacted the first nationwide wetlands regulatory program in Section 404 of the Clean Water Act and focused attention on coastal wetlands by enacting the Coastal Zone Management Act. In the 1970s and 1980s public concern led to the passage of tidal wetland protection laws in many coastal states and to stricter enforcement of federal laws. Each marine wetland is as individual as the estuary it borders, and the singular and distinctive nature of each is a part of the problem, for the difficulty of protecting a wetland is related to its legal definition. Under different legal interpretations and conditions some wetlands escape protection while other drier areas are included; in this case one size or definition does not fit all.

Biological Invaders 12.7

A combination of physical and biological barriers sets the geographical limits of a species. These limits have changed with time as climate patterns have altered and as plate tectonics has shifted the configurations of the oceans and continents. For example, the opening of the Bering Strait between Asia and North America allowed interchange between the marine organisms of the North Pacific and the Arctic basins, and the separation of South America from Antarctica allowed the currents associated with the Southern Ocean's West Wind Drift to carry organisms from one cold-water ocean to another.

When humans began to cross oceans they carried with them many organisms, plant and animal, intentionally and unintentionally. More than three and a half centuries ago settlers and traders coming to the North American continent brought with them large communities of widely varied organisms that had attached to and bored into the wooden hulls of their ships in the harbors or bays where their journeys began. Many of the organisms transported across the oceans in this way were certainly swept from the ships' sides by the waves and currents, but a few invaders were present when the ships anchored at their journeys' ends. Hundreds of years later there is no way of knowing when the first European barnacles or periwinkle snails began to colonize the eastern coasts of the United States and Canada; these two organisms are now considered typical of this region.

The steel hulls of today's ships do not carry such communities for they are protected by antifouling paints, and the ships move through the water at speeds sufficient to sweep away many of the organisms that wooden hulls carried with them. However, the ships of the present do carry ballast water that is loaded and unloaded to preserve the stability of the ship as it unloads and loads cargo. Tens of thousands of vessels with ballast tanks ranging in capacity from hundreds to thousands of gallons of water move across our oceans. These vessels rapidly transport this ballast water and its populations of small floating organisms across natural oceanic barriers; ballast water and living organisms may be released days or weeks later, thousands of kilometers from their point of origin.

J. T. Carlton and J. B. Geller sampled ballast water from 159 cargo ships from various Japanese ports that docked in Coos Bay, Oregon.* The ballast water contained members of all the major groups of floating organisms known as animal plankton: microscopic shrimplike copepods, marine worms, barnacles, flatworms, jellyfish, and shellfish. The copepod density was estimated to be greater than 1500 copepods per cubic meter; the density of juvenile forms or larvae of marine worms, barnacles, and shellfish was greater than 200 organisms per cubic meter. Whether organisms arrive in ballast water or in the packing of commercially harvested fish and shellfish, come attached to the floats of seaplanes, or are released from home aquariums, they all leave behind the natural controls of predators and disease found in their native environments. If these invaders are introduced into a hospitable new environment, they may flourish and severely disrupt its biological relationships, forcing out some species and destroying others.

Some biological invaders remain unnoticed for many years; others begin almost immediately to seriously disrupt the ecology of their adopted areas. In 1985 an Asian clam (*Potamocorbula amurensis*) (fig. 12.10) was discovered in a northern arm of San Francisco Bay. This clam was previously unknown in the area and appears to have arrived in its

*J. T. Carlton and J. B. Geller, 1993. Ecological Roulette: The Global Transport of Nonindigenous Marine Organisms. *Science:* 261 (5117): 78–82.

web link **Figure 12.10** Asian clam (*Potamocorbula amurensis*).

juvenile or larval form in the ballast water of a cargo ship from China. During the next six years the clam spread southward into the bay and formed dense colonies, as many as 10,000 clams per square meter. The Asian clam feeds on plantlike, single-celled diatoms and the larvae of crustaceans. The food requirements of these huge populations reduce amounts available for native species, stressing the system's species balance and food chains. In 1990 a second intruder, the European green crab, or green shore crab (*Carcinus maenas*) (fig. 12.11), was recognized in the southern part of San Francisco Bay. This crab moved from Europe to the East Coast of the United States in the 1820s and to Australia in the 1950s. The green shore crab is less than 6.5 cm (3 in) broad and is voracious and belligerent. It feeds on clams and mussels and may spawn several times from a single mating. It feeds enthusiastically on the Asian clam, but it can also feed on the native species and can outcompete them for food. *Carcinus maenas* is now moving northward along the West Coast of the United States. In 1997, it had reached Coos Bay, Oregon, and in 1998 is was discovered in Willapa Bay and Grays Harbor, Washington. Some shellfish exports from Willapa Bay were temporarily banned while processing facilities were checked by state agencies attempting to prevent the crabs from moving farther north into Puget Sound.

In 1982 the ballast water of a ship from the coast of America carried a jellyfish-like organism known as a comb jelly (*Mnemiopsis leidyi*) into the Black Sea. From the Black Sea it spread to the Azov Sea and has recently moved into the Mediterranean. It has no predator in these areas and devours huge quantities of plankton, small crabs, shrimp, fish eggs, and fish larvae. Russian scientists report that this small organism now dominates the Black Sea, devastating the fish catches of the 1990s.

Ballast water carrying the resting cells of Japanese organisms that cause red tides was responsible for shutting

 web link **Figure 12.11** European green crab (*Carcinus maenas*). The width of the crab's back (carapace) is about 4 cm (1.5 in). *Photo courtesy of Thomas M. Niesen, San Francisco State University.*

down the natural and aquaculture shellfish harvests of Tasmania and southern Australia in the 1980s; red tides are discussed in chapter 15. Fish can also be transported this way; for example, Japanese sea bass were introduced into Australia's Sydney Harbor region in 1982–83. A small crab (*Hemigrapsus sanguineus*) common in Japanese waters was identified in New Jersey in 1988 and has since been found as far south as Chesapeake Bay and as far north as Cape Cod. The North American razor clam (*Ensis directus*) was detected in Germany in 1979; and it has since spread to France, Denmark, the Netherlands, and Belgium.

Not all invaders are animals. In the 1980s, an attractive, fast-growing, bright green tropical seaweed (*Caulerpa taxifolia*) was introduced into European aquariums. In 1985 some plants escaped into the Mediterranean Sea during a routine tank-cleaning at Monaco's aquarium. The seaweed rapidly spread once released and grew to cover more than 44.5 km² (11,000 acres) of the northern Mediterranean coastline by 1997. It has recently been reported off the coast of North Africa. In June 2000 *Caulerpa taxifolia* was discovered in a coastal lagoon just north of San Diego, California, and later in Huntington Harbor in Orange County, California, just 125 km (75 mi) further north. In areas where it becomes well established it is capable of causing ecological and economic devastation by overgrowing and eliminating native species of seaweeds, seagrasses, reefs, and other organisms. In the Mediterranean *Caulerpa taxifolia* has had a negative impact on tourism, recreational boating, and diving and has harmed commercial fishing by causing changes in the distribution of fish and impeding net fishing. No effective method of removing it has yet been found.

As more and more alien species are recognized, biological oceanographers and marine biologists realize that an ecological revolution is taking place in estuaries and bays and along the rocky shores of all the continents. Many biologists refer to the introduction of these alien species as "biological pollution"; others are increasingly alarmed by the breakdown of natural barriers and the worldwide "biological homogenization" that may result.

In the case of ballast water, the introduction of zebra mussels into the Great Lakes and their subsequent invasion of

the rivers and streams of the central United States sounded the alarm for the freshwater environment. In 1990 the Nonindigenous Aquatic Nuisance Prevention and Control Act (NANPCA) was enacted in the United States. Under this law the United States adopted voluntary regulations for exchanging ballast water on the high seas for vessels bound for the Great Lakes; this provision became law in 1993.

It has been calculated that, at any one time, in excess of 3000 species must be in motion in the ballast water of ocean-going ships around the world.* As yet there is no law comparable to NANPCA that regulates ballast water in the marine environment, although both the United Nations and the International Council for the Exploration of the Sea have called for ballast water management. Exchange of ballast water at sea can be attempted, if there is no danger to ship or crew, but not all the organisms may be flushed out and the sediment in the bottom of the ballast tanks may not be removed. Proposals have included treating the ballast water by adding poisonous chemicals, heating the water, filtering the water, and exposing the water to ultraviolet radiation. However, none of the world's cargo vessels are designed for ballast management, and all these proposals require some redesign or refit of vessels.

Controlling ballast water will not close all doors to invading marine species, and it will be costly. However, it will lead to fewer foreign invasions, and it is well to keep in mind that "No introduced marine organism, once established, has ever been successfully removed or contained, or the spread successfully slowed" (James T. Carlton, Maritime Studies Program, Williams College).

Overfishing and Incidental Catch 12.8

Around the world, too many fishing boats are taking too many fish, too fast. Diminishing fish populations are victims of relentless overfishing and management that fails to acknowledge declining stocks. Changes in world fish catches between peak fishing years and 1992 are given in table 12.5. Fish are being taken faster than the populations can reproduce. A 1995 report from the UN Food and Agriculture Organization (FAO) estimates that 70% of fish stocks worldwide are now overfished and depleted, concluding that world fisheries cannot be sustained at their present levels. According to the U.S. Office of Fisheries Conservation and Management, 41% of the species in U.S. waters are overfished. The 200-year-old Newfoundland cod fishery was closed by Canada in 1992, and in 1993 the National Marine Fisheries Service (NMFS) closed large parts of the U.S. cod fishery. The failure of the cod fishery is discussed further in chapter 16. North Atlantic swordfish landings in the United States declined 70% from 1980–90, and the average weight of the swordfish fell from 52 to 27 kg (115 to 60 lb). In 1991, several countries worked to reduce the Atlantic

*J. T. Carlton and J. B. Geller, 1993. Ecological Roulette: The Global Transport of Nonindigenous Marine Organisms. *Science:* 261 (5117): 78–82.

Table 12.5 Changes in Fishing Catch Between Peak Year and 1992

Atlantic Ocean			Pacific Ocean			Indian Ocean[2]	
Area[1]	Peak Year	Percent Change	Area[1]	Peak Year	Percent Change	Area	Percent Change
NW	1973	−42	NW	1988	−10	West	+6
NE	1976	−16	NE	1987	−9	East	+5
WC	1984	−36	WC	1991	−2		
EC	1990	−20	EC	1981	−31		
SW	1987	−11	SW	1991	−2		
SE	1973	−53	SE	1989	−9		

1. NW, northwest; NE, northeast; WC, west-central; EC, east-central; SW, southwest; SE, southeast.

2. The increase in Indian Ocean catches was due to implementation of more-sophisticated fishing techniques. Average annual growth is between 1988 and 1992.

C. Safina, The World's Imperiled Fish *in Scientific American Presents the Oceans, 9(3): 58–63, 1998.*

swordfish catch; Spain and the United States complied by reducing catches 15%. However, Japan increased its catch 70%; Portugal's catch rose 120%, and Canada's 300%. The Atlantic bluefin tuna population dropped 80% between 1970 and 1993; the bluefin population that spawns in the Gulf of Mexico has dropped 90% since 1975; and the Mediterranean population has declined by 50%. Many observers believe that this species is doomed, for it is the world's most valuable fish, selling at $200/kg (nearly $91/lb) in Japan's specialty fish shops.

Fish populations are not the only casualties; other marine animals and marine birds are being affected as they compete for their share of the catch. In the Shetland Islands, nesting seabirds failed to breed in the mid to late 1980s, apparently in response to lack of food when the sand eels in the area were overfished. In Kenya overharvesting of the trigger fish on coral reefs allowed abnormal growth of sea urchins, thereby damaging the reef ecosystem. Alaska's Steller sea lion populations have plummeted. An estimated 140,000 Steller sea lions existed in 1960. By 1989 the count was 25,000, and by 1996 only 18,000 remained. Studies in the Gulf of Alaska and the Bering Sea show that more than 50% of the Steller sea lion's diet is pollock, but the heavily fished pollock stock in the Bering Sea was down from an estimated 12.2 million tons in 1988 to 6.5 million tons in 1995. The sea lions have to spend more time and energy obtaining the same amount of fish. The population of Steller sea lions in 2000 had grown to roughly 40,000, with about 500 living in California. The entire population has been listed as a threatened species since 1990, and the western U.S. stock was listed as endangered in 1997. Stellers are protected under the Marine Mammal Protection Act, which prohibits the killing, harming, or harassing of any marine mammal, as well as the Endangered Species Act. With this federal protection, there is hope for the recovery of the Steller sea lion population.

Two hundred million people make their living by catching, processing, or selling fish, and millions more depend on fish as their source of protein. In Southeast Asia more than 5 million people fish full time and contribute some $6.6 billion to the region's income. In northern Chile fishing accounts for 18,000 jobs and 40% of the national income; Iceland depends on fishing for 17% of its income and 12%–13% of its employment.

In 1977 the United States extended control over its fisheries to an Exclusive Economic Zone (EEZ) that extends 200 miles out to sea and within which only U.S. boats are permitted to fish. Today more than 122 other nations have claimed EEZs, and in most cases the declaration of an EEZ has been followed by an increase in the country's fishing capacity. The world's fishing industry today has about twice the capacity it needs to make its annual catch. The world's fishing fleet doubled between 1970 and 1999, from 585,000 to 1.2 million large boats. China's fishing fleet is the largest in the world, with an estimated 450,000 vessels, roughly a third of the global fleet. The European Union nations are estimated to have fleets that are more than 40% larger than required. Over-expansion is encouraged by economic and political pressures that push artificial government supports to modernize fleets, supply fuel as well as sophisticated fishing gear, and prevent industry collapse, but these programs often undermine management programs and encourage overfishing.

The U.S. fishing fleet (fig. 12.12) has grown dramatically through a federal loan program encouraging the building of new boats. The new boats are extremely efficient, equipped with sonar and depth recorders and computers that remember the sites of previous catches and home in on these sites at a later date. Planes, helicopters, and even satellite data are used to find the fish. In addition, the fishing boat and gear, the crew's wages, and the fuel needed are all increasing the cost of commercial fishing, requiring the vessels to put even greater pressures on the shrinking schools of fish as the fishers strive to maintain their incomes.

Some countries are responding to the situation by consolidating their fishing fleets. Taiwan no longer issues licenses to boats under 1000 tons and has started a buy-back program for boats more than fifteen years old. Malaysia is

web 🔗 link **Figure 12.12** Seiners gather in Petersburg, Alaska, waiting for the salmon season to open.

Table 12.6 Global Incidental Catch

Fish Type	Percent Discard of Landed Weight	Landed Weight (millions of metric tons)
Shrimp and prawns	520	1.83
Crabs	249	1.12
Flounder, halibut, sole	75	1.26
Redfish, bass, conger eel	63	5.74
Lobster, spiny-rock lobster	55	0.21

Data from D. M. Alverson et al., "A Global Assessment of Fisheries Bycatch and Discards," 1994. FAO *Technical Paper No. 339*.

cutting its number of fishers by 40% and favors more modern, higher-capacity boats. Iceland plans a 40% cut in its fishing capacity. These programs may decrease the number of fishing vessels, but they will not necessarily reduce the catch if the remaining vessels are larger and more efficient.

The oceanic drift-net fishery became a serious problem in the 1980s, laying out nets up to 65 km (40 mi) long each night. These gill nets of almost invisible nylon hang like walls in the water, trapping and killing nearly everything that swims into them. In 1990 the U.S. Marine Mammal Commission estimated the aggregate length of these nets at 40,000 km (25,000 mi), enough to ring the Earth. Three Asian nations—Japan, Taiwan, and South Korea—have used the nets to catch squid and fish in the Pacific and Indian Oceans; in the Atlantic several European nations have drift-netted, mainly for albacore tuna. These nets do not rot; sections torn free in storms float in the ocean for months, even years, as ghost nets catching everything they encounter; there is no estimate on the animal life they destroy. In the Pacific, Japan, Taiwan, and South Korea agreed to abide by a UN resolution to halt drift-net fisheries by the end of 1992. After four years of discussion, the European Union fisheries ministers voted to ban drift-net fishing in most of the northeast Atlantic beginning in 2002. The only countries to vote against the ban were France, which has the second biggest drift-net fleet in the European Union, and Ireland. Italy, which has the biggest, abstained while the other twelve member states voted for the ban. The ban will apply everywhere except in the Baltic waters where there are no dolphins. This concession was made in order to get the support of Denmark, Finland and Sweden. Fishermen in the Baltic use drift nets to catch salmon and sea trout. The decision ends a long campaign by environmentalists, including Greenpeace, which has been trying to convince the European Union to act for fifteen years.

Large numbers of marine animals die each year only because they are caught incidentally by people fishing for other species. **Incidental catch,** or **by-catch,** or what are often called "trash fish," represents a tremendous waste of marine resources. The FAO estimates that each year 30 million tons of fish, about 25% of all reported commercial marine landings, are caught as by-catch and discarded. The world's shrimp fishery is estimated to have an annual catch of 1.8 million tons; the associated discarded by-catch is 9.5 million tons. Shrimp trawlers are estimated to catch more than 45,000 sea turtles each year; more than 12,000 of them do not survive. Alaskan trawlers for pollock and cod throw back to the sea some 25 million pounds of halibut, worth about $30 million, as well as salmon and king crab because they are prohibited from keeping or selling this by-catch. Another 550 million pounds of bottom fish are discarded in Alaskan waters to save space for larger or more valuable fish. See table 12.6 for global incidental catch rates for the most severely affected fisheries.

The discard rate on by-catch varies from place to place; if incidental catch does not bring a high enough price, and if processors or markets are not available, these "trash fish" will be returned to the sea, usually dead. Whereas in the Gulf of Mexico 1 pound of shrimp results in 10.3 pounds of by-catch that is nearly all discarded, in Southeast Asia and other areas with local fisheries and fresh fish markets much of the by-catch is used.

Afterthoughts 12.9

The Earth and its environment are works in progress. Since its beginning the Earth has constantly reworked and modified itself; living things have interacted with their environments, making changes and achieving new balances. Humans, however, have acquired the power to accelerate natural change and make fundamental environmental alterations for their own purposes. Read the words of Rachel Carson at the beginning of this chapter. Few organisms compete with us successfully and few environments are able to resist our presence. Humans will never be a zero impact factor, but we can challenge ourselves to understand and consider the implications of our choices. Science can help us understand the consequences resulting from our choices, but each of us, individually, must carefully define and protect the process of making the "best" choices.

We must also remember that *there is no away*. What we dispose of on this planet remains on this planet—out of sight may be out of mind, but it is not out of our environment. In June 1995, after three years of deliberation, the Shell Oil Company planned to dump a decommissioned, floating 14,500-ton oil-storage tank 2000 m (6600 ft) down on the Atlantic's North Feni Ridge. Shell claimed that sea disposal of the tank, which contained an estimated 100 tons of heavy metal sludge, was less hazardous than land disposal. The plan was stopped by adverse publicity and a huge public outcry; the tank was interred at a land site. Was the best choice made? Should the public be as concerned about the land burial as it was about ocean disposal?

The more we understand our Earth system the better the choices that we can make. Each of us—by continuing our education (in school or out), by participating in the political process, by working with others—can help to make the intelligent, informed decisions that are required to maintain a healthy and productive planet.

Summary

Water quality is affected by dumping solid waste and liquid pollutants into coastal and offshore waters. Land runoff carries a mixture of toxicants, oil and gasoline residues, industrial wastes, pesticides, and fertilizers to estuaries and seacoasts. An expanding low-oxygen area in the Gulf of Mexico has been linked to increasing amounts of nitrates and phosphates flowing into the Gulf primarily from the Mississippi River. Toxicants are adsorbed onto silt particles and become concentrated in coastal sediments. Organisms further concentrate toxicants and pass them on to other members of marine food webs. Reducing the discharge of toxicants is showing success, with declines in lead, DDT, and PCBs.

Plastics are an increasing problem to marine life, killing thousands of fishes, birds, mammals, and turtles each year. Laws prohibit dumping of plastics at sea, but there is no international enforcement.

Because of the world's increasing quantities of solid wastes, the deep-ocean plains have been proposed for waste disposal by some scientists. It has also been suggested that long-lived nuclear wastes could be placed in deep-sea trenches for eventual recycling back into the Earth's mantle.

Oil spills are a special problem for inshore waters. Three significant oil spills are the sinking of the *Amoco Cadiz*, the grounding of the *Exxon Valdez*, and the wartime spills into the Persian Gulf. All have had devastating ecological effects that continue to demonstrate that there is no adequate cleanup technology at this time.

Wetlands border estuaries and coasts; they are important as areas of nutrients, food, and shelter for many marine species. Many wetlands have been filled, dredged, developed, and lost.

Organisms move across the oceans in the ballast water of thousands of cargo vessels, to be discharged into new environments far from their points of origin. Significant ecological disruptions have been found in San Francisco Bay, the Black Sea, the Azov Sea, coastal Australia, and coastal Europe. Many scientists refer to the introduction of alien species as biological pollution.

Overfishing in all the world's oceans is devastating the world's fisheries. People who make their living from fishing are losing their jobs, and the marine species that depend on the fish are declining as the fish populations decrease. The establishment of EEZs (Exclusive Economic Zones) has promoted overexpansion of fishing fleets. Incidental catch is the cause of a tremendous loss to fisheries and wildlife.

Key Terms

All key terms from this chapter can be viewed by term, or by definition, when studied as flashcards on this book's Online Learning Center at www.mhhe.com/sverdrup (click on this book's cover).

hypoxic, 339	incidental catch/
wetland, 347	by-catch, 354

Study Questions

1. If contaminated sediments are dredged from the floor of a harbor to use in a landfill, what hazards to the environment should be considered during the dredging, the transport, and the storage of the sediments?
2. Study table 12.1. What general trends can be seen in wastewater contaminants between 1975 and 1995? What do you think is responsible for these trends?
3. Nutrient concentrations in coastal waters are increasing. Why? What is the result?
4. Why is toxicant concentration in sediments higher than toxicant concentrations in the overlying water? How are organisms in a polluted area affected?
5. Why are plastics such a problem in marine waters?
6. Why do some scientists think that ocean disposal of hazardous waste materials is safer than land disposal?
7. Compare the oil spills from the Gulf War and the *Exxon Valdez*. Consider the geography and climate of each area, the dispersal of the oil, the effect of the oil on the beaches and the organisms, the effect on people who gain their living from the sea, and the effectiveness of the cleanup.
8. What happens to oil when it has been spilled at sea? Into coastal waters?
9. Why are marine wetlands ecologically valuable, and why are they decreasing in all ocean shore areas?
10. Explain the consequences of ballast water transport of marine organisms.
11. How did the establishing of Exclusive Economic Zones (EEZs) affect coastal fisheries?

12. What is incidental catch, or by-catch, and what is its effect on ocean fisheries?
13. Why can a biological invader such as *Caulerpa taxifolia* spread so rapidly in a new environment?
14. How could the incidental catch or by-catch of the world's fisheries be used effectively?
15. What would have to happen socially or economically in order to effectively use the world's incidental catch?

Links to Related Websites

Visit the book's Online Learning Center at www.mhhe.com/sverdrup (click on the book's cover) to find live Internet links for additional topics related to this chapter's content.

- Environmental preservation
- Coastal monitoring

- Toxic spills and cleanup
- Biological invaders
- Fish and fisheries

Visit the book's Online Learning Center at www.mhhe.com/sverdrup (click on this book's cover) to find these additional chapter tools: Suggested Readings; links to further information on boxed readings, selected figures, and related chapter topics; and additional study aids.

chapter 13

The Living Ocean

The world below the brine,
Forests at the bottom of the sea—the branches and leaves,
Sea lettuce, vast lichens, strange flowers and seeds—the thick tangle,
the openings, and the pink turf,
Different colors, pale gray and green, purple, white and gold—the play
of light through the water,
Dumb swimmers there among the rocks—coral, gluten, grass, rushes
and the aliment of the swimmers,
Sluggish existence grazing there, suspended, or slowly crawling close
to the bottom. . . .

Walt Whitman
From *The World Below the Brine*

A diver observes coral reef organisms in the Caribbean Sea.

The previous chapters combined basic principles of oceanography and contemporary information to provide an understanding of today's ocean sciences. In this chapter and those that follow we introduce the organisms that live in the oceans: in the water, on the sea floor, and along the shores. These organisms have much in common with organisms of the land, but their environment presents them with challenges and survival problems unknown to terrestrial organisms. In this marine environment some changes are gradual and extend over long distances; other changes occur quickly over small areas. The currents and waves, rising and falling tides, sandy sea floors, rocky shores, and dark cold deeps form a series of complex situations for living. To discuss the ocean environment and learn about the organisms that live in this environment one must also learn some principles of biological oceanography and marine biology. In this chapter we commence with the classifying of organisms and environments and then reconsider many of the topics in previous chapters but now in relationship to living organisms. An introduction to the many and various life forms of the oceans and their places in the marine environment is found in chapters 15, 16, and 17.

Ocean Biology 13.1

Marine biology is the scientific study of the organisms that live in the oceans. Biological oceanography or marine **ecology** is the scientific study of the interactions among marine organisms, the interactions between marine organisms and their environment (living and nonliving), and the effects of these interactions on the abundance and distribution of marine organisms. Those who study marine organisms and their environment may be called biological oceanographers, marine ecologists, or marine biologists. All these scientists study, with differing emphases, the abundance and distribution of marine organisms and the relationships between the organisms and their environment.

Groups of Organisms 13.2

The complexity of living systems requires that organisms and environments be subdivided into smaller units to provide an orderly framework for the classification of information. The terms *plankton*, *nekton*, and *benthos* are used to describe marine organisms with reference to their habitats. Plants and animals that live suspended in the water, floating or drifting with the water's movements, are known as **plankton** (fig. 13.1). Animals that swim freely and purposefully in the sea are the **nekton** (fig. 13.2), and animals that live on, or in, or attached to the sea floor are the **benthos** (fig. 13.3).

The scientific classification of different kinds of organisms is known as **taxonomy.** Taxonomic systems have been devised by biologists to sort organisms into groups and to provide names acceptable in all countries for the identification of living organisms. These taxonomic systems also help scientists identify and understand the relationships that exist among the organisms. More than 200 years ago, Carolus Linnaeus, a Swedish botanist, introduced a two-kingdom classification system that divided all living organisms into plants and animals. This Linnean system has since developed into the five-kingdom system, the most widely used taxonomic system at the present time. In this system all living organisms are placed in one of the five kingdoms: Monera, Protista, Fungi, Plantae, or Animalia. A new system (fig. 13.4) based on genetic and biochemical research proposes three categories above the kingdom level. These three large categories are known as domains. Members of the Monera kingdom (single-celled organisms without a membrane-bounded nucleus) are placed in either the Bacteria domain or the Archaea domain. The Eukarya domain (all nuclei-containing organisms) includes all the other kingdoms. For more information on members of the Archaea refer to the box titled "Extremophiles," in chapter 15. Consensus favoring the three-domain system is building among biologists, who are using it more and more frequently.

It is well to keep in mind that these systems are works in progress. Remember that any classification system is a human construct that remains useful only until additional information leads to its modification. Taxonomic categories for some common marine organisms are listed in table 13.1. Classification outlines for groups of organisms discussed in chapters 15, 16, and 17 are included in these chapters.

Figure 13.1 This plankton sample contains (*left* to *right*) shrimplike copepods, a stage in crab development, an arrow worm, and a fish egg.

Figure 13.3 The benthic sun star, *Solaster*, typically has ten arms and a diameter of 25 cm (10 in).

Figure 13.2 All fish are members of the nekton. Here a diver observes Sargent Major fish at Santo, Vanuatu.

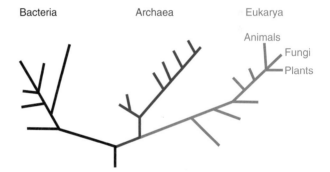

Figure 13.4 The new family tree of life is divided into three domains: Bacteria, Archaea, and Eukarya. The Bacteria and Archaea are single cells without nuclei. The Eukarya domain includes many single-celled organisms with nuclei as well as the animals, plants, and fungi.

Table 13.1 Taxonomic Categories of Some Marine Organisms

	Killer Whale	**Northern Fur Seal**	**Pacific (Japanese) Oyster**	**Giant Octopus**	**Sea Lettuce**	**Giant Kelp**
Kingdom	Animalia	Animalia	Animalia	Animalia	Plantae	Plantae
Phylum	Chordata	Chordata	Mollusca	Mollusca	Chlorophyta	Phaeophyta
Class	Mammalia	Mammalia	Bivalvia (Pelecypoda)	Cephalopoda	Chlorophyceae	Phaeophyceae
Order	Cetacea	Carnivora (Pinnipedia)	Anisomyaria	Octopoda	Ulvales	Laminariales
Family	Delphinidae	Otariidae	Ostreidae	Octopodidae	Ulvaceae	Lessoniaceae
Genus	*Orcinus*	*Callorhinus*	*Crassostrea*	*Octopus*	*Ulva*	*Macrocystis*
Species	*Orcinus orca*	*Callorhinus ursinus*	*Crassostrea gigas*	*Octopus dofleini*	*Ulva lactuca*	*Macrocystis pyrifera*

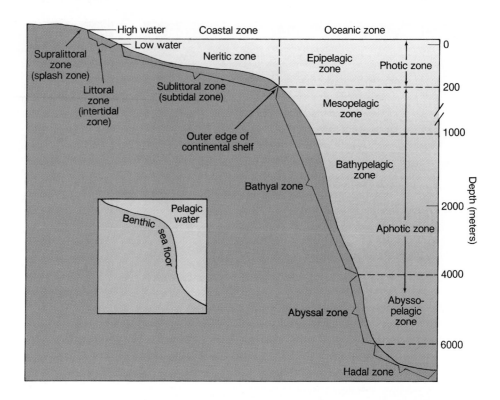

Figure 13.5 Zones of the marine environment.

Environmental Zones 13.3

Because the marine environment is so large and complex biological oceanographers, marine biologists, and marine ecologists divide the marine environment into subunits called zones. These zones are shown in figure 13.5. In this system the marine environment is divided first into two zones: the **pelagic zone** or water environment, and the **benthic zone** or seafloor environment. The pelagic zone is divided into the coastal or **neritic zone** (above the continental shelf) and the **oceanic zone** (open ocean away from the influence of land). Additional zones, for different increments of depth, are shown in figure 13.5. All oceanic subdivisions, except for the first 200 m (655 ft) in depth, are without sunlight.

The sea floor, or benthic environment, is also subdivided into zones. Tidal fluctuations at the shoreline define the **supralittoral zone** or **splash zone;** this zone is covered only by wave spray or during the highest spring tides. Here the organisms must cope with environmental extremes, for example, hot versus cold, wet versus dry, pounding surf versus exposure to air. The **littoral zone** or **intertidal zone** is covered and uncovered once or twice each day as the water level moves between high and low tides. As the seasons change, climatic conditions in the littoral change, depending on the latitude. Along different stretches of the same coast conditions in the littoral may vary greatly: from rock to sand, from crashing waves to gentle surf, from high-amplitude tides to low. Below the low-tide level the **sublittoral zone,** or **subtidal zone,** extends out over the continental shelf. The deepest portions of the subtidal zone are areas of perpetual darkness; these zones are known as the **bathyal** and **abyssal zones.**

Here conditions, worldwide, are more alike than different, and there is no seasonal change. The **hadal zone** lies below 6000 m (19,800 ft) and also shows no seasonal change. This zone is associated with the ocean's trenches and deeps (see again fig. 13.5).

Facts of Ocean Life 13.4

Buoyancy, Flotation, and Viscosity

Organisms that make their home on land require structural strength to support their bodies in air and against gravity. Trees must be able to hold up their canopies of leaves and animals need skeletons and muscles to give their bodies shape and the ability to move. Marine organisms, however, are supported by and move with the water. Except where breaking waves are encountered at the surface or along the shore, the mechanical stresses of the water are slight. In the following chapters you will see photographs and drawings of many very delicate and fragile organisms that survive and function well as long as they are surrounded by seawater.

The oceans supply marine animals with an environment that allows them to live and move on the sea floor and in three dimensions in the pelagic zone above the sea floor. Because the density of seawater is nearly the same as the density of many floating organisms, it provides the organisms with **buoyancy,** which maintains body support and flotation. The water helps to keep floating organisms at the surface, supports the bodies of the bottom-dwellers, and lessens the energy expended by the swimmers.

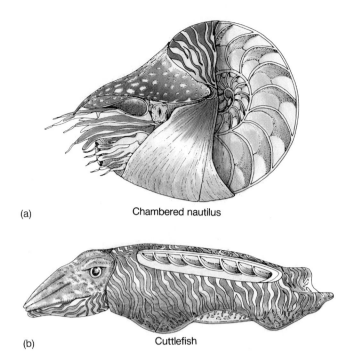

(a)
Chambered nautilus

(b)
Cuttlefish

Figure 13.6 (a) Chambered nautilus. (b) Cuttlefish. The chambered shell that provides buoyancy is shown in each organism.

Many organisms have ingenious adaptations to help them stay afloat. Some jellyfish-type animals—for example, the Portuguese man-of-war and the by-the-wind-sailor (see chapter 15, fig. 15.12)—secrete gases into a float that enables them to stay at the sea surface. Some seaweeds secrete gas bubbles and form gas-filled floats, which help them keep their fronds in the sunlit surface waters while they are anchored to the sea floor. One floating snail produces and stores intestinal gases; another forms a bubble raft to which it clings. The chambered nautilus (fig. 13.6*a*), a relative of the squid, continually adds chambers to its shell and moves to the last chamber as it grows. A specialized tissue removes the ions from the vacated chambers, causing water to diffuse out, then the chamber fills with gas, mainly nitrogen, from tissue fluids. The cuttlefish, another relative of the squid, has a soft, porous internal shell, or "bone," of calcium carbonate; this animal regulates its buoyancy by controlling the relative amounts of gas (also mainly nitrogen) and liquid within the chambers of the shell (fig. 13.6*b*).

Many species of fish have internal swim bladders filled with gas (mainly nitrogen and oxygen) that keep them neutrally buoyant. Some fill their swim bladders by gulping air at the surface; others release gas from their blood through a gas gland to the swim bladder. When a fish changes depth, it adjusts the gas pressure in its swim bladder to compensate for the pressure change in the water. If a fish with a swim bladder is forced suddenly to swim deeper, the increased external water pressure compresses the air, the bladder shrinks, and the fish sinks. If the fish is not able to readjust its system, it tires as it is forced to swim upward continually to compen-

sate for the loss of buoyancy. A deep-swimming fish brought quickly to the surface with a fishing line will show bulging eyes and a distended body, as the gas in the swim bladder expands with decreasing pressure. Fish living at great depths, 7000 m (23,000 ft) or more, and fish that migrate vertically have fat-filled swim bladders. Active, continuously swimming predatory species such as the mackerel, some tuna, and the sharks do not have swim bladders. For these fish this lack is advantageous because rapid vertical swimming would be difficult if the pressure on the swim bladder changed continually. Bottom fish also lack swim bladders.

Small members of floating plant and animal populations store their food reserves as oil droplets that decrease their density and retard sinking. Large surface-area-to-volume ratios that slow sinking are characteristic of small spheres, and single-celled organisms, particularly plants, benefit from this characteristic. Many have developed spines, ruffles, and feathery appendages that increase surface area and decrease their sinking rate, allowing them to more easily remain at or near the sea surface.

The tissues of most large marine animals are denser than seawater. To reduce the density of its body tissues, the giant squid excludes the denser ions from its body fluids and replaces them with less dense ions. Another squid species has one pair of its arms filled with low-density body fluids. Whales and seals decrease their density and increase their flotation by storing large quantities of blubber, which is mainly low-density fat. Sharks and some other varieties of fish store oil in their liver and muscle.

Seabirds float by using fat deposits in combination with light bones and air sacs developed for flight. Their feathers are waterproofed by an oily secretion called preen, which acts as a barrier to seal air between the feathers and the skin. This is important in keeping the birds warm, and it also helps to keep them afloat.

Viscosity of water is the internal friction of water that increases with decreasing temperature (see chapter 4, section 4.5). Because the increased viscosity of cold water increases the friction between the organism and the water, very small organisms float more easily in polar waters than in tropical waters. The tropical water organisms have adapted to the less viscous water by developing spines and frilly appendages to keep them afloat.

Salinity and Osmosis

Special problems are posed for living creatures if the salt content of their body fluids differs from the salinity of the water that surrounds them. The body fluids of living plants and animals are separated from the seawater by membrane boundaries that are semipermeable, allowing some molecules to move across the membrane boundaries while other molecules cannot. Molecules that move across these membranes do so along a gradient from a region of high concentration of a substance to a region of low concentration of that substance, a process called diffusion. Water molecules cross the membranes that separate body fluids from seawater by moving from a high concentration of water (low salinity) to a low concentration of water (high salinity). This special type of diffusion known as osmosis is discussed in chapter 5.

Most fish have body fluids with a salt concentration that is about halfway between that of fresh water and seawater. In salt water, their tissues tend to lose water as it moves along an osmotic gradient from an internal high water concentration and low salt concentration to the lower water concentration and higher salt concentration of the ocean environment. Fish must constantly expend energy to prevent dehydration and an increase in the salt concentration of their tissues. Fish stay in fluid balance by drinking seawater nearly continuously. They must excrete the excess salt, and because the outer skin of fishes is not completely permeable to seawater, this excretion of salt occurs at the gills. Sharks and rays do not have this problem, because their body fluids have the same approximate salt content as seawater, so there is no osmotic gradient. These fish maintain a high concentration of urea in their tissues. The urea allows the tissues to retain water and keeps body fluid at approximately the same salt content as seawater.

The body fluids of many bottom-dwelling organisms, such as sea cucumbers and sponges, are also at the same salt concentration as the seawater. There is no concentration gradient; the water diffuses equally in both directions across the membranes, and the salt content remains the same on both sides of the membranes. The fish, which does have to overcome an osmotic gradient, and the sea cucumber, which does not, are compared in figure 13.7.

Species may be limited in their geographic distribution by changes in salinity, for many organisms can maintain their salt-fluid balance over only limited salinity ranges. Because there is little change in salinity in deep water, species living below the surface layers are widely dispersed. The surface-dwelling forms are more likely to find salinity barriers in coastal waters than in the open ocean, because the salinity varies in bays and estuaries. Also, sudden severe storms may drop as much as 20–40 cm (8–16 in) of rain in a coastal area, lowering the water's salt content. Successful estuarine animals such as crabs are able to stabilize their osmotic processes by regulating the intake of salt ions. In some areas, a few species are able to survive in high-salinity lagoons and salt marshes, but they are unlikely to reproduce there, so the populations must be replenished by new recruits from the sea.

Some animals have an extraordinary ability to adapt to large changes in salinity over their life history. Salmon spawn in fresh water but move down the rivers as juveniles to live their adult lives in the sea. After several years (time depends on the species), the salmon return to their home streams. The American eel and the European eel reverse this process by migrating seaward from ponds, rivers, and estuaries to spawn in the Sargasso Sea. The new generation of eels drifts north and east with the Gulf Stream and returns to estuaries and rivers to live for up to ten years before migrating seaward. Other fish and crustaceans use the low-salinity coastal bays and estuaries as breeding grounds and nursery areas for their young, then as adults they migrate farther offshore into higher-salinity waters.

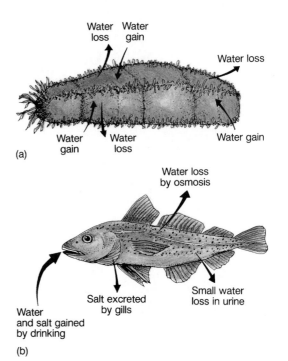

Figure 13.7 (a) The salt concentration of the seawater is the same as the salt concentration of the sea cucumber's body fluids (35‰). The water diffusing out of the sea cucumber is balanced by the water diffusing into it. (b) The salt concentration in the tissues of the fish is much lower (18‰) than that of the seawater (35‰). To balance the water lost by osmosis, the fish drinks salt water, from which the salt is removed and excreted.

Temperature

The temperature of the deep oceans is low and nearly constant. At the surface and close to shore the water temperature varies with seasonal climate changes and geographic latitude zones. Temperature, like salinity, affects the density of the seawater and also affects its viscosity; the density and viscosity of seawater are discussed in chapter 4. At polar latitudes, the surface water is cold, denser, and more viscous, and organisms float more easily. In tropical latitudes, the warm, less dense, less viscous water is home for species with more appendages, larger surface areas, and greater gas-bubble production, for the water offers less buoyancy (less resistance to sinking).

When surface conditions produce a warm, low-density surface layer overlying denser water, the water column is stable, and floating organisms are retained in the sunlit upper layers. Under these conditions, floating plantlike cells increase their rates of photosynthesis and reproduction. When surface waters cool and increase their density, they become unstable; the waters mix, overturn, and take the floating organisms with them into deeper water and away from the sunlight. The photosynthesis and reproduction of such displaced plantlike cells are decreased.

Marine animals other than birds and mammals do not have mechanisms to regulate their body temperatures. These organisms are **ectotherms,** and their body temperatures vary with environmental conditions. Their physiology is regulated by the temperature of the water, and within limits metabolic processes proceed more rapidly in warm water than in cold water. Cold-water forms frequently grow more slowly, live longer, and attain a larger size. To some species, changes in temperature act as signals to spawn or to become dormant. Many invertebrates spawn only when the water reaches a certain temperature. In various species environmental temperature may also influence the amount of egg and sperm production and the determination of sex. Although the heat capacity of the water reduces temperature fluctuations, the geographic distribution of seawater temperatures and their seasonal fluctuations are sufficient to affect the distribution of marine organisms.

Seabirds and mammals are **endotherms** and maintain nearly constant body temperatures that are well above the temperature of the seawater. Because these animals are less restricted by the temperature of the water, they often have wider geographic ranges; for example, whales annually migrate between polar and tropical waters. Some fish species are able to conserve heat in their swimming muscles, elevating muscle temperatures. Because their muscles work more efficiently at higher temperatures, these fish are able to swim rapidly and cruise long distances in the coldest water, making them efficient predators. Fish of this type include some tuna, mackerel-sharks, and the mako shark.

At the greater depths, the uniformity of temperature with latitude creates an oceanwide environment that is unaffected by seasonal changes. At the sea surface, the temperature changes with latitude much the same as the climate changes on land. Annual changes in open sea surface temperatures are small at the very high and very low latitudes; at the middle latitudes the annual changes in sea surface temperature are larger. Ocean areas close to land undergo still greater changes in surface temperature, owing to less water depth and the influence of the annual temperature changes over the landmasses. Seasonal fluctuation in ocean surface temperature at the middle latitudes is reflected in periods of spring and summer reproduction and growth and in winter dormancy.

On land, the general pattern of climate zones encountered by approaching the poles is also observed by increasing the altitude; the climate at sea level in polar zones is similar to the climate at the top of a high mountain peak at a lower latitude. In the ocean, conditions in surface water at polar latitudes are similar to those found at deeper depths at lower latitudes. Some shallow-water species of the polar seas are found at greater depths at the lower latitudes.

Pressure

Deep-living organisms such a worms, crustaceans, and sea cucumbers are unaffected by the great pressures of the deep ocean because they do not have gas-filled cavities or lungs that must be maintained at high pressure or mechanically protected against the pressure of the overlying water. Review "The Effect of Pressure" in chapter 4, section 4.6. It has been thought that pressure may lower metabolic rates and growth at greater depths, but these effects may also be adaptations

Table 13.2 Record Depths of Breath-Hold Divers

Species	Dive Depth (m)	(ft)
Human (Homo sapiens)	105	347
King penguin (Aptenodytes patagonicus)	240+	792+
California sea lion (Zalophus californianus)	250	825
Ridley sea turtle (Lepidochelys olivacea)	250	825
Common porpoise (Delphinus delphis)	260	858
Killer whale (Orcinus orca)	260	858
Emperor penguin (Aptenodytes forsteri)	>500	>1650
Bottle-nose dolphin (Tursiops truncatus)	535	1766
Pilot whale (Globicephala melanena)	610	2013
Beluga whale (Delphinopterus leucas)	650	2145
Weddell seal (Leptonychotes weddellii)	>700	>2310
Leatherback turtle (Dermochelys coriacea)	>1000	>3300
Elephant seal (Mirounga angustirostris)	>1500	>4950
Sperm whale (Physeter catodon)	>2000	>6600

Data from G. L. Kooyman, Diverse Divers, Physiology and Behavior, 1989; and G. L. Kooyman and P. J. Ponganis, The Challenges of Diving to Depth, 1997.

that reduce activity in response to poor quality and scarcity of food at deeper depths.

When humans descend into the sea, they need either protection from the pressure that will collapse their chest cavities and lungs or air supplied to the lungs at a pressure equal to the outside water pressure. Submarines and submersibles provide the first type of protection, and scuba (self-contained underwater breathing apparatus) and other commercial diving equipment supply the second type. After breathing gases under pressure, humans may experience the serious problems of decompression sickness, or "the bends," and nitrogen narcosis. Divers breathing air under pressure for extended periods take in quantities of gases, particularly nitrogen, that dissolve in their blood. When they return too quickly to shallow depths of less pressure, the oxygen is used up but not the nitrogen, which may form bubbles in their blood and body tissues. This causes extreme pain, paralysis, and sometimes death. Excess nitrogen dissolved in the bloodstream also has a narcotic effect producing nitrogen narcosis, which confuses the diver and restricts his or her ability to function normally. Record dives by humans in seawater require the divers to hold their breath for longer than two minutes.

Air-breathing marine mammals and seabirds make dives of spectacular depth and duration without encountering such difficulties. Record diving depths reported for humans and some marine animals are given in table 13.2. All the best diving mammals and birds have very streamlined shapes. This shape reduces the drag on their bodies, lessening their swimming effort and thereby lowering their oxygen consumption rate. The best divers show a significant difference in the distribution and concentration of the protein myoglobin when compared to terrestrial animals and birds. Myoglobin is found primarily in muscle tissue, and its main function is to bind oxygen. Diving animals and birds appear to rely on oxygen held in the myoglobin while diving and are able to regulate its flow as needed during the dive. The concentration of myoglobin in marine mammals and birds is three to ten times higher than that found in their terrestrial relatives.

People have a total oxygen capacity of 20 mL/kg of body mass; Weddell seals have a capacity of 87 mL/kg of body mass; and penguins have a capacity of 55 mL/kg of body mass. The lungs of seals collapse as they dive; most of their oxygen is stored in blood and muscle. Birds do not collapse their lungs, and, in addition, they have air sacs developed for flight. The emperor penguin stores ten times more oxygen than terrestrial birds, and its muscle oxygen storage exceeds its blood oxygen storage.

Gases

Life in the water requires carbon dioxide and oxygen, as does life on land. Carbon dioxide is required by plants for photosynthesis; it is contributed by animals and by decay processes, and it is absorbed by water from the atmosphere. There is no shortage of carbon dioxide in the ocean because seawater has the capacity to absorb it in large quantities. Additionally and very importantly, carbon dioxide acts as a buffer to limit the ocean's pH range, keeping the seawater a stable environment for living organisms (see chapter 5, section 5.1).

Oxygen is required by all organisms to liberate energy from organic compounds. Oxygen is available only at the ocean surface, as a by-product of photosynthesis and from the atmosphere. Life below the surface depends on the vertical circulation processes (discussed in chapter 7) to replenish oxygen at depth. Active species such as fish, squid, and crab require more oxygen than species like sponges and barnacles that stay in one place. Small organisms and larval forms with bodies only a few millimeters thick depend on diffusion for uptake of oxygen; larger worms, shellfish, and fish use gills and marine mammals have lungs.

The amount of oxygen in the water is influenced by the temperature, salinity, and pressure of the water; this also influences the distribution of organisms. On warm, quiet days, the temperature and salinity in shallow tidal pools and bays increase, lessening the ability of the water to hold oxygen. This situation forces motile animals out, limiting these areas to organisms that can successfully tolerate such changes. The slowly circulating water in the bottoms of deep, isolated basins (see fig. 5.6) may become anoxic, so low in oxygen that only non-oxygen-requiring, or anaerobic, organisms are able to survive.

Nutrients

Nitrate (NO_3^-) and phosphate (PO_4^{-3}) nutrients are required by the sea's plant and plantlike organisms. They are the fertilizers of the sea and are stripped from the surface layers by the plants, which incorporate them into their tissues. These nutrients are liberated at lower depths by the decay of plant as well as animal tissues, or they are returned to the water in the form of waste products of herbivores and carnivores (fig. 13.8). Vertical circulation and mixing transport

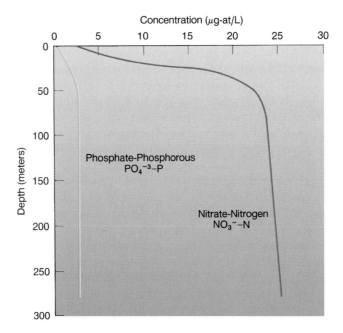

Figure 13.8 Nitrate and phosphate distribution in the main basin of Puget Sound in the late summer. The low surface values are the result of nutrient utilization by unicellular marine plants.

the nutrients back to the surface in upwelling areas where life is abundant. Estuaries and coastal waters, where nutrients are supplied by land runoff and mixing from the continental shelf's shallow sea floor, are also rich with organisms. Plant populations are limited by the lack of any essential nutrient; if the concentration of such a nutrient falls below the minimum required, the population's growth ceases until the nutrient is replenished. Small unicellular organisms known as diatoms (the grass of the sea) require dissolved silica to form their rigid outer coverings (see chapter 14). Iron, manganese, and zinc are necessary in some physiological systems, and zinc and copper are required in certain enzyme systems. Organic nutrients such as vitamins are also needed by some unicellular species. Nutrients were introduced in chapter 5 and are discussed further in chapter 14.

Light and Color

Sunlight Plant life is restricted to the **photic zone**, where there is sufficient light energy for the process of photosynthesis. The depth of the photic zone, about 200 m (660 ft) in clear ocean water, is controlled by factors discussed previously in chapters 1, 4, and 6, including (1) the angle at which the Sun's rays hit the Earth's surface, which is related to latitude and change of season; (2) the different rates at which the wavelengths of light are absorbed, which are determined by the properties of water; and (3) the suspended particulate material present, which affects the rate of absorption. Below the photic zone is the **aphotic zone,** the zone in which there is no photosynthesis.

However, the presence of light does not guarantee plant life; nutrients also must be available. It is because of

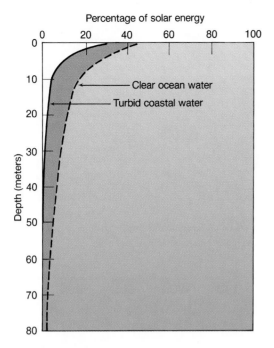

Figure 13.9 The percentage of solar energy available at depth in clear and turbid water.

lower concentrations of nutrients that so much of the open ocean, exposed to high-intensity sunlight, is less productive than the coastal ocean. Life is more abundant along the coasts and over the continental shelves because of the larger quantities of dissolved nutrients. Waves and currents of the coastal zone stir the bottom and mix up silt with its nutrients. The silt particles absorb and scatter light, reducing the depth to which it penetrates. As the single-celled plants reproduce, their increased numbers also act to limit light penetration, so that the photic depth may be reduced to less than 50 m (167 ft). The penetration of sunlight in clear and turbid seawater is compared in figure 13.9.

An interesting adaptation that allows photosynthesis to continue in an environment that has no direct sunlight has been discovered by Italian scientists working in the Antarctic. Inside sponges living 120 m (400 ft) below the ice, the researchers found tiny photosynthesizing green algae. These algae produce organic nutrients for their host sponges. Like many sponges, this species, *Rosella racovitzae*, contains silica **spicules,** small structures that support the body form of a sponge. But this species also has a light-guidance system; each spicule has a cross-shaped antenna at its top that captures the light and allows it to travel to the base of the spicules, where the algae grow. The scientists are looking for this ability in different sponges and in other light-reduced environments such as shore caves.

Bioluminescence Sunlight illuminates only the sea surface, but another source of light exists in the oceans. On a dark night, the wake of a boat may be seen as a glowing ribbon and disturbed fish may leave a trail of light; the water may flash with light as oars dip, or a person's hands may glow briefly when a net is hauled in. In all these cases, living organisms are producing the light. This light is **bioluminescence,** and it is produced by the interaction of the compound **luciferin** and the enzyme **luciferase.** This phenomenon is often incorrectly referred to as phosphorescence; it has nothing to do with phosphorus but is a chemical reaction that produces light with a 99% efficiency. The same phenomenon is seen on land in the flashing of a firefly or the ghostly glowing of a fungus in the woods.

Bioluminescence is triggered by the agitation of the water that disturbs microscopic organisms, causing them to flash and produce glowing wakes and wave crests. Animals that feed on these organisms often concentrate the chemicals in their tissues and also glow. So do one's hands if they come into contact with crushed tissue. Other bioluminescent organisms in the sea include squid, shrimp, and some fish. Many mid-depth and deep-water fish species carry light-producing organs, or **photophores,** some in patterns on their sides, possibly for identification. Others show photophores on their ventral surfaces, making them difficult to see from below against the light surface water, and still others have glowing bulbs dangling below their jaws or attached to flexible dorsal spines, acting as lures for their prey. Flashlight fish, living in the reefs of the Pacific and Indian Oceans, have a specialized organ below each eye that is filled with light-emitting bacteria. These fish use the light to see, communicate, lure prey, and confuse predators.

Color Some sea animals are transparent and blend with their water background: for example, jellyfish and most of the small floating animals in the surface layers. Other animals, particularly fish, use color in many ways. In the clear waters of the tropics, where light penetrates to deeper depths, bright colors play their greatest role. Some fish use bright colors as a disguise; they match the colors of the corals so well they become nearly invisible. Other fish conceal themselves with bright color bands and blotches that disrupt the outline of the fish and may draw the predator's attention away from a vital area to a less important spot; for example, a black stripe may hide the eye while a false eyespot appears on a tail or fin. Other fish use color to advertise; bold coloration increases during the breeding periods of some species and may be useful for sex recognition. Color also sends a warning; organisms that sting, taste foul, or have sharp spines or poisonous flesh are often striped and splashed with colors, for example, sea slugs and some poisonous shellfish. Among fish that swim near the surface in the well-lighted surface water—for example, herring, tuna, and mackerel—dark backs and light undersides are common. This color pattern, called countershading, allows the fish to blend with the bottom when seen from above and with the surface when seen from below (fig. 13.10). Fish living in deeper water tend to be much smaller. The average length of mezopelagic fishes is rarely more than 10 cm (4 in) in length, and they appear black at depths of 200–1000 m (700–3000 ft). These deep-water fish are discussed in chapter 16.

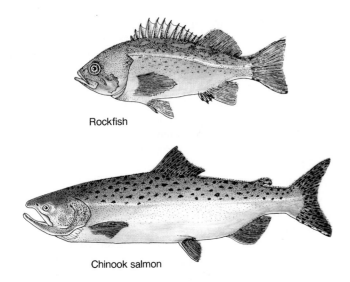

Rockfish

Chinook salmon

Figure 13.10 Viewed from above, the dark dorsal surface of the fish blends with the sea floor; viewed from below, the light ventral surface blends with the sea surface. This type of coloration is known as countershading.

Some species of deep-water shrimp are red when seen at the surface, but below the level of penetration of red wavelengths of light their red pigment absorbs blue and green light; little is reflected and the animals are dark and inconspicuous. In the deep ocean without light, color is of little importance except perhaps when combined with bioluminescence. Of course, we do not know how the animals see the colors, and there may be roles that color plays in the sea that we do not know or understand. Color is thought to be important in species recognition, courtship, and possibly in keeping a school of fish together.

In the turbid coastal water of temperate latitudes there is less light penetration and drab browns and grays conceal animals seen against kelp beds and rocks. Cold-water bottom fish are usually similar in color to the bottom or are speckled and mottled with neutral colors. The flatfish are well known for their ability to change their color, having skin cells that expand and contract to produce color changes (fig. 13.11). Their extraordinary color-changing ability enables them to conceal themselves by matching the bottom type on which they live (fig. 13.12). Squid may be the ocean's experts at color change; they are able to expand and contract pigment cells, producing rapid flashes and changing color patterns. Many squid also have bioluminescent cells, and some harbor colonies of light-emitting bacteria covered by a flap of skin. In combination these allow the squid to display hundreds of different and complex color patterns and sequences, enabling them to change color and appear to vanish almost instantaneously.

Circulation Patterns

The ocean's water is in constant motion, moved and mixed by currents (see chapter 8), waves (see chapter 9), and tides (see chapter 10). Below the surface layers, the ocean envi-

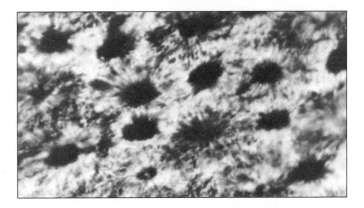

Figure 13.11 Pigment cells from a section of fish skin. The cells expand and contract to produce color change.

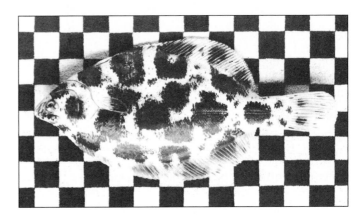

Figure 13.12 The winter flounder resting on a checkerboard pattern shows its use of camouflage.

ronment is very uniform, providing marine organisms with similar conditions of motion, temperature, and salinity in any ocean at any time. Oceanic circulation brings food and oxygen, replenishes nutrients, and removes wastes; it disperses floating organisms and scatters the reproductive stages of swimmers and attached forms.

Plants and animals that drift rather than swim are carried along by the currents and run the risk of being carried out of a suitable habitat by either vertical or horizontal water motion. However, analysis of their remains on the sea floor and observations of living populations show that this does not always happen. Populations of drifting animals appear to take advantage of their ability to move in the vertical direction either by swimming or by changing their buoyancy. They are able to maintain their place horizontally in ocean space by moving away from the surface during the day to depths where a current flows in a direction opposite to the surface current. They then move upward at night to be carried back to their starting position. Organisms also appear to maintain their position by adding to their population on the upstream side of a current to balance losses on the downstream side. This pattern is related to the new supply of food and nutrients brought into the population by the current upstream, in contrast to the food-depleted water downstream.

Vertical water motions in the sea are much slower than horizontal motions, but small vertical displacements can subject the organisms to substantial changes in light, salinity, temperature, and nutrient supply. If the vertical flow of the water is upward, it counteracts the tendency of organisms and other particulate matter to sink. In this way, light-dependent organisms are held in the photic zone. The upward motion of the water also supplies nutrients to the photic zone to promote plant growth. At the same time, these upward flows decrease the temperature of the surface waters and return water with a low oxygen content to the surface, where oxygen is replenished by photosynthesis and atmospheric exchange.

A downward vertical flow under an area of surface convergence accumulates a population of organisms as the surface flows move toward the area of downwelling. If the organisms cannot increase their buoyancy to compensate for the downward current, they are carried down to changes in light, temperature, salinity, nutrients, and gases. Areas of downwelling are usually regions of low plant growth, but at the surface convergence, the accumulation of organisms provides a rich feeding ground for carnivores.

Bottom Environments 13.5

Although seawater is critical for the survival of organisms that live on or in the sea floor, the type of ocean bottom—rock, mud, sand, or gravel—is equally important. The base on which an organism lives is its **substrate;** the substrate provides food, shelter, and a place of attachment. Each substrate type provides suitable living space for a different group of organisms. Benthic organisms that live attached to the surface of the sea floor or rocky shore areas (mussels) and those that move freely across it (crabs) belong to the **epifauna;** other benthic organisms belong to the **infauna,** animals that burrow and live buried in the sediment (clams).

Substrates show greatest variety along shallow coastal areas; sandbars, mudflats, rocky points, and stretches of gravel and cobble are frequently found along the same strip of coastline. Seaward and at increased depth the particle size of the sediments decreases and the environment becomes extremely homogeneous. Organic material that has descended from the sea surface is strewn patchily across the sea floor, providing food resources and life environments for many species of fish, squid, and small shrimplike organisms. Sediment grain size is also important to the distribution of benthic organisms. Seaweeds that require rocks or large cobbles for attachment are unable to live in sand, and a burrowing worm from a mudflat or a shrimp from a sand beach cannot survive on a rocky reef.

The infauna that live in soft sediments and derive at least some part of their nutrition from ingested sediments are **deposit feeders.** The sediment is a mixture of particulate organic material, dissolved substances, microorganisms, and inorganic substances. Different species of deposit feeders (clams, worms, sea cucumbers) feed in different ways: some ingest the sediments, some have tentacles to gather the particles, some sort out the inorganic particles (different species of clams and worms) before ingesting what remains.

The epifauna (barnacles, seaweeds, snails) live on hard surfaces including shell, coral, and rock. Some organisms capture food by sieving it from the water (corals and barnacles) or capturing particles on impact (sea anemones). Others are active feeders creating a current to take in food (some shellfish and worms); still others drill holes into the shells of other organisms (some snails).

Carnivorous animals, for example, crabs and some snails, hunt other animals. Some (sea stars) are able to detect soluble substances produced by their prey; others (crabs, fish, squid) use their vision, and still others (snails and worms) are able to immobilize and kill their prey by stinging. Some herbivorous animals (small shrimplike organisms) graze on single-celled algae in the water; others scrape the seaweed film from the rocks (some snails). Sea urchins and some fish tear off pieces from large seaweeds and marine plants (sea grasses).

Organisms living attached to the sea floor often modify their habitats, providing food, shelter, and additional surfaces for the attachment of still other organisms. Forests of large seaweed attached to rocky bottoms in 20 m (66 ft) of water and eel grass beds in shallow, quiet, sandy or muddy bays both provide such environments but for quite different populations of organisms. Some crabs moving across the sea floor carry sponges and barnacles attached to their shells. The most outstanding example of biological modification of a substrate is the tropical coral reef; here the organisms create a specialized environment over the calcareous skeletons of other organisms (see chapter 17).

Close Associations 13.6

Large numbers of close relationships between unrelated organisms are found in the oceans. Such a relationship is a form of **symbiosis,** a situation in which dissimilar organisms live together in a close association. If the relationship is one in which one partner benefits but the other partner is not affected, it is termed **commensalism.** For example, some worms find shelter living inside the shells of various shellfish, and certain barnacles live on whales. The worms benefit from shelter and a convenient food supply, and the barnacles have a surface for attachment. No harmful effect is known in either case. If both participants receive benefits from the relationship, the term **mutualism** is used. Various crabs carry sea anemones on their shells; the crab provides the anemone with a place of attachment and scraps of its prey, and the anemone protects the crab with its stinging tentacles. Anemonefish (clownfish) protect anemones from predatory fish, and the stinging tentacles of the anemone protect the fish and their eggs in another example of mutualism (fig. 13.13) Certain small fish remove parasites and diseased tissue from large fish; the larger fish do not attack the small fish, and both benefit. If one partner lives at the expense of the other, the relationship is **parasitism.** Parasites get food and shelter by damaging their hosts but not killing

Figure 13.13 Clownfish have a symbiotic relationship with the sea anemone.

them. Parasitic worms are found in most fish, and marine mammals are also heavily parasitized.

Barriers and Boundaries   13.7

Species and populations are isolated from one another by spatial changes in water properties. Near the surface these changes are often abrupt; examples are changes with depth in temperature (thermocline), density (pycnocline), and salinity (halocline) (see chapter 7). Light intensity also changes rapidly with depth (see chapter 4). These changes are barriers for marine organisms because the organisms may not survive if they are displaced through the boundary separating waters of different properties. As the water deepens, it becomes more homogeneous, and these effects lessen. In the horizontal direction, similar barriers exist at zones of surface convergences and divergences as well as between fresh water and seawater in coastal areas (see chapter 7). When one type of water is moving through or adjacent to another type, the boundaries are sharp; for example, populations that do well in the warm, salty waters of the Gulf Stream die if they are displaced into the cold, less salty water of the Labrador Current, which flows between the North American coast and the Gulf Stream.

Other boundaries are controlled by the topography of the sea floor (see chapter 3). Ridges may isolate one deep-ocean basin from another, separating deep waters of different characteristics with different populations. Or the water and the populations in the two basins may be similar, but the deep-water populations never mix because in order to cross the ridge the organisms are forced to move upward into water with properties that they cannot tolerate. Isolated

seamounts with their peaks in shallow water may support specific isolated communities of sea life in much the same way that mountaintops on land support their own specialized communities.

Near-surface and surface species of the tropical regions are prevented from moving between the Atlantic and Pacific Oceans by the land barrier of Central America. Tropical surface species such as sea snakes cannot migrate around the continental landmasses of North and South America because to do so they must pass through regions of much colder water. Africa also acts as a barrier, keeping the tropical species of the Indian Ocean from communicating freely with the tropical species of the Atlantic, because the water south of Africa is too cold for the tropical species of either ocean. Today the natural isolation of one biological community from another is being lost as species are transported across the oceans and released in areas they have never before inhabited. See chapter 12, section 12.7.

Practical Considerations: Modification and Mitigation   13.8

People have brought great and lasting changes to the ocean environment, especially in coastal bays and estuaries, where environments are typically small-scaled and varied. In many cases, people have altered an area to enhance populations of organisms considered "desirable" over those considered "undesirable." New artificial reefs are built using old ships and car bodies, bags of old shells, or chunks of fractured concrete; such reefs encourage the growth of organisms that are suited to recreational and sport fishing or commercial harvesting. In other areas sediment may be moved, or the slope of the bottom may be changed, or substrates may be altered. Such activities modify and often destroy marine habitats, substituting an altered habitat and its biological communities for the previous habitat complex. People can and do choose which habitats and which organisms are to be conserved and which are to be sacrificed.

Attempting to balance human and natural needs has led planners and developers to the coastal-management concept of **mitigation.** When development projects alter or destroy an environment, local, state, or federal law may require mitigation of these effects by requiring the developer to purchase an area of the same type as that to be developed and to arrange for it to be held in its natural state, or the developer may be required to reengineer an area to resemble the habitat that has been lost. In some cases the developer may bank mitigation credits by setting aside additional natural areas at one site to be used when development occurs at another site.

A successful mitigation project requires a thorough knowledge of the physical requirements needed to support the communities of plants and animals that the mitigation seeks to enhance. Any characteristics of the new environment that interfere with the mitigation process must be changed if the mitigated area is to sustain itself in the future.

Biodiversity in the Oceans

Biodiversity, or biological diversity, is defined as the number of species, or the variety of life forms, and their genes found in a defined area. Understanding the importance of the Earth's biodiversity as well as maintaining this diversity requires knowing what species are present in the oceans, since the oceans make up more than 90% by volume of the Earth's biosphere (that part of the Earth's environment in which organisms are found). The Earth has an estimated 10 million species, and between 1.5 million and 1.8 million have been identified and given scientific names. The total number of species in the oceans is still unknown, with only about 275,000 marine species having been described.

Census of Marine Life is an international project designed to fill in the gaps in our knowledge of marine life. In 2000, eight research groups in the United States were awarded $3.7 million to begin developing model Internet atlases to display information about marine organisms—from zooplankton to blue whales. This is a first step in identifying, counting, and mapping all marine life; it is expected to take in excess of ten years. The census intends to create computerized libraries to accommodate the results of intensive studies of selected ocean areas.

The more heterogeneous and complex the physical environment, the more varied the environments that are available for organisms to colonize and the higher the species diversity. Consider the oceans: coral reefs, coastal areas, estuaries, open-ocean waters, deep-sea floor, and hydrothermal vents are all distinctly different ecosystems. Nearly 90% of the ocean's water lies below 100 m (330 ft), where it is cold, dark, and one of the most homogeneous environments on Earth. However, we now realize that this deep-water environment is actually extremely patchy in space and time. Nutrient patches caused by sinking phytoplankton blooms, fish and marine mammal carcasses, and other kinds of organic material provide environments for several hundred species of fish, small shrimp-like organisms, and squids. Small differences in the otherwise homogeneous deep-sea environment allow for greater biological diversification than might be expected. Competition for food and space, predation by other species, and natural disturbances all help to control the biodiversity of an area.

Geographic barriers divide the oceans into a series of environments. The arrangement of the continents and oceans combined with latitude, topography, and related climatic zones organizes the oceans into a series of areas with different patterns of circulation and different water properties. Boundaries in this water world exist in both the horizontal and the vertical and include water mass borders, changes in temperature, salinity, light, and density as well as isolating currents. Coral reefs provide a large number of microhabitats and are found only in tropical waters. As latitude decreases from the poles to the tropics, ocean species diversity tends to increase, just as it does on land. The Pacific has a greater number of species than the Atlantic, because of the greater abundance of coral reefs. In the Pacific, coral reef diversity increases from all directions towards the Philippines and Indonesia.

In the last few decades rising concern over the loss of species and the importance of conserving the Earth's biodiversity have become a part of the discussion of human population growth and that growth's need for sustained development. The main cause of biodiversity loss is human-induced environmental degradation, and the most direct causes of biodiversity loss are habitat fragmentation, deterioration, and destruction as humans take up more and more of the planet. In addition the introduction of foreign species into local waters has led to extinctions as well as habitat homogenization (see chapter 12, section 12.7).

Yet to a very large degree the human species depends on preservation of biological diversity for its own survival. The richer the diversity of life the greater the opportunity for discovery of valuable drugs and other marine products. It has been said that conserving the variety of life is our insurance policy. In addition the appreciation of biological diversity keeps us aware of the continuity of life, and through our concern for our fellow species we gain an understanding of our biological heritage. However, our inadequate knowledge of biodiversity and its accurate measurement remain difficult problems. How much diversity is there? How fast is biodiversity being lost or degraded? What actions can slow or prevent an increasing rate of loss?

For centuries it was thought that humans could not drive any ocean-living species to extinction. The sea was too big and too deep; its inhabitants too numerous, prolific, and widespread. While in the last 200 years only one marine mammal (Stellar sea cow) and four marine shellfish have become extinct, estimates of extinctions among small organisms in coastal areas and on coral reefs range from 100 to over 1000. Think about the possibility of marine extinctions; consider the diversity of the ocean's food chains and remember that if you loose one component, you may upset an entire ecosystem and the other components suffer also. Think about loosing strands in the web of life as you read the next series of chapters.

For More Information About Biodiversity in the Oceans

Malakoff, D. 2000. Grants Kick Off Ambitious Count of All Ocean Life. *Science* 288 (5471): 1575–76.

Malakoff, D. 1997. Extinction on the High Seas. *Science* 277 (5325): 486–88.

Mills, C.E., and J.T. Carlton. 1998. Rationale for a System of International Reserves for the Open Ocean. *Conservation Biology* 12 (1): 244–47.

Solow, A.R. 1997. Biological Diversity in the Oceans. *Sea Technology* 38 (1): 50–52.

Internet References

Visit the book's Online Learning Center at www.mhhe.com/sverdrup (click on the book's cover) to explore links to further information on related topics.

After the area has been restructured it must be monitored, and changes must be made as needed to maintain the desired environment. If these preproject and postproject studies are not made and followed by necessary corrections, the mitigation effort will not succeed. Mitigation may preserve some habitats and species, but it only approximates a lost environment; it cannot duplicate it.

Pressures to develop deep-sea areas for energy and mining projects are likely to increase in the future. Tropical OTEC plants (ocean thermal energy conversion; see chapter 7) will bring cold, nutrient-rich water to the sea surface, changing the surface productivity and potentially interfering with the survival of tropical species. Manganese nodule mining (see chapter 3) will increase deep-sea turbidity, changing the environment for bottom organisms in mined areas. How mitigation might be undertaken in the open ocean has not yet been considered, but the threat of deep-sea exploitation is already causing some scientists to call for the setting aside of marine reserves in the open ocean.

Summary

Marine ecology is the study of interactions between marine organisms and their environment. The organisms of the sea are classified for identification and relationship. Marine organisms are commonly divided into plankton, nekton, and benthos. All organisms are classified for identification and relationship into kingdoms and a new system that includes domains.

The marine environment is subdivided into zones. The major environments are the benthic and pelagic zones; there are numerous subdivisions of each of these zones.

Organisms living in the sea are buoyed and supported by the seawater. Adaptations for staying afloat include low-density body fluids, gas bubbles, gas-filled floats, swim bladders, oil and fat storage, and extended surface areas and appendages.

Most marine fish lose water by osmosis. They drink continually and excrete salt to prevent dehydration. Sharks have the same concentration of salt in their tissues as there is in seawater. They therefore do not have a water-loss problem. Salinity is a barrier to some organisms; others can adapt to large salinity changes.

Temperature affects density, viscosity, and the water's buoyancy, as well as the stability of the water column. The body temperature and metabolism of all marine organisms, except for birds and mammals, are controlled by the sea temperature. Some fish conserve heat in their body muscles and elevate the temperature in these muscles. Temperatures at depth are uniform; sea surface temperatures change with latitude and seasons.

Changes in pressure affect organisms that have gas-filled cavities. Marine mammals have a unique ability to undergo large pressure changes due to their physiology and body chemistry. The swim bladders of fish are affected by pressure changes, and the fish must change depth slowly.

The carbon dioxide—oxygen balance in the oceans influences the distribution of all organisms. The availability of nutrients and light limits plant populations. The depth of light penetration in the oceans is controlled by the angle of the Sun's rays, the properties of the water, and the material in the water. Light limits plant life, but nutrients are also required. Some organisms produce chemical light, known as bioluminescence. Animals use color for concealment and camouflage and also to warn predators of poisonous flesh and bitter taste.

Winds, tides, and currents mix the water. Moving water carries food and oxygen, removes waste, and disperses organisms. Floating populations are not necessarily scattered but keep their place due to movements between surface currents and deeper currents. Upwellings supply nutrients and hold plants in the surface layers. Downwellings are regions of low plant growth.

Different substrates provide food, shelter, and attachment for different groups of organisms. Some live attached to the surface, some move across it, some burrow into the sea floor, and some modify the sea floor, providing habitats for others. Some animals are carnivores; others are herbivores. Large numbers of ocean organisms live in close associations such as commensalism, mutualism, and parasitism. Barriers for marine organisms include water properties, light intensity, zones of convergence and divergence, seafloor topography, and geography.

Development of marine areas, with the consequent loss of habitat and therefore populations, has led to the concept of mitigation. Mitigation requires developers to preserve or replace habitats in an effort to maintain and preserve species.

Key Terms

All key terms from this chapter can be viewed by term, or by definition, when studied as flashcards on this book's Online Learning Center at www.mhhe.com/sverdrup (click on this book's cover).

ecology, 358
plankton, 358
nekton, 358
benthos, 358
taxonomy, 358
pelagic zone, 360
benthic zone, 360
neritic zone, 360
oceanic zone, 360
supralittoral zone/splash
 zone, 360
littoral zone/intertidal
 zone, 360
sublittoral zone/subtidal
 zone, 360
bathyal zone, 360
abyssal zone, 360
hadal zone, 360
buoyancy, 360

ectotherm, 362
endotherm, 363
photic zone, 364
aphotic zone, 364
spicules, 365
bioluminescence, 365
luciferin, 365
luciferase, 365
photophore, 365
substrate, 367
epifauna, 367
infauna, 367
deposit feeder, 367
symbiosis, 367
commensalism, 367
mutualism, 367
parasitism, 367
mitigation, 368
biodiversity, 369

Study Questions

1. Describe and compare two different seafloor environments.
2. Explain why the substrate of the sea floor becomes less diversified as one moves from the shore to the deep ocean.
3. How do so many delicate and fragile organisms exist in the oceans without damage?
4. What will happen to the body fluids of a frog placed in seawater? A sea cucumber placed in fresh water?
5. Discuss the effect of temperature on the distribution of organisms. Consider changes with latitude and with depth.
6. Although seals and whales are mammals, they do not suffer from either decompression sickness or nitrogen narcosis during deep dives of long duration. Explain why.
7. How does the role of bioluminescence differ from the role of sunlight in the sea?
8. In what ways does upwelling contribute to increasing the populations of surface organisms? What properties of seawater act as barriers for marine organisms?
9. Compare the flotation problems of an organism with many frilly appendages to those of an organism without appendages but of the same density. Consider both organisms in 4°C water and in 20°C water.
10. Why is a neritic habitat at high-temperate latitudes more stressful to organisms than a deep-sea habitat?
11. The biodiversity of the oceans has recently become a topic of discussion and concern. Why is this happening now? What kinds of action are being taken? What should be done in the future?
12. How are commensalism and mutualism the same? How are they different?
13. Find an example of mitigation being used in your community and discuss its potential for success.
14. How does countershading aid the survival of fish in nearshore areas?
15. How do rapidly increasing populations of single-celled plants limit their own growth?

Visit the book's Online Learning Center at www.mhhe.com/sverdrup (click on this book's cover) to find these additional chapter tools: Suggested Readings; links to further information on boxed readings; and additional study aids.

chapter 14

Production and Life

One's first reaction to a close view of the life of the sea is confusion, unease. How can things be so beautiful, with such intricate balance and symmetry in their infinitely varied forms and, at the same time, seem so hostile, so threatening, ready to lunge, to snatch life from each other? So many of the moving parts of sea life seem to be designed, exquisitely tooled, for nothing but destruction and devouring. It is disconcerting, almost as though we have had the wrong idea about beauty and harmony: living things so evidently aimed at each others' throats should not have, as these things do, the aspect of pure, crystalline enchantment.

Perhaps something is wrong in the way we look at them. From our distance we see them as separate, independent creatures interminably wrangling, as a writhing arrangement of solitary adversaries bent on killing each other. Success in such a system would have to mean more than mere survival: to make sense, the fittest would surely have to end up standing triumphantly alone. This, in the conventional view, would be the way of the world, the ultimate observance of nature's law. It was to delineate such a state of affairs that the hideous nineteenth-century phrase "Nature red in tooth and claw," was hammered out.

What is wrong with this view is that it never seems to turn out that way. There is in the sea a symmetry, a balance, and something like the sense of permanence encountered in a well-tended garden.

Lewis Thomas
From *Sensuous Symbionts of the Sea*

Along the California coast blue rockfish swim in a forest of giant kelp (*Macrocystis pyrifera*) that reaches upward toward the sunlight.

I n all the environments of all the oceans, organisms prey on each other: big fish eat little fish, and little fish eat littler fish. These prey-predator relationships are called food chains. We are generally familiar with the predators at the upper ends of these food chains (for example, salmon, tuna, swordfish, and seals), but without the microscopic, planktonic organisms that begin these chains the large carnivores could not exist. Biological oceanographers devote much time and study to this first link in the food chain, for understanding the variation in the abundance of the plant plankton is the key to understanding the ocean's productivity. Productivity, or the rate at which organic material is produced in the sea, is directly related to the harvesting of the oceans on which millions of people depend for food and income.

Primary Production   14.1

Gross and Net

The single-celled plant plankton, or **phytoplankton,** that float or drift with the movement of the water, like all plants, require sunlight, nutrients or fertilizers, carbon dioxide gas, and water. The nutrients and carbon dioxide are dissolved in the water that surrounds and supports the phytoplankton. The phytoplankton cells contain the pigment **chlorophyll,** which traps the Sun's energy for use in **photosynthesis.** The photosynthetic process converts the carbon dioxide and water to high-energy organic compounds from which the cells form new plant material. The production of new plant material from photosynthesis and nutrients is termed **primary production.** The total amount or mass of organic material produced by photosynthesis is the **gross primary production** of the sea.

Photosynthesis is represented by the equation

$$6\,CO_2 \quad + \quad 6\,H_2O \quad \xrightarrow[\text{chlorophyll}]{\text{solar energy}} \quad C_6H_{12}O_6 \quad + \quad 6\,O_2$$

$$\underset{\substack{\text{molecules}\\ \text{carbon}\\ \text{dioxide}}}{6} \quad + \quad \underset{\substack{\text{molecules}\\ \text{water}}}{6} \quad \xrightarrow[\text{chlorophyll}]{\text{solar energy}} \quad \underset{\substack{\text{molecule}\\ \text{sugar}}}{1} \quad + \quad \underset{\substack{\text{molecules}\\ \text{oxygen}}}{6}$$

A more complex representation of photosynthesis is written

$$6\,CO_2 \quad + \quad 12\,H_2O \quad \xrightarrow[\text{chlorophyll}]{\text{solar energy}} \quad C_6H_{12}O_6 \quad + \quad 6\,H_2O \quad + \quad 6\,O_2$$

$$\underset{\substack{\text{molecules}\\ \text{carbon}\\ \text{dioxide}}}{6} \quad + \quad \underset{\substack{\text{molecules}\\ \text{water}}}{12} \quad \xrightarrow[\text{chlorophyll}]{\text{solar energy}} \quad \underset{\substack{\text{molecule}\\ \text{sugar}}}{1} \quad + \quad \underset{\substack{\text{molecules}\\ \text{water}}}{6} \quad + \quad \underset{\substack{\text{molecules}\\ \text{oxygen}}}{6}$$

The sugars produced by photosynthesis are broken down by the phytoplankton cells with the addition of oxygen to yield energy, carbon dioxide, and water in the process known as **respiration.** Respiration occurs in both plant and animal cells; this process provides all organisms with the energy required for their life processes.

Respiration is represented by the equation

$$C_6H_{12}O_6 \quad + \quad 6\,O_2 \quad \longrightarrow \quad 6\,CO_2 \quad + \quad 6\,H_2O \quad + \quad \underset{\substack{\text{support}\\ \text{energy}}}{\text{life}}$$

$$\underset{\substack{\text{molecule}\\ \text{sugar}}}{1} \quad + \quad \underset{\substack{\text{molecules}\\ \text{oxygen}}}{6} \quad \longrightarrow \quad \underset{\substack{\text{molecules}\\ \text{carbon}\\ \text{dioxide}}}{6} \quad + \quad \underset{\substack{\text{molecules}\\ \text{water}}}{6} \quad + \quad \underset{\substack{\text{support}\\ \text{energy}}}{\text{life}}$$

After deducting the part of primary production that is broken down by respiration to yield energy for life processes, the remaining gain in new organic material by a plant population is the population's **net primary production.** The net primary production is the gain in plant mass available for consumption by the plant eaters or herbivores and for decomposition by bacteria. Net and gross primary production

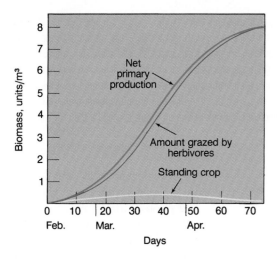

Figure 14.1 Net primary productivity is balanced by the grazing of herbivores. Both populations increase during the spring; the standing crop remains nearly constant throughout the year.

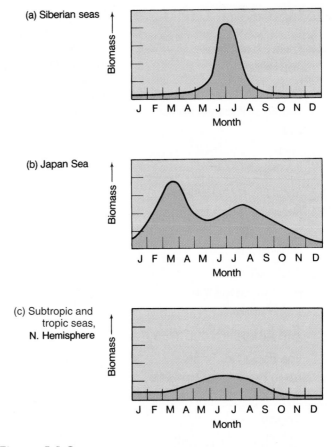

Figure 14.2 Phytoplankton growth cycles vary with the light available at different latitudes. At high latitudes (a), growth is limited to a brief period in midsummer. Seasonal light changes at the middle latitudes (b) increase growth in early spring and continue it through the summer. Sunlight levels vary little at tropic latitudes (c), where phytoplankton growth is nearly uniform throughout the year.

are usually reported for a unit volume of water or the volume of water under a fixed area of the sea surface. The organic matter produced is known as **biomass.** Because organic substances are based on the element carbon, primary production and biomass are most often given as the mass of dry weight organic carbon in grams per unit area or volume. The term **primary productivity** is used to express the rate at which biomass is produced by photosynthesis and is usually expressed as grams of carbon per day. In practice, the terms *production* and *productivity* are often used interchangeably.

Standing Crop

The total plant biomass under any area of sea surface at any instant in time is known as the **standing crop.** It is the result of growth, reproduction, death, and grazing. If the standing crop on two successive days shows an increase, then there is net production, but if the standing crop shows no change over the two days, the net production has not necessarily remained at zero but is the result of herbivores' grazing the phytoplankton population at the same rate as the net primary production increases it. This relationship is shown in figure 14.1.

Controls on Primary Production 14.2

Light

Phytoplankton growth is controlled in part by the light at the ocean surface. At polar latitudes the summer days are long and the intensity of daylight is low. Because the angle at which the Sun's rays strike the Earth is oblique, the light's penetration below the polar ocean surface is reduced. The result is a short period of light that promotes rapid plant growth during the midsummer, shown in figure 14.2a.

At the middle latitudes, the sunlight's intensity and duration vary with the seasons. The spring increase in solar radiation warms the surface and increases the density stability of the water column, helping the phytoplankton to re-

main at the surface. Figure 14.2b shows an initial increase in phytoplankton biomass with the lengthening of daylight hours in the spring. A second increase in the population occurs during the late summer.

In the subtropics and tropics (fig. 14.2c), the situation differs from those at polar and middle latitudes. Here abundant high-intensity solar energy is available year-round. Population increases associated with seasonal light changes are slight. Compare figures 14.2a and b to figure 14.2c.

To take advantage of the available sunlight, the phytoplankton must remain in the ocean's lighted surface layers. The warm surface layer of low-density water in the tropics exists all year long, and the phytoplankton are not displaced easily into the underlying denser water. At middle latitudes, a low-density, warm surface layer is found only during the summer. This layer appears gradually, and if the warm weather is late or is broken by cool periods or strong storms, the formation of the low-density layer is delayed and the active growth of the phytoplankton does not occur until the water is sufficiently stable and stratified to allow the plankton to remain in the upper layers. In the wintertime, the low-density surface layer of the middle latitudes is mixed with deeper water

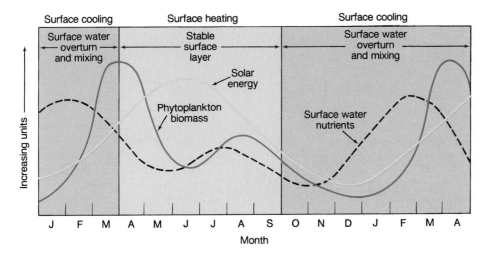

Figure 14.3 Phytoplankton biomass, nutrient supply, and surface water stability respond to solar energy changes at the middle latitudes in the Northern Hemisphere.

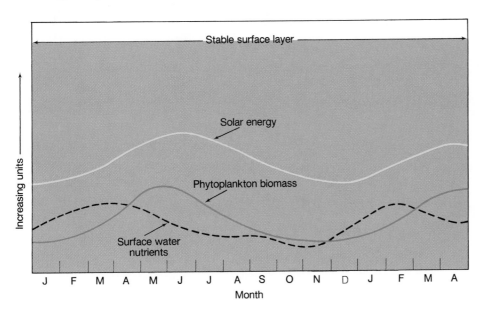

Figure 14.4 Lack of surface mixing and overturn at low latitudes results in a depressed phytoplankton biomass. This pattern is related to solar radiation levels that produce a year-round stable water column.

by surface cooling and winter storms. This vertical mixing combined with a decrease in available sunlight restricts winter growth of phytoplankton. At polar latitudes, the low level but nearly continuous sunlight during the summer warms the sea surface and forms a weakly stable density layer often associated with the fresh water that has returned from the melting sea ice. The meltwater helps to maintain a stable water column during the short polar growing season.

If the annual distribution of light were the sole factor influencing plant growth, the level of phytoplankton growth in the tropics (fig. 14.2c) would be expected to maintain a year-round biomass level approaching the summer values shown for the polar and middle latitudes (fig. 14.2a, b). However, nutrient supply must also be taken into account. Remember that land drainage, mixing, overturn, and up-

welling supply nutrients to the surface layers and that these processes occur seasonally and in specific areas of the oceans. Both light and nutrients must be available if the phytoplankton population is to increase.

Nutrients

In the polar seas, winter overturn resupplies the surface water with nutrients; the growing season is short and the nutrients rarely fall below the level that is necessary for population increase. At these latitudes, the availability of light controls phytoplankton growth. In the tropics, the phytoplankton remain in the surface layers because the water column is stable due to surface warming, but the upwelling and mixing processes for renewing the nutrient supply are weak and localized. This poor nutrient supply to the photic layers limits phytoplankton production in the tropics and subtropics despite the high levels of solar energy.

In temperate latitudes, if winter storms are severe, or if the cooling of the surface water produces a shallow overturn, the surface nutrients are replenished. Nutrients are available to the plant cells when the increase in spring sunlight warms the surface, decreases the surface water density, and increases the stability of the water column. Together the sunlight and nutrient supply trigger the first period of phytoplankton bloom. Grazing organisms consume the phytoplankton and release nutrients for a second bloom during midsummer. The rate at which nutrients are recycled or added to the surface layer controls the continued growth of the phytoplankton population. As autumn approaches, either the supply of nutrients or the level of sunlight becomes limiting.

The interactions between the sunlight, the stability of the surface water, the nutrients in the surface water, and the phytoplankton biomass are presented in figures 14.3 and 14.4. Note the replenishment of nutrients to Northern Hemisphere middle-latitude surface waters during periods of overturn and mixing and the decrease in nutrients during periods of increased sunlight and plant growth in figure 14.3. The lack of surface mixing and overturn at low latitudes is reflected in the low levels of nutrients and plant biomass despite the constant high level of solar energy; review figure 14.4.

Nutrient Cycles

When a plant or an animal dies, organisms of decay, known as **decomposers** (bacteria and fungi), release the energy

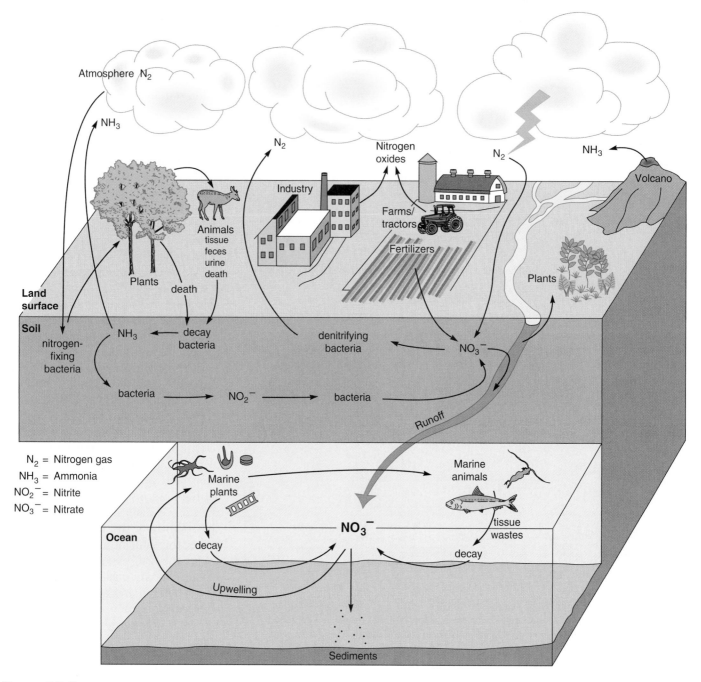

Figure 14.5 The nitrogen cycle. Although nitrogen makes up 78% of the Earth's atmosphere, only a few microorganisms are able to change nitrogen gas to the nitrate that is used by land and sea plants, which are in turn eaten and their nutrients recycled by animals.

within the organism's tissues as heat to the environment and break down the organism's complex organic molecules to basic molecules, such as carbon dioxide, nutrients, and water. Because no significant amount of new matter comes to the Earth from space, living systems must recycle inorganic molecules to form the organic compounds required for their systems and life processes. Organisms require many elements, but among the most important are nitrogen and phosphorus.

Nitrogen in the form of nitrate and phosphorus in the form of phosphate are two water-soluble nutrients required for life (refer to the discussion of nutrients in chapter 5). Nitro-

gen is essential in the formation of amino acids and proteins, and phosphorus is required in energy reactions, cell membranes, and nucleic acids. Nitrates and phosphates are removed from the water by the primary producers as the plant populations grow and reproduce. Both are cycled into the animal populations as the animals feed on the plants and are returned to the water as the organisms die and decay. Excretory products from the animals are also added to the seawater, broken down, and used again by a new generation of plants and animals. The cyclic nature of the nutrient pathways for nitrogen and phosphorus is illustrated in figures 14.5 and 14.6.

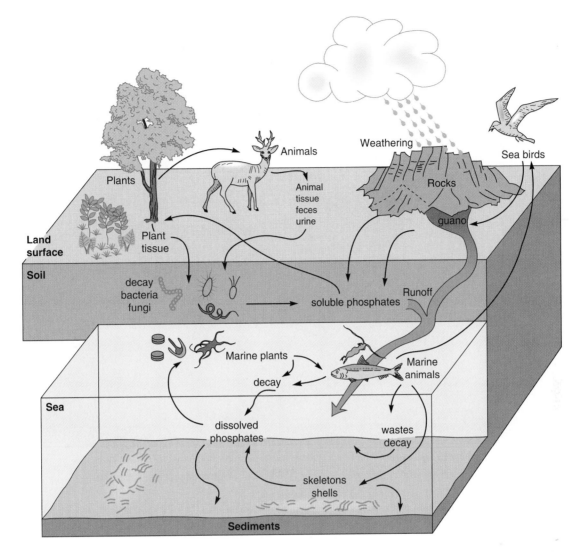

Figure 14.6 The phosphorus cycle. Dissolved phosphorus is carried to the sea by runoff from the land, where it is taken up by plants and recycled through animals until it is released from dead tissue by bacterial action.

To cycle nitrogen, a number of different microorganisms must participate. Nitrogen gas is useless to plants; it must be converted to nitrate to be incorporated into plant tissue. A few species of microorganisms convert nitrogen gas to ammonia, other species convert ammonia to nitrite, and still other species convert nitrite to nitrate, the form most easily absorbed by plants.

Increasing amounts of nitrogen from human activities such as fertilizer use and combustion of fossil fuels are entering the sea and degrading nearshore marine areas. Human activities have approximately doubled the global yearly rate of nitrogen fixation by terrestrial and aquatic bacteria between preindustrial times (140×10^9 kg) and the present (280×10^9 kg). Excess nitrogen increases phytoplankton blooms, and the decomposition of the resulting overabundant organic material decreases the oxygen in the water. Recently such problems have been reported from the Baltic and North Seas, the western Mediterranean, and Narragansett and Chesapeake Bays along the U.S. Atlantic coast.

In the Gulf of Mexico an oversupply of nitrates delivered by the Mississippi River has been linked to a low-oxygen zone along the Louisiana coast; this is the so-called dead zone discussed in chapter 12.

In the Pacific and Southern Oceans there is more light, more nitrogen, and more phosphorous than the plankton require, but the plankton are not more abundant in these oceans. This situation is related to the lack of iron in surface waters. Iron is delivered to the oceans as wind-blown dust and is the key limiting factor for many phytoplankton populations.

Experiments in 1994 and 1995 proved that it is possible to cause a surge in phytoplankton production by adding soluble iron to the surface water. However, after an initial spurt the growth slowed and leveled off as the iron reacted with other dissolved substances and then sank. In 2000 the Southern Ocean Iron Release Experiment (SOIREE) conducted a large-scale trial, releasing over 8500 kg (19,000 lb) of iron compound into an ocean area 8 km (5 mi) across.

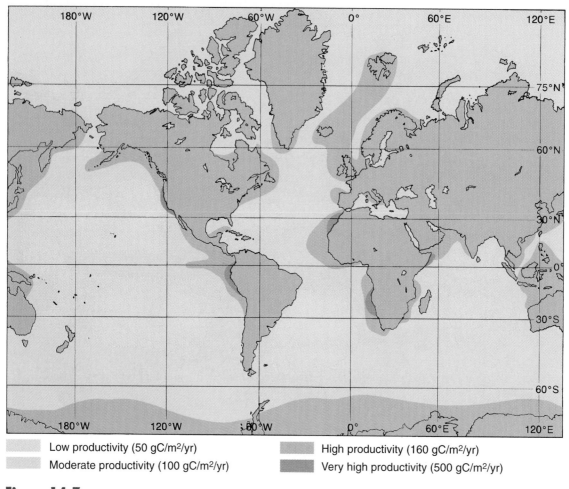

	Low productivity (50 gC/m²/yr)		High productivity (160 gC/m²/yr)
	Moderate productivity (100 gC/m²/yr)		Very high productivity (500 gC/m²/yr)

web link **Figure 14.7** The distribution of primary productivity in the world's oceans, based on many years of ocean sampling.

The results were the same; iron was limiting to phytoplankton growth.

Because the iron limits the growth of phytoplankton and because the plankton take up carbon dioxide when they photosynthesize, it has been suggested that the greenhouse effect could be mitigated by large-scale iron fertilization of the oceans. This theory proposes that the addition of iron and the resulting growth of phytoplankton populations will rapidly remove carbon from the seawater. The carbon required by the increasing phytoplankton populations will be replaced by absorption of excess carbon dioxide from the atmosphere, lessening the effects of fossil fuel burning. However, no one can predict the long-term effects of such massive experiments and there is always the possibility of creating more problems than are solved.

Global Primary Productivity   14.3

Figure 14.7 shows the distribution of the annual rates of global primary production and primary productivity. Refer to this figure while reading this section. Coastal areas are generally more productive than open ocean because rivers and land runoff supply nutrients along the coasts and in the estuaries. Also, currents cause mixing and turbulence, which supply nutrients from shallow depths to surface layers. The fresh water helps to create a stable, low-density surface layer that keeps the plant cells up where sunlight is plentiful.

Narrow areas of very high productivity are found against the west coasts of North and South America, the west coast of Africa, and along the west side of the Indian Ocean. These are the major upwelling zones on the sheltered sides of landmasses in the trade-wind belts and in areas of abundant sunlight. In these nutrient-rich zones, the populations of phytoplankton form the first step in the feeding relationships that have produced great schools of commercially valuable fish.

The equatorial Pacific demonstrates the influence of open-ocean upwelling on primary productivity that is caused by the equatorial divergence. The same is true for the surface divergence surrounding Antarctica. In contrast surface convergences at the centers of the large gyres are areas of low nutrient availability, which depresses primary productivity. Return to figure 8.13 and compare areas of surface divergence and convergence to the distribution of primary productivity in figure 14.7.

Table 14.1 World Ocean Primary Productivity

| Area | Primary Productivity (gC/m²/yr) | World Ocean Area | | Total Primary Productivity (metric tons carbon/yr) |
		(km²)	(%)	
Upwellings	640	0.36×10^6	0.1	0.23×10^9
Coasts	160	54×10^6	15.0	8.6×10^9
Open oceans	130	307×10^6	85.0	39.9×10^9
All ocean areas	135	361×10^6	100.0	48.73×10^9

Data from S. Smith and J. Hollibaugh. 1993 Coastal Metabolism and the Oceanic Organic Carbon Balance. *Review of Geophysics 31(1): 75–89.*

Table 14.2 Gross Primary Productivity Land and Ocean

Ocean Area	Range (gC/m²/yr)	Average (gC/m²/yr)	Land Area	Amount (gC/m²/yr)
Open ocean	50–160	130 ± 35	Deserts, grasslands	50
Coastal ocean	100–250	160 ± 40	Forests, common crops, pastures	25–150
Estuaries	200–500	300 ± 100	Rain forests, moist crops, intensive agriculture	150–500
Upwelling zones	300–800	640 ± 150	Sugarcane and sorghum	500–1500
Salt marshes	1000–4000	2471		

Data from S. Smith and J. Hollibaugh. 1993 Coastal Metabolism and the Oceanic Organic Carbon Balance. *Review of Geophysics 31(1): 75–89.*

On the average, upwelling areas are about four times more productive than coastal areas and five times more productive than the open ocean for the same units of area and time. See the first column of table 14.1. The second column shows that the total area of each type of oceanic region is inversely related to its primary productivity. There is more than 100 times more coastal area than upwelling area, and nearly six times more open-ocean area than coastal area. In the third column, the total primary productivity of each ocean area type is given. These values demonstrate that most of the organic carbon produced by the oceans' plants is scattered in low concentrations over large areas of the open sea. Smaller total amounts are produced along the coasts and in upwelling regions, but such amounts are concentrated in smaller areas, making these areas very rich in primary production on a unit area basis.

To compare primary productivity on land to primary productivity at sea, study table 14.2. Primary productivity per square meter in the open sea is about the same as that of the deserts on land; the vast expanses of open ocean are productive only because of their size, covering 85% of the ocean's total surface area. Upwelling areas are comparable to pastureland and lush forestland, and certain estuary systems approach the productivity of the most heavily cultivated land. Areas of intensively cultivated, fast-growing crops on land produce much more carbon than do most of the regions of the sea; however, on land people put large quantities of time and energy into raising their crops to produce the same high yields that natural processes provide in shallow estuaries. Keep in mind that, from the human standpoint, the plants on land are often used directly for food, whereas those of the sea are not.

Measuring Primary Productivity 14.4

Direct Methods

One method to determine the amount of plant material present in a water sample is to filter out the phytoplankton, count the cells, and multiply the number counted by the average mass per individual cell. A less-tedious method is to extract the chlorophyll from a sample of phytoplankton and determine the concentration of pigment present. Chlorophyll concentration can then be used to estimate the total quantity of plant material, or biomass. Repeated sampling and chlorophyll determination in a column of water or under a fixed area of sea surface over time yield data on changes in biomass with time, or net primary productivity. Another method exposes the chlorophyll in the phytoplankton cells to certain wavelengths of light and causes the chlorophyll pigment to fluoresce. The strength or intensity of the fluorescence is read electronically to give a direct measure of the chlorophyll and phytoplankton biomass present in a given volume of water. An instrument used to measure fluorescence, called a fluorometer, is often attached to a CTD (conductivity-temperature-depth sensor), or it may be carried by a robotic device such as a Seasoar (see chapter 7).

A common method of estimating the rate of photosynthetic production of organic matter in the oceans uses the radioactive isotope carbon-14. The uptake of dissolved

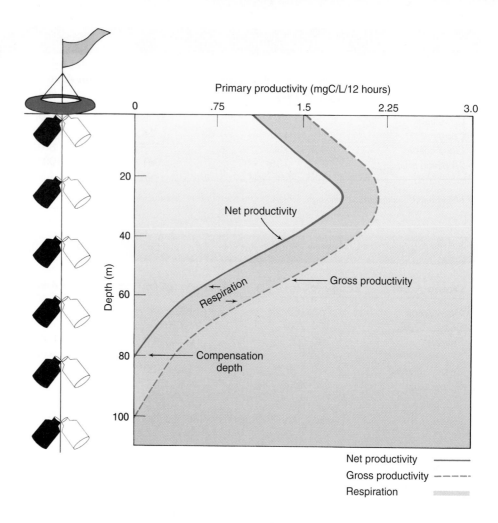

Figure 14.8 A light and dark bottle experiment provides values for respiration, net primary productivity, and gross primary productivity with depth. Respiration and productivity are measured in milligrams of carbon per liter per twelve-hour interval. The compensation depth occurs where net productivity is zero. In this example, light, nutrients, and stability of the water combine to provide conditions for the highest primary productivity at the depth of 30 m (100 ft).

inorganic carbon by the phytoplankton is measured and used to estimate the production of organic matter and the biological uptake of carbon. Each water sample from a specific depth is divided into three subsamples. One subsample is tested immediately as a control to assess the carbon present; the second is placed in a clear glass bottle or "light bottle," and the third is placed in a light-proof or "dark bottle." The light and dark bottles may be attached to a line suspended from a vessel and returned to the depths at which they were collected, or they may be attached to a free-floating system. A light and dark bottle experiment is illustrated in figure 14.8. Note that the values for phytoplankton respiration remain nearly constant with depth. Where net primary productivity is zero, gross productivity is equal to respiration. This is the compensation depth (see chapter 5, section 5.3).

If necessary the experiments can also be carried out onboard ship in a controlled temperature tank. Under these conditions the light bottles are partially covered to reduce the sunlight to levels similar to those at which the samples were taken. When the bottles have spent twelve to twenty-four hours in the water they are retrieved and daily organic carbon production is calculated. The light bottle provides a measure of net daylight organic carbon production; total daily organic carbon production is calculated from the difference between the light and dark bottles. These data can be used to compare carbon production in different ecosystems or between various communities.

When new plant material is produced by the phytoplankton, dissolved carbon dioxide is converted to organic carbon compounds; nitrogen from nitrate and phosphorus from phosphate (the nutrients dissolved in seawater) are also required. A ratio exists between the oxygen gas and organic compound produced and the nitrogen and phosphorus removed from the seawater. These fixed ratios by weight are

$$O_2 : C : N : P = 109 : 41 : 7.2 : 1$$

These ratios are used to estimate primary carbon production by measuring either the rate of nutrient uptake by the plant population or the rate of production of dissolved oxygen. If 10 mg of phosphorus is removed by the plants from a volume of water in a given time, then 410 mg of carbon have been produced and incorporated into the phytoplankton biomass. If the rate at which the nitrogen or phosphorus is carried into a region by upwellings and currents and the rate at which it is removed by other currents are known, the rate at which it is being incorporated into the phytoplankton can be calculated and primary productivity determined. This system allows the oceanographer to estimate primary productivity over large areas of the ocean and to associate it with large-scale water movements and chemical cycles.

Notice that 7.2 times more nitrogen than phosphorus is needed. If the mechanisms supplying these nutrients to the photic zone are similar, then the available nitrogen will normally be used up before the phosphorus. Depletion of the seawater's nitrate level tends to be the limiting nutrient factor for the ocean's primary productivity, because nitrogen is critical to living organisms. Without nitrogen from nitrates, phytoplankton growth and reproduction will lag, and primary productivity will decrease. The results of low iron concentrations on phytoplankton populations are discussed in section 14.2.

Remote Methods

Direct measurements of primary production using water samples collected by research vessels are limited and expensive. However, the continuous satellite measurements of sea surface chlorophyll made over the years can be used to prepare maps of the global distribution of the phytoplankton standing crop (fig. 14.9).

To produce these maps the satellite phytoplankton data must be calibrated to the actual abundance of phytoplankton in the top meter of seawater. The images in figure 14.9 were prepared from *SeaWIFS* satellite data collected during 1998 and 1999. Monthly average concentrations of phytoplankton are expressed as chlorophyll concentration per square meter of sea surface. The darker blue and magenta areas on the charts denote regions of low chlorophyll concentrations related to the mid-ocean gyres and convergence zones. Chlorophyll concentrations are shown increasing from lighter blue to green, to yellow, orange, and red, especially in nutrient-rich upwelling coastal waters. Follow the increase in chlorophyll in the Northern Hemisphere as spring advances into summer and the decrease in chlorophyll as fall changes to midwinter. Notice that seasonal changes in the chlorophyll of land plants can also be seen. Compare the distribution of chlorophyll and its abundance in these figures to the distribution of primary productivity from widely scattered direct measurements in figure 14.7.

Total Production 14.5

Food Chains and Food Webs

Plants are known as **producers** because they are able to manufacture organic matter from simple inorganic substances. Organisms such as animals that consume the producers are known as **consumers.** Primary production forms the first step in the **food chains** linking plants and animals. Where there is high primary production by phytoplankton, seaweeds, or other marine plants, there are large populations of animals. The **herbivores** eat plants directly; the **carnivores** feed on the herbivores or other carnivores. Notice that in either case the carnivores are still linked to the producers. The most numerous and the greatest biomass of herbivores are the plant-eating **zooplankton,** or animal plankton. These are the primary consumers that

convert plant tissue to animal tissue; in turn they become the food for other zooplankton, the flesh-eating carnivores, or secondary consumers. See figure 14.10 for the relationship between herbivorous zooplankton and their phytoplankton food source. Again, the response varies with latitude; the zooplankton peak lags behind the phytoplankton production at the polar latitudes (fig. 14.10a) and middle latitudes (fig. 14.10b). The zooplankton reproduce when sufficient phytoplankton are present to support an increase in herbivores, and as grazing by herbivores increases, the phytoplankton biomass decreases. The herbivore population then decreases in its turn as the phytoplankton population declines. The zooplankton population in the tropics (fig. 14.10c) remains nearly constant, with a lower but stable phytoplankton biomass.

Food chains may be long or short, but they are rarely simple and linear. Food chains are more likely to show complex interrelationships among organisms, in which case it is more appropriate to call the interconnecting feeding pattern a **food web.** See the herring food web diagram (fig. 14.11), in which organisms change their prey-predation levels as they mature, and the box titled "The Antarctic Food Web."

Trophic Pyramids

Food chains and food webs represent the pathways followed by nutrients and food energy as they move through the succession of plants, grazing herbivores, and carnivorous predators. These relationships are often demonstrated in the form of **trophic levels** representing links in the food chain; these levels form a **trophic pyramid** in which the trophic levels are numbered from the bottom of the pyramid to the top (fig. 14.12). The primary producers are always the first trophic level, the herbivorous zooplankton are the second trophic level, and the carnivores form the upper levels, up to the top carnivore, on which no other organisms prey (for example, sharks and killer whales). Figure 14.12 includes the Sun (the energy source directly necessary to the primary producers and indirectly necessary to all other trophic levels), the decay and decomposition processes that recycle nutrients to the primary producers, and energy loss.

In general, moving upward from the first trophic level, the size of the organisms increases and the numbers and biomass of organisms decrease. The larger numbers of small organisms at the lower trophic levels collectively have a much larger biomass than the smaller numbers of large organisms at the upper levels. Table 14.3 relates the abundance of organisms to their size.

The overall efficiency of energy transfer up each layer of an open-ocean trophic pyramid is estimated at about 10%. If, in order to add 1 kg of weight, a person ate 10 kg of salmon, to attain that weight the salmon had to consume 100 kg of small fish, and the fish needed to consume 1000 kg of carnivorous zooplankton, which in turn required 10,000 kg of herbivorous zooplankton, needing 100,000 kg of phytoplankton to supply the eventual 1 kg gain at the top of the pyramid. The 90% energy loss at each trophic level goes to the metabolic needs of the

web link **Figure 14.9** Charts of world phytoplankton abundance are based on *SeaWiFS* satellite measurements of chlorophyll concentration. *Dark blue* indicates a low abundance of chlorophyll; other colors (*light blue* to *green* to *yellow* to *orange*) indicate increasing abundance. Conditions are shown for April (a), July (b), October (c) in 1998, and January (d) in 1999.

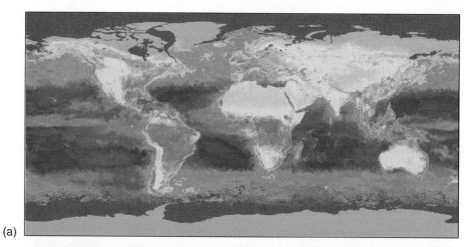

(a)

(b)

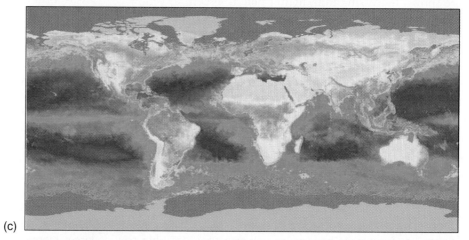

(c)

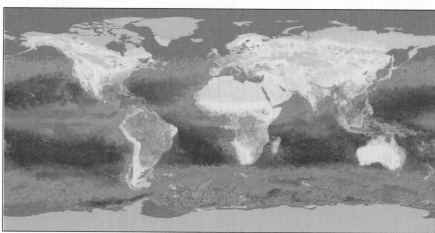

(d)

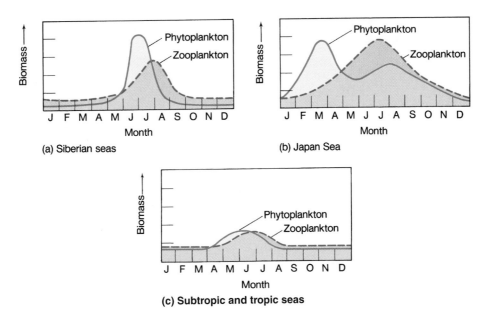

(a) Siberian seas

(b) Japan Sea

(c) Subtropic and tropic seas

Figure 14.10 The herbivorous zooplankton biomass varies directly with the phytoplankton biomass in (a) polar latitudes, (b) middle latitudes, and (c) tropical latitudes.

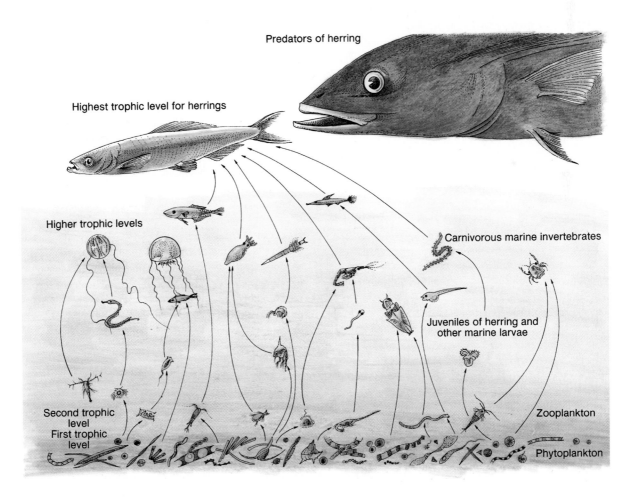

Figure 14.11 The food web of the herring at various stages in the herring's life. *Data from R. Buchsbaum and M. Buchsbaum, Basic Ecology 1957. Boxwood Press, Pacific Grove, Calif.*

The Antarctic Food Web

Almost all the life of Antarctica and the surrounding Southern Ocean depends on the sea. The fish, seals, whales, and birds are part of a food web that has at its base one zooplankton species, *Euphausia superba*, the Antarctic krill (box fig. 1). This shrimplike organism, only 6 cm (2.4 in) long, is the key to maintaining a balance of populations in the Southern Ocean. The krill are herbivorous shrimplike zooplankton feeding on the phytoplankton that bloom each summer as the ice pack melts back and primary productivity increases. Estimates of primary productivity vary, but at this time of year productivity is higher than the annual average of between 20 and 100 g of carbon per square meter per year (gC/m²/yr). Many of the food chains of Antarctica are very short and efficient, for krill are only one step removed from phytoplankton.

In summer, the ice pack melts back, the primary production increases, and the krill move toward the surface to graze the phytoplankton over large areas. In winter, primary production is very low in the surface waters because of reduced light, ice-pack cover, and increased turbulence; at this time, the krill appear to descend into deeper water and feed on the degraded and decomposing phytoplankton.

Squid are an important part of the Antarctic food web. There may be more than twenty species, some depending on krill as their most important food. The total annual consumption of squid by whales, birds, and seals is calculated at about 35 million tons. Some species of Antarctic fish stay in the Southern Ocean year-round, feeding on krill; other species migrate into Antarctic waters each summer to feed on krill. It is estimated that all Antarctic fish combined may consume as much as 100 million tons of krill each year.

There are few species of birds in Antarctica, but populations are usually very large. Birds of Antarctica feed principally on crustaceans (krill and large copepods), squid, fish, and carrion. Krill amounts to 78% of all the food they eat. Recent estimates indicate that about 115 million tons of krill are eaten annually by birds, either directly or indirectly. In winter, the birds either switch to a diet of squid or fish or they migrate northward. Most penguin species feed on krill supplemented, in some cases, by fish. The two largest species of penguin, the emperors and the kings, take fish and squid only.

Seven species of seals are found in the Southern Ocean; four species—the crabeater, leopard, Ross, and Weddell—are almost totally confined to the ice zones. Weddell seals eat mainly fish and squid; the crabeater's diet is about 94% krill; krill makes up 37% of the leopard seal's diet. The crabeater seal, now numbering about 30 million, is the most abundant seal in the world. The fur seals at South Georgia feed almost exclusively on krill. Stocks of Antarctic seals at present total about 33 million with a biomass of about 7 million tons. These seal populations annually consume over 130 million tons of krill (two or three times the current consumption by whales) and at least 10 million tons of squid.

Baleen, or plankton-feeding, whales of the Southern Ocean include the blue, finback, sei, minke, humpback, and southern right whales. Before whaling exploitation, they were probably about four times more abundant and had a biomass five times greater than at present. The major food of the baleen whales is krill. In 1904, the whale biomass was about 45 million tons, and whales consumed an estimated 190 million tons of krill. Competition for food probably limited their size and their numbers. By 1973, the whales had declined to a biomass of about 9 million tons, and they ate about 43 million tons of krill. The reduction in the whale population means that some 150 million tons of krill formerly eaten by whales have become available to the remaining whales, other predators, and human harvesters.

In the past thirty years, there has been an increase in the pregnancy rate of finback and sei whales and an apparent decrease in their age at maturity due, presumably, to the greater availability of food. The minke population may be double what it was before whaling; in 1930 the Antarctic minke became sexually mature when over 15 years of age; present minke populations are sexually mature at 7 years. The crabeater seal populations also have experienced a decrease in the age of sexual maturity (4 years in the 1950s to 2.5 years in the early 1960s), also thought to be the result of increased growth rates based on food abundance. The 14%–17% increase in the population of the South Georgia fur seals is unusually high, probably related to the abundance of krill.

The three most abundant penguins, the chinstrap, Adelie, and macaroni, have shown increases in population. Large increases in king penguins are thought to be due to their feeding on krill-eating squid. We do not know the response of the fish and squid stocks to the increase in available krill, but it is likely to be similar. The total quantity of krill taken by all predators in the Southern Ocean may be about 500 million tons per year. At one time scientists thought that the loss of the great whales would produce an overabundance of krill, but the smaller whales, the birds, and the seals now play larger roles in predation than they did a century ago and there is no oversupply of krill. The organisms have struck a new balance, but today as yesterday, directly or indirectly, the Antarctic food web depends on krill.

Internet References

Visit the book's Online Learning Center at www.mhhe.com/sverdrup (click on the book's cover) to explore links to further information on related topics.

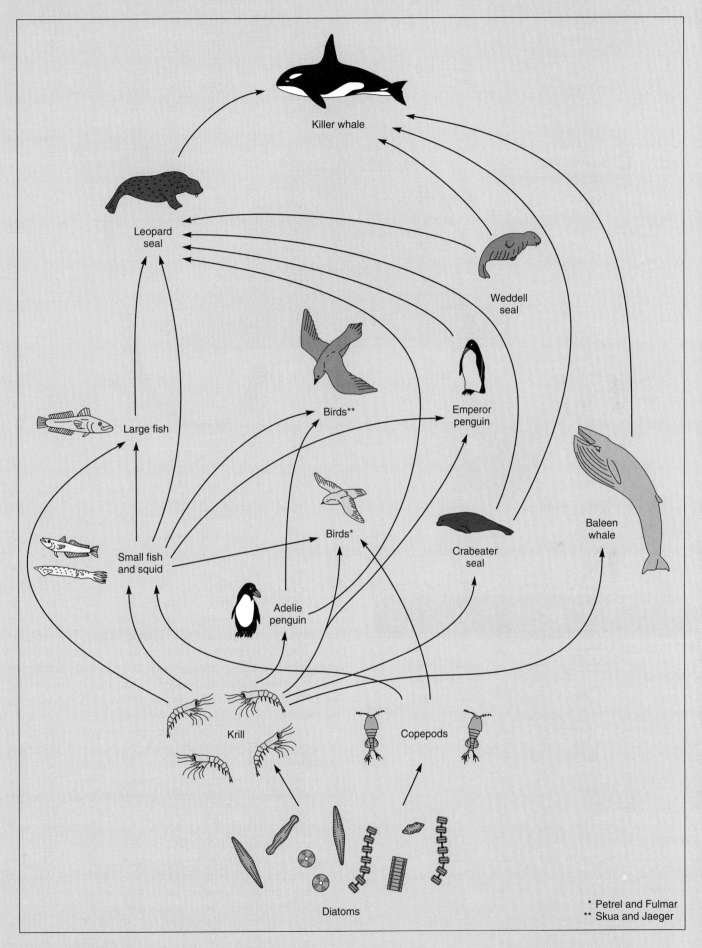

Killer whale

Leopard
seal

Weddell
seal

Birds**

Emperor
penguin

Large fish

Baleen
whale

Small fish
and squid

Birds*

Crabeater
seal

Adelie
penguin

Krill

Copepods

Diatoms

* Petrel and Fulmar
** Skua and Jaeger

web link **Box Figure 1** The Antarctic food web. The krill feed on the diatoms, and the other members of the food web depend on the krill.

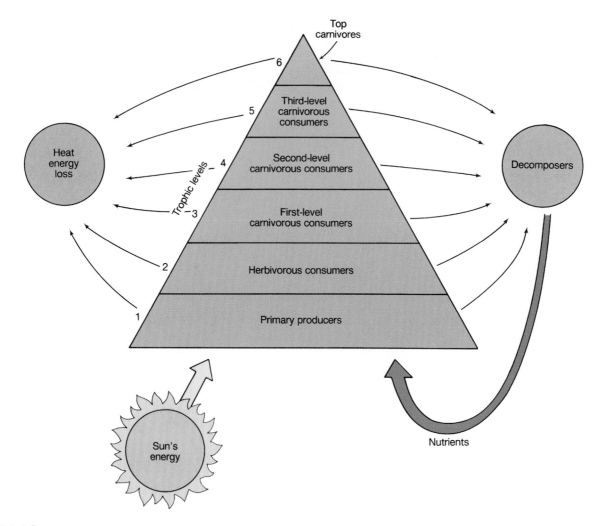

Figure 14.12 A trophic pyramid. Trophic levels are numbered from *base* to *top*. The first trophic level requires nutrients and energy. Nutrients are recycled at each level; energy is lost as heat at each level.

Table 14.3 Relative Abundance and Size of Marine Organisms

Organic Form	Size Range	Relative Abundance
Fish	10–100 cm	0.01
Zooplankton	1 mm–10 cm	1.0
Phytoplankton	0.001–0.02 mm	10.0

organisms—moving, breathing, feeding, and reproduction—and heat loss. In other words, an organism that consumes 100 units from the level directly below will use 90 units for its own metabolic needs and will convert only 10 units to body tissue available to predation from the level above. Therefore, feeding at high trophic levels is less energy efficient than feeding at low trophic levels.

Other Photosynthetic Systems

In some areas of the ocean, total production is not tied to the primary production of the phytoplankton. The shallow water above a tropical coral reef is clear, indicating a lack of phytoplankton. The reef, however, is a rich and varied community living in an independent, complex association that includes algae, herbivores, and carnivores. (Coral reefs are discussed in chapter 17.)

In shallow coastal areas and estuaries, active production of organic matter by primary producers occurs at all depths. In addition to the phytoplankton in the water column, masses of attached seaweeds, sea grasses, and unicellular algae are found where the sea bottom is exposed to sunlight. The rapid growth and relatively large size of many seaweeds provide a great amount of organic material for the animal population. Marshes that border low coastal areas also contribute. These areas are highly productive, in some areas yielding two crops of plants each year. Animals graze the living plants and feed from the fragments left after the plants have been battered by winter storms and natural seasonal die-off.

Table 14.4 Oceanic Food Production

Area	Plant Production (metric tons of carbon/yr)	Efficiency of Mass and Energy Transfer per Trophic Level	Trophic Level Harvested	Estimated Fish Production (metric tons/yr)
Open ocean	39.9×10^9	10%	5	4.0×10^6
Coastal regions	8.6×10^9	15%	4	29.0×10^6
Upwelling areas	0.23×10^9	20%	2	46.0×10^6

Data from S. Smith and J. Hollibaugh. Coastal Metabolism and the Oceanic Organic Carbon Balance. *1993.* Review of Geophysics *31(1): 75–89.*

In deep estuaries, most of the sea floor is below the photic zone, and attached seaweeds and marine plants play smaller roles. In a deep, partially mixed estuary system, the majority of the primary production comes from the phytoplankton in the upper water column. Although the concentration of organic carbon per cubic meter of water is highest near the surface, the water below the photic layer may contain a greater total amount of organic carbon because of active vertical mixing that displaces the plant cells downward. (Refer to table 14.2 to compare the primary productivity of estuaries and marshlands to that of the open ocean and coastal oceans.)

Primary Production and Chemosynthesis

In the deep-water communities of previously unknown organisms that surround hot-water vents on the deep-ocean floor, primary production is not dependent on solar energy but on the chemistry of the hot water flowing from cracks in the sea floor. Dense clouds of bacteria are the base for the trophic pyramids of these communities. These bacteria use dissolved chemicals in the hot vent water to obtain energy by a process known as **chemosynthesis.** Instead of being either directly or indirectly dependent on sunlight for energy to produce organic matter by photosynthesis, the ocean-bottom bacterial populations oxidize the sulfides, especially the dissolved hydrogen sulfide, in the hot vent water. These bacteria are able to do in the dark what plants do in the sunlight; they fix carbon dioxide into organic molecules such as simple sugars. The vent animals feed on the bacteria; no sunlight is necessary, and no food is needed from the sea surface. These self-contained vent communities are among the most productive in the world. In areas where the vents have become inactive, the animal communities have died when their energy source has been removed. Refer to chapter 17 for a discussion of the organisms in deep-ocean communities.

Practical Considerations: Human Concerns 14.6

When human activities supply additional nutrients, the primary production may exceed the ability of the local herbivores to consume it. A continuous and excessively high rate of primary production eventually results in an unnaturally high rate of decomposition, as unconsumed plant matter accumulates at depth, especially in an estuary. The decaying material consumes dissolved oxygen at a rapid rate, and a mat of organic debris may form over the sediments. The waters and the sediments become anoxic and adversely affect the organisms at the higher levels in the food webs and chains. The discussions of the Gulf of Mexico dead zone (see chapter 12) and Chesapeake Bay (see chapter 11) illustrate this problem.

Studies done in 1990 in the Southern Ocean around Antarctica indicate that springtime increases in ultraviolet radiation affect the phytoplankton at the sea surface. Higher levels of ultraviolet radiation directly under the ozone hole (a product of human activities, discussed in chapter 6) are estimated to reduce photosynthesis and phytoplankton reproduction by about 6%. When this reduction is adjusted for water and phytoplankton moving into and out of the high-ultraviolet-radiation area, the average reduction of annual phytoplankton production is estimated at 2%–4% of normal. This change may not seem large, but in the Antarctic the phytoplankton form the base of the food chain that supports fish, whales, seals, and birds; see the box titled "The Antarctic Food Web" in this chapter.

Custom, culture, economics, and availability influence the harvesting of the oceans by humans. Commercial harvests are made from both higher trophic levels (salmon, tuna, halibut, and swordfish) and lower levels (herring, shellfish, and anchovy). Harvesting high in the trophic pyramid is less energy efficient than harvesting at low trophic levels. Overharvesting any level endangers levels both above and below it by removing food resources from higher levels and preventing recycling of nutrients from higher to lower levels.

The yields of food resources for humans are highest when the harvest of marine species is conducted at the lowest possible trophic level. The relationship between total plant production and theoretical production of fish for three basic areas of the ocean is shown in table 14.4. The third column in this table gives the average efficiency of energy conversion between the trophic levels in each area, and the fourth column shows the trophic level at

CalCOFI—Fifty Years of Coastal Ocean Data

In the 1920s and 1930s the coast of Southern California, between Monterey and San Diego, was the site of a rich sardine fishery. During this period more sardines were caught off the California coast than any other fish in North America. This was the fishery that supplied the twenty-four canneries along Monterey's Ocean View Avenue, made famous in John Steinbeck's novel *Cannery Row*. But in the 1940s the fishery dwindled and collapsed. The 1945 catch of more than 550,000 metric tons dropped to just 100,000 metric tons in 1947. In response to the fishery's failure, California Cooperative Fisheries Investigations (CalCOFI) was formed and began a series of ocean research cruises in 1949. During the thirty-five years between 1949 and 1984 more than 23,000 stations were sampled between the southern reaches of Baja California, the California-Oregon border, and several hundred kilometers offshore. Since 1984 another 7000 stations have been routinely sampled between San Diego and Point Conception.

Although the original purpose of the program was to look at the distribution and abundance of the sardine, it became clear that to understand the fluctuations in the sardine population it was necessary to study the links between the ocean, its currents, and the atmosphere, especially with respect to El Niños, La Niñas, and the even longer-term cycles that affect the northeast Pacific.

Between 1958 and 1960 the warming of the California Current was related to an equatorial El Niño event; zooplankton and larval fish populations declined and the harvest of coastal fish dropped from 114,000 metric tons in 1956 to 79,000 metric tons in 1960. Nearshore seaweed forests near San Diego also declined during this period. In the late 1970s the California Current again became warmer and fresher and the long-term decline in the commercial fish catch accelerated. At the same time seabird counts decreased, seaweed forests declined, and changes were noted in the intertidal benthic and bird communities as southern species came to dominate communities. It is now recognized that an unexplained shift in northeast Pacific conditions occurred in 1977 on which later El Niños and La Niñas were imposed.

During the 1983–84 El Niño, zooplankton, kelp forests, and fish again declined. The range of many fish populations and invertebrates shifted northward in 1983; seabirds along the Oregon coast had a bad breeding season; and the numbers of pups born to California sea lions and Northern fur seals also dropped.

In 1985 sardine spawning reached 30,000 metric tons, by 1995 the sardine biomass reached 300,000 metric tons, and in 1999 the sardines surpassed 1 million metric tons for the first time since the mid-1940s. The stock has rebuilt and it has also extended its range northward, reaching Vancouver Island in 1998.

In the years 1998 and 1999 the area began a transition from warming to cooling, entering a La Niña cycle that peaked in 1999–2000. The phytoplankton population increased, with April 1999 chlorophyll measurements showing the highest values since 1984. It is thought that this may signal the end to the warming trend started in 1977. CalCOFI data banks now hold more than fifty years of coastal data concerning ocean currents, atmospheric conditions, and population levels of plankton, fishes, marine mammals, seabirds, and other marine organisms (box fig. 1).

web link **Box Figure 1** CalCOFI researchers use a plankton net known as a bongo net to collect duplicate samples of plankton.

In the California Current system warming episodes that last one to two years affect the strength of the current and also lower salinities—all linked to equatorial El Niños. Researchers investigating fish scales in sediment cores from the Santa Barbara basin have accumulated a 2000-year record that shows a pattern of alternating cycles of abundance between sardine and anchovy populations lasting not two but approximately thirty years. This cycle is thought to be related to the long-term northeast Pacific trend that started in 1977 (see previous). Sardines prefer warmer water and anchovies prefer cooler water, indicating some kind of long-term climatic oscillations that affect water temperature, productivity, and populations in this coastal area. Present indications are that the sardine is increasing in the California Current and the anchovy population is declining.

Another important use of CalCOFI data is associated with the interpretation of satellite data concerning phytoplankton. The *SeaWIFS* satellite measures phytoplankton absorption and reradiation of light, the changes in light leaving the water in response to concentrations of dissolved and suspended materials, and the fluorescence of photosynthetic pigments such as chlorophyll. These data are known as bio-optical data and are used to provide quantitative data on phytoplankton populations. However, to calibrate the satellite data one must have direct measurements of populations at sea. CalCOFI has been collecting bio-optical data at more than 300 stations since 1993, and this data provides a key to interpreting the data from the *SeaWIFS* satellites.

Internet References

Visit the book's Online Learning Center at www.mhhe.com/sverdrup (click on the book's cover) to explore links to further information on related topics.

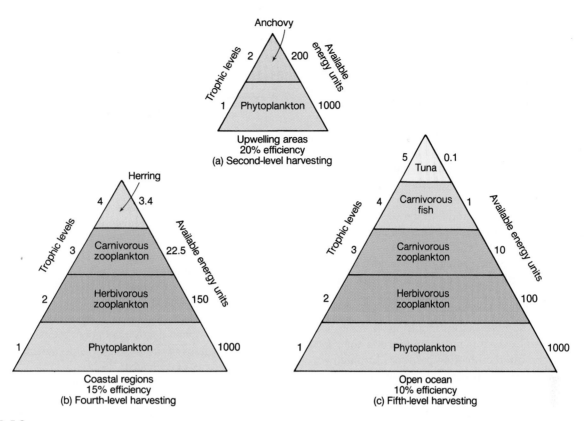

Figure 14.13 Trophic level efficiency varies among (a) upwelling areas, (b) coastal regions, and (c) the open ocean. The number of trophic levels and the level at which humans harvest differ with location.

which humans usually harvest their food. The well-mixed, nutrient-rich waters of the coasts and estuaries support short, efficient food chains, shown in pyramid form in figure 14.13a, b. Because of the high rates of plant production in these areas, food is easily obtained and energy is transferred efficiently to the next higher trophic level; consumers obtain food with less effort and fish production is high. Compare the efficiency of these food chains with the open-ocean food chain in figure 14.13c and the open-ocean fish production in table 14.4.

Although we need to know more about how to best harvest from one oceanic trophic level without depleting it and interfering with its transfer of energy to the next level, the more difficult task may be finding and then implementing agreements among harvesters and among nations that depend on the ocean's harvests to feed and employ their people. These topics are discussed in chapters 12, 16, and 17.

Summary

Plants use photosynthesis to produce organic compounds from carbon dioxide and water in the presence of sunlight and chlorophyll. Oxygen is formed as a by-product. Respiration breaks down organic compounds with the addition of oxygen to yield energy, water, and carbon dioxide.

Gross primary productivity is the total amount of organic material produced by photosynthesis per volume per unit of time; net primary productivity is the gain in organic material per volume per unit of time after the organic material used in plant respiration is deducted. The amount of organic carbon produced by primary production in a volume over a time period is measured as the rate of change in biomass. The biomass available at a location at a specific time is the standing crop.

Primary production is controlled by the interaction of sunlight, nutrients (nitrates and phosphates), and the stability of the surface water in an area. At polar latitudes, the availability of light controls phytoplankton growth; nutrients are not limiting. In the tropics, the sunlight is available year-round, and the stability of the water column helps hold the phytoplankton at the surface. However, the nutrient supply is poor and it limits production. At temperate latitudes, light, nutrients, and water column stability vary with the seasons. Nutrients cycle through the land and sea and through the plants and animals and are returned to the water by death and bacterial decomposition.

Where sunlight is not limiting, coastal waters are always more productive than the open ocean. Good mixing, the nutrients in the land runoff, and a water column made stable by the addition of fresh water combine to make shallow coastal regions very productive. Increasing nitrogen in the oceans is linked to high productivity followed by high rates of decomposition, which lowers the available oxygen and also future productivity. Lower productivity is also linked to lack of iron.

Upwelling areas are four times more productive than coastal water and six times more productive than the open ocean, but there is four times more open ocean than coastal water and more than 100 times more coastal water than upwelling area. Primary production is concentrated in coastal areas and at upwellings.

Primary productivity may be estimated by (1) counting the increase in plant cells; (2) determining the change in concentration of chlorophyll in a sample; (3) measuring the rates of nutrient supply and nutrient removal and using the known ratio of nutrients to calculate carbon production; (4) measuring the rate of incorporations of carbon-14 into the plant cells using light and dark bottles; or (5) using continuous satellite measurements to prepare maps of global phytoplankton production.

The phytoplankton, or primary producers, are preyed upon by herbivorous zooplankton. These are the primary consumers that are preyed upon by the secondary consumers, or carnivorous zooplankton. The relationship between the phytoplankton and the herbivorous zooplankton varies with latitude. The term *food web* is more appropriate than *food chain* to describe the interconnecting patterns. Trophic pyramids present the relationship between producers and consumers in terms of the transfer of biomass and energy. Open-ocean transfer efficiency is approximately 10% between trophic levels. Efficiency is higher in coastal and upwelling areas. Nutrients are recycled; energy is not.

Tropical coral reefs, very shallow coastal waters and estuaries, and deep-ocean vents are productive areas in which productivity is not keyed to the phytoplankton. Deep-water vent communities are highly productive and self-contained. They depend on chemosynthesis by bacteria to fix carbon into organic molecules.

If human activities supply additional nutrients to seawater, primary production increases, followed by decomposition and anoxia. Increasing solar radiation associated with the ozone holes may affect food webs. Harvested abundance is increased if the lower trophic levels are used, but care is required to conserve the stock for future harvests and for the requirements of other organisms.

Key Terms

All key terms from this chapter can be viewed by term, or by definition, when studied as flashcards on this book's Online Learning Center at www.mhhe.com/sverdrup (click on this book's cover).

Study Questions

1. Why is the efficiency of energy transfer between trophic levels only about 10% in the open sea? Compare this efficiency of energy transfer with that found in upwelling regions.
2. Explain the general relationship between the abundance and size of organisms shown in table 14.3.
3. Distinguish between the terms in each pair:
   a. Standing crop—biomass
   b. Photosynthesis—respiration
   c. Producer—consumer
   d. Food web—food chain
   e. Net productivity—gross productivity
4. Compare the productivity of polar, temperate, and tropical ocean regions. What factor or factors generally limit the productivity of each area?
5. The rate of primary production in the open ocean is less than the rate of primary production along the coasts, but the total primary production of the open ocean exceeds that of the coasts. Explain this apparent contradiction.
6. Which areas of the oceans are most productive? How does the productivity of these areas compare to the productivity of the land?
7. Why are estuaries less important than cultivated land as producers of human foods, although the primary production rate of estuaries approximately equals the most intensively cultivated land?
8. How do the surrounding land areas contribute to an estuary's productivity?
9. Draw a general diagram to explain the movement of (a) carbon dioxide or oxygen and (b) nitrate or phosphate through the ocean environment; include plant and animal populations.
10. Explain why the compensation depth is deeper in tropical areas with low productivity than it is in high temperate or subarctic areas with high primary productivity.
11. Consider figure 14.11 and explain how and why the herring food web changes with time.
12. Why is a shallow-water estuary more productive than either a deep-water estuary or an upwelling zone?
13. Discuss the factors that cause the changes in phytoplankton pigment concentrations in figure 14.9a–d.
14. Why does the respiration of phytoplankton appear nearly constant with depth in figure 14.8?
15. Why is nitrate-nitrogen limiting to production when the concentration of nitrates and phosphates are equal? Use the ratios of $O_2 : C : N : P$.

Study Problems

1. A sample of water from a depth of 5 m showed 8 mg of dissolved oxygen per liter. Part of this water was placed in a light bottle and part of it was placed in a dark bottle, and the bottles were returned to the 5 m depth. After six hours the oxygen content of the light bottle was found to be 8.9 mg of oxygen per liter; the dark bottle showed 7.4 mg of oxygen per liter. Calculate (a) the respiration rate, (b) net primary production, and (c) gross primary production. How many milligrams of new carbon were produced per liter in six hours?

2. If the nitrogen available as nitrate is removed from the water of an inlet at the rate of 3.6 mg of nitrogen per liter of water every eight hours, what is the rate of new carbon production by the phytoplankton?

Links to Related Websites

Visit the book's Online Learning Center at www.mhhe.com/sverdrup (click on the book's cover) to find live Internet links for additional topics related to this chapter's content.

- Primary productivity

Visit the book's Online Learning Center at www.mhhe.com/sverdrup (click on this book's cover) to find these additional chapter tools: Suggested Readings; links to further information on boxed readings, selected figures, and related chapter topics; and additional study aids.

chapter 15

The Plankton: Drifters of the Open Ocean

In a sudden awakening, incredible in its swiftness, the simplest plants of the sea begin to multiply. Their increase is of astronomical proportions. The spring sea belongs at first to the diatoms and to all the other microscopic plant life of the plankton. In the fierce intensity of their growth they cover vast areas of ocean with a living blanket of their cells. Mile after mile of water may appear red or brown or green, the whole surface taking on the color of the infinitesimal grains of pigment contained in each of the plant cells.

The plants have undisputed sway in the sea for only a short time. Almost at once their own burst of multiplication is matched by a similar increase in the small animals of the plankton. It is the spawning time of the copepod and the glassworm, the pelagic shrimp and the winged snail. Hungry swarms of these little beasts of the plankton roam through the waters, feeding on the abundant plants and themselves falling prey to larger creatures.

Rachel Carson
From *The Sea Around Us*

A drifting Scyphomedusa jellyfish, or sea jelly.

The word *plankton* comes from the Greek term *planktos,* meaning to wander, and the plant and animal plankton are the wanderers and drifters of the sea. They exist in vast swarms, limited in their mobility, moving with the currents. The diversity of planktonic organisms is so great that it is not possible to discuss all of their life forms here. Instead, representative animals, plants, and groups of animals and plants have been chosen for discussion. Because it is the plant life that is able to use the Sun's energy to form the basis of life for all the animals, we begin by considering the ocean's floating plant life, the primary producers of chapter 14, and then go on to describe the animals that drift with the plant plankton, grazing upon it and upon each other.

Kinds of Plankton 15.1

Although many plankton have a limited ability to move toward and away from the sea surface, they make no purposeful motion against the ocean's currents and are carried from place to place suspended in the seawater. Some plankton are quite large; jellyfish may be the size of a large washtub, trailing 15 m (50 ft) tentacles. But the phytoplankton and many zooplankton are generally too small for our unassisted vision and must be observed under a microscope.

Bacteria and very small phytoplankton cells are called **ultraplankton;** they are less than 0.005 mm in diameter and can be collected only by using special filtering systems. Slightly larger phytoplankton, called **nannoplankton,** have a size range between 0.005 and 0.07 mm. Zooplankton and phytoplankton between 0.07 and 1 mm are called **microplankton,** or **net plankton,** because they are usually captured in tow nets made of very fine mesh nylon.

The microscopic phytoplankton are the "grasses of the sea." Just as a land without grass and herbs could not support the insects, small rodents, and birds that serve as food for the larger, meat-eating carnivores, a sea without phytoplankton could not support the zooplankton and the other larger animals. As the British biological oceanographer Sir Alister Hardy has said, "All flesh is grass."

Phytoplankton

The phytoplankton are mainly unicellular (or single-celled) plantlike organisms known as algae. Each phytoplankton cell is **autotrophic** (or self-feeding) by the process of photosynthesis (see chapter 14). Each cell is an independent individual, and even in the species in which the cells attach together in **filaments** (long chains) or other aggregations, there is no division of labor between the cells. There is only one large, multicellular planktonic alga, the seaweed *Sargassum,* that is found floating in the area of the North Atlantic known as the Sargasso Sea. *Sargassum* reproduces vegetatively by fragmentation to form large mats, which provide shelter and food for a wide variety of organisms, including fish and crabs. The specialized organisms found living in the *Sargassum* mats occur nowhere else.

Groups of organisms belonging to the phytoplankton include the **diatoms, dinoflagellates,** and **coccolithophores;** silicoflagellates, cryptomonads, chrysomonads, green algae, and cyanobacteria or blue-green algae are present but less numerous. In the following discussion the emphasis is placed on the diatoms and the dinoflagellates because they are the most abundant and most important members of the marine phytoplankton.

Diatoms are sometimes called golden algae, because their characteristic yellow-brown pigment, fucoxanthin, masks their chlorophyll. Most diatoms have either **radial** or **bilateral symmetry.** Those that are radially symmetrical, sometimes called centric, are round and shaped like pillboxes; bilaterally symmetrical diatoms, sometimes called pennate, are elongate. Centric diatoms are more truly planktonic; they float better than pennate diatoms. Pennate diatoms are often found on the shallow sea floor or attached to floating objects. All are found in areas of cold, nutrient-rich water. Some common diatoms from temperate waters are shown in figures 15.1 and 15.2.

Around the outside of each diatom is a **frustule** (or cell wall) of pectin, a jellylike carbohydrate, impregnated with silica. The frustule is hard, rigid, transparent, and delicately marked with pores that connect the living portion of the cell inside to its outside environment

Figure 15.1 The diatoms dominate the phytoplankton from the temperate zone to the polar zone. (a) Two species of *Thalassiosira*. (b) The large centric diatom *Arachnoidiscus*. (c) An enlarged view of (b), showing the intricate details of the diatom's frustule. (d) The diatoms arranged in star-shaped groups are members of the genus *Asterionella*. (e) The large diatom, *Coscinodiscus*, lies between strands of *Ditylum*, another genus of diatom.

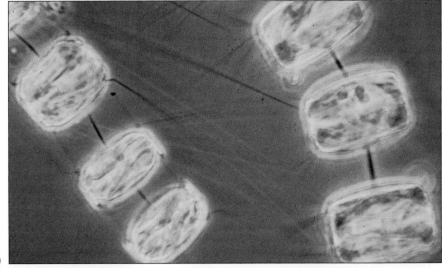

(a)

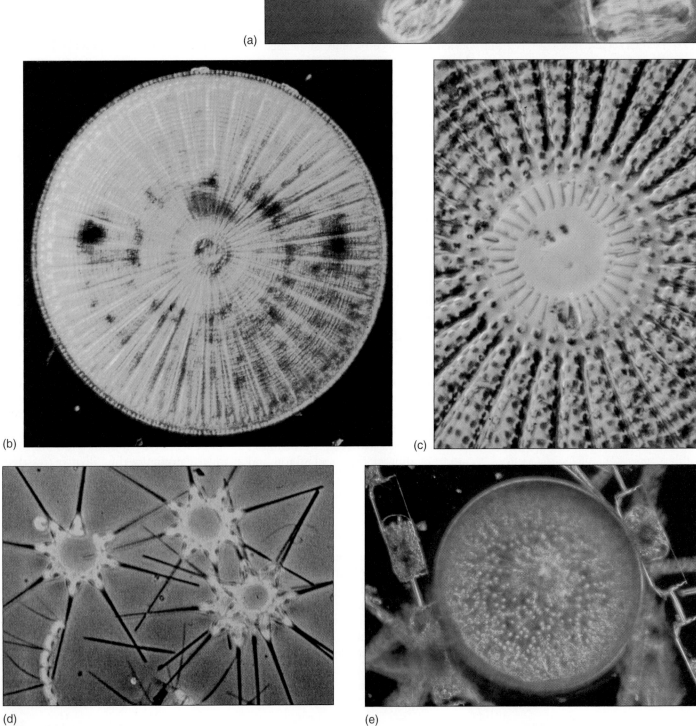

(b)

(c)

(d)

(e)

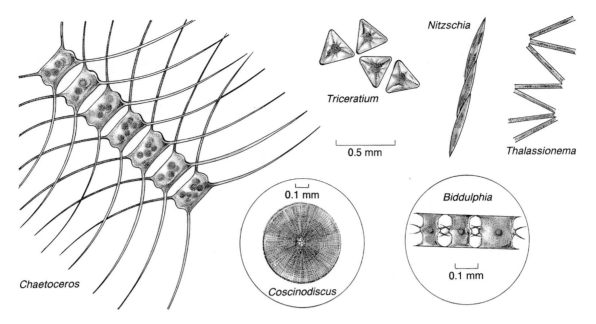

web link **Figure 15.2** Centric and pennate diatoms. Diatoms occur as single cells or form chains of cells.

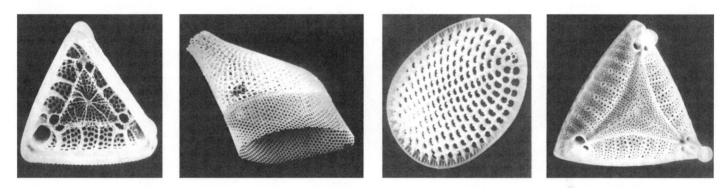

web link **Figure 15.3** Stereoscan micrographs of diatom frustules, showing the pores that connect the living portion of the cell to its outside environment.

(fig. 15.3). The two halves of a centric diatom's frustule fit together like a pillbox, and when the cell has grown sufficiently large, the cell inside divides, the two halves of the pillbox separate, a new inner half to each pillbox is formed, and two new daughter diatoms are produced. By this process, one of the new cells will be the same size as the parent; the other is always smaller, because its larger pillbox half is formed by the smaller bottom half of the parent pillbox. This process is shown in figure 15.4. When a cell reaches a size level of about 25% of the original parent's size, it stops dividing and begins a sexual cycle. In this cycle it produces a naked **auxospore,** which increases in size, forms a new frustule, and begins to divide again. Diatoms divide very rapidly, every twelve to twenty-four hours, under conditions of plentiful sunlight and nutrients. When this rapid division increases the population so that the water becomes discolored by the presence of millions and millions of cells, it is called a **bloom.**

The buoyancy of a diatom cell is increased by the low density of the cell's interior and the production of oil as a storage product. Fish that feed on large quantities of diatoms may have a distinctly oily taste. A diatom's small size also helps it to stay afloat, because a small spherical particle sinks slowly and has a large surface area compared to its volume or mass; this large surface-area-to-volume ratio reduces its sinking rate. The cell walls of some diatoms have spines or other extensions that further increase their surface area and retard their sinking. A diatom's large surface area also provides it with increased exposure to sunlight and to water containing the gases and nutrients necessary for photosynthesis and growth.

Diatoms that are not consumed by herbivores eventually die and sink to the ocean floor. In shallow areas, the cells reach the sea floor with some organic matter still locked inside, and they may, over a very long period of time, form petroleum deposits. Diatom frustules sinking to the

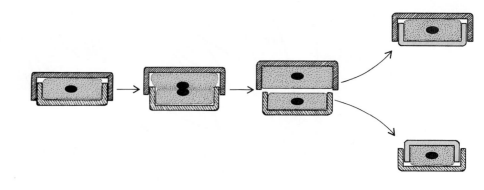

Figure 15.4 The division of a parent centric diatom into two daughter diatoms. The two halves of the pillboxlike cell separate, the cell contents divides, and a new inner half is formed for each pillbox. One daughter cell remains the same size as the parent; the other is smaller.

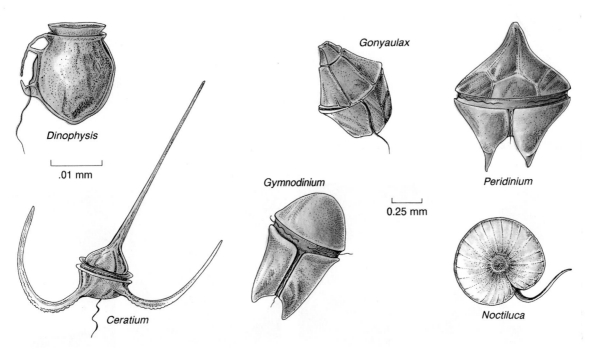

Dinophysis

.01 mm

Gonyaulax

Peridinium

Gymnodinium

0.25 mm

Ceratium

Noctiluca

web link **Figure 15.5** Dinoflagellates. *Noctiluca, Gymnodinium,* and *Gonyaulax* produce red tides. *Noctiluca* is a bioluminescent, nontoxic dinoflagellate. *Gymnodinium* and *Gonyaulax* produce toxic red tides and paralytic shellfish poisoning.

greater depths of the oceans build up siliceous sediments under surface areas of abundant diatom populations. (Refer to biogenous sediments in chapter 3.) Sometimes geological processes lift these silica-rich sediments above the sea, where they are mined as **diatomaceous earth,** used in industrial filtration systems (the filtering of wine as well as swimming pools) and as an abrasive in toothpaste and silver polish.

Dinoflagellates have red to green pigments and can exist at lower light levels than diatoms, because they can both photosynthesize like a plant and ingest organic material like an animal. They have both autotrophic and **heterotrophic** abilities. Heterotrophic organisms feed on other organisms or on organic substances. Their external walls do not contain silica but may be armored with plates of cellulose, giving them the appearance of helmets. Other dinoflagellates have a smooth, flexible outer surface showing no such plate structures.

Dinoflagellates usually have two **flagella,** or whiplike appendages, that beat within grooves in the cell wall. One groove encircles the cell like a belt, and the other lies at right angles to it. The beating of these flagella makes the cells motile and causes them to spin like tops as they move through the water. They also tend to migrate vertically in response to sunlight, but this ability to move is limited and they are still at the mercy of the waves and currents. Some dinoflagellates are called fire algae because they glow with bioluminescence at night. Bioluminescence is discussed in chapter 13. Representative dinoflagellates are shown in figure 15.5.

Dinoflagellates are found over most of the oceans but do not contribute to the bottom sediments, because both the cell walls and the soft parts decay completely. Although they do make up a substantial portion of the phytoplankton, dinoflagellates are not as important as the diatoms as a primary ocean food source. The cells reproduce by a division process similar to that found in diatoms but without the reduction in size. They can form blooms under favorable conditions, multiplying even more rapidly than diatoms.

Silicoflagellates and coccolithophores are relatives of the diatoms. Silicoflagellates are small autotrophic cells with flagella and an internal hollow skeleton of silica. They are less abundant than diatoms and are found in the Arctic, Antarctic, and open-ocean areas. The coccolithophores are single-celled, photosynthetic organisms with **coccoliths,** outer calcareous plates, that are deposited as sediment when the cells die. (See the section on sediments in chapter 3.) Like the dinoflagellates, coccolithophores possess two flagella, and like both diatoms and dinoflagellates, they may reproduce by simple division, although some species do have a form of sexual reproduction.

During the late summer–early fall of 1997 the water over most of the continental shelf in the eastern Bering Sea was colored aquamarine by a massive bloom of coccolithophores (fig. 15.6*a*). This bloom was so intense that it was clearly visible from space and was recorded in an image from the sensor SeaWiFs (sea-viewing wide-field-of-view sensor). The color of the water was due to light reflecting from the organisms' calcareous plates (fig. 15.6*b*). A bloom of this type has never before been recorded in this region. It was attributed to unusual climatic conditions; reduced cloud cover and few storms resulted in abundant sunlight and above-average surface water temperatures.

Zooplankton

The zooplankton are either grazers on phytoplankton (herbivores), feeders on other members of the zooplankton (carnivores), or feeders on both plants and animals (omnivores). Many of the zooplankton have some ability to swim and can even dart rapidly over short distances in pursuit of prey or to escape from predators. They may have some limited vertical motion, but, like the phytoplankton, they are transported by the currents and are therefore planktonic organisms.

Representatives of nearly every animal phylum are found in the zooplankton. The life histories of zooplankton types are varied and show many strategies for survival in a world where reproduction rates are high and life spans are short. These animals may produce three to five generations a year in warm waters, where food supplies are abundant and temperatures accelerate life processes. At high latitudes, where the season for phytoplankton growth is brief, the zooplankton may produce only a single generation in a year. The voracious appetites, rapid growth rates, and short life spans of the carnivorous zooplankton are responsible for the rapid liberation of nutrients to be recycled by the phytoplankton.

Zooplankton exist in patches of high population density between areas that are much less heavily populated. The high population patches attract predators, and the sparser populations between the denser patches preserve the stock, as fewer predators feed there. Turbulence and eddies disperse individuals from the densely populated patches to the intervening sparser areas. Convergence zones and boundaries between water types concentrate zooplankton populations, which attract predators.

Plankton accumulate at the density boundaries caused by the layering of the surface waters; variation of light with depth and day-night cycles play additional roles. Some zooplankton migrate toward the sea surface each night and re-

(a)

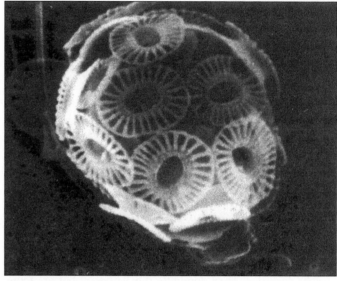

(b)

Figure 15.6 (a) A true-color image showing the extent of the coccolithophore bloom over the continental shelf of the eastern Bering Sea in September 1997. Image provided by the SeaWiFs Project, NASA/Goddard Space Flight Center. (b) A scanning electron microscope image of a coccolithophore. Light reflecting from the organisms' calcareous plates produces the color seen in (a).

turn to depth each day, either in an attempt to maintain their light level or in response to the movement of their food resource. This daily migration may be as much as 500 m (1650 ft) or less than 10 m (33 ft).

Accumulations of organisms in a thin band extending horizontally along a pycnocline or at a preferred light intensity or food resource level are capable of partially reflecting sound waves from depth sounders. The zooplankton layer is seen on a bathymetric recording as a false bottom or a deep scattering layer, the DSL (see chapter 4). Echo-sound studies are used to record the vertical migration of this layer of plankton and to estimate abundance by measuring the vertical and horizontal extent of the layer.

Among the most common and widespread zooplankton types worldwide are small **crustaceans** (shrimplike animals),

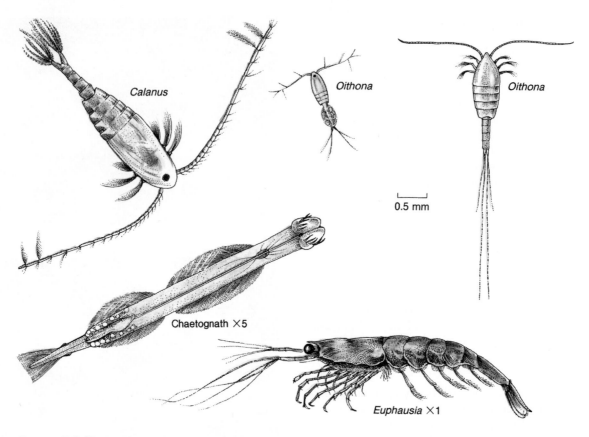

web [link] link **Figure 15.7** Crustacean members of the zooplankton and an arrowworm, or *chaetognath*. *Calanus* and *Oithona* are copepods. The shrimplike *Euphausia* is known as krill.

the **copepods,** and **euphausiids** (fig. 15.7). These animals are basically herbivorous and consume more than half their body weight each day. They dominate the Antarctic but are found throughout the world. There are more copepods than any other kind of zooplankton in the oceans and marginal seas. Copepods are a link between the phytoplankton, or producers, and first-level carnivorous consumers. Euphausiids are larger, move more slowly, and live longer than copepods. Euphausiids, because of their size, also eat some of the smaller zooplankton along with the phytoplankton that make up the bulk of their diet. Copepods and euphausiids both reproduce more slowly than diatoms, doubling their populations only three to four times a year. In the Arctic and Antarctic, the euphausiids are the **krill,** occurring in such quantities that they provide the main food for the **baleen** (or whalebone) whales. These whales have no teeth; instead they have netlike strainers of baleen suspended from the roof of their mouth. After the whales gulp the water and plankton, they expel the water through the baleen, leaving the tiny krill behind. Whales of this type include the blue, right, humpback, sei, minke, and finback whales; whales are discussed in chapter 16.

The Antarctic krill, *Euphausia superba* (fig. 15.8) is the most abundant of the eighty-five or so krill species and is the basis of the great Antarctic food web discussed in the box titled "The Antarctic Food Web" in chapter 14. The Antarctic krill are found over an enormous area, perhaps as large as 36×10^6 km^2, which is about four times the size of the United States.

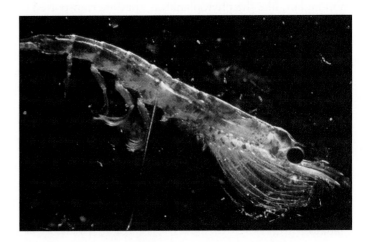

web [link] link **Figure 15.8** The Antarctic krill *Euphausia superba* dominates the zooplankton of the Antarctic Ocean.

The krill are circumpolar in distribution, but their concentration is not uniform; their greatest concentration occurs in the summer close to Antarctica. In summer the krill live near the surface in huge swarms made up of a billion or more individuals; in the cold, dark winter they are believed to live on the underside of the ice or perhaps dive down to the sea floor. Individuals are now known to live at least five years under natural conditions and up to nine years in the laboratory. In summer females lay up to 10,000 eggs at a time, several times a season.

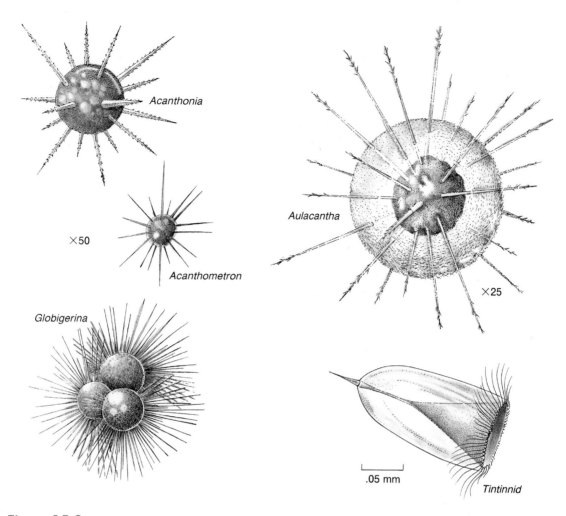

web ꙮ link **Figure 15.9** Selected radiolarians (*Acanthonia*, *Acanthometron*, and *Aulacantha*). A foraminiferan (*Globigerina*) is at *bottom left*; a tintinnid is at *bottom right*.

Estimates of the total biomass of Antarctic krill have varied from 5 million to 6 billion metric tons; the amount is still debated but is now thought to lie between 135 million and 1.35 billion tons. Because of these large biomass estimates the krill in the Southern Ocean have been considered a potentially valuable international fishery. The first harvests were made by the former Soviet Union in the 1960s, and in later years Japan, Korea, Poland, and Chile participated. The largest harvest, 529,000 metric tons, was made by Japan, Korea, Poland, and Chile during the Antarctic summer of 1980–81. During the next fifteen years fewer vessels fished the krill and the catch dropped: 81,000 metric tons in 1998 but increasing to 103,000 metric tons in 1999.

Marketing krill for human consumption has not been very successful, and the economic future of the krill fishery is questionable. Krill is sold whole, as peeled tail meat, minced, or as a paste; it is used as livestock and poultry feed in eastern Europe and as fish feed by the Japanese. Because the krill deteriorate rapidly, krill used for human consumption must be processed within three hours, limiting the daily harvest. The distance to the fishing grounds is long and the costs of vessels and fuel are high.

The Convention for the Conservation of Antarctic Marine Living Resources was established by treaty in 1981. The convention's goal is to keep any harvested population from dropping below levels that ensure the replenishment of the adult population. In 1991 krill harvesting limits in the South Atlantic were set at 1.5 million metric tons and in the Southern Indian Ocean at 390,000 metric tons. Although no harvests approaching these limits have been taken, the limits represent the first attempts made by any fishery to manage an entire ecosystem and maintain an equilibrium among all its species.

Arrowworms, or **chaetognaths** (see fig. 15.7), are abundant in ocean waters from the surface to the great depths. These macroscopic (2–3 cm, or 1 in), nearly transparent, voracious carnivores feed on other members of the zooplankton. Several species of arrowworms are found in the sea, and in some cases a particular species is found only in a certain water mass. The association between organism and water mass is so complete that the species can be used to identify the origin of the water sample in which it is found. In the North Atlantic, the arrowworm *Sagitta setosa* inhabits only the North Sea water mass and *Sagitta elegans* is found only in oceanic waters.

Foraminiferans and **radiolarians** are microscopic, single-celled, amoebalike protozoans; they are shown in figure 15.9.

Foraminiferans, such as the common *Globigerina,* are encased in a compartmented calcareous covering, or shell. Radiolarians are surrounded by a silica **test,** or shell. The radiolarian tests are ornately sculptured and covered with delicate spines. Openings in the test allow a continuity between internal protoplasm and an external layer of protoplasm. Pseudopodia (false feet), many with skeletal elements, radiate out from the cell. Radiolarians feed on diatoms and small protozoans caught in these pseudopodia. Both foraminiferans and radiolarians are found in the warmer regions of the oceans. After death, their shells and tests accumulate on the ocean floor, contributing to the sediments. Calcareous foraminiferan tests are found in shallow-water sediments; the siliceous radiolarian tests, which are resistant to the dissolving action of the seawater, predominate at greater depths, commonly below 4000 m (13,200 ft) (see chapter 3). **Tintinnids** (fig. 15.9) are tiny protozoans with moving hairlike structures, or **cilia.** These organisms are often called bell animals and are found in coastal waters and in the open ocean.

Pteropods (fig. 15.10) are mollusks; they are related to snails and slugs. They may or may not have a small calcareous shell, depending on the species, but all have a foot that is modified into a transparent and gracefully undulating "wing." Their hard, calcareous remains contribute to the bottom sediments in shallow tropical regions. Some pteropods are herbivores and some are carnivores.

Transparent, gelatinous, and bioluminescent, the **ctenophores,** or comb jellies (fig. 15.11), float in the surface waters. Some have trailing tentacles; all are propelled slowly by eight rows of beating cilia. The small, round forms are familiarly called sea gooseberries or sea walnuts; by contrast, the beautiful, tropical, narrow, flattened Venus' girdle may grow to 30 cm (12 in) or more in length. A group of Venus' girdles drifting at the surface and catching the sun-

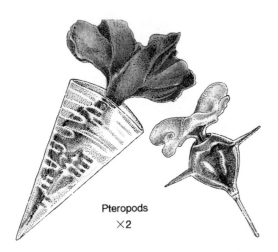

Pteropods
×2

Figure 15.10 The pteropods are planktonic mollusks.

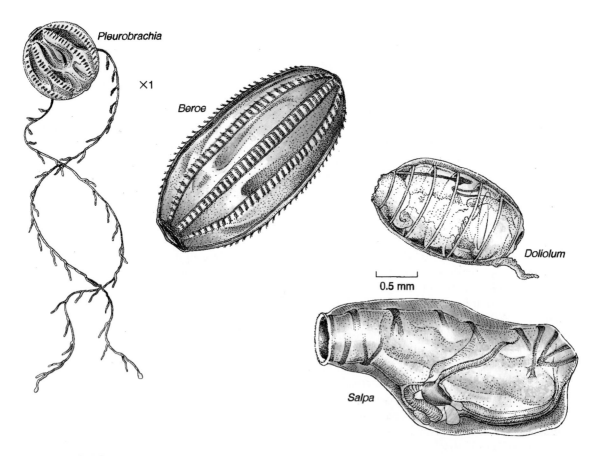

web link **Figure 15.11** The comb jellies (ctenophores) *Pleurobrachia* and *Beroe. Pleurobrachia* is often called a sea gooseberry. *Salpa* and *Doliolum* are tunicates.

light with their beating cilia is a spectacular sight from the deck of a ship. All ctenophores are carnivores, feeding on other zooplankton.

The tunicate, another transparent member of the zooplankton, is related to the more advanced vertebrate animals (animals with backbones) through its tadpolelike larval form. **Salps** (fig. 15.11) are pelagic tunicates that are cylindric and transparent; they are commonly found in dense patches scattered over many square kilometers of sea surface. Salps feed on phytoplankton and particulate matter.

Both ctenophores and pelagic tunicates, although jellylike and transparent, are not to be confused with jellyfish (fig. 15.12). True jellyfish, or sea jellies, come from another and unrelated group of animals, the **Coelenterata,** also

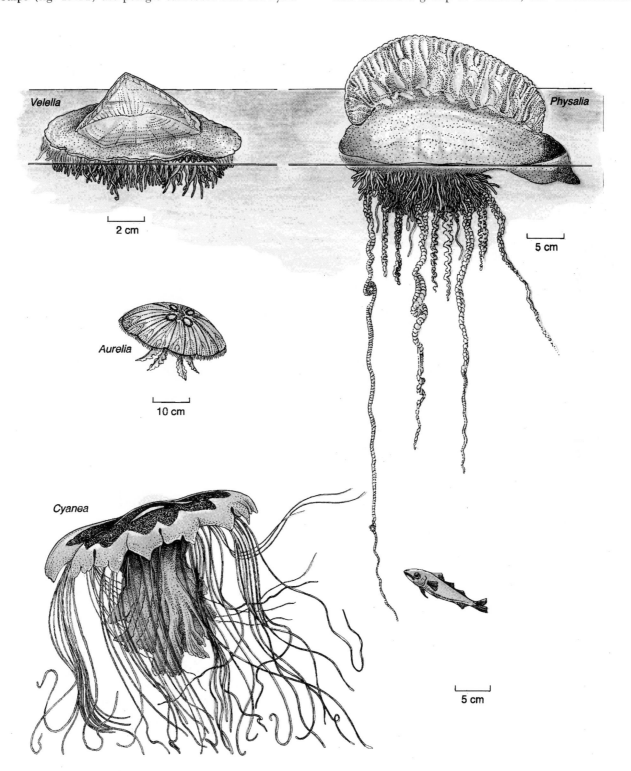

web link **Figure 15.12** Jellyfish belong to the Coelenterata, or Cnidaria. *Velella*, the by-the-wind-sailor (*top left*), and *Physalia*, the Portuguese man-of-war (*top right*), are colonial forms.

(a)

(b)

(c)

(d)

Figure 15.13 Animals that spend their entire lives in the zooplankton are known as holoplankton and include the single-celled (a) foraminiferans and (b) radiolarians and the more complex organisms such as (c) the jellyfish (*Polyorchis penicellatus*) and (d) a copepod.

called **Cnidaria.** Some jellyfish, such as the common *Aurelia* and the colorful *Cyanea*, with its trailing stinging tentacles that capture zooplankton, spend their entire lives as drifters. Others, such as *Gonionemus*, a small jellyfish of the Atlantic and Pacific Oceans, and *Aequorea*, found in many temperate waters, are members of the plankton for only a portion of their lives; they eventually settle and change to a bottom-dwelling, attached form similar to a sea anemone. Another group of unusual jellyfish are **colonial organisms,** including the Portuguese man-of-war, *Physalia*, and the small by-the-wind-sailor, *Velella*. Both are collections of individual but specialized animals. Some gather food, reproduce, or protect the colony with stinging cells, and others form a float.

All of the zooplankton discussed to this point, with the exception of certain jellyfish, spend their entire lives as plankton and are called **holoplankton** (fig. 15.13). However, an important portion of the zooplankton spends only part of its life as plankton; these are the **meroplankton.** The meroplankton include the eggs and the larval and juvenile stages of many organisms that spend most of their lives as either free swimmers (such as fish) or bottom-dwellers (such as crabs and starfish). For a few weeks, the **larvae** (or young forms) of oysters, clams, barnacles, crabs, worms, snails, starfish, and many other organisms are a part of the zooplankton. The currents carry these larvae to new locations, where they find food sources and areas to settle. In this way, areas in

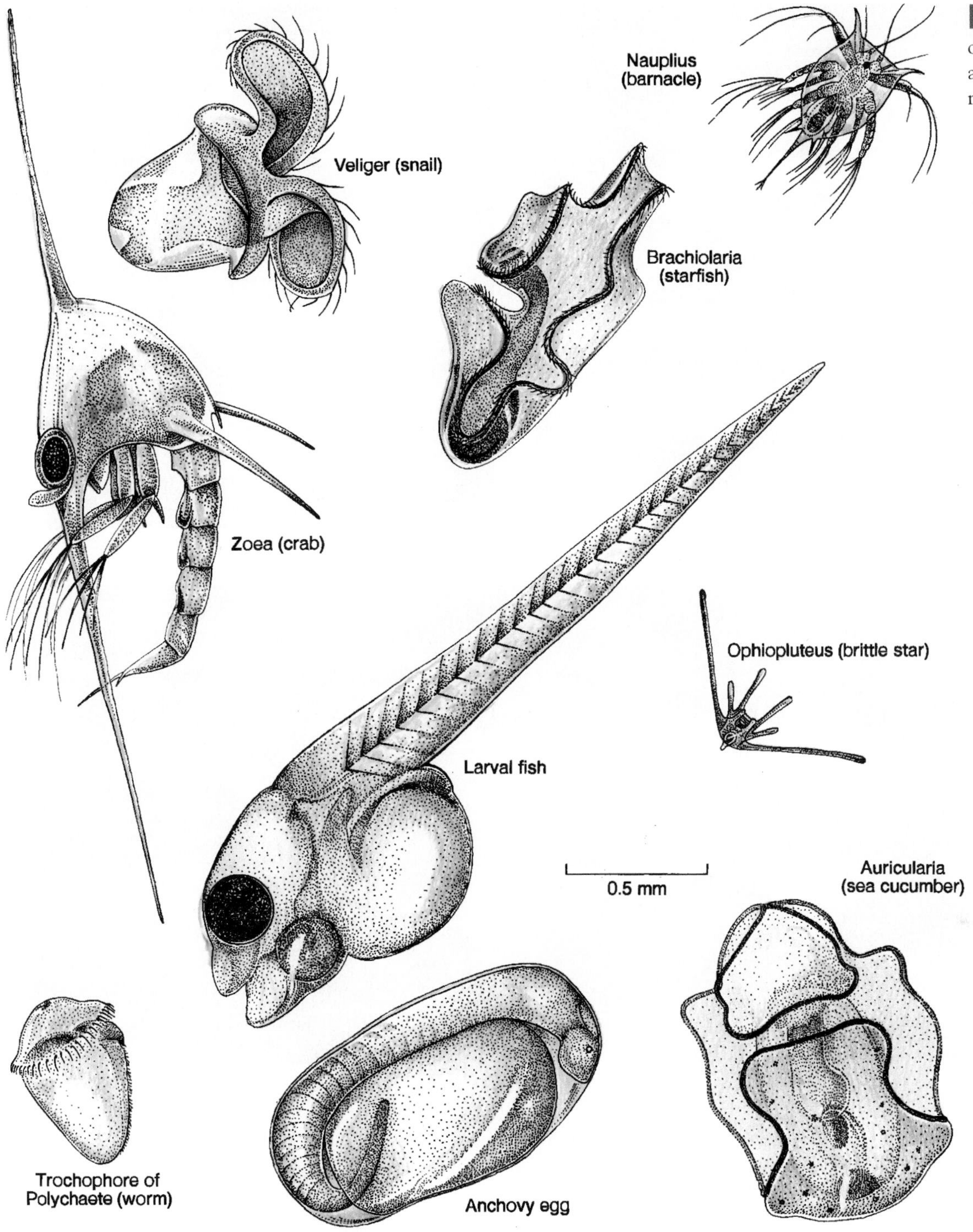

Figure 15.14 Members of the meroplankton. All are larval forms of nonplanktonic adults.

Veliger (snail)

Nauplius (barnacle)

Brachiolaria (starfish)

Zoea (crab)

Larval fish

Ophiopluteus (brittle star)

0.5 mm

Auricularia (sea cucumber)

Trochophore of Polychaete (worm)

Anchovy egg

which species may have died out are repopulated and overcrowding in the home area is reduced. Sea animals produce larvae in enormous numbers, so these meroplankton are an important food source for other members of the zooplankton and other animals. The parent animals may produce millions of spawn, but only small numbers of males and females must survive to adulthood to guarantee survival of the stock.

Larvae often look very unlike the adult forms into which they develop (figs. 15.14 and 15.15). Early scientists who found and described these larvae gave each a name,

thinking they had discovered a new type of animal. We keep some of these names today, referring, for example, to the trochophore larvae of worms, the veliger larvae of sea snails, the zoea larvae of crabs, and the nauplius larvae of barnacles.

Other members of the meroplankton include fish eggs, fish larvae, and juvenile fish. The young fish feed on other larvae, until they grow large enough to hunt for other foods. Some large seaweeds release **spores,** or reproductive cells, that drift in the plankton until they are consumed or settle out to grow attached to the sea bottom.

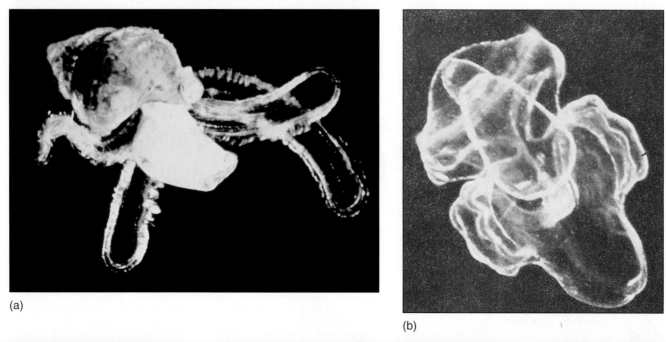

(a)

(b)

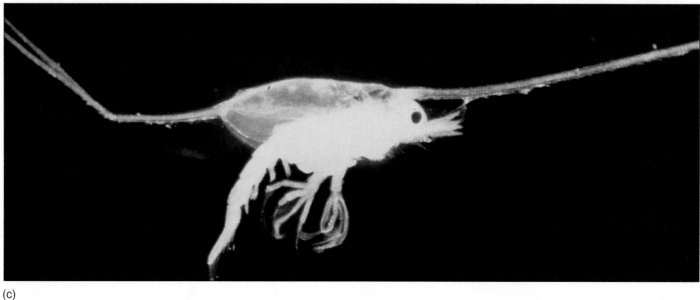

(c)

Figure 15.15 Larval forms of benthic marine organisms, or meroplankton, are an important part of the zooplankton. (a) The larva of a marine snail; (b) the larva of a starfish; (c) this zoea larva is a stage of crab development.

Bacteria 15.2

Bacteria are the smallest living organisms; they are microscopic, single cells without membrane-bounded nuclei and many are able to reproduce by cell division every few minutes when they inhabit a favorable environment. The bacteria are the most numerous organisms in the ocean. It is estimated that 1×10^{29} bacterial cells exist in marine environments; one-third occur in the upper ocean and two-thirds in deep water (2.5×10^{29} is the estimate for soils). Some marine environments produce 1×10^{30} generations of bacteria annually.

Autotrophic marine bacteria include the photosynthesizing **cyanobacteria,** most abundant in intertidal and estuarine areas and producing dense blooms in the warm water of tropical and subtropical latitudes, as well as the bacteria associated with the hydrothermal vents of the deep-sea floor, where they are the primary producers for the animal communities surrounding the deep-sea vents (see chapter 17). Heterotrophic marine bacteria are free-living in seawater and exist on every available surface including the sea floor, decaying material, the surface of organisms, and floating matter.

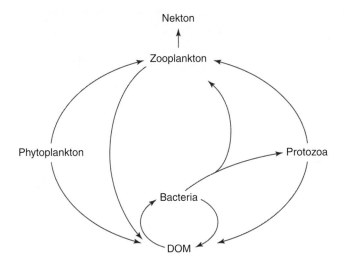

Figure 15.16 The microbial loop converts dissolved organic material (DOM) into particulate organic matter that is consumed by larger organisms.

Marine bacteria are a significant additional food source for planktonic larvae and a variety of single-celled protozoans. A film of bacteria is found on minute particles of floating organic material; the small size of these particles, with their attached bacterial population, makes them an ideal food for many small zooplankton. They are important in the food webs of the deep sea, where they are consumed by a large number of small animals. Bacterial activity on sinking plant remains decreases with depth and it is minimal below 2000 m (6600 ft), but once the remains have reached the seabed they are subject to vigorous bacterial activity. The deposit and decay of this plant material shortly after the spring phytoplankton bloom provide an important food resource for the deep-sea animal community. Bacteria play an important role in the decay and breakdown of organic matter, returning it to the sea as basic chemicals and compounds to be used again by new generations of plants and animals (see chapter 14).

The Microbial Loop

Recent studies emphasize that not all the organic matter produced by photosynthesis is eaten directly by the herbivorous members of food webs. Large quantities of dissolved organic material (DOM) are found in the oceans, and it was once thought that DOM was largely unused by other organisms. It is now known that DOM leaks out of phytoplankton cells and is also a product of the eating and excretion of zooplankton. While very few marine animals use DOM directly, bacteria are able to do so and they are the major consumers of DOM. Some phytoplankton in nearshore environments can also consume these dissolved organic nutrients.

A major pathway for organic matter entering marine food webs moves DOM through the **microbial loop** (fig. 15.16), which processes, on the average, half of oceanic primary production. In this system bacteria rapidly take up carbohydrates, amino acids, and other organic molecules and convert them to particulate organic matter; the organic matter is consumed by dinoflagellates, radiolarians, small zooplankton, and fish larvae. This is an energy-efficient transfer system that is of great importance to planktonic food webs.

Viruses 15.3

A **virus** is a noncellular particle made up of genetic material surrounded by a protein coat. Viruses are highly successful parasites; they infect plant, animal, and bacterial cells. Viruses are unable to carry on metabolic activities by themselves and can replicate themselves only inside a host cell.

Microbiologists have discovered an unexpected abundance of viruses in marine waters (10^6–10^9 viruses/mL). Most of the viruses appear to be free in the water, but some are associated with bacteria. Such high concentrations of viruses indicate that viral infection may be an important factor in the ecological control of planktonic bacteria. Estimates of total viral infection conclude that about 32% of the heterotrophic bacteria and about 15% of the cyanobacteria contain mature virus particles at any given time, indicating again that viral infection may be a significant mechanism of mortality in marine bacteria.

Laboratory experiments have shown that the addition of viral particles reduces primary productivity by as much as 78%. Because viruses can infect a variety of marine phytoplankton, including diatoms, they may affect ocean food webs by restricting or terminating phytoplankton blooms.

During replication within a host cell a virus may acquire some of the host's genes; it may then transport these genes to a new cell, leading to exchange of genetic material between organisms. The abundance of viruses in ocean waters indicates routine viral infection of aquatic bacteria, and this is likely to mean that genetic engineering in bacterial populations has been occurring naturally for a very long time.

Classification Summary of the Plankton 15.4

The types of plankton can be categorized in kingdoms, phyla, and classes, as in the following outline:

I. *Kingdom Monera*: cells, simple and unspecialized; single cells, without membrane-bounded nuclei, some in groups or chains.
   A. *Bacteria*: single cells, in chains or groups; autotrophic and heterotrophic, aerobic and anaerobic; important as food source and in decomposition.
   B. *Cyanobacteria*: blue-green algae; autotrophic single cells, in chains or groups, produce some red blooms in sea; phytoplankton.

II. *Kingdom Protista:* grouping of microscopic and mostly single-celled organisms; autotrophs (algae) and heterotrophs (protozoa).
   A. *Phylum Chrysophyta:* golden-brown algae; yellow to golden autotrophic single cells, in groups or chains; contribute to deep-sea sediments; phytoplankton.
      1. *Class Bacillariophyceae:* diatoms.
      2. *Class Chrysophyceae:* coccolithophores, silicoflagellates, and other flagellates.
   B. *Phylum Dinoflagellata:* fire algae; single cells with bioluminescence common, dinoflagellates; usually considered phytoplankton.
   C. *Phylum Sarcodina:* microscopic heterotrophic, moving with pseudopodia; radiolarians and foraminiferans.
   D. *Phylum Ciliophora:* microscopic heterotrophs, moving with cilia; ciliates; zooplankton.
III. *Kingdom Plantae:* plants; primarily nonmotile, multicellular, photosynthetic autotrophs.
   A. *Division Phaeophyta:* brown algae; *Sargassum* maintains a planktonic habit in the Sargasso Sea.
IV. *Kingdom Animalia:* animals; multicellular heterotrophs with specialized cells, tissues, and organ systems; zooplankton (holoplankton). For temporary members of the zooplankton (or meroplankton), see the Meroplankton listed in V.
   A. *Phylum Coelenterata,* or *Cnidaria:* radially symmetrical with tentacles and stinging cells.
      1. *Class Hydrozoa:* jellyfish (sea jellies) as one stage in the life cycle, including such colonial forms as the Portuguese man-of-war.
      2. *Class Scyphozoa:* jellyfish (sea jellies).
   B. *Phylum Ctenophore:* comb jellies; translucent; move with cilia; often bioluminescent.
   C. *Phylum Chaetognatha:* arrowworms; free-swimming, carnivorous worms.
   D. *Phylum Mollusca:* mollusks; the snaillike pteropod is planktonic.
   E. *Phylum Arthropoda:* animals with paired, jointed appendages and hard outer skeletons.
      1. *Class Crustacea:* copepods and euphausiids.
   F. *Phylum Chordata:* animals, including vertebrates, with dorsal nerve cord and gill slits at some stage in development.
      1. *Subphylum Urochordata:* saclike adults with "tadpole" larvae; salps.
V. *Meroplankton:* larval forms from the phyla Annelida (segmented worms), Mollusca (shellfish and snails), Arthropoda (crabs and barnacles), Echinodermata (starfish and sea urchins), and Chordata (fish). See also the classification summaries of the nekton (chapter 16) and the benthos (chapter 17).

Sampling the Plankton 15.5

The biological oceanographer needs to know what species of plants and animals make up the plankton in a given geo-

Figure 15.17 After a plankton tow, the net is rinsed to wash the plankton down into the sample cup at the narrow end of the net.

graphic area, their abundance, and where in the water column they are located. Traditionally, plankton are sampled by towing fine-mesh, cone-shaped nets (fig. 15.17) through the sea behind a vessel or by dropping a net straight down over the side of a nonmoving vessel and pulling it up like a bucket. After the net is returned to the deck it is rinsed carefully, and the catch is collected. The total water volume sampled is measured by placing a flow meter in the mouth of the net. Multiplying water flow times the cross-sectional area of the net yields the sampled volume. It is also possible to raise and lower the net as it is being towed horizontally. This motion allows sampling to be averaged over both horizontal distance and depth. If the sample is to be taken at a specific depth, the net may be lowered closed and opened only when the desired depth is reached. After the towing operation, the net is again closed before it is brought up through the shallower water.

Multiple-net systems may be mounted on a single frame; the nets are opened and closed on command from the ship. The frame also carries electronic sensors that relay data on salinity, temperature, water flow, light level, net depth, and cable angle to the ship's computer.

Plankton tows must be rapid enough to catch the organisms but slow enough to let the water pass through the net. If the tow is too fast, the water is pushed away from the mouth of the net, and less water than expected is filtered.

Extremophiles

Extremophiles are microorganisms that thrive under conditions that would be fatal to other life forms: extreme temperatures (hot and cold), high levels of acid or salt, no oxygen, no sunlight. Extremophiles not only flourish under these severe conditions but may actually require them to reproduce. Although some extremophiles have been known for forty years or more, scientists have been discovering more and more of these organisms in environments that were once thought lifeless.

These single-celled microorganisms resemble bacteria, for they have no membrane-bounded nucleus. However, when their genes were compared to the genes of bacteria, it was discovered that they are distinctly different. In fact, they appear to share a common ancestor with the eukaryotes, organisms such as ourselves with a membrane-bounded nucleus and cellular functional units, or organelles. This discovery opened for review the basic categories of all living organisms, and the result is a reorganization of life categories into three major domains: Bacteria, Archaea (the extremophiles), and Eukarya (all nuclei-containing organisms) (see fig. 13.4). This discovery was made by Professor Carl Woese at the University of Illinois; it has completely revised the way we think about microbial biology and the organization of life on Earth.

In the oceans, heat-loving members of the Archaea require temperatures in excess of 80°C (176°F) for maximum growth. *Pyrolobus fumarii* is an extreme example. It was found at a depth of 3650 m (12,000 ft) in a hot vent in the mid-Atlantic Ridge southwest of the Azores. Its name means "fire lobe of the chimney" from its shape and the black-smoker vent where it was found. *P. fumarii* stops growing below 90°C (194°F) and reproduces at temperatures up to 113°C (235°F). It uses hydrogen and sulfur compounds as sources of energy and can also use nitrogen gas. It is able to live with or without oxygen. Other Archaea are commonly found in the plumes of hot water that occur after an undersea eruption; it is unclear how deep into the Earth's crust these microorganisms are able to exist (box fig. 1).

Another surprise was finding that close relatives of the hot-vent Archaea are common and abundant components of the marine plankton, living in the cold, oxygenated waters off both coasts of North America. They have since been found at all latitudes in water below 100 m (330 ft), in the guts of deep-sea cucumbers, as well as in marine sediments. A microorganism found in Antarctic sea ice grows best at 4°C (39°F) and does not reproduce at temperatures above 12°C (54°F). And still other extremophiles have been found living in the salt ponds constructed for the evaporation of seawater.

Scientists in the United States, Japan, Germany, and other countries have a particular interest in the enzymes of extremophiles. Enzymes are required in all living cells to speed up chemical reactions without being altered themselves. Standard enzymes stop working when they are exposed to heat or other extremes, so those used in research and industrial processes must be protected during reactions and while in storage. Heat-loving extremophile enzymes are already being used in biological and genetic research, forensic DNA testing, and medical diagnosis and screening procedures for genetic susceptibility to some diseases. In industry, these enzymes have increased the efficiency of compounds that stabilize food flavorings and reduce unpleasant odors in medicines. Enzymes that work at low temperatures may be useful in food processing where

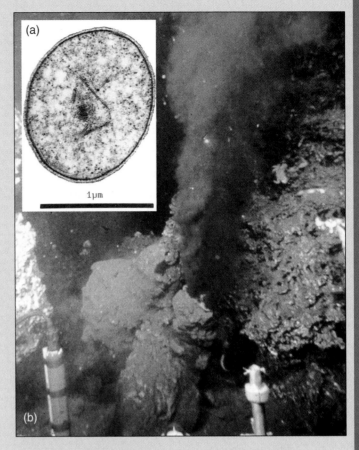

web link **Box Figure 1** The extremophile *Pyrococcus endeavorii* (a) was isolated from an East Pacific Rise black smoker (b). This microorganism grows at temperatures that exceed 100°C (212°F).

products must be kept cold to prevent spoilage. An important result of research with enzymes from extremophiles is learning how to redesign conventional enzymes to work under harsher conditions.

The discovery of microorganisms where none were assumed to exist and the recognition of a new branch of life are astonishing. It is likely that more discoveries and new technologies based on such discoveries will continue for some time to come.

To Learn More About Extremophiles

DeLong, E. 1998. Archaeal Means and Extremes. *Science* 280 (5363): 542–43.

Madigan, M., and G. Marrs. 1997. Extremophiles. *Scientific American* 276 (4): 82–87.

Pennisi, E. 1997. In Industry, Extremophiles Begin to Make Their Mark. *Science* 276 (5313): 705–6.

Internet References

Visit the book's Online Learning Center at www.mhhe.com/sverdrup (click on the book's cover) to explore links to further information on related topics.

The size of the mesh from which the net is made also influences the catch; a fine-mesh net clogs rapidly and a large-mesh net loses organisms. Plankton may also be filtered from samples taken by a water bottle or a submersible pump.

Whether the sample is taken by net, bottle, or pump, the next step is to determine the number and kinds of plankton in the sample. Because the numbers of organisms are so large, in most cases it is not possible to directly inspect or count the total catch. Instead the catch may be precisely subdivided, and a subsample may be checked under the microscope. If the amount of plankton is of greater interest than the kinds of organisms, an electronic particle counter may be used to find the total count (population density), or a subsample may be dried and weighed to determine its mass.

Sonar may be used to determine the quantity of zooplankton directly; the echo returned to the ship is related to the density of the zooplankton but does not determine the species present. The abundance of phytoplankton can also be determined by dissolving the chlorophyll pigment out of the sample and measuring the pigment concentration. New optical instrumentation has been designed to measure the natural fluorescent signal coming from chlorophyll pigment in phytoplankton cells. These measurements are made at sea, directly and instantaneously at depth, and then are related to phytoplankton abundance.

None of these methods tells us how the various plankton species behave. A new video plankton recorder designed at the Woods Hole Oceanographic Institution uses a strobe light with four video cameras at four different magnifications to photograph the plankton. The strobe flashes sixty times a second, capturing images of the organisms that researchers hope will enable them to learn about the swimming, feeding, and reproduction of the plankton. Eventually they hope to program the system to recognize different kinds of plankton, enabling them to acquire species and population information while the system is running at sea.

Practical Considerations: Marine Toxins 15.6

The number of toxic algal blooms reported in the world's coastal waters, the kinds of toxins present, and the species producing the toxins have increased dramatically in the last twenty to thirty years. Economic losses from toxins have surged, fishing, fish-farming, tourism, and recreation have all felt the impact of these blooms. Concerns over possible epidemics and the costs of testing programs to detect algal toxins have risen. The estimate of the overall impact of these organisms on the U.S. economy is over $40 million per year.

Harmful Algal Blooms

Harmful algal blooms (HABs) of single-celled organisms that discolor seawater are commonly known as **red tides.** HAB has become the preferred scientific term because these outbreaks have nothing to do with the tides, and they may or may not color the water red. Also, some species of algae

web link **Figure 15.18** This nontoxic red tide of the dinoflagellate *Noctiluca* extended for 10 km (6 mi) along the shore of Puget Sound, Washington, in 1996. The bloom persisted for a week.

may bloom and color the water but are not harmful to humans and other organisms (fig. 15.18). The Red Sea received its name because of dense blooms of nontoxic cyanobacteria that have large amounts of red pigment. The Gulf of California has been called the Vermillion Sea for the same reason.

About 6% of phytoplankton species are known to cause HABs, and most are produced by certain species of dinoflagellates (fig. 15.19). These blooms may or may not be poisonous to fish and other organisms and may or may not produce symptoms of paralytic shellfish poisoning (PSP), neurotoxic shellfish poisoning (NSP), diarrhetic shellfish poisoning (DSP), or amnesiac shellfish poisoning (ASP), also known as domoic acid poisoning (DAP). Humans are usually affected when they eat clams, mussels, or oysters that have ingested the toxic dinoflagellates.

In North American waters red tides are produced by several dinoflagellates (see fig. 15.5). *Gonyaulax, Alexandrium,* and *Gymnodinium* are toxic, but *Noctiluca* is not (see fig. 15.18). Species of *Gonyaulax* and *Alexandrium* found in temperate latitudes are generally nontoxic to the shellfish themselves, but the toxins produced by these dinoflagellates are concentrated in the tissues of the shellfish feeding on the bloom. In the warm waters of the Gulf of Mexico *Gymnodinium* produces spectacular fish kills, and *Gonyaulax* kills shrimp and crab as well.

The northeastern and western coasts of the United States are subject to outbreaks of paralytic shellfish poisoning (PSP) caused by the dinoflagellate *Alexandrium.* Cysts formed during the life cycle of *Alexandrium* are capsules with reduced metabolic activity that allow the organism to remain dormant in the bottom sediments through temperature extremes. The population increases only during periods of light and temperature favorable to growth. Humpback whales have been poisoned by feeding on mackerel that had

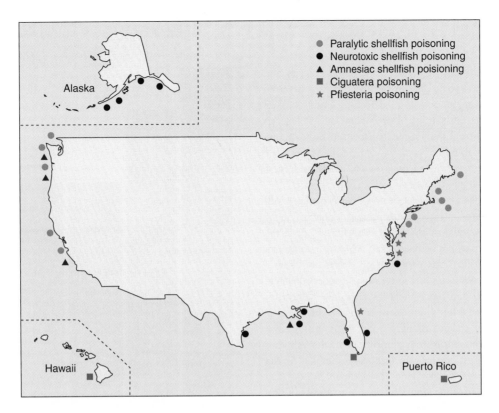

Legend:
- Paralytic shellfish poisoning
- Neurotoxic shellfish poisoning
- Amnesiac shellfish poisoning
- Ciguatera poisoning
- Pfiesteria poisoning

Alaska
Hawaii
Puerto Rico

web link **Figure 15.19** HAB events have increased in number and scale in the last twenty-five years. Also, more toxic algae species and more algal toxins have been identified.

Table 15.1 Phytoplankton Toxins

Toxin	Condition Produced	Phytoplankton Responsible	Characteristics
Brevetoxin	Neurotoxic shellfish poisoning (NSP)	*Gymnodinium breve*; dinoflagellate	Affects nervous system; respiratory failure in fish and marine mammals; food poisoning symptoms in humans
Ciguatoxin	Ciguatera fish poisoning	*Gambierdiscus toxicus*; dinoflagellate	Affects nervous system; human symptoms variable
Domoic acid	Domoic acid poisoning (DAP) or amnesic shellfish poisoning (ASP)	*Pseudo-nitzschia*; diatom	Acts on vertebrate nervous system
Okadaic acid	Diarrhetic shellfish poisoning (DSP)	*Dinophysis* and *Prorocentrum* dinoflagellates	Affects metabolism, membrane transport, cell division
Exotoxins		*Pfiesteria piscicida*; dinoflagellate	Mode of action unknown; produces mortality in fish, neurotoxic symptoms in humans
Saxitoxin	Paralytic shellfish poisoning (PSP)	*Alexandrium, Gonyaulax,* and *Gymnodinium*; dinoflagellates	Causes paralysis and respiratory failure in humans

web link *After J.T. Turner and P.A. Tester, Toxic Marine Phytoplankton, Zooplankton, Grazers and Pelagic Food Webs in Limnology and Oceanography 42 (5, part 2): 1203–14, 1997.*

fed on *Alexandrium*. Toxic events have also occurred frequently along the West Coast. The earliest reported cases of PSP in North America occurred in 1793 among Captain George Vancouver's crew; five crew members became ill and one died after eating mussels gathered along the British Columbia coast. At that time, Pacific Coast Indians were aware of toxic shellfish and their association with red tides.

A resident population of *Gymnodinium breve* is found in the Gulf of Mexico. The Florida Current and the Gulf Stream transport cells out of the Gulf of Mexico and into Atlantic waters. Eddy systems and Gulf Stream meanders bring the cells over the continental shelf. More than 150 Florida manatees died in 1996 due to toxins produced by *G. breve*. Reports of discolored water in the tropical Atlantic are found in sixteenth-century log books, and discolored water was linked to fish kills in 1844 reports.

The toxins associated with HABs (table 15.1) are powerful nerve poisons that can cause paralysis and death if

Rapid Detection of Algal Blooms

Algal blooms can pose serious health threats to human beings and marine organisms because some species of phytoplankton produce dangerous toxins. It is very important to detect the onset of algal blooms and to determine the species of phytoplankton involved as rapidly as possible so that, if the species is known to be toxic, the appropriate authorities can be notified. The rapid detection of toxic blooms can give authorities time to issue health alerts and close beaches and fisheries to minimize illness and possibly death.

Traditional methods of monitoring harmful algal bloom (HAB) species have required time-intensive sampling and identification by detailed microscopic analyses, as discussed in section 15.5. In the worst case, toxic blooms may not be recognized until their harmful impact on human or marine life is apparent in reported illness or death.

Scientists at the Monterey Bay Aquarium Research Institute (MBARI) have developed and successfully tested a new method for the rapid detection and identification of a variety of *Pseudo-nitzschia* phytoplankton species as they occur in natural assemblages in the water. The analysis is done both in the lab and aboard ship in real-time at present. An instrument that will do the complete analysis *in situ* is currently under development.

Pseudo-nitzschia species regularly occur in Monterey Bay and are often responsible for blooms along the coast (box fig. 1). Some of these species are known to be toxic; they produce domoic acid, which is responsible for amnesic shellfish poisoning (ASP). Other species are currently presumed to be nontoxic. It is important to note, however, that the specific environmental and genetic factors responsible for domoic acid production are not known. Variables including growth rate, light intensity, water temperature, and nutrient levels may all have a significant effect on the amount of toxin produced by an organism. Consequently, laboratory studies may not accurately predict a particular species' ability to produce toxins in the real world. Until a method for distinguishing species easily and accurately is found, it is safest to consider all species of *Pseudo-nitzschia* to be toxic.

The advantage of the new MBARI method is that identification and quantification of HAB species can be made with simple, sensitive, and rapid diagnostic tests in about one to two hours. These tests can provide early warning of the onset of HABs and the species of phytoplankton responsible. Traditional methods of identification typically take over a day, and the time required increases quickly with the number of samples analyzed.

Box Figure 1 Examples of six species of *Pseudo-nitzschia* commonly found in Monterey Bay, California. Included are the potential domoic acid producers *P. australis*, *P. pseudodelicatissima*, and *P. multiseries* as well as the presumed nontoxic *P. pungens*, *P. fraudulenta*, and *P. subpacifica*. Physical differences between species are subtle and difficult to detect without high-magnification electron micrographs.

Each species has its own unique genetic code. This code is contained in its nucleic acids (DNA and RNA). The process developed at MBARI uses new molecular-probe technologies to identify, and in some cases quantify, different species of plankton by analyzing the DNA or RNA in a sample. Molecular probes use synthetic molecules specially constructed to attach themselves to, or to pair

the breathing centers are affected. They are not single compounds but families of compounds with similar chemical structures. The saxitoxins that are produced by *Alexandrium* and *Gonyaulax* are fifty times more lethal than strychnine. They are not affected by heat, so cooking the shellfish does not neutralize the poison. Even after the colored water from a dinoflagellate bloom has disappeared, the shellfish can retain the toxin in their tissues for long periods, so the beaches must be kept closed to shellfish harvesting. Neurotoxic

shellfish poisoning (NSP) symptoms are similar to those of basic food poisoning; in addition wind-spread cells from dead organisms may cause irritated eyes, running noses, and persistent cough. In the past diarrhetic shellfish poisoning (DSP) may have been attributed to bacterial contamination rather than dinoflagellates.

Evidence shows that HABs are increasing. An initial outbreak of *Alexandrium* occurred along the New England coast in 1972, and it has persisted ever since. Outbreaks of

with, specific, or unique, complementary target sequences of DNA or RNA. The synthetic probes are relatively small molecules containing fifteen to fifty nucleotides; the DNA or RNA target molecules may consist of hundreds or thousands of nucleotides.

A probe must be "labeled" in order to measure its success in pairing with target molecules. Labels may include radioisotopes, fluorescent molecules, and enzymes that produce color changes or chemiluminescent reactions. Nonradioactive labels are the most desirable because they provide more flexibility in field and laboratory use. Thus, a successful pairing between probe and target molecules can be detected by the fluorescence of the species of phytoplankton being targeted while other species in the sample fail to react (box fig. 2). The creation of a species-specific probe requires (1) the identification of a nucleic acid sequence that is a unique signature of the organism and (2) the synthesis of a probe that recognizes the signature sequence. At this time, relatively little is known about the molecular sequence for HAB species as a whole, and much work is still to be done in enlarging the signature database for these organisms.

In the past decade, the hazards associated with domoic acid outbreaks have grown from localized events in eastern Canada to incidents along both coasts of North America, with the potential of spreading in many areas around the world. In May 1998 a number of sea lions became ill in Monterey Bay, suffering from seizures. At least 100, and probably more, died. Fourteen others survived after treatment at the Marine Mammal Center in Sausalito, California, but veterinarians there believe they may have suffered permanent brain damage in the form of short-term memory loss. The incident was the result of a toxic algal bloom in the bay that had been detected by scientists at MBARI a week before it hit its peak. The rapid detection of the bloom, and identification of the toxic species of phytoplankton responsible for it, allowed veterinarians to determine that the illness was domoic acid poisoning. The speed of the identification of the HAB was critical in the diagnosis of the sea lions' illness because the toxin stays in the bloodstream only for about four hours after it is ingested. Typical methods of identifying HABs would likely have taken so long that there would have been no chance to obtain blood samples from the animals while they were vomiting and having seizures, before the toxin was removed from the bloodstream. High levels of domoic acid were also found in anchovies and sardines, both common food sources for sea lions. Some shellfish were probably also contaminated, but they must have flushed the toxin out of their systems relatively quickly and no humans are known to have become ill. A similar toxic bloom was responsible for killing cormorants and pelicans in the area in 1991. Sport harvesting of shellfish is banned along the California coast each year from May through October because of the danger of toxic blooms.

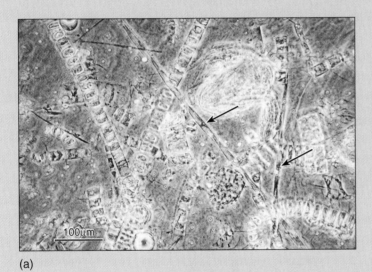

(a)   (b)

Box Figure 2 (a) Transmitted light micrograph of a mixed assemblage of phytoplankton. *Arrows* identify two chains of *Pseudo-nitzschia*. This plankton sample has been treated with a *P. australis* (a species of *Pseudo-nitzschia*) species-specific fluorescing probe. (b) Epifluorescent micrograph of the assemblage in which the fluorescence of the single chain identifies it as *P. australis*, while there is no label response on the other, nontarget chain.

G. *breve* have appeared along the Texas Gulf coast repeatedly during the last twenty-five years. In 1987–88 a bloom of G. *breve* occurred about the same time that menhaden fish began their annual summer migration northward; the menhaden fed on the G. *breve*, and the dolphins fed on the menhaden. More than 700 dolphins died along the Atlantic coast between Florida and New Jersey.

Brown tides have been reported along the Texas shore of the Gulf of Mexico and also along the shores of Rhode Island, New Jersey, and Long Island, New York. The organisms responsible are minute members of the Chrysophyta or golden-brown algae (see section 15.4): *Aureococcus anophagefferens* and *Aueoumbra lagunensis*.

HABs appear to be stimulated by the continuous addition of nutrients to coastal waters from sewage, agriculture, and aquaculture ponds. Extra nutrients mean extra growth, and the experience of the Japanese in the Inland Sea of Japan suggests that this may may be at least part of the problem. Red tide events in the Inland Sea increased from 44 in 1965 to more than 300 in 1973. Authorities introduced rigorous controls on industrial and sewage effluents, and the frequency of the red tides has been reduced by 50%. Also today's ships move algae from one place to another in their ballast water (see chapter 12, section 12.7), and young fish stock being transported to aquaculture pens may introduce HABs from other areas.

Pfiesteria-like Dinoflagellates

The dinoflagellate *Pfiesteria piscicida*, popularly known as the "cell from hell," was first identified in North Carolina in 1993. Since then, *P. piscicida* and other *Pfiesteria*-like organisms have caused major fish kills in estuaries and coastal waters of the mid-Atlantic and southeastern United States. Cells may undergo at least twenty-four transformations among amoeboid and flagellated swimming forms that stun and poison fish. They also form cysts on the bottom in an extraordinarily complex life history that is controlled by the availability of fish secretions, blood, and tissue. *P. piscicida* is a voracious predator on other marine microorganisms and is linked to human symptoms that include sores, severe memory loss, nausea, respiratory distress, and vertigo.

Ciguatera Poisoning

Ciguatera poisoning was first documented in the early sixteenth century, but it has probably been present in the tropics for much longer. It is estimated that each year between 10,000 and 50,000 people eating fish in tropical regions are affected by Ciguatera poisoning. More than 400 species of fish have been found to be affected, and there is no way to prepare affected fish to make it safe to eat. In the United States, there may be over 2000 cases of Ciguatera poisoning each year, clustered in Florida, Hawaii, the Virgin Islands, and Puerto Rico. Symptoms of Ciguatera poisoning are extremely variable; they may include headache, nausea, irregular pulse beat, reduced blood pressure, and, in severe cases convulsions, muscular paralysis, hallucinations, and death.

Several dinoflagellates are associated with Ciguatera poisoning but the dinoflagellate *Gambierdiscus toxicus* is most often associated with the problem. This dinoflagellate was first isolated in 1976 by scientists in Japan and French Polynesia. A toxin isolated at the University of Hawaii was named ciguatoxin; more toxic compounds have since been identified. Ciguatoxic dinoflagellates live attached to many types of seaweeds and appear to need nutrients exuded by the seaweeds. The seaweeds are eaten by herbivorous fish, and the ciguatoxins move through the food web stored in the fishes' liver. Initially only a few species are toxic, while at the peak of the outbreak almost all reef fish become toxic and in the final stages only large eels and certain snappers and groupers remain toxic. This cycle appears to take at least eight years and may last for as long as thirty years. Unfortunately there is, at present, no easy way to routinely measure ciguatoxin levels in any seafood. Outbreaks of Ciguatera seem to be preceded by a disruption of the reef system triggered by both natural phenomena such as monsoons and seismic events and also by human activities such as construction of piers and jetties. However, not all reef disruptions are followed by an increase in fish toxicity.

Toxic Diatom Blooms

In 1987, at Prince Edwards Island, Canada, three people died and more than a hundred others became sick from eating shellfish contaminated with domoic acid. Improved chemical detection allowed scientists to trace the toxin to a bloom of *Pseudo-nitzschia*, a diatom, not a dinoflagellate. Diatoms had not been known to produce toxins until this outbreak occurred. Domoic acid poisoning (DAP), also called amnesiac shellfish poisoning (ASP), may cause short-term memory loss in humans. Other symptoms include nausea, muscle weakness, disorientation, and organ failure. A second recorded case of DAP occurred in September 1991 in California's Monterey Bay, where more than 100 brown pelicans and cormorants died or showed unusual neurological symptoms after eating anchovies that had been feeding on *Pseudo-nitzschia*. That same year and the next razor clam and crab fisheries along the Oregon and Washington coasts were closed because of high levels of domoic acid. In 1998 more than 100 sea lions died off the California coast from feeding on anchovies and sardines that had fed on the diatoms. The reasons for the sudden appearances of toxic levels of dominic acid are unknown; for more information see the box titled "Rapid Detection of Algal Blooms."

Another diatom, *Chaetoceros*, has been a cause of mortality on pen-reared salmon since 1961. *Chaetoceros* cells are not toxic but have long, sharp spines that irritate the gills of the fish. This stimulates the production of excess mucus and leads to suffocation.

Cholera

Nutrient-rich coastal waters and their plankton blooms are associated with cholera epidemics. These areas are a reservoir for new environmentally hardy variants of the cholera bacterium (*Vibrio cholerae*) that adhere to members of the phytoplankton and zooplankton. In 1991 an epidemic of a new cholera variant broke out along the Peruvian coast; over a fifteen-month period more than half a million people became ill and nearly 5000 people died in nineteen Latin American countries. *V. cholerae* was isolated from phytoplankton in the upwelling zone along the Peruvian coast. The disease is thought to have crossed the Pacific in the bilge water of ships from Bangladesh (see chapter 12, section 12.7). Since 1960 seasonal outbreaks of cholera in Bangladesh have been related to costal phytoplankton blooms; the bacterium is found associated with cyanobacteria, diatoms, dinoflagellates, seaweeds, and most heavily with zooplankton egg sacs.

Summary

The plankton are the drifting organisms. The microscopic plankton are divided into groups by size. Phytoplankton are autotrophic single cells or filaments. *Sargassum* is the only large planktonic seaweed. The diatoms are found in cold, upwelled water; they are yellow-brown, with a hard, transparent frustule, and they store oil, which increases their buoyancy. Centric forms are round; pennate forms are elongate. Diatoms reproduce rapidly by cell division and make up the first trophic level of the open sea.

Dinoflagellates are single cells with both autotrophic and heterotrophic capabilities. Their cell walls are smooth or are heavily armored with cellulose plates. They, too, reproduce by cell division. These organisms are responsible for much of the bioluminescence in the oceans. Coccolithophores and silicoflagellates are very small autotrophic members of the phytoplankton.

Some herbivorous zooplankton reproduce several times a year; others reproduce only once, depending on the water temperature and phytoplankton food supply. Carnivorous zooplankton are important in the recycling of nutrients to the phytoplankton. Heavy concentrations of zooplankton are found at convergence zones and along density boundaries. Zooplankton migrate toward the sea surface at night and away from it during the day, forming the deep scattering layer.

Zooplankton that spend their entire lives in the plankton are called holoplankton. The copepods and euphausiids are the most abundant members of the holoplankton. Euphausiids are also known as krill; they form a basic food of the baleen whales. Krill is the zooplankton base for all the Antarctic food webs; recently it has been harvested for human consumption with mixed success.

Other small members of the holoplankton are the carnivorous arrowworms; the calcareous-shelled foraminiferans; the delicate, silica-shelled radiolarians; the ciliated tintinnids; and the swimming snails, or pteropods. Large zooplankton include the comb jellies, salps, and jellyfish; all are nearly transparent, but each belongs to a different zoologic group.

The meroplankton are the juvenile (or larval) stages of nonplanktonic adults. This group comprises fish eggs, very young fish, and the larvae of barnacles, snails, crabs, starfish, and many other nonplanktonic animals. The spores of seaweeds and the marine bacteria are also planktonic.

Great numbers of bacteria live free in seawater and coat every available surface. They are important as food sources and agents of decay and breakdown of organic matter. Dissolved organic material (DOM) moves through the microbial loop and is converted into particulate organic matter that is consumed by phytoplankton and zooplankton. Viruses are noncellular, parasitic particles that are abundant in seawater. Viruses can infect bacteria and other marine organisms.

Plankton sampling is usually done with a plankton net or with a water bottle. The kinds of organisms in a sample are determined microscopically; the numbers of organisms are counted, or samples are dried and weighed.

Heavy blooms of dinoflagellates and some other kinds of phytoplankton produce red tides. Some red tides are toxic; others are not. The toxin is concentrated in shellfish and produces paralytic and other types of shellfish poisoning in humans and sometimes in other animals. Red tides appear to be triggered by the particular combination of environmental factors, including the addition of increasing amounts of nutrients to coastal waters. *Pfiesteria*-like dinoflagellates have complex life cycles, cause fish kills, and affect humans. Domoic acid is produced by diatoms. Its presence causes shellfish-harvest closures and has affected birds as well as humans. Ciguatoxin is another dinoflagellate product that affects humans and hampers fishery development around the world. Warm, nutrient-rich coastal waters have been associated with outbreaks of new cholera variants.

Key Terms

All key terms from this chapter can be viewed by term, or by definition, when studied as flashcards on this book's Online Learning Center at www.mhhe.com/sverdrup (click on this book's cover).

ultraplankton, 393	chaetognath, 399
nannoplankton, 393	foraminiferan, 399
microplankton/net	radiolarian, 399
plankton, 393	test, 400
autotrophic, 393	tintinnid, 400
filament, 393	cilia, 400
diatom, 393	pteropod, 400
dinoflagellate, 393	ctenophore, 400
coccolithophore, 393	salp, 401
radial symmetry, 393	Coelenterata/Cnidaria, 400
bilateral symmetry, 393	colonial organism, 402
frustule, 393	holoplankton, 402
auxospore, 395	meroplankton, 402
bloom, 395	larva, 402
diatomaceous earth, 396	spore, 403
heterotrophic, 396	cyanobacteria, 404
flagella, 396	microbial loop, 405
coccolith, 397	virus, 405
crustacean, 397	extremophile, 407
copepod, 398	harmful algal bloom
euphausiid, 398	(HAB), 408
krill, 398	red tide, 408
baleen whale, 398	Ciguatera, 412

Study Questions

1. Why does a pycnocline located above the compensation depth promote a phytoplankton bloom?
2. Why are meroplankton produced in such large numbers?
3. When the discoloration has left the water after a toxic red tide, the shellfish may not be safe to eat. Explain why.
4. Patches with abundant populations of zooplankton are frequently found separated by patches with sparse populations. How does this pattern help to ensure survival from predators?
5. If the krill of the Southern Ocean were heavily harvested for human consumption, explain the possible effects on the rest of the organisms in that area.

6. Describe four ways to subdivide the plankton.
7. Why are centric diatoms more abundant in the plankton and pennate diatoms more abundant in the benthos?
8. Discuss what happens to a diatom population if no auxospores form.
9. Why is a planktonic stage important to a nonplanktonic adult?
10. Discuss how plankton may maintain themselves in a given region of the ocean even though there are currents flowing through that region.
11. When you are sampling plankton with a plankton net, how can you determine the quantity of the plankton in a volume of water? Assume that you know (a) the cross-sectional area of the net and (b) the length of time you towed the net and the speed at which you towed the net, or you know (a) the cross-sectional area of the net and (b) the distance you towed the net.
12. Distinguish between diatoms and dinoflagellates, between euphausiids and copepods. Which are heterotrophs and which are autotrophs?
13. What is the microbial loop, and why is it important?
14. Episodes of toxic phytoplankton blooms appear to be increasing along the world's coasts. (a) List possible reasons for this increase. (b) Distinguish among PSP, NSP, ASP, and Ciguatera.

15. Toxic materials, such as oil, may form a thin layer at the sea surface where many plankton also accumulate. How might this affect populations of the nekton and the benthos?

Links to Related Websites

Visit the book's Online Learning Center at www.mhhe.com/sverdrup (click on the book's cover) to find live Internet links for additional topics related to this chapter's content.

- The plankton
- Red tides and toxic blooms

Visit the book's Online Learning Center at www.mhhe.com/sverdrup (click on this book's cover) to find these additional chapter tools: Suggested Readings; links to further information on boxed readings, selected figures, and related chapter topics; and additional study aids.

chapter 16

The Nekton: Free Swimmers of the Sea

W e need another and a wiser and perhaps a more mystical concept of animals. Remote from universal nature, and living by complicated artifice, man in civilization surveys the creature through the glass of his knowledge and sees thereby a feather magnified and the whole image in distortion. We patronize them for their incompleteness, for their tragic fate of having taken form so far below ourselves. And therein we err, and greatly err. For the animal shall not be measured by man. In a world older and more complete than ours they move finished and complete, gifted with extensions of the senses we have lost or never attained, living by voices we shall never hear. They are not brethren, they are not underlings; they are other nations, caught with ourselves in the net of life and time, fellow prisoners of the splendour and travail of the earth.

Henry Beston
From *The Outermost House*

Steller sea lions on a buoy outside Valdez, Alaska.

The nekton are the free swimmers of the oceans, moving through the water independent of the motion of currents and waves. Approximately 5000 species of nekton swim freely through the pelagic and neritic regions of the oceans. Most of the nekton are fishes; others are marine mammals, ocean-living reptiles, and squid. Members of the nekton are able to move toward their food and away from their predators; many occupy the top trophic levels of the marine food webs, as either herbivores or carnivores. Sizes range from the smallest fish of the tropical reefs to the largest animal ever to have existed on Earth, the blue whale. This chapter examines representative swimmers in coastal waters and in the open sea; it contains additional information on their harvests, including quantities taken, methods used, and current population status.

Mammals 16.1

Marine **mammals** are endotherms and air-breathers. They may spend all of their lives at sea, or they may return to land to mate and give birth. In either case, the young are born live and are nursed by their mothers. Included in this group are large and small whales (including porpoises and dolphins), seals, sea lions, walruses, sea otters, and sea cows.

Whales and Whaling

Whales belong to the mammal group called **cetaceans**; see table 16.1 for information about representative whales. Some cetaceans are toothed, pursuing and catching their prey with their teeth and jaws (for example, the killer whale, the sperm whale, and the small whales known as dolphins and porpoises); others have mouths fitted with strainers of **baleen,** or whalebone, through which they filter the seawater and remove the krill. The blue, finback, right, sei, gray, and humpback whales are baleen whales. The mouths of toothed and baleen whales are compared in figure 16.1. The blue, finback, and right whales swim open-mouthed and engulf water and plankton. The tongue acts to push the water through the baleen, and the krill are trapped. The sei whale swims with its mouth partly open and uses its tongue to remove the organisms trapped in the baleen. The humpbacks circle an area rich in krill and expel air to form a circular screen, or a net of bubbles. The krill bunch together toward the center of this net, and the whales pass through the dense cloud of krill and scoop them up. The gray whale feeds mainly on small bottom crustaceans and worms.

Some whales migrate seasonally over thousands of miles; other whales stay in cold water and migrate over relatively short distances. The California gray whale and the humpback whale make long migratory journeys. In the summer, the California gray whale is found in the shallow waters of the Bering Sea and the adjacent Arctic Ocean, where they feed all summer, building up layers of fat and blubber. In October, when the northern seas begin to freeze, the whales begin to move south, and in December, the first gray whales arrive off the west coast of Baja California. Here they spend the winter in the warm, calm waters of sheltered lagoons, where the gray whales calve and mate but find little food; by the time they leave for their northward migration in February and March, they have lost 20%–30% of their body weight. The animals move north singly or in twos or threes, sometimes in groups of ten to twelve up the western coast of the United States, Canada, and Alaska. Moving at about 5 knots day and night, the whales make their annual 18,000 km (11,000 mi) migratory journey to link areas that provide abundant food with areas that ensure reproductive success (fig. 16.2a). Protection of the gray whales' calving and winter grounds in Baja California has allowed the population to build, and in 1994, the gray whale became the first marine mammal to be removed from the U.S. endangered species list.

The humpback whale also has well-defined migration patterns. Humpbacks are found in three geographically and reproductively isolated populations: in the North Pacific, North Atlantic, and Southern Ocean. The North Pacific humpback spends the summer feeding in the Gulf of Alaska, along the northern islands of Japan, and in the Bering Sea; in winter, these North Pacific humpbacks migrate to the Mariana Islands in the west Pacific, the Hawaiian Islands in the

Table 16.1 Principal Characteristics of the Great Whales

	Distribution	Breeding Grounds	Average Weight (tons)	Greatest Length (m)	Food
Toothed Whales					
Sperm	Worldwide; breeding herds in tropical and temperate regions	Oceanic	35	18	Squid, fish
Baleen Whales					
Blue	Worldwide; large north-south migrations	Oceanic	84	30	Krill
Finback	Worldwide; large north-south migrations	Oceanic	50	25	Krill and other plankton, fish
Humpback	Worldwide; large north-south migrations along coasts	Coastal	33	15	Krill, fish
Right	Worldwide; cool temperate	Coastal	50[1]	17	Copepods and other plankton
Sei	Worldwide; large north-south migrations	Oceanic	17	15	Copepods and other plankton, fish
Gray	North Pacific; large north-south migrations along coasts	Coastal	20	12	Benthic invertebrates
Bowhead	Arctic; close to edge of ice	Unknown	50[1]	18	Krill
Bryde's	Worldwide; tropical and warm temperate regions	Oceanic	17	15	Krill
Minke	Worldwide; north-south migrations	Oceanic	10	9	Krill

1. Estimate.

From K. R. Allen. 1980. Conservation and Management of Whales. *Washington Sea Grant Program. Reprinted by permission.*

(a)

(b)

Figure 16.1 (a) The killer whale (*Orinus orca*) is a toothed whale. (b) Bowhead whale (*Balaena mysticetus*) baleen. This dead whale was hauled onto the ice and is shown lying on its back.

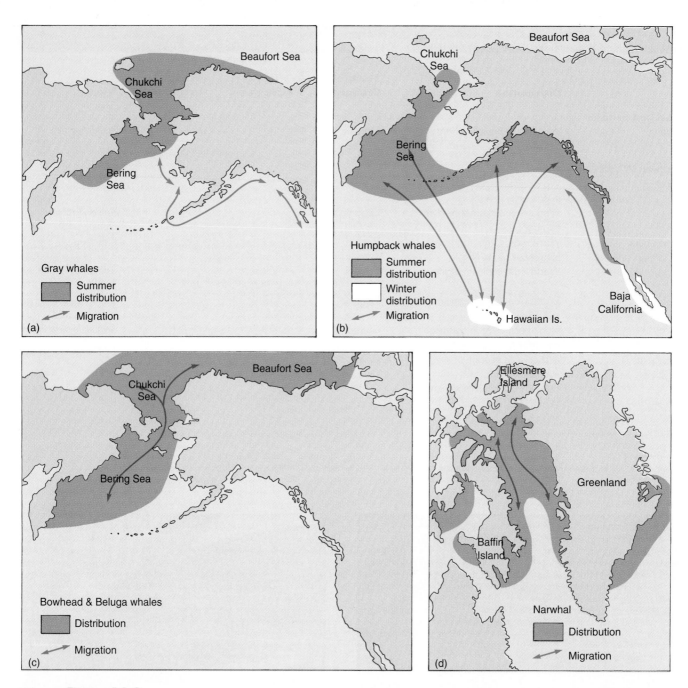

web link **Figure 16.2** Migration paths and seasonal distribution of whales. The California gray whale (a) and the humpback whale (b) travel long distances between cold-water feeding and warm-water calving and mating areas. The bowhead and beluga whales (c) and the narwhal (d) remain in cold water and migrate over short distances.

central Pacific, and along the west coast of Baja California in the eastern Pacific. At these warmer latitudes, calves are born and mating takes place (fig. 16.2b).

The bowhead, the beluga, and the single-tusked narwhal are whales that remain in cold water but still migrate over short distances each year. The largest population of bowhead whales is found in the Bering, Chukchi, and Beaufort Seas; this whale spends nearly all of its life near the edge of the Arctic ice pack. Bowheads, singly or in pairs, often accompanied by belugas, migrate north from the

Bering Sea to feed in the Beaufort and Chukchi Seas as the ice recedes in the spring, returning south to the Bering Sea in groups of up to fifty as the ice begins to extend in the winter. Their mating and reproductive cycles are not well known, but they probably mate during the spring migration and calve sometime during April and May of the following year (fig. 16.2c).

The narwhal is the most northerly whale and is found only in Arctic waters, most commonly on both sides of Greenland. In summer these whales move north along the

web link **Figure 16.3** Relative sizes of baleen and toothed whales.

coasts of Ellesmere and Baffin Islands, and in the autumn they return south to the waters along the Greenland coast (fig. 16.2d).

The whales of stories and songs are the "great whales": the blue, sperm, humpback, finback, sei, and right whales (fig. 16.3 and table 16.1); they are also the whales of the whaling industry. The earliest-known European whaling was done by the Norse between A.D. 800 and 1000. The Basque people of France and Spain hunted whales first in the Bay of Biscay, and then in the 1500s the Basque whalers crossed the

Mammals **419**

Atlantic to Labrador. They set up whaling stations along the Labrador coast to process the blubber of bowhead and right whales into oil for transport back across the Atlantic. In Red Bay, Labrador, the operation reached its peak in the 1560s and 1570s, when 1000 people gathered seasonally to hunt whales and produce 500,000 gallons of whale oil each year. By 1600, whaling had become a major commercial activity among the Dutch and the British, and at about the same time the Japanese independently began harvesting whales. In the 1700s and 1800s the whalers from the United States, Great Britain, the Scandinavian countries, and the other northern European countries pursued the whales far from shore, hunting them for their oil and baleen. Kills were made using handheld harpoons, and the whales were cut up and processed on land or onboard the ships at sea. Long voyages, intense effort, and dangerous combat between whalers and whales characterized these whale hunts.

In 1868, Svend Føyn, a Norwegian, invented the harpoon gun with its explosive harpoon and changed the character of whaling. Ships were motorized, and in 1925, harvesting was increased further by the addition of great factory ships, to which the small, high-speed whale-hunting vessels brought the dead whales for processing. This system freed the fleets, now centered in the Antarctic, from dependence on shore stations. These methods continued into the twentieth century, greatly increasing the efficiency of the hunt and rapidly depleting the whale stocks of the world.

In the 1930s the annual blue whale harvest reduced the population to less than 4% of its original numbers, threatening the species with extinction. In 1946 representatives from Australia, Argentina, Britain, Canada, Denmark, France, Iceland, Japan, Mexico, New Zealand, Norway, Panama, South Africa, the former Soviet Union, and the United States met in Washington, D.C., to establish the International Whaling Commission (IWC). Regulations prohibited the killing of the remaining blue, gray, bowhead, and right whales and of cows with calves. Opening and closing dates for whaling and minimum size data were set for each species harvested. Although each factory ship carried an observer, they could only report offenses and recommend disciplinary action; the government of the country registering the ship involved was responsible for any penalties. With IWC and its regulations in effect, 31,072 whales were killed in 1951, more than 50,000 in 1960, and in 1962 over 66,000 animals were killed. Annual harvests of whales from 1910–80 are given in figure 16.4. Keep in mind that it took New Bedford, Massachusetts, whalers over 150 years to kill 30,000 whales. The 1994 population status of the great whales is given in table 16.2.

As the 1970s drew to a close, the era of commercial whaling appeared to be closing as well. In 1979, the IWC placed a moratorium on all whaling in the Indian Ocean and outlawed the use of factory ships as floating bases from which to send out hunter vessels, but whaling continued from land bases in Antarctica. In the spring of 1982, the IWC voted a moratorium on commercial harvesting of whales, except dolphin and porpoise, in order to study the whale populations and assess their ability to recover. The moratorium began in 1985–86.

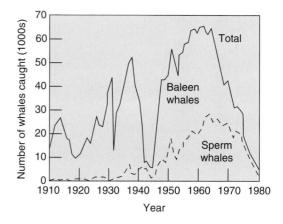

Figure 16.4 Total annual catches of baleen and sperm whales in all oceans from 1910–80.

Table 16.2 World Population Estimates of Great Whales

Species	Initial[1]	Present[2]	Status[1]
Right whale	100,000	8600	Severely depleted
Blue whale	>196,000	10,320	Severely depleted
Finback whale	>464,000	118,320	Severely depleted
Bowhead whale	>55,000	8000	Severely depleted
Humpback whale	>120,000	39,000	Severely depleted
Sei whale	>105,000	27,640	Depleted
Sperm whale	277,000	234,246	At or above optimum sustainable population
Gray whale	<20,000	26,900	Recovered
Minke whale	900,000	900,000	Sustained

1. From Endangered Whales: Status Update. 1994. *National Marine Mammal Laboratory, National Marine Fisheries Service, National Oceanic and Atmospheric Administration.*

2. Perry, S. L., D. P. DeMaster, G. K. Silber. 1999. The Great Whales: History and Status of Six Species Listed as Endangered Under the U.S. Endangered Species Act of 1973. *Marine Fisheries Review, Special Issue* 61(1):1–74.

In 1992 whaling nations reminded the IWC that the original intent of the organization was to "conserve" whales in order to protect the stocks as a harvestable natural resource. Citing the large population of North Atlantic minke whales, Norway chose to set its own national catch limits and resumed whaling in 1993 (226 whales); its 2001 harvest limit was 549 minkes. In 1994 the IWC voted, by a margin of twenty-three to one (Japan) and with several abstentions, to establish a whale sanctuary in Antarctic waters below 55°S. In 2000 Japan expanded its "scientific" hunt; six vessels took 160 minke whales, 50 Bryde's whales, and 10 sperm whales. This is the first time Japan has hunted Bryde's or sperm whales since 1987. Iceland left the IWC in 1999 with plans to resume whaling in 2001.

The IWC permits whaling by native peoples (in Alaska, Denmark [Greenland], and the former Soviet Union) in order to balance the conservation of whales with the cultural and subsistence needs of these peoples. In 1997 the Native American Makah tribe of Washington State, citing an 1895 treaty with the United States, received permission to harvest five gray whales from the annual limit of 140 gray whales harvested in the North Pacific for traditional aboriginal subsistence needs. One whale was harvested by the Makahs in 1999 and another was harvested in the early summer of 2001.

While populations of California gray whales have recovered and some populations of other whale species are slowly increasing, the majority of whale species are still present in low numbers when compared to their estimated original populations. Some researchers believe that the failure of the populations to recover is due to the difficulty of finding mates in such small populations; the possibility also exists that the noise produced by increasing ship traffic interferes with whale communication. Other scientists are concerned that krill harvesting and the global depletion of fish species are affecting the whale populations; pollution may also play a role.

The intertwined history of whales and people has not yet ended. In the words of Herman Melville from the pages of *Moby Dick*:

> The moot point is whether the Leviathan can long endure so wide a chase and so remorseless a havoc; whether he must not at last be exterminated from the waters, and the last whale, like the last man, smoke his last pipe, and then himself evaporate in the final puff.

Dolphins and Porpoises

Dolphins and porpoises are small, toothed whales (fig. 16.6e). They are curious, gentle, friendly, and easily trained, and they figure in many folktales, stories, movies, and television programs. In the open ocean, they are observed traveling at high speeds and in large schools; they occasionally leap clear of the water, in apparent fun and high spirits, and may even alter their course to keep a vessel company for hours, swimming easily just in front of the bow. Porpoises have been timed swimming at speeds in excess of 30 knots, a feat that interests scientists and researchers. Their ability to communicate and their intelligence, as demonstrated by their learning and recall abilities, are under study. They have also been trained to help divers and to act as messengers between those working underwater and surface vessels.

These smaller cetaceans are found in both tropical and temperate waters. Although marine, they will go up rivers and channels into shallow brackish waters. They have been seen moving across the very shallow lakes and canals of the Mississippi delta region in water barely deep enough to support their high-speed swimming.

The last two decades have placed great pressure on the world's dolphins and porpoises. A 1990 United Nations–sponsored symposium reported that more than a million of these mammals die each year in nets, usually the unwanted by-catch of those fishing for other species. Dolphin and porpoise populations are affected by certain fishing techniques used by the tuna industry; this situation is investigated further in the discussion of the tuna fishery later in this chapter. Two small species, the Mexican porpoise and the black dolphin of the Chilean coast, may be endangered. Only a few hundred Mexican dolphins are left in the Gulf of California, and the Chilean dolphins are said to have been killed in large numbers for use as fish and crab bait.

Seals and Sea Lions

Seals and sea lions belong to the **pinnipeds,** or "feather-footed" animals, so named for their four characteristic swimming flippers. Representative pinnipeds are shown in figures 16.5 and 16.6. These animals are marine mammals that still retain their ties to land, spending considerable time ashore on rocky beaches, on ice floes, or in caves. They are found from the tropics to the polar seas, ranging from the nearly extinct monk seal of the western Hawaiian Islands and Mediterranean area to the fur seals of the Arctic. The common harbor seal, the harp seal of the northwest Atlantic, the Weddell and leopard seals of the Antarctic, and the 2-ton male elephant seal, with its great pendulous snout, are all true seals, or seals without external ears and with torpedo-shaped bodies that require them to use a wriggling, wormlike motion to move on land. The northern fur seal and the sea lion (the seal of the circus and amusement park) are eared seals with longer necks and supple forelimbs tipped with broad flippers, which are used for walking and hold the animal's body in a partially erect position on land. Seals may undertake long sea migrations, congregating in spring and summer at specific locations for breeding. For example, the northern fur seal ranges the North Pacific, Bering, and Okhotsk Seas, coming ashore to breed in the Pribilof Islands. The habits and natural histories of each of the pinnipeds differ from those of the others, and much is still left to learn.

Some seals and sea lions are currently enjoying a period of relative peace, compared to the sealing days of the nineteenth and early twentieth centuries. Between 1870 and 1880 hunters for furs and oil reduced the northern elephant seal population to 100. The Guadalupe fur seal of Southern California was hunted to the brink of extinction, until, in 1892, only seven animals were thought to survive. This species' remarkable recovery to a current population of 1500 is due to protection by the U.S. and Mexican governments and to luck.

Whale Falls

The death of a whale suddenly sends a huge, localized source of food to the sea floor. In 1987, Craig Smith and colleagues from the University of Hawaii accidentally discovered the carcass of a blue whale on the floor of the Santa Catalina basin off California (box fig. 1a). Their studies show that a whale carcass, or whale fall, supports a large community of organisms. First the scavengers such as hagfish, crabs, and sharks reduce the body to bones in as little as four months, and then the bacteria take over. Whalebones are rich in fats and oils that give the animal buoyancy in life; after death they provide nutrition for anaerobic bacteria. These bacteria decompose the fatty substances and generate hydrogen sulfide and other compounds that diffuse out through the bone. Chemosynthetic bacteria then metabolize the sulfides and form bacterial mats over the skeleton. The bacterial mats are grazed by worms, mollusks, crustaceans, and other organisms. Other animals may be attracted, and they get their nutrition from the sulfides, the bacterial mats, the fatty substances in the bones, or other animals at the site.

The number of species found on a single skeleton is surprising: 5098 animals from 178 species were isolated from five vertebral bones recovered from one whale. Small mussels are shown on recovered whalebone in box figure 1b. The bone surface area totaled 0.83 m^2 (9 ft^2). Ten of these species, including worms and limpets, have been found only associated with whale skeletons. By comparison, the most fertile hydrothermal vent areas have yielded only 121 species and hydrocarbon seeps just 36.

How abundant are these whale falls? At this time that is difficult to say, for they can be anywhere on the ocean floor and are difficult to locate. In 1993 the U.S. Navy searched 20 km^2 (8 mi^2) of the Pacific Missile Range off California with side-scan sonar while seeking a lost missile. Eight whale falls were videotaped, and one of them was subsequently located and examined by Smith and his team using a submersible. Scientists in Japan, New Zealand, and Iceland are also looking for whale-fall sites. With permission from the National Marine Fisheries Service, Smith has taken two dead stranded whales out to sea and sunk them to observe the colonizing of the carcasses and learn more about the diversity of the organisms associated with whale falls.

As many as fifteen of the whale-fall species have also been found in other sulfide-rich habitats such as the deep-sea hydrothermal vents. Smith has suggested that the whale falls may serve as "stepping-stones" for the dispersal of organisms that depend on chemosynthesis, as from one hydrothermal vent to another. Objections to this theory include the small number of species that have been found to overlap whale falls and vents, that vents are found in more than 1500 m (5000 ft) of water while most whales live and die in shallower water along the edge of the continental shelf, and the increasing number of vents being found indicates that dispersal from vent to vent is not a problem. Whether or not the whale falls act as stepping-stones, it is of great interest that these whale-fall communities exist, that hydrothermal vents are not as isolated as was thought, and that another new kind of community has been discovered on the sea floor, a place once thought to be cold, dark, and unsuited to life.

(a)

(b)

Box Figure 1

(a) The skull, jawbones, and vertebrae of a 21 m (70 ft) blue whale on the floor of the Santa Catalina basin off Southern California, depth 1240 m (4000 ft). The skull is about 1.5 m (5 ft) long, and each vertebra is about 40 cm (16 in) long. (b) Sulfide-loving mussels (each about 1 cm [0.4 in] long) clustered on a recovered whalebone. Up to 178 species of animals have been found living on a single carcass.

To Learn More About Whale Falls

Deming, J., A. Reysenbach, S. Macko, and C. Smith. 1997. Evidence for the Microbial Basis of a Chemoautotrophic Invertebrate Community at a Whale Fall on the Deep Seafloor: Bone-Colonizing Bacteria and Invertebrate Endosymbionts. *Microscopy Research and Technique* 37: 162–70.

Life Among the Whale Bones. 1998. *Science* 279 (5355): 1302.

Showstock, R. 1998. Whale Falls May Provide Important Stepping Stone Habitat Between Deep Sea Vents and Seeps. *EOS* 79 (4): 45–46, 48.

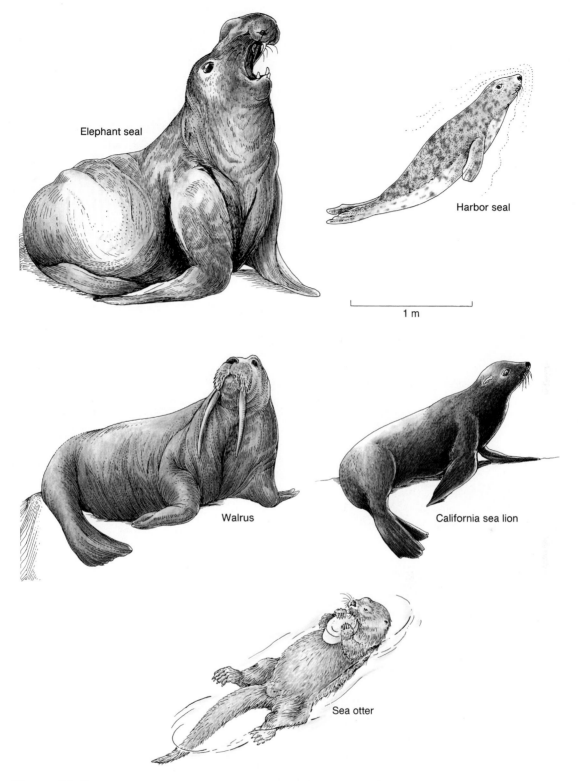

Labels in figure:
- Elephant seal
- Harbor seal
- 1 m
- Walrus
- California sea lion
- Sea otter

web link **Figure 16.5** Marine mammals. The elephant seal, harbor seal, walrus, and California sea lion are pinnipeds. The sea otter, related to river otters, feeds on clams and sea urchins while floating on its back.

Heavy hunting had reduced the population of northern fur seals in the Pribilof Islands to only 200,000–300,000 animals by 1910. In 1967–68, when commercial sealing and the killing of females ceased, the population was 2.5 million to 3 million. The minimum estimate of the annual native subsistence harvest over the years 1986–96 was 1600 seals (juvenile males only in the Pribilof Islands). The present total population is approximately 1.1 million. There has been a long-term downward population trend since the mid-1950s. This has been blamed on entanglement with nets,

Mammals **423**

(a)

(b)

(c)

(d)

(e)

Figure 16.6 Marine mammals. (a) The harbor seal (*Phoca vitulina*) is a friendly, curious animal that coexists well with humans. (b) This large male northern fur seal (*Callorhinus ursinus*) stands vigilant over his harem of females in the Pribilof Islands. (c) Sea otter (*Enhydra lutris*) and her pup dine on crab. (d) The walrus (*Odobenus rosmarus*) feeds from the bottom and relaxes on the Arctic ice floes. (e) Pacific white-sided dolphins (*Lagenorhynchus obliguidens*) keep pace with the bow wave of a research vessel.

lines, plastic strapping rings, and other debris, but recent studies suggest that the population's decline is primarily related to a reduced prey base, which in turn is related to the large commercial fish catches in the area. Seal food resources may also be subject to environmental changes related to the Pacific Decadal Oscillation (see chapter 8, section 8.6). For a discussion of the relationship between heavy commercial fishing and the Steller sea lion population in the Gulf of Alaska, see the section on overfishing and incidental catch in chapter 12 and the next section on the Alaskan groundfishery and its effect on sea otters.

In eastern Canada the white-fur pups of the Canadian harp seal were commercially harvested for 150 years. The trade was banned by the Canadian government after the harvest became the focus of public outcry against sealing in the 1970s and 1980s and the market collapsed. The loss of the Atlantic cod fishery in Newfoundland (see section 16.7) sent sealers back to the northern ice to fill the government quota of 250,000 in 1996, 1997, and 1998. The current harp seal population is estimated at 4 million.

Sea Otters

Sea otters (see figs. 16.5 and 16.6c) are related to river otters but are larger and live in salt water. This animal lives in coastal areas, taking shellfish and other food from the bottom in relatively shallow water. It differs from both seals and whales in having no insulating layer of blubber beneath the skin and so must depend entirely on its dense fur for warmth. This soft, thick fur became the sea otter's death warrant in the eighteenth and nineteenth centuries, when they were hunted nearly to extinction; prime pelts sold for more than $1000 each. The sea otter was brought under protection in 1911, and many scientists, at that time, doubted that the species could survive. The Alaskan population is now at a minimum of 100,000, and in California a small population of a minimum of 2300 animals exists, according to the National Fish and Wildlife Service.

In 1998 it was reported that killer whales, normally hunters of seals and sea lions, were now feeding on sea otters along the Aleutian Island chain of western Alaska. It appears that seal populations have declined because of groundfishing, and the lower numbers of seals force the killer whales to prey on sea otters. Not all researchers accept the link to groundfishing; they believe the situation to be more complex, including environmental factors such as warming ocean temperatures that have occurred in the area since the mid-1970s. In either case, the sea otters live in and around the kelp beds, feeding on the sea urchins that graze the kelp. As sea otter populations decrease, sea urchin populations increase and threaten to destroy the kelp beds that serve as nurseries for smaller fish and other animals; thus the entire coastal ecosystem may be affected.

Walrus

The walrus (see figs. 16.5 and 16.6d) is placed in a separate subgroup. It has no external ears and is able to rotate its hind flippers so that it can walk on a hard surface. Its heavy canine teeth (or tusks) are unique, and both males and fe-

males have them. These tusks help the walrus haul itself out of the water onto the ice and are probably used to glide over the bottom, like sled runners, while the animals forage for clams with their heavy muscular whisker pads.

The Pacific walrus occurs in the waters and on both coasts of the Bering and Chukchi Seas. It is thought that the population before the eighteenth-century exploitation was 200,000–250,000 animals. The 1950 population was 50,000–150,000, but since then it has increased dramatically. Counting the walrus requires international cooperation between the United States and Russia; the 1985 joint USA/USSR survey estimated the population at 234,000. No counts have been taken since then.

Over the last thirty-six years an average of 7000 walrus were harvested each year, but between 1992 and 1996 the average harvest dropped to 5000 walrus each year. This change is related to the ending of Russian ship-based harvesting and changing economic and social conditions among the hunters, as well as weather and ice conditions.

Sea Cows

Manatees and **dugongs** are also known as **sea cows** (fig. 16.7); they are members of the order **Sirenia** and are thought by some to be the source of mermaid stories. Manatees and dugongs are the world's only herbivorous marine mammals. Manatees are found in the brackish coastal bays and waterways of the warm southern Atlantic coasts and in the Caribbean, and dugongs are found in the seas of Southeast Asia, Africa, and Australia. At present, the growth of human populations and their need for protein is putting increasing hunting pressure on the dugongs in the southern Pacific. Manatees in the coastal waters of the Caribbean and South Atlantic are frequently injured and killed by collisions with the propellers of large and small vessels. Although they are protected along the Florida coast, the known manatee mortality between 1979 and June 1992 was more than 1700 animals, 26% from boat collisions. In 1996, 400 manatees died and at least 158 of their deaths were due to the neurotoxin produced by the red tide organism *Gymnodinium* (see chapter 15, section 15.6). In 1999 269 manatees died, in 2000 there were 273 deaths, and by April 2001 another 154 deaths had been recorded. Manatee deaths at or about the time of birth have increased from nine in 1977 to thirty in 1987 and fifty-eight in 2000. In addition the destruction of Florida's manatee habitat is accelerating, as salt marshes, sea grass beds, and mangrove areas are drained, reclaimed, or otherwise destroyed.

In former times the Steller sea cow existed in the shallow waters off the Commander Islands in the Bering Sea. These sea cows were slow-moving, docile, totally unafraid of humans, and present in only limited numbers. These characteristics, coupled with a low reproduction rate, made them unable to withstand the human hunting pressure. The last Steller sea cow was killed for its meat in about 1768.

Marine Mammal Protection Act

In 1972, the U.S. Congress established the Marine Mammal Protection Act, Public Law 92-522. This act includes a ban on the taking or importing of any marine mammals or marine mammal product. "Taking" is defined as the act of harvesting, hunting, capturing, or killing any marine mammal or attempting to do so. The act covers all U.S. territorial waters and fishery zones. It is also unlawful "for any person subject to the jurisdiction of the United States or any vessel or any convoy once subject to the jurisdiction of the United States to take any marine mammals on the high seas" except as provided under preexisting international treaty.

The act effectively removed the animals and their products from commercial trade in the United States. Only under strict permit procedures and with the approval of the Marine Mammal Commission can a few individual marine mammals be caught for scientific research and public display.

Native peoples are exempted from the act for purposes of subsistence hunting and of creating and selling authentic native articles of handicraft and clothing. If it is determined that a species or stock being hunted under this exemption is being depleted, further regulations may be established to conserve the animals.

In some cases, the Marine Mammal Protection Act has been very successful in its mission of protecting marine mammal populations; some would argue that it has been too successful. Interactions between people and marine mammals competing for the same resource or habitat, or both, are creating problems that are not easily solved. Since passage of the act, harbor seal and sea lion populations have increased 7%–10% per year along the U.S. West Coast. The seals and sea lions feed on fish that are harvested commercially and increasingly rob fishing nets directly. Under the provisions of the act, those who fish can only attempt to discourage the animals, and their attempts are not always successful. An additional problem accompanying the increasing populations of seals and sea lions is their deposit of fecal material in shallow beach areas. Populations of the bacterium *Escherichia coli*, or *E. coli*, associated with these deposits have led to closure of the areas to swimming and shellfish harvesting. In many cases, local people are sharply divided concerning these marine mammals. Some see the seals and sea lions as destructive raiders and polluters that no longer need protection; others enjoy watching them and consider them a natural part of the marine world.

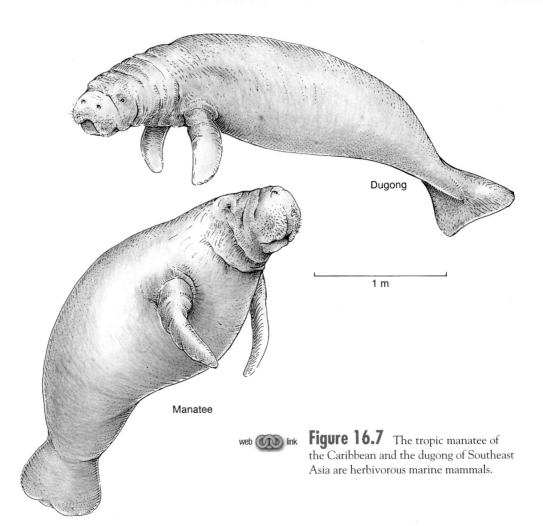

Figure 16.7 The tropic manatee of the Caribbean and the dugong of Southeast Asia are herbivorous marine mammals.

Communication

Many marine mammals use sound to communicate with each other and sound instead of sight to picture their underwater environment. The best-known communication between marine mammals are the "songs" of the male humpback whales. Different humpback populations have different songs, and the songs are transmitted from one individual to another within the population. Songs last up to thirty minutes and are changed and modified during each breeding season. These songs are thought to be announcements of presence and territory, although some scientists believe the singing is a secondary sexual characteristic of males in the breeding season. Female gray whales stay in contact with their calves by a series of grunts, and Weddell seals are known to communicate by audible squeaks.

Using sound to picture the environment is known as **echolocation.** The ability to make the sharp sounds required to produce the echoes that allow marine mammals to orient themselves and locate objects is suspected in all toothed whales, some pinnipeds (Weddell seal, California sea lion), and possibly the walrus. A few baleen whales—the gray, blue, and minke—also have this capacity. Although these animals can produce a range of sounds, the most useful sound for echolocation appears to be clicks of short duration released in single pulses or trains of pulses. The bottle-nose dolphin

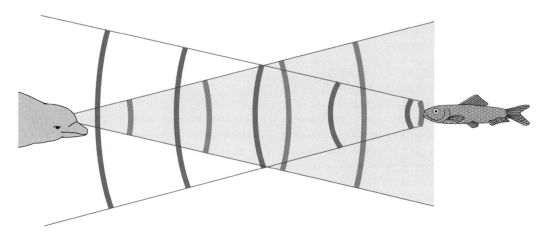

web link **Figure 16.8** The pattern of clicks produced by a porpoise for echolocation is shown in *red*. The reflected echoes by which the porpoise determines the speed, distance, and direction of the target are shown in *green*.

produces clicks in frequencies audible to the human ear and higher, each lasting less than a millisecond and repeated up to 800 times per second. When each click hits its target, part of the sound is reflected back; the animal continually evaluates the time and direction of return to learn the speed, distance, and direction of the reflecting target (fig. 16.8). Low-frequency clicks are used to scan the general surroundings, and higher frequencies are used for distinguishing specific objects.

Porpoises and dolphins move air in their nasal passages to vibrate the structures that produce the clicks; the whistles and squeaks are made by forcing air out of nasal sacs. The bulbous, fatty rounded forehead of the porpoise acts as a lens to concentrate the clicks into a beam and direct them forward. Sperm whales produce shorter, more powerful, long-range pulses at lower frequencies; these sounds may travel several kilometers. Each pulse from a sperm whale is compound, made up of as many as nine separate clicks. The sperm whale's massive forehead is filled with oil that may be used to focus the sound pulses.

These animals must be able to pick up the faint incoming echoes of their own clicks and screen out the louder outgoing clicks and other sea noises. Sounds enter through the lower jaw and travel through the skull by bone conduction. Within the lower jaw, fat and oil bodies channel the sound directly to the middle ear. Areas on each side of the forehead are also very sensitive to incoming sound. The hearing centers in the brains of marine mammals are extremely well developed, presumably to analyze and interpret returning sound messages. Their vision centers are less developed, and they are believed to have no sense of smell.

Marine Birds   16.2

About 3% of the estimated 8600 species of birds are marine species. Some are so well adapted to oceanic life that they come ashore only to reproduce; others move into coastal wa-

ters to feed; but all return to shore to nest. Most have definite breeding sites and seasons and they may migrate thousands of kilometers as they travel from feeding ground to breeding area.

Seabirds swim at the sea surface and underwater; they use their webbed feet, their wings, or a combination of wings and feet. They float by using fat deposits in combination with light bones and air sacs developed for flight. Their feathers are waterproofed by an oily secretion called preen, and the air trapped under their feathers helps keep the birds afloat, insulates their bodies, and prevents heat loss. When diving, the birds reduce their buoyancy by exhaling the air from their lungs and air sacs and pulling in their feathers close to their bodies to squeeze out the trapped air. The underwater swimmers such as cormorants and penguins have thicker, heavier bones, and penguins have no air sacs.

The eyesight of seabirds is highly developed, for they depend on their vision to locate fish in the water. Their senses of hearing and smell are less developed, and their least developed sense is taste; they have few taste buds and swallow quickly. They drink seawater or obtain water from their food. Excess salt is concentrated by a gland over each eye, and a salty solution drips down or is blown out through their nasal passages. To conserve water, the birds reduce and concentrate their urine, forming uric acid, a nearly nontoxic white paste that is mixed with feces and excreted. Because of their great activity and high metabolic rate, birds are huge eaters. Seabirds feed on fish, squid, krill, egg masses, and bottom invertebrates, as well as on carrion and garbage. They are most plentiful where food is abundant, and their presence is a good indicator of high productivity in surface water.

The wandering albatross of the southern oceans is the most truly oceanic of marine birds with the largest wing spread of all birds, 3.5 m (11 ft). These great white birds with black-tipped wings spend four to five years at sea before returning to their nesting sites. The smaller, North Pacific, black-footed albatross is a pelagic scavenger, searching the sea

The Sea Bear

The sea and sea ice are essential components of the polar bear's environment. The polar bear is the top predator of the Arctic's marine food chains, and its scientific name is appropriate, *Ursus maritimus*, or "sea bear" (box fig. 1). Polar bears are long-lived (fifteen to twenty years or more), late-maturing (four to five years) carnivores with dense fur and a blubber layer for insulation. Adult males weigh 250–800 kg (550–1700 lb) and measure 2.5–3.0 m (8–10 ft) from tip of nose to tail; females weigh 100–300 kg (200–700 lb) and measure 1.8–2.5 m (6–8 ft) (box fig. 2). In late October and November females make their dens in drifted snow. One or two cubs are born in December or January, and mother and cubs emerge in late March and early April. Most female bears keep their cubs with them until the cubs are about two and a half years old. The bears feed primarily on ringed seals and to a lesser extent on other seal species and whale and walrus carcasses. They use their keen sense of smell to locate seals and hunt by stalking the animals basking on the ice, lying in wait at breathing holes, and breaking into their birthdens. Polar bears travel long distances over the ice, 30 km (19 mi) or more each day; they are also strong swimmers, swimming continuously for 100 km (62 mi).

web ∞ link **Box Figure 1** A polar bear on the ice in the central Arctic Ocean.

Polar bear distribution is circumpolar but not continuous. Populations that return to feeding and denning areas are kept apart by the sea ice, ice movements, and land barriers. The bears move extensively between seasons, depending on regional patterns of freezing and breakup of the sea ice. Polar bears depend on sea ice as a platform from which to hunt and feed and as a base on which to seek mates, breed, and travel long distances. About 68% of the polar ice region appears to be suitable habitat for polar bears.

The world's total population of polar bears is estimated at between 25,000 and 40,000. The Canadian population is about 15,000, or roughly half of the world's polar bears. Polar bears are also found in Denmark (Greenland), Norway (Svalbard Islands), and Russia. There are two Alaska polar bear populations, estimated at 3000 animals; one population is in the vicinity of the Chukchi and Bering Seas and the other is in the Beaufort Sea area.

In the 1950s and 1960s there was growing international concern for the polar bear populations as the number of bears being killed for their hides increased. Russia prohibited all hunting of polar bears in 1956, but poaching increased dramatically. The U.S. Marine Mammal Protection Act of 1972 prohibited the killing of all marine mammals including polar bears, except by native people for subsistence purposes. Norway prohibited all polar bear hunting in 1973, and Canada increased regulations on native peoples' hunting and allowed a limited sport hunt of 500 bears per year. Greenland harvests 100–150 bears annually in a tightly regulated hunt for residents only.

In October 2000 the United States and Russia signed a treaty covering the polar bears in northwest Alaska and the Chukota region of northeast Siberia; the treaty includes the islands of the Chukchi and Bering Seas. New U.S. and Russian commissions were formed in which native tribes participate in setting harvest quotas. For the first time quotas were placed on the number of bears hunted in both countries. Den areas were ruled off-limits; commercial hunting and killing females and cubs under one year are prohibited. No poison, traps, snares, aircraft, or large motorized vehicles and vessels are permitted in the hunt.

The polar bears still face an uncertain future. If climate change increases ice-free periods, the bears cannot return to the ice to hunt and mate. Other potential problems include toxic contaminants, poaching, disruption of the food chain by overfishing, human

---

surface for edible refuse, including scraps from ships and fishing boats. Its primary foods are squid, crab, and surface fish. The smallest of the oceanic birds, Wilson's petrel, is a swallowlike bird that breeds in Antarctica and flies 16,000 km (10,000 mi) along the Gulf Stream to Labrador during the Southern Hemisphere winter, returning to the Southern Hemisphere for the southern summer, another 16,000 km.

Penguins (fig. 16.9a) do not fly but swim in flocks, using their wings for propulsion and steering with their feet. Their underwater swimming speed is almost 10 mph. They feed on fish, krill, squid, and shellfish. All but the Galápagos penguin are found in the Southern Hemisphere, and two species, the emperor and the Adelie, are found in Antarctica.

Pelicans and cormorants are large fishing birds with big beaks. They are strong fliers found mostly in coastal areas, but some venture far out to sea. A pelican (fig. 16.9b) has a particularly large beak from which hangs a pouch used in catching fish. White pelicans of North America fish in groups, herding small schools of fish into shallow water and then scooping them up in their large pouches. The Pacific's

activities related to oil and gas exploration and development, and increased tourism and human contacts.

To Learn More About the Sea Bear

Eliot, J. 1998. Polar Bears, Stalkers of the High Arctic. *National Geographic* 193(1): 52–71.

Internet References

Visit the book's Online Learning Center at www.mhhe.com/sverdrup (click on the book's cover) to explore links to further information on related topics.

Box Figure 2 This polar bear is being weighed, measured, and fitted with a tracking collar that will be followed by satellite. A measuring tape is lying on top of the bear. The scientists are members of the 1994 Arctic Section Expedition.

brown pelican does a spectacular dive from up to 10 m (30 ft) above the water to capture its prey. Cormorants are black, long-bodied birds with snakelike necks and moderately long bills that are hooked at the tip. They settle on the water and dive from the surface, swimming primarily with their feet but also using their wings. Because they do not have the water-repellent feathers of other seabirds, cormorants must return to land periodically to dry out.

Terns (fig. 16.9c) and gulls are found all over the world except in the South Pacific between South America and Aus-tralia. Gulls are strong flyers and will eat anything; they forage over beach and open water. The terns are small, graceful birds with slender bills and forked tails. The Arctic tern breeds in the Arctic and each winter migrates south of the Antarctic Circle, a round-trip of 35,000 km (20,000 mi).

Puffins, murres, and auks are heavy-bodied, short-winged, short-legged diving birds. They feed on fish, crustaceans, squid, and some krill. All are limited to the North Atlantic, North Pacific, and Arctic areas, where most nest on isolated cliffs and islands. The great auk was a large, slow,

Figure 16.9 (a) These Magellanic penguins are part of a 500,000-bird colony at Punta Tombo, Argentina. (b) A brown pelican takes off from the water. (c) Crested terns face the wind on the beach of Heron Island, a part of the Great Barrier Reef, Australia. (d) A mixed flock of dunlins and sanderlings take flight along a sandy beach. (e) Kittiwakes (gull-like birds) and common murres nesting on an island in Kachemak Bay, Alaska.

flightless bird, 0.6 m (2 ft) high, that provided food for generations of sea travelers. It was killed for both its meat and its feathers (for stuffing mattresses), and as its numbers dwindled museums and private collectors paid more and more for each bird. The last two were killed on a small island off Iceland on June 2, 1844.

Most shorebirds are migratory and are present in thousands and hundreds of thousands at certain times of the year (fig. 16.9d). There is no severe competition for food because different types of birds arrive at different times of the year, and different birds select different types of shore. Each bird exploits a different food resource related to the length of its neck, bill, and legs. Herons and egrets are waders; their long necks enable them to strike at small fish and organisms in the water. Sandpipers probe the sand and mud with their bills to locate worms and other organisms. Ducks, terns, and gulls swim and dive in the estuaries and rest on the beaches.

All over the world increasing numbers of people choose to live or vacation along coasts and beaches, and isolated shore areas available to birds are rapidly decreasing. Most recreational uses of shore areas are in the late spring and early summer, and this is also the time that most marine birds look for isolation to nest and rear their young (fig. 16.9e). Vacation homes, harbors, and resorts are built along once-lonely beaches; wetlands are diked and filled, removing food sources and nesting sites; runoffs from towns, industry, and agriculture contaminate estuaries, and accidental spills and industrial dumping pollute inshore waters. Also the intensity of fishing has increased in all the oceans, and as fish populations have diminished, there is less fish for both humans and birds. The flocks become smaller until at last they no longer visit beaches that once supported them by the hundreds of thousands.

Some birds do find ways to use the changes humans make, and these populations may thrive. The submerged portions of piers and pilings provide a home for many organisms that are an important part of some marine birds' diets. Jetties and docks are used as nesting sites and lookouts for prey, and artificially constructed and maintained beaches may become nesting and feeding areas.

Natural disasters such as volcanic eruptions and landslides disturb nesting areas, and events such as El Niño, with its disruption of the fish populations, depress the great bird populations along the west coast of South America (see chapter 6). However these disruptions are usually cyclical; a population decrease in one year builds up again over the next series of unaffected years. Human changes are generally more permanent, and the flocks of migratory seabirds that were described by naturalists of the seventeenth, eighteenth, and nineteenth centuries may never be seen again.

Marine Reptiles 16.3

Although many land reptiles visit the shore to feed, mainly on crabs and shellfish, few reptiles are found in today's seas. Examples of marine reptiles are shown in figure 16.10. The only modern marine lizard is the big, gregarious marine iguana of the Galápagos Islands. It lives along the shore and dives into the water at low tide to feed on the algae. This iguana has evolved a flattened tail for swimming and has strong legs with large claws for climbing back up on the cliffs. It regulates its buoyancy by expelling air, allowing it to remain underwater. The large monitor lizards of the Indian Ocean islands, known as the Komodo dragons, are capable of swimming but do so only under duress.

Alligators may enter shallow shore water, and crocodiles are known to go to sea. The estuarine crocodile of Asia is found in the coastal waters of India, Sri Lanka (Ceylon), Malaysia, and Australia. The Indian gavial and false gavial are slender-nosed, fish-eating crocodilians found in nearshore waters.

Sea Snakes

There are about fifty kinds of sea snakes found in the warm waters of the Pacific and Indian Oceans; there are no sea snakes in the Atlantic Ocean. Sea snakes are extremely poisonous; they have small mouths, flattened tails for swimming, and nostrils on the upper surface of the snout that can be closed when the snakes are submerged. The sea snakes' skin is nearly impervious to salt but it is permeable to gases, passing nitrogen gas as well as carbon dioxide and oxygen. The snake is able to lose nitrogen gas to the water, allowing it to dive as deep as 100 m (300 ft), stay submerged as long as two hours, and surface rapidly with no decompression problems. (See chapter 13 for a discussion of the way in which marine mammals handle diving and decompression.) Sea snakes eat fish, and most reproduce at sea by giving birth to live young.

Sea Turtles

The four species of large sea turtles are the green, hawksbill, leatherback, and loggerhead turtles. Green sea turtles are herbivores; they weigh 140 kg (300 lb) or more. The hawksbill is a tropical species found on reefs; it feeds primarily on sponges. The loggerhead weighs between 70 and 180 kg (150 and 400 lb); it is usually found around wrecks and reefs, feeding on crabs, mollusks, and sponges. The leatherback is the largest marine reptile, weighing up to 900 kg (2000 lb). This giant travels 5000 km (3100 mi) between nesting and feeding areas and dives to 1000 m (3300 ft) depth to feed on jellyfish. The Kemp's ridley turtle is the fifth sea turtle found in U.S. waters, the Gulf of Mexico, and the South Atlantic. It is a small sea turtle and the rarest and most endangered of all sea turtle species. There are three more species of sea turtle: the black turtle of the east Pacific (possibly a subspecies of the green), the Australian flatback, and the smaller olive ridley, which nests along Costa Rican beaches. The olive ridley is estimated to be the most abundant sea turtle.

All eight species of sea turtles, including the five found in U.S. waters, are endangered or threatened. The turtles live in the ocean but nest on land; they are migratory animals and make long sea journeys of many thousands of kilometers between their nesting sites and their foraging areas. The green turtles of the Brazil coast migrate 2250 km (1400 mi) to Ascension Island in the South Atlantic to lay their eggs and then return to Brazilian waters. Loggerhead turtles were known to migrate across the Atlantic, but their navigation mechanism was unknown until an experimental program carried out by scientists at the University of North Carolina showed that the turtles carry an internal compass sensitive to the Earth's magnetic field. Hatching occurs at night, and the compass is set by the direction of the night-light from the

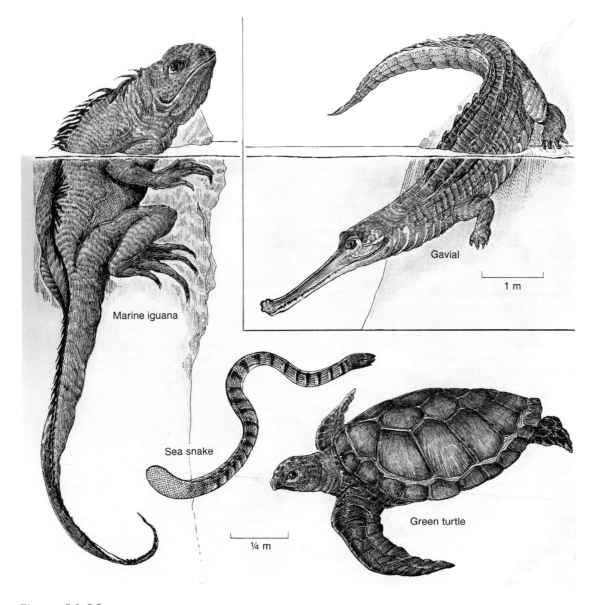

web ꞏ link **Figure 16.10** Marine reptiles. The marine iguana is the only modern marine lizard; it inhabits the Galápagos Islands. The gavial is a fish-eating crocodilian from India. Sea snakes are venomous; they are found only in the Pacific and Indian Oceans. The green turtle feeds on sea grasses and seaweeds in tropical coastal waters.

sea; the turtles orient themselves into the waves as they enter the water, and when they reach the Gulf Stream they move with the current until their internal compasses tell them to turn south into the Sargasso Sea. Later they use their compasses to bring them back to their home beaches in Florida and the Caribbean.

Each female turtle lays a hundred or more eggs in a scooped-out depression in the warm sand, covers them, and returns to the sea. While the eggs incubate, they are an easy prey for humans, dogs, rats, and other carnivores. In more and more cases no eggs survive to replenish the population. If the young turtles hatch, they must make their way across the sand while hungry birds attack and must then evade the waiting predatory fish when they do find the water.

Sea turtles are being driven to extinction by pollution, habitat degradation, fishing nets, and the demand for turtle products. Turtle eggs and turtle meat are prized by humans throughout the Pacific, and turtle nests are regularly raided by poachers. Hawksbill turtles are hunted for tortoiseshell, which is used to form combs, boxes, jewelry, and other ornaments. Turtle skins are used for leather and their fat and oils for cosmetics. The United States and 115 other countries have banned import or export of sea turtle products, and in 1992 Japan agreed to halt its imports. Eighty percent of all sea turtles in U.S. waters nest in Florida, where habitat degradation is the most serious problem. Beach lighting frightens nesting turtles and disorients hatchlings; seawalls and bulkheads destroy nesting areas; pumping sand onto

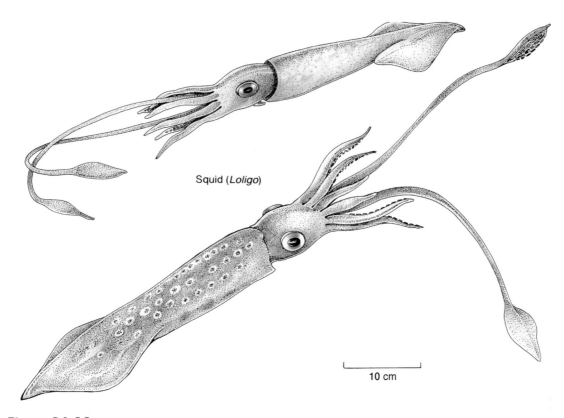

Squid (*Loligo*)

10 cm

web link **Figure 16.11** The squid is a swimming mollusk and a member of the nekton.

beaches to repair erosion buries nests and changes the quality of the sand; beach-cleaning machines crush the nests.

No one knew where Kemp's ridley turtles nested until 1947, when their only natural nesting site was found at Rancho Nuevo on Mexico's east coast. That year, 40,000 females came ashore to lay their eggs. Despite protection by the Mexican government, in 1992 there were fewer than 500 egg-laying females. The Kemp's ridley was under siege. Their meat was sold as a delicacy; their eggs were stolen to make aphrodisiac cocktails; their hides were tanned for leather; they suffocated in shrimp trawls; and their young were preyed on by raccoons and other scavengers. Scientists now transfer their eggs from the beach to a local hatchery, and the numbers are slowly rising.

Placing sea turtles on the U.S. endangered species list prevents the legal importing of turtle products into this country, but there is a fishery for these animals. In 1996, according to the United Nations Food and Agricultural Organization, 875 metric tons of sea turtles were harvested. There is also a vast international illegal trade in turtle products that rivals the illegal ivory trade in dollar value. The problem of turtles' drowning in shrimp nets plays a significant role in the depletion of sea turtle populations. Worldwide, ocean longline fisheries report an incidental catch of 40,000 sea turtles, with an estimated 42% mortality. In addition, shrimp trawlers catch more than 45,000 sea turtles each year, with an estimated mortality of 12,000. Regulations were passed in 1987 that required U.S. shrimp fishers to use turtle-excluding devices (TEDs) that allow the turtles but not the shrimp to escape the nets. Shrimpers initially resisted the TED requirement, claiming that it reduced shrimp yields, and they appealed enforcement of the regulation. Because of vacillating federal officials, lawsuits by shrimpers and environmental groups, and blockades by shrimpers, it was not until 1994 that the National Marine Fisheries Service (NMFS) required all shrimpers to install TEDs. At this time TEDs are used only in U.S. shrimp fisheries; see the discussion of fish by-catch in chapter 12.

Squid 16.4

Squid (fig. 16.11) are abundant in the world's oceans. A large world fishery exists for squid and their relatives, cuttlefish and octopus (discussed in chapter 17). The 1998 world harvest was 2.6 million metric tons. Until recently, little has been known about squid at all ocean depths. The present use of video cameras in submersibles and robots is providing researchers with a record of rarely seen, deep-water squids, many never before seen alive. They are elusive, swim rapidly, and are now known to not only float with neutral buoyancy but also to rest on the bottom. Their wide range of bioluminescence and coloration allows them to change color and disappear rapidly.

Squid have eight short arms and two thin, long tentacles; the tentacles have a number of suckers at their ends and, in some species, bioluminescent lures to attract prey. Squid can remain motionless or move backwards or forwards. They range from a few centimeters in length to the giant squid (*Architeuthis*), a deep-water species that may be 20 m (65 ft) long. Scientists know little about these huge organisms, for no giant squid has yet been captured alive.

Fish   16.5

The fish dominate the nekton. Fish are found at all depths and in all the oceans, but their distribution patterns are determined directly or indirectly by their dependency on the ocean's primary producers. Fish are concentrated in upwelling areas, shallow coastal areas, and estuaries. The surface waters support much greater populations per unit of water volume than the deeper zones, where food resources are sparser.

Fish come in a wide variety of shapes related to their environment and behavior. Some are streamlined, designed to move rapidly through the water (tuna and marlin); some are laterally compressed and swim more slowly around reefs and shorelines (snapper and tropical butterfly fish), and others are bottom-dwelling fish that are flattened for life on the sea floor (sole and halibut), while still others are elongate for living in soft sediments and under rocks (some eels). Fins provide the push or thrust for locomotion and occur in a variety of shapes and sizes. Fins are used to change direction, turn, balance, and brake. Flying fish use their fins to glide above the sea surface; mudskippers and sculpins walk on their fins. A sample of the variety of marine fishes is shown in figure 16.12.

Schooling is common among certain types of fish (herring, mackerel, menhaden). Schools may consist of a few fish in a small area, or they may cover several square kilometers; for example, herring in the North Sea have been seen in schools 15 km long and 5 km wide (9 × 3 mi). Usually the fish are all of the same species and similar in size. Fish schools have no definite leaders, and the fish change position continually. Schooling fish keep their relationship to one another constant as the school moves or changes direction. Most schooling fish have wide-angle eyes and the ability to sense changes in water displacement, which allow them to keep their place with respect to their neighbors. Schooling probably developed as a means of protection; each fish has less chance of being eaten in the school than alone. The school may also keep reproductive members of the population together.

Ocean fish are divided into two groups: (1) fish with skeletons of cartilage and (2) fish with skeletons of bone. Cartilaginous fish—sharks and their relatives, the skates, rays, and ratfish—are considered more primitive than bony fish. The fish that are used for food are mainly those with skeletons of bone.

Sharks and Rays

The shark is an ancient fish; it predates the mammals, having first appeared in Earth's oceans 450 million years ago.

Unlike most fishes, sharks bear live young. They also differ from other fish by their skeletons of cartilage rather than bone and by their toothlike scales. Shark scales have a covering of dentine similar to that on the teeth of vertebrates, and they are extremely abrasive; sharkskin has been used as a sandpaper and polishing material. The shark's teeth are modified scales; they are replaced rapidly if they are lost, and they occur in as many as seven overlapping rows. Sharks are actively aware of their environment through good eyesight; excellent senses of smell, hearing, and mechanical reception; and electrical sense.

A shark's vision is more important than previously thought; sharks see well under dim-light conditions. They have the ability to sense chemicals in their environment through smell, taste, a general chemical sense, and unique pit-organs distributed over their bodies. These pit-organs contain clusters of sensory cells resembling taste buds. The shark's sense of smell is acute; it has a pair of nasal sacs located in front of its mouth, and when water flows into the sacs as the shark swims, the water passes over a series of thin folds with many receptor cells. Sharks are most sensitive to chemicals associated with their feeding, and all are able to detect such chemicals in amounts as dilute as one part per billion.

Receptors along the shark's sides are sensitive to touch, vibration, currents, sound, and pressure. The movement of water from currents or from an injured or distressed fish are sensed by the shark's lines of receptors, which communicate with the watery environment by a series of tubes; water displacement stimulates nerve impulses along these systems. The shark is able to hear and uses its hearing in the location of its prey. Pores in the shark's skin, especially around the head and mouth, are sensitive to small electrical fields. Fish and other small marine organisms produce electrical fields around themselves, and the shark uses its electroreception sense to locate prey and recognize food. As a shark swims through Earth's magnetic field, an electrical field is produced that varies with direction, giving the shark its own compass.

We still do not understand completely how all these senses function or how they affect the behavior of sharks; we do know that the shark is extremely well tuned to its environment.

There are more than 350 known species of shark, and scientists are still discovering new species. Sharks are widely spread through the oceans and are found in rivers more than a hundred miles from the sea, and one species, the Lake Nicaragua shark, is found land-locked in fresh water. Some of these sharks are shown in figure 16.13. The whale shark is the world's largest fish, reaching lengths of more than 15 m (50 ft). This graceful and passive animal feeds on plankton and is harmless to other fish and mammals. The docile basking shark, 5–12 m (15–40 ft) long, is another plankton feeder. It is found commonly off the California coast and in the North Atlantic, where it has been harvested for its oil-rich liver. A third species of plankton-feeding shark, 4 m (14 ft)

 web link **Figure 16.12** The bony fishes dominate the world's aquatic environments. Their diversity is enormous, and they are found in almost every conceivable aquatic environment. Two of the thousands of species of reef fish: an emperor angel (a) and a butterfly fish (b). The wolf eel (c) is a carnivore with strong jaws and long teeth. The scales of the sea horse (d) are modified to form protective armor. Sea horse populations are declining, and their slow reproduction rate is not likely to support increasing harvests for aquariums, traditional medicine, and gourmet dishes. The red snapper (e) inhabits rocky coastal areas; it is used as a food fish.

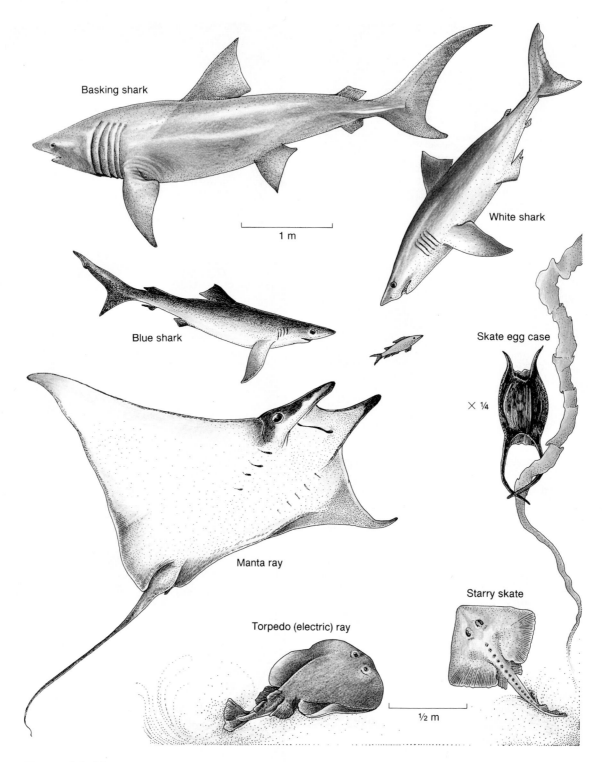

Basking shark

White shark

1 m

Blue shark

Skate egg case

× ¼

Manta ray

Starry skate

Torpedo (electric) ray

½ m

web link **Figure 16.13** Sharks, skates, and rays are all cartilaginous fishes. The leathery egg case of a skate is known as a mermaid's purse.

long and weighing approximately 680 kg (1500 lb), nick-named megamouth was discovered in 1976.

Many sharks are swift and active predators, attacking quickly and efficiently, using their rows of serrated teeth to remove massive amounts of tissue or whole limbs and body portions. They also play an important role as scavengers and, like wolves and the large cats on land, eliminate the diseased and aged animals. Sharks can and do attack humans, al-though the reasons for these attacks and the periodic fren-zied feeding observed in groups of sharks are not understood. A human swimming inefficiently at the surface may look like a struggling, ailing animal and be attacked; a diver swimming completely submerged may appear as a more nat-ural part of the environment and be ignored.

Skates and rays (fig. 16.13) are flattened sharklike fish that live near the sea floor; there are some 450–550 species of skates and rays in the world's oceans. They move by undulating their large side fins, which gives them the appearance of flying through the water. Their five pairs of gill slits are on the underside of their bodies, not along their sides. The large manta rays are plankton feeders, but most rays and skates are carnivorous, eating fish but preferring crustaceans, mollusks, and other benthic organisms. Their tails are usually thin and whiplike and in the case of stingrays, carry a poisonous barb at the base. Some of the skates and a few of the rays have shock-producing electric organs that can deliver shocks of up to 200 volts; these are located along the side of the tail in skates and on the wings of the rays; their purpose appears to be mainly defensive. Like the sharks, most rays bear their young live. Skates enclose the fertilized eggs in a leathery capsule called a sea purse or mermaid's purse that is ejected into the sea and from which the young emerge in a few months.

Commercial Species of Bony Fish

Most commercially valuable food fish are found between the ocean's surface layers and a depth of approximately 200 m (660 ft), or in the epipelagic zone. Most of these fish are streamlined, active, predatory, and capable of high-speed, long-distance travel. Among the most important species are the enormously abundant, small herring-type fish, such as sardine, anchovy, menhaden, and herring. These fish feed directly on plankton and are found in large schools in areas of high primary productivity. Other valuable fish that are commercially harvested include mackerel, pompano, swordfish, and tuna. These fish are caught at sea, out of sight of land.

Fish that live on or near the bottom, known as **demersal fish** or bottom fish, do not swim as rapidly as those that live in the water column above. The flounder, halibut, turbot, and sole are commercially valuable bottom fish. Perch and snapper tend to congregate along the sea floor in the shallower, nearshore areas. They are called rockfish because they hide among the rocks and live in the cracks of underwater reefs. Representative fish are shown in figure 16.14. In section 16.7, we discuss the commercial exploitation and economic significance of some of these food fishes.

Deep-Sea Species of Bony Fish

The fish of the deep sea are not well known, because the depths below the epipelagic zone are difficult and expensive to sample. No fish from these depths have been exploited commercially. Representative fish are shown in figure 16.15.

In the dim, transitional mesopelagic layer, between 200 and 1000 m (660 and 3300 ft), the waters support vast schools of small, luminous fish. Fishes of the mesopelagic are small in size, with large mouths, hinged jaws, and needlelike teeth. *Cyclothone* is believed to be the most common fish in the sea. Each of its species lives at a relatively fixed depth; the deeper-living species are black, and the shallower-living species are silvery to blend with the dim light. At this depth the lantern fish has a worldwide distribution; some 200 species are distinguished by the pattern of light organs along their sides. They are a major item in the diet of tuna, squid, and porpoises. *Stomias*, a fish with a huge mouth, long, pointed teeth, and light organs along its sides, and the large-eyed hatchet fish prey on the great clouds of euphausiids and copepods found at this depth.

In the perpetually dark bathypelagic zone there is little food; only about 5% of the food produced in the photic zone is available at these depths. Here the fish are small, between 2 and 10 cm (1 and 5 in), and have no spines or scales but are fierce and monstrous in appearance. They are mostly black in color with small eyes, huge mouths, and expandable stomachs. They breathe slowly and the tissues of their small bodies have a high water and low protein content. Floating at constant depth without using energy for swimming, these fish go for long periods between feedings by using their food for energy rather than for increased tissue production. Most have bioluminescent organs or photophores on their undersides that allow them to blend with any light from above. These photophores may show different patterns among species and between sexes; they may also identify predators and are used as lures to attract prey. Some species have large teeth that incline back toward the gullet so the prey cannot escape, and others have gaping mouths with jaws that unhinge to allow the catching and eating of fish larger than themselves.

Among the most famous of these predators are *Macropharynx longicaudatus* and the related *Saccopharynx*, which have funnel-like throats and tapering bodies ending in whiplike tails. When the stomach is empty the fish appears slender, but it expands to accept anything the great mouth can swallow. The female angler fish, *Ceratias halboelli*, has a dorsal fin modified into a fishing rod, which slides along a canal in the back of the fish; an illuminated lure dangles from the tip of the rod. Other fish are attracted to the lure, which the angler fish moves forward along her back until the bait is just above her jaws.

Classification Summary of the Nekton 16.6

The nekton, all members of the kingdom Animalia, can be classified in the following phyla and classes:

I. *Kingdom Animalia*: animals; multicellular heterotrophs with specialized cells, tissues, and organ systems.
   A. *Phylum Mollusca*: mollusks.
      1. *Class Cephalopoda*: cuttlefish, squid; the octopus is a member of the benthos.
   B. *Phylum Chordata*: animals with a dorsal nerve cord and gill slits at some stage in development.
      1. *Subphylum Vertebrata*: animals with a backbone of bone or cartilage.
      a. *Class Agnatha*: jawless fish; lampreys and hagfish.
      b. *Class Chondrichthyes*: jawed fish with cartilaginous skeletons; sharks and rays.

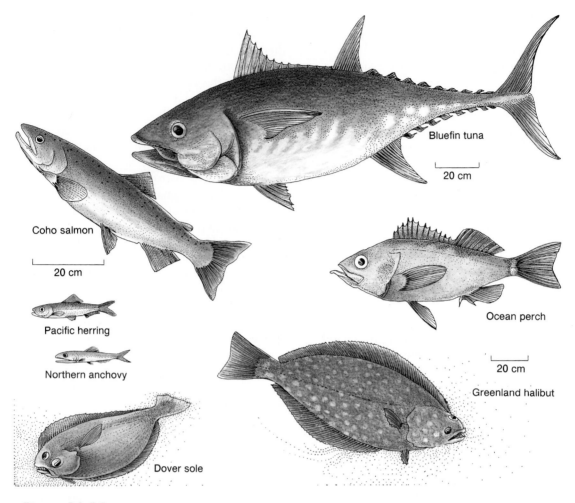

Bluefin tuna

20 cm

Coho salmon

20 cm

Pacific herring

Northern anchovy

Ocean perch

20 cm

Greenland halibut

Dover sole

web link **Figure 16.14** These bony fishes are all harvested commercially.

c. *Class Osteichthyes:* bony fish; all other fish, including commercial species such as salmon, tuna, herring, and anchovy.
d. *Class Reptilia:* air-breathers with dry skin and scales; young develop in self-contained eggs; turtles, sea snakes, iguanas, and crocodiles.
e. *Class Aves:* endotherms; skin covered with feathers, forelimbs are wings; embryo enclosed in shell; birds.
f. *Class Mammalia:* endotherms; body covering of hair; produce milk; young born live.
   (1) *Order Cetacea:* whales.
      (a) *Suborder Mysticeti:* baleen whales, including blue, gray, right, sei, finback, and humpback whales.
      (b) *Suborder Odonticeti:* toothed whales, including killer and sperm whales, porpoise, and dolphin.
   (2) *Order Carnivora:* sea otters, seals, sea lions, and walruses.
   (3) *Order Sirenia:* dugongs and manatees.

Practical Considerations: Commercial Fisheries 16.7

In 1950 the total world marine fish catch (as reported by the Food and Agricultural Organization of the United Nations) was approximately 21 million metric tons. During the next fifty years, as human populations exploded, the fishing effort by all nations intensified, and the technology and gear used to hunt and catch the fish improved dramatically (see chapter 12, section 12.8). By 1960 the catch was 40 million metric tons, and by 1970 it was 70 million metric tons. The tonnage continued to increase until, by 1989, the total world marine fish catch was 86 million metric tons. For the next five years the catch remained between 70 and 80 million metric tons, but in 1995 the marine catch rose again, reaching 86 million metric tons in 1996 and 1997. In 1998 there was a distinct drop to 78 million metric tons that was associated with the 1997–98 El Niño event.

The FAO relies on voluntary reporting of catches to estimate the amount of fish in the oceans. In 2001 an inde-

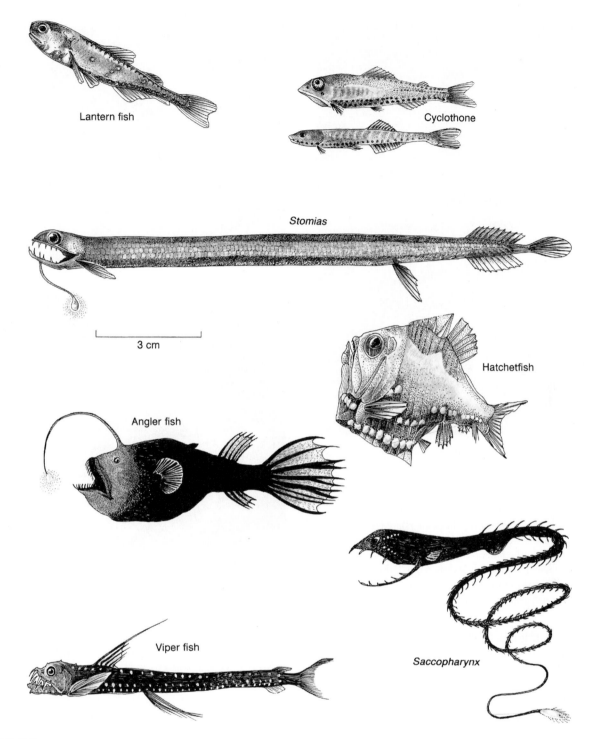

Figure 16.15 Fishes from the deep sea.

pendent analysis of FAO statistics by Canadian researchers showed world fish catches slowly increasing while local fisheries were decreasing. Some countries, particularly China, appear to have distorted their catch estimates. The independent analysis predicted a catch of 5.5 million metric tons for China in 1991; the official Chinese figure given to the FAO was 10.1 million metric tons. At present catch data is in some doubt and the ability of the FAO to assess world fish stocks may be compromised.

Over this period there has also been a shift in the use made of the catch. In 1950 90% of the world fish catch was consumed directly by humans and 10% was used to make fish-meal products to feed the poultry and livestock of the more-developed countries. Today, approximately two-thirds

(a)

(b)

(c)

Figure 16.16 (a) Freshly caught pollock are poured from a trawl net onto the deck of a seafood-processing vessel. (b) Fishes are cleaned and boned below decks in the processing plant. After the fishes have been washed to remove blood and oils and then dried, the fish meat becomes a flavorless, odorless fish product, surimi. (c) The surimi base is made into analogs such as crab shapes.

of the world catch is going to feed people, and one-third is being used as feed for domestic animals. The use of fish meal in this way is only about 20% efficient; that is, for every ton of fish meal fed, only 0.2 ton of additional animal protein is produced.

At the present time, total world marine fish catches increase only slowly despite the increase in fishing effort, the opening of new fisheries, and an increased consumer demand for fish and fish products. Alaskan pollock is an example of a more recent fishery that has responded to consumer demand. The pollock is a bottom fish; it is processed to remove the fats and oils that give the fish its flavor, and a highly refined fish protein called **surimi** is produced. Surimi is processed again and flavored to form artificial crab, shrimp, and scallops. The processing of pollock to form surimi is shown in figure 16.16. Surimi, a major fish product in Japan, is the fastest-growing product in the U.S. seafood market. Meanwhile, many traditional and valuable fisheries have declined, often as a result of overfishing. Five fisheries

are discussed in the following sections; each illustrates the difficulties faced by those fishing, the fishery managers, the fishing nations, and the fish consumers as the catches decline and the costs rise.

Anchovies

The anchovy fishery, concentrated in the upwelling zone off the coast of Peru, has produced the world's greatest fish catches for any single species. Anchovies are small, fast-growing fish that feed directly on the phytoplankton in the upwelling zone and travel in dense schools that are easy to net in large quantities. These fish are not caught for human consumption but are processed into fish meal that is exported as feed for domestic animals.

The fishery began in 1950 with a harvest of 7000 metric tons. By 1962, the harvest had risen to 6.5 million metric tons, and as the world demand for fish meal rose, the anchovy fishing intensified. In 1970, a peak fishing year, about 12.3 million metric tons were taken. As the fishing in-

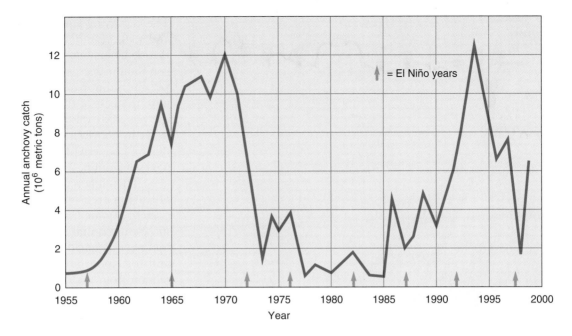

web link **Figure 16.17** The anchovy catch of Peru and Ecuador by calendar year, given in millions of metric tons. Years of El Niño are 1957, 1965, 1972, 1976, 1982–83, 1992–93 and 1997–98.

creased, the average size of the fish taken decreased and it took more and more fish (smaller and younger) to make up the same catch. During periods in which the Peruvian coast is visited by El Niños (see chapter 6 and fig. 16.17), the general wind pattern reverses, the winds become westerly, diminishing the upwelling and moving warmer, nutrient-poor surface water eastward into the areas of normally cold and productive upwelled water. The decrease in nutrients and the increase in temperature result in the destruction of the plant and animal populations; the decomposing organic material strips oxygen from the water and further reduces the populations. El Niño devastates the fishing, and large numbers of seabirds that feed on the schools of anchovies starve.

Follow the changes in the anchovy catch as related to El Niños in figure 16.17. Heavy fishing in 1970–71 combined with the severe effect of a 1972 El Niño resulted in only 2 million metric tons harvested in 1973. With government quotas in place the catch increased slowly in 1974 and 1975, but El Niño struck again in 1976 and in 1982–83. The catch stabilized in the years 1988–90 and continued to increase through the weak El Niño year of 1992–93. The 1994 catch of 12.5 million metric tons is the greatest catch yet taken from this area. Catches in 1995–97 steadied between 8.6 million and 7.6 million metric tons. The impact of the 1997–98 El Niño was dramatic; only 1.7 million metric tons were harvested in 1998. In this fishery overfishing combined with environmental changes intensifies the effect of overharvesting.

Tuna

The world tuna catch from 1988–98 has grown from 4.0 million metric tons to 5.5 million metric tons. This fishery is an expensive, sophisticated business that involves many na-

tions. The tuna boats take advantage of the tuna's schooling habit to catch the fish in huge seine nets 1100 m (3600 ft) long and 180 m (595 ft) deep; a single set of a net brings in over 150 tons of tuna. Although the modern open-ocean netting of tuna is an efficient method of harvesting, it has had a significantly undesirable incidental catch. Tuna are accompanied in their movement across the oceans by porpoises that swim just above the schools of tuna and are used by the tuna boats to locate the schools. The net is set in a circle around the school of tuna and pulled up into a bag. If porpoises are caught in the net they can drown. The total cost of the tuna harvest in porpoises is unknown, but since the passage of the 1972 Marine Mammal Protection Act, conscious and legally enforced efforts have been made to reduce the porpoise kill. It is estimated that the largest kill occurred in 1970, with a loss of more than 500,000 porpoises.

Today many nations, including the United States, ban the import of tuna caught by encircling dolphins with nets, and pressure is placed on other nations to abandon this technique. These efforts have decreased the annual dolphin kill to 20,000 from more than 150,000.

Salmon

Another fishery in which the United States has an important role and investment is the salmon industry of the Pacific Northwest and Alaska (fig. 16.18). These fish spawn in fresh water and remain there as juveniles for a year or less, depending on the species. Then they begin their long journey to the sea, where they will live and grow for one to four years. At the end of this time, the fully mature adults return to the same rivers, lakes, and streams in which they were born to spawn in their turn. As they return, the salmon are fished both commercially and for sport. The demand for

Figure 16.18 A salmon purse seiner hauling in its net.

The returning stocks of coho, chinook, and sockeye salmon along the coasts of British Columbia, Washington, and Oregon diminished greatly during the 1990s, yet the salmon runs for the year 2001 were the best of the previous decade. But the year 2001 was a drought year with low water in northwestern streams and rivers, raising concerns for the survival of the next generation.

The salmon depend on high-quality, pollution-free fresh water and clean, gravel-bottomed, shady, cool streams for successful natural spawning. People have routinely dammed rivers for power usage and for flood control, cutting off the salmon from their home streams; people have harvested trees up to the stream edge, removing shade and allowing mud and silt eroded from the exposed land to cover the streambed. People have used streams to carry away their wastes, degrading the water quality and damaging both the juvenile fish and the returning runs. In many areas, these activities are being corrected, but the naturally spawning fish are now few in number, and people must pay both the cleanup costs and the costs of hatchery programs to supplement the dwindling natural stocks.

Federal, state, and privately owned hatcheries harvest eggs from mature fish, raise the juveniles, and hold them until an optimal time for release. These juveniles go to sea and their survivors return to their natal streams at maturity. This practice can be called ocean ranching or sea ranching. Because the eggs and the juveniles are held in controlled environments, hatchery fish have a survival rate that is eight times that of wild fish. At sea the hatchery and wild fish mix together and return to the same streams. If commercial and sport fishing quotas are based on total numbers of returning fish, the wild stocks will become depleted as the production of hatchery fish increases.

salmon is high, and the fish bring a high price per pound, but like many other heavily exploited fish stocks, the salmon are becoming scarcer, and the fishing is now tightly controlled. Because fish that spawn in the rivers under one governmental jurisdiction are caught as they move through the coastal waters of other governmental jurisdictions, much friction occurs among those negotiating the length of fishing periods and the number of fish to be caught.

Each year's salmon fishery depends on the number of fish returning to spawn in previous years, and the numbers of returning fish can and do vary greatly from year to year.

Atlantic Cod
In 1992 Canada banned cod fishing in the waters off Newfoundland and in 1993 and 1994 extended the ban south into the Gulf of St. Lawrence, effectively shutting down its historic northwest Atlantic cod fishery. Iceland warned its fishing industry that the cod stock could collapse unless the annual catch was cut back by 40%. Also in 1994 the U.S. New England Fishery Management Council stopped the fishing for cod on Georges Bank.

Cod are bottom-dwellers or ground fish; they feed on small fish, crab, squid, and clams. They live twenty to twenty-five years, mature in three to seven years, and produce several million eggs per spawning fish. How could the fish that had fed the Pilgrims and produced catches of 50,000 metric tons by hand fishing a hundred years ago have failed so completely?

For nearly 400 years cod were fished with hand lines; these gave way to longlines with hundreds of hooks and net traps in the nineteenth century. By 1895 trawl nets 45–60 m (150–200 ft) long were in use, and in the 1950s huge factory trawlers from Great Britain, the former Soviet Union, Germany, and other nations joined the fishing. In 1968 the combined fleet caught 810,000 metric tons of cod, nearly three times more than had been caught in a single year before 1954. By the 1970s the cod catch had dropped to less than 200,000 metric tons, and both Canada and the United States extended their territorial waters out 320 km (200 mi) to exclude foreign vessels. Recommended cuts in harvest quotas were not made, and to maintain catch rates fishing effort was intensified. Between 1968 and 1992 there was a 69% decline in the catch, and between 1993 and 1996 another drop of the same magnitude occurred. More than 30,000 people lost their jobs with the closure of the Atlantic cod fishery. In the year 2000–2001 the Grand Banks area off Newfoundland remained closed to cod fishing.

The New England fishery followed the same course—enlarging the fleet from 825 boats to 1423 between 1977 and 1982. Although the catches had begun their decline, the New England Fishery Council was unwilling to control the fishing and eliminated catch quotas in 1982. In 1991 the National Marine Fisheries Service (NMFS) was sued to reinstate the regulations, and by 1993 the NMFS announced that 55% of the New England cod population had been caught and that only closing the fishery could save the stock. The 1999 cod catch was 9700 metric tons compared with 43,700 metric tons in 1990. The fishery remained tightly controlled in 2001 with fishing confined to certain areas and a limit of 10,000 metric tons.

Sharks

As overfishing takes its toll on familiar commercial pelagic species such as tuna and swordfish, the fishing pressure is transferred to other species. Until recently, sharks were fished for sport and for local markets. When U.S. commercial shark fishing began in the late 1970s, 21 metric tons of thresher sharks were brought to California docks; by 1982 the thresher catch was 1100 metric tons. In 1989 the catch fell to 300 metric tons, and the thresher fishery collapsed. The NMFS has reported a 75% decline in heavily fished shark species in U.S. waters.

Shark populations are being decimated in all the world's oceans. Shark catches have been increasing steadily since the 1940s; the 1990 world catch was 690,000 metric tons and the catches of 1996–99 exceeded 800,000 metric tons. The Asian demand for shark-fin soup has supported an expanding shark fishery that amputates the shark fins and

Figure 16.19 National Marine Fisheries Service scientists sample a shark population to evaluate the effects of commercial fisheries.

web link

throws the animals back into the ocean to die, a process known as "finning." Since 1970 there has been a 40% drop in shark caught in the northwestern Atlantic and a 50% drop in the southeastern Atlantic. The problem is intensified by sharks' low reproductive rate, their slow growth, and the long period they require for maturation.

In 1992–93, the NMFS set strict federal quotas on commercial shark fishing in the Atlantic, Gulf of Mexico, and Caribbean (fig. 16.19); recreational fishing was restricted and finning banned. South Africa declared a ban on fishing the great white shark, and Australia banned trade in white shark parts and set quotas for other species. Concern over the loss of the ocean's most widespread top predator is mounting; no one knows how many sharks are in the world's oceans and no one knows how their loss will affect the ocean food webs.

Fish Farming

The world's demand for seafood is increasing at the same time that the ocean harvest of wild fish is decreasing. An alternative way to increase the fish harvest is by fish farming, known as **aquaculture,** the growing or farming of animals and plants in a water environment. The term **mariculture** is used for the growing or farming of marine plants and animals. The UN Food and Agricultural Organization projects that by the year 2010 aquaculture will provide 35% of the world's fish supply for human consumption. Global aquaculture harvests more than doubled in weight between 1986 and 1996. At the present time more than 25% of all fish consumed by humans is raised in aquaculture ponds.

Farming the water began in China some 4000 years ago. The Chinese were culturing common carp in 1000 B.C., and in 500 B.C. a book was written giving directions on fish

Practical Considerations: Commercial Fisheries **443**

farming, including methods for building the pond, selecting the stock, and harvesting it. In China, Southeast Asia, and Japan, fish farming in fresh and salt water has continued to the present as a practical and productive method of raising large quantities of fish such as carp, milkfish, *Tilapia*, and catfish. Most of these farms are family-run, labor-intensive operations. Fish farmers may practice **monoculture,** raising fish as a single species, or **polyculture,** raising more than one species as surface feeders and bottom feeders to make use of the total volume of the pond.

In the United States fish farming produces only 2% of fishery products, but that amount includes 50% of the catfish and nearly all the trout. To be successful and profitable in the United States, fish farming requires a proven market and large-scale operation combined with the development of technology to reduce costs. The species chosen for farming must reproduce easily in captivity, have juvenile forms that survive well under controlled conditions, gain weight rapidly, eat cheap and available food, and fetch a high market price.

As commercial fishing has declined, mariculture has responded by nearly tripling its output in the last decade. In 1987 mariculture provided 4 million metric tons of fish and shellfish; in 1996 it provided 11 million metric tons. Marine fish are grown in floating fish cages or pens kept in calm, shallow, protected, coastal areas, which are becoming less and less available as shorelines are being increasingly built over and used for recreation. The lack of suitable sites, the impact of fish pens on water quality (increasing nutrients, decreasing oxygen), the release of antibiotics into the marine environment, and the possibility of the transmission of fish disease to wild populations are pushing the development of offshore, submersible systems that are more efficient and that distribute waste and excess food over larger areas. Additional research is going into high-energy feeds that promote faster growth and therefore decrease the time needed to bring the fish to market size.

Farming of Atlantic and Pacific salmon has surged past the catch of wild salmon. Sixty percent of the salmon consumed worldwide in 2000 was raised in pens (fig. 16.20). Norway produced 46% of this amount, Chile 23%, Scotland 12%, and Canada 7%. The combined output of Norway, Chile, and Scotland is more than 80% of the world salmon production. Farmed salmon depend on a diet that is 45% fish meal and 25% fish oil. In 1997 about 1.8 million metric tons of wild fish were required to produce the world salmon harvest, thus increasing the pressure on wild fisheries rather than decreasing it (see also shrimp aquaculture in chapter 17).

The Canadian government has adopted a national policy of encouraging salmon farms. In the United States, Maine, Washington, California, and Oregon allow salmon farming but the permit procedure is long and complex. Mariculture research is focusing on several other fish species for marketing by the year 2010. In Hawaii pen-reared mahi mahi now grow to 1.5 kg (3 lb) size in 150 days and are ex-

Figure 16.20 An aerial view of salmon pens, in which fish are raised to market size. *By Sea Farm Washington, Inc.*

pected to be the first to enter the market. Norway has the technology to grow Atlantic cod, but the cost is still too high for the commercial market. The United States and Canada are experimenting with the Atlantic halibut, and in Texas the craze for the cajun dish of blackened redfish has opened the way to redfish farming.

Summary

The nekton swim freely and independently. The marine mammals include whales, seals, sea lions, walruses, sea otters, and sea cows. The whalers hunted baleen whales and the sperm whales, which are toothed whales. As hunting techniques changed, harvesting became more intensive, and some species of whales have been threatened with extinction. The International Whaling Commission regulates whaling on a voluntary basis; a moratorium on commercial whaling is still in effect, but scientific whaling continues.

The fur seals and sea otters were hunted heavily during the nineteenth and early twentieth centuries; these animals are now feeling the effect of increasing groundfishery catches in the northeastern Pacific. Recent diminished walrus harvests are due to politics, economics, and weather conditions. The Stellar sea cow is extinct; manatees and dugongs are found in the warm waters of the Indian and Atlantic Oceans, where they are caught between increasing human contacts and loss of habitat. The U.S. Marine Mammal Protection Act was passed in 1972 to protect all marine mammals and to prohibit commercial trade in marine mammal products.

Only 3% of total bird species are marine; these birds are specialized for their life at sea. Many marine birds are migratory and travel long distances each year. Shorebirds and seabirds congregate in large numbers in shore areas for nesting; these areas are susceptible to human disturbance.

Sea snakes, the marine iguana, seagoing crocodiles, and sea turtles are the reptile members of the nekton. Sea turtle populations are under great pressure by hunters and poachers. The squid also belong to the nekton.

Fish are found at all depths in all oceans. Sharks, rays, and skates are primitive fish with cartilaginous skeletons. All other fish have bony skeletons, including the commercially fished species and the highly specialized types of the deeper ocean.

World fish catches have increased dramatically since 1950 due to increased fishing effort and improved technology. Although stock abundance has declined, new fisheries have developed, and the fishing effort has continued high to keep the catch stable. An independent analysis of FAO statistics in 2001 has questioned the reporting of catch estimates leading to doubts in the validity of some current data.

The problems of the traditional commercial fisheries include overfishing, increased regulation, and increased costs. The Peruvian anchovy fishery had produced the world's greatest fish catches until it was hit by the effects of overfishing and El Niño. The U.S. tuna fishery has had to cope with the killing of porpoises as a side effect of fishing technology, and the Pacific Northwest and Alaskan salmon fisheries are experiencing difficulty due to the degradation of the salmon's freshwater environment and the loss of natural salmon runs. The Atlantic cod fishery collapsed and remains closed due to overfishing. World shark fisheries are depleting shark populations in all the oceans.

Aquaculture and mariculture are increasing worldwide, and they have proved to be practical and productive in much of Asia. Norway, Chile, and Scotland produce most of the farmed salmon. In the United States, marine aquaculture is a small industry that requires large-scale operations, proven markets, and cost-cutting technology to succeed. Undesirable side effects include degraded water quality, possible disease transfer to the native stocks, and removal of wild juvenile fish to mature in pens.

Key Terms

All key terms from this chapter can be viewed by term, or by definition, when studied as flashcards on this book's Online Learning Center at www.mhhe.com/sverdrup (click on this book's cover).

mammal, 416
cetacean, 416
baleen, 416
pinniped, 421
manatee, 425
dugong, 425
sea cow, 425
Sirenia, 425

echolocation, 426
demersal fish, 437
surimi, 440
aquaculture, 443
mariculture, 443
monoculture, 444
polyculture, 444

Study Questions

1. The Marine Mammal Protection Act of 1972 has allowed the recovery of certain populations to such an extent that the marine mammals are now competing for food resources of economic value to humans. Under these circumstances should any adjustments be made to this act? Explain why.
2. What sensory abilities help a shark to find its prey?
3. In the last fifty years the world's marine fish catch has grown from 40 million metric tons to more than 80 million metric tons. During this period the catch from many large fisheries has declined. How do you explain this contradictory situation?
4. How have the problems of migratory seabirds changed during the last hundred years?
5. What uses do marine mammals make of sound?
6. Why has the sea otter population diminished in the area of the Aleutian Islands over the last five years?
7. How can the ability to predict El Niño help manage the Peruvian anchovy fishery?
8. Why are more toothed whales found at low latitudes and why do more baleen whales reside in temperate and polar latitudes? Consider food requirements only.
9. Compare the problems of the Atlantic cod fishery and the Pacific salmon fishery. Are there any factors in common to both fisheries? How do these fisheries differ?
10. Sea snakes are thought to have their origin in the land snakes of India and Southeast Asia. Why are there no sea snakes in the Atlantic Ocean? What might be the effect of a sea-level canal through the isthmus of Panama?
11. Explain the status of sea turtle populations at the present time.
12. Discuss the impact of humans on populations of walrus, fur seals, and sea otters.
13. Compare fishing as practiced in the United States and in less-developed countries.

14. Compare fish farming in the United States with fish farming in other countries.
15. What characteristics do deep-sea fish have in common? How do these characteristics enhance fish survival in this environment?

Links to Related Websites

Visit the book's Online Learning Center at www.mhhe.com/sverdrup (click on the book's cover) to find live Internet links for additional topics related to this chapter's content.

- Marine taxa
- Marine mammals

Visit the book's Online Learning Center at www.mhhe.com/sverdrup (click on this book's cover) to find these additional chapter tools: Suggested Readings; links to further information on boxed readings, selected figures, and related chapter topics; and additional study aids.

chapter 17

The Benthos: Dwellers of the Sea Floor

Bat stars cluster in a tide pool in the Queen Charlotte Islands, British Columbia, Canada.

The exposed rocks had looked rich with life under the lowering tide, but they were more than that; they were ferocious with life. There was an exuberant fierceness in the littoral here, a vital competition for existence. Everything seemed speeded-up; starfish and urchins were more strongly attached than in other places, and many of the univalves were so tightly fixed that the shells broke before the animals would let go their hold. Perhaps the force of the great surf which beats on this shore has much to do with the tenacity of the animals here. It is noteworthy that the animals, rather than deserting such beaten shores for the safe cove and protected pools, simply increase their toughness and fight back at the sea with a kind of joyful survival. This ferocious survival quotient excites us and makes us feel good, and from the crawling, fighting, resisting qualities of the animals, it almost seems that they are excited too.

John Steinbeck
From *The Log from the Sea of Cortez*

The Benthos: Dwellers of the Sea Floor

The animals and plants that live on the sea floor or in the sediments are members of the benthos. The plants are found in the sunlit shallow coastal areas and in the intertidal zones, while the benthic animals are found at all ocean depths. The benthos include remarkably rich and diverse groups of organisms, among them the luxuriant, colorful tropical coral reefs, the newly discovered organisms surrounding deep-sea hydrothermal vents, the great cold-water kelp forests, and the hidden life below the surface of the mudflat and the sandy beach. Benthic organisms are important food resources and provide valuable commercial harvests: for example, oysters, clams, crabs, and lobsters. In this chapter we first present an overview of the benthos by group and habitat, with special focus on the world's more intriguing benthic communities, and then we consider the harvesting of the benthos, its problems, and its potential. This chapter is intended to serve only as an introduction to this topic, for no single chapter can do justice to the profusion of organisms present in this group. If you wish to establish a greater appreciation and understanding of the benthos, we suggest additional readings in books and magazines devoted to marine biology and biological oceanography.

Algae and Plants 17.1

Seaweeds are members of a large group called **algae.** The unicellular planktonic algae were considered in chapter 15, and in this chapter we consider the large, benthic multicellular algae, or seaweeds, conspicuous members of coastal intertidal and subtidal communities. Because algae photosynthesize and contain chlorophyll pigments, they have been considered members of the plant kingdom, but most biologists view them as only plantlike, placing them in other categories. Algal body form, reproduction, accessory pigments, and storage products generally differ from those of land plants. These algae have simple tissues; they do not produce flowers or seeds, and their pigments and storage compounds vary from group to group.

General Characteristics of Benthic Algae

Seaweeds are benthic organisms; they grow attached to rocks, shells, or any solid object. Seaweeds are attached by a basal organ known as a **holdfast** that anchors the plant firmly to a solid base or substrate. The holdfast is not a root; it does not absorb water or nutrients. Above the holdfast is a stemlike portion known as the **stipe.** The stipe may be so short that it is barely identifiable, or it may be up to 35 m (115 ft) in length. It acts as a flexible connection between the holdfast and the **blades,** the alga's photosynthetic organs. Seaweed blades are thin and are bathed on all sides by water. They serve the same purpose as leaves, but they do not have the specialized tissues and veins of leaves because algae do not require that water be conducted from the ground, up a stem, and through the veins to leaf cells. The blades may be flat, ruffled, feathery, or even encrusted with calcium carbonate. The general characteristics of a benthic alga are shown in figure 17.1.

Seaweeds are not found in areas of mud or sand where their holdfasts have nothing to which they can attach. Sometimes during a storm the plants are dislodged, taking with them the rocks to which the holdfasts cling. The plants, carrying their rock anchors, may drift for some time before they sink back to the bottom. If they sink too deep for sufficient light to reach them, they die, but if their blades remain in the photic zone, they will continue to grow. Because the benthic algae are dependent on sunlight, they are confined to the shallow depths of the ocean, where they are surrounded by water, dissolved carbon dioxide, and nutrients. They are efficient primary producers, exposing a large blade area to both the water and the Sun.

In the sea, both the quality of light and the quantity of light change with depth (see chapter 4). The red end of the visible light spectrum is quickly absorbed at the surface, and the blue-green light penetrates to the greatest depths. On land and at the sea surface, plants receive the full spectrum of visible light, and it is the green algae with the same chlorophyll pigments as land plants, able to absorb both long and short wavelengths of light, that are found in shallow water. Algae found at moderate depths have a brown pigment that is more efficient at trapping the shorter wavelengths of solar light. At maximum growing depths, the algae are red, for the red pigment can

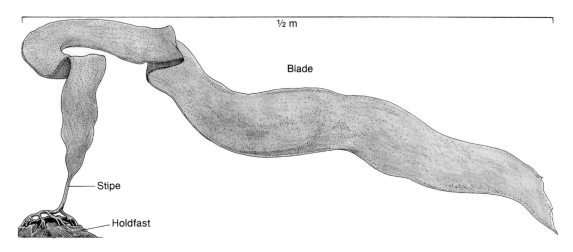

½ m

Blade

Stipe

Holdfast

Figure 17.1 Benthic algae are attached to the sea floor by a holdfast. A stipe connects the holdfast to the blade. *Laminaria* is a genus of kelp and a member of the brown algae.

best absorb the remaining blue-green light. Both the brown algae and the red algae possess the green chlorophyll, but the red and brown pigments mask its color, so that the seaweeds appear red or brown. Therefore, the characteristic pattern for seaweed growing on a rocky shore is the green algae in shallow water, then the brown algae, and last the red algae at greater depths. Few large red algae are seen at low tide, for they are primarily sublittoral species.

Seaweeds provide food and shelter for many animals. They act in the sea much as the forest and shrubs do on land. Some fish and other animals, such as sea urchins, limpets, and some snails, feed directly on the algae; other animals feed on shreds and pieces as they settle to the bottom. Some organisms use large seaweeds as a place of attachment; some of the smaller algae grow on the larger forms. Algae that produce calcareous outer coverings are important in the building of tropical coral reefs; these are discussed later in this chapter.

Kinds of Seaweeds

Algae can be divided into convenient groups based on their pigments. Green algae are moderate in size and may form fine branches or thin, flat sheets. They are mostly freshwater organisms, but a small number occur in the sea, including *Ulva*, the sea lettuce, and *Codium*, known as dead man's fingers. Green algae are most similar to the land plants; they have the same green chlorophyll pigments as land plants and also store starch as a food reserve.

All brown algae are marine and range from simple microscopic chains of cells to the **kelps,** which are the largest of all the algae. Kelps have more tissue structure than most algae but are much simpler than flowering land plants. The kelps have strong stipes and effective holdfasts that allow them to colonize rocky points in fast currents or heavy surf, a habitat favored by the sea palm, *Postelsia*. Other kelps grow with holdfasts well below the depth of wave action and float their blades at the surface supported by gas-filled floats; for example, the bull kelp, *Nereocystis*, is especially abundant

along the Alaskan, British Columbian, and Washington coasts, and the great kelp, *Macrocystis*, is found along the California coast. Other species are found off Chile, New Zealand, northern Europe, and Japan. Their brown color comes from fucoxanthin pigment, which masks the green chlorophyll. Storage products include the carbohydrates laminarin and mannitol but not starch.

The red algae are almost exclusively marine; they are the most abundant and widespread of the large algae. Their body forms are varied and often beautiful, flat, ruffled, lacy, or intricately branched. Their life histories are specialized and complex, and they are considered the most advanced of the algae. All contain the pigments phycoerythrin and phycocyanin, which mask their chlorophyll. Their storage product is floridean starch.

Although the algae are commonly classified by color, the visible color can be misleading. Some red algae appear brown, green, or violet, and some brown algae appear black or greenish. Representative algae are shown in figure 17.2.

Although the diatoms were discussed as part of the plankton (see chapter 15), there are also benthic diatoms. These diatoms are usually of the pennate type and grow on rocks, muds, and docks, where they produce a slippery brown coating.

Other Marine Plant Communities

There are a few flowering plants with true roots, stems, and leaves that have made a home in the sea. Sea grasses grow, often completely submerged, in patches along muddy beaches, where they help to stabilize the sediments. Eel grass, with its strap-shaped leaves, is found growing on mud and sand in the quiet waters of bays and estuaries along the Pacific and Atlantic coasts, and turtle grass is common along the Gulf Coast. Surf grass flourishes in more turbulent areas exposed to waves and tidal action. Sea grasses are important primary producers, and their decomposing leaves add large quantities of vegetative material and nutrients to shores and estuaries. They provide a place of attachment for sponges,

Algae and Plants **449**

web ○○ link **Figure 17.2** Representative benthic algae. *Ulva* and *Codium* are green algae. *Postelsia*, *Nereocystis*, and *Macrocystis* are kelps. The kelps and *Fucus* are brown algae. The red algae are *Corallina*, *Porphyra*, and *Polyneura*. *Corallina* has a hard, calcareous covering.

small worms, and tunicates and a feeding ground for many benthic organisms.

Temperate-area salt marshes are dominated by marsh grasses able to tolerate the brackish water. Marsh grasses are partly consumed by marsh herbivores, but much of the vegetation breaks down in the marsh and is washed into the estuaries by tidal creeks. There the vegetative remains are broken down by bacteria that release nutrients to the water to be recycled by the algae and the sea grasses. See the box on *Spartina* salt marshes in chapter 12.

The mangroves (see fig. 11.13) grow in the intertidal along humid, tropical coasts. They are salt-tolerant woody trees with special adaptations that allow them to thrive in oxygen-deficient muds. Their waxy leaves reduce water loss, and they excrete excess salt through glands located on their leaves. Some mangroves grow prop roots from overhanging branches to extend the root system above the water; others send up vertical roots from roots below the water's surface. Birds, insects, and other animals live in the mangrove's leafy canopy, and its intertwining prop roots provide shelter for small marine organisms. The roots also trap sediments and organic material, which help to build and extend the shore seaward.

Animals 17.2

Benthic animals, unlike benthic plants, are found at all depths and are associated with all substrates. More than 150,000 benthic species exist, as opposed to about 3000 pelagic species. About 80% of the benthic animals belong to the **epifauna;** these animals live on or attached to rocky areas or firm sediments. Animals that live buried in the substrate belong to the **infauna** and are associated with soft sediments such as mud and sand. There are only about 30,000 known species of infauna, as compared with 120,000 species of epifauna.

Some animals of the sea floor are **sessile,** or attached to the sea floor, as adults (for example, barnacles, sea anemones, and oysters), while others are motile all their lives (for example, crabs, starfish, and snails). Although most benthic forms produce motile larval stages that spend a few weeks of their lives as members of the plankton (refer to the discussion of the meroplankton in chapter 15), the planktonic larval stage is of particular importance to the sessile species. The mobility of the juvenile stages allows these species to relieve overcrowding and to colonize new areas. Sessile adults must wait for their food to come to them, either under its own power or carried in by waves, tides, and currents; motile organisms are able to pursue their prey, scavenge over the bottom, or graze on the seaweed-covered rocks.

The distribution of benthic animals is not governed by any single factor such as light or pressure. In fact, the increasing pressure that occurs with depth does not have much effect on animals that are more than 90% water and have no internal gas storage chambers or lungs. Animal distribution is controlled by a complex interaction of factors, creating living conditions that are extremely variable. The substrate may be solid rock, movable cobbles, shifting sand, or soft mud. Temperature is nearly constant in deep water but changes abruptly in shallow areas covered and uncovered by the daily tidal cycle; salinity, pH, exposure to air, oxygen content of the water, and water turbulence also change abruptly in the intertidal zone. Benthic animals exist at all depths and are as diverse as the conditions under which they live. Their lifestyles are related to their varied habitats, and the following sections provide an overview of habitats and lifestyles.

Animals of the Rocky Shore

The rocky coast is a region of rich and complex algal, plant, and animal communities living in an area of environmental extremes. It is a meeting place between the more variable

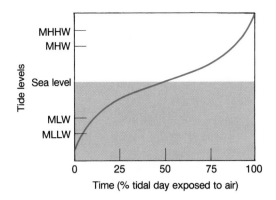

Figure 17.3 The time of exposure to air for intertidal benthic organisms is determined by their location above and below the sea level and by the tidal range. (MLLW = mean lower low water range; MLW = mean low water; MHW = mean high water; MHHW = mean higher high water.)

land conditions and the more stable sea conditions. As the water moves in and out on its daily tidal cycle, rapidly changing combinations of temperature, salinity, moisture, pH, dissolved oxygen, and food supply must be endured. At the top of the littoral zone, organisms must cope with long periods of exposure, heat, cold, rain, snow, and predation by land animals and seabirds as well as turbulence of waves and seaward water flow. At the littoral zone's lowest reaches, the organisms are rarely exposed but have their own problems of competition for space and predation by other organisms. The exposure endured by marine life at different levels in the littoral zone is shown in figure 17.3.

The distribution of the plants and animals is governed by their ability to cope with the stresses that accompany exposure, turbulence, and loss of water. Along the rocky coasts of such areas as North America, Australia, and South Africa, biologists have noted that patterns form as the algae and animals sort themselves out over the intertidal zone. This grouping is called zonation; **vertical zonation,** or **intertidal zonation,** is shown in figure 17.4. The distribution of the seaweeds, with the green algae in shallow water, the brown algae in the intertidal zone, and the red algae in the subtidal area, is an example of such vertical zonation.

The following is a general discussion of the more conspicuous life forms found in these rocky intertidal zones. It is not a guide to the marine life of rocky shores, as substantial differences occur from region to region. If you wish to explore the rocky shores in your area, obtain a manual written specifically for your location.

In the supralittoral (or splash) zone, which is above the high-water level and is covered with water only during storms and the highest tides, the animals and plants occupy an area that is as nearly land as it is ocean bed. The width of the zone varies with the slope of the rocks, variations in light and shade, exposure to waves and spray, tidal range, and the frequency of cool days and damp fogs. At the top of this area, patches of dark lichens and algae appear as crusts on the rocks; these crusts are often nearly indistinguishable

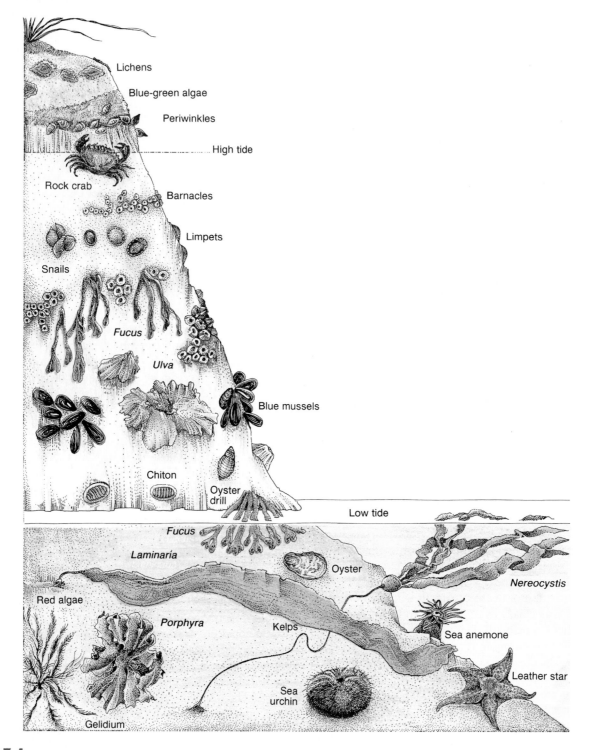

Figure 17.4 A typical distribution of benthic plants and animals on a rocky shore at temperate latitudes. Vertical zonation is the result of the relationships of the organisms to their intertidal environment.

from the rock itself. Scattered tufts of algae provide grazing for the small herbivorous snails and limpets of the supralittoral zone. The periwinkle snail *Littorina* is so well adapted to an environment that is more dry than wet that it is an air-breather, and some species will drown if caught underwater. At low tide snails withdraw into their shells to prevent moisture loss and seal themselves off from the air with a

horny disk called an operculum. Limpets are able to use their single muscular foot to press themselves tightly to the rocks to prevent drying out.

Just below the zone of snails and limpets, the small acorn barnacles filter food from the seawater and are able to survive even though they are covered with water only briefly during the few days of the spring tides each month. Barna-

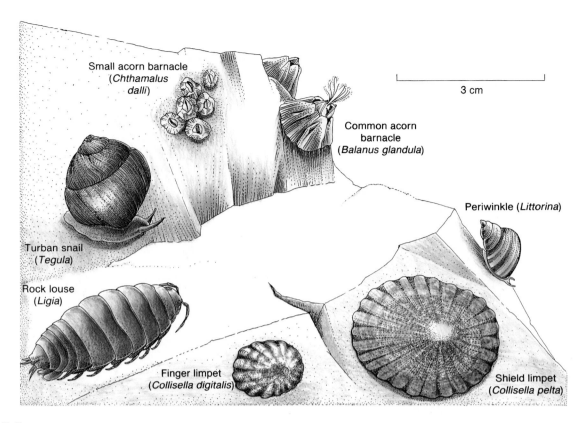

Figure 17.5 Organisms of the supralittoral zone. The limpet and the snail *Littorina* are herbivores. The barnacles feed on particulate matter in the water. *Ligia* is a scavenger.

Labels in figure:
- Small acorn barnacle (*Chthamalus dalli*)
- Common acorn barnacle (*Balanus glandula*)
- Periwinkle (*Littorina*)
- Turban snail (*Tegula*)
- Rock louse (*Ligia*)
- Finger limpet (*Collisella digitalis*)
- Shield limpet (*Collisella pelta*)
- 3 cm

cles are crustaceans, related to crabs and lobsters but cemented firmly in place. They have been described as animals that lie on their backs and spend their lives kicking food into their mouths with their feet. In some areas, the rocky splash zone is the home of another crustacean, the large (3–4 cm, or 1–2 in) isopod *Ligia*. Organisms of the supralittoral zone are shown in figure 17.5.

Conspicuous members of the upper midlittoral zone are illustrated in figure 17.6. These organisms include several other species of barnacles, limpets, snails, and two other **mollusks:** the bivalved (or two-shelled) mussels and chitons, which may appear to resemble limpets but on closer inspection will be seen to have shells of eight separate plates. Chitons, like limpets, are grazers that scrape algae from hard surfaces. Mussels are filter feeders; food strained from the water is trapped in a heavy mucus and moved to the mouth by liplike palps. A muscular foot anchors chitons and limpets; strong cement secures barnacles; and special threads attach mussels to the rocks. Tightly closed shells protect many of the organisms from drying out during periods of low tide, and their rounded profiles present little resistance to the breaking waves. Species of brown algae in this zone, typically rockweed *(Fucus)*, have strong holdfasts and flexible stipes.

Gooseneck barnacles are found attached to rocks where wave action is strong. They have evolved an interesting feeding style, facing shoreward and feeding by taking particulate matter from the runback of the surf rather than facing the sea, as might be expected. Mussel beds promote

shelter for less conspicuous animals, such as the segmented sea worm *Nereis* and small crustaceans. Shore crabs of varied colors and patterns are found in the moist shelter of the rocks. The crabs are active predators as well as important scavengers. Small sea anemones, huddled together in large groups to conserve moisture, are also found in the higher regions of the midlittoral zone.

The area is crowded; the competition for space appears extreme. The free-swimming juveniles (or larval forms) of these animals settle and compete for space, each with its own special set of requirements. New space in an inhabited area becomes available as the whelks, which are carnivorous snails, prey on some species of thinner-shelled barnacles, and the sea stars move up with the tide to feed on the mussels. This predation restricts the thinner-shelled barnacles to the upper levels of the littoral zone, which are too dry for the predator snails. Other species of barnacles inhabit the lower midlittoral zone in association with the predator snails, which cannot pierce the heavier plates of the mature individuals. Seasonal die-offs of algae and the battering action of strong seas and floating logs also act to clear space for newcomers.

A selection of organisms from the lower littoral zone is found in figure 17.7. The larger anemones are common inhabitants of the midlittoral and lower littoral zones. These delicate-looking, flowerlike animals attach firmly to the rocks and spread their tentacles, loaded with poisonous darts called nematocysts. The darts are fired when small fish, shrimp, or worms brush the tentacles. The prey is paralyzed,

Animals **453**

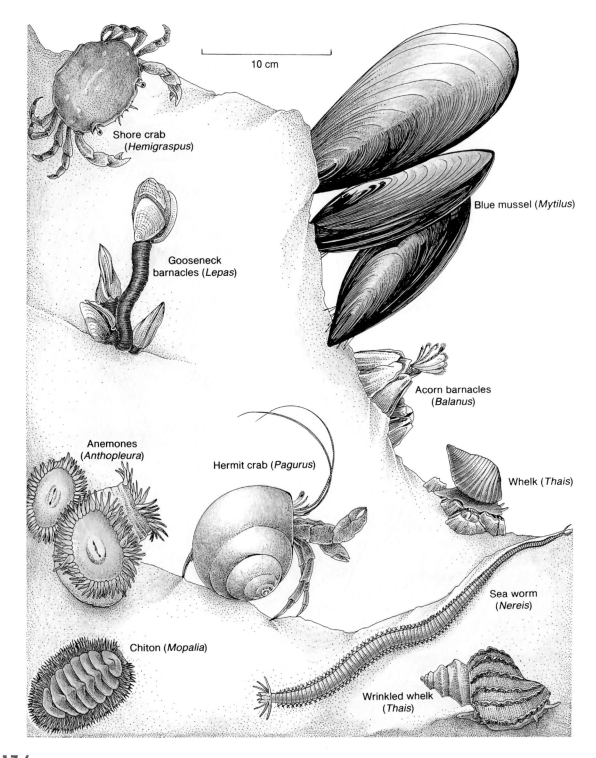

Figure 17.6 Representative organisms from the midlittoral zone. The mussels and barnacles filter their food from the water. The chitons graze on the algae covering the rocks. *Thais*, a snail, is a carnivore. Small shore crabs and hermit crabs are scavengers. *Balanus cariosus* is a larger and heavier barnacle than the barnacles of the supralittoral zone. *Nereis* is often found in the mussel beds. The anemones huddle together to keep moist when exposed.

the tentacles grasp, and the prey is pushed down into the anemone's central mouth. A finger inserted into the tentacles of most anemones will feel only a stickiness and a tingling sensation, but some large tropical species can produce a painful injury. Some snails and sea slugs are unaffected by

the nematocysts and prey on the anemones. Certain sea slugs store the anemone's nematocysts in their tissues and use them for their own defense.

Starfish of many colors and sizes make their home in the lower littoral zone; these slow-moving but voracious carni-

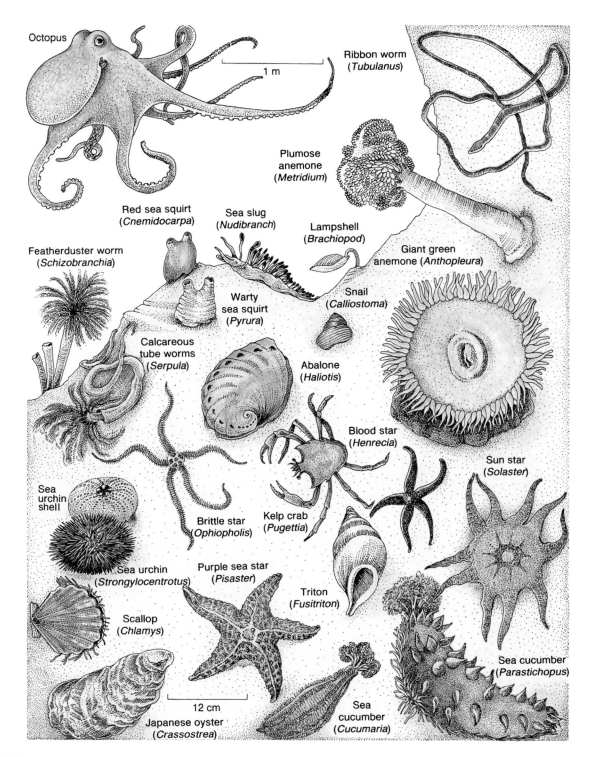

Figure 17.7 Lower littoral zone organisms. A variety of related organisms inhabit the area. The starfish feed on the oysters; the related sea urchins are herbivores; and the sea cucumbers feed on detritus suspended in the water. Among the mollusks are the oysters, scallops, snails, abalone, nudibranchs, and octopuses. The oysters and scallops are filter feeders. *Calliostoma* and the abalone are grazers; the nudibranchs, the octopus, and the triton snail are predators. Note the size of the anemones in this zone.

vores prey on shellfish, sea urchins, and limpets. Their mouths are on their undersides, at the center of their central disks, surrounded by strong arms that are equipped with hundreds of tiny suction cups, or tube feet. The tube feet are operated by a water-vascular system, a kind of hydraulic system that at-taches the animal very firmly to a hard surface. In feeding, the tube feet attach to the shell of the intended prey and, by a combination of holding and pulling, open the shell suffi-ciently to insert the starfish's stomach, which can be extruded through its mouth; enzymes are released, and digestion begins.

Shellfish sense the approach of a starfish by substances it liberates into the water, and some execute violent escape maneuvers. Scallops swim jerkily; clams and cockles jump away; even the slow-moving sea urchins and limpets move as rapidly as possible. Starfish are so effective as predators that oyster farmers must take care to exclude them from their oyster beds.

The filter-feeding sponges, some flat and some vase-shaped, encrust the rocks. On a minus tide, delicate, free-living flatworms and long **nemerteans,** or ribbon worms, armed with poison-injecting mouthparts, are found keeping moist under the mats of algae. Snails and crabs inhabit this zone; the scallop, another filter-feeding bivalve, and red algae are found as well, along the beds of kelp, eel grass, and surf grass. Calcareous red algae encrust some rocks and are seen as tufts on others. Occasionally **brachiopods,** or lampshells, are found in the lower littoral zone. They resemble clams but are completely unrelated to them. Their shells enclose a coiled ridge of tentacles used in feeding.

Beautiful, graceful, and colorful sea slugs, or **nudibranchs,** are active predators, feeding on sponges, anemones, and the spawn of other organisms. Although soft-bodied, they have few if any enemies, because they produce poisonous acid secretions. Herbivores are also present in the lower intertidal region; species of chitons and limpets as well as the sea urchin graze on the algae covering the rocks. Sea cucumbers are found wedged in cracks and crevices; some types are identifiable by their brightly colored tentacles, which act as mops to remove food particles from the water and thrust them into the animal's mouth. Tube worms secrete the leathery or calcareous tubes in which they live and extend only their graceful, feathered tentacles to strain their food from the water.

Octopuses are seen occasionally from shore on very low tides. These eight-armed carnivorous animals are soft-bodied mollusks. They feed on crabs and shellfish and live in caves or dens identifiable by the piles of waste shells outside. They are known for their ability to flash color changes and move gracefully and swiftly over the bottom and through the water. The world's largest octopus is found in the coastal waters of the eastern North Pacific. It commonly measures 2–3 m (10–16.5 ft) in diameter and weighs 20 kg (45 lb), but specimens in excess of 7 m (23 ft) and 45 kg (100 lb) have been observed. They are shy and nonaggressive, although they are curious and have been shown to have learning ability and memory.

Tide Pools

The zonation of benthic forms varies with local conditions; zones are generally narrow where the beach is steep and the tidal range is small, while in areas where the beach is flat and the range of tides is large the zones are wide. The orientation of the shore to sunlight and shade, wave action, and beach topography often results in the apparent displacement of benthic zones. A tide pool formed from water left by the receding tide in a rock depression or basin provides a habitat for animals common to the lower littoral zone. Some small, isolated tide pools provide a very specialized habitat of increased salinity and temperature due to evaporation and solar heating. On a summer day, the water in a tide pool may feel quite warm to the touch. Other tide pools act as catch basins for rainwater, lowering the salinity of the water and its temperature in the fall and winter. Isolated tide pools often support blooms of microscopic algae that give the water the appearance of pea soup, and in various parts of the world, a tiny, bright-red copepod, *Tigripus*, is also found.

The deeper the tide pool and the greater the volume of water, the more stable its environment during its isolation by the receding tide. The larger the tide pool, the more slowly it will change temperature, salinity, pH, and carbon dioxide–oxygen balance. Subtidal animals such as starfish, sea urchins, and sea cucumbers cannot survive in pools such as that described in the preceding paragraph. These animals do not tolerate significant changes in their chemical and physical environment; they require large, deep pools. A few fish species, such as the small sculpins, can be found in tide pools. These fish are patterned and colored to match the rocks and the algae within the pool. They spend much of their time resting on the bottom, swimming in short spurts from one resting place to another. Each tide pool is a specialized environment populated with organisms that are able to survive under the conditions established in that particular pool.

The bottom of the littoral (or intertidal) zone merges into the beginning of the sublittoral (or subtidal) zone extending across the continental shelf. If the shallow areas of the subtidal zone are rocky, many of the same lower littoral zone organisms will be found. When soft sediments begin to collect in protected areas or deeper water, the population types change, and animals of the rocky bottom are replaced by those of the mud and sand substrates.

Animals of the Soft Substrates

The distribution of life in soft sediments is shown in figure 17.8. A selection of animals from this region is found in figure 17.9. Along exposed gravel and sand shores waves produce an unstable benthic environment. Few algae can attach to the shifting substrate and few grazing animals are found. Sands and muds deposited in coves and bays with reduced water motion provide a more stable habitat. Here the size and shape of the sediment particles and the organic content of the sediment determine the quality of the environment. The size of the spaces between the particles regulates the flow of water and the availability of dissolved oxygen. Beach sand is fairly coarse and porous, gaining and losing water quickly, while fine particles of mud hold more water and replace the water more slowly. Sand beaches exchange water, dissolved wastes, and organic particles more quickly than mudflats. The finer the mud particles, the tighter they pack together and the slower the exchange of water; oxygen is not resupplied quickly and wastes are removed slowly. Digging into the mud will generally show a black layer 1 or 2 cm (0.5 or 1 in) below the surface. Above this layer, the water between the sediment particles contains dissolved oxygen; below the black layer, organisms (mainly bacteria) function without oxygen, producing hydrogen sulfide (the rotten egg smell).

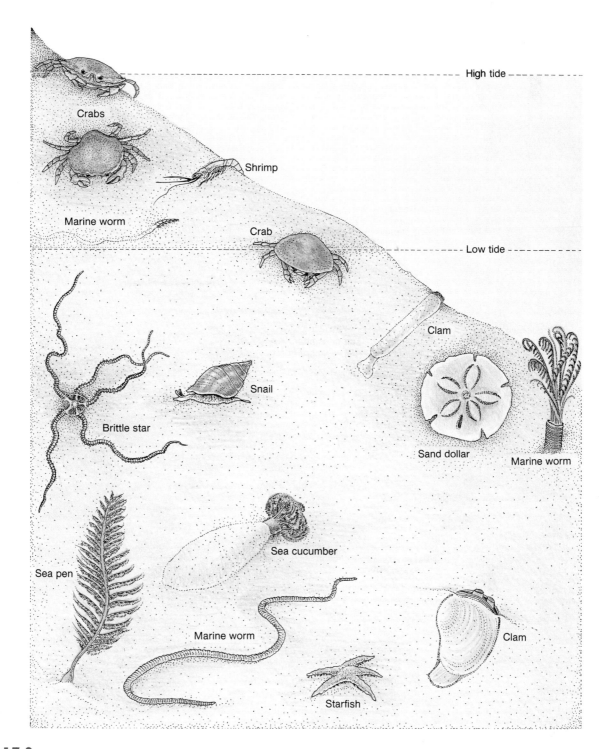

High tide

Crabs

Shrimp

Marine worm

Crab

Low tide

Clam

Snail

Brittle star

Sand dollar

Marine worm

Sea cucumber

Sea pen

Marine worm

Clam

Starfish

Figure 17.8 Zonation on a soft-sediment beach is less conspicuous than that found on a rocky beach. Animals living at the higher tide levels burrow to stay moist.

Lack of oxygen restricts the depth to which infauna species can be found, but some animals, such as clams, live below the oxygen level. Clams use long extensions called siphons to obtain food and oxygen from the water above the sediments.

In locations protected from waves and currents, eel grass and surf grass help stabilize the small-particle sediments and provide shelter, substrate, and food, creating a special community of plants and animals. (See "Other Marine Plant Communities" earlier in this chapter.)

Most sand and mud animals are **detritus** feeders. Most detritus is formed from plant material that is degraded by bacteria and fungi. The sand dollar feeds on detritus particles found between the sand grains. Clams, cockles, and some worms are filter feeders, feeding on the detritus and microscopic organisms suspended in the water. Other animals are deposit feeders that engulf the sediment and process it in their gut to extract organic matter in a manner similar to that of earthworms. These deposit feeders are usually

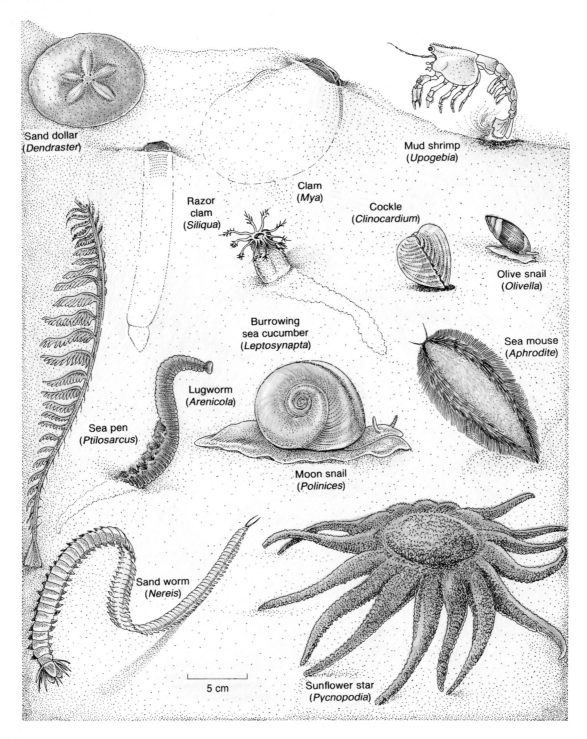

Figure 17.9 Organisms of the soft sediments. Infauna types include the shrimp (*Upogebia*), the lugworm (*Arenicola*), the clam, the cockle, and the burrowing sea cucumber. The sand dollar feeds on detritus; the Moon snail drills its way into shellfish; the sea pens feed from the water above the soft bottom. The sea mouse, like *Nereis*, is a polychaete worm.

found in muds or muddy sands that have a high organic content: for example, burrowing sea cucumbers and the lugworm *Arenicola*, which produces the coiled castings seen outside its burrow. Small crustaceans, crabs, and some worm species are scavengers, preying on any available plant or animal material, while still other worms and snails are carnivores. The Moon snail is a clam eater that drills a hole in the shell of its prey and then sucks out the flesh.

Bacteria not only play the major role in the decomposition of organic material; they also serve as a major protein source. It is estimated that 25%–50% of the material the bacteria decompose is converted into bacterial cell material, which is consumed by microscopic protozoans, which in turn serve as food for tiny worms, clams, and crustaceans. Areas of mud that are high in organic detritus produce large quantities of bacteria.

The intertidal area of a soft-sediment beach shows some zonation of benthic organisms, but it is not nearly as clear-cut as the zonation along a rocky cliff. In temperate latitudes, small crustaceans called sand hoppers are found at the high intertidal region; they are replaced by ghost crabs in the tropics. Lugworms, mole crabs, and ghost shrimp occupy the midbeach area, while clams, cockles, polychaete worms, and sand dollars are found in the lower intertidal region. The subtidal zone is home to sea cucumbers, sea pens, more crabs and clams, and some species of worms, snails, and sea slugs.

Animals of the Deep-Sea Floor

The deep-sea floor includes the flat abyssal plains, the trenches, and the rocky slopes of seamounts and mid-ocean ridges. The seafloor sediments are more uniform and their particle size is smaller than those of the shallower regions close to land sources. The environment of the bathyal, abyssal, and hadal zones is uniformly cold and dark.

Many of the animals of the deep-sea floor are very like their benthic relatives in the shallow waters of the continental shelf. But in general, it appears that, while the population of an organism decreases with depth, the number of species within that population increases. That is, diversity of species, which is determined from the ratio of number of species to number of organisms in a sample, increases with depth. Samples show that few species of burrowing copepods are found to be duplicated between samples. This diversity of species between samples is also evident in the protozoan foraminiferans. Thirty species of planktonic foraminiferans are known, whereas 1000 species of benthic foraminiferans have been described.

The stable conditions of the deep-sea floor appear to have favored deposit-feeding infaunal animals of many species. Many members of the deep-sea infauna are very small; they are known as the **meiobenthos** and measure 2 mm or less. The meiobenthos include nematode worms, burrowing crustaceans, and segmented worms. At 7000 m (23,000 ft), tusk shells lie buried in the ooze with tentacles at the sediment surface to feed on the foraminiferans. Acorn worms are found frequently in samples taken at 4000 m (13,000 ft). Hagfish burrow into the sediment at the 2000 m (6600 ft) depth. Detritus-eating worms and bivalved mollusks have been found on the sea floor in all the oceans.

Among the epifauna, protozoans are abundant and are widely distributed. Glass sponges attach to the scattered rocks on oceanic ridges and seamounts (fig. 17.10); so do sea squirts and sea anemones. Their stalks lift them above the soft sediments into the water, where they feed by straining out organic matter. Stalked barnacles attach to the stalks of glass sponges and sea squirts as well as to shells and boulders.

Tube worms are common, ranging in size from a few millimeters to 20 cm (8 in), and sea spiders with four pairs of very long legs that span up to 60 cm (27 in) are found at depths to 7000 m (23,000 ft). Snails are found to the greatest of depths; those in the deepest trenches frequently have no eyes or eye-stalks.

The beard worms, or **pogonophora,** are found in more-productive areas at depths to 10,000 m (33,000 ft). They secrete a close-fitting tube and stand erect, with only their lower portion buried in the sediment. They have no mouth, no gut, and no anus and absorb their needed molecules through their skin. The pogonophora associated with the vent communities are discussed later in this chapter.

Horny corals, or sea fans, which resemble plants more than animals, grow at depths of 5000–6000 m (16,000–20,000 ft); so do solitary stone corals, which grow larger at these depths than do the coral organisms in the surface waters. Sea lilies, or crinoids, which are related to starfish, are also found at this depth, as are brittle stars and sea cucumbers. Sea cucumbers live in areas of sediments that are rich in organic substances; they are a dominant and widely spread organism of the deep-sea floor.

The deep sediments are continually disturbed by sea cucumbers and worms in the process of extracting organic matter. This process, called **bioturbation,** reworks the sediment and destroys layers, resulting in the uniform sediments that cover vast areas of the ocean floor.

Fouling and Boring Organisms

Organisms that settle and grow on pilings, docks, and boat hulls are said to foul these surfaces. Fouling organisms include barnacles, anemones, tube worms, sea squirts, and algae. When these organisms grow on a vessel's hull, they slow its movement through the water and add to the costs in shipping time and haul-out for cleaning. Fouling organisms also make it difficult to operate equipment in the ocean environment, and much research goes into experiments with paints and metal alloys to discover methods of discouraging and controlling these fouling organisms.

Other organisms naturally drill or bore their way into the substrate. Sponges bore into scallop and clam shells; snails bore into oysters; some clams bore into rock. Organisms that bore into wood are costly, for they destroy harbor and port structures as well as wooden boat hulls. Two organisms are responsible for most of the wood boring: the shipworm (*Teredo*) and the gribble. *Teredo* is a wormlike mollusk with one end covered by a bivalved shell. These mollusks secrete enzymes that break down and at least partially digest the wood fibers, which are scraped away by the shell's rocking and turning movement, which bores deep and extremely destructive holes all through a piece of wood. The gribble is a crustacean; it gnaws more superficial and smaller burrows, but it, too, is very destructive (fig. 17.11).

(a)

(b)

(c)

Figure 17.10 (a) A vase-shaped glass sponge 640 m (2100 ft) under the surface on the Brown Bear Seamount in the northeastern Pacific Ocean. (b) A deep-sea crab photographed at a depth of 2000 m (6550 ft) on the Juan de Fuca Ridge, in an area of little life. (c) A group of deep-sea sponges and an anemone are seen at 684 m (2244 ft) on the Brown Bear Seamount.

Classification Summary of the Benthos 17.3

Benthic organisms include members of the following kingdoms and phyla:

I. *Kingdom Monera:* includes the bacteria, an important food source in mud and sand environments; also cyanobacteria or blue-green algae, especially abundant on coral reefs.

II. *Kingdom Protista:* convenience grouping of microscopic, usually unicellular plants and animals.
  A. *Phylum Chrysophyta:* golden-brown algae; mainly planktonic, but benthic diatoms are abundant.
  B. *Phylum Protozoa:* infauna include many members among sand and mud particles, including amoeboid and foraminiferan species.

III. *Kingdom Plantae:* plants; primarily nonmotile, multicellular, photosynthetic autotrophs.
  A. *Division Chlorophyta:* green algae; seaweeds such as *Ulva* and *Codium*.
  B. *Division Phaeophyta:* brown algae; coastal seaweeds, including the large kelps *Nereocystis*, *Macrocystis*, and *Postelsia*. *Sargassum* maintains a planktonic habit in the Sargasso Sea.
  C. *Division Rhodophyta:* red algae; littoral and sublittoral seaweeds of varied form.
  D. *Division Tracheophyta:* plants with vascular tissue; true roots, stems, and leaves present.
    1. *Class Angiospermae:* flowering plants; seeds enclosed in fruit; eel grass, surf grass, and mangrove trees.

Figure 17.11 Unprotected wood, such as driftwood, is subject to destruction by marine borers. (a) This beach log has been riddled by shipworms. (b) The small holes surrounding a shipworm hole are bored by a crustacean known as gribble.

IV. *Kingdom Animalia:* animals; multicellular heterotrophs with specialized cells, tissues, and organ systems.
   A. *Phylum Porifera:* sponges; simple, nonmotile filter feeders.
   B. *Phylum Coelenterata,* or *Cnidaria:* radially symmetric, with tentacles and stinging cells.
      1. *Class Anthozoa:* sea anemones, corals, and sea fans.
   C. *Phylum Platyhelminthes:* flatworms; free-living and parasitic forms.
   D. *Phylum Nemertina,* or *Nemertea:* ribbon worms.
   E. *Phylum Nematoda,* or *Aschelminthes:* round worms, or nematodes.
   F. *Phylum Brachiopoda:* lampshells.
   G. *Phylum Annelida:* segmented worms.
      1. *Class Polychaete:* free-living marine carnivores, including tube worms, *Nereis,* and clam worms.
   H. *Phylum Pogonophora:* deep-sea tube-dwelling worms; giant members found around hot vents on the sea floor.
   I. *Phylum Mollusca:* mollusks.
      1. *Class Polyplacophora:* chitons.
      2. *Class Scaphopoda:* tusk shells.
      3. *Class Gastropoda:* single-shelled mollusks; snails, limpets, and sea slugs.
      4. *Class Pelecypoda,* or *Bivalvia:* bivalved mollusks; clams, mussels, scallops, and oysters.
      5. *Class Cephalopoda:* octopus. The squid is a member of the nekton.
   J. *Phylum Arthropoda:* animals with paired, jointed appendages and outer skeletons.
      1. *Class Crustacea:* crabs, shrimp, and barnacles.
   K. *Phylum Echinodermata:* all marine, radially symmetric, spiny-skinned animals with water-vascular systems, including starfish, sea urchins, sand dollars, and sea cucumbers.
   L. *Phylum Hemichordata:* acorn worms.
   M. *Phylum Chordata:* animals, including vertebrates, with dorsal nerve cord and gill slits at some stage in development.
      1. *Subphylum Urochordata:* filter-feeding, saclike adults with "tadpole" larvae; sea squirt.

Classification Summary of the Benthos **461**

Deep-Sea Ice Worms

Gas hydrates are icelike, crystalline deposits that form at low temperatures and high pressures; they are composed primarily of methane gas with some hydrogen sulfide gas. Most gas hydrate deposits are found on the sea floor buried under heavy loads of sediments (see chapter 3, section 3.4). In 1997 a research cruise in the Gulf of Mexico found a gas hydrate mound free of sediments at a depth of 540 m (1800 ft). The mound showed two distinct color bands, yellow and white; the yellow band contained oil and the white band did not. More surprisingly the entire exposed surface of the hydrate was covered with worms, 2500 per square meter (box fig. 1). This is the first time nonmicrobial life has been seen on gas hydrates.

The worms live in individual oval depressions in the ice, and in more than 95% of the depressions only one worm was visible. The worms are 2–4 cm (about 1–1.5 in) in length, and each worm is pink with a red blood vessel down the back and a digestive tract. Its eyes are reduced, and the sexes were found to be separate. The researchers believe that the worms feed on the bacteria living on the surface of the ice. The worms can move their lateral extensions (parapodia), setting up water currents that raise the oxygen level for the worms and bacteria. This may also be the way in which the depressions are formed. The worms are a new species *(Hesiocaeca methanicola)* of the polychaete family of annelid or segmented worms. See the "Classification Summary of the Benthos," section 17.3.

The researchers also found the same worms living under as much as 10 cm (4 in) of sediment. This raises the question of how much deeper in the sediments the worms can colonize.

H. methanicola does not appear to suffer from predators although potential predators such as fish and large crustaceans were present on the hydrate surface. Revisiting the hydrate mound one month later researchers found no empty depressions, and the density of worms had increased to 3000 per square meter—perhaps the worms are unpalatable.

To Learn More About the Deep-Sea Ice Worms

Fisher, C.R. et al. 2000. Methane Ice Worms: *Hesiocaeca Methanicola* Colonizing Fossil Fuel Reserves. Naturwissenchaften 87: 184–187.

web link **Box Figure 1** Methane ice worms are found in depressions on the exposed surface of a gas hydrate. The worms are believed to create the depressions.

Tropical Coral Reefs   17.4

Coral reefs are the most diverse and complex of all marine communities (fig. 17.12). The largest coral reef in the world, the Great Barrier Reef, stretches more than 2000 km (1200 mi), from New Guinea southward along the eastern coast of Australia. Reef-building corals require warm, clear, shallow, clean water and a firm substrate to which they can attach. Because the water temperature must not go below 18°C and the optimal temperature is 23°–25°C, their growth is restricted to tropical waters between 30°N and 30°S and away from cold-water currents. Waters at depths greater than 50–100 m (150–300 ft) are also too cold for significant secretion of calcium carbonate. Most Caribbean corals are found in the upper 50 m (150 ft) of lighted water; Indian and Pacific corals are found to depths of 150 m (500 ft) in the more transparent water of those oceans. Reefs usually are not found where sediments limit water transparency.

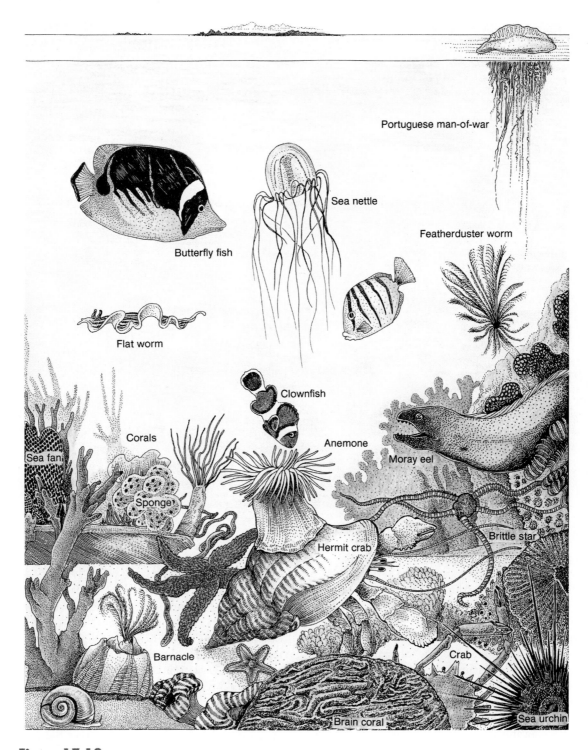

Portuguese man-of-war

Sea nettle

Featherduster worm

Butterfly fish

Flat worm

Clownfish

Corals

Anemone

Sea fan

Moray eel

Sponge

Hermit crab

Brittle star

Barnacle

Crab

Brain coral

Sea urchin

web link **Figure 17.12** A coral reef is a complex, interdependent but self-contained community. Members include corals, clams, sponges, sea urchins, anemones, tube worms, algae, and fish.

The Corals

Corals are colonial animals, and individual coral animals are called **polyps** (fig. 17.13). A coral polyp is very similar to a tiny sea anemone with its tentacles and stinging cells, but, unlike the anemone, a coral polyp extracts calcium carbonate from the water and forms a calcareous skeletal cup. The corals of the curio shop and jeweler are the calcareous skeletons of the coral polyps. Large numbers of these polyps grow together in colonies of delicately branched forms or rounded masses.

Clear, shallow water is required by the reef-building coral, because within the tissues of the polyps are masses of single-celled dinoflagellate algae called **zooxanthellae** that require light for photosynthesis and therefore are limited to

Figure 17.13 A colony of star coral polyps on a reef in the Caribbean Sea, Bonaire, Lesser Antilles.

the photic zone. Polyps and zooxanthellae have a symbiotic relationship in which the coral provides the algal cells with a protected environment, carbon dioxide, and nitrate and phosphate nutrients, and the algal cells photosynthesize, returning oxygen and removing waste. The zooxanthellae supply the corals with substantial amounts of their photosynthetic products; some coral species receive as much as 60% of their nutrition from the algae. Zooxanthellae also enhance the ability of the coral to extract the calcium carbonate from the seawater and increase the growth of their calcareous skeletons. The degree of interdependence between zooxanthellae and coral is thought to vary from species to species. The polyps feed actively at night, extending their tentacles to feed on zooplankton, but during the day, their tentacles are contracted, exposing the outer layer of cells containing zooxanthellae to the sunlight.

The Reef

The corals require a firm base to which they can cement their skeletons. The classic reef types of the tropical sea—fringing reefs and barrier reefs—are attached to existing islands or landmasses. Atolls are attached to submerged

seamounts. (For a review of reef formation around a seamount, see chapter 3.) Corals are slow-growing organisms; some species grow less than 1 cm (0.4 in) in a year, and others add up to 5 cm (2 in) a year. The same corals may be found in different shapes and sizes, depending on the depth and the wave action of an area. Environmental conditions vary over a reef, forming both horizontal and vertical zonation patterns that are the product of wave action and water depth, as shown in figure 17.14. On the sheltered (or lagoon) side of the reef, the shallow **reef flat** is covered with a large variety of branched corals and other organisms. Fine coral particles broken off from the reef top produce sand, which fills the sheltered lagoon floor. On the reef's windward side, the reef's highest point, or **reef crest,** may be exposed at low tide and is pounded by the breaking waves of the surf zone. Here the more massive rounded corals grow. Below the low-tide line to a depth of 10–20 m (35–65 ft) on the seaward side is a zone of steep, rugged buttresses, which alternate with grooves on the reef face. Masses of large corals grow here, and many large fish frequent the area. The buttresses dissipate the wave energy, and the grooves drain off fine sands and debris, which would smother the coral

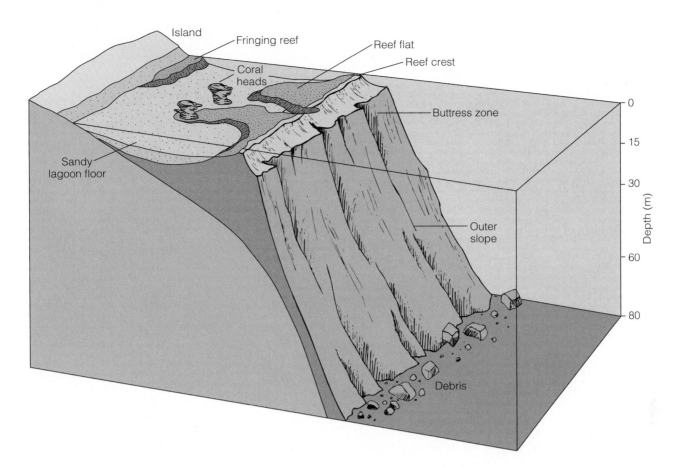

Figure 17.14 Coral reef zonation. Water depths and wave action vary over the reef. The reef crest may be exposed at low tide. Sand from coral debris fills the lagoon and drains off down the grooves of the outer slope, where buttresses dissipate wave energy from the open sea.

colonies. At depths of 20–30 m (65–100 ft), there is little wave energy, and the light intensity is only about 25% of its surface value; still it is adequate to support reef algae and corals. The corals are less massive at this depth, and more delicately branched forms are found here. Between 30 and 40 m (100 and 130 ft), the slope is gentle and the level of light is very reduced; sediments accumulate at this depth, and the coral growth becomes patchy. Below 50 m (165 ft) the slope drops off sharply into the deep water. The reef exists as a balance between the growth of the organisms on the reef surface, as they build up on top of old, dead, calcareous remains, and the wearing away of the reef by mechanical and biological forces.

Coral reefs are complex assemblages of many different types of algae and animals (see figs. 17.12 and 17.15), and competition for space and food is intense. Algae, sponges, and corals are constantly overgrowing and destroying each other. Some species are active only at night: some fishes, snails, shrimp, the octopus, fireworm, and moray eel. During the day, other species depend on color and vision to make their way. It has been estimated that as many as 3000 animal species may live together on a single reef. The giant reef clam *Tridacna* can measure up to a meter in length and weighs over 150 kg (330 lb). It also possesses zooxanthellae in large numbers in the colorful tissues that line the edges of the shell. Crabs, moray eels, colorful reef fish, poisonous stonefish, long-spined sea urchins, seahorses, shrimp, lobsters, sponges, and many more organisms are all found living here together. The total reef community can be considered to be an autotrophic association based on the Sun's energy as transformed by its photosynthetic members. On some reefs, the zooxanthellae have been shown to produce several times more organic material per unit of space than the phytoplankton, probably owing to the rapid recycling of nutrients between the corals and the zooxanthellae.

The reefs are not formed exclusively from the calcium carbonate skeletons of the coral. Encrusting algae that produce an outer calcareous covering also contribute; so do the minute shells of foraminifera, the shells of bivalves, the calcareous tubes of polychaete worms, and the spines and plates of sea urchins. All are compressed and cemented together to form new places for more organisms to live. At the same time, some sponges, worms, and clams bore into the reef; some fish graze on the coral and the algae; and the sea cucumbers feed on the broken fragments, reducing them to sandy sediments.

Bleaching

Oceanographers, marine biologists, and all those who have enjoyed the experience of a tropical coral reef have become

web link **Figure 17.15** Coral reefs are rich communities of organisms. (a) Corals and a sea fan at Mana Island, Fiji. (b) A butterfly fish (left) and a moorish idol (right) at Soma Soma Straits, Taveuni Island, Fiji.

increasingly concerned with the health of these biologically rich but delicately balanced regions of the oceans. Coral bleaching episodes, in which corals expel their zooxanthellae and turn white, have occurred from time to time, but unless the episodes were especially severe the corals regained their algae and recovered. Between 1876 and 1979 there were only three bleaching events, but over sixty bleach events were documented between 1979 and 1990. In the last two decades this bleaching has become more frequent and more severe, and not all reefs have recovered. Large amounts of bleaching occurred at the time of the 1982–83 El Niño, during which shallow reefs of the Java Sea lost 80%–90% of their living coral cover. In 1990–91 there was large-scale bleaching in both the Caribbean and French Polynesia. During the 1997–98 El Niño event, which lasted for over a year, surface waters reached the highest temperatures on record throughout the Indian and western Pacific Oceans.

Fifteen hundred scientists from more than fifty countries met at the 2000 International Coral Reef Symposium in Bali and concluded that "coral reefs face a bleak future" without immediate and drastic action. At the same meeting the Global Reef Monitoring Network reported bleaching from the Indian Ocean to the coast of Brazil during the 1997–98 El Niño. The Network estimated 11% of the world's coral reefs had been destroyed by human activities before 1998 (table 17.1), and since that time it is estimated that another 16% have been "severely damaged." Scientists studying the death of coral reefs in the Indian Ocean report that 50%–95% have died in the past two years.

Predation and Disease

Periodically there is a dramatic population increase of the sea star *Acanthaster* known as the crown of thorns; this sea star feeds on the coral polyps. Fossil evidence points to outbreaks

Table 17.1 The Effect of El Niño on Coral Reefs

Region	Pre-1998 Destruction (%)	1998 Destruction (%)	Total (%)
Indian Ocean	13	46	58
Arabia	2	33	35
South, East Asia	16	18	34
Caribbean	21	1	22
Pacific Ocean	4	5	9
Australia Papua New Guinea	1	3	4
Global total	11	16	27

Source: *Global Coral Reef Monitoring Network, 2000, "The El Niño Effect." Reprinted with permission from* Science: *Vol. 290, No. 5492, p. 682. Copyright 2000 American Association for the Advancement of Science.*

having occurred along the Great Barrier Reef for at least the last 8000 years. Why the *Acanthaster* population increases so rapidly is unknown, but evidence points to a correlation between rainy weather (low salinity) and runoff (increased nutrients), allowing large numbers of *Acanthaster* larvae to survive. The clearing of land for agriculture and the development of coastal areas may also be related. Concern also exists that the harvesting of the large conchs that prey on the starfish may upset the population balance. A recent outbreak began in 1995 and by 1997 was affecting 40% of the reef. It takes ten to fifteen years for reef areas to recover from an *Acanthaster* outbreak, but the opening up of areas on the reef by the starfish may also allow slower-growing coral species to expand.

Reef-building algae as well as corals have recently been found under attack by several previously unknown diseases. One of these is coralline lethal orange disease, known as CLOD. CLOD is caused by a bright orange bacterial pathogen that is lethal to the encrusting red algae (corallines) that deposit calcium carbonate on the reefs. These algae cement together sand, dead algae, and other debris to form a hard, stable substrate. The disease was initially found in 1993 in the Cook Islands and Fiji; by 1994 it had spread to the Solomon Islands and New Guinea; by 1995 it was found over a 6000 km (3600 mi) range of the South Pacific. No one knows whether CLOD has been recently introduced from some obscure location or whether it has been present on the reefs but has now evolved into a more virulent form.

Between 1996 and 1997 black-band disease on Florida's coral reefs increased by nearly 300%. Scientists do not know why such an unprecedented change occurred, but many think that poor water quality and polluted runoff into Florida Bay may have been contributing factors.

Human Activities

Humans and their activities are among the greatest threats to the reefs. The first global survey of coral reef health was carried out in 1997, and it was reported that 95% of the world's coral reefs have been damaged by overfishing, dynamiting, poisoning, and pollution. Reefs are damaged by careless sport divers trampling delicate corals and by boats grounding or dragging their anchors. As the tropical forests are cleared, the trees are replaced by farms, mines, resorts, and other human enterprises. The result is increased runoff, carrying sediments, agricultural fertilizers, and human sewage into coastal waters. Additional nutrients favor the growth of algae; the algae outcompete the corals for space, smother the existing corals, and prevent new coral colonies from starting.

Coral reefs are despoiled by shell collectors and mined for building materials. In French Polynesia, Thailand, and Sri Lanka, tons of coral are used as construction material each year. Low-lying coastal areas that lose their coral reefs have no protection against storm surges that accompany extreme weather events. This is particularly important to island nations in the Pacific.

Where catching reef fish with a net is difficult those who fish use sodium cyanide and explosives. Whether it is cyanide squirted into the water or explosives detonated in the water, the fish are temporarily stunned, making them easy to capture. These methods also kill many other reef organisms, including the reef-building organisms. Although banned in Indonesia and the Philippines, these methods are still the first choice of those in the very lucrative international aquarium and restaurant businesses.

Ensuring the continued existence of these beautiful and productive areas requires both an increased understanding of the complex nature of reef communities and the development of policies designed to protect them from human interference. The United Nations Environment Program, the Intergovernmental Oceanographic Commission, the World Meteorological Organization, and scientists from many different countries are involved in monitoring stations that cover the world's major reefs. The NOAA program, Coral Reef Watch, is developing a long-term coral reef monitoring system able to predict coral bleaching in all U.S. coral reef areas. Plans call for the installation of meteorological and oceanographic monitoring stations in the Bahamas and the U.S. Virgin Islands. The data from these stations will be used with satellite sea temperature data and biological monitoring to provide a Coral Reef Early Warning System (CREWS). Although it is not possible to halt or reverse stressful conditions, there may be time to reduce impacts from other factors.

High-Energy Environments 17.5

Recent work has shown that intertidal communities constantly battered by the waves are more productive than the world's lush, green rain forests. On the average, waves deliver 0.335 watts of energy per square centimeter of coastline, about fifteen times more energy than comes from the Sun. Even during calm periods, wave energy is 100% greater than solar energy. The algae have little woody tissues; instead, the kelps of the rocky intertidal have 2.5 times more photosynthetic area per square meter of growing surface and are from two to ten times more productive than rain forest vegetation. The mussels, which are consumers, have been found to match or exceed the rain forest productivity when growing in areas of high wave action.

The method of harnessing the energy is indirect. Wave action reduces predators, such as starfish and sea urchins, allowing more mussels and more kelp to live in a unit area. The moving water brings a constant supply of nutrients to the algae and keeps their blades in motion, so they are never in the shade for long. Last, the waves can dislodge mussels, allowing more kelp to move in. The waves allow the primary producers to become highly concentrated, so the consumers can grow and expand their populations as well.

Deep-Ocean Chemosynthetic Communities   17.6

Hot Vents

In March 1977, an expedition from Woods Hole Oceanographic Institution using the research submersible *Alvin* discovered densely populated communities of animals living around hydrothermal vents along the Galápagos Rift of the East Pacific Rise. Prior to that time deep-sea benthic communities were assumed to consist of small numbers of deposit-feeding, slow-growing animals living on and in the soft sediments of the sea floor and depending for nutrition on the slow descent of decayed organic material from the surface layers. Since that first discovery vent communities have been found scattered throughout the world at seafloor spreading centers, and nearly 300 new species have been identified. Vent communities are known between 9°N and 21°N latitude along the East Pacific Rise, in the Mariana and Okinawa troughs and North Fiji basin of the western Pacific, along the Mid-Atlantic Ridge, in the east Pacific from 49°–22°S, and on the Gorda and Juan de Fuca Ridges of the northeast Pacific.

The animals living in these deep-sea vent areas include filter-feeding clams and mussels, in addition to anemones, worms, barnacles, limpets, crabs, and fish. The clams are very large and show the fastest growth rate of any known deep-sea animal, up to 4 cm (2 in) per year. The tube worms (fig. 17.16) are startling in size, up to 3 m (10 ft). They have been placed in the phylum Vestimentifera (a phylum is the most basic of taxonomic categories). These crowded communities have a biomass per unit area that is 500–1000 times greater than the biomass of the usual deep-sea floor.

The primary producers in the communities surrounding vent areas are bacteria that use chemosynthesis to fix carbon dioxide into organic molecules. See section in chapter 14 on "Primary Productivity and Chemosynthesis" for more information.

The tube worms in the vent areas have no mouths and no digestive systems; the soft tissue mass of their internal body cavity is filled with bacteria. The tube worms have the ability to transport sulfide made nontoxic by a binding pro-

Figure 17.16 Tube worms crowd around deep-sea hydrothermal vents. The internal body cavity of each worm is filled with bacteria that synthesize organic molecules by chemosynthesis.   © *Richard A. Lutz.*

tein, carbon dioxide, and oxygen to the bacteria held in their body tissues; the synthesis of organic molecules is done within the bacteria. Despite their size and abundance these tube worms are completely dependent on an internal symbiosis with the bacteria for their nutrition. Both clams and mussels have large numbers of bacteria in their gills. The mussels have only a rudimentary gut, and the clams and mussels have red flesh and red blood; the color is due to the oxygen-binding molecule hemoglobin. The oxygen is needed for oxidation of the hydrogen sulfide and to maintain body tissues and high growth rates.

The vent plankton and the bacteria in the water provide nutrition for a variety of filter feeders. Other bacteria form mats that surround the vents, and these bacterial mats are the food for snails and other grazers. The bacteria are the base of a broad-based, self-contained trophic system in which nutrition passes from animal to animal by symbiosis, grazing, filtering, and predation.

In December 2000, researchers using the submersible *Alvin* and the imaging vehicle *Argo II* discovered a new type of hydrothermal vent field. This field, called the Lost City, was discovered on a terrace 700–800 m (2300–2600 ft) down in the North Atlantic, just off the Mid-Atlantic Ridge. Lost City has no black smokers but instead has steep-sided, white carbonate pinnacles 10–30 m (33–99 ft) tall. These carbonate chimneys support dense bacterial populations of both Archaea and Eubacteria. There are no large populations of larger organisms surrounding the vents; crabs and sea urchins are rare, sponges and corals more abundant. See chapter 2, section 2.7 for more information about the Lost City.

During studies of the large shrimp populations surrounding some vent areas, researchers noticed a reflective spot just behind the head of the shrimp. The reflective spot consists of two lobes, each connected to the brain by a large nerve cord; light-sensitive pigment related to visual pigment is also present. This system allows the shrimp to detect the radiation associated with hot vents (see fig. 2.38) and may allow the shrimp to orient themselves with the vents and their food supply.

In 1991 a submersible monitoring vents on the East Pacific Rise, west of Acapulco and at a depth of 2500 m (1.5 mi), came upon a vent area immediately after a volcanic eruption that killed the larger organisms and left a white carpet of bacteria over the sea floor. Returning in 1992, researchers found the bacterial mats being grazed by large groups of fish, crabs, and other species. By 1993 giant tube worms were more than a meter (3.3 ft) high. Growing at nearly 1 m per year, these tube worms are the fastest growing of all known marine invertebrates. At this time, researchers found that metal-rich sulfide deposits had grown into 10 m (30 ft) chimneys. Previously, geologists had believed that such formations took decades to form. Between 1995 and 1997 the number of species more than doubled, from twelve to twenty-nine. Mussels and small worms called serpulids moved in and so did clouds of tiny crustaceans, but no giant white clams had yet arrived.

How vents are first colonized by various animals is still not fully understood. Initially the distances between vent areas appeared to be a barrier, but as more and more vents are discovered distance seems less of an obstacle when combined with the fast maturation of the organisms and the large numbers of free-swimming larvae they produce. For another way in which organisms might move between vents see the box titled "Whale Falls" in chapter 16.

Cold Seeps

Off the coasts of Florida, Oregon, and Japan communities based on chemosynthesis have been found associated with cold seepage areas. Bacteria are the primary producers of the communities, using methane and hydrogen sulfide. Along the continental slope of Louisiana and Texas in the Gulf of Mexico, faulting has fractured the sea bottom and produced environments where oil and gas seep up and onto the surface of the sea floor. Clams, mussels, and large tube worms were first collected from these sites (fig. 17.17) in 1985. A variety of other animals including fishes, crustaceans, and mollusks commonly found along the continental shelf are abundant around these seeps, attracted by the food supply.

In 1990 salt seeps were reported on the floor of the Gulf of Mexico. The escape of gas through surface sediments has formed depressions that filled with salt brine more than 3.5 times the usual salinity of seawater. These extremely dense brine lakes also contain methane and are surrounded by large communities of mussels.

Sampling the Benthos   17.7

At low tide, beaches all over the world are sampled by researchers armed with buckets and shovels. To understand the relationships between the organisms and their environment, one needs to know where the plants and animals are found on the beach with respect to the tide levels, or the vertical zonation pattern. This pattern is established by determining the slope of the beach and marking locations relative to mean sea level; this is done by surveying a **beach**

transect line directly up the beach from low-tide level to a point above high tide. Trenches paralleling the transect line are dug on sandy beaches to reveal the infauna populations below a measured surface area, and on a rocky shore, surface counts of individuals within an area of specific size are made along the transect line. Once the transect line is established, it is possible to return to the same place, season after season or year after year, to study seasonal changes in the populations or to determine the number of juveniles added yearly. Another type of study compares a natural area to an area stripped of all benthic organisms. The rate of repopulation and the sequence in which the plants and animals return give information on the relationships that exist among species as well as the recovery rates in the event of a catastrophe due to human error or natural occurrences.

The classic method of obtaining information on subtidal species is by using a bottom dredge towed by a slowly moving ship. A dredge is a metal frame to which a heavy net bag is attached; it is dragged over the sea floor, scraping up the organisms in its path (fig. 17.18). This is a strictly qualitative sampling method, as the number and kinds of organisms cannot be accurately related to the area sampled. It is possible to attach a measuring wheel to the dredge frame, so that when the dredge is on the bottom, the wheel measures the distance over which it is dragged; then, knowing the distance and the width of the dredge, researchers can compute the approximate size of the area sampled, assuming the dredge did not bounce or skip over a rough bottom.

Soft bottoms can be sampled with a bottom grab or a box corer (see chapter 3). The sediment collected is washed through a series of mesh screens of varying sizes, and the organisms are collected and counted.

Divers wearing scuba gear can sample and photograph bottom populations directly in depths to approximately 35 m (105 ft). A diver can place a frame of a specific area on the bottom, identify the species within it, and count individuals of each type. If the bottom is not disturbed, it is possible to return to the sample plot and check it again. Other methods of observing the bottom without disturbing it include underwater photography and television cameras that are operated remotely from a ship or from a submersible.

In the past sampling the benthos of the bathyal, abyssal, and hadal zones required large research vessels and extended periods of time at sea to sample small, widely separated areas. Today submersibles, ROVs (remotely operated vehicles), and AUVs (autonomous underwater vehicles) provide new ways to reach previously unvisited parts of the deep ocean and view its inhabitants, undisturbed and in their natural world.

Practical Considerations: Harvesting the Benthos   17.8

The Animals

Benthic animals are a valuable part of the seafood harvest; they include crustaceans (crabs, shrimp, prawns, and lobsters) and mollusks (shellfish—clams, mussels, and oysters—

as well as squid). The world's wild harvests for 1999 were catches of 5 million metric tons of shellfish, including squid, and another 5 million metric tons of crustaceans. Because of the demand for shellfish and crustaceans, the catches are more important in dollar value than the weight of the catches would suggest. In the United States, oysters are harvested in the Gulf of Mexico, southern New England, Chesapeake Bay, and Puget Sound; lobsters are fished in New England; crabs and clams are caught along all U.S. coasts; and shrimp are caught in the gulf area and off other coasts as well. In many cases, these fisheries make important contributions to the local economy.

Many of the problems of the finfish fisheries are repeated in the benthic fisheries. For example, between 1975 and 1980 more and more boats entered the king crab fishery of the Bering Sea (fig. 17.19). The catch of 1979 (70,000 metric tons) had dwindled to 5000 metric tons in 1994. The rapid decrease in catch was apparently due to overfishing combined with an insufficient knowledge of the king crab's natural history. King crab migrate across the floor of the Bering Sea, but without better knowledge of crab populations and migration patterns effective fishery conservation programs are difficult to implement. Bristol Bay is the most productive region of the Bering Sea, but in 1994 and 1995 the fishery was closed be-

Figure 17.17 Photographs from Gulf of Mexico oil and gas seeps. (a) Mussels and tube worms. (b) Tube worms. (c) Seeping gas and oil, bacterial mats, and chemosynthetic communities in the Green Canyon oil lease area. Note sampling devices and gas bubbles.

Figure 17.18 A biological dredge used to collect epibenthic organisms from rocky substrate.

cause of low numbers of female crabs in the Bristol Bay population. This Bristol Bay fishery reopened in 1996 and by 1998 the catch had reached 7500 metric tons, but the catch declined in succeeding years and the 2001 quota was set at 3200 metric tons.

Attempts to increase the harvest of crustaceans and mollusks have focused on expanding world aquaculture or mariculture efforts. In 1999 the combined mariculture harvest of crustaceans and mollusks reached 11.5 million metric tons. Oysters, mussels, scallops, and clams are raised on mariculture farms around the world. In Asia and Europe, raft culture of mollusks is popular. The larvae attach to ropes trailing below rafts, and this attachment keeps the shellfish in the water column, with abundant food and few predators. Mariculture in Japan, the Republic of Korea, France, and Spain is dominated by oysters and mussels. Raft culture in Japan is responsible for a harvest of about 250,000 metric tons of oysters each year. Spain produces over 150,000 metric tons of mussels by raft culture, the Japanese also culture scallops in hanging net cages as well as on the bottom, yielding a yearly harvest of 250,000 metric tons. Scallop culture in China has increased to more than 1 million metric tons. Mussel and oyster farmers in the United States (fig. 17.20) have many of the difficulties encountered by fish farmers. Costs, licensing policies, technology to replace hand labor, and research to improve diet and disease control require attention if the U.S. seafood harvest is to increase in the same way.

World shrimp and prawn production from mariculture ponds reached 1 million metric tons in 1999, but shrimp production declined at the end of the year due to disease problems. Thailand was the world's top shrimp producer and in Latin America production was up 10%, with Ecuador as the

web link **Figure 17.19** Alaska king crab being unloaded from a crab pot.

web link **Figure 17.20** Commerical raft culture of bay mussels in a cove on Whidbey Island in Puget Sound. This mussel farm was the first mussel aquaculture venture in the United States.

main producer. Crustaceans contribute 4% of the world's aquaculture production by weight, but their value is 18% of the total value. Shrimp feed is about 30% fish meal and 3% fish oil; intensive shrimp aquaculture adds to the pressure on wild fisheries (see also the discussion of salmon aquaculture in chapter 16). The United States is the world's largest market for shrimp, approaching 50% of all seafood imports, and U.S. corporations are taking advantage of the lower costs in Latin America with large-scale investments in shrimp farming.

The Algae

Seaweeds are gathered from the wild in northern Europe, Japan, China, and southeast Asia; they are also an important part of the Japanese aquaculture/mariculture industry. The 1999 world estimate for total seaweed harvests, mainly from mariculture, was 9 million metric tons. The 1999 algal mariculture harvest in metric tons was 5.0 million for brown algae, 1.8 million for red algae, and 1.0 million for green algae.

Algae are good sources of vitamins and minerals but not of food calories, because most of the cellular material is indigestible. Certain species of green algae, known as sea lettuces, are used in seaweed soups, in salads, and as a flavoring in other dishes. A species of kelp, *Laminaria japonica*, is the **kombu** of Asia. Its blade is used in soups and stews, and it is used fresh, dried, pickled, and salted. It is also sweetened and shredded for use in candy and cakes. Another species of *Laminaria* is used in Europe in the same way. Historically along coastal areas, kelp was used as winter fodder for sheep and cattle and to mulch and fertilize the fields. Along the northwest coast of the United States and Canada, the stipes of bull kelp are made into pickles. The red alga *Porphyra* is the **nori** of Japan and the **laver** of the British Isles. It has been cultivated in Japan since 1700; in 1996, 850,000 metric tons were harvested. Nori is used in soups and stews and is rolled around portions of rice and fish for flavor. Laver is fried or used in salads.

There are also important industrial uses for some algal products. **Algin** is extracted from brown algae, and **agar** and **carrageenan** are obtained from red algae. In California more than 160,000 tons of kelp have been harvested yearly for the production of algin. Algin derivatives are used as stabilizers in dairy products, paints, inks, and cosmetics; and to strengthen ceramics, improve the consistency of plaster, and thicken jams as well as put a longer-lasting head on beer. Agar-producing algae are harvested in Japan, Africa, Mexico, and South America. Agar is used as a medium for bacterial culture in laboratories and hospitals, in addition to serving as an ingredient in desserts and in pharmaceutical products. Algae rich in carrageenan are gathered in the wild in New England and northeastern Canada; it is a stabilizer and emulsifier used to prevent separation in ice creams, salad dressings, soups, puddings, cosmetics, and medicines. About 455 metric tons (1 million pounds) of agar and 4550 metric tons (10 million pounds) of carrageenan are used in the United States each year.

Biomedical Products

Many benthic organisms produce biologically active compounds that have therapeutic use. Extracts of some sponges yield anti-inflammatory and antibiotic substances, and an anticoagulant has been isolated from red algae. Some corals produce antimicrobial compounds, and the sea anemone *Anthopleura* provides a cardiac stimulant. Extracts of abalone and oyster act as antibacterial agents, and a muscle relaxant has been isolated from the snail *Murex*. The adhesive secreted by mussels is used to secure dental fillings and crowns, and good results have been obtained using this substance to repair the cornea and retina of the eye.

Thousands of active compounds from marine organisms have been screened since the mid-1980s for their anticancer, anti-inflammatory, antitumor, immune suppressant, and antiviral possibilities. The tools of molecular biology have now made it possible to screen substances rapidly: for example, to test natural compounds for their effect on enzymes responsible for the growth of cancer cells. Discodermolide is an anticancer compound derived from a marine

(a)

(b)

(c)

(d)

Figure 17.21 The benthos is a large, varied group of animals living on or in the sea floor, especially the organisms of the tide pools and the intertidal areas of the rocky coasts. (a) The anemone *Tealia crassicornis* is a sessile carnivore. The graceful and beautiful nudibranchs, or sea slugs, include (b) the white *Dirona albolineata*, (c) the orange-flecked *Triopha carpenteri,* and (d) *Hermissenda crassicornis* with its orange-and-white striped tips.

sponge. The sponge was first collected in 1987; the compound was isolated in 1990; and a license for development and manufacture was granted in 1998.

Most of the substances studied come from soft-bodied invertebrate organisms that rely on toxins for defense; many also live in dense populations and under very crowded conditions: for example, on coral reefs and pier pilings. These animals have developed an array of compounds that prevent predation and give them an advantage in the competition for space. Toxic compounds poison predators or adjacent organisms, providing space for growth, and nontoxic substances make the animals unpalatable. Since the mid–1980s more than 5000 compounds of this type have been reported.

Collecting, extracting, identifying, testing, and evaluating active natural compounds from the world's benthos are time-consuming and expensive endeavors. From discovery to pharmacy shelf takes ten to fifteen years and may cost

hundreds of millions of dollars. If the process is successful, then there is the additional problem that the demand for these organisms might exceed their supply until the substance is successfully synthesized.

The benthos is a remarkably diverse grouping of plants and animals (fig. 17.21). It is easily accessible in the shallow littoral zone, where it has been studied by scientists and students for hundreds of years, and today submersibles, ROVs, and AUVs survey previously unsampled marine habitats. As we continue to explore, we may expect to make discoveries as surprising and as unexpected as the vent communities in the rift areas. When we consider the uses we make of the oceans at present and the uses for which we need the oceans in the future, we must remember their impact on the benthic environment and its inhabitants. We gain little if we use one resource at the expense of another. The balance within the marine environment is fragile, easy to degrade,

(e)

(f)

(g)

(h)

(i)

Figure 17.21 continued Starfish come in a remarkable diversity of shapes and sizes: (e) the bright orange bloodstar (*Henrica levinscula*), the slender-armed *Evasterias troschelli*, the many-armed *Solaster dawsoni*, *Mediaster aequalis* with its wide disk and broad arms, the leather star (*Dermasterias imbricata*), and the purple, rough-skinned *Pisaster ochraceus*. (f) All starfish are carnivores and use their tube feet to hold and open the shellfish on which they feed. (g) The purple sea urchin (*Strongylocentrotus purpuratus*) and the green urchin (*S. droebachiensis*) are closely related to the sea stars but they are herbivores, clipping off the algae with their especially constructed mouthparts (h). (i) The pink sea scallop (*Chlamys hastata hericia*) lies open as it filters organic particles from the seawater.

Genetic Manipulation of Fish and Shellfish

Fisheries scientists in North America, Japan, and northern Europe are using gene-transfer and chromosome manipulation techniques to keep coastal waters and fish farms stocked with species of rapid development and high growth rate. Chromosome manipulation techniques are being used to produce sterile fish, fish of selected gender, and fish known as **triploid,** which carry an extra set of **chromosomes** (box fig. 1).

Triploid salmon and trout are produced by exposing the eggs to temperature, pressure, or chemical shock shortly after fertilization. The shock interferes with the division of the egg nucleus; the egg retains an extra, or third, set of chromosomes. The third set is retained throughout the fishes' development and inhibits their sexual maturation. Salmon with normal numbers of chromosomes grow, mature sexually, spawn, and die, while triploid fish do not mature sexually and reach a larger size since they are not spending energy on egg or sperm production. The improved growth and survival of these fish has led to widespread production of farmed triploid trout in the United Kingdom.

This same technique allows hybridization of salmon and trout. Whereas normal chromosome-number hybrids between different species of salmon and trout do not survive, triploid hybrids do. In Idaho, a hybrid between rainbow trout and coho salmon has been shown to resist a common viral disease that causes serious economic loss to trout producers.

The gender of fish can be determined by chromosome manipulation and hormone treatment. Using a technique called gynogenesis, researchers expose fish sperm to ultraviolet light, denaturing its chromosomes. Eggs fertilized with the inactivated-chromosome sperm are briefly chilled, causing the eggs to develop with only the female's chromosomes. The result is all female offspring. Many female fish grow bigger and live longer than males. Female flounder grow to twice the size of males; female coho salmon have firmer and more flavorful flesh; and an all-female brood of sturgeon would bring higher profits to caviar producers. An alternative to gynogenesis is androgenesis, in which the egg chromosomes are inactivated and the sperm chromosomes are doubled.

Genes for a specific characteristic can be extracted from a species of fish, or an artificial gene that codes for a specific trait can be developed to enhance fish stocks. Either can be transferred into an egg or sperm of a different species of fish to produce a **transgenic** fish that may have faster growth, greater disease resistance, or more efficient food conversion than its parents. The Food and Drug Administration has been asked to approve the marketing of farmed Atlantic salmon that grows to market size in about half the time it takes under usual conditions. The freezing temperatures in parts of Canada have been a major obstacle to the farming of Atlantic salmon. An antifreeze protein (AFP) that inhibits ice crystal formation in fish has been successfully transferred to Atlantic salmon stock and their offspring.

Summer mortality and fluctuations in the marketability of oysters have kept the oyster industry seasonal. Normal oysters enter a summer reproductive phase during which their meat is poor in quality and unmarketable. Triploid oysters, or all-season oysters, do not usually produce sperm or eggs but continue to grow steadily, becoming significantly larger than normal oysters and ready for the summer market. Triploid Pacific oysters have become an important part of U.S. West Coast aquaculture, where they account for one-third to one-half of the total production. Tetraploid oysters have also been produced. These oysters have four sets of chromosomes, and they are fertile. When they are mated with normal diploid oysters (two sets of chromosomes), only triploid oysters are produced.

The era of commercial transgenic fish and shellfish has arrived. There are no regulations in either the United States or Canada concerning transgenic fish, and researchers go to great pains to prevent their experimental stock from escaping to the wild environment, rigging pens with fences, screens, and alarm systems. However occasional pen failures do allow transgenic fish to escape into the wild. If mixing occurs between experimental and normal fish, can it endanger the spawning population? Will the rapid inbreeding that is possible produce populations more susceptible to disease or environmental problems? How will the higher growth rate of sterile fish affect a mixed population? Although these problems and questions must be addressed, chromosome and gene manipulation as a useful tool in aquaculture and fish management has arrived and is in use.

Internet References

Visit the book's Online Learning Center at www.mhhe.com/sverdrup (click on the book's cover) to explore links to further information on related topics.

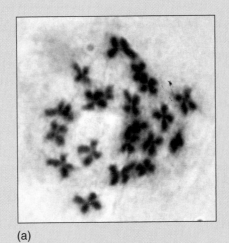
(a)

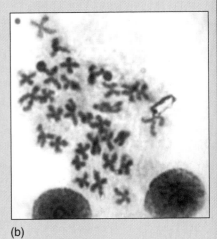
(b)

web link **Box Figure 1** (a) Chromosomes of the Pacific oyster (*Crassostrea gigas*) seen through a compound microscope. The diploid oyster has two sets of ten chromosomes. (b) The triploid oyster has three sets of ten chromosomes.

and difficult (probably impossible) to reconstruct. The health of the ocean's plants and animals is important to us as a sign of the health of our planet. We are all linked and each of us affects the others.

Summary

Benthic algae are anchored to firm substrates. These algae have a holdfast, a stipe, and photosynthetic blades but no roots, stems, or leaves. Algal growth along a rocky beach ranges from green algae at the surface through brown algae at moderate depths to red algae, which are found primarily below the low-tide level. Each group's pigments trap the available sunlight at these depths. Algae are generally classified by their principal pigment. The brown algae include the large kelps. Seaweeds provide food, shelter, and substrate for other organisms in the area. There are also benthic diatoms and a few seed plants, including sea grasses and mangroves.

Benthic animals are subdivided into the epifauna, which live on or attached to the bottom, and the infauna, which live buried in the substrate. Animals that inhabit the rocky littoral region are sorted by the stresses of the area into a series of zones. Organisms that live in the supralittoral (or splash) zone spend long periods of time out of water. The animals of the midlittoral zone experience nearly equal periods of exposure and submergence. These animals have tight shells or live close together to prevent drying out. The area is crowded, and competition for space is great. The lower littoral zone is a less stressful environment. It is home to a wide variety of animals. The organisms of the littoral zone are herbivores and carnivores, and each has its specialized lifestyle and adaptations for survival.

The zonation of the organisms in the benthic region varies with local conditions. Tide pools provide homes for lower littoral zone organisms; they can also become extremely specialized habitats.

Mud, sand, and gravel areas are less stable than rocky areas. The size of the spaces between the substrate particles determines the porosity of the sediments. Some beaches have a higher organic content than others. Few algae can attach to soft sediments, and thus few grazers are found here. Eel grass and surf grass provide food and shelter for specialized communities. Most organisms that live on soft sediments are detritus feeders or deposit feeders. Zonation patterns are not conspicuous along soft bottoms. Bacteria play an important role in the decomposition of plant material and its reduction to detritus. The bacteria themselves represent a large food resource.

The environment of the deep-sea floor is very uniform. The diversity of species increases with depth, but the population density decreases. The microscopic members of the deep-sea infauna are the meiobenthos. Larger burrowers such as sea cucumbers continually rework the sediments. The organisms of the epifauna are found at all depths.

Some organisms specialize in attaching to surfaces and others bore into them. Wood-borers are very destructive.

Tropical coral reefs are specialized, self-contained systems. The coral animals require warm, clear, clean, shallow water and a firm substrate. Photosynthetic dinoflagellates, called zooxanthellae, live in the cells of the corals and the giant clams. The reef exists in a complex but delicate biological balance, which can easily be upset. Reefs have a typical zonation and structure associated with depth and wave exposure. Coral reefs are presently under great stress from coral bleaching, predation, and disease. Human activities are among the greatest threats; damage is due to overfishing, pollution, and the use of dynamite and poison. An international network of monitoring stations is being implemented.

High-energy coastal benthic environments are two to ten times more productive than rainforest vegetation. Self-contained deep-ocean benthic communities made up of large, fast-growing animals depend on chemosynthetic bacteria for the first step in their food chains. Communities are associated with hot-water vents and cold seeps. Dense bacterial communities have been found associated with carbonate chimneys in the North Atlantic.

Benthic organisms are sampled by hand in the intertidal zone; deeper samples are obtained with dredges, grabs, corers, ROVs, AUVs, and submersibles. Sampling is both qualitative and quantitative.

Shellfish and crustacea are valuable world food resources. Large-scale aquaculture is being used to increase harvests of oysters, mussels, and shrimp.

Algae are gathered in many countries and are cultivated in Japan. Some are used directly as food; others are used as stabilizers and emulsifiers in foods and other products. Biologically active substances with potentially practical uses have been isolated from benthic organisms. Thousands of active compounds are being screened for anticancer, antitumor, and other properties.

Key Terms

All key terms from this chapter can be viewed by term, or by definition, when studied as flashcards on this book's Online Learning Center at www.mhhe.com/sverdrup (click on this book's cover).

Study Questions

1. Explain the relationship between sulfides, bacteria, tube worms, and clams in a hydrothermal vent environment.
2. In what ways are the benthic algae (seaweeds) adapted for life in the littoral and sublittoral zones? Consider their structure, pigments, and life requirements.
3. In what ways are the benthic algae important in the ocean environment?
4. Discuss the food-gathering strategies of motile and sessile organisms in the littoral and sublittoral zones.
5. Discuss the factors that are responsible for the littoral zonation of marine organisms along a rocky shore.
6. Design an original organism to inhabit the supralittoral, the littoral, or the sublittoral zone. Consider its requirements for food, shelter, and protection from predators, its adaptations to its environment, and its life history.
7. Why are some subtidal forms found in a tide pool high on a rocky beach, while other subtidal forms are not?
8. Why are there few benthic organisms on a beach made up of noncohesive sediments in a wave and surf area?
9. Discuss the importance of bacteria to benthic organisms.
10. Compare a square-meter area of deep-sea floor with a square-meter area of the rocky intertidal zone. What differences do you expect to find? Consider biomass, species abundance, and substrate.
11. How are coral reefs able to support a rich and varied population, when the water surrounding the reef is clear and devoid of planktonic primary producers?
12. What is coral bleaching? Why does it happen? What is its result?
13. Compare the organisms found growing around deep-ocean hot-water vents and the organisms found around cold gas and oil seeps.
14. Discuss the genetic manipulation of fish and shellfish. Do the advantages of such techniques outweigh the possible disadvantages?
15. Compare photosynthesis and chemosynthesis; how are they similar and how are they different?

Links to Related Websites

Visit the book's Online Learning Center at www.mhhe.com/sverdrup (click on the book's cover) to find live Internet links for additional topics related to this chapter's content.

- Tide pools
- Algae
- Corals and coral reefs

Visit the book's Online Learning Center at www.mhhe.com/sverdrup (click on this book's cover) to find these additional chapter tools: Suggested Readings; links to further information on boxed readings, selected figures, and related chapter topics; and additional study aids.

Scientific (or Exponential) Notation

The writing of very large and very small numbers is simplified by using exponents, or powers of 10, to indicate the number of zeroes required to the left or to the right of the decimal point. The numbers that are equal to some of the powers of 10 are as follows:

$$1,000,000,000. = 10^9 = \text{one billion}$$
$$1,000,000. = 10^6 = \text{one million}$$
$$1000. = 10^3 = \text{one thousand}$$
$$100. = 10^2 = \text{one hundred}$$
$$10. = 10^1 = \text{ten}$$
$$1. = 10^0 = \text{one}$$
$$0.1 = 10^{-1} = \text{one tenth}$$
$$0.01 = 10^{-2} = \text{one hundredth}$$
$$0.001 = 10^{-3} = \text{one thousandth}$$
$$0.000001 = 10^{-6} = \text{one millionth}$$
$$0.000000001 = 10^{-9} = \text{one billionth}$$

149,000,000 is rewritten by moving the decimal point eight places to the left and multiplying by the exponential number 10^8, to form 1.49×10^8. In the same way, 605,000 becomes 6.05×10^5.

A very small number such as 0.000032 becomes 3.2×10^{-5} by moving the decimal point five places to the right and multiplying by the exponential number 10^{-5}. Similarly, 0.00000372 becomes 3.72×10^{-6}.

To add or subtract numbers written in exponential notation, convert the numbers to the same power of 10. For example,

$$
\begin{array}{ccc}
1.49 \times 10^3 & 14.90 \times 10^2 & 1.490 \times 10^3 \\
+6.05 \times 10^2 = & 6.05 \times 10^2 = & 0.605 \times 10^3 \\
\hline
& 20.95 \times 10^2 & 2.095 \times 10^3
\end{array}
$$

$$
\begin{array}{ccc}
2.36 \times 10^3 & 23.60 \times 10^2 & 2.360 \times 10^3 \\
-1.05 \times 10^2 = & -1.05 \times 10^2 = & -0.105 \times 10^3 \\
\hline
& 22.55 \times 10^2 & 2.255 \times 10^3
\end{array}
$$

To multiply, the exponents are added and the numbers are multiplied:

$$
\begin{array}{r}
4.6 \times 10^3 \\
\times 2.2 \times 10^2 \\
\hline
10.12 \times 10^5 = 1.012 \times 10^6
\end{array}
$$

To divide, subtract the exponents and divide the numbers:

$$\frac{6.0 \times 10^8}{2.5 \times 10^3} = 2.4 \times 10^5$$

The following prefixes correspond to the powers of 10 and are used in combination with metric units:

Exponential Notation

Exponential Value	Prefix	Symbol
10^{18}	exa	E
10^{15}	peta	P
10^{12}	tera	T
10^9	giga	G
10^6	mega	M
10^3	kilo	k
10^2	hecto	h
10^1	deka	da
10^{-1}	deci	d
10^{-2}	centi	c
10^{-3}	milli	m
10^{-6}	micro	μ
10^{-9}	nano	n
10^{-12}	pico	p
10^{-15}	femto	f
10^{-18}	atto	a

appendix B
SI Units

The *Système international d' unités*, or International System of Units, is a simplified system of metric units (known as (SI units) adopted by international convention for scientific use.

Basic SI Units

Quantity	Unit	Symbol
length	meter	m
mass	kilogram	kg
time	second	s
temperature	Kelvin	K

Derived SI Units

Quantity	Unit	Symbol	Expression
area	meter squared	m^2	m^2
volume	meter cubed	m^3	m^3
density	kilogram per cubic meter	kg/m^3	kg/m^3
speed	meter per second	m/s	m/s
acceleration	meter per second per second	m/s^2	m/s^2
force	newton	N	$(kg)(m)/s^2$
pressure	pascal	Pa	N/m^2
energy	joule	J	(N)(m)
power	watt	W	J/s; (N)(m)/s

Length: The Basic SI Unit Is the Meter

Unit	Metric Equivalent	English Equivalent	Other
meter (m)	100 centimeters 1000 millimeters	39.37 inches 3.281 feet	0.546 fathom
kilometer (km)	1000 meters	0.621 land mile	0.540 nautical mile
centimeter (cm)	10 millimeters 0.01 meter	0.394 inch	
millimeter (mm)	0.1 centimeter 0.001 meter	0.0394 inch	
land mile (mi)	1609 meters	5280 feet	0.869 nautical mile
nautical mile (nm)	1852 meters	1.151 land miles 6076 feet	1 minute of latitude
fathom (fm)	1.8288 meters	6 feet	

Area: Derived from Length

Unit	Metric Equivalent	English Equivalent	Other
square meter (m²)	10,000 square centimeters	10.76 square feet	
square kilometer (km²)	1,000,000 square meters	0.386 square land mile	0.292 square nautical mile
square centimeter (cm²)	100 square millimeters	0.151 square inch	

Volume: Derived from Length

Unit	Metric Equivalent	English Equivalent	Other
cubic meter (m³)	1,000,000 cubic centimeters 1000 liters	35.32 cubic feet 264 U.S. gallons	
cubic kilometer (km³)	1,000,000,000 cubic meters	0.239 cubic land mile	0.157 cubic nautical mile
liter (L) or (l)	1000 cubic centimeters	1.06 quarts, 0.264 U.S. gallon	
milliliter (mL) or (ml)	1.0 cubic centimeter		

Mass: The Basic SI Unit Is the Kilogram

Unit	Meteric Equivalent	English Equivalent
kilogram (kg)	1000 grams	2.205 pounds
gram (g)		0.035 ounce
metric ton, or	1000 kilograms	2205 pounds
tonne (t)	1,000,000 grams	
U.S. ton	907 kilograms	2000 pounds

Time: The Basic SI Unit Is the Second

Unit	Metric and English Equivalent	
minute	60 seconds	
hour	60 minutes; 3600 seconds	
day	86,400 seconds 24 hours	} mean solar day
year	31,556,880 seconds 8756.8 hours 365.25 solar days	} mean solar year

Temperature: The Basic SI Unit Is the Kelvin

Reference Point	Kelvin (K)	Celsius (°C)	Fahrenheit (°F)
absolute zero	0	−273.2	−459.7
seawater freezes	271.2	−2.0	28.4
fresh water freezes	273.2	0.0	32.0
human body	310.2	37.0	98.6
fresh water boils	373.2	100.0	212.0
conversions	$K = °C + 273.2°$	$°C = \dfrac{(°F - 32)}{1.8}$	$°F = (1.8 \times °C) + 32$

Speed (Velocity): The Derived SI Unit Is the Meter per Second

Unit	Metric Equivalent	English Equivalent	Other
meter per second (m/s)	100 centimeters per second 3.60 kilometers per hour	3.281 feet per second 2.237 land miles per hour	1.944 knots
kilometer per hour (km/h)	0.277 meter per second	0.909 foot per second	0.55 knot
knot (kt)	0.51 meter per second	1.151 land miles per hour	1 nautical mile per hour

Acceleration: The Derived SI Unit Is the Meter per Second per Second

Unit	Metric Equivalent	English Equivalent
meter per second per second (m/s²)	12,960 kilometers per hour per hour 100 centimeters per second per second	8048 miles per hour per hour 3.281 feet per second per second

Force: The Derived SI Unit Is the Newton

Unit	Metric Equivalent	English Equivalent
newton (N)	100,000 dynes	0.2248 pound force
dyne (dyn)	0.00001 newton	0.000002248 pound force

Pressure: The Derived SI Unit Is the Pascal

Unit	Metric Equivalent	English Equivalent	Other
pascal (Pa)	1 newton per square meter 10 dynes per square centimeter		
bar	100,000 pascals 1000 millibars	14.5 pounds per square inch	0.927 atmosphere 29.54 inches of mercury
standard atmosphere (atm)	1.013 bars 101,300 pascals	14.7 pounds per square inch	29.92 inches of mercury or 76 cm of mercury

Energy: The Derived SI Unit Is the Joule

Unit	Metric Equivalent	English Equivalent
joule (J)	1 newton-meter 0.2389 calorie	0.0009481 British thermal unit
calorie (cal)	4.186 joules	0.003968 British thermal unit

Power: The Derived SI Unit Is the Watt

Unit	Metric Equivalent	English Equivalent
watt (W)	1 joule per second 0.2389 calorie per second 0.001 kilowatt	0.0569 British thermal unit per minute 0.001341 horsepower

Density: The Derived SI Unit Is the Kilogram per Cubic Meter

Unit	Metric Equivalent	English Equivalent
kilogram per cubic meter	0.001 gram per cubic centimeter (g/cm^3)	0.0624 pound per cubic foot

To Learn More about Metrics U.S. Metric Association
(USMA), Inc. **http://lamar.colostate.edu/~hillger/**

appendix C
Equations and Quantitative Relationships

Proportion Ratios

Modeling: A / B = A′/B′

Scaling and vertical distortion: $(D/L) \times V$. Distortion = D′/L′

Distance, Rate, Time Problems

Distance = speed × time

Area = area spreading rate × time

Volume = volume flow rate × time

Mass transferred = mass transfer rate × time

Rotation = rotation rate × time

Volume flow rate = current speed × area

Equilibrium, Steady-State Problems

Volumes, masses, and energy are held constant in time

All inflows equal ouflows. $(rate \times time)_{in} = (rate \times time)_{out}$

Residence time = mass or volume / supply or removal rate

Exponential Problems

$X = Y e^{\pm(rz)}$ Requires math tables or calculator

In time-dependent problems z = time, $r = (A/time)$, a rate

X is the value of Y after a given time. If rz is +, then X > Y

If rz is −, then X < Y

When X = 0.5 (Y), (rz is negative) and (rz) = 0.693

Radioactive decay example. Half-life calculation:

$$0.5 \,(Y) = 1 \,(Y)\, e^{-(rt)}$$

Light attenuation example:

$$I_{depth\,z} = I_{z=0}\, e^{-(xz)}$$

where I = light intensity, x = light attenuation rate, z = depth or distance over which light is attenuated. The attenuation rate of light with depth, x, is about $1.7/D$, where D is the Secchi disk depth. x can be determined for total available light or for individual wavelengths of light.

Population example:

$$P_{time=t} = P_{t=0}\, e^{+(rt)}$$

P is the number of individuals or mass. r in this case is a combination of reproduction rate minus the (death rate + grazing rate).

Three-Dimensional Problems

Volume = length × width × depth

Area = length × width

Slope Problems

Slope = rise/run

Horizontal distance × slope = elevation or height change

Small-Particle Settling Velocity; Stokes Law

$V = (2/9)\, [g\, (\rho_1 - \rho_2)/\mu] r^2$

V = settling speed, terminal velocity in cm/s

2/9 = shape factor constant for all small spheres

g = acceleration due to earth's gravity, 981 cm/s^2

ρ_1 = quartz particle density, g/cm^3

ρ_2 = seawater density, g/cm^3

μ = viscosity of seawater

r = radius of particle in cm

$$V_{cm/s} = (2.62 \times 10^4)\, (r_{cm})^2$$

Used to determine settling rate of small particles with diameters less than 0.125 mm. Can be used in reverse to determine size of small particles by measuring settling rates.

Heat Problems

Latent heat of fusion of water at 0°C = 80 cal/g

Latent heat of vaporization of water at 100°C = 540 cal/g

Heat capacity of water = 1 cal/g/°C

Calories = mass in g × heat capacity × temperature change, °C

Interpolation of Data Tables

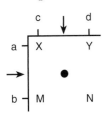

Find the table values at axis value between a and b, and c and d.

(b − a) e + a = axis value between a and b
(d − c) f + c = axis values between c and d
Solve for e and f.

Solve for appropriate table values intermediate to X and M, and Y and N, and then X and Y, and M and N.

(M − X)e + X = 0 = interpolated value between X and M
(N − Y)e + Y = P = interpolated value between Y and N
(P − 0)f + P = interpolated value of table number at chosen axis values between a and b, and c and d

Seawater, Constancy of Composition

$Cl_1‰/Cl_2‰ = S_1‰/S_2‰ = ionA_1/ionA_2 = ionB_1 / ionB_2$

Subscripts 1 and 2 refer to two different concentrations.

Hydrostatic Pressure

$$P = \rho g z$$

P is pressure, ρ is fluid density, g is Earth's gravity, and z is the height of the fluid.

Progressive Surface Water Waves

Simple sinusoidal wave theory:

Wave speed, C, = L/T, where L = wavelength and T, the wave period, is conserved.
$C^2 = (g/2\pi) L \, Tanh[(2\pi/L)D]$, where g = Earth's gravity, D = water depth, L = wavelength, and $Tanh[\varnothing]$ = the hyperbolic tangent of angle$[\varnothing]$. Here $\varnothing$ equals $(2\pi/L)D$ in radians. See math tables or use calculator.

Deep-water approximation:

When $D > L/2$, $Tanh[(2\pi/L)D] \approx 1.0$. $C^2 = (g/2\pi) L$

Shallow-water approximation:

When $D < L/20$, $Tanh[(2\pi/L)D] \approx (2\pi/L)D$. $C = \sqrt{gD}$

Internal Waves

The celerity, C, or wave speed of internal waves along a boundary of a two-layer system is determined by a relationship between the density of the layers and their thicknesses. When the wavelength, L, is long compared with the water depths:

$$C^2 = (g/2\pi)L\left[\frac{\rho - \rho'}{\rho \cotanh (2\pi h/L) + \rho' \cotanh (2\pi h'/L)}\right]$$

h and ρ are the thickness and density of the denser lower layer; and h' and ρ' are the thickness and density of the less-dense upper layer. Cotanh is the hyperbolic cotangent. When L is small compared to the water depths, cotanh $(2\pi h/L)$ and cotanh$(2\pi h'/L)$ approach the value of 1.0 and

$$C^2 = (g / 2\pi)L\left[\frac{\rho - \rho'}{\rho + \rho'}\right]$$

This equation relates to the deep-water wave equation at the sea surface which is the boundary of two fluids—air and water. In this later case ρ' of air is about 1/1000 the value of ρ of water and is neglected.

Standing Waves, Seiches

Based on $L/T = \sqrt{gD}$, shallow-water wave condition.
Closed basin:

The period of oscillation is $T = (1/n) (2l/\sqrt{gD})$, where n = the number of wave nodes, $L = 2 \times$ basin length, l; $L = 2l$, and D = basin depth.

Open-end basin:

The period of oscillation is

$T = (1/n) (4l/\sqrt{gD})$. $L = 4 \times$ basin length, $L = 4l$

Refraction of Light, Sound, and Waves

Snell's Law:
The change of speed of propagation causes refraction, bending, or wave rays.

$$C_1 \sin(a_2) = C_2 \sin(a_1)$$

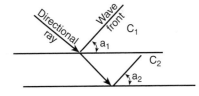

Tide-Raising Forces

The force between two masses due to their mutual gravitational attraction.

$$F = G (M_1 M_2)/R^2$$

where M_1 = mass 1, M_2 = mass 2, R = distance between centers of masses M_1 and M_2, and G = Newton's constant of gravitation = 6.67×10^8 cm^3/g/s^2.

The gravitational force per unit mass on the Earth, M_1, caused by the Sun or the Moon, mass M_2, is,

$$F/M_1 = G (M_2 / R^2)$$

where R is the distance between the unit Earth mass M_1 and the Sun or the Moon center.

The tide raising force, $\Delta F/M_1$, is the difference between the gravitational force acting on a unit mass at the Earth's surface directly in line with the tide-raising body and a unit mass at the Earth's center. (F/M_1)sur. $-$ (F/M_1)cen. $= \Delta F/M_1$

$$\Delta F/M_1 = [G\,M_2/(R-r)^2] - [GM_2/(R)^2]$$

where R is the distance between the center of the Earth and center of the Sun or Moon and r is the Earth's radius.

$$\Delta F/M_1 = (-GM_2/R^2)\,[1 - (1/(1-r/R)^2)]$$
$$= (-GM_2/R^2)\,[-2r/R + (r/R)^2]/(1-r/R)^2$$

If the magnitude of r/R and $(r/R)^2$ is calculated, then $(r/R)^2 \ll 2r/R$ and $(r/R)^2$ is neglected in the numerator; $(r/R) \ll 1.0$ and (r/R) is neglected in the denominator.

$$\Delta F/M_1 = (-GM_2/R^2)\,[(-2r/R)/1] = 2GM_2\,r/R^3$$

Water and Salt Budgets for Partially Mixed Estuaries

Water in = water out = water budget, volume constant

Salt in = salt out = salt budget, salt content constant

T_o = volume transport of mixed water out, vol/time

T_i = volume transport of seawater in, vol/time

R = river flow into estuary, vol/time

$\bar{S}_o$ = average salt content of T_o water, mass/vol

$\bar{S}_i$ = average salt content of T_i seawater, mass/vol

Water budget: $T_o = T_i + R$

Salt budget: $T_o\,\bar{S}_o = T_i\,\bar{S}_i$

Combined budgets: $T_o = [\bar{S}_i/(\bar{S}_i - \bar{S}_o)]\,R$

Estuary flushing time or estuary water residence time equals estuary volume / T_o.

$(\bar{S}_i - \bar{S}_o)/\bar{S}_i$ = fraction of fresh water in T_o type flow

$(\bar{S}_i - \bar{S}_a)/\bar{S}_i$ = fraction of fresh water in estuary volume
where $\bar{S}_a$ = average salt concentration of estuary water

$[(\bar{S}_i - \bar{S}_a)/\bar{S}_i] \times$ estuary volume = freshwater volume stored in the estuary

(Freshwater volume of estuary / R) = residence time of fresh water in the estuary

Plant Production of Carbon by Phytoplankton

Production of carbon (C), release of oxygen (O_2), and use of the nutrients nitrogen (N) and phosphorus (P) by phytoplankton in photosynthesis occurs with fixed ratios on a mass basis. If the use of or production rate of one of the terms is known, the others are calculated from mass-to-mass ratios.

$$O_2{:}C{:}N{:}P = 109{:}41{:}7.2{:}1$$

glossary

A

absorption taking in of a substance by chemical or molecular means; change of sound or light energy into some other form, usually heat, in passing through a medium or striking a surface.

abyssal pertaining to the great depths of the ocean below approximately 4000 m.

abyssal clay is lithogenous sediment on the deep-sea floor composed of at least 70% clay-sized particles by weight.

abyssal hill low, rounded submarine hill less than 1000 m high.

abyssal plain flat ocean basin floor extending seaward from the base of the continental slope and continental rise.

abyssopelagic oceanic zone from 4000 m to the deepest depths.

accretion natural or artificial deposition of sediment along a beach, resulting in the buildup of new land.

acoustic profiling the use of seismic energy to measure sediment thickness and layering on the sea floor.

active margin *see* leading margin.

adsorption attraction of ions to a solid surface.

advection horizontal or vertical transport of seawater, as by a current.

agar substance produced by red algae; the gelatin-like product of these algae.

algae marine and freshwater organisms (including most seaweeds) that are single-celled, colonial, or multicelled, with chlorophyll but no true roots, stems, or leaves and with no flowers or seeds.

algin complex organic substance found in or obtained from brown algae.

alluvial plain flat deposit of terrestrial sediment eroded by water from higher elevations.

amphidromic point point from which cotidal lines radiate on a chart; the nodal, or low-amplitude, point for a rotary tide.

amplitude for a wave, the vertical distance from sea level to crest or from undisturbed sea level to trough, or one-half the wave height.

anaerobic living or functioning in the absence of oxygen.

andesite type of volcanic rock that is intermediate in composition between basalt and granite; associated with subduction zones.

anion negatively charged ion.

anoxic deficient in oxygen.

Antarctic Circle *see* Arctic and Antarctic circles.

antinode portion of a standing wave with maximum vertical motion.

aphotic zone that part of the ocean in which light is insufficient to carry on photosynthesis.

aquaculture (mariculture) cultivation of aquatic organisms under controlled conditions.

Arctic and Antarctic Circles latitudes 66 1/2°N and 66 1/2°S, respectively, marking the boundaries of light and darkness during the summer and winter solstices.

armored beach a beach that is protected from wave and water erosion by being covered with lag deposits.

asthenosphere upper, deformable portion of the Earth's mantle, the layer below the lithosphere; probably partially molten; may be site of convection cells.

atmospheric pressure pressure, at any point on Earth, exerted by the atmosphere as a consequence of gravitational force exerted on the column of air lying directly above the point.

atoll ring-shaped coral reef that encloses a lagoon in which there is no exposed preexisting land and which is surrounded by the open sea.

attenuation decrease in the energy of a wave or beam of particles occurring as the distance from the source increases; caused by absorption, scattering, and divergence from a point source.

autotrophic pertaining to organisms able to manufacture their own food from inorganic substances. *See also* chemosynthesis and photosynthesis.

autumnal equinox *see* equinoxes.

auxospore naked cell of a diatom, which grows to full size and forms a new siliceous covering.

B

backshore beach zone lying between the foreshore and the coast, acted on by waves only during severe storms and exceptionally high water.

baleen whalebone; horny material growing down from the upper jaw of plankton-feeding whales; forms a strainer, or filtering organ, consisting of numerous plates with fringed edges.

bar offshore ridge or mound of sand, gravel, or other loose material that is submerged, at least at high tide; located especially at the mouth of a river or estuary or lying a short distance from and parallel to the beach.

barrier island deposit of sand, parallel to shore and raised above sea level; may support vegetation and animal life.

barrier reef coral reef that parallels land but is some distance offshore, with water between reef and land.

basalt fine-grained, dark igneous rock, rich in iron and magnesium, characteristic of oceanic crust.

basin large depression of the sea floor having about equal dimensions of length and width.

bathyal pertaining to ocean depths between approximately 1000 and 4000 m.

bathymetry study and mapping of seafloor elevations and the variations of water depth; the topography of the sea floor.

bathypelagic oceanic zone from 1000–4000 m.

beach zone of unconsolidated material between the mean low-water line and the line of permanent vegetation, which is also the effective limit of storm waves; sometimes includes the material moving in offshore, onshore, and longshore transport.

beach face section of the foreshore normally exposed to the action of waves.

Beaufort scale scale of wind forces by range of velocity; scale of sea state created by winds of these velocities.

Benioff zone dipping patterns of earthquake activity that descend into the mantle along convergent plate boundaries.

benthic of the sea floor, or pertaining to organisms living on or in the sea floor.

benthos organisms living on or in the ocean bottom.

berm nearly horizontal portion of a beach (backshore) with an abrupt face; formed from the deposition of material by wave action at high tide.

berm crest ridge marking the seaward limit of a berm.

bilateral symmetry having right and left halves that are approximate mirror images of each other.

biodiversity the number of species in an area compared to the number of individuals.

biogenous sediment sediment derived from organisms.

biological pump photosynthetic transfer of carbon as CO_2 from the atmosphere to the ocean in the form of organic molecules; carbon is transferred to intermediate and deep-ocean water when organic material sinks and decays.

bioluminescence production of light by living organisms as a result of a chemical reaction either within certain cells or organs or outside the cells in some form of excretion.

biomass the total mass of all or specific living organisms, usually expressed as dry weight in grams of carbon per unit area or unit volume.

bioturbation reworking of sediments by organisms that burrow into them and ingest them.

blade flat, photosynthetic, "leafy" portion of an alga or a seaweed.

bloom high concentration of phytoplankton in an area, caused by increased reproduction; often produces discoloration of the water. *See also* red tide.

breaker sea surface water wave that has become too steep to be stable and collapses.

breakwater structure protecting a shore area, harbor, anchorage, or basin from waves; a type of jetty.

buffer substance able to neutralize acids and bases, therefore able to maintain a stable pH.

bulkhead structure separating land and water areas; primarily designed to resist Earth sliding and slumping or to reduce wave erosion at the base of a cliff.

buoy floating object anchored to the bottom or attached to another object; used as a navigational aid or surface marker.

buoyancy ability of an object to float due to the support of the fluid the body is in or on.

by-catch *see* incidental catch.

C

caballing mixing of two water types with identical densities but different temperatures and salinities; the resulting mixture is denser than its components.

calcareous containing or composed of calcium carbonate.

calcareous ooze fine-grained deep-ocean biogenous sediment containing at least 30% calcareous tests, or the remains of small marine organisms.

calorie amount of heat required to raise the temperature of 1 g of water 1°C.

calving breaking away of a mass of ice from its parent glacier, iceberg, or sea ice formation.

capillary waves waves with wavelengths less than 1.5 cm in which the primary restoring force is surface tension.

carbonate a sediment or rock formed from the accumulation of carbonate minerals ($CaCO_3$) precipitated organically or inorganically.

carbonate compensation depth (CCD), also known as the calcite compensation depth, is the depth at which the amount of calcium carbonate ($CaCO_3$) preserved falls below 20% of the total sediment. This is also commonly defined as the depth at which the amount of calcium carbonate produced by organisms as skeletal material in the overlying water column is equal to the rate at which it is dissolved in the water. No calcium carbonate will be deposited below this depth.

carnivore flesh-eating organism.

carrageenan substance produced by certain algae that acts as a thickening agent.

cation positively charged ion.

cat's-paw patch of ripples on the water's surface, related to a discrete gust of wind.

centrifugal force outward-directed force acting on a body moving along a curved path or rotating about an axis; an inertial force.

centripetal force inward-directed force necessary to keep an object moving in a curved path or rotating about an axis.

chaetognaths free-swimming, carnivorous, pelagic, wormlike, planktonic animals; arrowworms.

chemosynthesis formation of organic compounds with energy derived from inorganic substances such as ammonia, sulfur, and hydrogen.

chloride atom of chlorine in solution, forming an ion with a negative charge.

chlorinity (Cl‰) measure of the chloride content of seawater in grams per kilogram.

chlorophyll group of green pigments that are active in photosynthesis.

chromosome one of the bodies in a cell that carries the genes in a linear order.

chronometer portable clock of great accuracy used in determining longitude at sea.

ciguatera toxin found in fish of tropical regions; produced by dinoflagellates.

cilia microscopic, hairlike projections of living cells that beat in coordinated fashion and produce movement.

cluster a group of galaxies. A cluster may contain thousands of galaxies.

coast strip of land of indefinite width that extends from the shore inland to the first major change in terrain that is unaffected by marine processes.

coastal circulation cell (drift sector, littoral cell) longshore transport cell pattern of sediment moving from a source to a place of deposition.

coccolithophore microscopic, planktonic alga surrounded by a cell wall with embedded calcareous plates (coccoliths).

cohesion molecular force between particles within a substance that acts to unite them.

colonial organism organism consisting of semi-independent parts that do not exist as separate units; groups of organisms with specialized functions that form a coordinated unit.

commensalism an intimate association between different organisms in which one is benefited and the other is neither harmed nor benefited.

compensation depth depth at which there is a balance between the oxygen produced by algae through photosynthesis and that consumed by them through respiration; net oxygen production is zero.

condensation process by which a vapor becomes a liquid or a solid.

conduction transfer of heat energy through matter by internal molecular motion; also heat transfer by turbulence in fluids.

conservative constituent component or property of seawater whose value changes only as a result of mixing, diffusion, and advection and not as a result of biological or chemical processes; for example, salinity.

consumer animal that feeds on plants (primary consumer) or on other animals (secondary consumer).

continental crust crust forming the continental land blocks; mainly granite and its derivatives.

continental drift the movement of continents; the name of Alfred Wegener's theory, preceding plate tectonics.

continental margin zone separating the continents from the deep-sea bottom, usually subdivided into shelf, slope, and rise.

continental rise gentle slope formed by the deposition of sediments at the base of a continental slope.

continental shelf zone bordering a continent, extending from the line of permanent immersion to the depth at which there is a marked or rather steep descent to the great depths.

continental shelf break zone along which there is a marked increase of slope at the outer margin of a continental shelf.

continental slope relatively steep downward slope from the continental shelf break to depth.

contour line on a chart or graph connecting the points of equal value for elevation, temperature, salinity, or other property.

convection transmission of heat by the movement of a heated gas or liquid; vertical circulation resulting from changes in density of a fluid.

convection cell circulation in a fluid, or fluidlike material, caused by heating from below. Heating the base of a fluid lowers its density, causing it to rise. The rising fluid cools, becomes denser, and sinks, creating circulation.

convergence situation in which substances come together, usually resulting in the sinking, or downwelling, of surface water and the rising of air.

convergent plate boundary a boundary between two plates that are converging or colliding with one another.

copepod small, shrimplike member of the zooplankton; in the class Crustacea.

coral colonial animal that secretes a hard outer calcareous skeleton; the skeletons of coral animals form in part the framework for warm-water reefs.

corange lines in a rotary tide, lines of equal tidal range about the amphidromic point.

core vertical, cylindric sample of bottom sediments, from which the nature of the bottom can be determined; also the central zone of the Earth, thought to be liquid or molten on the outside and solid on the inside.

corer device that plunges a hollow tube into bottom sediments to extract a vertical sample.

Coriolis effect apparent force acting on a body in motion, due to the rotation of Earth, causing deflection to the right in the Northern Hemisphere and to the left in the Southern Hemisphere; the force is proportional to the speed and varies with the latitude of the moving body.

cosmogenous sediment sediment particles with an origin in outer space; for example, meteor fragments and cosmic dust.

cotidal lines lines on a chart marking the location of the tide crest at stated time intervals.

covalent bond chemical bond formed by the sharing of one or more pairs of electrons.

cratered coast a primary coast formed when the seaward side of a volcanic crater located at the water's edge is eroded away or is blown away by a volcanic eruption; opening the interior of the crater to the sea and creating a concave bay.

cratons large pieces of Earth's crust that form the centers of continents.

crest *see* berm crest, reef crest, wave crest.

crust outer shell of the solid Earth; the lower limit is usually considered to be the Mohorovicic discontinuity.

crustacean member of a class of primarily aquatic organisms with paired jointed appendages and a hard outer skeleton; includes lobsters, crabs, shrimps, and copepods.

ctenophore transparent, planktonic animal, spherical or cylindrical with rows of cilia; comb jelly.

Curie temperature temperature at which the magnetic signature of molten rock is frozen into it during cooling.

current horizontal movement of water.

current meter instrument for measuring the speed and direction of a current.

cusp one of a series of evenly spaced, crescent-shaped depressions along sand and gravel beaches.

cyclone *see* typhoon, hurricane.

D

deadweight ton (DWT) capacity of a vessel in tons of cargo, fuel, stores, and so on; determined by the weight of the water displaced.

declinational tide *see* diurnal tide.

decomposer heterotrophic; microorganisms (usually bacteria and fungi) that break down nonliving organic matter and release nutrients, which are then available for reuse by autotrophs.

deep exceptionally deep area of the ocean floor, usually below 6000 m.

deep scattering layer (DSL) layer of organisms that move away from the surface during the day and toward the surface at night; the layer scatters or returns vertically directed sound pulses.

deep-water wave wave in water, the depth of which is greater than one-half the wavelength.

degree an arbitrary measure of temperature. Temperature is measured in degrees using one of three different scales; Fahrenheit (°F), Celsius (°C), and Kelvin (K). See appendix B for conversions between these scales.

delta area of unconsolidated sediment deposit, usually triangular in outline, formed at the mouth of a river.

demersal fish fish living near and on the bottom.

density property of a substance defined as mass per unit volume and usually expressed in grams per cubic centimeter or kilograms per cubic meter.

depth recorder *see* echo sounder.

desalination process of obtaining fresh water from seawater.

detritus any loose material, especially decomposed, broken, and dead organic materials.

dew point temperature to which air must be cooled for its water vapor to condense.

diatom microscopic unicellular alga with an external skeleton of silica.

diatomaceous ooze sediment made up of more than 30% skeletal remains of diatoms.

diffraction process that transmits energy laterally along a wave crest.

diffusion movement of a substance from a region of higher concentration to a region of lower concentration (movement along a concentration gradient); may be due to molecular motion or turbulence.

dinoflagellate one of a class of planktonic organisms possessing characteristics of both plants and animals.

dipole a magnetic field like the Earth's, with two opposite poles.

dispersion (sorting) sorting of waves as they move out from a storm center; occurs because long waves travel faster in deep water than short waves.

diurnal inequality difference in height between the two high waters or two low waters of each tidal day; the difference in speed between the two flood currents or two ebb currents of each tidal day.

diurnal tide (declinational tide) tide with one high water and one low water each tidal day.

divergence horizontal flow of substances away from a common center, associated with upwelling in water and descending motions in air.

divergent plate boundary a boundary between two plates that are diverging or moving apart from one another.

Dobson unit Ozone unit, defined as 0.01 mm thickness of ozone at Standard Temperature and Pressure (0°C and 1 atmosphere pressure). If all the ozone over a certain area is compressed to 1 atmosphere pressure at 0°C, it forms a slab of a thickness corresponding to a number of Dobson units.

doldrums nautical term for the belt of light, variable winds near the equator.

downwelling zone sinking of water toward the bottom, usually the result of a surface convergence or an increase in density of water at the sea surface.

dredge cylindric or boxlike sampling device made of metal, net, or both, which is dragged across the bottom to obtain biological or geological samples.

drift bottle bottle released into the sea for use in studying currents; contains a card identifying date and place of release and requesting the finder to return it with date and place of recovery.

drift sector *see* coastal circulation cell.

dugong *see* sea cow.

dune wind-formed hill or ridge of sand.

dune coast a primary coast formed by the deposition of sand in dunes by the wind.

dynamic equilibrium state in which the sum of all changes is balanced and there is no net change.

E

Earth sphere depth uniform depth of Earth below the present mean sea level, if the solid Earth surface were smoothed off evenly (2440 m).

ebb current movement of a tidal current away from shore or down a tidal stream as the tide level decreases.

ebb tide falling tide; the period of the tide between high water and the next low water.

echolocation use of sound waves by some marine animals to locate and identify underwater objects.

echo sounder (depth recorder) instrument used to measure the depth of water by measuring the time interval between the release of a sound pulse and the return of its echo from the bottom. *See also* precision depth recorder (PDR).

ecology the study of interactions between organisms and their environment and of interactions among organisms.

ecosystem the organisms in a community and the nonliving environment with which they interact.

ectotherm organism that maintains body temperature by taking in heat from or returning heat to the environment—fish, reptiles, and invertebrates.

eddy circular movement of water.

Ekman spiral in a theoretical ocean of infinite depth, unlimited extent, and uniform viscosity, with a steady wind blowing over the surface, the surface water moves 45° to the right of the wind in the Northern Hemisphere. At greater depths the water moves farther to the right with decreased speed, until at some depth (approximately 100 m) the water moves opposite to the wind direction. Net water transport is 90° to the right of the wind in the Northern Hemisphere. Movement is to the left in the Southern Hemisphere.

electrodialysis separation process in which electrodes of opposite charge are placed on each side of a semipermeable membrane to accelerate the diffusion of particles across the membrane.

electromagnetic radiation waves of energy formed by simultaneous electrical and magnetic oscillations; the electromagnetic spectrum is the continuum of all electromagnetic radiation from low-energy radio waves to high-energy gamma rays, including visible light.

electromagnetic spectrum *see* electromagnetic radiation.

El Niño wind-driven reversal of the Pacific equatorial currents resulting in the movement of warm water toward the coasts of the Americas, so called because it generally develops just after Christmas.

endotherm organism that maintains body temperature by internal mechanisms—birds and mammals.

entrainment mixing of salt water into fresh water overlying salt water, as in an estuary.

epicenter point on the Earth's surface directly above an earthquake location, specified by identifying the latitude and longitude of the earthquake. *See also* focus, hypocenter.

epifauna animals living attached to the sea bottom or moving freely over it.

epipelagic upper portion of the oceanic pelagic zone, extending from the surface to about 200 m.

episodic wave abnormally high wave unrelated to local storm conditions.

equator 0° latitude, determined by a plane that is perpendicular to the Earth's axis and is everywhere equidistant from the North and South Poles.

equilibrium tide theoretical tide formed by the tide-producing forces of the Moon and Sun on a nonrotating, water-covered Earth.

equinoxes times of the year when the Sun stands directly above the equator, so that day and night are of equal length around the world. The vernal equinox occurs about March 21, and the autumnal equinox occurs September 22–23.

escarpment nearly continuous line of cliffs or steep slopes caused by erosion or faulting.

estuary semi-isolated portion of the ocean that is diluted by freshwater drainage from land.

euphausiid planktonic, shrimplike crustacean. *See also* krill.

eustatic change global change in sea level that affects all of the world's coastlines.

evaporation process by which liquid becomes vapor. Returns moisture to the atmosphere by the hydrologic cycle; water warmed by the Sun's heat reaches its vapor point, becomes vapor or gas, and rises into the atmosphere.

evaporite deposit formed from minerals left behind by evaporating water, especially salt.

evapotranspiration the combined effect of evaporation and transpiration.

extremophile microorganism that thrives under extreme conditions of temperature, lack of oxygen, or high acid or salt levels; conditions that kill other organisms.

F

fast ice sea ice that is anchored to shore or the sea floor in shallow water.

fathom a unit of length equal to 1.8 m (6 ft); used to measure water depth.

fault break or fracture in the Earth's crust, in which one side has been displaced relative to the other.

fault bay a bay formed by faulting along a primary coast.

fault coast a primary coast formed by tectonic activity and faulting.

fetch continuous area of water over which the wind blows in essentially a constant direction.

filament chain of living cells.

fjord narrow, deep, steep-walled inlet of the ocean formed by the submergence of a mountainous coast or by the entrance of the ocean into a deeply excavated glacial trough after the melting of the glacier. A deep, small-surface-area estuary with moderately high river input and little tidal mixing.

flagellum long, whiplike extension from a living cell's surface that by its motion moves the cell; plural, flagella.

floe discrete patch of sea ice moved by the currents or by the wind.

flood current movement of a tidal current toward the shore or up a tidal stream as the tide wave increases.

flood tide rising tide; the period of the tide between low water and the next high water.

flushing time length of time required for an estuary to exchange its water with the open ocean.

focus the location of an earthquake. Focus is specified by identifying latitude, longitude, and depth of the earthquake. *See also* epicenter.

fog visible assemblage of tiny droplets of water formed by condensation of water vapor in the air; a cloud with its base at the surface of the Earth.

food chain sequence of organisms in which each is food for the next member in the sequence. *See also* food web.

food web complex of interacting food chains; all the feeding relations of a community taken together; includes production, consumption, decomposition, and the flow of energy.

foraminifera minute, one-celled animals that usually secrete calcareous shells.

foraminifera ooze sediment made up of 30% or more skeletal remains of foraminifera.

forced wave wave generated by a continuously acting force and caused to move at a speed faster than it freely travels.

foreshore portion of the shore that includes the low-tide terrace and the beach face.

fouling attachment or growth of marine organisms on underwater objects, usually objects that are made or introduced by humans.

fracture zone large, linear zone of irregular bathymetry of the sea floor, characterized by asymmetric ridges and troughs; commonly associated with fault zones.

free wave wave that continues to move at its natural speed after its generation by a force.

friction resistance of a surface to the motion of a body moving (for example, sliding or rolling) along that surface.

fringing reef reef attached directly to the shore of an island or a continent and not separated from it by a lagoon.

frustule siliceous external shell of a diatom.

G

gabbro a coarse-grained, dark igneous rock, rich in iron and magnesium, the slow-cooling equivalent of basalt.

galaxy a huge aggregate of stars held together by mutual gravitation.

generating force disturbing force that creates a wave, such as wind or a landslide.

geomorphology study of the Earth's land forms and the processes that have formed them.

geostrophic flow horizontal flow of water occurring when there is a balance between gravitational forces and the Coriolis effect.

glacier mass of land ice, formed by the recrystallization of compacted old snow, flowing slowly from an accumulation area to an area of ice loss by melting, sublimination, or calving.

Global Positioning System (GPS) a worlwide radio-navigation system consisting of twenty-four navigational satellites and five ground-based monitoring stations. GPS uses this system of satellites as reference points for calculating accurate positions on the surface of the Earth with readily available GPS receivers.

Gondwanaland an ancient landmass that fragmented to produce Africa, South America, Antarctica, Australia, and India.

graben a portion of Earth's crust that has moved downward and is bounded by steep faults; a rift.

grab sampler instrument used to remove a piece of the ocean floor for study.

granite crystalline, coarse-grained, igneous rock composed mainly of quartz and feldspar.

gravitational force mutual force of attraction between particles of matter (bodies).

gravity Earth's gravity; acceleration due to Earth's mass is 981 cm/s^2; g is used in equations.

gravity wave water wave form in which gravity acts as the restoring force; a wave with wavelength greater than 2 cm.

great circle the intersection of a plane passing through the center of the Earth with the surface of the Earth. Great circles are formed by the equator and any two meridians of longitude 180° apart.

greenhouse effect the gradual increase in average global temperature caused by the absoprtion of infrared radiation from Earth's surface by "greenhouse gases" in the atmosphere such as water vapor and carbon dioxide.

Greenwich Mean Time (GMT) solar time along the prime meridian passing through Greenwich, England; also known as Universal Time or Zulu Time.

groin protective structure for the shore, usually built perpendicular to the shoreline; used to trap littoral drift or to retard erosion of the shore; a type of jetty.

group speed speed at which a group of waves travels (in deep water, group speed equals one-half the speed of an individual wave); the speed at which the wave energy is propagated.

guyot submerged, flat-topped seamount.

gyre circular movement of water, larger than an eddy; usually applied to a larger system.

H

habitat place where a plant or animal species naturally lives and grows.

hadal pertaining to the greatest depths of the ocean.

half-life time required for half of an initial quantity of a radioactive isotope to decay.

halocline water layer with a large change in salinity with depth.

harmful algal bloom (HAB) *see* red tide.

harmonic analysis process of separating astronomical tide-causing effects from the tide record, in order to predict the tides at any location.

heat a measure of total kinetic energy of atoms and molecules in a substance.

heat budget accounting for the total amount of the Sun's heat received on Earth during one year as being exactly equal to the total amount lost because of radiation and reflection.

heat capacity the quantity of heat required to produce a unit change of temperature in a unit mass of material. *See also* specific heat.

herbivore animal that feeds only on plants.

heterotrophic pertaining to organisms requiring preformed organic compounds for food; unable to manufacture food from inorganic compounds.

higher high water higher of the two high waters of any tidal day in a region of mixed tides.

higher low water higher of the two low waters of any tidal day in a region of mixed tides.

high water maximum height reached by a rising tide.

holdfast organ of a benthic alga that attaches the alga to the sea floor.

holoplankton organisms living their entire life cycle in the floating (planktonic) state.

hook spit turned landward at its outer end.

horse latitudes regions of calms and variable winds that coincide with latitudes at approximately 30°–35°N and S.

hot spot surface expression of a persistent rising jet of molten mantle material.

hurricane severe, cyclonic, tropical storm at sea, with winds of 120 km (73 mi) per hour or more; generally applied to Atlantic Ocean storms. *See also* typhoon.

hydrogen bond in water, the weak attraction between the positively polar hydrogen of one water molecule and the negatively polar oxygen of another water molecule.

hydrogenous sediment sediment formed from substances dissolved in seawater.

hydrologic cycle movement of water among the land, oceans, and atmosphere due to vertical and horizontal transport, evaporation, and precipitation.

hydrothermal vent seafloor outlet for high-temperature groundwater and associated minerals; a hot spring.

hypocenter *see* focus.

hypoxic having low oxygen levels in the water; organisms may find a hypoxic environment difficult or impossible to survive.

hypsographic curve graph of land elevation and ocean depth versus area.

I

iceberg mass of land ice that has broken away from a glacier and floats in the sea.

igneous rock rock formed by congealing rapidly or slowly from molten magma.

incidental catch the portion of any catch or harvest taken in addition to the targeted species.

inertia the property of matter that causes it to resist any change in its motion.

infauna animals that live buried in the sediment.

inner core the innermost region of the Earth. It is solid and consists primarily of iron with minor amounts of lighter elements that likely include nickel, sulfur, and oxygen.

internal wave wave created below the sea surface at the boundary between two density layers.

international date line an imaginary line through the Pacific Ocean roughly corresponding to 180° longitude, to the east of which, by international agreement, the calendar date is one day earlier than to the west.

intertidal see littoral.

intertidal volume in an embayment, the volume of water gained or lost owing to the rise and fall of the tide.

inverse estuary an embayment, often located in arid climates, with high evaporation and little freshwater input. Circulation is seaward at depth and inward at the surface, opposite that of a typical estuary. Salinity is generally higher than average seawater salinities.

ion positively or negatively charged atom or group of atoms.

ionic bond the electrostatic force that holds together oppositely charged ions formed by a transfer of electrons from a metal to a nonmetal atom.

island arc system chain of volcanic islands formed when plates converge at a subduction zone.

isobar line of constant pressure.

isobath contour of constant depth.

isohaline having a uniform salt content.

isopycnal having a uniform density.

isostasy mechanism by which areas of the Earth's crust rise or subside until their masses are in balance, "floating" on the mantle.

isothermal having a uniform temperature.

isotope atoms of the same element having different numbers of neutrons.

J

jellyfish or sea jelly semitransparent, bell-shaped pelagic organism, often with long tentacles bearing stinging cells.

jet stream (polar) a stream of air, between 30° and 50°N and S and about 12 km above Earth, coming from the west at an average speed of 100 km/h.

jetty structure located to influence currents or to protect the entrance to a harbor or river from waves (U.S. terminology). *See also* breakwater, groin.

K

kelp any of several large, brown algae, including the largest known algae.

kinetic energy energy produced by the motion of an object.

knot a unit of speed equal to 0.51 m/s or 1 nautical mile per hour.

krill small, shrimplike crustaceans found in huge masses in polar waters and eaten by baleen whales.

L

lag deposits large particles left on a beach after the small particles are washed away.

lagoon shallow body of water that usually has a shallow, restricted outlet to the sea.

Langmuir cells shallow wind-driven circulation; paired helixes of moving water form wind rows of debris along convergence lines.

La Niña condition of colder-than-normal surface water in the eastern tropical Pacific.

larva immature juvenile form of certain animals.

latent heat of fusion amount of heat required to change the state of 1 g of water from ice to liquid.

latent heat of vaporization amount of heat required to change the state of 1 g of water from liquid to gas.

latitude distance north or south of the equator. Latitude is the angle between the equatorial plane and a line drawn outward from the center of the Earth to a point on the surface of the Earth. Latitude varies from 0° to +90° north of the equator and 0° to −90° south of the equator. Together with longitude it specifies the location of a point on the surface of the Earth.

Laurasia an ancient landmass that fragmented to produce North American and Eurasia.

lava magma, or molten rock, that has reached Earth's surface; the same material solidified after cooling.

lava coast a primary coast formed by active volcanism producing lava flows that extend to the sea.

leading margin (active margin) the edge of the overriding plate at a trench or subduction zone.

lee shelter; the part or side sheltered from wind or waves.

light-year the distance light travels in one year. A light-year is equal to 9.46×10^{12} km, or 5.87×10^{12} mi.

lithification the conversion of loose sediment to solid rock.

lithogenous sediment sediment composed of rock particles eroded mainly from the continents by water, wind, and waves.

lithosphere outer, rigid portion of the Earth; includes the continental and oceanic crusts and the upper part of the mantle.

littoral area of the shore between mean high water and mean low water; the intertidal zone.

longitude distance east or west of the prime meridian. Longitude is the angle in the equatorial plane between the prime meridian and a second meridian that passes through a point on the surface of the Earth whose location is being specified. Longitude may be specified in one of two ways; either from 0° to 360° east of the prime meridian, or 0° to +180° east and 0° to −180° west. Together with latitude it specifies the location of a point on the surface of the Earth.

longshore current current produced in the surf zone by the waves breaking at an angle with the shore; runs roughly parallel to the shoreline.

longshore transport (littoral drift) movement of sediment by the longshore current.

loran navigational system in which position is determined by measuring the difference in the time of reception of synchronized radio signals; derived from the phrase "long-range navigation."

lower high water lower of the two high waters of any tidal day in a region of mixed tides.

lower low water lower of the two low waters of any tidal day in a region of mixed tides.

low-tide terrace flat section of the foreshore seaward of the sloping beach face.

low water minimum height reached by a falling tide.

lunar month time required for the Moon to pass from one new Moon to another new Moon (approximately twenty-nine days).

lysocline depth at which calcareous skeletal material first begins to dissolve.

M

magma molten rock material that forms igneous rocks upon cooling; magma that reaches Earth's surface is referred to as lava.

magnetic pole either of the two points on Earth's surface where the magnetic field is vertical.

manatee see sea cow.

manganese nodules rounded, layered lumps found on the deep-ocean floor that contain, on average, about 18% manganese, 17% iron, with smaller amounts of nickel, cobalt, and copper and traces of two dozen other metals; a hydrogenous sediment.

mantle main bulk of Earth between the crust and the core; increasing pressure and temperature with depth divide the mantle into concentric layers.

mariculture *see* aquaculture.

maximum sustained yield maximum number or amount of a species that can be harvested each year without steady depletion of the stock; the remaining stock is able to replace the harvested members by natural reproduction.

meander turn or winding curve of a current that may detach from the main stream as an eddy.

mean Earth sphere depth the depth below sea level of the surface of the solid Earth if it was perfectly smooth with no variation in elevation. This is 2440 m (8003 ft) below present sea level.

mean ocean sphere depth the depth of the ocean if the solid Earth was perfectly smooth with no variation in elevation. This is 2686 m (8810 ft).

mean sea level average height of the sea surface, based on observations of all stages of the tide over a nineteen-year period in the United States.

meiobenthos very small animals living buried in the sediments of the sea floor.

Mercator projection a map projection in which the surface of the Earth is projected onto a cylinder. Distortion is great at high latitudes and the poles cannot be shown. Mercator projections are frequently used for navigation because a straight line drawn on them is a line of true direction or constant compass heading.

meridian circle of longitude passing through the poles and any given point on Earth's surface.

meroplankton floating developmental stages (eggs and larvae) of organisms that as adults belong to the nekton and benthos.

mesopelagic oceanic zone from 200–1000 m.

mesosphere either the layer of the atmosphere above the stratosphere extending from about 50–90 km, or the region of the mantle beneath the asthenosphere.

metamorphic rock a rock created by the alteration of a preexisting rock by high temperature and/or pressure.

microbial loop component of marine food webs in which dissolved organic material cycles through bacteria and nanoplankton then back to small members of the zooplankton.

microplankton net plankton, composed of individuals from 0.07–1 mm in size but large enough to be retained by a small mesh net.

minus tide low-tide level below the mean value of the low tides or the zero tidal depth reference.

mitigation coastal management concept requiring developers to replace developed areas with equivalent natural areas or to reengineer other areas to resemble areas prior to development.

mixed tide type of tide in which large inequalities between the two high waters and the two low waters occur in a tidal day.

Mohorovicic discontinuity (Moho) boundary between crust and mantle, marked by a rapid increase in seismic wave speed.

mollusks marine animals, usually with shells; includes mussels, oysters, clams, snails, and slugs.

monoculture cultivation of only one species of organism in an aquaculture system.

monsoon name for seasonal winds; first applied to the winds over the Arabian Sea, which blow for six months from the northeast and for six months from the southwest; now extended to similar winds in other parts of the world; in India, the term is popularly applied to the southwest monsoon and also to the rains that it brings.

Moon tide portion of the tide generated solely by the Moon's tide-raising force, as distinguished from that of the Sun.

moraine glacial deposit of rock, gravel, and other sediment left at the margin of an ice sheet.

mutualism an intimate association between different organisms in which both organisms benefit.

N

nannoplankton plankton that passes through an ordinary plankton net but can be removed from the water by centrifuging water samples.

nautical mile unit of length equal to 1852 m, or 1.15 land miles or 1 minute of latitude.

neap tides tides occurring near the times of the first and last quarters of the Moon, when the range of the tide is least.

nebula a large dense cloud of gas and dust in space.

nekton pelagic animals that are active swimmers; for example, adult squid, fish, and marine mammals.

neritic shallow-water marine environment extending from low water to the edge of the continental shelf. *See also* pelagic.

net circulation the long term transport of water out of an estuary at the surface and into the estuary at depth averaged over many tidal cycles.

net plankton *see* microplankton.

node point of least or zero vertical motion in a standing wave.

nonconservative constituent component or property of seawater whose value changes as a result of biological or chemical processes as well as by mixing, advection, and diffusion; for example, nutrients and oxygen in seawater.

nudibranchs soft-bodied, gastropod mollusks; sea slugs.

nutrient in the ocean, any one of a number of inorganic or organic compounds or ions used primarily in the nutrition of primary producers; nitrogen and phosphorus compounds are examples.

O

oceanic pertaining to the ocean water seaward of the continental shelf; the "open ocean." *See also* pelagic.

oceanic crust crust below the deep-ocean sediments; mainly basalt.

offshore direction seaward of the shore.

offshore current any current flowing away from the shore.

offshore transport movement of sediment or water away from the shore.

onshore direction toward the shore.

onshore current any current flowing toward the shore.

onshore transport movement of sediment or water toward the shore.

oolith small, rounded accretionary body of calcium carbonate. Created by precipitation of calcium carbonate in shallow, warm seawater due to a change in water temperature or acidity.

ooze fine-grained deep-ocean sediment composed of at least 30% calcareous or siliceous remains of small marine organisms, the remainder being clay minerals.

orbit in water waves, the path followed by the water particles affected by the wave motion; also, the path of a body subjected to the gravitational force of another body, such as Earth's orbit around the Sun.

orographic effect precipitation patterns caused by the flow of air over and around mountains.

osmosis tendency of water to diffuse through a semipermeable membrane to make the concentration of water on one side of the membrane equal to that on the other side.

osmotic pressure pressure that builds up in a confined fluid because of osmosis.

outer core a region surrounding the inner core. It is liquid and consists primarily of iron with minor amounts of lighter elements that likely include nickel, sulfur, and oxygen.

overturn sinking of denser water and its replacement by less dense water from below.

oxygen minimum zone in which respiration and decay reduce dissolved oxygen to a minimum, usually between 800 and 1000 m.

ozone a form of oxygen that absorbs ultraviolet radiation from the Sun; O_3.

P

paleoceanography the study of ocean characteristics and processes in the past.

paleomagnetism study of ancient magnetism recorded in rocks; includes study of changes in location of Earth's magnetic poles through time and reversals in Earth's magnetic field.

Pangaea an ancient landmass that consisted of all of the present-day continents; it fragmented into Laurasia and Gondwanaland.

parallel circle on the surface of the Earth parallel to the plane of the equator and connecting all points of equal latitude; a line of latitude.

parasitism an intimate association between different organisms in which one is benefited and the other is harmed.

partially mixed estuary is one in which there is a strong net seaward flow of fresh water at the surface and a strong inward flow of seawater at depth.

passive margin see trailing margin.

pelagic primary division of the sea, which includes the whole mass of water subdivided into neritic and oceanic zones; also pertaining to the open sea.

period see tidal period, wave period.

pH measure of the concentration of hydrogen ions in a solution; the concentration of hydrogen ions determines the acidity of the solution; $pH = -\log_{10}(H^+)$, where H^+ is the concentration of hydrogen ions in gram atoms per liter.

phosphorite a sedimentary rock composed largely of calcium phosphate and largely in the form of concretions and nodules.

photic zone layer of a body of water that receives ample sunlight for photosynthesis; usually less than 100 m.

photo cell a device that converts light energy to electrical energy; used to determine solar radiation below the sea surface.

photophore luminous organ found on fish.

photosynthesis manufacture by plants of organic substances and release of oxygen from carbon dioxide and water in the presence of sunlight and the green pigment chlorophyll.

physiographic map portrayal of Earth's features by perspective drawing.

phytoplankton microscopic algal and plant forms of plankton.

pinniped member of the marine mammal group that is characterized by four swimming flippers; for example, seals and sea lions.

plankton passively drifting or weakly swimming organisms.

plate one of roughly a dozen segments of lithosphere that move independently on the asthenosphere. Plates consist of oceanic and/or continental crust fused to the rigid upper mantle overlying the asthenosphere.

plate tectonics theory and study of Earth's lithospheric plates, their formation, movement, interaction, and destruction; the attempts to explain Earth's crustal changes in terms of plate movements.

polar easterlies winds blowing from the poles toward approximately 60°N and 60°S; winds are northeasterly in the Northern Hemisphere and southeasterly in the Southern Hemisphere.

polar molecule a molecule with unbalanced electrical charge. One end of the molecule has a slight positive charge and the other end has a slight negative charge.

polar reversal the periodic reversal of the Earth's magnetic field where the north magnetic pole becomes the south magnetic pole and vice versa.

polar wandering curve a plot of the apparent location of the Earth's north magnetic pole as a function of geologic time.

Polaris also known as the North Star, it is located less than 1° from the celestial pole, a line corresponding to the extension of the Earth's axis of rotation into the sky from the north geographic pole. The elevation of Polaris above the horizon corresponds to the latitude of an observer in the northern hemisphere.

polychaetes marine segmented worms, some in tubes, some free-swimming.

polyculture cultivation of more than one species of organism in an aquaculture system.

polyp sessile stage in the life history of certain members of the phylum Coelenterata (Cnidaria); sea anemones and corals.

potential energy energy that an object has because of its position or condition.

precipitation falling products of condensation in the atmosphere, such as rain, snow, or hail; also the falling out of a substance from solution.

precision depth recorder (PDR) instrument used to obtain a continuous pictorial record of the ocean bottom by timing the returning echoes of sound pulses. *See also* echo sounder.

primary coast coastline shaped primarily by land forces rather than sea forces.

primary production the amount of living matter, or biomass, synthesized by photosynthetic or chemosynthetic organisms, usually expressed in grams of carbon. In practice, the terms *production* and *productivity* are often used interchangeably.

primary productivity the rate at which biomass is produced by photosynthesis and chemosynthesis, usually expressed as grams of carbon produced per day. In practice, the terms *production* and *productivity* are often used interchangeably.

prime meridian meridian of 0° longitude, used as the origin for measurements of longitude; internationally accepted as the meridian of the Royal Naval Observatory, Greenwich, England.

producer an organism that manufactures food from inorganic substances.

progressive tide tide wave that moves, or progresses, in a nearly constant direction.

progressive wind wave wave that moves, or progresses, in a certain direction.

projection a system of projecting lines of latitude and longitude onto a plane surface to create a map with specific physical properties; see Mercator projection.

protozoa minute, mostly one-celled animals.

pseudopodium flowing, temporary extension of the protoplasm of a cell, used in locomotion or feeding.

pteropod pelagic snail whose foot is modified for swimming.

pteropod ooze sediment made up of more than 30% shells of pteropods.

P-wave (or primary wave) a type of seismic wave in which material is alternately compressed and stretched in the direction of propagation of the wave.

pycnocline water layer with a large change in density with depth.

R

radar system of determining and displaying the distance of an object by measuring the time interval between transmission of a radio signal and reception of the echo return; derived from the phrase "radio detecting and ranging."

radial symmetry having similar parts regularly arranged around a central axis.

radiation energy transmitted as rays or waves without the need of a substance to conduct the energy.

radiolarian ooze sediment made up of more than 30% skeletal remains of radiolarians.

radiolarians single-celled protozoans with siliceous skeletons.

radiometric dating determining ages of geological samples by measuring the relative abundance of radioactive isotopes and comparing isotopic systems.

rafting transport of sediment, rock, silt, and other land matter out to sea by ice, logs, and the like, with the deposition of the rafted material when the carrying agent disintegrates.

rain shadow an area of low precipitation on the leeward (or sheltered) side of an island or mountain. Precipitation occurs on the windward side as the air is forced to ascend the side of the island or mountain so the descending air on the leeward side is dry.

range *see* tidal range.

red clay red to brown fine-grained lithogenous deposit, of predominantly clay size, which is derived from land, transported by winds and currents, and deposited far from land and at great depth; also known as brown mud and brown clay.

red tide red coloration, usually of coastal waters, caused by large quantities of microscopic organisms (generally dinoflagellates); some red tides result in mass fish kills, others contaminate shellfish, and still others produce no toxic effects.

reef offshore hazard to navigation, made up of consolidated rock or coral, with a depth of 20 m or less.

reef coast a secondary coast formed by reef-building corals in tropical waters.

reef crest highest portion of a coral reef on the exposed seaward edge of the reef.

reef flat portion of a coral reef landward of the reef crest and seaward of the lagoon.

reflection rebounding of light, heat, sound, waves, and so on, after striking a surface.

refraction change in direction, or bending, of a wave.

relict sediments sediments deposited by processes no longer active.

reservoir a source or place of temporary residence for water, such as the oceans or atmosphere.

residence time mean time that a substance remains in a given area before being replaced, calculated by dividing the amount of a substance by its rate of addition or subtraction.

respiration metabolic process by which food or food-storage molecules yield the energy on which all living cells depend.

restoring force force that returns a disturbed water surface to the equilibrium level, such as surface tension and gravity.

reverse osmosis the process of producing fresh water from seawater by forcing seawater through a semipermeable membrane that traps salt ions and other impurities.

ria coast a primary coast formed when rising sea level, caused by the melting of glaciers and ice sheets following the last ice age, flooded coastal river valleys.

ridge long, narrow elevation of the sea floor, with steep sides and irregular topography.

rift valley trough formed by faulting along a zone in which plates move apart and new crust is created, such as along the crest of a ridge system.

rift zone a region where the lithosphere splits and separates allowing new crustal material to intrude into the crack or rift.

rip current strong surface current flowing seaward from shore; the return movement of water piled up on the shore by incoming waves and wind.

rise long, broad elevation that rises gently and generally smoothly from the sea floor.

rotary current tidal current that continually changes its direction of flow through all points of the compass during a tidal period.

rotary tide tide that is the result of a standing wave moving around the central node of a basin.

S

salinity measure of the quantity of dissolved salts in seawater. It is formally defined as the total amount of dissolved solids in seawater in parts per thousand (‰) by weight when all the carbonate has been converted to oxide, all the bromide and iodide have been converted to chloride, and all organic matter is completely oxidized.

salinometer instrument for determining the salinity of water by measuring the electrical conductivity of a water sample of a known temperature.

salt budget balance between the rates of salt addition to and removal from a body of water.

salt marsh a relatively shallow coastal environment populated with salt-tolerant grasses. These are often found in temperate climates. They are extremely productive biologically and extend the shoreline seaward by trapping fine sediment.

salt wedge intrusion of salt water along the bottom; in an estuary, the wedge moves upstream on high tide and seaward on low tide.

sand spit *see* spit.

satellite body that revolves around a planet; a Moon; a device launched from Earth into orbit around a planet of the Sun.

saturation value the maximum amount of any substance that can be held in solution without any of the substance coming out of solution. When a dissolved substance has reached its saturation value, the rate at which it is dissolved into the solution will equal the rate at which it precipitates out of the solution.

scarp elongated and comparatively steep slope separating flat or gently sloping areas on the sea floor or on a beach.

scattering random redirection of light or sound energy by reflection from an uneven sea bottom or sea surface, from water molecules, or from particles suspended in the water.

sea same as the ocean; subdivision of the ocean; surface waves generated or sustained by the wind within their fetch, as opposed to swell.

sea cow (dugong, manatee) large herbivorous marine mammal of tropical waters; includes the manatee and the dugong.

seafloor spreading movement of crustal plates away from the mid-ocean ridges; process that creates new crustal material at the mid-ocean ridges.

sea level height of the sea surface above or below some reference level. *See also* mean sea level.

seamount isolated volcanic peak that rises at least 1000 m from the sea floor.

sea smoke type of fog caused by dry, cold air moving over warm water.

sea stack isolated mass of rock rising from the sea near a headland from which it has been separated by erosion.

sea state numerical or written description of the roughness of the ocean surface relative to wave height.

Secchi disk white, or white and black, disk used to measure the transparency of the water by observing the depth at which the disk disappears from view.

secondary coast coastline shaped primarily by marine forces or marine organisms.

sediment particulate organic and inorganic matter that accumulates in loose, unconsolidated form.

sedimentary rock a rock formed by the accumulation and cementation of mineral grains by wind, water, or ice transportation to the site of deposition or by chemical precipitation at the site.

seiche standing wave oscillation of an enclosed or semienclosed body of water that continues, pendulum fashion, after the generating force ceases.

seismic pertaining to or caused by earthquakes or Earth movements.

seismic sea wave *see* tsunami.

seismic tomography the use of seismic data to produce computerized, detailed, three-dimensional maps of the boundaries between Earth's layers.

seismic waves elastic disturbances, or vibrations, that are generated by earthquakes.

semidiurnal tide tide with two high waters and two low waters each tidal day.

semipermeable membrane membrane that allows some substances to pass through it but restricts or prevents the passage of other substances.

sessile permanently fixed or sedentary; not free-moving.

set direction in which the current flows.

shallow-water wave wave in water whose depth is less than one-twentieth the average wavelength.

shingles flat, water-worn pebbles, or cobbles, found in beds along a beach.

shoal elevation of the sea bottom comprising any material except rock or coral (in which case it is a reef) and that may endanger surface navigation.

shore strip of ground bordering any body of water and alternately exposed and covered by tides and waves.

sidereal day time period determined by one rotation of Earth relative to a far-distant star, about four minutes shorter than the mean solar day.

sigma-t abbreviated value of the density of seawater, neglecting pressure and at a given temperature and salinity: $\sigma_t = (\text{density} - 1) \times 1000$.

siliceous containing or composed of silica.

siliceous ooze fine-grained deep-ocean biogenous sediment containing at least 30% siliceous tests, or the remains of small marine organisms.

sill shallow area that separates two basins from one another or a coastal bay from the adjacent ocean.

slack water state of a tidal current when its velocity is near zero; occurs when the tidal current changes direction.

slick area of smooth surface water.

sofar channel natural sound channel in the oceans in which sound can be transmitted for very long distances; the depth of minimum sound velocity; derived from the phrase "sound fixing and ranging."

solar constant rate at which solar radiation is received on a unit surface that is perpendicular to the direction of incident radiation just outside Earth's atmosphere at Earth's mean distance from the Sun; equal to 2 cal/cm^2/min.

solar day time period determined by one rotation of Earth relative to the Sun; the mean solar day is twenty-four hours.

solstices times of the year when the Sun stands directly above 23 1/2°N or 23 1/2°S latitude. The winter solstice occurs about December 22, and the summer solstice occurs about June 22.

sonar method or equipment for determining, by underwater sound, the presence, location, or nature of objects in the sea; derived from the phrase "sound navigation and ranging."

sorting *see* dispersion.

sounding measurement of the depth of water beneath a vessel.

sound shadow zone area of the ocean into which sound does not penetrate because the density structure of the water refracts the sound waves.

Southern Oscillation a periodic reversal of the low and high pressure areas that typically dominate the eastern and western Equatorial Pacific, respectively.

specific gravity ratio of the density of a substance to the density of 4°C water.

specific heat ratio of the heat capacity of a substance to the heat capacity of water.

spectrum the regular ordering of oscillatory wave phenomena such as electromagnetic radiation by wavelength or frequency. A familiar spectrum is the regular ordering of visible light from the long-wavelength red to the short-wavelength violet when it passes through a prism.

sphere depth thickness of a material spread uniformly over a smooth sphere having the same area as Earth.

spicules small calcareous or siliceous skeletal structures found in sponges.

spit (sand spit) low tongue of land, or a relatively long, narrow shoal extending from the shore.

splash zone *see* supralittoral.

spoil dredged material.

spore minute, unicellular, asexual reproductive structure of an alga.

spreading center region along which new crustal material is produced.

spreading rate the rate at which two plates move apart. Spreading rates are generally between about 2 and 10 cm (0.4 and 8 in) per year.

spring tides tides occurring near the times of the new and full Moon, when the range of the tide is greatest.

standing crop biomass present at any given time.

standing wave type of wave in which the surface of the water oscillates vertically between fixed points called nodes, without progression; the points of maximum vertical rise and fall are called antinodes.

steepness of wave *see* wave steepness.

stipe portion of an alga between the holdfast and the blade.

storm berm *see* winter berm.

storm center area of origin for surface waves generated by the wind; an intense atmospheric low-pressure system.

storm tide (storm surge) along a coast, the exceptionally high water accompanying a storm, owing to wind stress and low atmospheric pressure, made even higher when associated with a high tide and shallow depths.

stratosphere the layer of the atmosphere above the troposphere where temperature is constant or increases with altitude.

subduction zone plane descending away from a trench and defined by its seismic activity, interpreted as the convergence zone between a sinking plate and an overriding plate.

sublimation transition of a substance from its solid state to its gaseous state without becoming a liquid.

sublittoral benthic zone from the low-tide line to the seaward edge of the continental shelf; the subtidal zone.

submarine canyon relatively narrow, V-shaped, deep depression with steep slopes, the bottom of which grades continuously downward across the continental slope.

submersible a research submarine, designed for manned or remote operation at great depths.

subsidence sinking of a broad area of the crust without appreciable deformation.

substrate material making up the base on which an organism lives or to which it is attached.

subtidal *see* sublittoral.

summer berm a seasonal berm that is built by low energy waves during the summer and removed by high energy waves in the winter.

summer solstice *see* solstices.

Sun tide portion of the tide generated solely by the Sun's tide-raising force, as distinguished from that of the Moon.

supersaturation when the concentration of a dissolved substance is higher than its normal saturation value.

supralittoral benthic zone above the high-tide level that is moistened by waves, spray, and extremely high tides; also called splash zone.

surf wave activity in the area between the shoreline and the outermost limit of the breakers.

surface tension tendency of a liquid surface to contract owing to bonding forces between molecules.

surimi refined fish protein used to form artificial crab, shrimp, and scallop meat.

swash the up-rush of water onto the beach from a breaking wave.

swash zone beach area where water from a breaking wave rushes.

S-wave (or secondary wave) a type of seismic wave in which material is sheared from side to side, perpendicular to the direction of propagation of the wave.

swell long and relatively uniform wind-generated ocean waves that have traveled out of their generating area.

symbiosis living together in intimate association of two dissimilar organisms.

T

taxonomy scientific classification of organisms.

tectonic pertaining to processes that cause large-scale deformation and movement of Earth's crust.

tektites particles with a characteristic round shape, derived from cosmic material.

temperature a measure of the rate of atomic or molecular motion in a substance.

terranes fragments of Earth's crust bounded by faults, each fragment with a history distinct from each other fragment.

terrigenous of the land; sediments composed predominantly of material derived from the land.

test the shell of an organism

thermocline water layer with a large change in temperature with depth.

thermohaline circulation vertical circulation caused by changes in density; driven by variations in temperature and salinity.

thermosphere the layer of the atmosphere above the mesosphere; extends from 90 km to outer space.

tidal bore high-tide crest that advances rapidly up an estuary or river as a breaking wave.

tidal current alternating horizontal movement of water associated with the rise and fall of the tide.

tidal datum reference level from which ocean depths and tide heights are measured; the zero tide level.

tidal day time interval between two successive passes of the Moon over a meridian, approximately twenty-four hours and fifty minutes.

tidal period elapsed time between successive high waters or successive low waters.

tidal range difference in height between consecutive high and low waters.

tide periodic rising and falling of the sea surface that results from the gravitational attractions of the Moon and Sun acting on the rotating Earth.

tide wave long-period gravity wave that has its origin in the tide-producing force and is observed as the rise and fall of the tide.

tombolo deposit of unconsolidated material that connects an island to another island or to the mainland.

topography general elevation pattern of the land surface (or the ocean bottom). *See also* bathymetry.

toxicant substance dissolved in water that produces a harmful effect on organisms, either by an immediate large dose or by small doses over a period of time.

trace element an element dissolved in seawater at a concentration less than one part per million.

trade winds wind systems that occupy most of the tropics and blow from approximately 30°N and 30°S toward the equator; winds are northeasterly in the Northern Hemisphere and southeasterly in the Southern Hemisphere.

trailing margin (passive margin) the continental margin closest to the mid-ocean ridge.

transform fault fault with horizontal displacement connecting the ends of an offset in a mid-ocean ridge. Some plates slide past each other along a transform fault.

transform plate boundary a boundary between two plates that are sliding past one another. This boundary is marked by a transform fault.

transgenic describing an organism that contains hereditary material from another organism incorporated into its genetic material.

transpiration process by which plants return moisture to air. Plants take up water through roots and lose water through pores in their leaves. An actively growing plant daily transpires five to ten times as much water as it holds at one time.

transverse ridge ridge running at nearly right angles to the main or principal ridge.

trench long, deep, and narrow depression of the sea floor with relatively steep sides, associated with a subduction zone.

triploid condition in which cells have three sets of chromosomes.

trophic relating to nutrition; a trophic level is the position of an organism in a food chain or food (trophic) pyramid.

Tropics of Cancer and Capricorn latitudes 23 1/2°N and 23 1/2°S, respectively, marking the maximum angular distance of the Sun from the equator during the summer and winter solstices.

troposphere the lowest layer of the atmosphere, where the temperature decreases with altitude.

trough long depression of the sea floor, having relatively gentle sides; normally wider and shallower than a trench. *See also* wave trough.

T-S diagram graph of temperature versus salinity, on which seawater samples taken at various depths are used to describe a water mass.

tsunami (seismic sea wave) long-period sea wave produced by a submarine earthquake or volcanic eruption. It may travel across the ocean for thousands of miles unnoticed from its point of origin and build up to great heights over shallow water at the shore.

tube worm any worm or wormlike organism that builds a tube or sheath attached to a submerged substrate.

turbidite sediment deposited by a turbidity current, showing a pattern of coarse particles at the bottom, grading gradually upward to fine silt.

turbidity loss of water clarity or transparency owing to the presence of suspended material.

turbidity current dense, sediment-laden current flowing downward along an underwater slope.

typhoon severe, cyclonic, tropical storm originating in the western Pacific Ocean, particularly in the vicinity of the South China Sea. *See also* hurricane.

U

ultraplankton plankton that are smaller than nannoplankton; they are difficult to separate from water.

Universal Time solar time along the prime meridian passing through Greenwich, England; also known as Greenwich Mean Time (GMT) or Zulu Time.

upwelling rising of water rich in nutrients toward the surface, usually the result of diverging surface currents.

V

vernal equinox *see* equinoxes.

vertebrates animals with backbones, or a spinal column.

virus a noncellular infectious agent that reproduces only in living cells.

viscosity property of a fluid to resist flow; internal friction of a fluid.

W

water bottle device used to obtain a water sample at depth.

water budget balance between the rates of water added and lost in an area.

water mass body of water identified by similar patterns of temperature and salinity from surface to depth.

water type body of water identified by a specific range of temperature and salinity from a common source.

wave periodic disturbance that moves through or over the surface of a medium with a speed determined by the properties of the medium.

wave crest highest part of a wave.

wave height vertical distance between a wave crest and the adjacent trough.

wavelength horizontal distance between two successive wave crests or two successive wave troughs.

wave period time required for two successive wave crests or troughs to pass a fixed point.

wave ray line indicating the direction waves travel; drawn at right angles to the wave crests.

wave steepness ratio of wave height to wavelength.

wave train series of similar waves from the same direction.

wave trough lowest part of a wave.

well-mixed estuary is one in which there is strong tidal mixing and low river flow. The salinity of the water in the estuary is relatively constant with depth and decreases from the ocean to the river.

westerlies wind systems blowing from the west between latitudes of approximately 30°N and 60°N and 30°N and 60°S; southwesterly in the Northern Hemisphere and northwesterly in the Southern Hemisphere.

wind wave wave created by the action of the wind on the sea surface.

winter berm a relatively permanent berm that is formed by high energy waves in the winter.

winter solstice *see* solstices.

Z

zenith point in the sky that is immediately overhead.

zonation parallel bands of distinctive plant and animal associations found within the littoral zones and distributed to take advantage of optimal conditions for survival.

zooplankton animal forms of plankton.

zooxanthellae symbiotic microscopic organisms (dinoflagellates) found in corals and other marine organisms.

Zulu Time solar time along the prime meridian passing through Greenwich, England; also known as Greenwich Mean Time (GMT) or Universal Time.

credits

Text

Prologue

Pro I Poem Excerpt: From "The Sea is History" from *Collected Poems: 1948–1984* by Derek Walcott. Copyright © 1986 by Derek Walcott. Reprinted by permission of Farrar, Straus and Giroux, LLC. **Pro. V; Pro. VII; Pro. XXIII:** From *Fundamentals of Oceanography*, 4th edition, Duxbury, Duxbury, and Sverdrup. Copyright 2000 The McGraw-Hill Companies. All rights reserved.

Chapter 1

Figure 1.18; Figure 1.19; Figure 1.22: From *Fundamentals of Oceanography*, 4th edition, Duxbury, Duxbury, and Sverdrup. Copyright 2000 The McGraw-Hill Companies. All rights reserved.

Chapter 2

Figure 2.1: From Plummer/McGeary/Carlson, in *Physical Geology—Earth Revealed*, 4th edition, Copyright 2001 The McGraw-Hill Companies. All rights reserved. **Figure 2.2; Figure 2.20; Figure 2.22; Figure 2.23; Figure 2.24; Figure 2.25; Figure 2.26; Figure 2.29; Figure 2.30; Figure 2.31; Figure 2.33:** From *Fundamentals of Oceanography*, 4th edition, Duxbury, Duxbury, and Sverdrup. Copyright 2000 The McGraw-Hill Companies. All rights reserved. **Figure 2.11:** Earthquake Epicenters, 1961–1967. From Barazangi and Dormar, *Bulletin of Seismological Society of America*, 1969. Two maps depicting earthquake epicenters Seismological Society of America, El Cerrito, CA. Reprinted by permission. **Figure 2.17:** From D. H. Tarling and J. C. Mitchell, "The Earth's Magnetic Polarity as a Function of Millions of Years Before Present" in *Geology*, 4.3(1976). The Geological Society of America. Reprinted by permission.

Chapter 3

Figure 3.12; Figure 3.14; Figure 3.18: From *Fundamentals of Oceanography*, 4th edition, Duxbury, Duxbury, and Sverdrup. Copyright 2000 The McGraw-Hill Companies. All rights reserved. **Figure 3.24a & b:** From M. N. Hill, ed., *The Sea*, Vol. 3. Copyright © 1963. This material is used by permission of John Wiley & Sons, Inc.

Chapter 4

CO Chapter 4 Poem Excerpt: From "Boats in a Fog," Copyright 1925 and renewed 1953 by Robinson Jeffers, *The Selected Poems of Robinson Jeffers*, by Robinson Jeffers. Used by permission of Random House, Inc.; Electronic rights granted by Stanford University Press, Stanford, CA. Used by permission. **Figure 4.16:** From Kinsler and Frey, *Fundamentals of Acoustics*, 2nd edition. Copyright 1962 John Wiley & Sons, Inc., New York.

Chapter 5

Figure 5.8: From *Climate Change* 1994, Intergovernmental Panel on Climate Change.

Chapter 7

CO Chapter 7 Poem Excerpt: Reprinted with permission of Simon & Schuster from *Kon-Tiki* by Thor Heyerdahl. Copyright 1950, 1960, 1984 by Thor Heyerdahl.

Chapter 9

Figure 9.4; Figure 9.13; Figure 9.29: From *Fundamentals of Oceanography*, 4th edition, Duxbury, Duxbury, and Sverdrup. Copyright 2000 The McGraw-Hill Companies. All rights reserved.

Chapter 10

CO Chapter 10 Excerpt: From *Spring Tides* by Samuel Eliot Morison. Reprinted by permission of Curtis Brown Ltd. Copyright 1965 by Samuel Eliot Morison.

Chapter 12

Figure 12.2: Based on data from N. Rabalais, R. Turner, and W. Wiseman. "Keeping the Stygian Waters at Bay," *Science* 291:968–973, February 9, 2001. **Figure 12.3:** Based on data from N. Rabalais, R. Turner, and W. Wiseman. "Keeping the Stygian Waters at Bay," *Science* 291:968–973, February 9, 2001.

Chapter 14

CO Chapter 14 Excerpt: From "Sensuous Symbionts of the Sea" by Lewis Thomas in *Natural History*, Vol. 80, No. 7. Reprinted by permission of the American Museum of Natural History, New York, NY. **Box Fig. 1:** From *Fundamentals of Oceanography*, 4th edition, Duxbury, Duxbury, and Sverdrup. Copyright 2000 The McGraw-Hill Companies. All rights reserved.

Chapter 17

CO Chapter 17 Excerpt: From *The Log From the Sea of Cortez* by John Steinbeck. Copyright 1941 by John Steinbeck and Edward F. Ricketts. Copyright renewed © 1969 by John Steinbeck and Edward F. Ricketts, Jr. Used by permission of Viking Penguin, a division of Penguin Putnam, Inc.

Photo

Prologue

Opener: © Ed Young/SPL/Photo Researchers, Inc.; **1a-c:** Courtesy of the Burke Museum of Natural History & Culture, Seattle, WA (cat 1-11433, 1-794, 1-158); **2:** © Stephen D. Thomas; **3:** Courtesy of Sea & Shore Museum, Port Gamble, WA; **4:** From Ptolemy's *Geographia*, 15th century Italy; **5:** From Johannes van Keulen's *Sea Atlas*, 1682-84; **6:** National Maritime Museum, London; **7:** Franklin-Folger, 1769; **8:** Courtesy of NASA/JPL/Caltech; **9:** U.S. Navy Photo; **10:** From Matthew F. Maury's *Physical Geography of the Sea*, 1855; **11:** Vol. 25 *Faunna & Flaura dus Golfes von Neaple*, 1899; **12a-i:** *Challenger Expedition*, Dec. 21, 1872 - May 24, 1876. Engravings from *Challenger Reports*, Vol.1, 1885; **13,14:** Courtesy Norwegian Maritime Museum; **15:** Scripps Institution of Oceanography; **16:** © D. Weisman/Woods Hole Oceanographic Institution; **17:** Courtesy Joe Creager, University of Washington; **18:** Ocean Drilling Program/Texas A&M; **19:** CZCS Project/NASA/Goddard Space Flight Center and the Goddard Earth Science Distributed Active Archive Center; **Box 1:** © Mary Rose Trust; **Box 2:** Hans Hammerskiold for Vasa Museum, Stockholm, Sweden; **Box 3:** Henry Thomason, Texas Historical Commission; **Box 4:** © Quest Group Ltd.; **Box 5:**

NOAA/Inter Network, Inc.;
Box 6: U.S. Navy/Inter Network, Inc.;
Box 7: Jet Propulsion Lab/Inter Network, Inc.; **Box 8, 9:** European Space Agency.

Chapter 1
Opener: Keith Sverdrup; **1.1:** NASA; **1.2:** AP/Wide World Photos; **1.3:** Goddard/NASA; **1.9:** Alyn & Allison Duxbury; **1.13:** Courtesy Stephen P. Miller, UC-Santa Barbara; **1.14:** From Ortelius, *Theatre of the World,* 1570; **1.17a:** Department of Defense; **Box 1:** © James A. Sugar/CORBIS; **Box 2, 3:** L.A. Frank/University of Iowa.

Chapter 2
Opener: © Kevin Shafer/Tom Stack; **2.5:** From Dziewonski & Anderson, *American Scientist,* 1984, 72:483-94; **2.8:** Paleogeographic Maps by Christopher R. Scotese, PALEOMAP project, University of Texas at Arlington (www.Scotese.com); **2.9:** B. Heezen & M. Tharp, *World Ocean Floor.* © Marie Tharp, 1977, South Nyack, NY 10960. Reproduced by permission; **2.12:** Dr. Robert M. Kieckhefer, Chevron Overseas Petroleum, Inc., San Ramon, CA; **2.15:** Courtesy Ocean Drilling Program/Texas A&M University, College Station, TX; **2.27:** OAR/National Underseas Research; **2.32:** P.W. Lipman, U.S. Geological Survey; **2.35:** Paleogeographic Maps by Christopher R. Scotese, PALEOMAP project, University of Texas at Arlington (www.Scotese.com); **2.37:** News Office/©Woods Hole Oceanographic Institution, Photo by Rod Catanech; **2.38a-d:** Courtesy John Delaney/School of Oceanography, University of Washington; **2.39a,c, 2.40:** Courtesy John Delaney/School of Oceanography, University of Washington; **Box 1:** Peter J. Auster; **Box 2:** ©Woods Hole Oceanographic Institution; **Box 3:** Peter J. Auster; **Box 4:** ©Woods Hole Oceanographic Institution; **Box 5:** Alyn & Alison Duxbury; **Box 6:** Courtesy Deborah Kelley, University of Washington; **Box 7:** Courtesy Deborah Kelley, University of Washington; **Box 8:** Alyn & Alison Duxbury; **Box 9:** Courtesy Deborah Kelley, University of Washington.

Chapter 3
Opener: NASA/Image Works; **3.1:** Alyn & Alison Duxbury; **3.2:** Courtesy John Delaney/School of Oceanography, University of Washington; **3.3:** Courtesy Walter H.F. Smith/ NOAA; **3.9:** Alyn & Alison Duxbury; **3.10:** Official U.S. Navy Photo by R.F. Diel; **3.19a,b:** Courtesy Williamson & Associates, Seattle, WA; **3.21a,b:** Ken Adkins, School Fisheries, University of Washington; **3.21c:** Courtesy of Steve Nathan and Dr. R. Mark Leckie, University of MA;

3.22: Courtesy of Dr. Virgil E. Barns; **3.23:** Courtesy Joe Creager, University of Washington; **3.25a:** Courtesy Dean McManus, University of Washington; **3.25b:** Courtesy Dawn Wright, Oregon State University; **3.26:** Courtesy Kathy Newell/School of Oceanography, University of Washington; **3.27c:** Courtesy Sara Barnes/School of Oceanography, University of Washington; **3.27d:** Alyn & Alison Duxbury; **3.27e:** Courtesy Dr. Richard Sternberg/School of Oceanography, University of Washington; **3.28:** Courtesy Chevron Corporation; **Box 1:** Courtesy Williamson & Associates, Seattle, WA.

Chapter 4
Opener: © Superstock; **4.3:** Sandra Hines, News and Information/University of Washington; **4.11:** Alyn & Alison Duxbury; **4.12:** Dennis K. Clark, MOBY Project Team Leader, NOAA; **4.14:** Courtesy EPC Laboratories, Inc., Danvers, MA; **4.17a:** Courtesy Kathy Newell/School of Oceanography, University of Washington; **4.17b:** Courtesy Roger Anderson, Polar Science Center, Applied Physics Laboratory, University of Washington; **4.17c:** Captain Lawson W. Brigham, retired, USCG; **4.18a,b:** J. Zwally /Goddard Space Flight Center; **4.18c,d:** D.J. Cavaneri and P. Gloersen/Goddard Space Flight Center; **4.19a:** Edward Joshberger/U.S. Geological Survey, Dept. of Interior; **4.19b:** Courtesy Kathy Newell/School of Oceanography, University of Washington; **4.20:** Reproduced by permission of Earth Observatory Satellite Company, Lanham, MD. ©1991Discover Magazine; **4.21:** © William Rosenthal/Jeroboam, Inc.; **Box 1:** K. Heumann, Alfred-Wagner-Institute for Polar and Marine Research.

Chapter 5
Opener: © Randy Morse/Tom Stack; **5.5:** Kathy Newell/School of Oceanography, University of Washington; **5.9:** Courtesy John Conomos, U.S. Geological Survey; **Box 1:** Peter Grootes/Quarterly Research Center, University of Washington; **Box 2:** Courtesy Christine Massey, University of Washington, **Box 3:** Courtesy Christine Massey, University of Washington; **Box 4:** Peter Grootes/Quarterly Research Center, University of Washington.

Chapter 6
Opener: © Superstock; **6.7a,b:** M. Chahine/Jet Propulsion Laboratory, J. Sussking/Goddard Space Flight Center; **6.12:** Richard McPeters, Ozone Processing Team Head/GSFC, NASA; **6.20:** W. Timothy Liu, NASA, JPL, California Institute of Technology; **6.27:**

NOAA satellite photo; **6.29a-c;** Scientific Visualization Studio, NASA, Goddard Space Flight Center; **6.31:** National Hurricane Center, NOAA, U.S. Dept. of Commerce; **6.32:** National Weather Service; **Box 1:** © Hubertus Kanus/Photo Researchers; **Box 2:** Alyn and Alison Duxbury;

Chapter 7
Opener: © Bob Pool/Tom Stack; **7.10:** Alyn and Alison Duxbury; **7.11:** James R. Postel/School of Oceanography, University of Washington; **7.12:** Chelsea Technologies Group; **7.13:** News Office, WHOI; **7.14:** Dana Swift/School Oceanography, University of Washington; **7.15a-c:** Courtesy Lockheed Missiles and Space Co., Sunnyvale, CA; **Box 1, 2:** Captain Lawson W. Brigham, retired, USCG; **Box 3:** Sandra Hines, University of Washington News Office; **Box 4, 5:** James Johnson/Applied Physics Lab, University of Washington; **Box 6, 7:** Charles C. Eriksen/School of Oceanography, University of Washington.

Chapter 8
Opener: Jeff Rotman; **8.8:** Courtesy Otis Brown; **8.9:** NASA; **8.13a,b:** W.W. Hseih, Dept. of Oceanography, University of British Columbia, Vancouver, CANADA; **8.17:** School of Oceanography, University of Washington; **8.18a:** Kathy Newell/School of Oceanography, University of Washington; **8.18b:** Aanderaa Instruments; **8.20:** Courtesy of Mark Holmes/School of Oceanography, University of Washington; **Box 1:** Bauer, Moynihan & Johnson, Seattle, WA; **Box 2:** Rowland Studio, Seattle, WA.

Chapter 9
Opener: EDA Rogers; **9.1:** Alyn and Alison Duxbury; **9.2:** Eda Rogers; **9.6:** U.S. Army Corps; **9.9:** Exxon Mobil, Leonard Wolhuter; **9.17:** Alyn & Alison Duxbury; **9.20 a-b:** Alyn & Alison Duxbury; **9.21:** © Superstock; **9.22:** Richard Sternberg, School of Oceanography/University of Washington; **9.23a-c:** National Geophysical Data Center/NOAA, **9.24:** Courtesy Dr. Yoshinobu Tsuji; **9.25:** Courtesy Dr. Yoshinobu Tsuji, Tohoku University/Japan National Defense Academy; **9.26:** © 1993 Discover Magazine; **9.32:** Japan Marine Science & Technology Center (JAMSTEC); **Box 1:** Courtesy Steve Niland/Datasonics, Cataumet, MA.

Photo Essay "Going to Sea"
1,2: Courtesy School of Oceanography, University of Washington; **3:** Courtesy Thomas Chaplin, School of Oceanography/University of Washington; **4, 5:** Kathy Newell, School of Oceanography/University of Washington; **6:** Curtis Martin, Enigma Communications; **7:** Courtesy Mark

Holmes, USGS, School of Oceanography/ University of Washington; **8:** Kathy Newell, School of Oceanography/University of Washington; **9, 10:** Courtesy School of Oceanography/University of Washington; **11:** Courtesy Mark Holmes, School of Oceanography/University of Washington; **12, 13:** Kathy Newell, School of Oceanography /University of Washington; **14:** School of Oceanography/University of Washington; **15:** Mark Holmes, USGS, School of Oceanography, University of Washington; **16:** Jan Newton, School of Oceanography, University of Washington; **17:** School of Oceanography, University of Washington; **18:** Alyn & Alison Duxbury.

Chapter 10

Opener: Alyn & Alison Duxbury; **10.15a,b:** Tourism New Brunswick, Canada; **10.16:** Tourism New Brunswick, Canada; **10.18:** Nova Scotia Power Plant; **Box 1:** From Egbert, G.D. & Foreman, M.G.G, "TOPEX/POSEIDON Tides estimated using a global inverse model," *J. Geophys. Res.* 99 (1994), pp. 24821-24852. © American Geophysical Union.

Chapter 11

Opener: Courtesy Joe Creager, University of Washington; **11.1a-c, 11.2a-c, 11.3:** Alyn & Alison Duxbury; **11.4:** Joe Creager, University of Washington; **11.5:** Chesapeake Bay Foundation; **11.6:** Alyn & Alison Duxbury; **11.7:** Courtesy Dr. Sherwood Maynard, Univ. of Hawaii; **11.8:** Courtesy © Terra-Nova International; **11.9:** Courtesy Dr. Richard Sternberg/School of Oceanography, University of Washington; **11.10:** Alyn & Alison Duxbury; **11.11:** © Alex S. McClean/ Landslides; **11.12-11.17:** Alyn & Alison Duxbury; **11.20:** Carr Clifton; **11.24:** Courtesy Dr. Richard Sternberg/ School of Oceanography, University of Washington; **11.25:** Alyn & Alison Duxbury; **11.26:** Courtesy Dr. Richard Sternberg/ School of Oceanography, University of Washington; **11.27:** Alyn & Alison Duxbury; **11.29:** © Bob Evans/La Mer Bleu Productions; **11.30:** U.S. Army Corps of Engineers; **11.38:** John Conomos, U.S. Geological Survey; **11.39:** NASA; **Box 1:** © Francois Gohier/Photo Researchers, Inc.; **Box 2:** © Larry Lipsky/Tom Stack; **Box 3:** U.S. Army Corps of Engineers/NASA/ EOSAT, Geological Survey, School of Oceanography, University of Washington.

Chapter 12

Opener: Hazardous Materials Response Branch; **12.4a-c:** Courtesy Joe Lucas/Marine Entanglement Research Program/Nat Marine Fisheries Service, NOAA; **12.4d:** Courtesy

R. Herron/Marine Entanglement Research Program/Nat Marine Fisheries Service, NOAA; **12.6a-e:** Courtesy Hazardous Materials Response Branch/Jerry Gault/ NOAA; **12.6f:** Alyn & Alison Duxbury; **12.7:** Courtesy Clean Sound Cooperative, Seattle, WA; **12.8:** Courtesy Port of Seattle; **12.9:** © Alex MacLean/Landslides; **12.10:** Dr. Fred Nichols/U.S. Geological Survey, Menlo Park, CA; **12.11:** Thomas Niesen, San Francisco State University; **12.12:** Alyn & Alison Duxbury; **Box 1:** Courtesy P. Flanagan; **Box 2:** U.S. Army Corps of Engineers.

Chapter 13

Opener: © Aldo Brando/Peter Arnold; **13.1:** A. Hardy, *The Open Sea*, Harper Collins, Ltd., Hammersmith, London, 1956; **13.2:** Kathy Newell/School of Oceanography, University of Washington; **13.3:** © Leo J. Shaw, The Seattle Aquarium; **13.11:** Courtesy James Sumich; **13.12:** Field Museum, Z82860; **13.13:** Courtesy Leo J. Shaw, Seattle Aquarium.

Chapter 14

Opener: © Chuck Davis/Tony Stone Images; **14.9a-d:** Image provided by ORBIMAGE. © Orbital Imaging Corporation and processing by NASA Goddard Space Flight Center; **Box 14.2:** Scripps Institution of Oceanography.

Chapter 15

Opener: © William Amos/Star Brook Photography; **15.1a-d:** Alyn & Alison Duxbury; **15.1e:** School of Oceanography, University of Washington; **15.3a-d:** Enge's Equipment Co.; **15.6a,b:** Tiffany Vance, NOAA; **15.8:** Courtesy Kendra Daly/School of Oceanography, University of Washington; **15.13a-c:** Alyn & Alison Duxbury; **15.13d:** Courtesy Paulette Brunner, Seattle Aquarium; **15.15a,b:** Alyn & Alison Duxbury; **15.15c:** Courtesy Deborah Penry, University of Washington; **15.17:** Courtesy School of Oceanography, University of Washington; **15.18:** Alyn & Alison Duxbury; **Box 1a,b:** Dr. John Baross/School of Oceanography, University of Washington; **Box 2a-f:** © Monterey Bay Aquarium Research Institute © MBARI 1998. Photo by Peter Miller; **Box 3a,b:** Miller, P.E. & Scholin, C.A. *J. Phycol.* 34 (1998) p. 377, fig.3a-b. Courtesy *Journal of Phycology*. Photo by Peter Miller, © MBARI 1998.

Chapter 16

Opener: Alyn & Alison Duxbury; **16.1a:** Courtesy Dr. Albert Erickson/Fisheries Research Institute, University of Washington; **16.1b:** Courtesy David Whitrow/National

Marine Fisheries Service, NOAA; **16.5a:** Courtesy Leo J. Shaw, Seattle Aquarium; **16.5b:** Courtesy Kemper, Seattle Aquarium; **16.5c:** Courtesy Leo J. Shaw, Seattle Aquarium; **16.5d:** Courtesy Kemper, Seattle Aquarium; **16.5e:** Alyn & Alison Duxbury; **16.9a:** Courtesy P. Dee Boersma/Dept. of Zoology, University of Washington; **16.9b-e:** Alyn & Alison Duxbury; **16.12a-c:** Courtesy Leo J. Shaw, Seattle Aquarium; **16.12d:** Courtesy Paulette Brunner, Seattle Aquarium; **16.12e:** Courtesy Gail Scott, Seattle Aquarium; **16.16a,b:** Courtesy Wayne A. Palsson, Washington State Dept. of Fisheries; **16.16c:** Courtesy Uni Seafoods, Inc., Richmond, WA; **16.18:** Bruce W. Buls, Seattle, WA; **16.19:** H. Wes Pratt, Apex Predator Investigation/NOAA-NMFS Narragansett Laboratory, RI; **16.20:** Courtesy Northwest Sea Farm Inc.; **Box 1a:** © Dr. Craig R. Smith/Dept. of Oceanography, University of Hawaii; **Box 1b:** © H. Kukert, courtesy Craig R. Smith; **Box 2, 3:** Captain Lawson W. Brigham, Retired, USCG.

Chapter 17

Opener: © Art Wolfe/Tony Stone/Getty Images; **17.10a-c:** Courtesy Verena Tunnicliffe/University of Victoria, B.C.; **17.11a,b:** Alyn & Alison Duxbury; **17:13:** © David Hall/Photo Researchers, Inc.; **17.15a,b:** Courtesy Kathy Newell/School of Oceanography, University of Washington; **17.16:** Richard A. Lutz; **17.17a-c:** Courtesy Dr. James Brooks/Geochemical & Environmental Research Group, College Station, TX; **17.18:** Courtesy Kathy Newell/ School of Oceanography, University of Washington; **17.19:** Brad Matsen, Seattle, WA; **17.20:** Courtesy Ken Chew/School of Fisheries, University of Washington; **17.21a:** Courtesy Seattle Aquarium; **17.21b:** Courtesy Leo J. Shaw, Seattle Aquarium; **17.21c:** Courtesy Paulette Brunner, Seattle Aquarium; **17.21d,e:** Courtesy Seattle Aquarium; **17.21f,g:** Courtesy Leo J. Shaw, Seattle Aquarium; **17.21h:** Courtesy Tina Link, Seattle Aquarium; **17.21i:** Courtesy Leo J. Shaw, Seattle Aquarium; **Box 1:** Dr. Charles Fisher, Prof. of Biology, 208 Mueller Lab, The Pennsylvania State University, University Park, PA 16802; **Box 2a,b:** Reprinted by permission from Washington Sea Grant Program, University of Washington, January 2000.

index

A

Aanderaa current meter, 244
abalone, 455
absorption, light, 142
abyssal clay, 114
abyssal hills, 107–9
abyssal plains, 106–7
abyssal zone, 360
Acanthaster, 466–67
acceleration, SI units and, 480
accretion, 312
acid rain, 186, 187
acorn barnacles, 453
acorn worms, 459
Acoustically Navigated Underwater
    Survey System (ANGUS), 94
acoustic Doppler current profiler, 217
acoustic profiling, 121
Acoustic Thermometry of Ocean Climate
    (ATOC), 145, 146–47
active continental margins, 79, 101, 104
Adelie penguins, 384
adsorption, 162
Advanced Environmental Orbiting
    Satellite (ADEOS), 193
Advanced Television Infrared
    Observation Satellite (TIROS),
    23, 180
Advanced Unmanned Search System
    (AUSS), 89
advanced very high-resolution radiometer
    (AVHRR), 192
advective fog, 152
Adventure, 12
Aegean Sea, 218
Aequorea, 402
Africa
    continental motion, 81, 85
    dune coast, 304
African plate, spreading rate, 81
agar, 472
Agassiz, Alexander, 18

agriculture
    Chesapeake Bay, damage from
        industrial and domestic, 330
    runoff from, 337–39
    San Francisco Bay, damage from, 330
Agulhas Current, 232, 254
air
    benthic organisms and exposure to, 451
    composition, 182
    convection cells in, 186–88
    movement (*see* winds)
    pollution (*see* pollution)
    pressure (*see* atmospheric pressure)
airborne particles, 120, 121
Alaska
    coast, 301
    earthquake (1964), 263
    Exxon Valdez in Prince William Sound,
        335, 344–46
    terranes, 86
    tidal bores in Cook's Inlet, 290
Alaska, Gulf of, 353
Alaska Current, 231, 237
albacore tuna, 354
albatross, 427–28
Albatross, 18
Aleutian Trench, 110, 111
Alexander the Great, 4
Alexandrium, 408–9, 410
algae
    aquaculture, 472
    biomedical products and, 472
    blooms, 340, 408–12
    brown (*see* brown algae)
    characteristics, 455, 456
    characteristics of benthic, 448–49,
        451–52, 453
    in coral reefs, 463–64, 465, 467
    green (*see* green algae)
    harvesting, 472
    photosynthesizing green, 365
    planktonic, 393–97
    red (*see* red algae)
algin, 472

Allen, Hurricane (1980), 200
alligators, 431
alluvial plain, 304
Alvin (submersible), 88, 94
    black smoker research, 92–93
    hydrothermal vent research, 90, 91,
        94–95, 468
    Juan de Fuca Ridge, research on,
        80–81, 92
    Lost City, discovery of, 468
    Project FAMOUS, 87
Amazon River (South America),
    289, 304
Amery Ice Shelf, green icebergs from, 153
amnesiac shellfish poisoning (ASP),
    408, 412
Amoco Cadiz, 343–44, 345
amphidromic point, 287
amplitude, wave, 249
Amundsen, Roald, 17, 19
anaerobic, 165
anchovies, 388, 389
    commercial fishing, 437, 438, 440–41
andesite, 78
Andes Mountains (South America), 78
Andrew, Hurricane (1992), 205
androgenesis, 475
anemones. *see* sea anemones
angel fish, emperor, 435
angler fish, 437, 439
anions, 157, 159–60
Annapolis River (Canada), tidal power
    plant on, 295
Anne, (queen, England), 11
anoxic water, 165
Antarctica
    continental movement, 122, 124
    food web, 384–85
    icebergs, 147–49
    ice core research, 168–69
    oozes in, 117
    ozone depletion, 185–86
    sea ice in, 148–49
Antarctica Treaty (1959), 128

E

France, 354
    Amoco Cadiz oil spill, 343–44, 345
    aquaculture, 471
    commercial tidal plant on La Rance
        River, 294
    tidal bores in Mont-Saint-Michel, 290
    whaling, 419
Frank, Louis, 33
Franklin, Benjamin, 12
Franklin-Folger map of the Gulf
    Stream, 12
free waves, 251, 285
freshwater lid, 212
fringing reefs, 109, 110
Frisius, Gemma, 11, 41
Frobisher, Martin, Sir, 9
frustule, 393, 394, 395
fucoxanthin, 393
Fucus, 450, 452
Fundy, Bay of, 289, 290, 294–96
fur seals, 341, 342, 423, 424, 425

G

gabbro, 74
Galápagos Islands, 227, 431, 432
Galápagos Rift, 87, 90, 94–95, 468
galaxies, 30–31
Galilei, Galileo, 10
Gambierdiscus toxicus, 412
Ganges River (Bangladesh, India), 113, 304
gas, as resource, 125–26
gases
    atmospheric, 181–82
    greenhouse (*see* greenhouse gases)
    in seawater, 164–66, 167, 364
gas hydrates, 126, 462
gavial, 431, 432
Geller, J.B., 351
general circulation models (GCMs), 195
genetic manipulation, 475
Geochemical Ocean Sections
    (GEOSECS), 212
Geodynamic Experimental Ocean
    Satellite, US Navy (GEOSAT),
    22, 23, 24, 291
Geographia, 5
geological long-range inclined asdic
    (GLORIA), 108–9
geological oceanography, 2
Geological Service, US
    National Tsunami Hazard and
        Mitigation Program, 264–65
Geological Survey, US
    gas hydrates, data on, 126
geologic time, 34–35, 36

geomorphology, 302
Georgia, barrier islands, 307
Georgia, Strait, 119, 324
geostationary operational environmental
    satellites (GOES), 23, 199
geostrophic flow, 229–30
Germany, cod fishing, 443
ghost crabs, 459
giant reef clams, 465
giant squid, 361
Gjoa, 17, 19
glaciation
    coastal formation from, 302–4, 317
    oceanic sedimentation and, 114
    sea level changes and, 302, 303–4
Glacier Bay, 301
glass sponges, 459, 460
glider, oceanographic, 220, 222–23
Global Ocean Atmosphere-Land System
    (GOALS), 24
Global Ocean Observing System
    (GOOS), 25
Global Positioning System (GPS),
    42, 44–45
    plate movements, measurement of, 81
    seaglider research and, 223
Global Reef Monitoring Network, 466
global warming, 184–85
Globigerina, 301, 400
Glomar Challenger, 21, 64, 120
GLONASS (Soviet Union, former), 45
Golden Hind, 10
Gondwana, 84–85
Gondwanaland, 60
Gonionemus, 402
Gonyaulax, 396, 408, 410
gooseneck barnacles, 453, 454
Gorda Ridge, 468
graben, 73
grab samplers, 121, 122
granite, in continental crust, 56, 58
grasses, benthic, 449–50, 457
gravels, 112, 124
gravity
    seafloor topography and, 100–101
    tides and, 279–82
gravity corer, 123
gravity waves, 248, 249
gray seals, 342
Great Barrier Reef (Australia), 308, 430,
    462, 467
Great Britain
    cod fishing, 443
    Dorset coast, 301, 305
    groins at Hastings, 317
    Land's End, Cornwall, 299
    whaling, 420

great circles, 39–40
Great Lakes, biological invaders, 352
*Great New and Improved Sea-Atlas
    or Water-World*, 8
great white sharks, 443
Greeks, ancient oceanographers, 2, 4
green algae, 449, 450, 452, 455
    harvesting, 472
    photosynthesizing, 365
Green Canyon oil lease area, 470
greenhouse effect, 183–85
greenhouse gases
    carbon dioxide and greenhouse effect,
        182–85
    ozone, 185–86
green icebergs, 153
Greenland Current, 213
Greenland Ice Core Project (GRIP),
    168–69
green turtles, 431, 432
Greenwich Mean Time (GMT), 42, 44
Gregorian calendar, 37–38
gribble, 459, 461
groins, 316–17
gross primary production, 373
grunts, 319
Guadalupe fur seal, 318, 421
Guembelitria, 124
Gulf Stream, 213, 218, 231
    eddies, 233–34, 235
    episodic waves, 254
    Franklin-Folger map of, 12
    red tides, 408, 409
    western intensification, 232–33
Gulf War, 346
gulls, 429, 430
guyots, 83, 107, 108
Gymnodinium, 396, 408, 409, 411, 425
gynogenesis, 475
gyres, ocean, 228–29

H

hadal zone, 360
Hadean eon, 34, 36
hagfish, 459
half-life, 34, 35
halibut, 437, 438
    as by-catch, 354
    fish farming, 444
Halley, Edmund, 10
halocline, 210, 211
Hannu, 2–3
Hansa Carrier, 236–38
harbor seals, 423, 424
Hardy, Alister, Sir, 393